CELEBRATING THE MEGASCALE

Proceedings of the Extraction and Processing Division Symposium on Pyrometallurgy in Honor of David G.C. Robertson

Cover Photographs: Images on front cover, clockwise from top: (1) Ladle pouring. Photo courtesy of Rodney Jones, Mintek; (2) Photo of David G.C. Robertson; (3) Metal tapping. Photo courtesy of Isabel Geldenhuys, Mintek; (4) Molten silver cast from a levitation coil (1961). Photo courtesy of School of Materials Science and Engineering, University of New South Wales; (5) Slag tapping. Photo courtesy of Anglo American Platinum. Additional image on back cover; (6) Metallurgical plant near green countryside at the Polokwane Smelter. Photo courtesy of Anglo American Platinum.

TMS2014

143rd Annual Meeting & Exhibition

CELEBRATING THE MEGASCALE

Proceedings of the Extraction and Processing Division Symposium on Pyrometallurgy in Honor of David G.C. Robertson

Proceedings of a symposium sponsored by
the Extraction and Processing Division of
The Minerals, Metals & Materials Society (TMS)

held during

TMS2014
143rd Annual Meeting & Exhibition

February 16-20, 2014
San Diego Convention Center
San Diego, California, USA

Edited by:

Phillip J. Mackey | Eric J. Grimsey
Rodney T. Jones | Geoffrey A. Brooks

Editors
Phillip J. Mackey
Eric J. Grimsey
Rodney T. Jones
Geoffrey A. Brooks

ISBN 978-3-319-48591-1
ISBN 978-3-319-48234-7 (eBook)
DOI 10.1007/978-3-319-48234-7

Chemistry and Materials Science: Professional

Published by Springer International Publishers, Switzerland, 2016
Reprint of the original edition published by John Wiley & Sons, Inc., 2014, 978-1-118-88961-9

Printed on acid-free paper

This Springer imprint is published by Springer Nature
The registered company is Springer International Publishing AG
The registered company address is: Gewerbestrasse 11, 6330 Cham, Switzerland

TABLE OF CONTENTS
Celebrating the Megascale: Proceedings of the Extraction and Processing Division Symposium on Pyrometallurgy in Honor of David G.C. Robertson

Keynote Session

Ferro-Alloys

Non-Ferrous Smelting, Converting, and Refining

Iron and Steel Production

Process Modeling and Simulation

Metallurgical Education

Pyrometallurgy Process Fundamentals I

Pyrometallurgy Process Fundamentals II

Pyrometallurgy Process Fundamentals III

Poster Session

PREFACE

This volume contains the proceedings of the symposium, *Celebrating the Megascale: Proceedings of the Extraction and Processing Division Symposium on Pyrometallurgy in Honor of David G.C. Robertson*, held at the 2014 TMS Annual Meeting & Exhibition in San Diego, California. Dr. David G.C. Robertson, Professor Emeritus of Metallurgical Engineering at the Missouri University of Science and Technology, has devoted his career to the education of metallurgical professionals and to the science of all types and sizes of pyrometallurgical processes.

During the 2011 TMS Annual Meeting in San Diego, an informal gathering of colleagues and ex-students of Dr. Robertson began discussing the idea of holding a special symposium as a tribute to his significant contributions in pyrometallurgy. Support was strong, and Dr. Adrian Deneys played a key role in piloting this idea through TMS and identifying the venue, selected to be the TMS 2014 Annual Meeting. It was during these early discussions that Dr. Mark Schlesinger of Missouri University of Science and Technology coined the working title of the symposium: "Celebrating the Megascale". This was in recognition of Dr. Robertson's work across different fields that have contributed to the understanding of pyrometallurgical processes, in this way contributing to the successful scale-up of ferrous and non-ferrous metallurgical processes to the unprecedented levels seen today.

As discussed in one of the papers at the symposium[1], when David Robertson began his career in metallurgy (in the 1960s), the output of a well-running iron blast furnace was about 2,000 tonnes of pig iron/day, while the typical equivalent of the best copper reverberatory furnace was about 300 tonnes of Cu/day. Today, single furnaces run at up to 4 to 5 times these rates, that is, of the order of 10,000 tonnes of pig iron/day, and for copper, the equivalent of about 1,500 tonnes of Cu/day – truly the "mega-scale".

Dr. Robertson's own contribution to the development of realistic theories for describing the kinetics of processes has been particularly influential, and has assisted understanding to enable the scale-up of processes to what is referred to now as the "mega-scale". Through his work, Dr. Robertson has contributed to the understanding of many of the big developments in extractive metallurgy over the last half century; these include the large, modern blast furnace which remains a critical component of the world iron and steel industry, steel making technologies, the Noranda Process for copper smelting, the Outokumpu Flash Furnace, Top Submerged Lance technology and the large electric furnace for ferro-alloy production, to name a few.

With a symposium topic and venue identified, an organizing committee consisting of Kent Peaslee, Phillip Mackey, Eric Grimsey, and Rodney Jones was established. Sadly, Kent Peaslee passed away on May 17, 2013. We express condolences for the untimely

[1] P.J. Mackey, "Evolution of the Large Copper Smelter - 1800s to 2013", this symposium volume.

loss of our colleague, who was also a Past President of the Association for Iron & Steel Technology, a sister organization to TMS. Subsequently, the organizing committee invited Professor Geoff Brooks, AIST 2013 John F. Elliott Award recipient, to join the organizing committee.

The contribution of Dr. Brooks is gratefully acknowledged as is the guidance and help provided by Dr. Robertson throughout the planning process. The organizing committee developed an outline of topics both reflecting the interests of Dr. Robertson and which would also enable a review of the present state and practice of extractive metallurgy, covering both ferrous and non-ferrous metallurgy.

The outcome is the present symposium volume which contains almost 70 papers covering the important topics and issues in pyrometallurgy today including papers as follows: eight keynote papers including a tribute to Dr. Robertson, workforce skills needed in the profession going forward, copper smelting, ladle metallurgy, process metallurgy and resource efficiency, new flash iron making technology, ferro-alloy electric furnace smelting and on horizontal single belt casting for aerospace applications. Topics then covered in detail in this volume include ferro-alloys (seven papers), non-ferrous metallurgy (nine papers), iron and steel (eight papers), modeling (eight papers), education (eight papers, including papers related to sustainability in the metals industry, with particular reference to pyrometallurgy), and fundamentals (17 papers); there are also two papers as posters. We are indebted to all authors who have contributed to this volume.

The editors would like to express their appreciation to all the TMS staff for their help and dedication to this project. In particular, we would like to thank Louise Wallach, Trudi Dunlap, Patricia Warren, Joni Zychowski Minehart, and Diana Grady for their contributions.

Thanks are also given to all authors for their contributions and thanks to the session chairs who contributed to the reviewing of the manuscripts in their respective sessions.

Phillip J. Mackey
Eric J. Grimsey
Rodney T. Jones
Geoffrey A. Brooks

ABOUT THE EDITORS

LEAD EDITOR

Phillip J. Mackey is a consulting metallurgical engineer and specialist in non-ferrous metals with more than 40 years of international experience in all aspects of the non-ferrous and ferrous metals business. Dr. Mackey is originally from Australia where he received his Ph.D. in metallurgical engineering from the University of New South Wales studying under Professor N.A. Warner, one of the innovative leaders of his time. He carried out his graduate work at the same time as David Robertson. With his degrees, Dr. Mackey was armed to take on his first challenge at Noranda Mines in Canada. There he played a leading role in the development of the Noranda Process, the world's first commercial continuous copper smelting and converting process and one of the important copper technologies developed in the twentieth century. He was later responsible for the marketing of this technology to a number of other companies worldwide. His role in introducing the Noranda Converter, a new continuous converting process, was recognized by the Noranda Inc. Technology Award given in 1998. Active in the copper world, he co-founded the Copper/Cobre series of international conferences, with the first one held in Chile in 1987. He was also involved in a number of major nickel sulphide and nickel laterite projects around the world. Dr. Mackey worked in senior positions for many years with Xstrata (formerly Falconbridge/Noranda) before retiring at the end of 2009 to start his own consulting company. He presently acts in a consulting role for a number of Canadian and international mining and metallurgical companies. He has authored and co-authored more than 100 publications covering many aspects of nonferrous metallurgy. Dr. Mackey is a Metallurgical Society of CIM Past-President (1984–1985) and a fellow of both CIM and TMS. A recipient of several professional awards in Canada and the United States, he also received the 2007 TMS Distinguished Service Award, the Selwyn G. Blaylock Medal of the CIM in 2010, and the Airey Award of the Metallurgical Society of CIM in 2012.

Eric J. Grimsey is an extractive metallurgist who obtained a Ph.D. from the University of Queensland in 1975. He has extensive experience as a university educator, administrator, researcher, and industry consultant. His publications and expertise are mainly in the field of non-ferrous pyrometallurgical processes, encompassing heat and mass balances, thermodynamics, slag/metal interactions, and process modelling. He has also written extensive materials for the courses he has taught in pyrometallurgical fundamentals and pyrometallurgical processes.

Dr. Grimsey is a fellow of the Institution of Engineers Australia and the Australasian Institute of Mining and Metallurgy. He is a former Vice President of The AusIMM, a Director of Curtin University's Western Australian School of Mines (WASM), and a recipient of the AusIMM President's Award from his contribution to the institute and minerals education. He is currently an emeritus professor of minerals engineering with Curtin University and has a continuing involvement with AusIMM as chair of the institute's Ethics Committee.

Dr. Grimsey is pleased to have the opportunity to serve on the organizing committee to celebrate the work of Dr. David Robertson, who has been a friend and colleague since they met at the University of Missouri-Rolla in 1989.

Rodney T. Jones has worked in the Pyrometallurgy Division at Mintek since 1985. He holds a B.Sc. (Eng) degree in chemical engineering from the University of the Witwatersrand (Wits) in Johannesburg, a B.A. degree in logic and philosophy from the University of South Africa, and a M.Sc.(Eng) degree in metallurgy from Wits University. He is a registered Professional Engineer, a fellow and senior vice president of the Southern African Institute of Mining and Metallurgy (SAIMM), and a fellow of the South African Institute of Chemical Engineers (SAIChE). He was a visiting professor at the Center for Pyrometallurgy, University of Missouri-Rolla, during July and August 1996, and in 2002 and 2003 also lectured in pyrometallurgy at Murdoch University, Perth, as an adjunct associate professor. The National Research Foundation in South Africa rated him in 2009 as an "Internationally Acclaimed Researcher."

The SAIMM awarded him an Honorary Life Fellowship in 2010. Rodney is also a member of the Board of Trustees for OneMine.org.

Geoffrey A. Brooks has been a professor in the Faculty of Engineering and Industrial Sciences at Swinburne University of Technology since 2006, where he leads the High Temperature Processing research group. Previously, he was a senior principal research scientist at CSIRO (2004–2006), an associate professor in materials science and engineering at McMaster University (2000–2004), and a senior lecturer at the University of Wollongong (1993–2000). In the 20 years since completing his Ph.D. at University of Melbourne, he has published more than 140 papers and run many research projects with funding from companies and government agencies. Dr. Brooks is currently active in steelmaking, aluminum production and refining, and solar thermal research. He has been a key reader for *Metallurgical and Materials Transactions* since 1998 and is a fellow of the Institute of Engineers (Australia). In 2013, he received the John Elliott Lectureship from the Association for Iron & Steel Technology (AIST), acknowledging his contribution to research and training in steelmaking. Dr. Brooks has been a TMS member since 1990.

ORGANIZING COMMITTEE

Phillip J. Mackey (Chair)
P.J. Mackey Technology Inc.
Montreal, Canada

Eric J. Grimsey
Curtin University
Kalgoorlie, Western Australia

Rodney T. Jones
Mintek and SAIMM
Randburg, South Africa

Geoffrey A. Brooks
Swinburne University of Technology
Melbourne, Australia

The committee received guidance and help from Adrian Deneys of Praxair, Lloyd R. Nelson of Anglo American Platinum Limited, Quinn Reynolds of Mintek, and Dr. David Robertson.

TECHNICAL REVIEWERS

Nic Barcza

Ian Barker

Geoffrey A. Brooks

Hurman Eric

Eric J. Grimsey

Peter C. Hayes

Rodney T. Jones

Shin-ya Kitamura

Phillip J. Mackey

Michael S. Moats

Quinn Reynolds

M. Akbar Rhamdhani

David G.C. Robertson

Mark E. Schlesinger

Katie Schumacher

Merete Tangstad

SESSION CHAIRS

Keynote Session
Phillip J. Mackey, P.J. Mackey Technology Inc
Eric J. Grimsey, Curtin University

Ferro-Alloys
Rodney T. Jones, Mintek
Lloyd R. Nelson, Anglo American Platinum Limited

Non-Ferrous Smelting, Converting, and Refining
Michael S. Moats, Missouri University of Science and Technology
Katie Schumacher, Stillwater Mining Corporation

Iron and Steel Production
Roderick I.L. Guthrie, McGill University
Gordon A. Irons, McMaster University

Process Modeling and Simulation
Geoffrey A. Brooks, Swinburne University
Mark W. Kennedy, ProVal Partners SA

Metallurgical Education
Peter C. Hayes, University of Queensland
Merete Tangstad, Norwegian University of Science & Technology (NTNU)

Pyrometallurgy Process Fundamentals I
Mansoor Barati, University of Toronto
P. Chris Pistorius, Carnegie Mellon University

Pyrometallurgy Process Fundamentals II
Shin-ya Kitamura, Tohoku University
M. Akbar Rhamdhani, Swinburne University of Technology

Pyrometallurgy Process Fundamentals III
Ken S. Coley, McMaster University
In-Ho Jung, McGill University

CELEBRATING THE MEGASCALE

Proceedings of the Extraction and Processing Division Symposium on Pyrometallurgy in Honor of David G.C. Robertson

Keynote Session

Celebrating the Megascale: Proceedings of the Extraction and Processing Division
Symposium on Pyrometallurgy in Honor of David G.C. Robertson
Edited by: Phillip J. Mackey, Eric J. Grimsey, Rodney T. Jones, and Geoffrey A. Brooks
TMS (The Minerals, Metals & Materials Society), 2014

DAVID GORDON CAMPBELL ROBERTSON: A BIOGRAPHICAL SKETCH

J.B. See[1]

[1] Consultant, Millfield, NSW, 2325, Australia

Keywords: Biography, Education, Research, Pyrometallurgy

Emeritus Professor David Robertson of the Missouri University of Science and Technology was born in Dublin Ireland on 29 December 1941. His father was a merchant navy Captain who served during WWII and during David's early years his family lived in Dublin and Donegal where David went to the local elementary school. In 1954 he moved to London with his parents and attended Highgate School before commencing metallurgy at the Royal School of Mines, Imperial College, London in 1960.

David graduated in 1963 with a BSc(Eng) with first class Honors in Metallurgy having also been awarded the Imperial College Charles Salter Prize for excellence in metallurgy. The distinguished faculty at Imperial College included Professors Richardson, Bradshaw, Alcock, Hills and Dr Julian Szekely. Rod Guthrie, now of McGill University, was among David's fellow classmates. Upon graduation, on Professor Richardson's recommendation, David decided to study for his doctorate at the University of New South Wales (UNSW) in Sydney, Australia arriving in Sydney just before Christmas 1963.

David's thesis advisers at UNSW were Dr Alex Jenkins and Dr Noel Warner. David was a Teaching Fellow in 1964 and a Broken Hill Proprietary (BHP) Scholar from 1965 to 1968 and hence had extensive contact with then Director of BHP Research, Dr Robert G Ward, author of the textbook *The Physical Chemistry of Steelmaking*.

UNSW was then a major world centre for university research in pyrometallurgy. Alex Jenkins, and his students Bruce Harris and Les Baker, had perfected the use of levitation melting to study the reactions of metal drops in gases whilst Noel Warner, a chemical engineer, was a pioneer in the application of heat and mass transfer principles to pyrometallurgical processes. In addition David's fellow PhD students at UNSW included Dr Phillip Mackey, a co-editor of this volume and a well-known and respected industrial metallurgist in North America.

David's PhD thesis topic was "Heterogeneous Reactions between Liquid Metals and Gases," an investigation of the kinetics of the reaction of iron alloy drops in pure oxygen and a topic of great interest in steelmaking research even today. On graduating with his PhD from UNSW in 1968 David accepted an appointment as a Lecturer back at Imperial College. As David has since bemoaned; "So I left one of the world's most beautiful cities for cold grey London - that was before the days of global warming!" Professors Richardson, Bradshaw and Hills became very supportive colleagues together with Jim Jeffes who had joined the faculty in the interim. David also interacted with then graduate students the late Professor Keith Brimacombe and Dr Amit Chatterjee, now an international authority in the area of iron and steel. David also worked closely with Professor Paul Grieveson, who joined Imperial College in 1977.

David remained a faculty member at Imperial College from 1969 to 1985 after being promoted to Reader in 1982 and supervised 17 PhD theses and 6 other higher degree theses - M Phil and DIC - on a wide range of topics in pyrometallurgy. In 1985 he accepted the position of Professor of Metallurgical Engineering at the University of Missouri-Rolla (now Missouri S&T) and, from 1985 to 1996, was also the Director of the Generic Mineral Technology Center for Pyrometallurgy. He worked closely with the first Director, Dr. Art Morris, and was responsible for all aspects of running this national center for university-based research, funded through the United States Bureau of Mines and with a budget of about $1,000,000 per annum. This appointment allowed David to interact closely with pyrometallurgical researchers from all over the US when their projects were funded by the Center.

Dr Robertson organized two successful international conferences sponsored by the Center - in 1986 at Purdue University on "Gas-Solid Reactions in Pyrometallurgy", and in 1988 at the University of Utah on "Flash Reaction Processes."

When the Center for Pyrometallurgy closed in 1996, with the demise of the United States Bureau of Mines, David remained at the University as a Professor. He retired in 2008 and was awarded the title of Professor Emeritus.

In his role as Professor, David supervised 7 PhD theses and 10 MS theses. One of his most distinguished graduate students in this period was the late Dr. Kent Peaslee, initially an organiser of this conference, who passed away recently.

David has travelled frequently to be a thesis examiner, give lectures, teach short courses, take part in conferences, and visit companies as a consultant. In 1995, he spent a period of six months on development leave at Mintek in South Africa, working on a major project on fuming zinc from lead blast furnace slags. In 2006 he worked at BHP Billiton's Technical Marketing Group in Australia for six months and he was a Visiting Professor at Tohoku University for four months in 2008 and at IIT Kanpur for five months in 2010. He is currently a Visiting Professor at Chongqing University.

David is a Member of TMS and AIST. In 2008 he was the recipient of the EPD Distinguished Lecturer Award from TMS, and in 2010 he was the recipient of the J.F. Elliott Lecturer Award from AIST.

Professor Robertson has a wide variety of teaching, research and administrative experience over more than 40 years since receiving his doctorate in 1968. His teaching duties have covered courses from 'Introductory Materials for Engineers' to 'Transport Phenomena in Extractive Metallurgy'. His major emphasis has been in teaching the application of engineering principles to metallurgical processes for smelting, refining, and solidification.

His research work, embodied in over 100 technical publications and many presentations, has been concentrated in the same area, including research into the fluid mechanics of processes such as gas injection into melts, gas jet atomization of liquid metals, and the behavior of bubbles rising through melts. Dr Robertson has also worked on thermodynamic measurements in melts, gas-solid reactions, and mathematical modelling.

David married Ruth Docker from Goulburn, New South Wales, Australia, in London in 1970. They have two daughters, Anne and Vera, and four grandchildren.

Over a period of more than 40 years, encouraged and supported by his wife Ruth, Professor David Robertson has made a significant contribution to teaching, research and consulting in pyrometallurgy. He has worked and interacted with the best and brightest in his field and taught and inspired a generation of graduate and undergraduate students, many of whom hold senior positions in industry and elsewhere. David is distinguished by his incisive intellect and dedication but, even more importantly, by his integrity and commitment to the highest standards in his personal and professional life. This symposium is a fitting tribute to David and his achievements.

APPENDIX
Dr. David G.C. Robertson References

1. D. G. C. Robertson, "Heterogeneous Reactions Between Liquid Metals and Gases," Ph.D. Thesis, U.N.S.W., 1968.

2. A. E. Jenkins and D. G. C. Robertson, "Some Recent Studies in the Technique of Dispersed Phase Refining," Research in Chemical and Extraction Metallurgy, Aus. I.M.M., Melbourne, 1-5, 1967.

3. O. C. Roberts, D. G. C. Robertson, and A. E. Jenkins, "The Electromagnetic Levitation of Liquid Metal Sulfides and Their Reaction in Oxygen," Trans. A.I.M.E., 245, 2413-2420, 1969.

4. D. G. C. Robertson and A. E. Jenkins, "The Reaction of Liquid Iron and Its Alloys in Pure Oxygen," Heterogeneous Kinetics at Elevated Temperature, University of Pennsylvania, Philadelphia, 393-408, 1970.

5. D. G. C. Robertson, N. A. Warner, and A. E. Jenkins, "Basic Oxygen and Spray Steelmaking, Some Research on the Mechanism of the Reactions," Chemeca 70 Proceedings, Butterworth, London, 54-69, 1972.

6. D. G. C. Robertson, D. P. O'Shaugnessy, and N. A. Molloy, "The Mechanism of Sheath Formation by Plunging Jets," Chem. Eng. Sci., 28, 1635-1637, 1973.

7. D. G. C. Robertson and B. B. Staples, "Model Studies on Mass Transfer Across a Metal-Slag Interface Stirred by Bubbles," Process Engineering of Pyrometallurgy, I.M.M., London, 51-59, 1974.

8. S. Mizoguchi, D. G. C. Robertson, and A. V. Bradshaw, "Stream Degassing," Physical Chemistry of Process Metallurgy: The Richardson Conference, I.M.M., London, 135-142, 1974.

9. D. G. C. Robertson, "Oxidation of Copper and Nickel Drops by Carbon Dioxide," Trans. I.M.M., 85, C60-C62, 1976.

10. F. D. Richardson, D. G. C. Robertson, and B. B. Staples, "Mass Transfer Across Metal-Slag Interfaces Stirred by Bubbles," Physical Chemistry in Metallurgy: The Darken Conference, U.S. Steel, Pittsburgh, PA, 25-48, 1976.

11. N. H. El-Kaddah and D. G. C. Robertson, "Homogeneous Nucleation of Carbon Monoxide Bubbles in Iron Drops," J. Colloid and Int. Sci., 60, 349-360, 1977.

12. T. Deb Roy, N. H. El-Kaddah, and D. G. C. Robertson, "Mixed Transport Control in Liquid Metal Reactions," Metall. Trans. B, 8B, 271-277, 1977.

13. S. Mizoguchi, D. G. C. Robertson, and A. V. Bradshaw, "Bubble Growth During Stream Degassing," Second Australasian Conference of Heat and Mass Transfer, Sydney, Australia, 1977.

14. N. H. El-Kaddah and D. G. C. Robertson, "Equilibria in Reactions of CO and CO_2 with Dissolved Oxygen and Carbon in Liquid Iron," Metall. Trans. B, 8B, 569-579, 1978.

15. N. H. El-Kaddah and D. G. C. Robertson, "The Kinetics of Gas-Liquid Metal Reactions Involving Levitated Drops. Carburization and Decarburization of Molten Iron in CO-CO_2 Mixtures at High Pressures," Metall. Trans. B, 9B, 191-199, 1978.

16. T. Deb Roy and D. G. C. Robertson, "A Mathematical Model for Stainless Steelmaking. Part I. Argon-Oxygen-Steam Mixtures," Ironmaking and Steelmaking, 5, 198-206, 1978.

17. T. Deb Roy and D. G. C. Robertson, "A Mathematical Model for Stainless Steelmaking. Part II. Application to AOD Heats," Ironmaking and Steelmaking, 5, 206-210, 1978.

18. S. Mizoguchi, D. G. C. Robertson, and A. V. Bradshaw, "Nozzle Pressure During Stream Degassing," Trans. I.S.I. Japan, 18, 177-180, 1978.

19. T. Deb Roy, S. R. Chandrashekar, and D. G. C. Robertson, "Fuming of Stannous Oxide from Slags," Trans. I.M.M., 86, C225-C230, 1979.

20. D. S. Conochie and D. G. C. Robertson, "A Ternary Interfacial Energy Diagram," Trans. I.M.M., 89, C61-C64, 1980.

21. D. S. Conochie and D. G. C. Robertson, "The Behavior of the Third Phase Produced in Gas Bubble-Liquid Reactions," Trans. I.M.M., 89, C54-C60, 1980.

22. M. Nilmani and D. G. C. Robertson, "Model Studies of Gas Injection at High Flow Rates Using Water and Mercury," Trans. I.M.M., 89, C42-C53, 1980.

23. D. G. C. Robertson, D. S. Conochie, and A. H. Castillejos, "Model Studies on Gas-and-Solid Injection and Related Phenomena in Liquid Metal Baths," Scaninject II, Second International Conference on Injection Metallurgy, T. Lehner, Editor, MEFOS, Lulea, Sweden, 4:1-4:36, 1980.

24. D. G. C. Robertson, "Research in Chemical Process Metallurgy in the John Percy Group," Tisco Tech. J., January-October, 153-170, 1980.

25. M. Nilmani, T. T. Maxwell, D. G. C. Robertson, and D. B. Spalding, "Prediction of Initial Motion of a Gas Bubble in Liquids," Appl. Math. Modeling, 5, 24-28, 1981.

26. C. J. Hallet, A. J. Monhemius, and D. G. C. Robertson, "Oxygen Mass Transfer in Pachuca Tanks," Extraction Metallurgy 1981, I.M.M., London, 308-320, 1981.

27. J. Herbertson, D. G. C. Robertson, and A. V. Bradshaw, "Critical Experiments on the Role of Surface Tension Driven Flow in the Kinetics of Oxygen Transfer Between Gases and Molten Silver," Can. Met. Q., 22, 1-8, 1983.

28. L. Farias and D. G. C. Robertson, "Physical Modeling of Gas-Powder Injection into Liquid Metals," Third Process Technology Conference Proceedings, J. W. Hlinka, Editor, Iron and Steel Society, Pittsburgh, PA, 206-222, 1982.

29. L. Farias and D. G. C. Robertson, "Analysis of Pressure Traces in Submerged Injection Systems," Injection Phenomena in Extraction and Refining, A. E. Wraith, Editor, Univ. of Newcastle-upon-Tyne, E1-E25, 1982.

30. S. Ohguchi and D. G. C. Robertson, "Heat Transfer Calculations for Accretion Formation Around a Tuyere During Submerged Gas Injection into Molten Steel," Injection Phenomena in Extraction and Refining, A. E. Wraith, Editor, Univ. of Newcastle-upon-Tyne, J1-J37, 1982.

31. S. Ohguchi and D. G. C. Robertson, "Formation of Porous Accretions Around a Tuyere During Gas Injection," Ironmaking and Steelmaking, 10, No. 1, 15-23, 1983.

32. D. G. C. Robertson, S. Ohguchi, B. Deo, and A. Willis, "Theoretical and Laboratory Studies of Injection Phenomena," Scaninject III, Third International Conference on Refining of Iron and Steel by Powder Injection, G. Carlsson, Editor, MEFOS, Lulea, Sweden, 8:1-8:47, 1983.

33. D. G. C. Robertson, B. Deo, and S. Ohguchi, "A Multicomponent Mixed Transport Control Theory for the Kinetics of Coupled Slag-Metal and Slag-Metal-Gas Reactions: Application to Desulphurization of Molten Iron," Ironmaking and Steelmaking, 11, 41-55, 1984.

34. S. Ohguchi, D. G. C. Robertson, B. Deo, P. Grieveson, and J. H. E. Jeffes, "Simultaneous Dephosphorization and Desulphurization of Molten Pig Iron," Ironmaking and Steelmaking, 11, 202-213, 1984.

35. D. S. Conochie, C. A. Ebiogwu, and D. G. C. Robertson, "Oxygen Solubility in Molten Lead," Trans. I.M.M., 93, C45-C48, 1984.

36. S. Ohguchi and D. G. C. Robertson, "A Kinetic Model for Refining Using Submerged Injection," Ironmaking and Steelmaking, 11, 262-273, 1984.

37. S. Ohguchi and D. G. C. Robertson, "A Model of Bulk Phase Mixing and Its Application to the Kinetics of Powder Injection Refining," Ironmaking and Steelmaking, 11, 274-282, 1984.

38. T. Okazaki and D. G. C. Robertson, "Removal of Tramp Elements: Mathematical Modeling," Ironmaking and Steelmaking, 12, 295-298, 1985.

39. L. M. Keilman and D. G. C. Robertson, "Natural Convection in Gas-Solid Packed Beds," Gas-Solid Reactions in Pyrometallurgy, H. Y. Sohn and D. G. C. Robertson, Editors, Center for Pyrometallurgy, Rolla, MO, 187-194, 1986.

40. D. G. C. Robertson and H. Y. Sohn, "Gas-Solid Reactions in Pyrometallurgy," Journal of Metals, 38, No. 9, 33, 1986.

41. D. G. C. Robertson, "Computer Models in Materials Processing," Tenth International CODATA Conference on the Computer Handling and Dissemination of Data, P. Glaeser, Editor, North Holland, Amsterdam, Netherlands, 159-164, 1986.

42. A. Unal and D. G. C. Robertson, "Pilot Plant Gas Atomizer for Rapidly Solidified Metal Powders," International Journal of Rapid Solidification, 2, 219-229, 1986.

43. S. Taneka and D. G. C. Robertson, "Reaction Kinetics of Friable Chromites Between 1350°C and 1750°C," Pyrometallurgy 1987 Proceedings, I.M.M., London, England, 937-976, 1987.

44. S. Taneka and D. G. C. Robertson, "Carbothermal Reduction of Friable Chromite in a Small-Scale Transferred-Arc Furnace," Pyrometallurgy 1987 Proceedings, IMM, London, England, 977-1024, 1987.

45. S. T. Johansen, D. G. C. Robertson, K. Woje, and T. A. Engh, "Fluid Dynamics in Bubble Stirred Ladles: Part 1. Experiments," Metall. Trans. B, 19B, 745-754, 1988.

46. N. H. El-Kaddah and D. G. C. Robertson, "The Nucleation of CO Bubbles in Molten Iron-Carbon Drops Reacting with Oxidizing Gases," Metall. Trans. B, 19B, 831-837, 1988.

47. S. P. Beutner, R. D. Hagni, and D. G. C. Robertson, "Applied Mineralogy of Refinery Slags," Journal of Metals, 41, No. 4, 46-49, 1989.

48. D. G. C. Robertson, S. Wei, and Anand Raman, "Process Dynamics of Metal/Slag and Metal/Slag/Gas Reactions," Chemical Process Metallurgy: The Elliott Symposium, G. R. St. Pierre, and P. J. Koros, Editors, Iron and Steel Society, Warrendale, PA., 413-445, 1990.

49. A. K. Agrawal and D. G. C. Robertson, "Modeling of a Counter-Current Reactor for Metal Refining," 1991 Steelmaking Proceedings, Iron and Steel Society, Warrendale, PA, 845-851, 1991.

50. R. Immele-Meziou, O. A. Pringle, D. G. C. Robertson, and G. J. Long, "Mossbauer Effect Study of Reduction of Ilmenite," Trans. I.M.M., 101, C57-C59, 1992.

51. H. Y. Sohn, D. G. C. Robertson, A. K. Agrawal, and K. M. Iyer, "Fluid Flow and Mixing in Two-Phase Channel Reactors with Bottom Gas Injection," Savard-Lee International Symposium on Bath Smelting, J. K. Brimacombe et al., Editors, TMS, Warrendale, PA, 377-412, 1992.

52. K. D. Peaslee, D. K. Panda and D. G. C. Robertson, "Physical Modeling of Metal/Slag/Gas Interactions and Reactions in Steelmaking," 1993 Steelmaking Proceedings, Iron and Steel Society, Warrendale, PA., 637-644, 1993.

53. B. S. Terry, C. L. Harris and D. G. C. Robertson, "The Decoppering of Liquid Lead Bullion by Elemental Sulfur Additions. Part 1: The Sulphidation of Lead by Elemental Sulfur," Trans. I.M.M., 102, C57-C62, 1993.

54. B. S. Terry, C. L. Harris and D. G. C. Robertson, "The Decoppering of Liquid Lead Bullion by Elemental Sulfur Additions. Part 2: A Mechanism and Model for Decoppering," Trans. I.M.M., 102, C63-C69, 1993.

55. D. G. C. Robertson, A. E. Morris, L. R. Nelson and K. Narayana Swamy, "Process Modeling Software from the Center for Pyrometallurgy," Second International Symposium on Computer Software in Chemical and Extractive Metallurgy, C.W. Bale and G. A. Irons, Editors, CIM, Montreal, Canada, 127-142, 1993.

56. C. Vercruyssen, P. Wollants, J. Roos, D. G. C. Robertson, and L. Bertels, "The Mathematical Modeling of the Refining of Stainless Steel in the MRPA Converter," Ironmaking and Steelmaking, 21, (4), 287-296, 1994.

57. D. Panda and D. G. C. Robertson, "A Simple Apparatus for Producing Single Liquid Drops," Review of Scientific Instruments, 65, (7), 2388-2394, 1994.

58. K. D. Peaslee and D. G. C. Robertson, "Fluid Dynamics of Inclined Jetting on a Slag-Metal Bath," EPD Congress 1994, G. W. Warren, Editor, TMS, Warrendale, PA, 1129-1145, 1994.

59. K. D. Peaslee and D. G. C. Robertson, "Modeling of Fluid and Particle Flow to Study Metal Powder Production Through Atomization," Powder Processing Education for the Year 2000, K. Vedula, W. Frazier, G. Jankowski, Editors, TMS, Warrendale, PA, 111-124, 1994.

60. L. R. Nelson, D. G. C. Robertson, and K. Narayana Swamy, "Thermodynamic and Kinetic Simulation of a Novel Counter-Current Reaction Launder Process for the Production of Refined Low Carbon Ferromanganese," EPD Congress 1994, G.W. Warren, Editor, TMS, Warrendale, PA, 713-737, 1994

61. D. K. Panda and D. G. C. Robertson, "Model Studies of Drop Reactions with Gas Evolution," EPD Congress 1994, G. W. Warren, Editor, TMS, Warrendale, PA, 1109-1128, 1994.

62. K. D. Peaslee and D. G. C. Robertson, "Model Studies of Splash, Waves, and Recirculating Flows within Steelmaking Furnaces," 1994 Steelmaking Conference Proceedings, Iron and Steel Society, Warrendale, PA, 713-722, 1994.

63. L. R. Nelson, D. G. C. Robertson, and K. Narayana Swamy, "Kinetic Simulation of the Zinc Fuming of Lead Blast Furnace Slag in Center-Fed, Hollow Graphite Electrode, D.C. Plasma-Arc Furnaces," Metallurgical Processes for the Early Twenty-First Century, H. Y. Sohn, Editor, TMS, Warrendale, PA, 283-306, 1994.

64. L. R. Nelson and D. G. C. Robertson, "Modeling of a Counter-Current Reactor Launder (CCRL) Process for Metal Refining," 1994 Steelmaking Conference Proceedings, Iron and Steel Society, Warrendale, PA, 665-680, 1994.

65. D. K. Panda and D. G. C. Robertson, "Physical Modeling of Metal Drop-Slag Reactions with Gas Evolution," 1994 Steelmaking Conference Proceedings, Iron and Steel Society, Warrendale, PA, 443-456,1994.

66. L. R. Nelson, D. G. C. Robertson, and K. Narayana Swamy, "Kinetic Simulation of the Control of the Nitrogen Content of Low Carbon Ferromanganese in a Novel Counter-Current Reduction Launder Process," INFACON 7, International Ferroalloys Conference, Trondheim, Norway, 1995.

67. D. G. C. Robertson, "The Computation of the Kinetics of the Reactions Between Multiple Phases," EPD Congress 1995, G. W. Warren, Editor, TMS, Warrendale, PA, 347-361, 1995.

68. L. R. Nelson, D. G. C. Robertson, and K. Narayana Swamy, "Physical Modeling of Three-phase Mixing in a Counter-Current Reaction Launder (CCRL) Process for Metal Refining," EPD Congress 1995, G. W. Warren, Editor, TMS, Warrendale, PA, 321-346, 1995.

69. K. Narayana Swamy and D. G. C. Robertson, "Mathematical Modeling of the Kinetics of Reactions Between Multiple Phases: Applied to the de-S and de-P of Molten Iron," 13th Process Technology Conference Proceedings, Iron and Steel Society, Warrendale, PA, 357-362, 1995.

70. K. D. Peaslee and D. G. C. Robertson, "The Application of Computers to Study the Effects of Jetting During Steelmaking," 1995 Steelmaking Conference Proceedings, Iron and Steel Society, Warrendale, PA, 689-696, 1995.

71. N. A. Barcza, D. G. C. Robertson, A. F. S. Schoukens, F. Shaw, G. M. Denton, T. W. Worcester, and D. J. Bailey, "Enviroplas Technology for the Recovery of Lead and Zinc From Lead Blast Furnace Slags," 6th International Conference of the Lead and Zinc Study Group, Recycling Lead and Zinc into the 21st Century, Madrid, Spain, 1995.

72. K. N. Swamy, D. G. C. Robertson, and K. D. Peaslee, "The Effect of High Gas Flowrates on the Performance of a Counter-Current Reaction Launder (CCRL) Process for Metal Refining," EPD Congress 1996, G. W. Warren, Editor, TMS, Warrendale, PA, 95-108, 1996.

73. Hongjie Li, A. E. Morris, and D. G. C. Robertson, "The Development of a Multi-stage Model for a Ferromanganese Furnace and Simulation Results", 1996 Electric Furnace Conference Proceedings, Iron and Steel Society, Warrendale, PA, 227-235, 1996.

74. A. C. Deneys, D. G. C. Robertson, and N. Schupp, "A Two-Liquid Slag Mixing Technique for Studying the Oxidative Pre-fuming Treatment of Lead Blast Furnace Slag," EPD Congress 1997, B. Mishra, Editor, TMS, Warrendale, PA, 621-633, 1997.

75. H-J. Li, D. G. C. Robertson, and A. E. Morris, "Process Modeling by Using the METSIM Computer Program", 9th International Conference on High Temperature Materials Chemistry, Electrochemical Society Proceedings, Volume 97-39, K. H. Spear, Editor, Penn State University, 171-178, May 19-23, 1997.

76. A. C. Deneys, D. G. C. Robertson, R. T. Jones, and A. W. Worcester, "Recovery of Cobalt from a Viburnum Trend Lead Blast Furnace Slag", Nickel-Cobalt 97, Vol 2, Pyrometallurgy Fundamentals and Process Development, C.A. Levac and R. A. Berryman, Editors, CIM, Montreal, Canada, 23-37, 1997.

77. H-J. Li, and D. G. C. Robertson, "Effect of Moisture, Combined Water, and Volatile Elements on Fixed Carbon Requirement in a Ferro-manganese Smelting Furnace, 1997 Electric Furnace Conference Proceedings, Iron and Steel Society, Warrendale, PA, 269-279, 1997.

78. A. C. Deneys and D. G. C. Robertson, "Fluid Flow in a Laboratory-Scale DC Arc Furnace for Slag Cleaning", 1998 Electric Furnace Conference Proceedings, Iron and Steel Society, Warrendale, PA, 423-432, 1998.

79. H-J. Li, A. E. Morris and D. G. C. Robertson, "Thermodynamic Model for MnO-Containing Slags and Gas-Slag-Metal Equilibrium in Ferromanganese Smelting", Metall. Trans. B, Vol. 29B, 1181-1191, 1998.

80. A. C. Deneys and D. G. C. Robertson, "Flow Visualization and Temperature Measurements Close to the Arc Attachment Zone of a Laboratory Scale DC Arc Furnace for Slag Cleaning", Fluid Flow Phenomena in Metals Processing, N. El-Kaddah, Editor, The Metallurgical Society, Warrendale, PA, 127-137, 1999.

81. D. G. C. Robertson and S. Kang, "Model Studies of Heat Transfer and Flow in Slag-Cleaning Furnaces", Fluid Flow Phenomena in Metals Processing, N. El-Kaddah, Editor, The Metallurgical Society, Warrendale, PA, 157-168, 1999.

82. L. S. Todd, D. G. C. Robertson, and H-J. Li, "A Direct Reduction Zinc Process". EPD Congress 1999, B. Mishra, Editor, TMS, Warrendale, PA, 865-879, 1999.

83. K. N. Swamy, D. G. C. Robertson, P. Calvert, and D. S. Kozak, "Factors Effecting Carbon Consumption in the Production of High Carbon Ferromanganese", INFACON 9 Proceedings, The Ferroalloy Association, Washington DC, 293-301, 2001.

84. D. G. C. Robertson, "The Kinetics of Reactions in Steelmaking", Siderurgia 2001, Avances Recientes en la Produccion y Manufactura del Hierro y el Acero, A Conejo, Editor, Instituto Tecnologico de Morelia, Mexico, 1-1 to 1-12, 2001.

85. D. G. C. Robertson, "The Fuming of Zinc from Oxide Slags", Siderurgia 2001, Avances Recientes en la Produccion y Manufactura del Hierro y el Acero, A Conejo, Editor, Instituto Tecnologico de Morelia, Mexico, 11-1 to 11-10, 2001.

86. D. G. C. Robertson, D. K. Panda, and A. C. Deneys, "The Chemistry of Fuming Zinc from Oxide Slags using Coke", Yazawa Conference Proceedings, H. Y. Sohn and Florian Kongoli, Editors, TMS, Warrendale, PA, 2003, page 87.

87. J. J. Peter, K. D. Peaslee and D. G. C. Robertson, "Study of Current Steelmaking Practices to Evaluate the Viability of Continuous Steelmaking", AISTech 2004 Conference, Nashville, AIST, Warrendale, PA, 2004.

88. M. Velez, M. Karakus, W.L. Headrick, R.E. Moore, B. Varghese, R, Zoughi, S. Van Key, H. Fan, R.A. Rossow, D.G.C. Robertson and J. M. Almanza, "Sensor development for the glass industry", in *Advances in Fusion and Processing of Glass III; Rochester,* N.Y., Ceramic Transactions, 141, pp. 413-420, 2004

89. J. J. Peter, K. D. Peaslee and D. G. C. Robertson, "Review of Progress in Developing Continuous Steelmaking", AISTech Conference 2004, Nashville, AIST, Warrendale, PA, 2004. Also in Iron and Steel Technology, February 2005, 53-60.

90. J. J. Peter, K. D. Peaslee and D. G. C. Robertson, "Introduction of a Novel, Scrap-based, Fully Continuous Steelmaking Process", AISTech 2005 Conference, Charlotte, AIST, Warrendale, PA, 2005.

91. J. J. Peter, K. D. Peaslee and D. G. C. Robertson, "Experimental Study of Kinetic Processes during the Steel Treatment at Two LMF's", AISTech 2005 Conference, Charlotte, AIST, Warrendale, PA, 2005.

92. S. N. Lekakh, D. G. C. Robertson and C. R. Loper, "Thermochemistry of Nodulization and Inoculation of Irons", AFS Cast Iron Inoculation Conference, AFS, Schaumberg, IL, 2005.

93. G. Gupta, D. G. C. Robertson and M. E. Schlesinger, "Tellurium Thermodynamics in Austenitic Iron", Can. Metall. Quarterly, 44, 351-356, 2005.

94. S. N. Lekakh, D. G. C. Robertson and C. R. Loper, "Thermochemistry and Kinetics of Iron Melt Treatment", World Foundry Conference, Birmingham, England, 68/1-68/10, 2006.

95. J. J. Peter, K. D. Peaslee and D. G. C. Robertson, "Simulations of a New Continuous Steelmaking Process", AISTech 2006 Conference, Cleveland, AIST, Warrendale, PA, 2006.

96. S. Lekakh, D. G. C. Robertson, S. Rimoshevsky and V. Tribushevsky, "Recycling Nonferrous Oxides in Iron and Steel Melting", Sohn Symposium, F. Kongoli and R. Reddy, Editors, TMS, Warrendale, PA, 429–434, 2006.

97. G. S. Rao and D. G. C. Robertson, "A Fundamental Study of Oxygen-Melt Reactions in the AOD Process", Sohn Symposium, F. Kongoli and R. Reddy, Editors, TMS, Warrendale, PA, 325-342, 2006.

98. S. N. Lekakh, C. H. Rawlins, D. G. C. Robertson, V. L. Richards, and K. D. Peaslee, "Kinetics of Aqueous Leaching and Carbonization of Steelmaking Slag," Metall. and Materials Trans. B, 39B, 125-134, 2008.

99. S. N. Lekakh, D. G. C. Robertson, C. H. Rawlins, V. L. Richards, and K. D. Peaslee, "Two-Stage Reactor for Geological Sequestration of Carbon Dioxide Using Steelmaking Slag," Metall. and Materials Trans. B, 39B, 2008.

100. D. G. C. Robertson, "Smelting and Refining – Cheaper, Faster, Smoother", TMS Distinguished Lecturer Award paper, TMS Annual Meeting, New Orleans, 2008

101. D. G. C. Robertson, "The Computation of the Kinetics of the Reactions between Multiple Phases in Chemical and Metallurgical Processes," Advanced Steel Refining Technology Symposium, IMRAM, Tohoku University, Sendai, 1-15, 2008

102. D. G. C. Robertson, 'The Kinetics of CO Formation During Coupled Reactions Involved in the Refining of Steel – Understanding G(CO)," Bulletin of the Advanced Materials Processing Building, (SOSAIKEN IHO), IMRAM, Tohoku University, 64, No. 12, 45-57, 2008

103. S. N. Lekakh, D.G.C. Robertson, S. Rimoshevsky, V. Tribyshevsky and B. Tribyshevsky, "Dynamic processes for recycling of low-quality charge materials" Light Metals 2008, pp. 1103-1106, 2008

104. S. Kitamura, S. Saito, K. Utagawa, H. Shibata, and D. G. C. Robertson, "Mass Transfer of P_2O_5 between Liquid Slag and Solid Solution of $2CaO.SiO_2$ and $3CaO.P_2O_5$," ISIJ International, 49, No.12, pp. 1838-1844, 2009

105. D. G. C. Robertson, A. C. Ebiogwu, and D. S. Conochie, "The Behavior of Slags in Rising Bubbles in Liquid Metals, Using the Open Top Bubble Technique", International Conference on the Advances in Theory of Ironmaking and Steelmaking, ATIS 2009, Govind S. Gupta, Editor, IISc, Bangalore, 2009

106. S. N. Lekakh, D.G.C. Robertson, S. Rimoshevsky, V. Tribyshevsky, and N. Bestuzev, "Optimization of ductile iron treatment by computing the refining reactions in liquid irons". EPD Congress 2009 , pp. 583-590, 2009

107. V. Thapliyal, S. N. Lekakh, K. D. Peaslee and D. G. C. Robertson, "Novel modeling concept for vacuum tank degassing", AISTech 2012 Conference, Atlanta, GA, AIST, Warrendale, PA, pp.1143 – 1149, 2012.

108. L. Zhang, J.W. Gao, L.N.W. Damoah and D. G. C. Robertson, "Removal of iron from aluminum: A review", Mineral Processing and Extractive Metallurgy Review, 33, No 2, pp. 199-157, 2012.

109. A. J. Shukla, B. Deo, and D.G.C. Robertson, "Modeling of scrap dissolution in molten iron containing carbon for the case of heat transfer controlled process", Submitted to ISIJ International, May 2013.

110. S. N. Lekakh, D. G. C. Robertson, "Analysis of melt flow in metallurgical vessels of steelmaking processes", ISIJ International, 53, No 4, pp. 622-628, 2013.

111. D. G. C. Robertson, "The Kinetics of Metal-slag-gas Reactions: Developments and Applications of the Coupled Reaction Model", Ralph Lloyd Harris Memorial Symposium, 61-69, Conference of Metallurgists, Canadian Institute of Mining Metallurgy and Petroleum, Montreal, October 2013.

112. A. J. Shukla, B. Deo, and D.G.C. Robertson, "Role of Air Gap in the Scrap Dissolution Process", Met. Trans B, 44, No. 6, pp. 1398-1406, 2013.

113. A. J. Shukla, B. Deo, and D.G.C. Robertson, "Scrap Dissolution in Molten Iron Containing Carbon for the Case of Coupled Heat and Mass Transfer Control", Met. Trans B, 44, No. 6, pp. 1407-1427, 2013.

Celebrating the Megascale: Proceedings of the Extraction and Processing Division Symposium on Pyrometallurgy in Honor of David G.C. Robertson
Edited by: Phillip J. Mackey, Eric J. Grimsey, Rodney T. Jones, and Geoffrey A. Brooks
TMS (The Minerals, Metals & Materials Society), 2014

EVOLUTION OF THE LARGE COPPER SMELTER – 1800s TO 2013

P.J. Mackey[1]

[1]P.J. Mackey Technology Inc., Kirkland, QC, Canada
E Mail: pjmackey@hotmail.com

Keywords: Copper smelting, technology, Top Submerged Lance, ISASMELT™, flash smelting, bath smelting, Noranda Process, El Teniente Converter, Continuous converting, SKS, future trends.

Abstract

Over the course of Dr. Robertson's career, the ferrous and non-ferrous plants have seen enormous changes in technology and increases in plant capacity, essentially amounting to a "technological revolution". In iron and steel, the "mega" blast furnace of some 6,000 m^3 working volume is now standard (~10,000 tonnes (mt) of pig iron/day). Similar huge changes in process technology and plant size have occurred in the non-ferrous industry. As an example, the fuel-fired reverberatory furnace, once the mainstay of the copper industry, has disappeared - replaced by large capacity flash and bath smelting technologies. The energy consumption per unit mass of metal produced has also been reduced considerably. Our understanding of the thermodynamics and mechanisms of metallurgical reactions, a field to which Dr. Robertson has significantly contributed, has made great strides. This paper reviews these changes with particular reference to the copper smelting industry, providing also comments on expected future trends.

Introduction

When Dr.David Robertson and the writer began their careers in metallurgy (1960s), the best iron blast furnace produced about 2,000 mt/day of pig iron, while in copper smelting, a well-running reverberatory furnace produced the equivalent of about 300 mt/day of Cu, corresponding to a concentrate smelting rate for a single furnace of about 50 mt/hr. When the writer first worked at the then Noranda Mines Limited in the early 1970s, roasted concentrate (calcine) was delivered to the "hot charge" reverberatory furnaces from roasters in the adjacent building using a number of 8-tonne rail-mounted "calcine cars" driven by the calcine operator. The calcine car driver used a hand-operated brass bell (Figure 1) as both a safety alert and also to notify furnace feeders. The plant's reverberatory furnaces - amongst the largest and most efficient in the world at the time - each required about 170 such cars/day. While at University in the UK, Dr. Robertson visited a steel plant at Brymbo in Wales on a field trip in 1962, and he recalls the blast furnace being fed by a man with a wheelbarrow [30] and the whole town red with fume. The annual production of the Brymbo plant in 1960 was 143,000 long tons, some 430 mt/day.

Today, a modern blast furnace can produce 10,000 t/day of pig iron [1]. Large, computer-controlled BOFs can refine this pig iron in about 30 minutes, giving the 50 heats a day required to match the above blast furnace output. Copper concentrate, now handled by a computer controlled advisory system replacing the former motor-driven calcine car and brass bell, is today smelted at rates in the range of 165 to 175 mt/hr [2], over 3 times the rate of about 50 years ago[1]. David Robertson has indeed witnessed and participated in unprecedented developments and growth in the metallurgical industry, with plants now truly at the "mega scale" as celebrated at this conference in his honor.

Figure 1. Hand-operated brass bell used to signal arrival of calcine cars at the hot charge reverberatory furnaces as a safety alert and to advise furnace feeders at the former Noranda Horne smelter. The bell stands about 13 cm high.

It is not only in the recent decades, but throughout the long history of metallurgy, individual plant units such as the blast furnace, roasters, smelting furnaces, converters and the like have progressively become larger and larger in size. This trend has also typically been accompanied by an increase in unit productivity such as tonnes of metal per unit furnace volume, etc. This trend to larger units as noted is not new and has increased metal output per furnace – essentially leading to improved overall efficiency and lower operating costs since fixed costs are spread over a greater amount of metal output – in effect achieving the "economies of scale" plus operating fewer furnace units for a given plant output.

Beginnings of the Modern Era in Copper Smelting

The technical foundation of the modern era in copper smelting technology might be considered to have been laid down during the period of the 1930s to 1960s. During these three decades, the following new technology developments began for the smelting of filter cake or bone dry copper concentrate to a medium to high grade matte (and eliminating the need for roasting), producing an off-gas suitable for acid making [2, 3].

1. Initial testing of flash smelting – first in Canada and France - later first adopted commercially in Canada and Finland (1930s-1940s);

---

[1] Copper electrorefining does not have quite the long sweep of history as is the case with copper smelting. While electrorefining is not part of the present paper, it is of interest to note that significant plant size increases occurred soon after the first copper electrorefinery opened in 1869 [29]. Tanks became larger and anode size increased fairly quickly in size in the first 30 to 40 yrs (from about 12 kg), with the typical weight today being 400 kg. Increases in current density also occurred (from about 75 A/m^2), a typical value today is about 300 to 340 A/m^2, which is about the present practical limit with conventional tanks and present electrolyte flows. The key recent developments have been in the use of stainless steel cathodes, increased automation and mechanized handling.

2. Continuous smelting and converting testwork using a top lance system and stationary furnace in Czechoslovakia (present day Slovakia), 1960s;
3. Noranda and Mitsubishi bath smelting in Canada and Japan, later El Teniente in Chile (1960s-early 1970s);
4. Top Submerged Lance tests began in Australia (1960s);
5. The use of tonnage oxygen in smelting and converting was tested, along with improvements in the Peirce-Smith converter at a number of plants around the world; other process developments such as Vanyukov in Russia (1960s) began.

Several factors contributed to these beginnings: (i) The desire to simplify the copper smelting process; (ii) Finding effective ways to utilize the fuel value of the sulphide concentrate; and (iii) Energy usage issues and environmental concerns (principally SO_2 gas emissions).

The above developments together would usher in a new set of technologies, leading to the demise of the reverberatory furnace which had been the mainstay of the copper industry for centuries. The smelting rate in the first commercial applications of the new technologies in the period 1950s to 1970s was modest (~20 to 30 mt/hr) - in fact, far less than the smelting rate of a single reverberatory furnace of the era. A later wave of developments and the widespread use of tonnage oxygen allowed large increases in capacity at the new plants, leading to the "mega scale" plants seen today. The technologies were subsequently classified as bath or flash smelting depending on the nature of the new smelting mechanisms involved.

This paper puts these developments into perspective. The paper addresses primary copper smelters (as opposed to secondary or scrap smelters), and smelting rate, generally given as tonnes/hour (mt/hr) of concentrate, is used to discuss furnace capacity (or size) in the present paper (in some cases, the estimated corresponding copper output is also given). Originally this paper was planned to examine changes since about 1950, but it was soon realized that change is continuous, and that important changes had taken place before that time, hence the scope of the paper was widened. The actual smelting mechanism in the intensive processes used today is quite different to that in smelting say 50 years ago. In the past, big capacity smelters employed a large number of furnaces to provide for total plant capacity; there are multiple furnaces at some of the larger plants today, but typically it may be only two large, modern smelting units. While large capacity plants have an economy of scale with attendant lower unit costs, cost issues are not covered in the present paper. Some thoughts on future directions are also discussed in the paper.

The Large Copper Smelter of Yesteryear

The trend to large metallurgical plants as noted is not new. An early example from the last century illustrates similar changes were occurring. M. Eissler [4] discussed the trend to the large "modern" smelter of 1903 - exactly 110 years ago - in Chapter 2, "The Production of Copper from the Ore" of his text," Hydrometallurgy of Copper" as follows:

> *There is a striking difference in a modern copper-smelting plant, when compared with those which were in operation 20 or 30 years ago. A 30 to 50 tons cupola was then considered a large furnace, whereas, if the notable copper mines of to-day are passed in review, it will be found that the up-to-date water jackets are smelting, year-in and year-out, from 200 to 250 tons and even 300 tons daily. The same comparison applies to reverberatory smelting furnaces, whose ancestors can be seen in Swansea, where tradition and custom have adhered*

to the original size of 9 feet by 15 feet internal dimensions, whereas a modern reverberatory furnace has a laboratory of 16 feet by 35 feet smelting 50 tons of ore daily; and even these are small when compared with those lately erected at the Boston and Montana Company's works, where regenerating reverberatory furnace are smelting from 125 to 150 tons of roasted ore into matte daily.

It is not only in the direction of capacity and size that advances have been made, but also in the simplification of the processes of extraction. Let any student on the subject take up old textbooks and compare the many and varied manipulations through which the ores, matte and the great nomenclature of intermediate products had to pass before fine copper was obtained, with the simplicity of manipulations of the modern metallurgist, obtaining the same results in two or three operations. These resolve themselves, practically, into (i) converting the raw or unroasted ores into a liquid condition – by smelting; and (ii) the conversion of the resulting matte, by bessemerising, into metallic copper.

The most notable advance in smelting has been the utilization of the sulphur, iron, and other oxidisable substances in the ores as generators of heat, so that they practically smelt themselves without the aid of extraneous heat; the application of this principle being known as "pyritic smelting".

The reverberatory furnace discussed above by Eissler [4] advanced considerably over the next 60 years, becoming the mainstay of copper smelting worldwide (handling filter cake concentrate or calcine). The early reverberatory furnaces were designed with a fire grate burning lump coal introduced by hand, refer Figure 2 [5]. The operation of these furnaces required hard manual labour particularly for handling coal and the furnace charge, as well as for matte and slag tapping. Gradually, as larger and larger furnaces were built (equipped with natural gas, fuel oil or pulverized coal firing), mechanized means of feeding and tapping were also introduced – still, manual labour was always part of reverberatory furnace operations. Smelting occurred on the charge banks heated by the flame.

The trend in the increase in reverberatory furnace size (using hearth area as the size parameter) from the early 1800s to about 1960 (after which little change in size occurred) is shown in Figure 3 [6]. Of interest is the small change in furnace size over the initial 80 year period – no doubt reflecting Swansea preference [4] for maintaining the "original size" of 9 ft by 15 ft (2.74 m by 4.57 m) with an L/W ratio of 1.7 and a hearth area of 12.5 m^2, compared as Eissler [4] suggested above, to the newer, larger furnaces in the US. The huge increase in furnace size from about 1890 to say 1910 coincided not surprisingly with the large increase in copper demand as Europe, the US and elsewhere truly entered the "electrical age".

Not only were larger reverberatory furnaces built with the capability for much higher smelting rates, but improvements in firing efficiency, feeding techniques and later, adoption of process control features led to significant improvements in fuel efficiency (fuel being a major cost item) and higher specific smelting rates. Expressed as mt/hr m^2 hearth area, specific smelting rate increased from about 0.03 mt/hr m^2 around the mid-1800s, to about 0.06 mt/hr m^2 in ~1900, and then to about 0.22 mt/hr m^2 as achieved at the Noranda reverberatory furnace discussed below. Figure 3 also shows the decrease in furnace fuel consumption over the same time period.

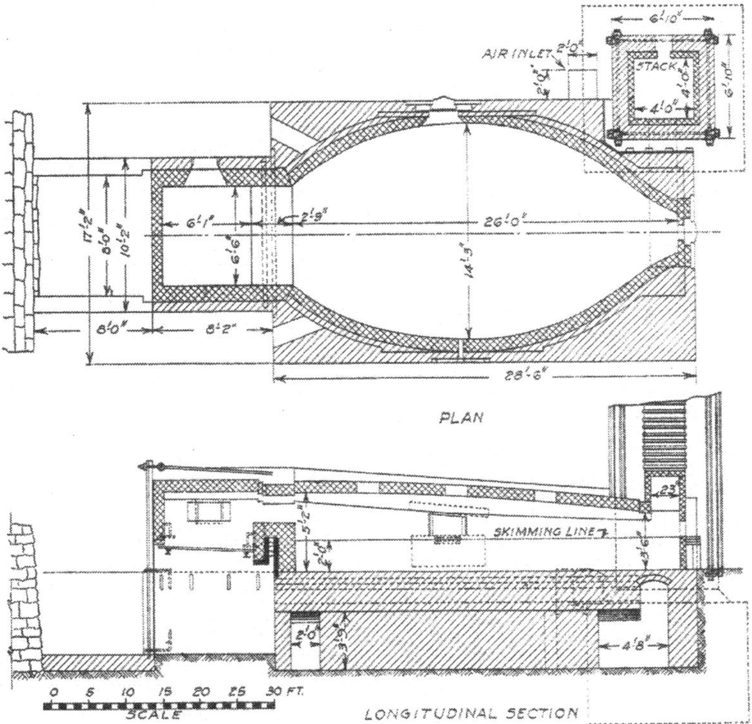

Figure 2. Plan and section of the reverberatory furnace at the Anaconda smelter, Butte, 1893 (Taken from Ref. [5]). This furnace smelted about 30 mt/day, or a little over 1 mt/hr. Multiple furnaces such as this were typically employed at the copper smelters of the day to provide for sufficient plant capacity.

The former Noranda smelter in Canada (now the Glencore Horne smelter) achieved record smelting rates for a single reverberatory furnace in the 1940s [7, 8]. The plant included 10-multi-hearth roasters, 2-large "hot charge" or calcine charged, coal-fired reverberatory furnaces, 5-Peirce-Smith converters and 1-reverberatory type anode furnace plus anode casting facilities (a total of 18 furnace units).

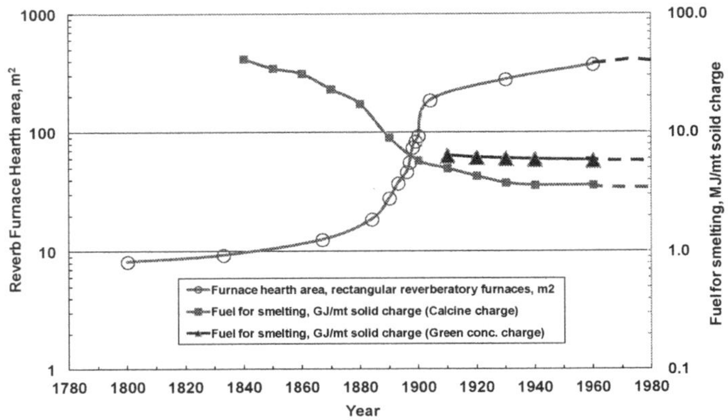

Figure 3. Trend in size of copper reverberatory furnaces, 1800 to 1960 (Based in part on Ref. [6]).

During the period of high tonnage operation (June 1944 to July 1949), only one furnace was in operation on account of a shortage of miners at the adjacent Horne mine [7, 8]. Likely an all-time record for a single copper reverberatory furnace, the No. 1 furnace at Noranda smelted a record 1,960 mt/day of solid charge or 82 mt/hr (2,160 short tons/day) in April 1946 (average daily rate). The coal firing rate corresponded to 10.4 t coal/t solid charge, or for the coal then in use, a fuel ratio of 3.1GJ/mt of solid charge (2.7 million BTU/short ton of solid charge).

More than 30 years would elapse before the above record smelting rate would be surpassed in one of the new bath and flash smelting technologies as discussed below. As indicated in Figure 3, furnace size had started to plateau by about the 1960s. A plan and section of the No. 2 reverberatory furnace at Noranda (essentially identical to the No. 1 furnace mentioned above), which was close to the largest size indicated in Figure 3, is shown in Figure 4 [8]. The calcine car mentioned above can be seen near the firing end of the furnace shown in longitudinal section; calcine was delivered along each side for almost the entire length of the furnace. Factors contributing to the successful increase in reverberatory smelting rate included: (i) Burner design and allowance for sufficient combustion space needed for coal consumption (coal was burnt at the rate of about 8 mt/hr - the use of oxygen enrichment to improve combustion efficiency only came later) – the charge banks formed by calcine additions along each side of the furnace melted to form the bath, (ii) Control of hearth build-up, and, (iii) The ability to handle the required slag and matte tapping rates with the existing design. It is considered however, that the plateau indicated in Figure 3 was likely close to the maximum practical size with existing designs.

It is noted that it was common for large smelters of the day to employ multiple furnaces to provide for sufficient plant capacity (for example, refer [9], where 5 reverberatory furnaces were used). Thus, by 1957, a new furnace ("green" or filter-cake concentrate charge furnace) had been built at Noranda; the 3 reverberatory furnaces then treated about 840,000 mt of solid charge/yr, producing some 132,000 mt/yr of anode copper [8]. The world's largest smelter at the time was Chuquicamata in Chile. In the mid- to late 1970s, this plant (an early mega-scale plant) was producing of the order of 390,000 mt/yr of new copper from the smelting of over 960,000 mt/yr of partly dried local concentrate (41% Cu) in 4 oxy-fuel fired reverberatory furnaces. The concentrate smelting rate per furnace was about 34 mt/hr. Chuquicamata operated 6 Peirce-Smith converters, with 6 anode refining furnaces and 3 casting wheels (18 furnace units including two dryers). Chuquicamata remains a large plant today. Not far behind was the Rhokana smelter in Zambia (later the Nkana smelter and which was subsequently closed in the early 2000s). Equipped with 5 large (green charge) reverberatory furnaces (also using some level of oxygen enrichment in the combustion air), 6 Peirce-Smith converters and 3 anode furnaces (a total of 14 furnace units), the plant produced 299,945 mt of copper in the 12 months from mid-1969 to mid-1970 from 1,118,384 mt of furnace charge [9]. It is noted that the actual smelting rate per furnace given as mt/hr was not that particularly high - about 26-28 mt/hr of charge - relative to the above noted record achieved at Noranda (albeit on calcine feed).

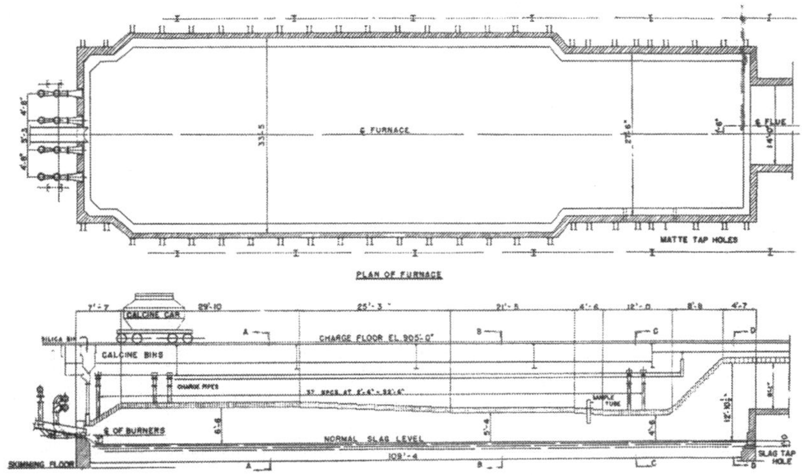

Figure 4. Plan and longitudinal section of the No. 2 coal-fired reverberatory furnace at the former Noranda smelter in the 1950s [8]. The L/W ratio of this furnace was about 3.4. In April 1946, the No. 1 reverberatory furnace at this plant (identical to the furnace shown here) set a record for the smelting rate for a single furnace (1,960 mt/day of charge or 82 mt/hr).

In the event, soon the new flash and bath technologies would be optimized and these plants expanded, bringing copper smelters to unprecedented capacity levels as will be described.

Including both the new technologies and the old, Figure 5 shows the average trend in hourly feed smelting rate over the period from about 1860 up to the present day (as noted, covering a range of feed types and different technologies)[2].

The steady rise in smelting rate in the early period shown on the graph is evident. The increase in furnace smelting rate from the 1900s to the 1940s (an increase by a factor of about 10) was due in part to productivity improvements and also due to the above noted increase in furnace size (furnace size had increased by a factor of about 4 over this period).

In virtually all fields of metallurgy, unit furnace or unit equipment size has generally increased over time. The following examples are given to illustrate that the above trends in copper technology are generally consistent with developments in other metallurgical areas. The trend in iron blast furnace size and blast furnace productivity in Japan from the 1950s to the early 2000s is illustrated in Figure 6 [1], while Figure 7 presents blast furnace output as tonnes of pig iron/day [1]. The trend in Figure 7 from the 1950s to the mid-1970s corresponded to similar trends worldwide [10]. It is seen here that by about the mid-1970s, pig iron output per furnace had reached about 10,000 mt/day, and this figure seems to have reached a plateau. This in part reflects physical limitations of the maximum penetration of the blast air in the furnace; a "dead zone" could occur with a further increase in furnace diameter.

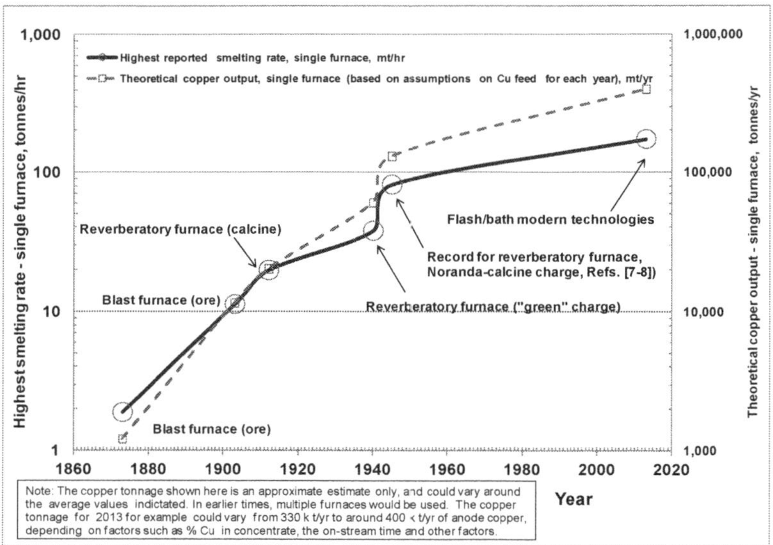

Figure 5. Highest reported smelting rate of copper feed over time, mt/hr.

[2] The data up to about 1930 are considered approximate. It is also noted that a smooth curve through the available data points may represent somewhat of a simplification. The sharp increase from 1940 to 1946 reflects the high tonnages achieved at Noranda as discussed above.

Figure 8 shows the trend in the size of fluid bed roasters (given as bed area in m^2) treating sulphide feed materials [11]. There was almost a 10,000 times increase in bed area (~100 times increase in diameter) from the original pilot unit first used in 1945 to the commercial units designed and built in the early 1950s. Thereafter it is seen there was a general plateauing of the bed size.

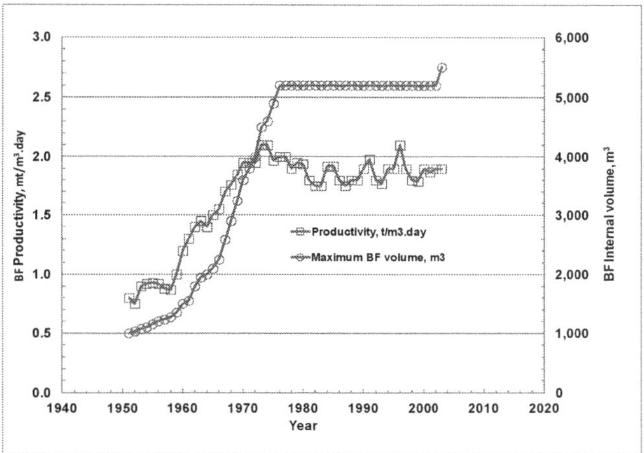

Figure 6. Increase in iron blast furnace volume and furnace productivity in Japan, 1950 to 2010 [1].

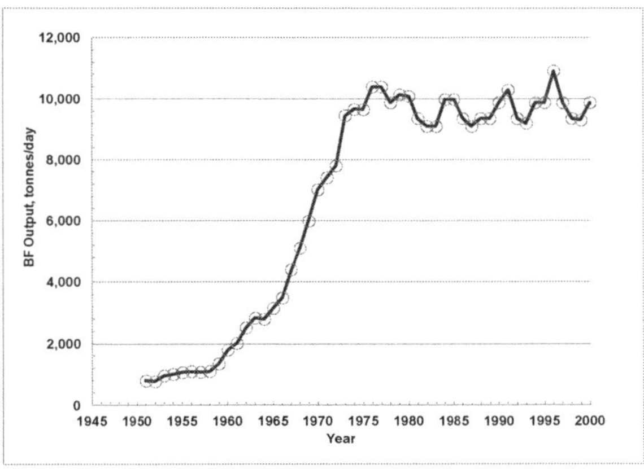

Figure 7. Trend showing the increase in pig iron output per blast furnace at blast furnaces in Japan [1].

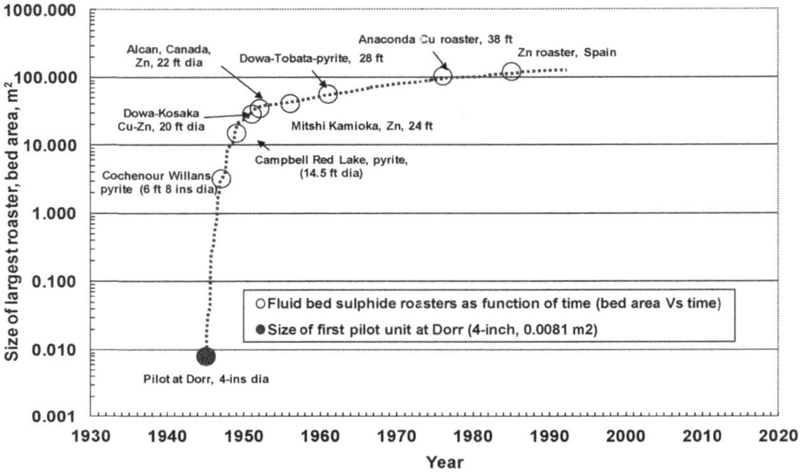

Figure 8. Trend in the increase in size of fluid bed roasters (shown as bed area) treating sulphide feed [11].

The Large Copper Smelter Today and Beyond

World copper smelter production reached a record 15.8 million mt of copper in 2011 (Figure 9 [12]). World demand for copper is increasing and by 2025, based on assumptions for world growth, it is expected to approach 30 million mt/yr (smelted copper would be approximately 24 million mt/yr). Much of this copper as concentrate will come from a number of recently announced copper projects in Chile and Peru. Certainly more smelters and refineries will be required to transform this material into copper cathode.

Copper smelting technology in use today can be broken into three types as shown in Table I, which also gives the proportion of world copper produced by each type (2010-2011 data).

Table I – Technology type and proportion of world smelted copper [2]

No.	Technology type	Proportion of world smelted copper, %
1	Flash smelting	49
2	Bath smelting	40
3	Other (reverberatory, electric furnace, blast furnace, etc)	11

Bath smelting includes: Noranda, El Teniente, TSL including ISASMELT™ and Outotec/Ausmelt, Mitsubishi and Vanyukov technologies. Production by the new SKS process in China [2] would also be included in this category

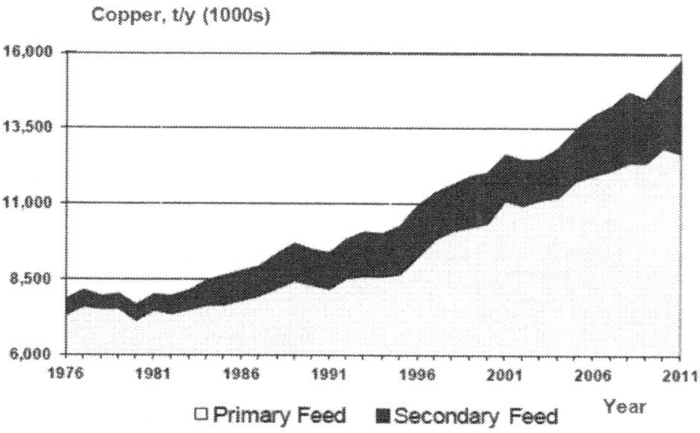

Fig 9. World smelter production, 1976 to 2011, units are 1000s mt/yr [12].

The two largest copper smelters today are Guixi in China (740,000 mt/yr of Cu) and Onsan, Republic of Korea (520,000 mt/yr of Cu); these plants each operate with two smelting furnaces. There are about 10 smelters with capacities between 350,000 and 450,000 mt/yr of Cu and some of these plants also operate with two smelting units.

The present large bath and flash smelting furnaces are now reaching concentrate smelting rates in the range of 165 to 210 mt/hr for a single furnace. Depending on the copper grade of the concentrate and other factors, this corresponds to a copper output in the range 330,000 to 400,000 mt/yr of Cu[3] - certainly approaching the "mega scale" (single furnace). It is understood that larger plants are under commissioning or are on the drawing board. It is of interest to note that reaching these high smelting rates with the new technologies occurred over a period of time - in this way, it allowed plant components to be safely and effectively adapted to higher capacities – the feed system, smelting furnace itself, off-gas cooling and cleaning, acid plant along with converter and anode departments. Not only were the original flash and bath smelting units thoroughly piloted [2], but the capacity of the first commercial plants were modestly scaled-up based on the pilot plant capacity - typically by a factor of about 10 based on feed rate. Thereafter, steady tonnage increases were generally incremental, the driver for this typically being market conditions and copper demand at the time, as well as the constant need to capture the economy of scale. Importantly, technical development work conducted by the technology developers, operating plants and a number of institutes and universities contributed significantly to the capacity increases. David Robertson's work across many fields was an example of these supporting activities.

Figure 10 [2] illustrates the tonnage increase from 1973 to 1988 for the Noranda Process reactor at the Glencore smelter in Canada (formerly the Noranda Horne smelter). This shows about a 3 times increase over these 15 years (as indicated, tonnage increases continued after 1988). A similar trend was observed for the Mt. Isa ISASMELT™ plant in Australia [2] as shown in

[3] The annual copper output from concentrate depends on the concentrate grade, the number of days per year, the on-stream time and the copper recovery. A number of assumptions were made to develop these estimates of copper output.

Figure 11 for the period 1992 to 2000. Since that time, the technology elsewhere has achieved higher smelting rates.

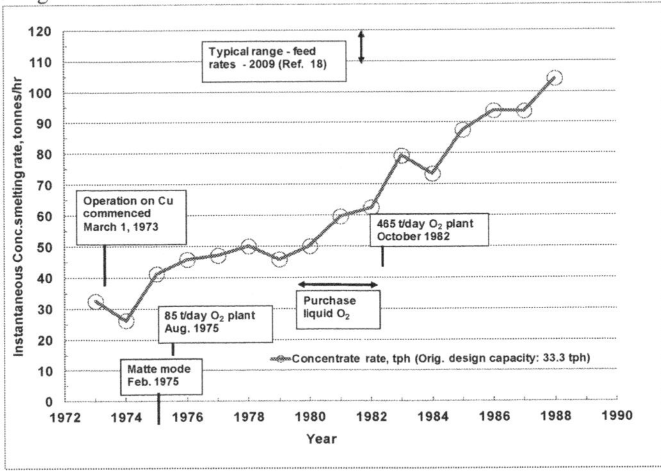

Figure 10. The increase in the instantaneous smelting rate for the Noranda Process - 1973 to 1988 [2].

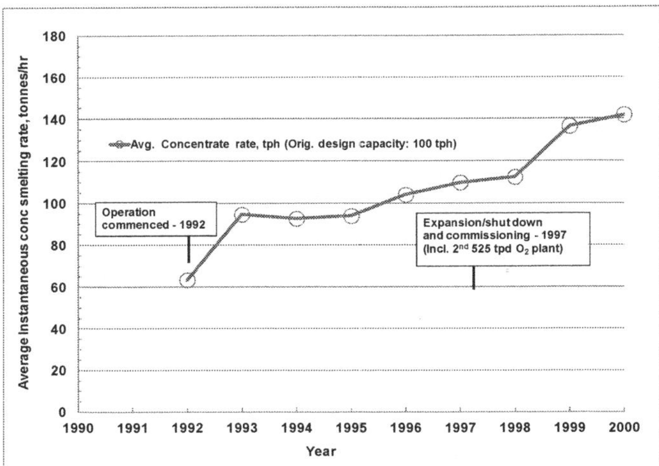

Figure 11. The increase in the instantaneous smelting rate for the Mt. Isa ISASMELT™ plant in Australia, 1992 to 2000 [2].

The trend in the average highest smelting rate over time in installed Outotec flash furnaces (formerly Outokumpu flash furnace) by year is shown in Figure 12. That is, the data in Figure 12 present the highest nominal smelting rate of the installed plants for the years indicated. In the beginning, the smelting rate is seen to be quite low – around 20 to 30 mt/hr of concentrate – and of a similar range to the initial tonnages for the Noranda Process shown in Figure 10, and certainly less than the best smelting rate in the then dominant technology – the fuel-fired reverberatory furnace (Figure 5). Flash furnace smelting rates now exceed about 210 mt/hr, approximating to a specific smelting rate of about 5.3 mt/hr m^2 of shaft area. The sharp recent increase in smelting rate in Figure 12 reflects the Outotec plant installed by Guangxi Jinchuan Non-Ferrous Metals Co.,Ltd and built at the Fangchenggang port area near Bei Hai City in Guanxi Zhuang autonomous region in China. The plant has a design capacity of the order of 400,000 t/yr of Cu, requiring concentrate smelting rates of over 175 mt/hr, possibly up to 250 mt/hr (with lower grade copper concentrate) [18]. It is expected that new plants based on the new SKS process [2] will be built, however initially the capacity of these plants will likely be less than about 200,000 to 250,000 mt/yr of Cu.

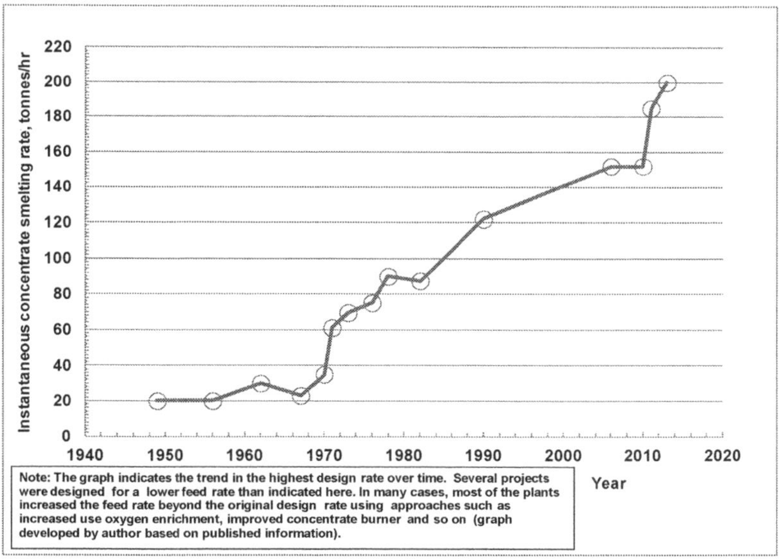

Figure 12. Estimated trend in the average highest smelting rate in the installed Outotec flash furnaces over time (1949 to the present day).

Thus, the gradual change to ever larger capacity plants permitted the new plants (covering all technologies) to be designed and built using the proven experience of earlier plants. This, combined with a considerable body of excellent technical work, ensured in most cases that a good start-up profile would be achieved (for example, refer [13]), and typically approaching a

"Type 1" or a "Type 2" start-up[4] [14]. Interestingly, this notable start-up record of new, large copper smelters, benefiting from the successful technology history, contrasts with the situation observed for example in nickel laterite plants.

In the nickel laterite case, several recent large (or even, "mega" scale) projects using new technology involving both pyrometallurgy and hydrometallurgy, combined with high scale-up factors (on feed rate) - in some cases over 700 to 800 times that of the pilot testing - are presently in the start-up phase. These plants now appear [15, 16] to be showing a "Type 4" [14] start-up performance, which amongst other factors, might be linked in part to the large scale-up factors involved. It is of interest to note that this situation in nickel laterite metallurgy today is reminiscent of the outcome of plants referred to as the "Ultra large single stream chemical plants" in post-war UK in the 1950s and 1960s [17]. Holroyd [17], in a Presidential Address to the Society of Chemical Industry, showed that what could have been a robust 915 mt/day ammonia project with a 1-year start-up profile, nevertheless, became a poor performing project for the owners with the 3-year start-up that occurred. As discussed by Mackey and Nessett [14], Holroyd's analysis of the causes of the slow start-up of these "Ultra large" plants in general remains valid today (faults in design due to technology stretch (= large size), faulty equipment/erection and/or materials of construction, and a component due to operator related error). It is thus noteworthy that the relatively slower capacity increase trend for copper smelters in effect allowed effective start-up of the later so-called mega-scale copper smelters.

Copper Converting

While important tonnages of matte are now converted by the new, flash or bath continuous converters, at the present time the Peirce-Smith converter remains the dominant technology for copper matte converting. As discussed below, the trend to even larger "mega scale" smelting plants could change this situation, in that it may be found that continuous converting could have the edge in large part on account that only a single or perhaps two continuous converter units may be required, whereas for a large plant (>400,000 mt/yr Cu), 4 large Peirce-Smith converters would be required (3 hot, with 1 on repair, and 2 blowing at a time).

Converter size (using vessel internal volume as the parameter) increased greatly over the first 40 years of pneumatic converting, Figure 13 [19] (by a factor of nearly 50 times). The so-called "13 by 30" Peirce-Smith converter (13 ft dia. by 30 ft or 3.96 m dia. by 9.14 m) introduced in the US in the late 1920s, soon became the "standard size" and remained so for some 50 years until larger units were eventually introduced (to increase copper output, multiple converters were installed). Since the 1920s, vessel size has more than doubled, Figure 13 [19].

Today, the typical large size Peirce-Smith converter is of the order of 4.2 m to 4.5 m dia. by 12.2 m to 13.4 m long, with the capability to operate at a blowing rate of over 55,000 Nm^3/hr, corresponding to a matte conversion rate of over 80 to 100 mt/hr (depending on the actual matte grade and the levels of oxygen enrichment).

As an example, the Ilo smelter in Peru was designed to treat some 1,200,000 mt/yr of concentrate in the ISASMELT™ plant installed in 2007 (smelting rate approximately165 t/hr conc.), [20-21].

[4] The reference to "Type 1" etc start-up refers to the four plant start-up types as classified by McNulty and discussed further in [14]. A Type 1 start-up is considered a good start-up (up to about one year to reach full production), while Type 4 would be a poor start up.

There are 4 Peirce-Smith converters (somewhat smaller than the above noted sizes - 3.96 m dia. by 10.67 m long - 13 ft by 35 ft), with 3 hot (and 1 on repair) and 2 blowing at any one time to handle the 62.5% Cu matte. The design copper output was 328,000 mt/yr of anode copper. It is estimated that a similar configuration (4 converters), but with vessels sized 4.5 m dia. by 13.4 m long would handle well over 400,000 mt/yr of Cu output, Table II.

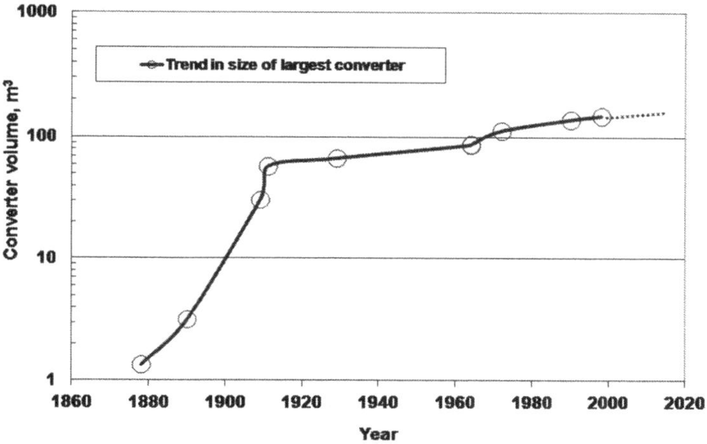

Figure 13. Growth in Peirce-Smith converter size (vessel internal volume), 1880 to the present day [19].

The newest "double flash" smelter in China (one flash smelting furnace and one Kennecott-Outotec flash converter) was designed to produce 400,000 mt/yr of Cu [18]. It may be that as the capacity of the "mega-plants" increases further, continuous converting, which can require fewer vessels and could generate a lower off-gas volume, may have an edge over the modern version of the workhorse of the copper industry - the Peirce-Smith converter. Nevertheless, the jury is considered still out on this issue, since the great flexibility and simplicity of the Peirce-Smith converter in the ability to handle scrap copper (old and new) coupled with modern design for hooding, and likely cost, still makes a compelling case for this converting technology even for large projects.

Some Factors Affecting Further Increases in Smelting Rate/Smelter Capacity

The key equipment options considered for a mega smelter (>400,000 mt/yr of Cu in one smelting furnace) are summarized in Table II.

Modern bath or flash furnaces can equally handle the smelting of copper concentrate to produce of the order of 400,000 mt/yr of Cu; modern converters can also handle the resulting converting load. In looking at capacities significantly beyond this, some factors as regards the smelting unit itself would include the following.

i. Vessel Cooling

High smelting rates (say >185 to 200 mt/hr) involve more intense smelting with tonnage oxygen and as such will tend to require some additional degree of heat removal from the smelting zone. It is recalled that in the 1970s to 1990s, flash furnaces were installed with a falling water film for cooling the exterior shell of the smelting shaft [23-24]; this type of shell cooling was once effectively used on the Noranda Process reactor [25] on the area opposite the tuyeres during a period requiring the use of a large natural gas burner (~85 to 105 GJ/hr - before large scale use of tonnage oxygen).

As shown in Figure 14, this type of cooling increases the wall heat flux over that achieved by natural convection by about a factor of about 2 or 3, but film cooling has now been replaced with more effective finger and waffle coolers (Figure 14). External shell cooling however will continue to remain important for certain situations. As smelting rates increase much beyond 200 mt/hr, it is expected that there will be more use of wall/shell cooling on the smelting furnace itself.

Table II - Key Equipment – Mega Smelter (>400,000 mt/yr of Cu in one smelting furnace)

Unit	Main Features		
Smelting unit	1 large bath smelter (TSL) or flash furnace (and O_2 plant)		
Converter	Peirce-Smith (4.5 m dia. by 13.4 m) 4 vessels (3 hot) 2 blowing/time	Continuous Bath (Noranda or TSL type) 1 or 2 units	Continuous Flash (Kennecott- Outotec) 1 or 2units
Anode	3 (or 4) - 4.5 m dia. by 11 m furnaces 2 - M18 twin-casting wheels		
Slag	Slag milling assumed		
Acid plant	Double absorption acid plant		

Typically this represents between 6 to 8 furnace units including dryer – about half the number at large plants in the 1960s-1970s and for significantly higher copper output.

Note: As shown by Coursol at al. [22], the energy requirement for the production of anode copper from copper concentrate by modern smelting technologies was almost independent of technology type. The variation in total calculated energy requirement [2, 22] amongst four modern processes (Outotec-Flash Smelt - Flash Convert, ISASMELT^{TM} - PS Converters, Mitsubishi Process, Noranda-Teniente - PS Converters) was about 8%, with the average total energy being 11,403 GJ/mt of anode Cu. The electrical consumption was 8,691 GJ/mt of anode Cu, while fossil fuel was 2,712 GJ/mt of anode copper (an average power plant efficiency of 38% was adopted by Coursol et al. [22] in computing energy for electrical requirements).

ii. Matte and slag tapping

Higher smelting rates requires both more frequent matte and slag tapping, and also tapping at a higher flow rate, along with improved taphole design and maintenance. Figure 15 presents calculated slag and matte tapping rates as a function of taphole diameter. Measured and calculated tapping rates for a number of Ni-Cu electric furnaces are included.

The required tapping rates of slag and matte for a 400,000 mt/yr Cu smelter are also indicated. It is seen that a matte taphole of the order of 6 cm diameter and a slag taphole of the order of 8 to 9 cm diameter would be required.

Even at the tapping rates indicated, tapping is required for 20 hr/day for both slag and matte. It is anticipated that beyond the above copper output, multiple tap holes, already standard practice on some furnaces, will become more common.

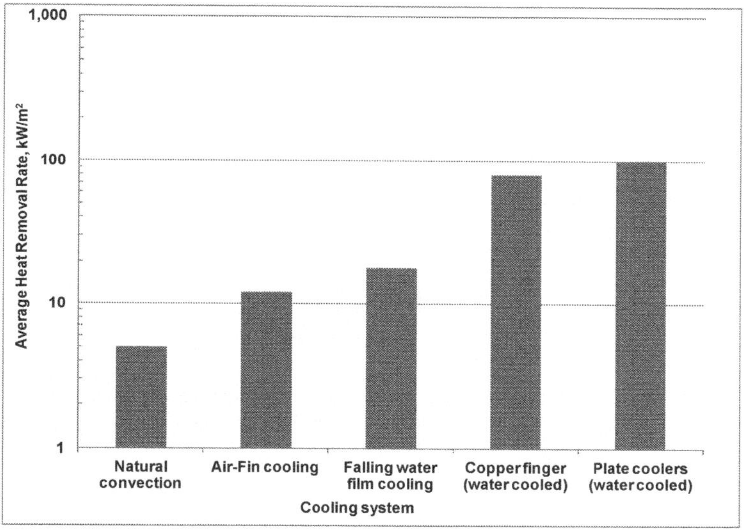

Figure 14. Range of average heat flux levels for different furnace wall cooling systems generally applicable to copper smelting furnaces. It is noted that depending on design, waffle type coolers (not included here) can provide heat fluxes over 500 kW/m^2 for more intense cooling, such as now used on high temperature Ni-Cu smelting or ferro-nickel electric furnaces. Based in part on information in [26].

Efficient material handling (copper concentrate, flux, copper, slag, dust, etc.) is always important at a copper smelter of any size, but becomes even more so for a mega-scale plant. Handling large tonnages of copper concentrate, sulfuric acid and slag in particular becomes a key issue. Effective disposal of large tonnages of slag or slag tailings is another important aspect. Mega plants operate most efficiently with a high on-stream time; hence a strong preventative maintenance program would be essential (and in any case, this is also true for a smaller plant).

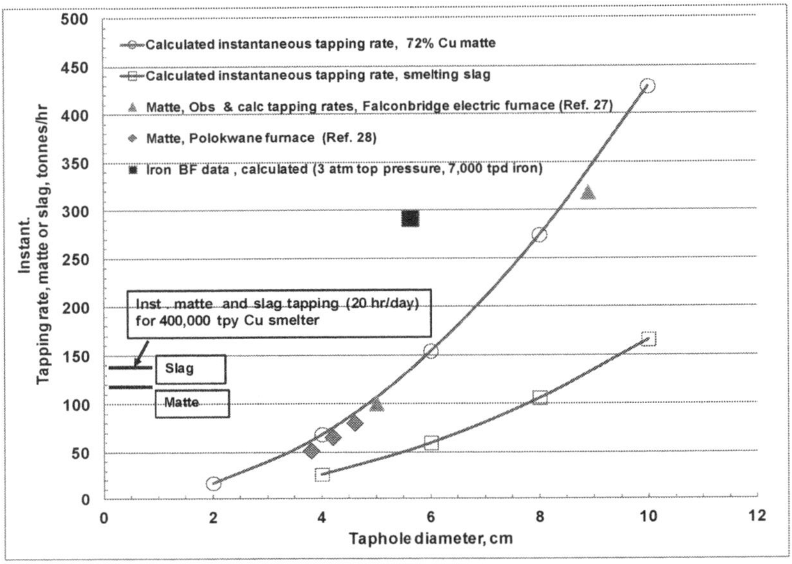

Figure 15. Calculated instantaneous matte and slag tapping rates as a function of tap hole diameter. Operating and calculated data for Ni-Cu electric furnaces [27, 28] and the iron blast furnace included for comparison.

Given the right design, there seems no physical reason why concentrate smelting tonnages over 200 mt/hr per furnace cannot be achieved. For example, a rate of 215 mt/hr of 30% Cu concentrate would correspond to a copper output of 500,000 mt/yr, while a rate of 260 mt/hr of 25 % Cu concentrate is needed for the same copper output. Plant tonnage could be doubled with two lines. While tonnages higher than this may be possible, other aspects such as logistics and economic factors would also become important.

Conclusions

The paper examined the steady trend towards higher furnace smelting rates as seen over a long period of time in the primary copper smelting industry. Today's technologies smelt copper concentrate in an intense process with oxygen enriched air by completely different mechanisms to that common 50 years ago. It was shown that the present large bath and flash smelting furnaces are reaching concentrate smelting rates in the range 165 to 210 mt/hr in a single furnace; for a 30%Cu concentrate, it is estimated that this corresponds to a copper output between 380,000 mt/yr and over 400,000 mt/yr. It is expected that the trend towards higher smelting rates will continue, not only at the highest capacity plants but covering all plant capacity levels. Converting and anode furnace technology and other plant areas will adapt to the needs of the higher capacity plants. Peirce-Smith converter technology will continue to dominate copper smelting for the time being, however there may be inroads by the new continuous converting technologies particularly for the largest capacity plants.

It was reported [2] that for the year 2010, some 22% of world primary copper smelting capacity was at plants having capacities greater than 300,000 mt/yr of Cu, while 37% of capacity was at plants having the capacity range of 200,000 to 300,000 mt/yr of Cu; the balance (41%) was at plants in the capacity range below 200,000 mt/yr of Cu. As noted, the trend to larger plants across the whole capacity range will continue and in future, there will be a greater proportion of smelting capacity in the range above 300,000 mt/yr of Cu. It is also expected that plants, in particular custom smelting plants, with capacities greater than 400,000 mt/yr of Cu and likely single furnace plants, will be built in future.

Still, there will continue to be a place for the smaller, highly efficient specialized or niche primary plants as well as secondary copper smelters; however secondary copper smelting was not covered in the present paper.

While single furnace plants with capacities larger than say 450,000 to 500,000 mt/yr of Cu are considered technically feasible, a trend to employing two smaller capacity furnaces for plant capacities beyond 500,000 mt/yr of Cu may also develop.

Acknowledgements

The author thanks David George of Rio Tinto, David Robertson, Emeritus Professor at the Missouri University of Science and Technology, Kim Fagerlund of Outotec and Gerardo Alvear of Xstrata Technology for useful discussions concerning large furnaces and smelters and for information and ideas they generously provided. The author also expresses his appreciation for the opportunity to discuss smelter projects big and small with numerous colleagues over the years.

References

1. M. Naito, "The Development of Iron Making Technology" (Nippon Steel Technical Report, No. 94, June 2006), http://www.nssmc.com/en/tech/report/nsc/pdf/n9402.pdf

2. P .J. Mackey, "Copper Smelting Technologies in 2013 and Beyond"(Paper to be presented at Copper 2013, Santiago, Chile, December 2013, The Chilean Institute Mining Engineers, 2013).

3. C. Díaz, H. Schwarze and J. Taylor, "The changing landscape of copper smelting in the Americas", in *Copper 95-Cobre 95, Vol. IV: Pyrometallurgy of Copper,* W.J. (Pete) Chen et al., Eds., (Montreal, Québec, Canada: MetSoc of CIM, 1995) 3-28.

4. M. Eissler, *The Hydro-Metallurgy of Copper* (New York: D.Van Nostrand Company, 1902), 9-10.

5. F. Laist, "History of Reverberatory Smelting in Montana 1879 to 1933", *Trans. TMS-AIME (Special Volume on Copper Metallurgy)*, 105 (1933), 23-87.

6. P.Tarassoff and P.J. Mackey, "The Changing Face of Copper Pyrometallurgy" (Paper presented at the 22nd Chemical Engineering Conference, CSChE, Toronto, Ontario, Canada, September, 1972), 17-20.

7. W.B. Boggs, J.N. Anderson and W.J. Stevens, "The Noranda Smelter from 1939 to 1945", *Trans. Canadian Institute Mining and Metallurgy,* XLVIII (1945), 402-428.

8. J.N. Anderson, "Reverberatory Furnace Practice at Noranda", *Trans. AIME,* 200 (1954), 745-758.

9. W.B. Eastwood, J.S.Thixton and T.M. Young, "Recent Developments in the Smelting Practice on Nchanga Consolidated Copper Mines Rokana Smelter" (Paper presented at the AIME Annual Meeting, 1975), Paper No. A71-75.

10. W.F. Cartwright, "The Growth of Unit Output and its Effect on Works Planning and Management", *Journal of the Iron and Steel Institute (London),* 207 (Part 6) (June 1969), 729-738.

11. A. Barnes, P.J. Mackey and G.R.F. Alvear F., "Process Development and Growth of Non-Ferrous Metals Production -The Role of Pilot Plants" (Plenary Paper delivered to EMC 2009, June 28 to July 1, 2009, Innsbruck, Austria).

12. International Copper Study Group (ICSG), *The World Copper Fact Book,* (Lisbon, Portugal: The ICSG, 2012). Web site: www.icsg.org

13. T. Lehner and J. Wiklund, "The New Ronnskar Smelter", in *Proceedings of EMC 2001,* Friedrichshafen, Germany, 18 to 21 September, (Clausthal-Zellerfeld, Germany: GDMB, 2001), 49-59.

14. P.J. Mackey and J.E. Nesset, "The Impact of Commissioning and Start-up Performance on a Mining/Metallurgical Project", *paper presented to the 35th Annual Meeting of The Canadian Mineral Processors,* S. Wilson, Ed., Ottawa, Ontario, Canada, 21-23 January (The Canadian Institute of Mining, Metallurgy and Petroleum, 2003), 331-347.

15. Glencore Inc., Presentations at Investor Day, 10 September, 2013, http://www.glencorexstrata.com/assets/Investors/InvestorDay2013/GLEN-Investor-Day-2013-Complete-full.pdf

16. Sherritt Inc., Company presentations, www.sherritt.com, also, http://www.sherritt.com/getattachment/cd41b967-c0de-4566-b8cc-1328119dcc0f/Sherritt-Investor-Day-Presentation-Avail-868055.

17. R.Holroyd, "Ultra Large Single Stream Chemical Plants: Their Advantages and Disadvantages", *Chemistry and Industry,* August 5 (1967), 1310-1315.

18. Outotec Oy Press Release, Outotec to Deliver Copper Flash Smelting and Flash Converting Technology to China, 2011.

19. A.E. Wraith, C.L. Harris and P.J. Mackey, "On Factors Affecting Tuyere Flow and Splash in Converters and Bath Smelting Reactors", in *From Agricola to the Present - A Meeting Between East and East: Responsibility for the Future - Metallurgy I,* EMC '94, Freiberg - Dresden, Germany, June 1994, (Clausthal-Zellerfeld, Germany: GDMB, 1994) 50-78.

20. L. Mariscal S. and W. Torres P., "Modernization and Start-up of the SPCC Ilo Smelter", in *Copper 2007- Cobre 2007, The Carlos Diaz Symposium on Pyrometallurgy,* Vol. III, Book 2, Eds. A.E.M. Warner, C.J. Newman, A.Vahed, D.B. George, P.J. Mackey and A.Warczok, Toronto, Ontario, Canada, August (Montreal, Quebec, Canada: The Metsoc of CIM, 2007), 193-202.

21. G.R.F. Alvear F., P. Arthur and P. Partington, "Feasibility to Profitability with Copper ISASMELTTM", in Proceedings of Copper 2010, June 6 to 10, Hamburg, Germany, Volume 2-Pyrometallurgy I (Clausthal-Zellerfeld, Germany: GDMB, 2010), 615 to 630.

22. P. Coursol, P.J. Mackey and C. Díaz, "Energy Consumption in Copper Sulphide Smelting", *Proceedings of Copper 2010,* June 6-10, Hamburg, Germany, Volume 2-Pyrometallurgy I, (Clausthal-Zellerfeld, Germany: GDMB, 2010), 649-668.

23. S. Wallner, A. Filzwieser and J. Kleicker, "Some Aspects for the Use of Water Cooled Furnace Walls" (Report by RHI aG, Vienna, Austria).

http://www.mettop.com/mediacache/22watercooled_furnace_walls__water_the_best_ref_153004.pdf

24. A. Weidner, "Refractory Lining of Flash-Smelting Furnaces", (Paper presented at the 104th AIME Annual Meeting, 1975), Paper No. A75-69.

25. J.B.W. Bailey and G.C. Balfour, "Refractory Practice for the Noranda Process", (Paper Presented at the 109th AIME Annual Meeting, Las Vegas, Nevada, USA, Feb.24-28 1980).

26. N. Voermann, et al., "Furnace Cooling Design for Modern High-Intensity Pyrometallurgical Processes", in *Proceedings of the Copper 99 – Cobre 99 International Conference*, October, Vol. V, Phoenix, Arizona, U.S.A., (Metsoc of CIM-TMS, 1999) 573-582.

27. I.A. Cameron, R. Sriram and F. Ham, "Electric Furnace Matte Tap Hole Developments at Falconbridge Inc.", *CIM Bulletin,* June (1995), (Montreal, Quebec, Canada :CIM), 102-108.

28. C.A. Snyders, J.J. Eksteen and A.Moshokwa, "The Polokwane Smelter Matte Tapping Channel Model", (Paper presented at Fifth International Conference on CFD in the Process Industries, CSIRO, Melbourne, Australia, 13-15 December, 2006).

29. P.J. Mackey and A.E. Wraith, "The Pembrey Copper Cathode", *Mineral Processing and Extractive Metallurgy (Trans. Inst. Min. Metall. C),* 121, No. 3 (2012), 166-172.

30. www.wrexham.gov.uk/english/heritage/brymbo_steelworks/making_steel_1900.htm

Celebrating the Megascale: Proceedings of the Extraction and Processing Division Symposium on Pyrometallurgy in Honor of David G.C. Robertson
Edited by: Phillip J. Mackey, Eric J. Grimsey, Rodney T. Jones, and Geoffrey A. Brooks
TMS (The Minerals, Metals & Materials Society), 2014

EVOLUTION OF THE MEGA-SCALE IN FERRO-ALLOY ELECTRIC FURNACE SMELTING

Lloyd Robert Nelson[1]

[1]Anglo American Platinum Ltd, 55 Marshall Street, Johannesburg, 2107, South Africa
lloyd.nelson@angloamerican.com

Keywords: Ferro-alloy, Electric Furnace, Smelting

Abstract

The historical progression in installed electric furnace smelting capacity of some 1650 calcium, chromium, manganese, nickel and silicon ferro-alloys furnaces is reviewed. Key increases in the inherent installed electrical capacity, often achieved through uprating furnaces, are identified.

Possible factors facilitating such advances are explored, including: specific process engineering and pyrometallurgical enhancements (e.g., improved control over the consistency, size and chemical form of raw materials, extents of preheating and pre-reduction delivered into the furnace; and patterns of feed distribution); furnace electrical configuration (AC or DC; immersed, submerged-, shielded- or open-arc); form of electrode (graphite, self-baking or composite); furnace configuration (circular or rectangular; closed or open and 1-, 3- or 6-electrodes); better engineering and equipment designs (e.g., high-intensity cooling) and state of furnace electrical and metallurgical control systems.

A view is presented on the prevailing state of evolution of the Mega-scale in ferro-alloy smelting and opportunities for still further improvements.

Introduction

Review of ferro-alloy electric furnace applications and their evolution presents a daunting array of furnaces and applications worldwide characterized by, with the possible exception of ferro-nickel [1], limited collation into a coherent historical progression based on installed electric furnace size. This is further complicated by the sheer number of furnaces worldwide, with over 1500 ferro-alloy companies quoted in China alone [2]. Clearly the some 1670 furnace applications (some furnaces are repeated, as they perform swing duty between the various ferro-alloy commodities depending on market conditions - Table 1), involving some 1360 "identified furnace start dates" (i.e., can include the same furnace more than once, if uprated - Table 2) isolated here is woefully incomplete as a true record. Nonetheless, collection of these furnaces into a historical progression of installed electric furnace MVA according to commodity (despite obvious potential shortcomings related to transformer MVA being over-sized, sometimes grossly, relative to actual applied furnace active power, MW) has proven instructive to identify some trends that inform on the current state of the ferro-alloy industry and the future potential for further development of the electric furnaces it adopts for primary smelting.

The historical progression of calcium ferro-alloys (predominantly calcium carbide, CaC_2) is first presented, because the initial evolution of the furnace capacity appears fastest there. This is followed by review of the chromium ferro-alloys (various high, medium and low carbon ferro-chrome, charge chrome and silico-chrome); manganese ferro-alloys (various high, medium and low carbon ferro-manganese and silico-manganese); ferro-nickel and finally silicon alloys (metallurgical grade silicon, and the suite of ferro-silicons).

Table 1: Breakdown of furnaces used per commodity

Source	Cr	CaC_2	CaSi	Mn	SiMn	Ni	FeSi	Si
Number of Furnaces	345	142	6	282	218	90	411	177
Grand Total	1671							

Table 2: Number of Furnaces with "identified start dates" per ferro-alloy commodity

Ferro-alloy Commodity	Cr	Ca	Mn	Ni	Si	Total
Number of Furnaces	345	148	309	90	474	1366

Calcium Ferro-alloy Furnaces

A notable feature is rapid progression of furnace capacity above 40 MVA by the 1950s (Figure 1), associated with an increase of about 1 MVA/annum installed furnace size until 1980 to a maximum furnace of 108 MVA. Thereafter, there is noticeable drop in the number of furnaces, with identified furnaces being typically 30 MVA or smaller. Clearly a declining market for an acetylene precursor to plastics, coupled with difficulty transporting a reactive product, rather than technological limits of furnace design specifically, dominated this period. Equally remarkable though is the surge in China post-2010, with numerous mega-scale furnace applications greater than 81 MVA, including five planned 190 MVA (90 MW) 6-in-line (6iL) rectangular furnaces.

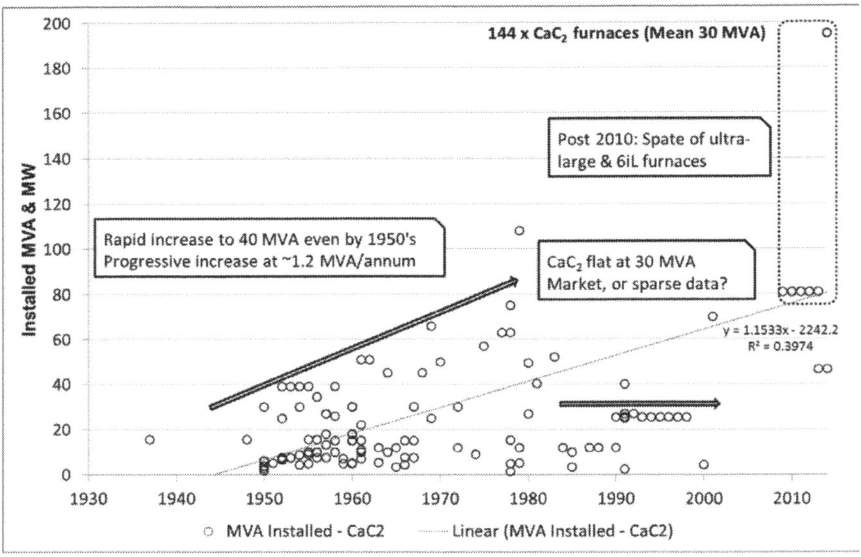

Figure 1: Historical progression of installed MVA of calcium ferro-alloys furnaces

Chromium Ferro-alloy Furnaces

Following a slower progression in furnace capacity to above 40 MVA by only the 1970s [3], there has been a fairly steady progression of the largest units at about 1.5 MVA/annum to an installed maximum of 132 MVA (100 MW demonstrated - Figure 2). Again there is evidence of a period of limited activity (possibly market related) between mid-1980 to 2000. Certainly in this period there is documented evidence of a plethora of smaller (1-10 MVA) furnaces installed in India [4], matched by China, and more recently in Zimbabwe [5]. Even in South Africa, with a more concentrated larger-scale furnace capacity to traditionally dominate world ferro-chromium production (~40% market share [6]), there is very recent installation of small electric furnaces for specialty low and medium carbon ferro-chromium, with downstream argon oxygen decarburization (AOD) and vacuum oxygen decarburization (VOD) refining [7].

Another striking feature since the 1980s is the emergence of an alternative furnace technology in the form of generally larger (mean 45 MW) DC-arc furnaces, with furnace capacity increasing from an initial 12 MW to an imminent 72 MW on four furnaces; a rate of about 1.4 MW/annum.

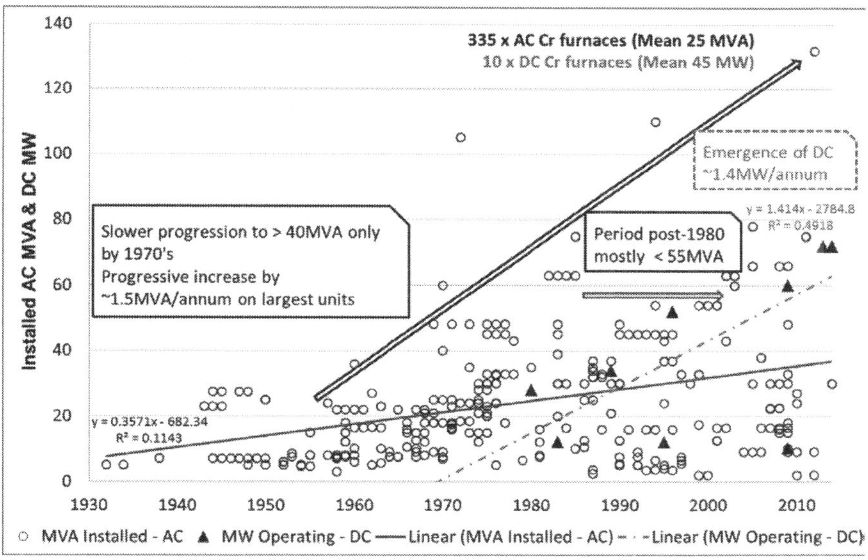

Figure 2: Historical progression of installed MVA of chromium ferro-alloys furnaces

Manganese Ferro-alloy Furnaces

Noticeable is the frequent dual smelting duty for a given furnace to swing between high carbon ferro-manganese (HCFeMn) and silico-manganese (SiMn) production (Figure 3). There is often little distinction when viewing furnace MVA; operating differences only really becoming apparent when looking at active power setpoint, and the near double specific energy consumption [8] of the latter product (as low as 2 MWh/t HCFeMn compared to 4 MWh/t SiMn).

There appears a moderate progression to above 30 MVA by the 1960s, followed by a more rapid progression of about 2.5 MVA/annum to 81 MVA in the 1970s [9,10]. Then again, there is a significant drop to 30 MVA units and numerous significantly smaller furnaces for a period of two decades following 1980, which dramatically lowers the mean installed overall furnace capacity to about 20 MVA. These smaller units again represent new capacity in China and India [4,9]. With the exception of the largest 102 MVA HCFeMn furnace, a repeated rapid growth at about 2.5 MVA/annum in furnace capacity to 81 MVA has transpired, and this appears to present somewhat of a capacity limit.

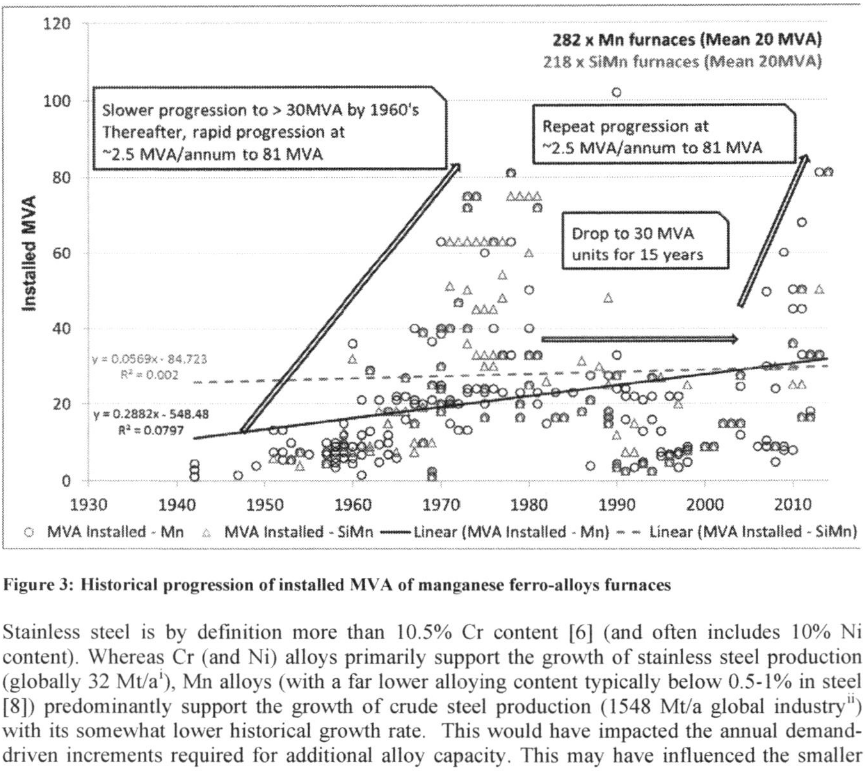

Figure 3: Historical progression of installed MVA of manganese ferro-alloys furnaces

Stainless steel is by definition more than 10.5% Cr content [6] (and often includes 10% Ni content). Whereas Cr (and Ni) alloys primarily support the growth of stainless steel production (globally 32 Mt/a[i]), Mn alloys (with a far lower alloying content typically below 0.5-1% in steel [8]) predominantly support the growth of crude steel production (1548 Mt/a global industry[ii]) with its somewhat lower historical growth rate. This would have impacted the annual demand-driven increments required for additional alloy capacity. This may have influenced the smaller size of furnaces installed in the 1980s and 1990s – before demand from China started to have a pronounced effect on the market in the succeeding period.

[i] "Global stainless steel production reaches new record in 2011", http://www.worldstainless.org/Files/issf/non-image-files/PDF/media_releases/ 120321ISSFFY2011productionstatisticsmediarelease_EN.pdf.

[ii] "World crude steel output increases by 1.2% in 2012", http://www.worldsteel.org/media-centre/press-releases/2012/12-2012-crude-steel.html.

Ferro-nickel Furnaces

Striking features of the progression in ferro-nickel furnace capacity (Figure 4) are:

- Most consistent overall progression of about 1.3 MVA/annum to an impressive installed mean of 50 MVA, with power only marginally lagging at 1 MW/annum, in somewhat fewer furnaces (only 90 listed, uncertainty mainly associated with status of furnaces in China).
- A distinct period of more rapid evolution of maximum furnace power rating of about 1.7 MW/annum from the 1970s, associated with technical development of the "shielded-arc" in preference to the "immersed" electrode mode of operation, with accompanying design of high-intensity water-cooled copper coolers and improved furnace controls; the latter culminating in sophisticated thyristor-based electrical controllers capable of stabilizing power input despite the incidence of calcine feed collapse into the shielded-arc.

The ferro-nickel industry is also characterised by both large (75-94 MW) circular furnace designs [1], as well as large-scale rectangular designs [1,11] (80-85 MW, with a 102 MW design in execution). Also in ramp-up is a novel twin-graphite electrode DC furnace (80 MW design) that, like many of the rectangular furnaces, has applied a 3-dimensional binding system to provide a contant furnace compressive load to the furnace to improve structural brick and steel lining integrity, and to provide some relief against the onset of inevitable furnace ratcheting.

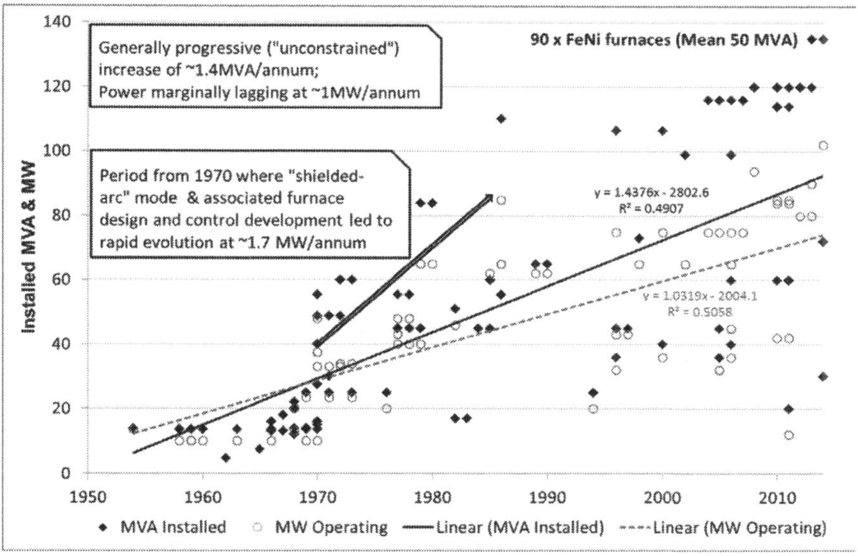

Figure 4: Historical progression of installed MVA of ferro-nickel furnaces

Silicon Ferro-alloy Furnaces

While there is a fairly quick evolution through the 1960s to 30 and even 40 MVA capacity furnaces, often dual commodity metallurgical grade silicon and ferro-silicon in nature (Figure 5), the peculiar process requirements of the former metallurgical grade silicon production (including "rotating" furnaces to help secure the required crucible and burden porosity process conditions) effectively have limited furnace capacity to under 45 MVA, with one notable exception at 75 MVA (45 MW). The specific requirement to avoid iron contamination has also predicated graphite electrodes, or later composite Söderberg electrodes (with a graphite core). For reasons of commodity flexibility, this has also precipitated a plethora of tiny (< 5 MVA) furnaces, capable of swing production [2,4]. This to the extent that China advocated phasing out furnaces smaller than 6.3 MVA by 2011, with approval only of furnaces larger than 12.5 MVA, as a means of protection against effectively "exporting" scarce and increasingly more costly electricity [1] (given high specific energy consumption up to 13 MWh/t Si).

Progression in FeSi capacity was rapid at about 2 MVA/annum until the late 1970s; whereafter again far smaller furnace installations dominated [1,4]. This impacted on the mean furnace capacity of identified furnaces being as low as 20 MVA; in reality far smaller because of the multitude of tiny Chinese and Indian furnaces for which data are scarce, if not entirely unavailable.

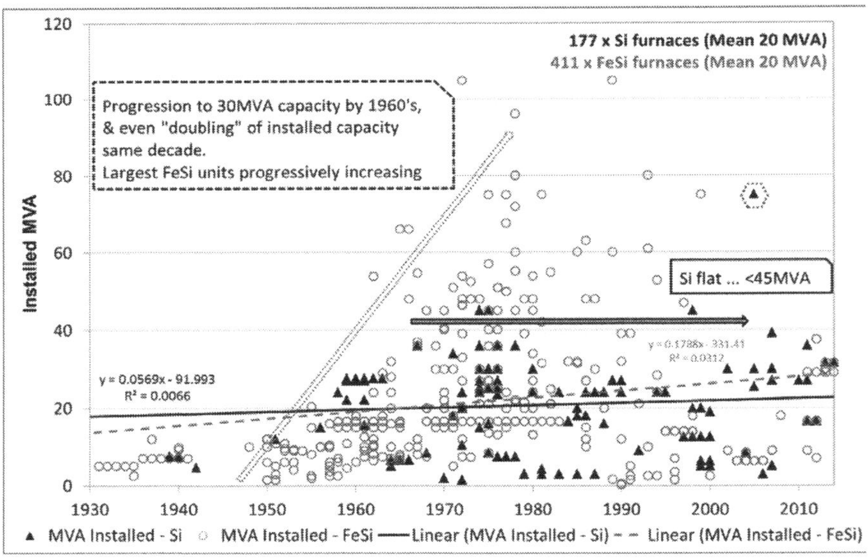

Figure 5: Historical progression of installed MVA of silicon ferro-alloys furnaces

Furnace Design Evolution and Characteristics

Using the SMS Siemag references [10] as an example, one might find it remarkable that almost all common ferro-alloy furnace configurations, AC versus DC, circular versus rectangular, and one versus multiple electrodes had been adopted by around the beginning of the last century

(Table 3). The "rotating"[iii] circular furnaces, developed to move the location of the smelting zones to break up crusts and improve burden porosity, emerged commercially in 1950. This configuration found broad application across the ferro-alloys, and is still deemed a near pre-requisite for silicon smelting with its associated high specific energy consumption (< 13 MWh/t Si), associated high CO/CO_2 gas generation, and a fundamental aspect of its pyrometallurgical processing crucially involving intermediate SiO(g) and control of its generation, back-reaction and extent of bypass (e.g., to make microsil), and intermediate "refractory"[iv] SiC and its control.

Initially AC three electrode-in-line (3iL) rectangular furnaces found application, but difficulties even on small scale with control leading to "live" and "dead" electrodes [12], prompted switch to lower reactance "knap-sack" designs [13], with 3 single phase instead of single 3-phase transformers becoming increasingly popular. Rectangular six-in-line (6iL) furnaces were considered [14], but concerns seemingly on the costs of near double the electrode equipment largely dissuaded their application in ferro-alloys smelting (despite successful installation in other smelting industries).

Unfortunately the data integrity is poor on the quantitative distribution of open, semi-closed and closed furnaces. Qualitatively, of positively identified furnace types, closed furnaces totally dominate in nickel, represent almost 60% in calcium and manganese, and almost 40% of chromium applications (Table 3). Open submerged-arc furnaces (SAFs), permitting access for furnace stoking to improve bed porosity, dominate silicon alloy applications (82%), largely for reason of flexibility in operation to address the process metallurgy duty listed above.

Table 3: Distribution of furnace designs and dates per commodity

	"Rotating"	Rectangular	DC	AC Circular	Type		
Year installed	1950	1913*	1906* (CaC_2)	1911* (FeCr)	Open	Semi-	Closed
Cr	5%	-	3%	97%	34%	29%	36%
Ca	54%	7%	-	93%	39%	4%	57%
Mn	25%	6%	-	94%	26%	17%	57%
Ni	-	30%	4%	66%	0%	0%	100%
Si	94%	1%	0.2%	99%	82%	10%	8%
*SMS Siemag Reference List	Only 26% Si+FeSi furnaces "positively" identified rotating + 1x 3-pair, 6-electrode	Historically 3iL & now usually 6iL	6MW pilot plant for Si	Mige[v] 1-electrode furnaces = 3%	Data incomplete, as Type only "positively" identified in list for: 67% of Cr, 92% of Ca, 64% of Mn, 73% of Ni & 28% of Si/FeSi furnaces		

[iii] "Rotating" can be either full 360° perpetual rotation, or periodic back and forth rotation around the central axis.

[iv] "Refractory" used in the context here of elevated melting temperature.

[v] Described by Shkirmontov [15] as "one-electrode Mige furnaces installed at Zaporozhie Metallurgical Plant prior to 1941 to make ferrosilicon", whereas Robiette [16] describes "development in France between the wars of single-phase Miguet furnaces".

With the exception of silicon alloys (owing to their peculiar process metallurgy), closed furnaces are likely to predominate for reasons of:

- Being cooler top (less vapour/dust losses) and less oxidized, for improved hygiene, environmental capture and control (including negligible hexavalent chromium production);
- Lower off-gas temperature and thus slightly lowered specific energy consumption;
- Less infiltration air requirements for combustion and cooling, thus less direct energy losses through a smaller volume of off-gas, requiring lower capacity and less energy for operating downstream off-gas handling systems;
- Less dilution of the combustible chemical energy potential (i.e., CO and H_2) suitable for later energy recovery and co-generation [11].

By way of example, easily the sensible and chemical energy of the off-gas of a closed DC ferro-chrome furnace equates to over one-third of the furnace electrical power input. There are many examples of commercial energy recovery and co-generation plants being integrally operated on ferro-alloy smelting sites to great benefit [11].

Furnace Integrity

An aspect of large-scale furnaces neglected at one's peril is furnace design integrity; especially on mega-scale units exceeding 15 m wide × 40 m long rectangular [17] or 22.2 m diameter circular [18]. Especially for a thermally (and electrically, for DC) conducting hearth design, even the minuscule $\sim 1\times10^{-6}$mm/mm.°C linear thermal expansion of graphites can lead to a not insignificant thermal expansion of 40mm; and significantly greater for magnesia-based bricks typical of DC conductive hearth designs. In the high-intensity platinum group metal (PGM) base metal sulfide smelting industry (68 MW), matte superheat can exceed 650°C, or 5 times that of ferro-alloy systems [19]. This makes it near impossible to secure a stable bottom hearth freeze lining, so requiring total reliance on loaded brick-to-brick contact to prevent infiltration and leakage. Given this experience, one cannot but wonder if the ferro-alloy industry is adequately sighted on furnace lining and structural design requirements of future ultra-high power, mega-scale furnaces, if protective metal heel freeze linings cannot always be positively secured.

Flexibility and Economies of Scale

The plethora of small electric furnaces producing ferro-alloys world-wide is testimony to certain advantages to flexibility to quickly enter and exit production. Smaller furnaces will undeniably be more amenable to swing production between the different ferro-alloy grades and even commodities, which can provide ability to respond profitably to short-term commodity cycles. Also some variable costs may well be cheaper for smaller furnaces [10]: need for less classified raw materials (negating screening and associated costs, and costly fines agglomeration); ability to use higher proportions of non-metallurgical coke reductants such as coals; lower Söderberg electrode costs, and reduced risk of production loss due to electrode breaks; improved utilization through quicker ramp to power set point and less complicated process metallurgical, electrical and electrode control issues to facilitate better adherence to that set point.

However, the claim that: "M/s. Electro kemish A/S of Norway who supplied most of the SAFs to India … had expressed that there will be about 10% reduction in specific power while smelting in larger size furnaces compared to small size furnaces. But it is surprisingly not found to be correct in India! Smelting power and other usage norms reported by producers in small furnaces are found to be almost equal to those obtained in large furnaces!" appears overstated

[4]. It is precisely because smaller furnaces are less efficient that China has deemed it appropriate to regulate and curtail smaller furnace operations in their attempt to "limit exporting of electricity" [2].

Moreover, there is ample evidence that large furnaces [1], or the same furnace, but operated at higher 68 MW power input [20], with only a marginal increase in heat loss with applied power and so a lower proportion of heat losses relative to power input, are indeed more energy efficient. In addition on larger furnaces the greater throughput and productivity per FTE (full time equivalent) can more effectively dilute significant fixed cost components [10].

So while even larger-scale furnaces may not be the panacea, there is certainly a compelling argument to pursue even larger capacity furnaces if they can deliver a position at the bottom of the cost curve, especially with secured product offtake, electricity and raw materials supply.

Division of the Smelting Function

Emi [21] provided a most illuminating description of the role of division of the refining function in effectively decoupling the more tedious quality and process requirements essential to steelmaking, so permitting the essential bulk metal-making process units to be focused on their core activity, of generating upwards of 10 000 tpd hot metal. At a significantly smaller metal production rate, but in an analogous manner, in ferro-alloys smelting there have been numerous pre- and post-treatment processing developments that have been pivotal in increasing ferro-alloy smelting productivity, regardless of furnace size. Indeed many of these developments such as consistency of charge composition, size, dried, preheated and even pre-reduced state, coupled with enhanced more stable furnace and off-gas control and post-treatment for reliable energy recovery, have proven crucial to permitting furnace size, active power and ultimately ferro-alloys smelting rates to be increased significantly. Of course it is essential to ensure that the resultant coupling of unit processes indeed improves overall productivity, guarding against the risk that it inadvertently and adversely impacts overall equipment effectiveness (OEE).

It is difficult to do justice to the fundamental importance of these aspects of division of the smelting function [22] and their role in securing and sustaining ongoing evolution in electric furnace smelting capacity and throughput. Nonetheless some important milestones are listed.

Ore Grade and Quality

Chromite ore grade is fundamental to the ferro-chromium grade produced, as well as its efficiency of production. Compared to many chromites, the dominant reserve of South African ores is generally poor quality, being characterized by low contained 38-45% Cr_2O_3 content (lump versus fine concentrate, respectively) & low Cr/Fe ~ 1.5 (even ~ 1.35 for abundant UG2, as a byproduct of PGM processing [23]). It is friable and less "refractory" in nature owing to low Cr_2O_3 content and so high relative presence of gangue (of low < 0.8 MgO/Al_2O_3 ratio). However, being a spinel mineral (and so even adopted for production of high-temperature refractory), reduction kinetics are frequently limiting, and so smelting is strongly influenced by temperature.

It is reported [24] for submerged-arc furnace smelting of high-grade Kazakhstan chromite (> 50% Cr_2O_3 & Cr/Fe > 3) that a "1% decrease in Cr_2O_3 content increases power consumption 4% and slag chromium losses by 2%". Extrapolated to conventional smelting of far lower grade South African chromites, this provides a remarkably accurate projection of chromium recoveries actually achieved. When coupled with appropriate pre-treatment practices (e.g., ore fine agglomeration & sintering), benchmark Japanese, Russian and Kazakhstan smelters can achieve Cr recoveries approaching 90-95%, compared to Cr recoveries as low as 70% when smelting directly 70% by mass chromite South African chromite fines (-6mm) in semi-closed submerged-arc furnaces. Remarkably, with these very same fine ores, DC open-bath smelting can deliver upwards of 90% Cr recovery, but at an ~15% penalty to specific energy consumption.

Manganese ores can be far more variable in nature, with Mn oxidation states ranging from +2 to +4, and with more complex mineralogy (e.g., in addition to oxides, now hydroxides and carbonates). Carbonate ores frequently undergo sinter pre-treatment to produce a more uniform classified feed of reduced in-furnace decrepitation and gas evolution potential (less H_2O, CO_2 and higher oxides of Mn). Another feature is that the reduction of all higher oxides down to MnO by CO is exothermic. This has prompted careful blending of pyrolusite (MnO_2), Mn sinters and high-grade Mn ores to an optimum Mn^{4+}/Mn^{2+} ratio in benchmark Japanese operations, to be able to achieve high slag practice specific energy consumption approaching 2 MWh/t alloy (even lower if coupled with preheating). Selective smelting of "refractory" (more alumina-rich) Mn ores is believed to provide improved smelting conditions and performance in other instances.

Ni laterites are also extremely variable in nature, making accurate blending of over 10 ore types onto stacker-reclaimers frequently essential to achieve stable high productivity downstream smelting. Coupled with gangue SiO_2/MgO (that largely dictates slag liquidus and therefore smelting process temperature), Ni ore grade and Ni/Fe ratio are important determinants of the final ferro-alloy Ni grade achievable after reduction, at acceptable recoveries [25].

Raw material purity is especially crucial for metallurgical grade silicon production, to deliver 96-99+ % Si content; especially relevant for its function as a precursor feedstock to some high-purity downstream silicone and ultra-high (99.9999999% Si) polysilicon applications. In addition to selection of the purest quartz sources, this often entails washing to remove traces of gangue. This provides an important secondary function of removing fines and so helping to secure a well-classified ore crucial to good burden permeability.

Reductant Selection and Quality

Always important is adequate availability of high quality reductant (low volatiles, ash, sulfur and phosphorus), of appropriate strength in-service in a hot furnace (e.g., burden in submerged-arc furnaces), of an appropriate size distribution for optimum burden resistivity and reactivity, and "type" of reactivity suitable for the metallurgical reduction duty at hand. This is even more applicable to a high capacity furnace.

Often scarce, metallurgical coke is always the preferred reductant for large-scale submerged-arc furnaces to secure adequate burden permeability and favourable burden resistivity. Much has been written on the importance of coke to efficient submerged-arc smelting (to the extent of describing a particular reductant-rich zone in proximity under the electrodes as a "coke" bed). A stable "coke" bed is therefore sought for optimal metallurgical and energy dissipation in the furnace [26]. Borrowed from the steel industry, the coke reactivity index (CRI) and coke strength after reduction (CSR) provide "some" guidance on coke reactivity applicable to ferro-alloy submerged-arc furnaces.

In the case of silicon and silicon-rich ferro-alloy production, wood chips are frequently added to promote burden porosity. Charcoal and coals are also used. Their fundamental less-ordered carbon structure (with many free carbon atoms available for reaction because of only short-range order of crystallites) also promotes high "gaseous" reactivity; crucial to promote early reduction reactions in the colder upper zones of submerged-arc furnaces, but also for efficient reaction with any gaseous SiO(g) intermediate. Frequently the amenability of reductants to the latter reaction is evaluated through a dedicated SiO(g) reactivity test [27].

When iron is substantially involved in the reduction process (as in most ferro-alloys), in zones where temperature has progressed through that for onset of reduction of FeO to incipient melting of the product iron (lower melting temperatures possible for iron with increasing dissolved carbon content), another type of "liquid" reductant reactivity emerges. This is explained through the preferred dissolution of certain reductants in iron as $\underline{C}$ (the "liquid" reactivity, and facilitated by a liquid iron solvent), and specifically the finding that the Boudouard reaction largely responsible for generating the primary reductant CO involved in the classical coupled "gas-ferrying" iron oxide (and other ferro-alloy metal oxide [28]) reduction reactions, is about an order of magnitude faster [29,30] when through dissolved $\underline{C}$, than through solid reductant C.

Carbon dissolution reaction:	$C(s) = \underline{C}_{\text{dissolved in iron}}$	(1)
Boudouard reaction:	$CO_2 + \underline{C} = 2CO$	(2)
Iron oxide gas-solid reduction:	$2CO + 2FeO = 2Fe + CO_2$	(3)

Simplistically, more ordered (higher rank [31]) reductants (better approaching a graphitic-type planar hexagonal ring structure) resist gaseous reaction, because they are reasonably strongly bonded in the structure. However, in a liquid metal environment, especially if the solid reductant possesses good wettability [30], once a carbon atom has been pulled out of the planar structure by dissolution into metal, the bond force holding adjacent carbon atoms is considerably weakened. Now the higher ordering of the reductant facilitates more rapid dissolution of carbon atoms of that crystal plane into the metal, in a more rapid "unzipping" mechanism. So paradoxically, less ordered reductants (like bituminous coal) of high "gaseous reactivity", are typically of lower "liquid reactivity" than more ordered reductants (e.g., anthracite), and vice versa [29,30].

Therefore, in all probability a suite of reductants with differing gaseous and liquid reactivities, will best optimize reduction to ferro-alloys in the multi-zone burden of submerged-arc furnaces.

Certainly strong evidence of the effects of reductants of differing "gaseous" (coal) and "liquid" (anthracite) reactivities has been experienced in rotary kiln electric furnace (RKEF) ferro-nickel [32] and ferro-cobalt [33] production. And this was similarly a key issue in development of the DC-arc open-bath chromite fines smelting process, where "liquid" reactivity is more appropriate. This was found to require not only anthracite, but specific anthracites for good reduction. The key learning in all instances is that a seemingly innocuous change in reductant (e.g., anthracite) source can lead to dramatically different reduction behavior. The maxim "a carbon is not a carbon is not a carbon", unless proven positively to the contrary, is so true!

Finally reductant size also plays a very important role in submerged-arc furnace burden resistivity. Fundamental modeling [34] suggests that both small reductants (many point contacts) and large reductants (continuous paths of low resistivity) will display enhanced electrical conductivity. The modeling also identified a surprisingly significant potential contribution to overall resistivity of the size of "insulating" phases (ores and fluxes) located between reductant conductors. Certainly all ferro-alloys submerged-arc furnace operators know of the ability of quartz to "kill the current" when recovering from an overly conductive electrode break situation. As reaction proceeds, particle sizes diminish continually, so this again suggests that some optimum blend of reductant sizes and flux sizes may be required to best match the particle sizes of the specific ore available, in order to optimize burden resistivity.

Charge Classification and Furnace Resistance

The importance of a well-classified charge especially to submerged-arc operations cannot be overstated. Most truly high-productivity smelter processes feature a prepared and well-classified charge. Excessive charge fines (usually considered -6 or -10 mm) can be extremely deleterious to burden permeability, promote charge bridging and even sintering, and so impede free egress of off-gases. This can lead to hot gas and charge blows, cause furnace equipment damage, and in the extreme cause significant and hazardous furnace eruptions. Lower energy efficiency is inevitable with the best of (hazardous) stoking practices used in open furnaces to retain burden porosity.

For this reason benchmark smelters with large high-productivity furnaces, screen lump raw materials to remove fines to better guarantee smelting efficiency[10]. Resulting ore fines are typically recovered for some form of agglomeration and smelting.

In 1971 Selmer-Olsen [35] reported for low Cr/Fe ratio chromites "furnace resistance with a pelletised charge is very high and 2.5 mΩ seems well within reach, which could make electric furnaces of the 100 MW range possible". Some 4 decades later, 100 MW power input is indeed being achieved in Finland, but with about half the furnace resistance and so facilitated through the largest 2.1 m diameter Söderberg electrodes delivering up to 170 kA current (Table 4). This highlights the (then) little appreciated magnitude of the role of increasing electrode diameter (often ascribed to power factor), in lowering resistance with increasing furnace capacity.

Case Study – Oxidatively-sintered chromite pellets

With the large-scale roll out of oxidative sintering of pellets (~13 mm diameter) in South Africa since the late 1990s, similar benefits of increased resistance were also experienced. However, far less well appreciated were the benefits of such a well-classified ore to overall furnace performance. This was probably best captured by an operator who, while watching a 70% fine chromite ore by mass furnace charge blowing with red top conditions, asked "why was the adjacent furnace off". With only 30% by mass ore pellets it was blacktop and so stable by comparison, yet actually operating at a slightly higher power input! Explanations developed for this included the following:

Fundamentally, the pellets are near uniform in size (~13 mm diameter) so their permeability is theoretically predicted to be significantly enhanced. A Rosin-Rammler plot (Figure 6) illustrates the improvement in particle size distribution for increased pellet additions; namely more: to the *R*ight (larger size), up*R*ight (narrower range) and st*R*aight (more uniform distribution with less fine, or too large, tails). This 3 *R*'s classification is useful to assess the permeability and thus likely furnace burden aerodynamic performance of the submerged-arc furnace, which critically needs to "breathe easily" to perform optimally.

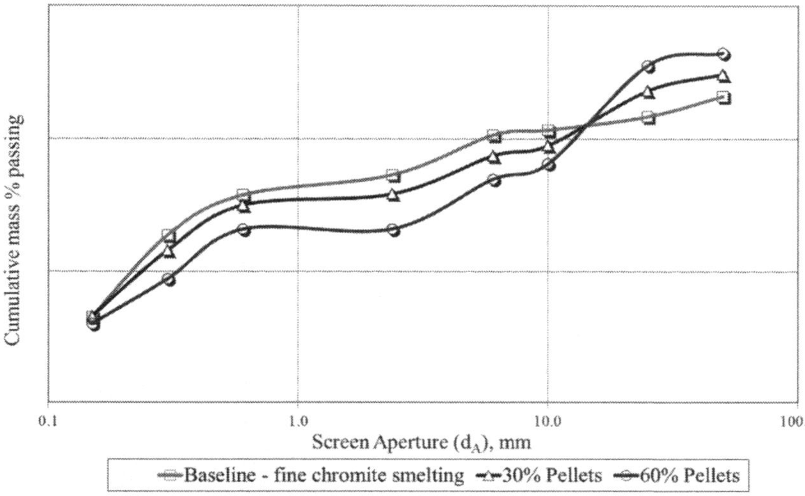

Figure 6: Rosin Rammler particle size distribution illustrating an improved 3*R*'s and more permeable charge with increasing pellet content to 60% by mass ore

Next, the d_{32}, volume-surface (Sauter), mean particle size for i mass fractions, x_i, of that mean particle size, i, can be calculated according to [36]:

$$\frac{1}{d_{32}} = \sum_i \frac{x_i}{\bar{d}_i}$$

Remarkably, well-classified pellets (with little fines) yield a $d_{32} \approx 10$ mm, similar to that of lump ore with top size as large as 70mm. Further, when the improved shape factor [36] of the pellets (P) relative to lump (L) ore ($\lambda_L \approx 1.5\lambda_P$) and 1.75 times greater voidage predicted [37] for the pellets ($\omega_L \approx 0.57\omega_P$) are considered, in the context of the Ergun equation for pressure drop [36], for the lump particle size to be equivalent to that of the pellets assuming purely laminar flow, it can be estimated that: $d_{32,L} \approx \{(\lambda_L / \lambda_P)^2(\omega_P / \omega_L)\}^{1/2}.d_{32,P}$. This suggests that lump size would need to be $d_{32,L} \sim 3 \times d_{32,P}$ to be equivalent in pressure drop effect to pellets.

Estimated similarly, for purely turbulent flow: $d_{32,L} \approx \{(\lambda_L / \lambda_P)(\omega_P / \omega_L)^2\}.d_{32,P}$. This implies lump size would need to be $d_{32,L} \sim 4.6 \times d_{32,P}$ to be equivalent in pressure drop effect to pellets.

This larger "effective" pellet d_{32} size delivered into the furnace, even when compared to that of ostensibly larger lump ore, helps to explain why, despite the seemingly retrogressive step of oxidizing a spinel before introducing it into the furnace to reduce it (so actually raising the thermodynamic energy requirement) results in a more stable and colder blacktop operation (remember the operator's "furnace off" comment). The far more permeable burden of considerably improved product gas aerodynamics, coupled with improved reduction kinetics of oxidized spinel (Fe^{2+} now oxidized to Fe^{3+}, so sesquioxide mineralogical form) leads to overall lowering in specific energy consumption of ~80 kWh/t/10% pellets by mass ore being realized!

Additionally, an improvement of ~1 % point in Cr recovery/10% pellets by mass ore is also realized. While improved kinetics of reduction can partially explain this due to less potential for tapping partially reduced chromite spinels, there is another remarkable consequence of pellet smelting. Not only do the pellets deliver a more resistive charge, but part of the reason they do so is because they considerably raise the refractoriness of the pellets (Figure 7, compared to more molten standard ore charges - Figure 8), so significantly raising temperatures achievable around the electrode tips. These hotter conditions favourably shift the primary electrical conduction path deeper to the electrode tips and the higher temperatures are crucial to driving improved Cr_2O_3 reduction thermodynamics and kinetics, promoting overall Cr recovery.

Figure 7: Chromite pellet charge – individual "refractory" pellets still evident at 1500°C. Promotes higher temperatures for enhanced Cr_2O_3 reduction.

Figure 8: Typical lump chromite ore charge – substantially molten and reduced at 1500°C. Poorer overall extent of Cr_2O_3 reduction will result.

Drying

Dry charge is a critical component of high-intensity smelting. Not only does it directly lower the specific energy requirement of water removal in the furnace as steam, but its removal prior to electric smelting also helps lower off-gas volume and associated energy losses; limits otherwise unwanted carbon loss (of reductant and expensive electrode carbon) through endothermic water-gas shift reaction according to:

$$C + H_2O = CO + H_2 \ (\Delta H^o \sim 2 \text{ MWh/t water}) \quad (4)$$

and eliminates potential risks of batching errors based on equivalent dry raw materials used in most furnace charge calculations.

Two strategies adopted are either: to store raw materials under cover, especially common in colder climates where wet raw materials can freeze to a solid block, becoming problematic to handle especially in spring as they thaw; or to provide active drying, in the anticipation that active management of dry feed provides sufficient process stability and productivity benefit.

Calcination

Chemical waters of crystallisation, hydroxides and carbonate ores and fluxes can be advantageously decomposed by calcination ahead of electric furnace smelting. This is routinely performed on nickel laterites (typically in rotary kilns) [1,11]. It is a particularly useful consequence of manganese ore sintering [10,38]. Sintering helps to produce a more competent classified charge less susceptible to decrepitation, and of lower in situ gas release and resulting lower propensity to cause furnace blows and eruptions (with Mn carbonates now substantially calcined in a feed charge of up to ~65% sinter by mass Mn ore). It is also a beneficial consequence of stainless steel belt oxidative sintering of chromite pellets [39], and rotary kiln pre-reduction of chromite and flux charges (either SRC [40] or Premus [41]). Especially where considerable limestone is used as a slag-modifying flux (such as Chrome Direct Reduction-DC furnace), the benefits of calcination include saving electrical energy of over 0.8 MWh/t lime calcined, and beneficially leading to lower electrode consumption (up to one-third less, to as low as 1 kg electrode/MWh) due to less oxidation of the electrode carbon by H_2O (Reaction 4) and CO_2, as:

$$C_{electrode} + CO_2 = 2CO \quad (5)$$

Pre-heating and Hot Transfer

Generally a saving of 30-80 kWh/t_{feed}/100°C in specific energy consumption is realizable for preheating dependent on specific ferro-alloy feeds (lower range for ferro-nickel calcines; upper range for treated chromite ores [39]). In ferro-manganese, rotary kiln preheating is beneficially reported (< 2 MWh/t alloy in EF [42]). In conventional nickel laterite RKEF smelting [1,11] (<0.45 MWh/t dry ore); frequently in oxidatively-sintered chromite pellet smelting (<3 MWh/t alloy in EF with >86% Cr recovery [39]) and in chromite RK pre-reduction processes [40,41], delivery of hot charge into the electric furnace at temperatures above 600°C (typically 800°C) is effected, substantially improving the productivities of the primary smelting furnaces.

Pre-reduction

Given the exothermic nature of the CO reduction of higher oxides of manganese, pre-reduction other than sintering is not an option practised. In nickel laterite smelting substantial reduction of Fe^{3+} to Fe^{2+} is sought together with some extent of NiO reduction to promote downstream EF smelting productivity. Over-reduction can lead to less stable smelting conditions with bath foaming which, although beneficial to arc shielding and enhanced heat transfer, can prove decidedly detrimental if not well-controlled, considerably complicating control of electrode tip position.

A similar situation prevails in chromite pre-reduction. High degrees of 75% Cr reduction (requiring low volatile coke breeze to limit hot pellet "bursting") substantially lower specific electrical energy consumption to 1.7 MWh/t alloy in the original SRC process [39]. The newer Premus process [41] typically targets somewhat lower extents of Cr reduction (50% at 2.4 MWh/t alloy), seeking instead to make use of cheaper anthracites fines, matching a higher RK capacity to an increased furnace smelting capacity, and a change in operating philosophy of maximizing metallization to maximizing energy output from the kiln to achieve a more reliable coupled RKEF equivalent. Both processes achieve a credible ~90% overall Cr recovery on South African chromites.

In the context of evolution of the mega-scale in ferro-alloy electric furnace smelting, the productivity benefits of such technological advances, while potentially facilitating even larger scale smelting, may equally have diminished the need for their immediate implementation. As the primary aim of smelting is actually to make "cash", not alloy, such advances are extremely valuable in their own right. However, the apparent steady progression of ferro-nickel smelting to the largest installed mean capacity suggests that technological advances for a commodity in general demand, did not act to slow down advance to larger installed smelting unit capacities.

Hydrometallurgical Upgrading

An emerging trend is hydrometallurgical upgrading to produce intermediates amenable to smelting. Higher commodity price nickel (USD/t Ni) is amenable to this, providing a means to more effectively utilize available smelting capacity to deliver additional saleable Ni units (effectively higher grade) per unit alloy produced [43].

Deleterious "Volatile" Impurities

Trace elements that are volatile under smelting conditions can be particularly deleterious to otherwise efficient submerged-arc furnace smelting, in that they can enter a concentrating recycle of volatilization in hot regions near the electrode tips; followed by condensation in colder upper portions of the burden; potentially forming bridges and lowering burden permeability; only to be returned back in descent of the burden to the hotter primary smelting zone to repeat the volatile impurity concentration cycle.

In the extreme, such recycle in the case of volatile Zn can lead to serious furnace eruptions and associated hazard to personnel and damage to equipment, as experienced in ferromanganese smelting [44]. Another similar cycle of volatile impurity recycle, this time of K, has similarly been modeled [45] to lead to furnace energy inefficiency and deleterious unwanted fixed carbon consumption. Therefore, tight monitoring of raw materials quality to limit the introduction of deleterious volatile impurities is an important requirement to efficient and safe smelting.

Sulfur and silicon are generally considered as impurities with deleterious impact on the quality of ferro-chromium. One unexpected consequence of development of the open-bath DC smelting of chromite was the significantly lower alloy silicon and sulfur contents (as low as 1% Si and 0.007% S, respectively); to the extent that the DC-arc furnace alloy can even attract a market premium. Technically an explanation for this is believed to be the absence of a charge burden, so permitting direct removal of volatile SiO(g) and SiS(v) species from the open-bath, without the cooling condensation, capture and return of volatile impurity species, and so recycle associated with traditional submerged-arc operation. This is an example of an alternative process metallurgy obviating a ferro-alloy product quality challenge.

Feed Distribution and Gas Permeability

When operating at the limits of electric furnace power input not only are feed quality and form important, but also its distribution within the furnace. Appropriate distribution of adequate feed maximizes capture efficiency of energy dissipated by the electrode/s and avoids local feed/power imbalances and inappropriate metallurgy. This is also core to maintenance of an effective "shielded-arc" in ferro-nickel production [32] (but also DC-arc ferro-cobalt production [33]) where it can otherwise drive unwanted endothermic fuming reactions and even lead to silicon reversion and CO boil events that challenge both process stability and furnace integrity.

A related challenge at increasing power density and associated increased reduction rate is to limit local product gas generation (mainly CO reduction product) and its impact on the permeability of any calcine or burden surrounding the arc. This is to avoid excessive accumulations of gas so generated being released by inefficient hot furnace blows and jets, which concurrently cause arc instability due to ejecting and collapsing feed. While it has been well-argued that the "rate of generation of off-gas should be roughly proportional to the power dissipation ... electrode cross-sectional area ... and hence area of burden through which gas must escape" [46], especially when process metallurgy is shifted into a more reducing regime (such as low-grade ferro-nickel), limitations to calcine burden permeability and "shielded-arc" stability may well develop with attempt to increase power input [1].

The value of optimal feed distribution to best capture energy dissipated by the electrode tip, is illustrated by the (unfortunately slow) evolution of DC-arc fine chromite smelting to a more distributed "trickle charge" feeding of fines. Despite installation, in a 1988 furnace upgrade from 16 to 40 MVA, of roof side feeders specifically to supplement hollow electrode centre-feeding (with its initial decided operational challenges), only 100% centre-feeding was ever practiced. This was then copied onto the next DC furnace (without any side feeders). Evidence for the limited efficiency of open-arc, open-bath energy capture in this centre-fed hollow electrode configuration, was the need to retrofit water-cooled copper plate coolers into the furnace freeboard, not the bath crucible sidewall, and their adoption on a still further DC furnace [47].

Later, and with the hindsight of favorable experiences on 100% side feeding with solid graphite electrodes in ferro-cobalt production [33], side feeding was finally attempted; and with such success that it was designed into the further furnace [47]. With improved energy capture by side-fed fines from the roof, a ~10% lowering in specific energy consumption is estimated. These benefits of "halo" feeding of fines were identified early in the development of the novel 80 MW twin-electrode DC furnaces for ferro-nickel, now in commissioning in New Caledonia.

Taphole and Post- Taphole Practices

A fundamental prerequisite to higher productivity electric furnace ferro-alloy smelting is the ability to tap alloy and slag unimpeded. With substantial absence of a head of overlying slag ($\rho g h_{slag}$), being a "dry" process, metallurgical grade silicon tends to exhibit slower and more difficult taphole and tapping conditions. Even with multiple metal tapholes (often numbering 4) this may contribute to its more limited capacity and productivity relative to other ferro-alloys.

Given the ability of the largest blast furnaces to tap at rates in excess of 10 000 tpd hot metal, there would appear considerable scope to further increase ferro-alloy tapping rates, coupled with appropriate post-taphole logistics.

Possibly slightly different for ferro-alloys, especially high-grade ferro-nickel with metal fall lower than 15%, is the requirement to tap high volumes of slag. Fortunately application of multiple slag tapholes (up to 6 is not uncommon), coupled with slag water granulation systems, provides high capacity ... provided that first pass metal recovery is adequate. If inadequate, slow cooling would instead be required to permit secondary metal recovery from slag by means of gravity and mechanical jigs.

Slag Energy Recovery

Given the quantity of slag tapped (especially in ferro-nickel) and its considerable sensible energy, it is not surprising that numerous slag energy recovery schemes have been devised. Up to 140 tpd systems were even operated for some time on ferro-nickel slags [48]. However, the technology was eventually discarded. Results of a pilot plant using a dry approach (important in regions of scarce water supply) involving casting slags into metal balls, followed by breakup of the solidified hot slag and recovery of energy by counter-current heat transfer are awaited [11].

Off-gas Handling and Energy Recovery

Given the scarcity and cost of electricity in most regions, informed by the environmental imperative, energy recovery systems for utilization of sensible energy (all furnaces), coupled with co-generation systems for better utilization of chemical energy (closed furnaces) will become increasingly common [11]. This will be especially true on mega-scale electric furnace smelters of the future. Of course to realize their full recovery benefit, they must be operated at the highest possible power input. Given the increasing tendency of electricity supply regulators to seek more flexible demand-side supply agreements and tariffs [49], this is a factor to consider.

Furnace and Electrical Control

The first direct consequence of scale up in power is that processes happen faster, requiring quicker corrective control action. The second consequence is that if the required control is not achieved effectively, conditions deteriorate more rapidly and to a far worse state than possible on a smaller scale.

Also, a fundamental truth of scaling the capacity of all AC electric smelting furnaces is that an increase in furnace power is accompanied by lowered furnace resistance owing to larger electrode sizes and lower power factor [15,50,51,52], because the circuit inductances at typical supply frequencies typically increase with furnace size [13,49,46]. A slightly different, and more quantitative interpretation of this effect, is that electrode resistance (R_E) decreases according to: $R_E \alpha D_E^{-1.5}$, where D_E = electrode diameter. This can be inferred from the Andreae formula [53] in metric form [46] for maximum electrode current (I_E): $I_E = 55D_E^{1.5}$ and the general validity of the proportionality has been demonstrated in analysis of innumerable industrial electrode dip tests, together with application of more robust design formulae.

So especially when the electrodes get sufficiently large, and electrode resistance drops, power factor: $\cos\varphi = P/S$ falls below 0.7 (where P = real or active power (MW) and S = apparent power (MVA)), and the "interaction effect" magnifies and can begin to significantly complicate electrode movement and control [46,49].

On one large commercial silico-manganese furnace, good old-fashioned pyrometallurgy uncovered a statistical difference in the tapping temperatures of ~85°C between the slag tapped through the dedicated slag taphole opposite a "live" electrode compared to slag tapped opposite a "dead" electrode. By exploring the phase rotation on that particular submerged-arc furnace and developing the appropriate phasor diagram, it was possible to confirm the existence of a significant "interaction effect", matching the "live" (hotter) and "dead" (colder) electrodes observed. Barker et al [54] discuss further theoretical implications of the "interaction effect" on such a large furnace.

To circumvent this, increasingly advanced and specific furnace control has proven essential, largely based upon the principles of resistance [55] or impedance control to control individual electrode positions, with primary measurements usually being used to improve accuracy and infer secondary electrical conditions. More recently [11,56], predictive thyristor-based control has been applied on the furnace electrical reactance to stabilize fluctuations in furnace power at the speeds of the inherent electrical supply frequency (e.g., 60 Hz), to great advantage of improved control.

Another emerging trend driven by demand-side electricity supply, is the need and benefit to operate the furnace at varying electrical load (either seasonally or even daily) depending on the availability and price of electricity [49], and market demand [57]. This has been practised for over two decades in Japan [58], but could present some control and electrode management challenges on larger furnaces. In the extreme, it has recently proven more beneficial for some South African ferro-chromium smelters to shut down entirely for extended periods to help lower grid demand in power buy-back deals with the local electricity supply utility [60], rather than incur high pricing or risk uncertain supply.

DC Versus AC

DC power [46,49] offers the potential to obviate the affliction of AC reactance issues of low power factor and the "interaction effect", leaving the property of DC parallel current-carrying conductors to attract their arcs to each other, as opposed to the less desirable repulsion and arc flare associated with AC (with local asymmetric increase in sidewall and roof heat fluxes).

Additionally, in DC with zero AC frequency, the "skin effect" that afflicts all AC electrodes by impeding effective AC electrical conduction across the full cross-sectional area of the electrode, is entirely absent. Typically for AC, the skin depth (δ) is: $\delta = (\rho/\pi f \mu_0 \mu_r)^{1/2}$ (where, ρ = specific resistivity of the conductor, f is AC frequency in Hz, μ_0 is magnetic permeability of free space and μ_r is relative permeability of the conductor). It is estimated that most of the current flow is concentrated to a skin depth of about 0.3 m in Söderberg electrodes and 0.15 m in graphite electrodes [46], with negative connotations to electrode capacity and operability, and consequently scale up.

Good electrode performance is fundamental to having reliable power input, with the consequences of an electrode break being more dire with larger furnaces given the larger electrodes required (owing to electrode current density limitations). Fundamentally, electrode mechanical strength deteriorates with size [46,49], which can present a limitation to furnace scale. Electrodes are also particularly susceptible to thermo-mechanical stresses. While freeboard energy transfer, dependent on processing conditions, is important to electrode baking and condition (especially in lower current density ferro-nickel furnace electrodes), the role of electrode current and associated ohmic I^2R heating dominates in all other ferro-alloy electrodes, leading to complex time-dependent tensile, compressive and thermal stresses.

In shutdowns (I = 0), modeling [59,60] shows that the cooling of the electrode leads initially to development of tensile stresses at the electrode periphery and compressive stresses in the core, that first grow and then start to reverse and increase further with time (magnified by electrode dimension and shutdown duration). So that by the time the furnace is restarted, the core may be subjected to tensile stresses exceeding 3.5 N/mm^2 (with compressive stresses calculated to exceed 5 N/mm^2 in the electrode periphery [60]). Especially if prior furnace operation has been cyclical, these stresses can magnify still further [59] and, given a lower electrode tensile than compressive strength, a hard break of the electrode simply results if the tensile limit is exceeded. Clearly, therefore, the maldistribution in electrode current induced by the AC "skin effect", with greater local current densities for an equivalent current, has potential to worsen the situation, compared to DC.

In comparing DC with AC, one must balance these distinct advantages of DC against some more challenging aspects of DC operation, including the need to:

- Install a complex power supply unit, which includes rectifiers and current harmonic mitigation equipment.
- Be able to secure electrical contact to the conducting hearth through an overlying slag whose resistivity is strongly influenced by temperature. This is more problematic after extended period of downtime, especially for single electrode units.
- Continually battle the natural tendency of electricity to "take the easiest" path and, promoted by a strong potential driving force, risks of roof arcing and rapid skewing of the arc attachment to the detriment of equipment and furnace integrity (either towards more conductive zones, e.g., metal tapholes; or away from less conductive zones, e.g., oxidized carbon-based hearth refractories, or hydrated refractories). Basically it is easy to skew 1- or 2-electrode DC arcs that are inherently unstable!
- Mitigate the fundamental consequence of the Wiedemann-Franz Law; namely: $\kappa/\sigma = LT$ (where κ = thermal conductivity, σ = electrical conductivity, L = Lorenz number and T = absolute temperature in K), that thermal conductivity and electrical conductivity are proportional, so a good electrical conductor (essential to bottom hearth conduction) is by corollary also a good thermal conductor. So DC current imposes additional load on the furnace cooling design requirements to dissipate energy generated directly through current flow, in addition to normal crucible heat transfer through the bottom conductive hearth.

Electric Furnace Capacity – Current Limits and Future Potential

Guided by best demonstrated available technologies and operating practices, an attempt will be made to project realistic limits to furnace capacity in the different ferro-alloy commodities.

Chromium Furnaces

The largest AC, circular submerged-arc furnace, recently commissioned, smelts high resistivity oxidatively sintered chromite pellets at 100 MW/132 MVA (Table 4). The benefits of presenting to the furnace such a consistent, more permeable, pretreated and higher resistivity ore charge as a fundamental enabler to such mega-scale operation should be clear. Scope probably still exists to increase this operating power further through increased resistance per electrode by either:

- Further optimizing charge burden resistivity (100% oxidatively sintered pellets yet?), or
- Marginally increasing arc resistance [15] (sometimes described as, e.g., "a 10% increase in secondary voltage" [24]) and as realized in many South African chromite smelters [46].

Converting this to an AC rectangular, 6iL furnace design, 202 MW power input should be entirely plausible (assuming electrode resistance remains unchanged for a 2.1 m diameter electrode, and furnace reactance substantially unchanged). Indeed the "great flexibility in voltage and in power in each pair of electrodes" and advantages of less-cluttered, more symmetrical roof access for "charging advantages" were recognized as long ago as 1974 [14], and as realized in ferro-nickel, manganese and 68 MW PGM 6iL furnaces [20]. The 6iL configuration permits decidedly more flexible control of individual electrode pairs. So in the event of an electrode break, unlike the 3-electrode configuration where the entire furnace power input is adversely impacted, fully two-thirds of the power input can be maintained unaffected (crucial to high OEE). Demonstrated commercial multi-taphole operations (4 metal and 6 slag tapholes) should also eliminate furnace tapping as a constraint.

Table 4: Best ferro-chromium furnace power inputs are based on AC in Finland and DC in Middelburg [47], and future projected power increases are based on best demonstrated limits

	FeCr	FeCr **Max**	FeCr **Max**	FeCr **Max**	FeCr *	FeCr **Max**	FeCr **Max**
Electrical Supply	AC	AC	DC	DC	DC	DC	DC
Mode of operation[a]	SAF	SAF	SAF	SAF	Open	Open	Open
Furnace shape[b]	Circ.	Rect.	C/R	C/R	Circ.	Circ.	Circ.
Number of electrodes	3	6	6	6	1	1	3
Electrode type[c]	Söd.	Söd.	Gr.	Gr.	Gr.	Gr.	Gr.
Electrode diameter[d], m	2.1	**2.1**	0.75	0.75	0.6	**0.75**	0.75
Electrode current[e], kA	167	167	78	28	71	**150**	71
Resistance per electrode[f], mΩ	1.2	1.2	5.6	5.6	12	12	12
Current density, A/cm^2	4.8	4.8	17.7	6.3	25.1	**34.0**	16.1
Power per electrode, MW	34	34	**34**	4	60	270	**60**
Electrode power density, MW/m^2	10	10	77	**10**	214	611	137
Total active power, MW	101	**202**	**205**	**26**	60	**270**	**181**
Reported operation/ *Potential benefit*, MW	100	*+102*	*+105*	*-74*	60	*+170*	*+81*

Max = Maximum projected power, subject to **constraint in bold**

[a] SAF = "submerged-arc"; Open = "open-arc"; Shield = "shielded-arc"[1]. Notice "immersed" electrode absent in all highest power furnaces.

[b] Circ. = circular; Rect. = rectangular; C/R = optional circular or rectangular for DC

[c] Söd. = Söderberg. Gr = graphite

[d] Söderberg maximum existing = 2.1 m. Graftech ALX [62] maximum graphite = 0.75 m

[e] DC maximum = 150 kA Graftech ALX [62]. AC maximum = metric Andreae electrode current [46], of $I_E = 55D_E^{1.5}$

[f] Assumes electrode resistance α $D_E^{1.5}$ (Andreae's [53] metric formula [46]), applied to AC and DC SAFs.

A DC, 6 × 0.75 m diameter graphite electrode furnace should be capable of operating at a higher equivalent 5.6 mΩ electrode resistance. For it to run in "submerged-arc" mode, the closest industrial precedent would be enforced DC "immersed-arc" 40 MW operation in commercial ferro-cobalt production [33]. This could increase operating power to 205 MW, if the present "submerged-arc" 34 MW per electrode proves to be limiting. However, if the prevailing ferro-chromium electrode power density (10 MW/m^2) proves to be limiting to scale-up, given the far smaller diameter of even the largest commercial graphite electrodes, this would impose a probably intolerable and significant 7.7 times increase. To retain any 10 MW/m^2 electrode power density process constraint, only a paltry 26 MW total power input would be projected.

Clearly it is incumbent on furnace designers to resolve the real nature of this process (not electrical) constraint to scale up!

The lower electrode current-carrying capacity of ~150 kA of graphite (determined by prevailing graphite size and quality limitations [62]) is nowhere near being approached; and any calls, at least in DC "submerged-arc" ferro-chromium smelting, for novel DC Söderberg submerged-arc operation (conceived [63], but still to be demonstrated commercially) may be somewhat premature, unless purely for reasons of lowering operating cost.

The largest DC "open-bath" single graphite electrode furnaces currently operate at 60 MW with conducting hearth designs on chromite smelting of 75% higher power per electrode than the largest AC submerged-arc furnace in operation (at 100 MW, with 67% higher power). This highlights the significant benefits to:

- "Open-bath" mode more effectively decoupling the power input from the process than "submerged-arc" mode of processing
- Using arcing to increase resistance and power input (a core feature of the more rapid evolution in capacity of ferro-nickel furnaces)
- More intensely cooled, "thermally" conducting hearth designs.

Uprating this to the electrode-carrying limit of a single graphite electrode may theoretically yield a furnace power input of some 270 MW. Yet a resulting excessive near 185% increase in electrode power density (611 MW/m^2), while not all bath power (with associated resistive I^2R_{bath} heating) that directly imposes increased sidewall heat flux, may still provide challenges to:

- The integrity of the furnace infrastructure (roof and sidewalls) and ability to withstand such a concentrated arc energy source
- The current-carrying capacity of existing electrically conducting hearth designs; e.g., if the conducting hearth were to contribute 1% to overall resistance, at a current of 150 kA a not insubstantial additional 2.7 MW of $I^2R_{conducting\ hearth}$ energy needs to be dissipated.
- Securing roof electrical isolation to avoid stray arcs (implies fully 1.7 times the highest operating voltages prevailing in ferro-nickel shielded-arc smelting [1])
- Maintaining the process feed/power balance locally under the arc without excessive local bath overheating and a risk of driving unwanted energy intensive fuming reactions (including SiO(g) and Mg(g)).

Therefore, a more measured approach, may be to consider 3 graphite electrodes to deliver proven power and voltage per electrode, while inherently constricting the arc inwards with DC, rather than flaring it outwards as would be the case of AC. Three graphite electrodes, such as operated as early as in the 1980s for DC steelmaking at Societe Metallurgique de l'Escaut [64], as opposed to twin-electrode DC implemented both in steelmaking and more recently in ferro-nickel, would also inherently provide a more stable arc (like a tripod), without the need for complicated (and expensive) arc deflection correction equipment, while yielding an impressive scale up to 181 MW power input. This is within demonstrated power per electrode (60 MW) with a calculated 36% lower electrode power density (137 MW/m^2) than prevailing on the current commercial DC furnace.

The engineering scalability of the DC power input projected for 3 graphite electrode sources in near proximity to any conducting hearth design, may warrant more detailed investigation to corroborate the potential for its viability in sustained operation.

Manganese Furnaces

Given the frequency with which swing production between high carbon ferro-manganese and silico-manganese is practised (positively identified for at least one-third of all manganese furnaces), the furnace capacity limits are treated together (although it is well appreciated that not all furnaces perform equally efficiently in swing production duty). It has been argued that AC, circular submerged-arc furnaces, some of late 1970s vintage, are close to their limit [46] at 50 MW/81 MVA, due to their very low resistance (and susceptibility to excessive fuming of volatile of manganese under more arcing conditions), resulting low power factor < 0.7 and excessive associated "interaction effect" [49].

Table 5: Best manganese – circular (Ukraine [50] and South Africa) and rectangular (Ukraine [50]) and ferro-silicon alloy furnace power inputs, and future projected power increases are based on best demonstrated limits

	Mn	Mn, **Max**	Mn, **Max**	Mn §	Mn, **Max**	FeSi	FeSi, **Max**	FeSi, **Max**	FeSi, **Max**
Electrical Supply	AC	AC	DC	AC	AC	AC	AC	AC	DC
Mode of operation[a]	SAF	SAF	SAF	SAF	SAF	SAF	SAF	SAF	SAF
Furnace shape[b]	Circ.	Circ.	C/R	Rect.	Rect.	Circ.	Circ.	Rect.	C/R
Number of electrodes	3	3	6	6	6	3	3	6	6
Electrode type[c]	Söd.	Söd.	Gr.	Söd.	Söd.	Söd.	Söd.	Söd.	Gr.
Electrode diameter[d], m	1.9	**2.1**	0.75	1.8	**2.1**	1.8	**2.1**	1.8	0.8
Electrode current[e], kA	144	167	72	110	167	133	167	133	70
Resistance per electrode[f], mΩ	0.8	0.7	3.2	0.7	0.6	1.0	0.8	1.0	3.7
Current density, A/cm^2	5.1	4.8	16.3	4.3	4.8	5.2	4.8	5.2	15.8
Power per electrode, MW	17	19	**17**	8	16	18	22	**18**	**18**
Electrode power density, MW/m^2	6	6	**38**	3	4	7	6	**7**	**41**
Total active power, MW	50	**58**	100	51	**93**	53	**67**	**106**	**109**
Reported operation/ *Potential benefit*, MW	50	*+8*	*+50*	50	*+43*	55	*+12*	*+51*	*+54*

Max = Maximum projected power, subject to **constraint in bold**
See Table 4 footnotes for definitions and clarifications
§ Equivalent diameter, rectangular electrode 3.2m × 0.85m; actual operating current; estimated cosφ ≈ 0.79 (50 MW/63 MVA) [50]

Little seems published on the operating power of the largest installed 102 MVA furnace in France. However, even if equipped with the largest available 2.1 m diameter Söderberg electrodes (and projected still lower 0.7 mΩ resistance) and operated on a well-classified, high resistivity and more refractory (e.g., alumina-rich ore) charge, only marginal improvement to 58 MW operation would result, primarily constrained based on maximum available electrode diameter (Table 5). While still larger Söderberg electrodes are conceivable, they would come at the expense of lower electrode resistance, poorer power factor and greater "interaction effect".

Remarkably, 50 MW Russian-design, AC, 6iL rectangular furnaces (also proposed by the Japanese for silico-manganese production in the 1960s [65]) equipped with rectangular electrodes [50], were concurrently installed along-side the standard circular 81 MVA Tanabe furnaces, but achieving the power input with a lower installed 63 MVA. Yet no further commercialization of this furnace type, with a significantly improved power factor of almost 0.8, occurred in manganese alloys.

Before addressing scale-up it may be prudent to consider a view [50] that a 3-electrode delta configuration be considered for silico-manganese, "which requires that energy be concentrated in the space between the electrodes and the surface of the bath". To make ferro-manganese, the 6iL configuration is "recommended to avoid overheating and excessive oxidation losses of manganese due to its high vapor pressure" [50]. Coincidently, it is believed that the actual operating practice may be that the circular furnaces are in fact as frequently applied to ferro-manganese production. The 6 MW/m^2 electrode power densities estimated for commercial AC, circular submerged-arc Mn alloy furnaces are some 40% lower than for commercial ferro-chromium (Table 4), which supports a real process metallurgy requirement to limit electrode power density and resultant vaporization of manganese (and potentially SiO(g) for silico-manganese).

Modernising the basic 6iL design concept with circular 2.1 m diameter electrodes to deliver more uniform current density and lower thermo-mechanical stresses should improve resilience to electrode breakage. Despite a resultant estimated lower 0.6 mΩ electrode resistance, this favourably yields marginally lower electrode power and power density (16 MW and 4 MW/m^2), crucial to avoiding unwanted manganese fuming. Scale up to an impressive 93 MW capacity should be realizable.

This would appear a less radical conceptual technological scale up for roughly doubling furnace power input, than adopting an entirely novel DC, 6-graphite electrode, submerged-arc furnace configuration of calculated higher 3.2 mΩ electrode resistance. This is projected capable of delivering 100 MW power input, within the current demonstrated operating constraint of 17 MW power per electrode, but with concomitant risk of a more than 6-fold increase in electrode power density (to 38 MW/m^2), with significant risk to increased manganese fuming local to the electrode.

Silicon Furnaces

The specific process chemistry, energy density and arc requirements for intermediate SiO(g) fuming and recapture; requirements of the burden and electrode crater for optimum (not excessive) presence of "refractory" SiC, together with a need for permeability typically enhanced through full, or periodic back and forth, furnace rotation, to secure energy-efficient smelting of metallurgical grade silicon, impose distinct limitations to further furnace scale up. This is compounded by stringent product quality requirements, only partly ameliorated by "iron-free" composite Söderberg electrode designs.

Benchmark operation often further demands the unusual flexibility for furnace PCD to be modified periodically, to best match burden resistivity dependent on quartzite ore and reductant qualities. The latter, when coupled with an innovative control approach to seek continually the apex of the equivalent of the theoretical C/SiO_2 stoichiometric ratio of 2, secures demonstrated minimum specific energy consumption (up to 2 MWh/t alloy lower than conventionally controlled operation on the same units). The latter is achieved by deliberately slightly under-coking, and then progressively raising the reductant ratio past the "optimal" apex, before reverting back to the initial slightly under-coked condition, and repeating the control cycle. This proves to be more effective than trying instead to perpetually control exactly on (and so invariably miss) the optimum metallurgical condition.

While 3 pairs of electrodes (for a total 6 electrodes) in a rotating cylindrical furnace configuration has been advocated [11], and indeed attempted commercially [12], significant advantages and sustained performance of the design have still to be demonstrated.

While ferro-silicon is often produced in silicon furnaces, its production is generally more forgiving. The largest commercial unit operates around 55 MW without furnace rotation. Adopting the largest proven Söderberg electrode diameter of 2.1 m (and associated lower 0.8 mΩ electrode resistance) would only marginally increase this to 67 MW power (Table 5).

The calculated prevailing operational limit of 18 MW power per electrode (7 MW/m^2 electrode power density) surprisingly only marginally exceeds that of the largest commercial ferro-manganese and silico-manganese furnaces, yet is some 30% lower than for commercial ferro-chromium (Table 4). While the considerably higher specific energy requirement of smelting silicon and its ferro-alloys is undeniable, it is less clear that this imposes any real requirement in design for increased electrode power density compared to other ferro-alloys.

However, given residual potential sensitivity to SiO(g) and fuming conditions, an opportunity may remain for further scale up by simply introducing additional electrodes in a still stationary furnace design. In a simpler (more roof access for symmetrical feed port placement) and potentially actively bound and more structurally robust rectangular configuration, an AC rectangular 6iL could potentially deliver double the prevailing power input of 106 MW.

Conceptually this could also be achieved in a novel DC, 6-graphite electrode furnace configuration of calculated higher 3.7 mΩ electrode resistance. This is projected to be capable of delivering 109 MW power input, within the prevailing demonstrated operating constraint of 18 MW power per electrode, but with concomitant risk of an almost 6-fold increase in electrode power density (to 41 MW/m^2). Possibly the addition of still more electrodes, either DC without risk of lowered power factor and the "interaction effect" [49], or even AC (the author has witnessed operation of 3 × 3-in delta, 9-electrode operations) may even still be possible.

To increase furnace power still further without more electrodes, increased arc power (higher working voltage) remains as an option. It has been suggested [51] that to "alleviate resultant risk of short-circuiting between the electrodes, the design include partitions made of a non-conducting refractory to separate the electrodes". Charging "refractory" and "insulating" quartzite fines "into the space between the electrodes over the height of the bath" was proposed. While innovative, this would present some challenge in limiting participation of the partition (of incorrect C/SiO_2 stoichiometry for reduction) with the rest of the furnace charge.

Ferro-nickel Furnaces

While the higher resistance nature of electric furnace smelting to ferro-nickel has simplified furnace control and scale up issues, there are unquestionably more innovative technological advancements that have supported the progressive increase to fewer larger capacity ferro-nickel furnaces than the other ferro-alloys. These include:

- Process metallurgy development – a shift from a traditional "immersed-arc" to a higher intensity "shielded-arc" mode of operation [1]
- Development of high-intensity water-cooled copper elements for greater crucible integrity
- Enhanced thyristor-based arc and power control to effect improved furnace electrical stability

- With higher-intensity, rationalizing the number of furnaces in operation [1], with now redundant electrical transformer pairs being innovatively coupled in series to double MVA.

A fundamental enabler, is that almost all ferro-nickel operations practise a maximum of "division of the smelting function", by rigorous ore blending (up to 100 kt prepared stockpiles), preheating (usually rotary kiln), calcination (typically rotary kilns, as long as 185 m), hot calcine transfer (typically at 800°C) ahead of large electric furnaces, followed by near-continuous slag tapping and granulation, and sometimes multi-stage crude metal refining (combination of decarburization, desiliconisation, dephosphorisation and desulfurization) to deliver a final saleable granulated ferro-nickel product. The process metallurgical stability so enforced permits the electric furnace to function as a bulk smelting unit - effectively a "slag making" machine with typically 15% metal fall by mass. The frequent selection and reliance on a "single line" highly-coupled production process, also focuses great attention on the only working line and was probably pivotal to the fundamental development of effective large-scale RKEF production of ferro-nickel.

Three large-scale variants exist (Table 6); the twin-electrode DC variant is under commissioning so not yet considered a best demonstrated practice to warrant much further analysis here. Certainly the paradox has been recognized, that ferro-nickel circular furnaces still dominate seven out of the eight largest furnaces reliably exceeding 75 MW [1], despite key furnace designers consistently advocating the decided advantages of rectangular furnaces with robust 3-dimensional binding systems [17]. Both furnace variants require "shielded-arc" mode of operation to provide increased effective total power input (P_{total}) = P_{arc} + P_{bath}, through the application of an increased ratio of arc power (P_{arc})/bath power (P_{bath}). Such use of arc resistance to drive power input ($P_{arc} = I^2R_{arc}$), helps effectively decouple the negative increased furnace sidewall heat fluxes and associated crucible lining wear imposed directly by bath power ($P_{bath} = I^2R_{bath}$).

Another critical process metallurgy factor is nickel grade: high-grade (40% Ni) requires far less reduction of iron oxide (alloy lower in dissolved $\underline{C}$ and $\underline{Si}$) and associated CO(g) generation, than low-grade (20% Ni) ferro-nickel. The requirement for adequate calcine permeability to permit controlled egress of the product gas is frequently underestimated. Excessive furnace "blows" are known to drive unwanted highly endothermic reduction of silica to $\underline{Si}$ dissolved in alloy and SiO(g) through fuming reactions (often including even more endothermic reduction and fuming of Mg(g) [32]); destabilize calcine supporting the "shielded-arc"; lower the arc stability and effective power input despite application of fast-reacting thyristor-based electrical controls; exacerbate off-gas heat losses by periodic blast of superheated gas and hot solids, in the extreme, directly damaging furnace freeboard and roof equipment.

Given uncertainties to further scale up under certain metallurgical conditions of the electrode power (high-grade alloy 25 MW limit?), and given the decidedly improved furnace structural capability of a rectangular furnace configuration, only a mega-scale 6iL is contemplated based on best demonstrated practice. The resulting projected power is 150 MW (Table 6); not too much beyond the 102 MW rectangular furnace already in construction [11]. The performance of this latter unit will be watched with great interest, to help inform on the potential for still further capacity increase in ferro-nickel electric furnace smelting.

Table 6: Best ferro-nickel furnace power inputs [1,17] and future projected power increases are based on best demonstrated limits

	FeNi # *Low-Grade*	FeNi # High-Grade #	FeNi # High-Grade	FeNi # High-Grade **Max**	FeNi *##*	FeNi ## **Max**
Electrical Supply	AC	AC	AC	AC	DC	DC
Mode of operation[a]	Shield	Shield	Shield	Shield	Open	Open
Furnace shape[b]	Circ.	Circ.	Rect.$^{\Sigma}$	Rect.$^{\Sigma}$	Circ.$^{\Sigma}$	Circ.$^{\Sigma}$
Number of electrodes	3	3	6	6	2	3
Electrode type[c]	Söd.	Söd.	Söd.	Söd.	Gr.	Gr.
Electrode diameter[d], m	1.8	1.8	1.4	1.8	0.711	0.711
Electrode current[e], kA	40	25	14	25	*45*	45
Resistance per electrode[f], mΩ	20	40	75	40	*20*	20
Electrode current density, A/cm^2	1.6	1.0	0.9	1.0	*11.3*	11.3
Power per electrode, MW	31	25	14	**25**	*40*	**40**
Electrode power density, MW/m^2	12	10	9	**10**	*101*	**101**
Total active power, MW	94	75	84	**150**	***80***	**120**
Reported operation/ *Potential benefit*, MW	94	75	84	*+56*	*80*	*+26*

Max = Maximum projected power, subject to **constraint in bold**
See Table 4 footnotes for definitions and clarifications
\# Operating electrode currents and resistances used [1]. *## Still commissioning estimates.*
$^{\Sigma}$ 3-dimensional binding system

Conclusions

There is a long tradition of evolution in ferro-alloy power input and furnace size. As "making cash" is ultimately the objective, ferro-alloy productivity takes precedence over sheer electric furnace size, as dictated by market conditions and the sustainability of environmental factors (supply of raw materials, electricity, water, CO_2-equivalent, etc.). Efforts are likely to be directed first towards achieving key improvements and efficiencies through further development of the division of the smelting function. The resultant process stability and optimization, together with still further improved process control, are likely key enablers to facilitate even more efficient mega-scale ferro-alloy electric smelters. A comparison of current best demonstrated operations and processing and electrical limits, suggests that considerable scope remains to achieve still further progression to even more productive units.

Acknowledgements

The contributions of past and current colleagues who stimulated debates on aspects of ferro-alloy smelting and electric furnace design, and specifically valuable discussions with and assistance from Nic Barcza, Ian Barker, Johan Basson, Rolf Degel, Kim Fagerlund, Laszlo Kadar, Mark Kennedy, Bjørnar Larsen, Thomas Magnussen, Johannes Nell, Bruce Nourse, Dawid Theron, Roger Urquhart, Nils Voermann and Bert Wasmund are gratefully acknowledged.

References

1. C. Walker et al., "Nickel Laterite Rotary Kiln-Electric Furnace Plant of the Future", Proceedings of COM 2009, Sudbury, Canada, August 23-26, 2009. 19.

2. *Silicon and Ferrosilicon: Global Industry Markets and Outlook*, (Thirteenth edition, Roskill Information Services Ltd., 2011), ISBN 978 0 86214 573 6.
3. N.A. Barcza and R.A. Featherstone, "The growth of ferro-alloy production in South Africa", Ferroaleaciones '82, ILAFA, 1982. 25.
4. C.N. Harman & N.S.S. Rama Rao, "Indian Ferro Alloy Industry - Present Status and Future Outlook". Indian Institute of Metals. 18. http://www.iim-delhi.com/upload_events/FerroAlloys_FACOR.pdf
5. J. Chirasha, "Ferrochrome smelting and smelting capacity investment in Zimbabwe', Southern African Pyrometallurgy 2011, Southern African Institute of Mining and Metallurgy, Johannesburg, 6–9 March, 2011. 77-82.
6. *The Economics of Chromium*, (Eleventh edition, Roskill Information Services Ltd., 2009), ISBN 978 0 86214 551 4.
7. "SA firm set to become world's largest low and medium carbon ferrochromium producer", Business Day, Monday 16 September, 2013.
8. Manganese and Manganese Alloys, (5th edition, Ullman's Encyclopedia of Industrial Chemistry, VCH Verlagsgesellschaft mbH, Weinheim, 1990). 77-121.
9. *The Economics of Manganese*, (Eleventh edition, Roskill Information Services Ltd., 2008), ISBN 978 0 86214 536 1.
10. W.A. Gericke, "A techno-economic assessment of large and small furnaces", INFACON 3, Tokyo, Japan, 8-11 May 1983. 91-102.
11. R. Degel et al., "Innovative electric smelter solutions of the SMS Group for the ferroalloys and Si-metal industry", Paper and Presentation, INFACON 13, Almaty, Kazakhstan, 9-13 June, 2013. 291-305.
12. R.C. Urquhart, *Private Communication*, 2013.
13. I.J. Barker and A.B. Stewart, "Inductive reactance and the operation of large submerged-arc furnaces", J. SAIMM, Vol. 80, No. 3, Mar 1980, 123-128.
14. S.G. King, "Electric Smelting Furnaces in Southern Africa", INFACON 74, Johannesburg, 1974, 135-142.
15. A. Shkirmontov, "Establishing the theoretical foundations and energy parameters for the production of ferroalloys with a larger-than-normal gap under the electrode", Metallurgist, Vol. 53, Nos. 5–6, 2009, 300-308.
16. A.G.E. Robiette, *Electric melting and smelting practice*, (London, Griffen,1955), 286-287.
17. *SAF Bulletin 201*, SMS Siemag.
18. L. Rodd et al., "SNNC: a new ferronickel smelter in Korea", INFACON XII, Helsinki, Finland June 6 – 9, 2010, 697-708.
19. A. Shaw et al., "Challenges and solutions in PGM furnace operation: high matte temperature and copper cooler corrosion", Platinum 2012, Sun City, South Africa, SAIMM, 193-216.
20. R.J. Hundermark, L.R. Nelson et al., "Redoubling platinum group metal smelting intensity – operational challenges and solutions", Celebrating the Mega-scale: An EPD Symposium in Honor of David G.C. Robertson, TMS Annual Meeting 2014, San Diego.
21. T. Emi, 'Trends in steel refining technology', SCANINJECT VII, Part I, Luleå, Sweden, 7-8 June, 1995. 225-248.
22. L.R. Nelson, "Production of Cr and Mn ferro-alloys", SAIMM Pyrometallurgy School, Randburg, 20-21 May 1997. 51.
23. L.A. Cramer, J. Basson and L.R. Nelson, "The Impact of Platinum Production from UG2 Ore on Ferrochrome Production in South Africa", INFACON X, SAIMM, Cape Town, South Africa, 1-4 Feb., 2004. 47-59.

24. A.V. Suslov et al., "Ferrochrome production at Aksu ferroalloy plant", Steel in Translation, 2009, Vol. 39, No. 12, 1088–1092. © Allerton Press, Inc., 2009. Original Russian Text © 2009, published in "Stal," 2009, No. 12, 34–38.
25. M.Y. Solar and I. Candy, "Selection of Optimum Ferronickel Grade", CIM Bulletin, Vol. 101, No. 1107, Mar.-Apr., 2008. 8.
26. R.C. Urquhart, "The role of the coke bed in the electric-furnace production of ferroalloys", Elec. Fce. Conf. Spec. Arc Pre-prints, 1977, 79-84.
27. T. Lindstad et al., "Improved SINTEF SiO-reactivity test" INFACON XI, New Delhi, India, 2007, 414-423.
28. P.C. Hayes, "Aspects of SAF smelting of ferrochrome", INFACON X, Cape Town, South Africa, 1–4 February 2004. 1-14.
29. S. Teasdale and P.C. Hayes, "Mechanisms and kinetics of reduction of FeO from slags by graphite, coke and coal char under iron blast furnace conditions", James M. Toguri Symposium, Ottawa, Aug. 2000, CIM, pp.467-479.
30. V. Sahajwalla, M. Dubikova and R. Khanna, "Reductant characterisation and selection: implications for ferroalloys processing", INFACON X, Cape Town, South Africa, 1–4 February 2004. 351-362.
31. R.M.S. Falcon and C.P. Snyman, "An introduction to coal petrography: Atlas of petrographic constituents in the bituminous coals of Southern Africa", The Geological Society of South Africa, Review Paper No. 2, February, 1986.
32. L.R. Nelson, et al., "Role of operational support in ramp-up of the FeNi-II furnace at PT Antam in Pomalaa", INFACON X1, New Delhi, India, 2007, 798-813.
33. L.R. Nelson et al., "Application of a high intensity cooling system to dc-arc furnace production of ferrocobalt at Chambishi," J. SAIMM, vol. 104, (2004), 551-561.
34. H.M. Dijs and D.J. Smith, "Factors affecting the resistivity and reactivity of carbonaceous reducing agents for the electric-smelting industry", J. SAIMM, Aug 1980, 286-296.
35. S. Selmer-Olsen, "Trends in ferro-alloy production", J. SAIMM, May, 1971, 210-214.
36. G.H. Geiger and D.R. Poirier, "Transport phenomena in metallurgy", Addison-Wesley, 1980, 92-102.
37. C.C. Furnas, "Grading Aggregates - I. - Mathematical Relations for Beds of Broken Solids of Maximum Density", Ind. & Eng. Chem., 23 (6), 1931, 1052-1058.
38. A.H. Olsen, "Two decades of manganese alloy smelting in Tasmania, Australia", INFACON 3, Tokyo, Japan, 8-11 May 1983. 103-113.
39. H. Tuovinen et al., "Recent and future developments in ferroalloy production", ILAFA Ferroalloy Congress – 82, Rio de Janeiro, Brazil, 19-21 May, 1982. 53.
40. T. Uemura, "Progress of high carbon ferrochrome by the SRC Process", Chromium Association Split Conference, May, 1990. 36.
41. O. Naiker, "The development and advantages of XSTRATA'S PREMUS process", INFACON X1, New Delhi, India, 2007. 112-119.
42. F. Yoshida, T. Honma and T. Sasaki, "Automation and reduction of labor in the operation of electric arc furnace for HC FeMn production", INFACON 8, Beijing, 1998, 337-342.
43. A. Oxley and N.A. Barcza, "Hydro-Pyro Integration in the processing of Nickel Laterites", Mineral Engineering, Elsevier, 6 April 2013. http://www.sciencedirect.com/science/article/pii/S0892687513000691
44. Y.E. Lee and D.S. Kozak, "The role of zinc in the eruption of high carbon FeMn smelting furnace", 51st Elec. Furn. Conf., Washington DC, 1993, 145-150.
45. K.N. Swamy, D.G.C. Robertson et al., "Factors affecting carbon consumption in the production of high carbon ferromanganese", INFACON 9, 293-301.

46. N.A. Barcza et al., "The application and scale-up of a.c. and d.c. smelting furnaces for ferro-alloys", Proc. 60th Elect. Furn. Conf., Vol. 60, Warrendale, PA: Iron & Steel Society, 2002. 425-437.
47. D. Sager et al., "Low cost ferroalloy extraction in DC-arc furnace at Middleburg Ferrochrome", J. SAIMM, Vol. 110, December 2010, 717-724.
48. J. Andō et al., "Development of slag blast granulating plant characterized by innovation of the slag treatment method, heat recovery and recovery of slag as resources", Technical Review, Mitsubishi Heavy Industries, Ltd., June 1985. 7.
49. I.J. Barker, "Some considerations on future developments in ferro-alloy furnaces", Southern African Pyrometallurgy 2011, SAIMM, Misty Hills, South Africa, 6–9 March 2011. 1-14.
50. A. Shkirmontov, "Determination of the energy parameters for the smelting of manganese ferroalloys with increases in the electrode gap and electrode spacing", Metallurgist, Vol. 53, Nos. 7–8, 2009, 512-517.
51. A. Shkirmontov, "Determination of the energy parameters for the smelting of ferrosilicon with an increased electrode gap. Limitation on the electrode–electrode current", Metallurgist, Vol. 53, Nos. 7–8, 2009, 434-438.
52. A. Shkirmontov, "Energy parameters of ferrosilicon production with larger-than-normal values for the electrode gap and electrode spacing under factory conditions", Metallurgist, Vol. 53, Nos. 9–10, 2009. 642-647.
53. F.V. Andreae, "Design and control of ferroalloy furnaces", Trans. American Inst. Elec. Engrs., Vol. 69, 1950, 557-562.
54. I.J. Barker et al., "The "interaction effect" in submerged-arc furnaces", 49th Elec. Furn. Conf., Toronto, Canada. ISS-AIME, 1991, 305–310.
55. A.L. Moolman, M.S. Rennie and P. Brereton-Stiles, "Experiences in Operation of Various Electric-Arc Furnaces under Resistance Control", INFACON 9, Quebec City, Canada, 3-6 Jun. 2001. 103-107.
56. M. Sedighy, T Ma, and N. Voermann, "New developments in furnace power stabilization with SPLC", INFACON XII, Helsinki, Finland, 6-9 June, 2010, 867-878.
57. G. Folmo et al., "Furnace management in ERAMET manganese during the 2009 crisis", INFACON XII, Helsinki, Finland, 2010. 521-530.
58. M. Morimoto et al., "Modification of operating parameters for ferroalloy electric furnaces to allow extension of intervals to tap", INFACON 7, Trondheim, Norway, 1995. 565-572.
59. R. Innvær, K. Fidje and R. Ugland, "Effect of current variations on material properties and thermal stresses in Søderberg electrodes", INFACON 86, Rio de Janeiro, Brazil, 1986, 321-330.
60. R. Innvær, L. Olsen and A. Vatland, "Operational parameters for Søderberg electrodes from calculations, measurements, and plant experience", MINTEK 50, Johannesburg, 1984, 787-794.
61. "Samancor cuts ferro-chrome output in Eskom deal", Metal Bulletin, January 11, 2013. http://www.metalbulletin.com/Article/3140406/Raw-materials/Samancor-cuts-ferro-chrome-output-in-Eskom-deal.html.
62. Graftech, ALX Graphite Electrode, http://www.graftech.com/PRODUCTS/Graphite-Electrodes/ALX%e2%84%a2-Graphite-Electrodes.aspx, ALX™ Sell Sheet.pdf. 2.
63. K. Bergman and B. Kjellberg, "DC arc furnace technology applied to smelting applications", INFACON 9, Quebec City, Canada, 3-6 June 2001, 80-89.
64. M. Hamy, and J.M. Thebault, "Three years production in the DC arc furnace in the Societe Metallurgique de l'Escaut mini-mill", Metallurgical Plant & Technology, May 1989, 54-60.
65. S. Tomioka et al., "A six electrode, silicomanganese smelting furnace", 5th Int. Cong. on Electroheat, No. 103, Wiesbaden, Germany, 30 Sep-5 Oct., 1963. 8.

Celebrating the Megascale: Proceedings of the Extraction and Processing Division
Symposium on Pyrometallurgy in Honor of David G.C. Robertson
Edited by: Phillip J. Mackey, Eric J. Grimsey, Rodney T. Jones, and Geoffrey A. Brooks
TMS (The Minerals, Metals & Materials Society), 2014

FROM SULFIDE FLASH SMELTING TO A NOVEL FLASH IRONMAKING TECHNOLOGY

Hong Yong Sohn

Department of Metallurgical Engineering, University of Utah, Salt Lake City, Utah 84112

Keywords: carbon dioxide, coke, concentrate, energy, flash ironmaking, flow sheet, hydrogen, magnetite, natural gas, process simulation, reformerless, taconite

Abstract

A novel flash ironmaking technology is under development at the University of Utah under the support of U.S. Department of Energy (DOE) and American Iron and Steel Institute (AISI). The history of the development from the conception of the idea to the current status will be discussed. The flash ironmaking process produces iron directly from fine iron ore concentrates without requiring cokemaking and pelletization/sintering, which will enable the technology to significantly reduce energy consumption and carbon dioxide emissions compared with blast furnace ironmaking. Unlike other gas-based ironmaking processes, this technology will not suffer from the problems of solid sticking and fusion. Current work is focused on the method of supplying the energy required to maintain the necessary temperature, as an intermediate step to determine the scalability for larger, industrial-scale pilot trials.

Introduction

Development of sustainable metallurgical processes and the fusion of ideas from different disciplines for technology innovation are two most important factors in modernizing the metallurgical industry. One of the core technical problems facing the steel industry is to develop a new technology to produce iron in the future from the viewpoint of energy consumption and greenhouse gas emissions. This paper describes the development of a novel ironmaking process, which is based on a process used extensively in copper and nickel production. This also represents an appropriate example of linkage between ferrous and nonferrous technologies that is highly beneficial and thus should take place to greater extents for the growth of both industries.

Ironmaking is an energy intensive process and is one of the largest producers of greenhouse gas emissions, specifically CO_2. Unfortunately, after decades of refinement, the traditional technology is approaching its practical efficiency limits. In addition, the existing plants are in need of replacement. A typical steel mill produces in excess of 3 million tons of iron per year and thus requires immense capital investment.

An ideal process would replace the blast furnace and coke oven; would directly use iron ore concentrates for reasons presented below; and would greatly reduce energy requirements. It should also require much less capital investment than the blast furnace/coke oven combination, and must be capable of producing 5,000 – 10,000 tons of metal per day so that it can support existing steel mills.

The steel industry contributes around 5~7 % to total anthropogenic emission of CO_2 [1]. The novel Flash Ironmaking Technology, conceived by Sohn [2], has the potential of significantly reducing energy consumption by up to 20-35% and lowering carbon dioxide emissions by 36-97% compared with the average blast furnace ironmaking route, depending on whether natural gas or hydrogen is used as the fuel/reductant. These benefits come largely from the elimination of cokemaking and sintering/pelletization, by using gaseous reduction of iron oxide concentrate in a flash reactor. It takes advantage of the large specific area of fine particles of the concentrate, much like the flash smelting of copper or nickel sulfide concentrate.

More than 90% of iron is currently produced by the Blast Furnace process, while the balance is produced by Direct Reduction processes [3]. The Blast Furnace process requires sinters or pellets of iron ore and coke made from high-grade coking coal, which consume much energy and are problematic from an environmental perspective. Also, the BF process is highly capital intensive, requiring large-scale infrastructure and operation. Those constraints limit the flexibility of the BF process in operation and choice of materials [4].

There has been a great deal of effort to develop alternate ironmaking technologies such as direct gaseous reduction and smelting reduction processes [5, 6]. Most of these processes, however, are not sufficiently intensive to replace the blast furnace because they cannot be operated at high temperatures due to the sticking and fusion of particles. The shaft furnace processes, which are the dominant types among these, require pelletization of concentrate with the accompanying costs and environmental problems, and also suffer from pellet disintegration problems.

The process under development uses a reductant gas such as natural gas, hydrogen or syngas, while bypassing pelletization/sintering and cokemaking steps required in other ironmaking processes [2, 7, 8]. Unlike other gas-based ironmaking processes based on shaft furnaces or fluidized-bed reactors, the flash reduction technology will not suffer from the sticking and fusion of particles. The disintegration problems encountered in the use of pellets are also eliminated.
The availability of large quantities of iron ore concentrate produced in the U.S. and elsewhere will also be a backbone for the possible development of this technology. 60-70% of the U.S. iron production is based on taconite concentrates (50-55 million tons/year) [9]. Newer reserves will also need to be ground down to concentrate sizes in order to beneficiate them and make them acceptable feed materials for ironmaking.

Description of Technology

Natural gas or hydrogen will be partially combusted with industrial oxygen through a burner on top of the flash reactor to generate a reducing gas stream at 1600 – 1900 K. Iron ore concentrate will be injected in the vicinity of the burner and will undergo reduction reaction as it flows down. In one version of the process the reduced iron will be collected as a molten bath with the possibility of direct steelmaking in a single unit, but the reduced product may also be collected as solid particles and briquetted to be used in separate steelmaking furnaces.

The Flash Ironmaking Technology is initially expected to be operated using natural gas as a fuel and reducing agent. Natural gas is an abundant readily available resource in the U.S. and elsewhere in the world. According to the 2011 Annual Energy Outlook [10], the estimations on the U.S. production rate of natural gas predict a steady increase; there is a big push to increase natural gas production in the U.S. from its considerable reserves. The recently discovered reserves in Marcellus Shale [11] represent a good example of this trend.

Kinetic Feasibility - Reduction Kinetics of Magnetite Concentrate Particles

A critical condition for the proposed process is for iron oxide concentrate to be reduced to a high metallization degree within the few seconds of residence time available in a typical flash furnace. Hydrogen is the main reducing agent in the new process that takes place at temperatures above 1473 K, even when partial combustion of natural gas producing an H_2+CO mixture is used. Most previous work was done on the reduction of iron ore particles at temperatures below 1373 K, which is much lower than the temperature at which the new process is expected to operate [2, 7]. What is more, none of the rate equations reported in the literature contained all the information needed for reactor design relevant to the new process. Thus, it was necessary to develop a comprehensive kinetics expression applicable to the new conditions.

Such reduction kinetics of the gaseous reduction of magnetite concentrates was measured and a rate equation that contains the effects of process conditions in a flash ironmaking reactor was developed. These conditions include temperature, partial pressures of hydrogen and water vapor, and particle size. Furthermore, data covering the entire fractional reduction degree vs. time, which had not been available for the conditions applicable to the proposed ironmaking process, were generated in the author's laboratory. An additional emphasis on the formulation was that such a rate expression be readily applicable to the development of an overall process model for the design and simulation of a reactor for the proposed ironmaking technology.

Details of the materials, experimental apparatus and procedure, and approach to kinetic analysis can be found elsewhere [12, 13]. Here, only the pertinent results will be summarized.

The nucleation and growth kinetics expression at a constant temperature given by the following equation was found to best describe the reduction rate of magnetite concentrate particles:

$$[-Ln\ (1-X)]^{1/2} = [k \cdot f_p(p_{H_2}, p_{H_2O}) \cdot f_d(d_p)] \cdot t = k_{app} \cdot t \quad (1)$$

where X = fractional reduction degree, t = reaction time, k = rate constant, d_p = particle size, $f_p(p_{H_2}, p_{H_2O})$ = concentration driving force, $f_d(d_p)$ = rate dependence on particle size, and k_{app} = apparent rate constant.

The reduction of magnetite to iron proceeds through the formation of FeO. However, for small particles, it is extremely difficult to follow the irregular evolution of morphology involving FeO formation and the overlapping regions of different reduction extents. Furthermore, the reduction of Fe_3O_4 to FeO accounts only for the first 25 % of conversion, whereas the main interest in this work and the new ironmaking technology is the attainment of high degrees of reduction. Thus, equation (10) was developed as a global rate equation for the overall reduction process.

The following expression for pressure dependence, which satisfies the equilibrium condition as a special case when the net rate is zero, was found applicable with K being the equilibrium constant of the reaction (FeO + H_2 = Fe + H_2O):

$$f_p(p_{H_2}, p_{H_2O}) = p_{H_2}^{1/2} - (p_{H_2O}/K)^{1/2} \quad (2)$$

Surprisingly, the reduction rate increased with the particle size at 1423 and 1473 K within the size range tested. However, the effect of particle size on reduction rate was negligible after temperature reached 1573 K. We believe the major factor in this size dependence is the variation in the contents of gangue materials. The contents of the main gangue materials CaO, MgO and

SiO_2 in iron oxide concentrate particles were higher for larger particles. Previous investigations have shown that the addition of CaO [14] or MgO [15] increases the rates of gaseous reduction of wustite. What is more, cracks tend to be formed in the large particles during the reduction more readily than in the smaller particles, and the dimension of solid between cracks can be smaller than the size of the smaller particles. The rate dependence on the particle size was assumed to take the form of $(d_p)^s$. Thus,

$$f_d(d_p) = (d_p)^s \tag{3}$$

with

$$s = \frac{2.73\times10^8}{T^2} - \frac{345930}{T} + 109.5 \qquad T = 1423{\sim}1573\ \mathrm{K} \tag{4}$$

$$s = 0 \qquad T \geq 1573 \tag{5}$$

After the functional dependence of the rate on partial pressures of gaseous species (f_p) and particle size (f_d) were determined, the reaction rate constant k was obtained as follows:

$$k = (9.9\times10^{14})\cdot exp(-\frac{463000}{RT}) \tag{6}$$

resulting in the following conversion-vs.-time relationship:

$$[-Ln\ (1-X)]^{1/2} = (9.9\times10^{14})\cdot e^{-\frac{463000}{RT}}\cdot\left[p_{H_2}^{1/2} - (p_{H_2O}/K)^{1/2}\right]_{am}\cdot(d_p)^s\cdot t \tag{7}$$

where R is 8.314 J/mol/K, T is in K, p is in atm, d_p is in μm, and t is in seconds.

The instantaneous rate of reaction, which is valid under varying temperature and gas composition, can now be expressed as

$$\frac{dX}{dt} = 2\,k\cdot\left[p_{H_2}^{1/2} - (p_{H_2O}/K)^{1/2}\right]\cdot(d_p)^s\cdot(1-X)\cdot[-Ln\ (1-X)]^{1/2} \tag{8}$$

This equation satisfactorily represents the experimentally determined rate of magnetite concentrate, as a sample set of data given in Fig. 1 indicates [13].

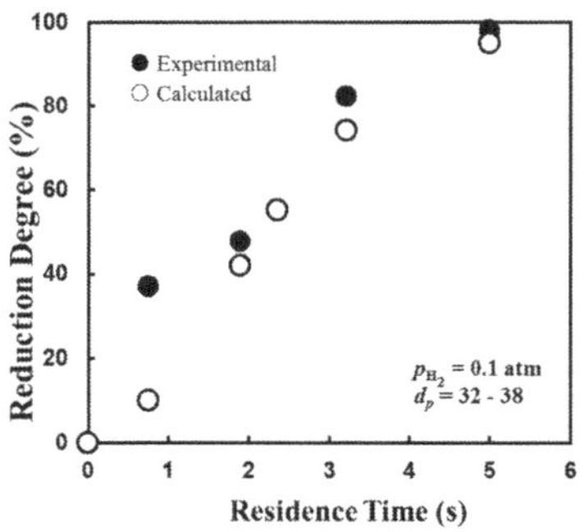

Figure 1. Comparison of experimental and calculated results. p_{H2} = 0.1 atm; d_p = 32 - 38 μm; 1400°C. (Adapted from Wang and Sohn [13].)

The rate measurements indicated that magnetite concentrate particles can be reduced to a sufficiently high degree of metallization within 2~7 seconds of residence time available in a flash reactor.

Tests in a Bench Flash Reactor

Based on the above results, larger scale suspension reduction tests were conducted by Sohn et al. [7]. The maximum temperature obtained with the set-up was 1420 K in 76 cm of isothermal zone. The average residence time varied from 3.5 to 5.5 seconds.

As shown in Fig. 2(a), the reduction extent was determined at three different average residence times; 3.5, 4, and 4.5 seconds in which the reduction extents were 21, 29 and 43%, respectively, with 0% excess H_2 and approached over 90% with large excess H_2 at 1423 K which was the highest temperature that could be generated in this reactor. The reduction extents in longer residence times were always higher and only 0.5 second difference made a notable difference in the reduction rate. Figure 2(b) shows that the extent of reduction significantly depends on the excess hydrogen. As shown in Fig. 2(c), additional experiments conducted with moderate % excess H_2 from 50 to 100% and in longer residence times from 5.0 to 5.5 seconds at 1423 K supported the above statement with the results that 64% and 71% reductions were reached in 5.0 and 5.5 seconds, respectively, with 100% excess H_2.

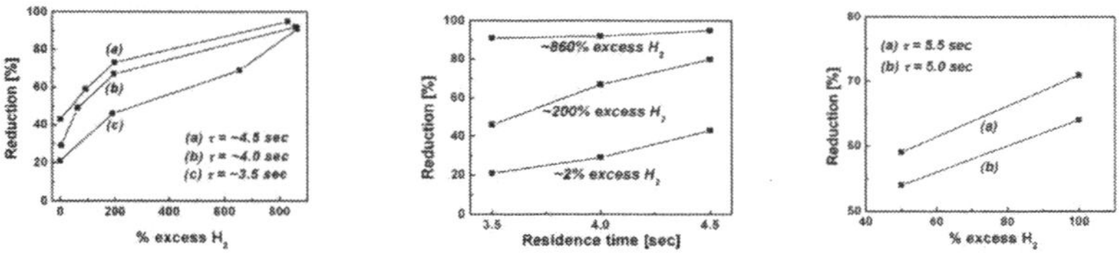

(a) (b) (c)

Figure 2. Reduction rate of iron oxide concentrate vs. (a) % excess H_2 in 3.5~4.5 seconds (b) residence time with 2~860% excess H_2 and (c) % excess H_2 in 5.0 and 5.5 seconds. (Adapted from Sohn et al. [7].)

Process Design and Simulation

Detailed flow sheets for different versions of the new process depending of the fuel type were developed by Pinegar et al. by using commercially available software METSIM [16-18]. Although hydrogen is the cleanest fuel/reductant for the proposed ironmaking process, it is still expensive to produce. Thus, the results for using natural gas in a "reformerless" mode will be summarized here [17]. The reformerless process involves an ironmaking step and off-gas treatment/hydrogen recycling steps. Water gas shift (WGS) reaction and pressure swing adsorption (PSA) were incorporated in the off-gas treatment steps. A block diagram of the reformerless process is shown in Fig. 3. Descriptions of unit operations and streams as well as other details are given elsewhere [17]. The annual iron production rate was set to be 1 million tons with 300 days of annual operation. Major results of the process simulation can be summarized as follows:

- CO_2 emissions from the new process: 1.0 ton/ton Fe vs. 1.6 ton/ton Fe from an average blast furnace.

- Energy requirement for the new process: 14.7 GJ/ton Fe vs. 18.0 GJ/ton Fe for an average blast furnace. See Table 1 for further details, including the case of using hydrogen. (Note: These numbers are those calculated based on approach used in DOE reports. For a discussion on different items included in 'energy requirements' for ironmaking, the reader is referred to a paper on the subject presented at this meeting [19].)

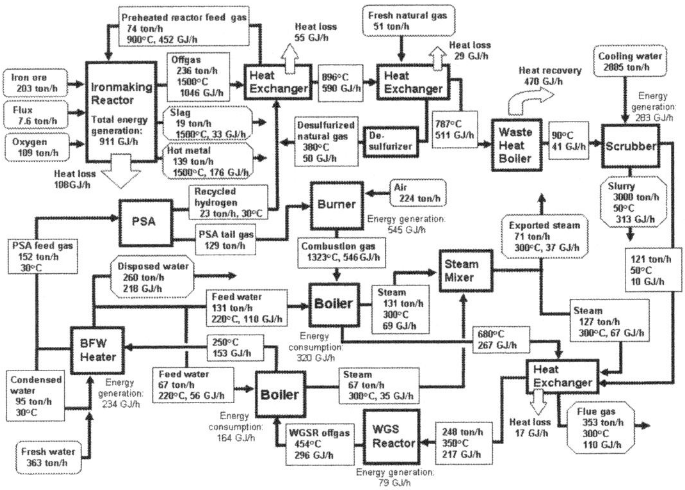

Figure 3. Material and energy flows for a reformerless ironmaking process producing 1 million tons of iron annually (300 days operation in 1 year). (Adapted from Pinegar et al. [17].)

Table 1. Energy requirement for commercial-scale Flash Ironmaking Process using natural gas or hydrogen compared with that for the average blast furnace process.

		Reformerless Natural Gas	**Hydrogen**	**Blast Furnace**[a]
INPUT[b] **(GJ/mtFe)**	Fuel combustion[c]	19.22	14.05	13.6
	Pelletizing/ Sintering/ Cokemaking			3.01/0.65/2.02
	Limestone calcination	0.26	0.26	
	Total	**19.48**	**14.31**	**19.28**
OUTPUT[d] **(GJ/mtFe)**	Heat recovery	4.77	2.80	1.32
	Total	**4.77**	**2.80**	**1.32**
NET = INPUT – OUTPUT (GJ/mtFe)		**14.7**	**11.5**	**18.0**

[a] Calculated based on material balance following the same procedure as that for flash ironmaking.
[b] All raw materials are considered to enter the system at room temperature (25°C).
[c] Fuel combustion energy input in this case is the heat of combustion of the entire fuel + reductant.
[d] Energy output is the useful heat recovered (plus any heating values which in this case are zero because we considered fully oxidized species) in both flash ironmaking processes and blast furnace.

Economic analysis based on the simulation results indicated that the new process has favorable economic feasibility when operated with natural gas, and the capital investment would be significantly lower than for a new plant based on blast furnace and coke oven. Any carbon penalty would make the new technology even more attractive [20].

Further Development Work

The development work is continuing toward the design and construction of a large-scale bench reactor and, beyond that, a 50,000-100,000 ton of iron per year industrial pilot testing [21]. The large-scale reactor test program aims to overcome several technical difficulties before an industrially viable technology can be finalized, such as the method of supplying energy required to maintain the necessary temperature, the installation of proper refractory and insulation linings, and the safe control of the entire operation. The large-scale bench reactor will produce iron directly from iron ore concentrate (10~100 μm) using natural gas or hydrogen as the fuel as well as the reductant in a closed cylindrical reactor. Of the specific design and functional features necessary for the bench facility, the reactor heating system, which has been the major limitation in the preliminary scale-up tests, is a key issue.

The bench reactor under construction consists of a refractory-lined vertical vessel (80 cm ID, 150 cm long) in which iron oxide concentrate is to be reduced in flight by hot reducing gases produced by partial combustion of natural gas or hydrogen. The reducing gas is to be introduced through a swirl burner installed at the top part of the reaction chamber. The concentrate powder will be fed using a pneumatic feeding system through the burner and auxiliary ports located on the vessel roof. A typical feed rate of iron ore concentrate in this scale will be 5 kg/h. Concentrate particles are to be supplied through the burner or in the vicinity of the hydrogen /oxygen flame. The average temperature of the reaction shaft will be varied in the range of 1650~1850 K. The energy for the sensible heat of solid and gas feed materials and for the reduction reaction of iron oxide is to be supplied by a single burner (1.27 GJ/h) in combination with a 100 kW plasma torch as a supplementary heat source. The entire system will be monitored by several safety measures and controlled by proper emergency shut-off procedures.

This system to be installed at the University of Utah has many features that resemble an eventual industrial reactor. They include a burner as the main heat source and a refractory-lined vessel. Upon completing this project phase, the engineering data necessary for industrial pilot-plant operations will be determined, such as heat loss, product collection, raw materials feeding, process control strategy, burner design, and off-gas treatment. In addition, a risk mitigation plan addressing the critical technical hurdles to be overcome has been devised, as follows:

1. Heat supply. If the necessary temperature cannot be attained because of the large heat loss typically present in a small test reactor, plasma can be used to augment the heating. Calculations indicated that the heat loss in the bench reactor could reach more than 30% of total heat input to the reactor. Similar calculations for an industrial-scale reactor (15 m ID, 15~45 m height) showed that the heat loss would drop to ~2% of total input.
2. Particle residence time. The burner is designed to provide different levels of swirl flow which allows the control of the particle residence time.
3. Refractory lining. Different types of refractory liners may be tested in the course of the experiments.

Safety will be the most important issue in this bench testing. A great deal of attention is paid to safety-related details such as the Sequence of Operation and the Emergency Shut-Off Procedures for a safe and sound operation as well as a Hazards and Operability (HAZOP) study.

References

1. K.-H. Tacke and R. Steffen, *Stahl und Eisen.*, 124 (4) (2004), 45.

2. H. Y. Sohn, *Steel Times International*, May/June (2007), 68.

3. IISI, Steel Statistical Yearbook (2007). http://www.worldsteel.org/pictures/publicationfiles/SSY2007.pdf.

4. C. P. Manning and R. J. Fruehan, *JOM*, 53 (10) (2001), 36.

5. Lockwood Greene Technologies, "Ironmaking Process Alternatives Screening Study" (Report LG Job No. 010529.01, U.S. Department of Energy, 2000).

6. D. Macauley, *Steel Times International*, 21 (1) (1997), 20.

7. H. Y. Sohn et al., *Iron & Steel Technology (AIST Trans.)*, 6(8) (2009), 158.

8. H. Y. Sohn and M. E. Choi, *Collected Proceedings: Supplemental Proceedings: Vol. 1 Materials Processing and Properties, International Symposium on High-Temperature Metallurgical Processing*, 139th TMS Annual Meeting, Seattle, WA, ed. J. Drelich et al. (Warrendale, PA: TMS, 2010), 347.

9. USGS, "2005 Minerals Yearbook - Iron Ore" (Report U.S. Department of the Interior, U.S. Geological Survey, 2007).

10. U.S. Energy Information Administration, *Annual Energy Outlook 2011*, http://www.eia.gov/forecasts/aeo/index.cfm

11. Pittsburgh Post-Gazette (2011), http://www.post-gazette.com/pg/11202/1161933-503.stm; http://www.eia.gov/analysis/studies/usshalegas/

12. M. E. Choi and H. Y. Sohn, *Ironmaking and Steelmaking*, 37 (2010), 81.

13. H. Wang and H. Y. Sohn, *Metall. Mater. Trans. B*, 44B (2013), 133.

14. N. Shigematsu and H. Iwai, *ISIJ, International*, 29(6) (1989), 486.

15. P. W. Tasker, E. A. Colburn, and W. C. Mackrodt, *J. Amer. Ceram. Soc.*, 68 (1985), 74.

16. H. K. Pinegar, M. S. Moats, and H. Y. Sohn, *Steel Res. Int'l.*, 82 (8) (2011), 951.

17. H. K. Pinegar, M. S. Moats, and H. Y. Sohn, *Ironmaking Steelmaking*, 39 (2012), 398.

18. H. K. Pinegar, M. S. Moats, and H. Y. Sohn, *Ironmaking Steelmaking*, 40 (2013), 32.

19. H. Y. Sohn and M. Olivas-Martinez, "Determination of Energy Requirements for Ironmaking Processes: It's Not That Straightforward," (Paper presented in Energy Education, Energy Technologies and Carbon Dioxide Management at TMS Annual Meeting, San Diego, CA, 2014).

20. H. K. Pinegar, M. S. Moats, and H. Y. Sohn, *Ironmaking Steelmaking*, 40 (2013), 44.

21. H. Y. Sohn, M. E. Choi and M. Olivas-Martinez, *Proceedings of 6th Intl Congress on Sci. and Tech. of Ironmaking (ICSTI '12)*, Oct. 14 – 18, 2012, Rio de Janeiro, Brazil, ISSN 2176-3135, Paper No. 22117 (2012), 460-468.

Celebrating the Megascale: Proceedings of the Extraction and Processing Division
Symposium on Pyrometallurgy in Honor of David G.C. Robertson
Edited by: Phillip J. Mackey, Eric J. Grimsey, Rodney T. Jones, and Geoffrey A. Brooks
TMS (The Minerals, Metals & Materials Society), 2014

FOSTERING MINERALS WORKFORCE SKILLS OF TOMORROW THROUGH EDUCATION AND TRAINING PARTNERSHIPS

Gavin Lind[1]

[1]Minerals Council of Australia; 8th Floor; 10-16 Queen Street; Melbourne; Victoria; 3000; Australia
E Mail: gavon.lind@minerals.org.au

Keywords: Higher Education, Academic Collaboration, Industry Engagement, Minerals Industry

Abstract

The Minerals Council of Australia (MCA), through its Minerals Tertiary Education Council (MTEC), builds capacity in higher education in the core disciplines of mining engineering, metallurgy and minerals geoscience. Over the past fourteen years, this all-of-industry approach in securing the long-term supply of these critical skills (which remain a chronic skills shortage for the Australian minerals industry) through nationally collaborative programs across sixteen Australian universities delivers spectacular and sustainable results for the industry. These unique, world-first programs are built on a healthy platform of dedicated industry funding and in-kind support and forms part of the MCA's broader uninterrupted, sustainable education and training pathway to increase workforce participation, workforce diversity and workforce skills, regardless of the business cycle in the industry. This paper will highlight the origins, iterations and current successful programs of MTEC, including its future vision, and presents a mechanism for industry and academia to collaborate to address future professional skills needs in the minerals industry globally.

Introduction

The Minerals Council of Australia (MCA) represents Australia's exploration, mining and minerals processing industry, nationally and internationally, in its contribution to sustainable economic and social development. MCA member companies produce more than 85 per cent of Australia's annual mineral output, contributing some A$163 billion in 2011-12 (52 per cent) of Australia's total exports [1].

Minerals-related higher education courses (mining engineering, metallurgy and earth sciences) have been chronically underfunded in Australia, requiring the industry to subsidise these courses heavily. The Minerals Tertiary Education Council (MTEC) – the higher education arm of the Minerals Council of Australia (MCA) – engages with the tertiary education sector in Australia to help ensure the sustainability of this sector and so influence the supply of suitably educated and trained professionals for the minerals industry.

Since 1999, over A$35 million of industry funds have been allocated to assist MTEC in the development of industry-focused courses and in the employment of academic staff and educational specialists [2]. MTEC fosters the partnership between industry, government and academia to promote and provide opportunities in the tertiary education arena by working with a network of selected university partners (Figure 1). These partners are dedicated to achieving world-class education by cooperating in the development and delivery of undergraduate and postgraduate programs in the specialist disciplines of mining engineering, metallurgy and earth science (specifically minerals geoscience). MTEC works with sixteen partner universities across Australia – four in mining engineering, three in metallurgy and nine in minerals geoscience in building capacity in the higher education arena to deliver high quality graduates to industry. Through industry engagement, Australia not only has the capability to produce high quality graduates but also the opportunity of delivering tertiary minerals education to the global minerals industry.

Minerals Geoscience	Mining Engineering	Metallurgy
University of Tasmania	The University of New South Wales	The University of Queensland
The University of Western Australia	The University of Queensland	Curtin University (WA School of Mines)
James Cook University	Curtin University (WA School of Mines)	Murdoch University
The University of Adelaide	The University of Adelaide	
The University of Melbourne		
Monash University		
The Australian National University		
Curtin University (WA School of Mines)		
University of Newcastle (provisional)		

Figure 1. Minerals Education Tertiary Education Council (MTEC) partner universities.

Importantly for the minerals industry in Australia, the MCA reports final-year student number increases at MTEC-supported universities over the years 2007-2011 of 94 per cent in minerals geoscience (Honours), 84 per cent in mining engineering and 50 per cent in extractive metallurgy [3,4]. These increases are encouraging but, considering that employment in the Australian minerals industry has doubled over the same period, the gap in delivering professional skills is not only widening, but these professions are falling behind in their ability to meet industry's demand for skilled engineers and geoscientists.

Australian minerals industry involvement in minerals education

Simply, without direct industry financial and in-kind support, mining engineering, extractive metallurgy and minerals geoscience programs at Australian universities would be under severe threat of closure, predominantly due to low historic enrolments as described previously and universally high teaching costs associated with these programs. Some core contributors to the increasing teaching costs of metallurgy degrees include:

- The failure of past Australian Governments to index higher education funding which has resulted in a growing inability of university departments to be viable under the student numbers-based system – especially when student numbers are low;
- A failure by the Australian Government to recognise metallurgy as a discipline of national interest and afford it appropriate levels of funding;
- Universities investing in teaching and learning capabilities for high enrolment, low teaching cost courses (i.e. volume-driven programs);

- Shortages of skilled academic staff, compounded by the ageing profile of academics.

The Australian Government committed significant reforms in higher education in 2011 [5] through the reintroduction of indexation and a demand-driven student funding scheme. However, these reforms (realistically) are of little consequence or benefit to the extractive metallurgy departments in Australia as they do not address the chronic structural underfunding of this core discipline.

In addition, to secure a future supply of professionals for the minerals industry direct investment by the minerals industry, has been and is still a requirement. This investment ensures that universities can sustainably deliver high-cost/low-student number programs by increasing their capacity to attract and retain high quality academics and researchers. Industry involvement also facilitates the delivery of undergraduate programs that equip students with relevant high quality technical and decision making skills that take account of the social, environmental and financial aspects of development. Graduates who have these skills can actively play key roles in delivering on the national innovation agenda, by accounting for the rights and interests of both current and future generations.

Figures 2, 3 and 4 show the actual graduate and enrolment numbers (and predicted enrolments to 2016) for the collective final years at MTEC partner universities for mining engineering, metallurgy and minerals geoscience respectively. Obvious from these graphs are the increases in graduates in these core disciplines (some more impressive than others).

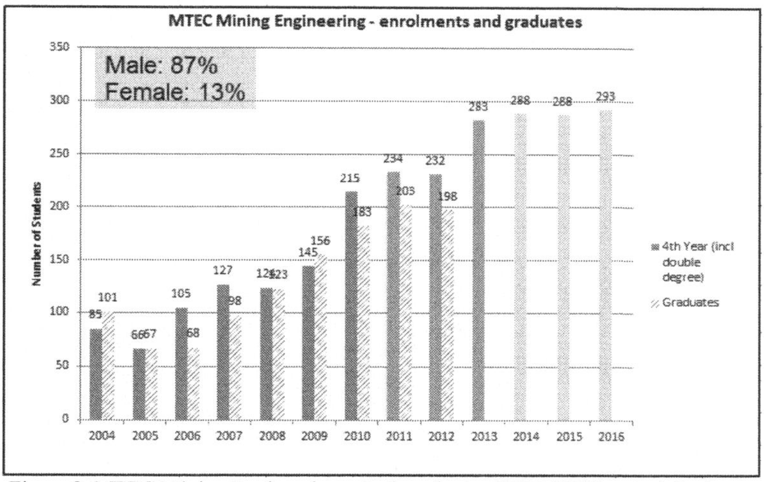

Figure 2. MTEC Mining Engineering actual graduates (2004-2012), actual final year enrolments (2004-2013) and predicted final year enrolments (to 2016).

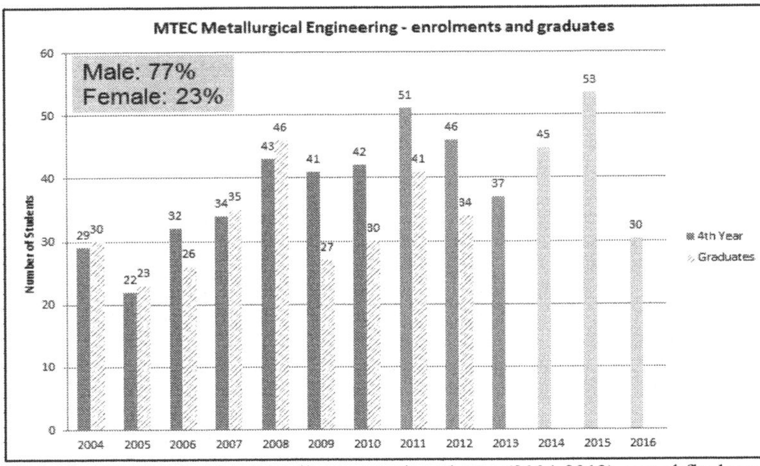

Figure 3. MTEC Extractive Metallurgy actual graduates (2004-2012), actual final year enrolments (2004-2013) and predicted final year enrolments (to 2016).

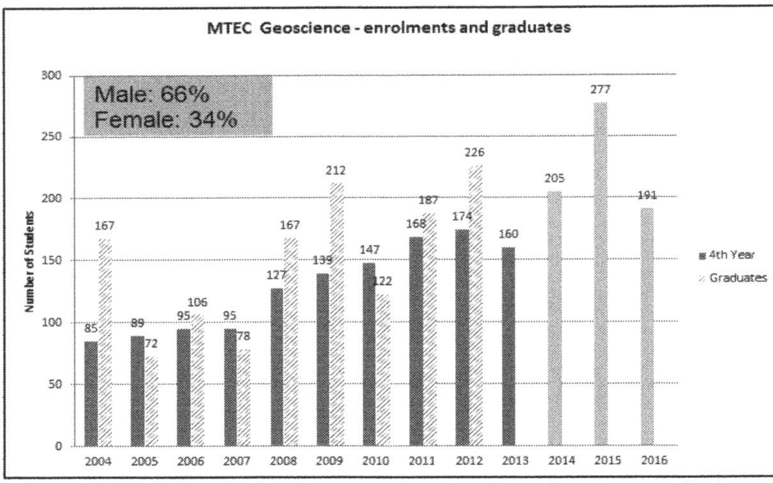

Figure 4. MTEC Minerals Geoscience actual graduates (2004-2012), actual final year enrolments (2004-2013) and predicted final year enrolments (to 2016).

A higher education sector with the institutional capacity and flexibility to ensure an ongoing supply of minerals industry professionals is critical to the future of this industry and to the prosperity of Australia. The minerals sector is concerned at the decline in sector-specific higher education funding as first reported in the MCA's *'Back from the Brink'* report of 1998 [6]. The Minerals Tertiary Education Council (MTEC) was established in 2000 by the Minerals Council of Australia specifically to address this capacity constraint. MTEC is funded by voluntary levy from MCA member companies and its efforts are directed towards teaching programs that are:

- collaborative in nature;
- national in scope;
- tapping sources of recognised teaching and research expertise in universities; and
- delivered by institutions with robust links to industry.

Over the period 1998 - 2011, key MTEC achievements have included:

- Development of the Industry Experience for Undergraduates (IEU) Program - a coordinated initiative for undergraduate students seeking work experience in order to develop their careers in the industry;
- The cross-institutional development and delivery of new postgraduate courses and modules. This is supported by *PATHWAYS* - a postgraduate coursework program to assist industry professionals with skills development and/or re-training;
- Employment of new academic staff. Twelve mostly industry experienced academic staff have been employed at participating universities to assist with course development and delivery;
- Establishment of and contribution of funding to centres of excellence - including the National Centre for Mine Ventilation; Centre for Ore Deposit Research (CODES) as well as specific Cooperative Research Centres (CRC for Landscape Environments and Mineral Exploration, Predictive Mineral Discovery CRC and the Centre for Sustainable Resource Processing);
- Development of a social competencies postgraduate course to provide formalised training for community relations practitioners working in the industry. The course builds core competencies in cultural studies, community consultation and engagement, and community sustainability and development;
- Establishment of Mining Education Australia; a joint venture with the Universities of New South Wales, Queensland and Curtin University (through the Western Australian School of Mines – WASM); and the University of Adelaide since January 2011, in which the teaching resources of the three partner universities are combined to deliver nationally consistent third and fourth years of a mining engineering degree;
- Development of a formal partnership for collaborative teaching of extractive metallurgy (Metallurgical Education Partnership) for final year engineering undergraduate students of the only three universities in Australia that teach this discipline – University of Queensland, Murdoch University and Curtin University/WASM [4];
- Development of a collaborative, national advanced level undergraduate (Honours) program in earth sciences (Minerals Geoscience Honours) delivering industry-relevant skills to final-year students at eight participating universities – Universities of Tasmania, Western Australia, Melbourne, Adelaide and James Cook University, Monash University, Australian National University and Curtin University /WASM, and
- Development of a collaborative national geoscience Masters by coursework program (Minerals Geoscience Masters) for practising professionals to up-skill using the collective resources of the Universities of Western Australia and Tasmania, Curtin University/WASM and James Cook University.

- Winning two major national awards for its collaborative programs.
    - 2001 - An Australian Museum Eureka Prize for Industry in the Research and Innovation category, for the creation of the Minerals Tertiary Education program which encourages cooperation between universities and the minerals industry to provide enhanced education opportunities for students and graduates of geoscience, mining engineering and metallurgy; and
    - 2010 - A prestigious Australian Teaching and Learning Council (ALTC) award for Educational Partnerships and Collaborations with Other Institutions MTEC's flagship mining program – Mining Education Australia (MEA).
- Award of a Science Lectureships Initiatives (SLI) grant in 2001 to assist in building and delivering collaborative courses in earth science, mining engineering and extractive metallurgy, and
- Receipt of a Collaboration and Structural Reform (CASR) grant in 2007 to establish Australia's first and only 4 university national undergraduate program in mining engineering - Mining Education Australia (MEA), and
- MEA was successful in obtaining a Council on Australia and Latin American Relations (COALAR) grant to explore the possibility of taking MEA to a global audience based on the teaching of sustainable mining practices.

MTEC into the future

From 2012, MTEC formed an integral part of the MCA Education and Training portfolio and policy framework, focusing on initiatives to develop and implement national strategies to ensure the adequate supply of skills to the Australian mineral industry labour productivity by advocating public policy and institutional capacity building for improved delivery in the higher education university sector in minerals related areas, and is encapsulated in the MCA Workforce Skills, Education and Training Policy Brief [7]. The MCA advocates building an uninterrupted, sustainable education and training pathway to increase workforce participation, workforce diversity and workforce skills, regardless of the business cycle in industry. Significantly, the MCA policy position is viewed as being well-established internationally [8].

Also in 2012, MTEC gained a significant grant (A$700,000) from the Australian Commonwealth Government to undertake a feasibility study and prepare for implementation national Associate Degrees in Mining Engineering and Geoscience, known as the Minerals Industry National Associate Degree (MINAD) Project. The MCA is working with industry and the tertiary education sector to develop new, nationally consistent Associate Degrees in Mining Engineering and Mining Geoscience. Associate degrees in minerals engineering will provide the industry with efficient, appropriately skilled para-professional workers as part of their workforce profile. Associate Degree graduates will complement the work of traditional-4 year university graduate engineers, facilitating a more flexible yet targeted approach to engineering in the resources sector.

The Minerals Industry National Associate Degree (MINAD) Project aims to help address the shortage of mining engineers and geoscientists [9, 10] by developing, piloting and evaluating nationally consistent Associate Degrees to be delivered in VET, dual sector and Higher Education institutions. The Project seeks to be a true partnership between industry, government and academia to reshape minerals education in Australia, and will provide the minerals industry with efficient, appropriately skilled paraprofessional workers as part of their workforce profile.

MCA has undertaken extensive consultation with industry, recent graduates, operational supervisors and technical experts from a broad range of commodity groups and mining operations to develop and disseminate a piloted Associate Degree program model. Industry participants are defining new workforce roles for paraprofessionals, including the requisite generic qualities and technical knowledge required of engineering para-professionals. Through this extensive consultation exercise, the MINAD Pilot Model has been developed, which outlines the graduate capabilities framework and fields of knowledge to be gained within the Associate Degree program. The MIND Model is being piloted nationally in 2014.

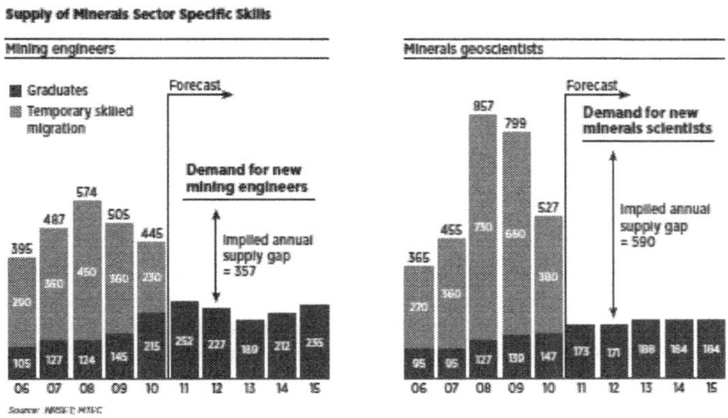

Figure 5. Supply of minerals sector specific skills [9].

From 2014 onwards, MTEC will be funded as a compulsory part of MCA membership, rather than the opt-in approach taken since MTECs formal constitution in October 2000. This is a signal moment for MTEC from the industry in that MTEC is viewed as part of the long-term vision of the industry in successfully securing the supply of quality graduates in the core disciplines of mining engineering, metallurgy and minerals geoscience.

Conclusions

The Australian minerals industry – through the Minerals Tertiary Education Council (MTEC) – is a proud initiator and supporter of national collaborative initiatives in higher education to build capacity in the core disciplines of mining engineering, metallurgy and minerals geoscience to deliver high quality graduates which industry desires. The Australian minerals industry continues to experience professional skill shortages in its core disciplines and recognizes that its investment in higher education is one way to ensure the future employment pipeline.

This paper has highlighted how direct industry investment in tertiary education on a national collaborative basis delivers benefits to students, universities and industry. The all-of-industry approach delivers results on the uninterrupted education and training pathway, and presents a mechanism for industry and academia to collaborate to address future professional skills needs in the minerals industry globally.

Acknowledgements

The author thanks the Minerals Council of Australia for permission to publish this paper and for the financial support in attending this esteemed gathering.

References

1. Minerals Council of Australia (MCA) 2012, *2013-14 Pre-budget submission*, viewed on 4 September 2013, < http://www.minerals.org.au/news/2013_14_pre_budget_submission/>.

2. Minerals Tertiary Education Council (MTEC) 2013, *MTEC 2013 key performance measures report*, viewed on 4 September 2013,
< http://www.mtec.org.au/resource/item/mtec_key_performance_measures_report_2013/>.

3. Minerals Tertiary Education Council (MTEC) 2011, *Higher education base funding review submission*, viewed on 4 September 2013,
< http://www.minerals.org.au/news/higher_education_base_funding_review/>.

4 Lind, G 2012, Minerals industry engagement in metallurgical education in Australia, *Minerals Industry Education and Training: A collection of papers from the Special Symposium on Human Resource Development, held during XXVI International Mineral Processing Congress (IMPC 2012), September 24-28, 2012, New Delhi, India*, (The Indian Institute of Mineral Engineers, 2012), pp. 87-104.

5. Department of Education, Employment and Workplace Relations (DEEWR) 2012, viewed on 6 September 2013, <http://www.deewr.gov.au/HigherEducation/Pages/default.aspx>.

6. Minerals Council of Australia (MCA) 1998, Discussion Paper, *Back From the Brink – Reshaping Minerals Tertiary Education*, February,
< http://www.minerals.org.au/focus/mtec>.

7. Minerals Council of Australia (MCA) 2012, Policy Brief, Workforce skills, education and training – The policy principles of the Australian minerals industry, December,
< http://www.minerals.org.au/news/workforce_skills_education_training>.

8. National Research Council 2013. *Emerging Workforce Trends in the U.S. Energy and Mining Industries: A Call to Action*. Washington, DC: The National Academies Press,
< http://www.nap.edu/catalog.php?record_id=18250>.

9. Department of Education, Employment and Workplace Relations (DEEWR) 2010, *Resourcing the Fut*ure: National Resources Sector Employment Taskforce (NRSET), June,
<http://www.innovation.gov.au/skills/SkillsTrainingAndWorkforceDevelopment/NationalResourcesSectorWorkforceStrategy/Pages/default.aspx>

10. Minerals Council of Australia (MCA) 2012, Monograph, *Opportunity at risk – Regaining our competitive edge in minerals resources*, September,
<http://www.minerals.org.au/news/opportunity_at_risk_-_regaining_our_competitive_edge_in_minerals_resources>.

Celebrating the Megascale: Proceedings of the Extraction and Processing Division
Symposium on Pyrometallurgy in Honor of David G.C. Robertson
Edited by: Phillip J. Mackey, Eric J. Grimsey, Rodney T. Jones, and Geoffrey A. Brooks
TMS (The Minerals, Metals & Materials Society), 2014

MODELING OF LADLE METALLURGY IN STEELMAKING

Gordon Irons[1], Krishnakumar Krishnapisharody[2] and Kevin Graham[3]

[1]Steel Research Centre, McMaster University, Hamilton, Canada
[2]Saarstahl Innovation, Volklingen, Germany
[3]Strategic & Business Planning, Vale, Toronto, Canada

Keywords: Modeling, steelmaking, ladle metallurgy, injection, slag, reactions

Abstract

Ladle or secondary metallurgy in steelmaking is an essential process step for the production of high-quality steel. Professor David Robertson was a pioneer in the modeling of these processes, in particular gas-metal fluid dynamic interactions and the rates of slag-metal reactions. This is now a highly active area of research, and some of the recent developments in this area will be reviewed.

Introduction

Ladle or secondary steelmaking is the process step after primary steelmaking (BOF or EAF steelmaking) and before casting, so it is the final opportunity to modify the steel chemistry. Therefore, it is essential that the chemistry in terms of dissolved elements, and inclusion chemistry and size distribution be correct.

Gas injection is central to most Ladle Metallurgy operations. During ladle processing, the steel melt is usually stirred by an inert gas like Argon through porous plug(s) fitted in the bottom of the ladle or a lance inserted deep into the melt. The purging serves several purposes *viz.*, homogenizing the melt, flotation of inclusions, enhancing the rates both of refining reactions and dissolution of alloying elements and promoting inter-phase interactions in the ladle.

The other central issue to Ladle Metallurgy is the refining reactions between the synthetic top slag and the steel. Sulphur removal from the steel is the usual purpose of refining, but because this reaction is electrochemical coupled, other elements are affected. More recently, steelmakers have recognized that there is a direct link between the slag refining conditions and the inclusion composition.

Professor Robertson has made important contributions to both the areas of gas injection and refining reactions. Recent work following from his work in the authors' group will be briefly reviewed.

Gas Injection

Gas injection is widely used in metallurgical operations, and has been studied extensively. Figure 1a is taken from the work of Farias and Robertson [1]. In this work they injected different gases to investigate the effect of the gas to liquid density ratio on the gas behavior; at high jet momentum this is an important parameter. In Ladle Metallurgy, the flow rates are much lower, so the gas plume appears to be more like the water model in Figure 1b, taken from the work of Anagbo and Brimacombe [2].

Figure 1(a). Nitrogen injection into water at high modified Froude number (Fr_m = 41000, d_o = 3.18 mm, Q = 8.3 x 10^{-3} $m^3 s^{-1}$, Farias and Robertson [1])

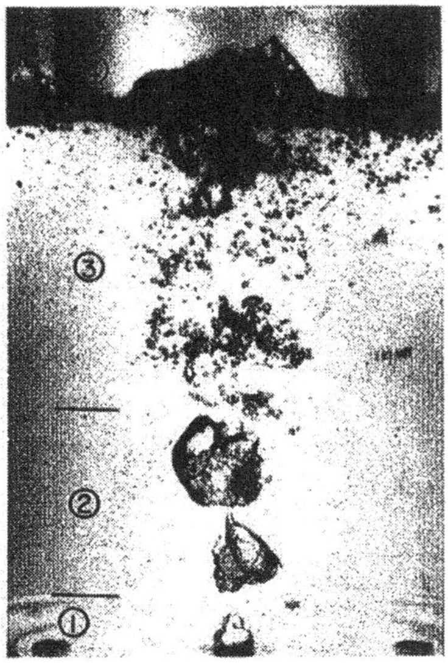

Figure 1(b). Air injection into water at low modified Froude number (Fr_m = 8, d_o = 6.35 mm, Q = 6.5 x 10^{-4} $m^3 s^{-1}$): the regimes of gas dispersion in liquid at low modified Froude numbers (1) Primary bubble (2) free bubble (3) plume and (4) spout, as represented by Anagbo *et al.* [2].

In a previous work by the present authors on the dynamics of two-phase plumes, [3] it was proven that only two dimensionless quantities are required to characterize free-rising plumes as in ladle metallurgy applications. These are the non-dimensionalized gas flow rate (Q^*) and axial height (z^*), defined, respectively, as:

$$Q^* = \frac{Q}{g^{0.5} H^{2.5}} \tag{1}$$

$$z^* = \frac{z}{H} \tag{2}$$

where Q is the gas flow rate, H is the metal height, z is the vertical position and g is the gravitational acceleration. It was also proven that the proper form of the Froude number relevant to Ladle Metallurgy is the plume Froude number, based on the liquid or plume velocity as the relevant velocity and the bath height as the characteristic dimension:

$$Fr_P = \frac{\overline{U}_l^2}{\overline{\alpha} g H} \tag{3}$$

The plume Froude number arose from the non-dimensionalization of the equation for momentum in the free-rising part of the plume. In that study the plume Froude number was mathematically related to Q^* [3]:

$$Fr_P = \Phi\left(Q^*, z^*\right) \tag{4}$$

Therefore, dynamic similarity is achieved by matching either Fr_P or Q^* and z^* in the plume. It is more convenient to use Q^* because it contains variables that are set by the operator.

Further, by using the mathematical descriptions of two-phase systems, *viz.*, (1) the well-known Eulerian-Lagrangian or bubble tracking models and (2) momentum and continuity for the phases, a similarity framework was developed to correlate all the major plume parameters to Q^* and z^* on a fundamental basis [3]. In this framework, the similarity parameters for the void fraction, plume radius, liquid and gas velocities, respectively are:

$$\overline{\alpha} = 1.13\left(Q^*\right)^{0.63}\left(z^*\right)^{-1.57} \tag{5}$$

$$R_p^* = \frac{R_p}{H} = 0.65\left(Q^*\right)^{0.2}\left(z^*\right)^{0.5} \tag{6}$$

$$\overline{U}_l^* = \frac{\overline{U}_l}{\sqrt{gH}} = 1.16\left(Q^*\right)^{0.32}\left(z^*\right)^{-0.28} \tag{7}$$

$$\overline{U}_g^* = \frac{\overline{U}_g}{\sqrt{gH}} = 3.52\left(Q^*\right)^{0.37}\left(z^*\right)^{-0.43} \tag{8}$$

where the overbar indicates that the quantity is averaged over the plume cross-section and the superscript * indicates that that the variable is non-dimensional.

The model was also extended to mixing times, t_m (the time to achieve a particular degree of chemical homogeneity, 95% in this case) as a function of gas injection rate [4]:

$$t_m^* = t_m\sqrt{\frac{g}{H}} = 82.4\,(Q^*)^{-0.31}(R^*)^{2} \tag{9}$$

Where R^* is the dimensionless vessel radius:

$$R^* = \frac{R}{H} \tag{10}$$

The equations in this model, Equations 5 to 9, have been compared with all available data in Figure 2. Further details are contained in references [4] and [5]. It is clear that there is remarkably good agreement between the model and data over a wide range of conditions. It should also be noted that the model is entirely self-consistent which means that the variables scale properly, according to plume Froude number similarity. Those references [4, 5] further argue that because the *modified* Froude number (widely used in gas injection studies) and the specific energy input should not be used in ladle metallurgy applications because they do not conform to plume Froude number similarity.

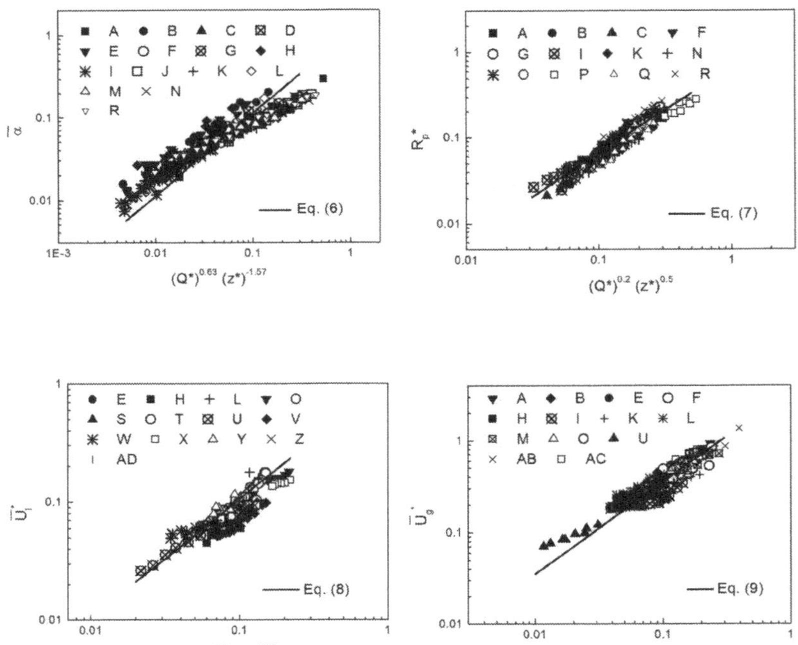

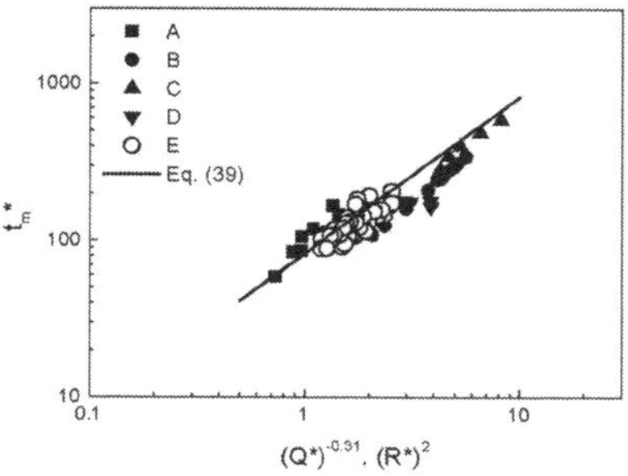

Figure 2: Similarity representation of the plume void fraction (a), plume radius (b), liquid velocity (c), gas velocity (d) and mixing time (e). The letters in the figures refer to the source of the data, explained in the original references [4, 5]. Equation 39 in Figure 2e is Equation 9 in this text.

Slag-Metal Reactions

It has been known since the work of King and Ramachandran [6] that slag-metal reactions are electrochemically coupled, but it has been difficult to develop a process model for Ladle Metallurgy reactions. The work of Robertson, Deo and Ohguchi [7] was a breakthrough in this regard, and it has been adopted by the present authors for a model of slag metal reactions in a full-scale Ladle Metallurgy station at ArcelorMittal Dofasco [8, 9]. The model is currently being extended to the transformation of inclusions.

The mixed mass transfer control model of Robertson *et al.* [7] is based on diffusion-controlled reactions in slag and metal, assuming equilibrium at the slag-metal interface. The generalized interfacial slag-metal reaction is:

$$x\underline{M} + y\underline{O} = M_xO_y \tag{11}$$

For which the equilibrium constant can be given as:

$$K_M = \frac{a_{M_xO_y}}{h_M{}^x h_O{}^y} \tag{12}$$

where K_m is the deoxidation constant, a_{MxOy} the activity of the interfacial oxide component, and h_O and h_M the respective interfacial Henrian activities for oxygen and selective metallic components in steel.

Assuming conservation of mass, flux density equations for each interfacial slag-metal reaction can be stated as:

$$k_m^M C_{Vm}\left(X_M^b - X_M^*\right) = k_{sl}^{M_xO_y} C_{Vs}\left(X_{M_xO_y}^* - X_{M_xO_y}^b\right) \tag{13}$$

where $k_m{}^M$ and $k_{sl}{}^{MxOy}$ are the mass transfer coefficients in the metal and slag phases for the respective metal and oxide components, C_{Vm} and C_{Vs} represent the molar volumes of metal and slag, and X^b and X^* the bulk and interfacial mole fractions of metal and oxide components in either the slag and metal phases.

The basic approach proposed by Robertson et al. [7] involves first rearranging equation (12) for each slag-metal equilibrium reaction in terms of an expression for the interfacial oxide concentration. These derived expressions are then substituted into flux density equations (13) and manipulated in order to obtain relationships for the interfacial concentration of metallic components in terms of mass transfer coefficients and bulk metal and slag phase concentrations. Exact solution of the interfacial slag and metal concentration at any instant in time requires first coupling the various slag-metal reactions using an overall oxygen mass balance equation as:

$$k_m^O C_{Vm}\left(X_O^b - X_O^*\right) = \sum_{i=1}^{n} y_i C_{Vs} k_{sl}^{M_xO_y}\left(X_{M_xO_y}^* - X_{M_xO_y}^b\right) \tag{14}$$

Equation (14) expresses the fact that the total amount of oxygen change in the metal phase equals the total oxygen change in the slag phase. Substitution of the metallic and oxide interfacial concentration expressions into equation (14) yields a single non-linear equation containing only interfacial oxygen concentration as an unknown. A numerical solver based on the Newton-Raphson method was developed by the present authors for finding the root of the oxygen mass balance equation *(i.e.* the interfacial oxygen content). Having solved for the oxygen concentration at the interface, all other interfacial values can be determined and composition trajectories can be projected forward by numerical integration of first order differential equations of the form:

$$-\frac{dX_M}{dt} = k_m^M\left(\frac{A}{V_m}\right)\left(X_M^b - X_M^*\right) \tag{15}$$

$$\frac{dX_{M_xO_y}}{dt} = k_{sl}^{M_xO_y}\left(\frac{A}{V_{sl}}\right)\left(X_{M_xO_y}^* - X_{M_xO_y}^b\right) \tag{16}$$

The model was applied to results obtained at ArcelorMittal Dofasco. Figure 3 shows the way in which the gas flow rate through porous plugs, total electrical power input and alloy additions are made in a typical heat. No modifications were made to operating practice; the model inputs were made to change as they did in the plant.

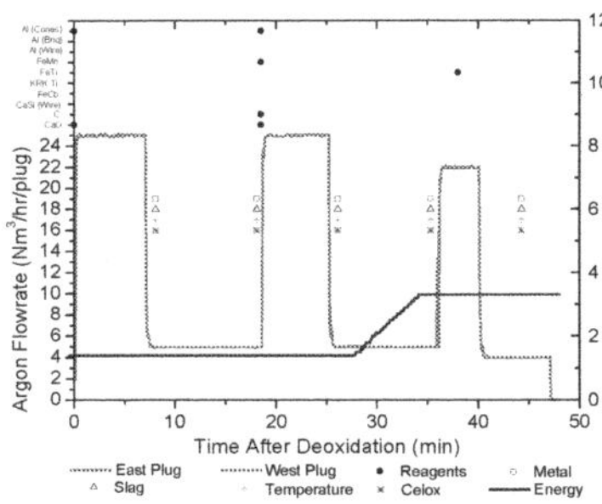

Figure 3: Summary of stirring, heating, alloy addition, and sampling procedure for one typical heat at ArcelorMittal Dofasco.

The model requires known values for the mass transfer coefficients. These were deduced from analysis of the rate of desulphurization. The rate was found to be dependent on the gas flow rate as shown in Figure 4.

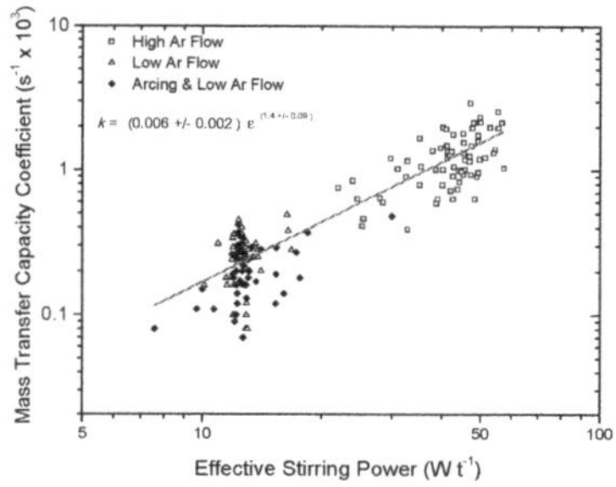

Figure 4: Relationship between mass transfer capacity coefficient for sulphur and effective stirring power which is proportional to the gas flow rate.

The results were validated against 41 heats at ArcelorMittal Dofasco. Figure 5 shows the agreement between the model and experiment for heat 40. At the 18 minute mark aluminum was added to the steel and the stirring rate was increased which lead to a rapid decrease in the sulphur content. These conditions were so reducing that silica and manganese oxide were reduced from the slag into the melt. As can be seen, these changes are well-described by the model.

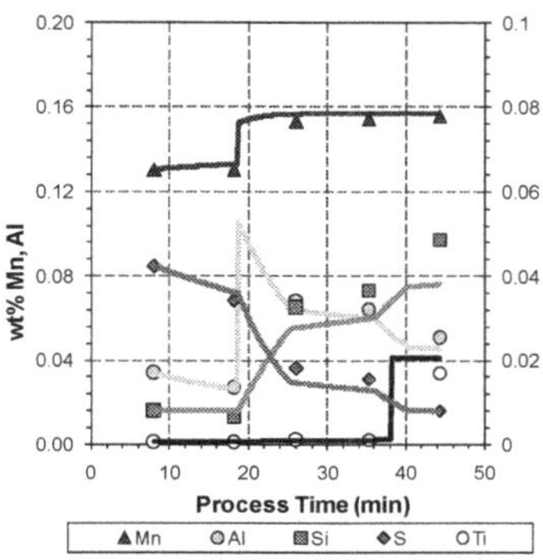

Figure 5a: Comparison of the model and experimental data for heat 40. At 18 minutes aluminum was added, and the gas stirring rate was increased. At 38 minutes titanium was added.

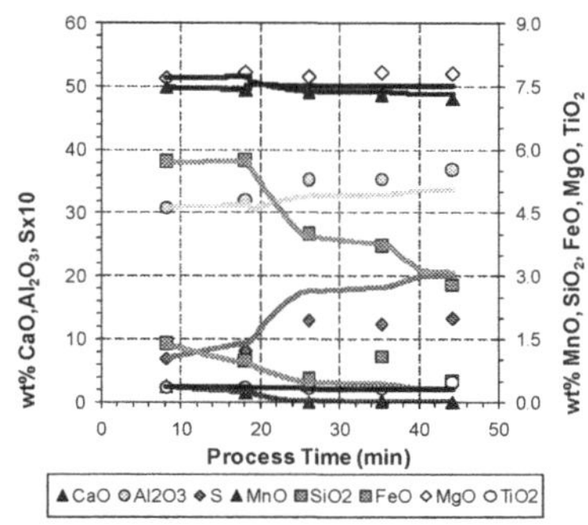

Figure 5 b: Comparison of slag compositions for the model and data for Heat 40.

Summary

The hallmark of Professor Robertson's work is that he has applied fundamental approaches to solving significant problems in metals processing. His works cited in this paper are just two examples of the many fine contributions that he has made. On a personal note, it has always been a pleasure to discuss our mutual interests, and he has been very gracious with my students as well.

References

1. L.R. Farias and D.G.C. Robertson, Private Communication, 1982.

2. P.E. Anagbo, J.K. Brimacombe and A.H. Castillejos, A Unified Representation of Gas Dispersion in Upwardly Injected Submerged Gas Jets, *Can. Metall. Q.*, 1989, 28, 323-30.

3. K. Krishnapisharody and G. A. Irons, "A Unified Approach to the Fluid Dynamics of Gas-Liquid Plumes in Ladle Metallurgy", *ISIJ International*, 50, (10), 1413-1421.

4. K. Krishnapisharody and G. A. Irons, "An Analysis of Recirculatory Flow in Gas-Stirred Ladles", *Materials Processing Towards Properties, Seetharaman Seminar*, June 14 - 15, 2010, Stockholm, Jernkontoret, 317 – 326.

5. K. Krishnapisharody and G.A. Irons, "A Critical Review of the Modified Froude Number in Ladle Metallurgy", in press, *Metallurgical and Materials Transactions B*, 2013.

6. S. Ramachandran, T.B. King, and N.J. Grant: Trans. AIME, 206 (1956) 1549-53.

7. D.G.C. Robertson, B. Deo, and S. Ohguchi, "A Multicomponent Mixed Transport Control Theory for the Kinetics of Coupled Slag/Metal and Slag/Metal/Gas Reactions: Application to Desulphurization of Molten Iron", *Ironmaking and Steelmaking*, 1984, 11, 41-55.

8. K.J. Graham and G.A. Irons, Towards Integrated Ladle Metallurgy Control, *Proceedings AISTech2008*, Pittsburgh, Pa, May 5-8, 2008, AIST, 13.

9. K.J. Graham and G.A. Irons, , Coupled Kinetic Phenomena in Ladle Metallurgy, *International Symposium on Highly Innovative Novel Operations "Future Steelmaking Metallurgy"*, Tokyo, May 23-25, 2010, ISIJ and JSPS, 65 - 74.

Celebrating the Megascale: Proceedings of the Extraction and Processing Division Symposium on Pyrometallurgy in Honor of David G.C. Robertson
Edited by: Phillip J. Mackey, Eric J. Grimsey, Rodney T. Jones, and Geoffrey A. Brooks
TMS (The Minerals, Metals & Materials Society), 2014

PROCESS METALLURGY AN ENABLER OF RESOURCE EFFICIENCY:
LINKING PRODUCT DESIGN TO METALLURGY IN PRODUCT CENTRIC RECYCLING

Markus Reuter[1], Antoinette van Schaik[2]

[1]Outotec Oyj, Puolikkotie 10, Espoo, Finland
[2]MARAS, The Hague, The Netherlands
(corresponding author: markus.reuter@outotec.com)

Keywords: Design for Recycling, Design for Sustainability, Resource Efficiency

Abstract

In this paper the link between process metallurgy, classical minerals processing, product centric recycling and urban/landfill mining is discussed. The depth that has to be achieved in urban mining and recycling must glean from the wealth of theoretical knowledge and insight that have been developed in the past in minerals and metallurgical processing. This background learns that recycling demands a product centric approach, which considers simultaneously the multi-material interactions in man-made complex 'minerals'. Fast innovation in recycling and urban mining can be achieved by further evolving from this well developed basis, evolving the techniques and tools that have been developed over the years. This basis has already been used for many years to design, operate and control industrial plants for metal production. This has been the basis for Design for Recycling rules for End-of-Life products. Using, among others, the UNEP Metal Recycling report as a basis (authors are respectively Lead and Main authors of report), it is demonstrated that a common theoretical basis as developed in metallurgy and minerals processing can help much to level the playing field between primary processing, secondary processing, recycling, and urban/landfill mining and product design hence enhancing resource efficiency. Thus various scales of detail link product design with metallurgical process design and its fundamentals.

Introduction– Classical Minerals and Metallurgical Processing

Minerals processing and extractive metallurgy have evolved to a high level of sophistication through the past decades in which for example:

- particle systems are described in terms of detailed particle descriptions, their properties and their liberation [1-3];
- fundamental phenomena, physical and chemical separation and concentration of minerals have been modelled to various degrees of detail [4-5];
- complete plants and systems can been simulated [6] and optimised [7];
- systems can be characterised with sophisticated measurement such as Mineral Liberation Analysis (MLA) [8-10] to help calibrate simulation models and understand the mineralogy of materials;
- sophisticated sampling of materials can be performed as described by Gy's sampling theory to assist in characterising materials as well as provide statistical relevance to analysis of streams [11-12]; and
- ore bodies can be characterised and quantified by for example Kriging [13-14].

These techniques have been applied in various ways in commercial software [15]. For example Figure 1 shows software that can model by the use of the aforementioned theory and simulate and control complete systems e.g. from rock to copper metal as depicted.

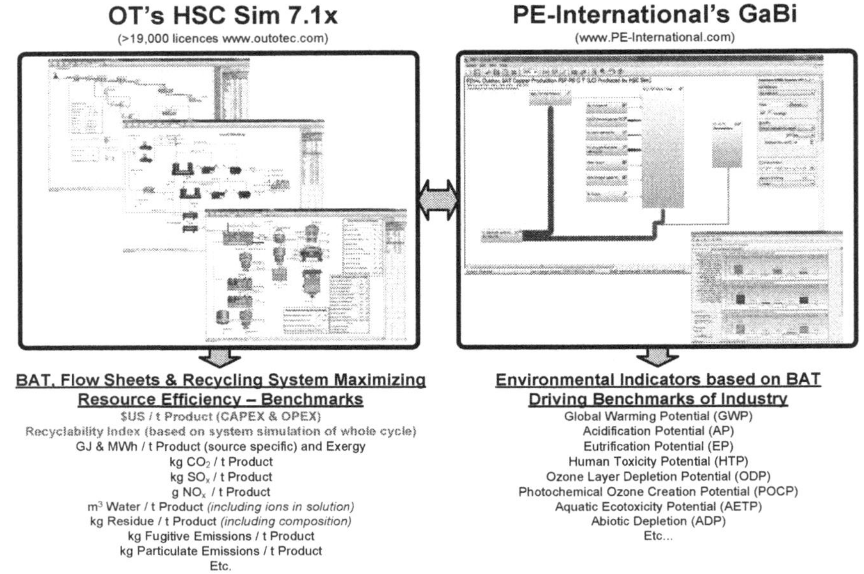

Figure 1: A complete rock to metal flowsheet for copper as provided by Outotec (www.outotec.com) [15], showing the link of physical separation with metallurgical processing as represented by the HSC Sim process simulator with a link to GaBi to estimate the environmental footprint (www.pe-international.de) [16].

Furthermore these systems are supported and linked by sophisticated environmental software [16]. In addition Gibbs free energy minimisation may be used to quantify equilibrium systems to assist in the simulations and understanding of the behaviour of species if they undergo chemical change in hydro- and pyrometallurgical reactors and complete plants [15, 17].

It is self-evident that there is much to be learnt and used from these techniques in classical minerals and metals processing. Figure 2 reflects the similarities between classical minerals and metallurgical processing of complex primary ores in relation to the recycling of even more complex designed 'mineralogies' of multi-material consumer products. This reveals that a Product Centric approach, which is capable of modelling and optimising the recovery/losses of all combined materials of the modern ore at the same time, similar to what is daily practice in minerals and metallurgical processing, is crucial to optimise recovery in recycling and urban/landfill mining. This reveals the need for a rigorous basis and simulation tools to describe and optimise these 'new' mineral systems.

This paper will describe briefly how classical minerals and metallurgical processing theory and insights have been used to:

- describe, simulate and optimise recycling and residue processing systems, while highlighting a Product Centric approach to recycling of materials and EoL (End-of-Life) products;
- simulate complete systems on this fundamental basis with existing well developed tools and the challenges thereof for recycling theory, product data requirements and hence urban mining;
- develop Design for Recycling rules based on this detail; and
- thus link the various disciplines on a consistent basis i.e. linking product design with metallurgical design and hence providing the basis to quantify sustainability.

The paper then concludes how the aforementioned may be of use to urban mining and product design and thus touching on the various scales of the complete system.

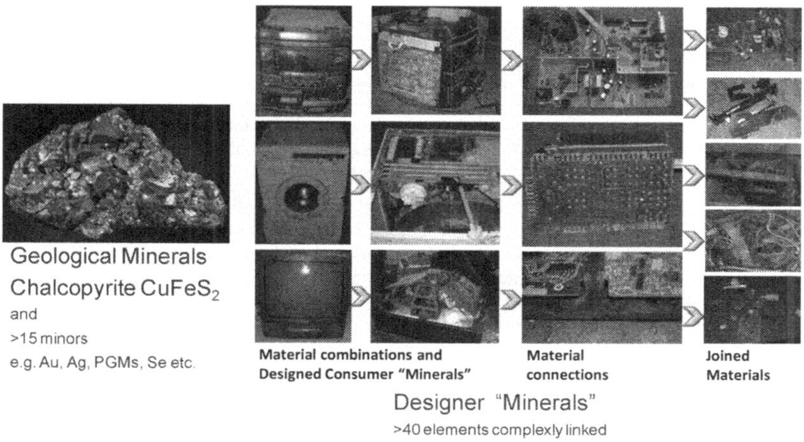

Figure 2: Learning from the simulation of complex primary minerals processing and metallurgical plants to process complex "designer" minerals, which may have even more complex "mineralogies", revealing the detail and resolution of product and recyclate data to become available to simulate and optimise recycling/resource efficiency.

Theory of Product Centric Recycling and Residue Processing

Understanding the existent minerals and metallurgical processing fundamental basis well, has been the basis for the authors to use this theory to describe recycling and residue processing systems [18-22]. We have linked End-of-Life (EoL) products and their "mineral" structure by simple "mineral" liberation data structures while at the same time linking this to metallurgical processing.

Crucial to this understanding of recycling and residue processing is for example liberation modelling in which parallels were drawn between classical minerals processing and EoL products [21]. As an example Table 1 compares various aspects of relevance between crushing and grinding from classical minerals processing and EoL shredding and cutting. Understanding this well makes it possible to better quantify dispersion and losses of materials, compounds and elements in the system, either through physical phenomena or through dissolution of metals in inappropriate phases. This is the true basis for quantifying sustainability. This table shows that significant work still has to be done on the description on how particles are liberated in residue and EoL processing systems, although some work is available that has attempted to better understand these phenomena [21-22].

Table 1: Comparative elements between comminution modelling in minerals processing and EoL shredding modelling with often non-brittle materials and complex joining.

Minerals processing comminution	**EoL Shredding/cutting**
Mineralogical Texture: Size of mineral grains in the particles, the grade or analysis, their shape, orientation and position, liberation (may be > 15).	*Design:* Materials choices, size and shape of the elements of the product, the joints between these different elements, complexity of product, the detailed analysis of many elements (may be >40).
Minerals properties: Minerals are generally considered as brittle and hard materials.	*Materials properties:* Since the large range of elements composing cars, WEEE and other complex multi-material products, different materials behaviours have to be considered, from brittleness to very ductile in a wide range of different combinations.
Liberation: Detachment by comminution (particle size reduction) of one or more species of mineral from the complex in which it is bound in a larger fraction of gangue material of little economic value.	*Liberation in End-of-Life products, waste, residues:* To be further developed on the basis of the complexity of modern products, where all materials can be valuable. Focus of liberation/sorting on the moment lies on commodity streams due to weight driven recycling targets for cars/WEEE. Emphasis on critical/scarce elements/minors and their Resource Efficiency changes this.

Table 1 reflects some of the complexity of EoL materials and the possible issues that describe and affect the liberation behaviour and thus the losses from the system. Figure 1 suggests that the deportation of all elements in chalcopyrite can be simulated well and a process can be designed on this simulated basis. This is common practice in for example Outotec (www.outotec.com) when we design plants as shown in this figure. However, the complexity of EoL products and often the lack of good sampling data for scrap, lack of detailed "mineralogies" of EoL products, lack of understanding of liberation behaviour etc. make the simulation of recycling systems significantly more difficult. Nevertheless, by the systematic use and also description of recycling systems on the basis of well-established minerals processing and further developments of this theory, some steps have been made to better understand EoL recycling systems.

Figure 3 shows, that by the use of the same tool as shown in Figure 1 for classical copper rock to metal simulation, EoL systems can be simulated and scrap particle systems appropriately calibrated. This fundamental rigour provides a basis for further developing of sampling systems, data acquisition and storage, data resolution of both designs and recycling streams and depth in terms of analysis and composition (grade), etc. Above all it provides the relative basis to which things are measured, much as has already been developed in minerals and metals processing. This standardisation of data will help much to provide the detail that can help enable Resource Efficiency. This is common practice for example in minerals and metals processing system design and optimisation. It is also common practice

in metallurgy in which all data is measured in such a way that it can calibrate standard polynomials for heat capacities (c_p), solution models etc. The recycling field still has much to learn from this.
In order to be able to fundamentally describe the system requires a key understanding that EoL products, waste materials etc. and should be considered very much like minerals as shown in Figure 3. Reuter coined the word Product Centric (i.e. equivalent to Mineral Centric) [23] recycling. This is identical way to considering a mineral centric description of systems, which is standard in minerals processing. This then by default brings into play the richness of all the theory and tools available through classical minerals and metals processing.

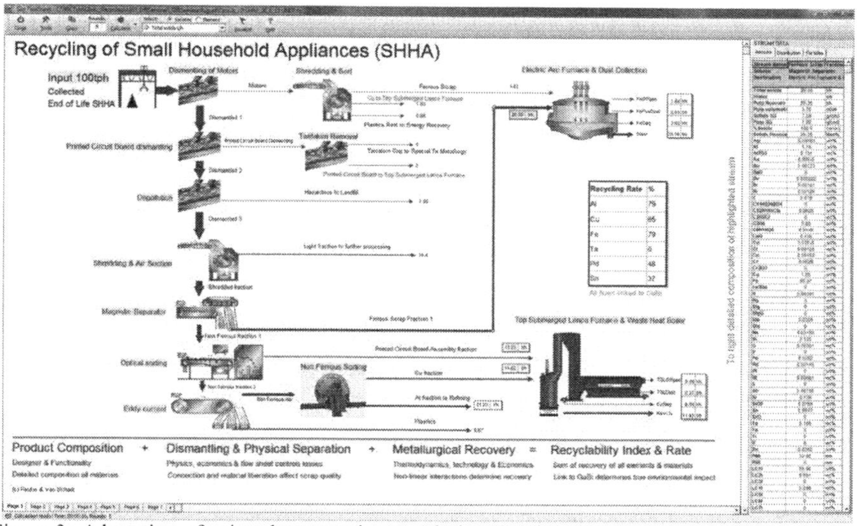

Figure 3: Adaptation of minerals processing to "designer" minerals and simulating and providing Recyclability Index for products on a rigorous simulated basis – thus true quantified sustainability and recycling rate calculations for the entire product, but also for all materials, compounds and elements, while addressing all the scales of the system – from theory to product design.

Through this Product-Centric approach as illustrated by Figure 3 [23], which at the same considers all metals, alloys, materials etc. in an EoL product, recycling has the potential to capture much greater value from waste streams, extracting the metals now possibly lost. Thus resource efficiency is captured, much in the way as we capture the losses of all metals in classical minerals and metals processing system with our models.

This contrasts with a rather more simplistic Material (& Metal)-Centric approach, which is favoured more by the waste processing field and has its root from the bulk recycling industry that would try to increase recycling of a specified metal, focuses on this while forgetting all other connected materials and the related thermodynamics and physics. This Material Centric approach would be similar if one would just consider ONLY the copper in the chalcopyrite in Figure 2 and neglect all non-linear effects of all other included elements, compounds etc. It is therefore extremely important for resource efficiency to step away from the Material Centric perspective to the Product Centric perspective when considering EoL products, waste streams, residues etc. This requires a deep understanding of extractive metallurgy, slag chemistry, separation physics and systems engineering as well as the techno-economic (CAPEX & OPEX) aspects of processes and systems.

Figure 3 furthermore provides a brief glimpse of some of the factors and aspects that have to be understood in a Product-Centric systemic and physics based manner in order to optimise resource efficiency. Also a clear understanding of the various losses that occur is imperative (many governed by product design, physics, chosen technology and linked economics), which also requires a deep compositional understanding of all residues, but also the understanding of unaccounted flows (poor statistics, data as well as collection) and the economics of the complete system are critical.
This Product-Centric approach, with its roots in minerals and metals processing, is an example of the practical application of systems thinking to a complex issue – optimising it requires Design for Resource Efficiency. It is an approach which leads to a set of conclusions that have strong implications for public recycling policy. It reveals when we will hit a "brick wall", suggests change, prompts innovation and not just continuing research without a bigger perspective. Furthermore, it gives a true physics-based meaning to the so often and easily fuzzily used word "holistic".

It suggests that, to optimise recycling and the recovery for various metals and materials from secondary resources such as sludges, slags, residues, EoL products, scrap, waste dumps, the following conditions would have to be in place (also please refer to Figure 3):

- treatment should happen in Best Available Techniques (BAT) by a certified system of operators in order to level the playing field;
- economic drivers and KPIs capture the performance of Figure 3 best and should be the driver;
- thermodynamically underpinned recycling policy should be supported by suitable simulation and rigorous fundamental calibration data;
- incentives should be present for all stakeholders;
- technology and system innovation should consider the complete system in fundamental rigour;
- recyclate data and its resolution should be on the compositional level similar as those describing minerals;
- collection, sampling of the urban mine and landfills, thus knowing the product "mineralogy" is key and should be based on rigorous sampling theory and Kriging; and
- above all, this Design for Resource Efficiency of products, systems, waste etc. should be based on rigorous process simulation. Simple material flow analysis is not good enough.

The next paragraph describes various rules that have resulted from the aforementioned to possibly provide some guidance to describing and design resource processing systems – thus learn from design for recycling to design systems for resource efficiency.

Design for Recycling

Recently Van Schaik, Reuter and NVMP/Wecycle have presented a set of 10 Design for Recycling Rules [24]. These are based on several of the aforementioned and similar other deliberations and divided into 5 fundamental design rules and 5 derived guidelines. Please read these with the glasses of both urban/landfill Mining and Design for Resource Efficiency (DfRE) and consider the questions in the next section to possibly assist in the innovation of Resource Efficiency also in the field of landfill mining as well as recycling.

Fundamental Design for Recycling (DfR) Rules

1. DfR rules are product and recycling system specific; oversimplification of recycling by defining general DfR rules will not produce the intended goal of resource efficiency.
   Figure 1 to Figure 3 shows the different facets and complexity of recycling complex multi-material products such as WEEE and reveals all details which should be considered in DfR (Design for Recycling) & DfRE (Design for Resource Efficiency).
   - Due to its functional and unique mix of materials each product has a unique recyclability profile and fingerprint. This implies that every product has a unique set of DfR guidelines which are product and recycling system specific.
   - These DfR guidelines are to be derived and iteratively refined for each product/product group separately by the application of recycling simulation models that can map any (BAT - Best Available Technique) recycling system and its opportunities and limits.

2. DfR needs model and simulation based quantification.
   - DfR demands a tool (process simulation models Figure 1 & Figure 3) that is capable of quantifying the product's recycling profile and performance to pinpoint DfR issues of importance and to give priority within Design adjustments to be implemented and insight in the effects of improved design on Resource Efficiency (Recycling Rate, toxicity, scarce material recovery/losses, environmental impact, etc.).

3. Design data should be accessible and available in a consistent format, which is compatible with the detail required to optimise and quantify recycling performance of products for all metals, materials and compounds present.
   - Detail and format of the product data on product material composition (including chemical compounds) and construction should have the resolution to quantify, identify and localise the commodity/critical/disturbing materials (see Figure 4) very much as is usual in minerals processing (Figure 2).

4. Economically viable technology infrastructure and rigorous tools must be in existence for realising industrial DfR rules and methodology.
   - Design must be based on a robust physical separation and sorting infrastructure that is minimally capable of producing economically valuable recyclates.
   - A robust metallurgical infrastructure and system must be in place to ensure maximum recovery of all "critical" materials from complex recyclates and dismantled functional sub-units of a product. Figure 3 illustrates the role of the metallurgical processing infrastructure which takes up the different recyclates from dismantling and sorting. The Metal Wheel [25] shows the destination (recovery/losses) of the different elements contained in an EoL product and different recyclates of physical recycling (where the quality is design and sorting determined). The carrier metallurgical infrastructure (segment in the Metal Wheel) in which the recyclate will end up determines the recovery, losses and fugitive emissions of product recycling.
   - Environmental footprint and eco-design should include the whole chain of processing to ensure that all materials, residues, fugitive emissions, losses are tracked (Figure 1) in their true compositional "mineralogy". This requires suitable global policy to be in place.

5. CAD, Process and System Design tools must be linked to recycling system process simulation tools to realise technology based, realistic and economically viable DfR.
Linking of existing and industry applied Process Simulation tools to CAD/Design tools is a necessary step forward to realise realistic and economically viable DfR. This is a rigorous basis for industrial useful DfR rules and methodology. The examples in Figure 1 & Figure 3 shows that based on rigorous recycling process simulation and detailed product compositional data derived from CAD/Design tools Recycling performance indicators and Recycling Indices including Environmental Analyses can be derived. The simulated Recovery rates for commodity and minor/critical materials and Recycling indicators provide the basis for DfR. Ecolabels for a product can be derived on this basis, which provide the depth to rigorously distinguish differences in product designs and identify the most Resource Efficient designs. This approach rigorously incorporates the detailed product design and recycling system dependent composition (grade) of all streams, residues, effluents etc. into the evaluation for DfR.

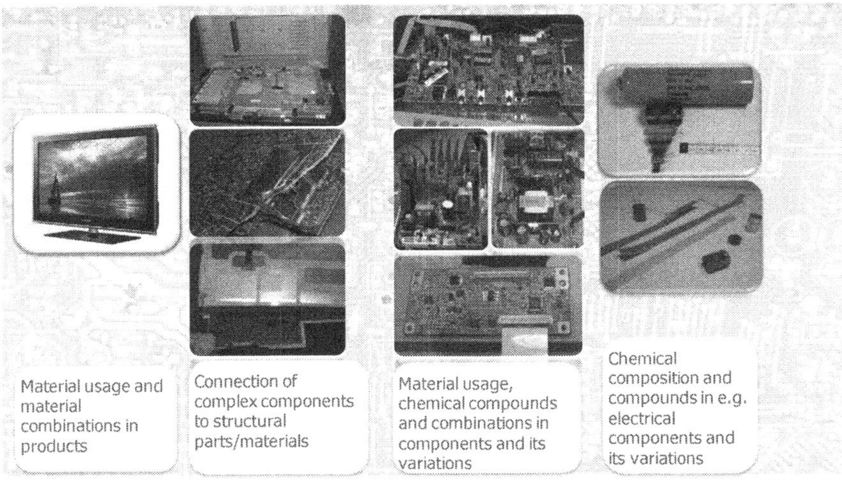

Figure 4: Examples of different levels and details of data required (from Bill of Materials/Full Material Declaration/chemical content analyses etc.) that capture the recycling profile of a product and also pin point critical DfR and DfRE issues and hence prompt innovation.

Possible Design for Recycling Guidelines–Derived from the Fundamental DfR Rules

Various Design for Recycling Guidelines can be derived by applying the above listed Fundamental DfR rules and principles (Iteratively checked by simulation and for validity, while being subject to a mindful consideration of product/component functionality). Recycling process simulation tools are used to define, validate and quantify the set of guidelines per product (of which the list below shows some possible guidelines). This physics based approach also can set priorities between the different guidelines and quantify the necessity and potential result of DfR.

Examples of the below guidelines can be found in general DfR approaches. However, by the implementation of the Fundamental Rules and simulation as a basis to derive and refine these, unique sets of guidelines are derived per product as a function of material mix and (BAT) recycling systems, including a mindful consideration of product functional demands. In contrast a fixed set of all possible guidelines as applied in general DfR approaches will leave no room for the designer to design/construct a product. Input from recyclers has also been used in defining examples for the certain guidelines.

6. Identify and minimise the use of materials, which will cause losses and contaminations in recycling due to material characteristics and behaviour in sorting.
   Some examples are:
   - Use and construction of concrete in washing machines, which is difficult to dismantle and distributes over different recyclate streams.
   - Use of glass plates in fridges, which is (currently) considered as process waste and hence decreases recyclability.
   - Application of insulation materials which are difficult to sort and will contaminate different recyclate fractions and hence lead to losses.
   - Use of coloured (other than brown, green) Printed Wire Boards, which will lead to contamination of PWB fraction with e.g. coloured plastics or losses of PWB to other recyclates.

7. Identify components/clusters in a product, which will cause problems and losses in recycling due to combined and applied materials.
   - A compatibility matrix can be derived per product based on the knowledge behind the Metal Wheel and is useful for a quick first screening. It is important to realise that compatibility tables are no DfR tool on their own, since this gives no indication of resource efficiency, material recovery, losses and fugitive emissions.

8. Design clusters or sub-units in products that can be easily removed and which match with the final treatment recycling options (i.e. Metal Wheel) e.g.:
   - Design products in such a way that critical components are present on separable or removable parts or components.
   - Examples are the use or removal of Al applied as heat sinks on a Printed Wire Board; separate PWBs for different functions e.g. power boards (high Fe content, which will be lost in PWB processing route), control boards; removal of Ta capacitors, etc.

9. Labelling (including carefully considered standardisation) of products/components based on recovery and/or incompatibility so that they can be easily identified from recyclates and waste streams. Thus Design for Waste stream sorting or Design for (Automated) Dismantling/Sorting are important.
   Examples are:
   - Use of colour or identification based labelling & easy to break connections for Ta capacitors. Of crucial importance is metallurgical knowledge as captured in Process Simulations tools to understand what the quality requirements are for the existing industrial Ta production/recycling infrastructure.
   - Labelling/identification based Waste Stream sorting of CFL from LED lamps.
   - Standardisation of marking & identification (e.g. type of marking and position) of cooling liquid & gas and marking of type of liquid & gas applied in fluid system and foam (these might be different).
   - Marking of tapping point on compressors.

10. Be mindful of liberation of materials in design (Design for Liberation)
Simulation and knowledge on liberation behaviour in relation to design, particulate and recyclate quality and recycling (metallurgical) efficiency is crucial. Examples are:
   - Avoid bolts/rivets of dissimilar materials (e.g. Fe bolts to Al, PWB, plastics, etc.) as these generally produce a liberation problem, therefore creating cross-contamination of the different recyclate fractions.
   - Minimise the use of non-reversible adhesives for incompatible & undesired material combinations.
   - Examples are gluing of glass on steel mask of CRT TV, shrink films on tube lamps, sealed batteries, PUR foam glued to steel/aluminium/plastic, wood glued to plastic, etc.

The Metal Wheel indicates the consequences in a simple overview manner, whereas simulation quantifies these more precisely in terms of recoveries, losses and emissions capturing the complexities and depth of recycling and metallurgical technology, its physics, chemistry and economics.

Policy targets for Recycling and Ecodesign
Some policy targets can be derived from these rules with regard to recycling and eco-design:

- should remain within what is technologically (physically and thermodynamically) and economically possible and be based on recycling process simulation tools and knowledge;
- must be set in ways that account for the inevitable losses of materials due to mixing in products for product functional specifications;
- should stimulate Best Available Technology/Techniques (recycling and metallurgical infrastructures)
- should reflect the interest and expertise of all stakeholders in the product and recycling system and stimulate interaction;
- should respect the dynamic (time-varying) product characteristics and recycling profile over time.

What can urban/landfill mining learn from Product Centric Recycling, DfR and DfRE?

The recent UNEP Metal Recycling report [26] gives a detailed overview over metal recycling and also links recycling to Urban Mining and (Enhanced) Landfill Mining, with the latter being defined as "the safe conditioning, excavation and integrated valorisation of (historic and/or future) landfilled waste streams as both materials (Waste-to-Material, WtM) and energy (Waste-to-Energy, WtE), using innovative transformation technologies and respecting the most stringent social and ecological criteria" [27]. In view of the brief discussion above and the UNEP report, some points can be kept in mind when advancing the field urban/landfill mining while focusing on Design for Resource Efficiency and its suggested more product centric view. Thus with reference to the so very diverse forms of materials and spatially distributed diverse, heterogeneous, reacted/decayed, fines containing excavated waste from the landfills:

- develop as far as possible the description of the system on the basis of the theory that is already available from minerals and metals processing in order to optimise the systems as rapidly as possible, while also further innovating and advancing the theory and deeper understanding;
- base the DfRE efforts and descriptions on rigorous process simulation as done for recycling;
- this will help also to further develop a Product Centric approach to mining landfill, very much like mining classical minerals and processing these in a minerals (product) centric manner;

- ensure that data structures for Urban/Landfill mining are developed in accordance with standard minerals processing standards and sampling is done in accordance with rigorous available techniques, which enables the calibration of rigorous simulation models;
- ensure that the complete system is modelled from landfill to metallurgy to ensure a decent economic CAPEX and OPEX analysis can be made to ensure the developed ideas make economic sense;
- the rigour suggested by the DfR rules also ensures that research is focussed on the issues that are of true value and will lead to true innovation;
- understand the thermodynamics of the system as well as kinetic issues that may affect the outcome;
- understand the distributed nature of all variables in the system in order to obtain an indication of the risk involved in the urban mine "ore body", thus the use of Kriging techniques and sampling theory are of paramount importance; and
- understand the dynamics of the market as well as the available metallurgical infrastructure as these are key to extract metal value.

Obviously this list can be extended or perhaps even shortened, but is a basis for a rigorous discussion on resource efficiency, which is ultimately the goal of DfR, DfRE and urban/landfill Mining. Fundamental rigour and theory is the common language, which is important to harmonise discussion and bridge the gap between the waste processing, metallurgical, OEM, resource recovery, materials science, recycling and other silos. This should be a pressing outcome for our exciting but urgent innovation ahead of us. Society and the taxpayers demand that of us.

References

1. J.R.G Andrews and T.S. Mika, 1976. Comminution of heterogeneous material. Development of a model for liberation phenomena, In: Proceedings 11th International Mineral Processing Congress, 59-88.
2. R.K. Mehta, G.T. Adel and R.H. Yoon, 1989. Liberation Modeling and Parameter Estimation for Multicomponent Mineral Systems, Powder Technology, 58, 195-209.
3. K. Heiskanen, 1993. Particle Classification, Chapman & Hall, London, Great Britain. ISBN 0 412 49300 4.
4. R.P. King, 1994. Linear stochastic models for mineral liberation, Powder Technology, 81(3), 217-234.
5. S.L. Gay, 2004. A liberation model for comminution based on probability theory, Minerals Engineering, 14, 525-534.
6. R.P. King, 2001. Modeling and Simulation of Mineral Processing Systems, Butterworth-Heinemann Publications, Oxford, Great Britain. ISBN 0750648848.
7. M.A. Reuter and J.S.J. Van Deventer, 1990. The Use of Linear Programming in the Optimal Design of Flotation Circuits Incorporating Regrind Mills", International Journal of Mineral Processing, 28, 15-43.
8. D.N. Sutherland and P. Gottlieb, 1991. Application of automated quantitative mineralogy in mineral processing, Minerals Engineering, 4(7-11), 753-762.
9. G. Bonifazi, 1995. Digital multispectral techniques and automated image analysis procedures for industrial ore modelling, Minerals Engineering, 8(7), 779-794.
10. W.R. Goodall, 2008. Characterisation of mineralogy and gold deportment for complex tailings deposits using QEMSCAN®, Minerals Engineering, 21(6), 518-523.

11. P. Gy, 1981. Proportional sampling — A new philosophy of metallurgical accounting, International Journal of Mineral Processing, 8(3), 279-286.
12. P.M. Gy, 1976. The sampling of particulate materials - A general theory Review Article, International Journal of Mineral Processing, 3(4), 289-312.
13. G. Matheron, 1963. Principles of geostatistics, Economic Geology 58 (8), 1246–1266.
14. N.A.C. Cressie, 1993. Statistics for Spatial Data, Revised Edition, Wiley, New York.
15. HSC and HSC Sim 7.1, 1974-2012. Thermochemical and process simulation, Outotec Research Center, www.outotec.com.
16. PE: GaBi 6, 1992-2013. Software and System Databases for Life Cycle Engineering", Stuttgart-Echterdingen, www.pe-international.com.
17. FACT Sage 6.4, 1976-2013. CRCT-ThermFact Inc. & GTT-Technologies, www.factsage.com.
18. M.A. Reuter, 1998. The simulation of industrial ecosystems", Minerals Engineering, 11(10), 891-917.
19. A. van Schaik and M.A. Reuter, 2004. The time-varying factors influencing the recycling rate of products", Resources, Conservation and Recycling, 40(4), 301-328.
20. A. van Schaik, M.A. Reuter and K. Heiskanen, 2004. The influence of particle size reduction and liberation on the recycling rate of end-of-life vehicles", Minerals Engineering, 17, 331-347.
21. A. Richard, A. van Schaik and M.A. Reuter, 2005. A comparison of the modelling and liberation in minerals processing and shredding of passenger vehicles", In: EPD Congress 2005 (Edited by M.E. Schlesinger TMS (The Minerals, Metals & Materials Society)), 1039-1052.
22. B. Castro, H. Remmerswaal, H. Brezet A. van Schaik and M.A. Reuter, 2005. A simulation model of the comminution-liberation of recycling streams: Relationship between product design and the liberation of materials during recycling, International Journal of Minerals Processing, 75(3-4), 255-281.
23. M.A. Reuter and A. van Schaik, 2012. Opportunities and Limits of recycling – A Dynamic-Model-Based Analysis", MRS Bulletin, 37(4), 339-347.
24. A. van Schaik, M.A. Reuter and NVMP/Wecycle, 2013. 10 Design for Recycling Rules", Presented 29th August 2013, www.producenten-verantwoordelijkheid.nl.
25. M.A. Reuter, K. Heiskanen, U. Boin, A. van Schaik, E. Verhoef and Y. Yang, 2005. The Metrics of Material and Metal Ecology, Harmonizing the resource, technology and environmental cycles, Elsevier BV, Amsterdam, 706p. (2005) (ISBN: 13 978-0-444-51137-9).
26. UNEP Metal Recycling: Opportunities, Limits, Infrastructure, 2013. United Nations Environmental Programme UNEP (Lead author M.A. Reuter with principal contributors: C. Hudson, A. van Schaik, K. Heiskanen, C. Meskers, C. Hagelüken,) 316p., http://www.unep.org/resourcepanel/Publications/MetalRecycling/tabid/106143/Default.aspx.
27. P.T. Jones, D. Geysen, Y. Tielemans, S. Van Passel, Y. Pontikes, B. Blanpain, M. Quaghebeur and N. Hoekstra, 2013. Enhanced Landfill Mining in view of multiple resource recovery: a critical review, Journal of Cleaner Production, 55, 45-55.

Celebrating the Megascale: Proceedings of the Extraction and Processing Division
Symposium on Pyrometallurgy in Honor of David G.C. Robertson
Edited by: Phillip J. Mackey, Eric J. Grimsey, Rodney T. Jones, and Geoffrey A. Brooks
TMS (The Minerals, Metals & Materials Society), 2014

HORIZONTAL SINGLE BELT CASTING (HSBC) of Ca-BASED BULK METALLIC GLASS (BMG) STRIPS

Roderick I.L. Guthrie, Mihaiela Isac, Donghui Li and Luis Calzado[1],

[1]McGill Metal Processing Center, McGill University, 3610 University Street, Montreal, QC H3A 2B2, Canada

Keywords: Horizontal Single Belt casting (HSBC), Ca-Mg-Al-based bulk metallic glass, mathematical modeling, strip product.

Abstract

The HSBC is an ideal process for the production of BMG strips. The purpose of the present work was to investigate the casting of Ca-based, bulk amorphous strips. The possibility of casting three Ca-based alloys (60Ca-20Mg-20Al, 60Ca-15Mg-10Al-15Zn and 55Ca-15Mg-10Al-15Zn-5Cu (at.%).) was investigated, using the HSBC simulator. Numerical simulation and mathematical modeling were used to predict temperature variations and cooling rates, under different interfacial gaseous atmospheres (helium and air) near the meniscus region. Given the critical cooling rates and glass temperatures needed to be reached for the vitrification of the respective alloys, this leads to predictions of the critical thicknesses possible with this novel casting process. To validate our predictions, these alloys were then prepared and cast onto the sand-blasted, solid copper substrate mould of the HSBC simulator. The possibility of producing Ca-based metallic glassy strips via the HSBC processes was confirmed, using optical microscopy, X-ray diffraction and Scanning Electron Microscopy, to characterize the glassy properties of the cast strips. It was found that the HSBC process would be able to provide sufficient cooling to form amorphous strips for the 60CA-15Mg-10Al-15Zn and 55Ca-15Mg-10Al-15Zn-5Cu alloys, but not the 60Ca-20Mg-20Al alloy.

Introduction

In preparing this paper, the authors firstly wish acknowledge the seminal contributions of Professor David Robertson to the field of Process Metallurgy over the past forty odd years. We also acknowledge his review of our HSBC casting work that appears in the Treatise of Process Metallurgy [1]. The present work represents a small part of that overall casting program, and was carried out to investigate the possibility of casting Ca-Mg-Al based bulk amorphous strip using this new horizontal single belt strip casting process, or HSBC. Similar attempts to form BMG's had previously failed using a vertical twin roll strip casting process, at IMI, Boucherville.

The Horizontal Single Belt Strip Casting (HSBC) process is a near net-shape strip casting technology, producing thin strips directly from liquid metal. This is accomplished by depositing a layer of liquid, iso-kinetically, onto a moving, water-cooled belt. The HSBC process also has the potential to provide the very high cooling rates for forming bulk metallic glassy strips. On a commercial scale, it is a process that potentially enjoys a very high productivity, together with relatively low operating and investment costs.

Recently, calcium based bulk metallic glasses have attracted the interest of a number of research groups, thanks to their unique properties [2-11]. Ca-Mg-Zn-based bulk metallic glasses (BMG) have a low density (about 2000kg/m^3) and their Young's modulus values (17~35GPa) are comparable to the modulus of human bones [2-4]. The elements of Ca, Mg and Zn are all biocompatible, which makes the Ca-Mg-Zn alloys attractive for use in bio-medical applications. By adding aluminum, this will improve the corrosion resistance of these alloys [5]. The amorphous structure provides unique mechanical properties, such as a high elastic strain, high fracture strength and high fatigue resistance. Numerous Ca-based BMGs have been produced and studied, and many of these BMGs have high glass forming ability. Fully amorphous plates and rods, with thicknesses or diameters larger than 15mm, can be produced by casting these BMG's [6-11]. It was reported that the 60Ca-15Mg-10Al-15Zn alloy was able to form an amorphous matrix of 4 mm thickness in a water-cooled copper mould [9], and that its glass transition temperature was 130°C.

Most of the Ca-Mg-Al-based bulk metallic glasses in previous research studies have been cast in small scale, wedge-like, copper moulds, in order to test their glass forming ability. In this paper, Ca-Mg-Al-based alloys were cast on a flat copper substrate mould, using an HSBC simulator, in order to obtain large sheets of thin metallic glassy strips. Three different Ca-Mg-Al-based alloys with chemical composition: 60Ca-15Mg-10Al-15Zn; 55Ca-15Mg-10Al-15Zn-5Cu, and 60Ca-20Mg-20Al were selected for the BMG alloy compositions. The temperature variation and related cooling rates during solidification were predicted by mathematical modeling. These alloys were then cast on the HSBC simulator to form thin metallic strips. Later, the strip samples were analyzed using x-ray diffraction and scanning electron microscope.

Theory and Numerical Method

Until now, information on the critical cooling rates for these three Ca-Mg-Al based alloys to form fully amorphous matrices has not been available. However, it is possible to predict the temperature variations and cooling rates inside the thin metallic strips, when the melt is deposited onto a moving copper substrate, or onto a rapidly moving, water chilled, thin belt of steel. So mathematical modeling using numerical simulation techniques was adopted for the present work.

General Governing Equation

In the horizontal single belt strip casting process, the flow of the liquid metal adjacent to the belt surface region was taken to be one of plug flow, parallel to the casting direction. Because of the high ratios of the metallic strips lengths and widths to the strip's thickness, heat transfer in the transverse and longitudinal directions can be ignored, in comparison with the heat flow in vertical direction. As such, the transfer of heat from the liquid metal to the water-cooled belt below can be simplified to being a one dimensional, unsteady state, heat conduction problem through a finite medium.

According to these simplifications, the relevant partial differential equation for heat flowing through the BMG alloy is given as follows:

$$\frac{\partial(\rho C_p T)}{\partial t} = k\frac{\partial^2 T}{\partial x^2} + S_T \qquad (1)$$

where: T is the temperature of the liquid metal, t is the time, and k, ρ and C_p represent thermal conductivity, density and specific heat of the melt, and solidified melt, respectively. The x term represents the distance within the sump of liquid metal from the bottom surface of the forming strip. S_T is the source term related to the latent heat released during solidification. In the case of forming a fully amorphous matrix in the metallic strips, no latent heat is released during solidification. This is equivalent to making the source term zero. i.e. $S_T = 0$.

Given that the physical properties of k, ρ, C_p of the melt vary with temperature, the analytical solution to describe the temperature in the metallic strip becomes inadequate, and Equation (1) has to be solved using the techniques of numerical simulation.

In order to solve the governing equation (1) by numerical simulation techniques, the boundary condition has to be applied for the top and bottom boundaries within the computational domain. The overall heat transfer coefficient values for strip casting process [12], h, on the bottom of the thin melt layer, was adopted. i.e. $h = 17.3\ V_{cast}^{\ 0.65}$, where V_{cast} stands for the casting speed. On the top of the thin melt layer, the boundary condition of radiation and natural convection into the ambient environment were considered in the modeling. The initial temperature of the melt was set as 530°C, which represented the casting temperature of the alloys, while the calculations are stopped when the temperature of all the melt reaches the glass transition point, T_g.

It has been established that the main interfacial thermal resistance to heat transfer in HSBC casting derives from the entrapped air layer at melt/mould interface. Given that the thermal conductivity of helium is more than five times greater than that of air, it is possible to enhance interfacial heat fluxes and alloy cooling rates by using a helium atmosphere rather than default air, during the casting of Ca-Mg-Al based metallic glassy strips.

Physical Properties

The thermal properties of the liquid metals, such as density, thermal conductivity and specific heat, were needed for solving the governing equation (1). The density and thermal conductivity of these Ca-Ma-Al-based alloys are temperature dependent, and are given in Figure 1 and Figure 2. Specific heat of these alloys 60Ca-20Mg-20Al, 60CA-15Mg-10Al-15Zn and 55Ca-15Mg-10Al-15Zn-5Cu were considered as constants, 0.917, 0.78 and 0.758, kJ/kg/°C respectively.

Predicted Results for the Casting Ca-Mg-Al Based Strips

Firstly, a typical casting speed of 1m/s and a casting temperature of 530°C were chosen, producing a 3mm thick strip in the modeling. Figure 3 gives the typical temperature profiles, in the melt, and the cooling rates of 60Ca-15Mg-10Al-15Zn alloy at different moments. Because of the good thermal conductivity of the alloy, the temperature gradient in the thin liquid melt layer is low and the temperature profiles display smooth and flat curves. After 0.50s from initial contact with the mould, all temperatures within the melt are predicted to become lower than the glass transition temperature of 130°C (T_g). Similarly, the minimum cooling rate in the thin liquid metal film during the rapid cooling process is about 320°C/s. When the alloys of 60Ca-20Mg-20Al and 55Ca-15Mg-10Al-15Zn-5Cu were considered in the modeling, similar results were obtained.

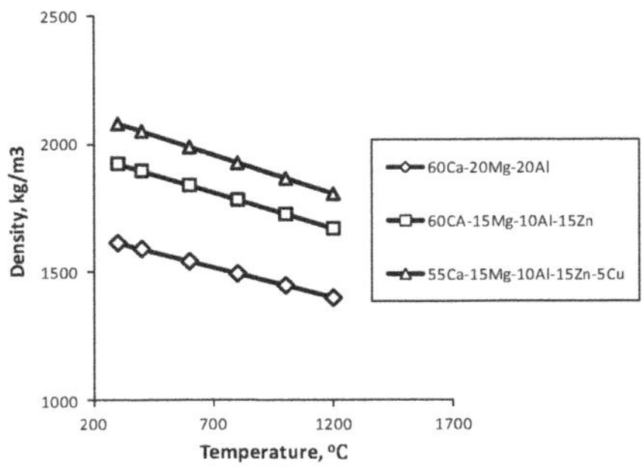

Figure 1. Computed density of the Ca-Mg-Al alloys.

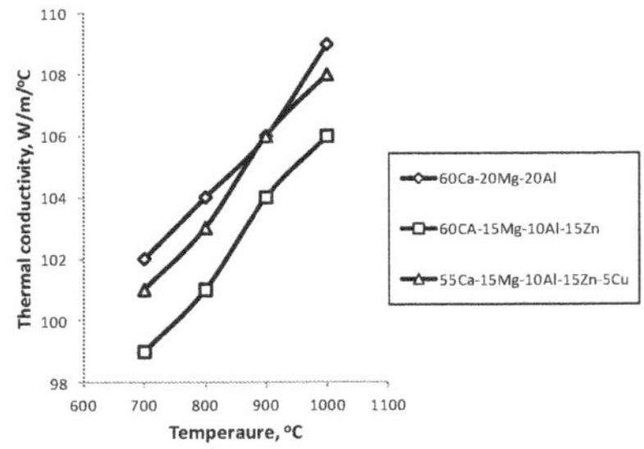

Figure 2. Thermal conductivity of the Ca-Mg-Al- alloys

When the melt thickness was reduced to 2mm and all the other casting parameters were kept the same, predicted results are given in Figure 4, for typical solidification times of t=0.1s, t=0.2s and t=0.3s. Because the thickness of the thin liquid metal layer decreases, it only takes about 0.30~0.33s for all the three liquid alloys to cool down rapidly to a temperature lower than 130°C. Due to the difference in the physical properties of these three Ca-Mg-Al-based alloys, the temperature profiles of 55Ca-15Mg-10Al-15Zn-5Cu are always a little bit higher than the other two alloys', resulting in having lower cooling rates as a consequence. The minimum cooling rate is 584°C/s during solidification, for the case of casting 60Ca-20Mg-20Al alloy. In the case of casting 60CA-15Mg-10Al-15Zn alloy and 55Ca-15Mg-10Al-15Zn-5Cu alloy, the minimum predicted cooling rates achieved by HSBC process are 586 °C/s and 601 °C/s respectively.

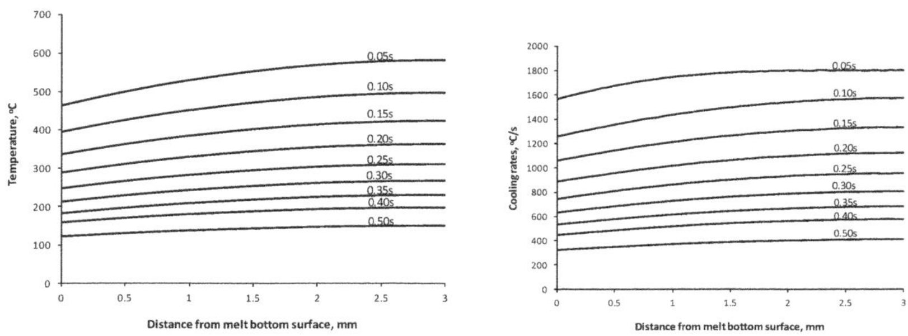

Figure 3. Typical temperature profiles and cooling rates of 60Ca-15Mg-10Al-15Zn alloy, strip thickness=3mm.

Until now, the critical cooling rates for these Ca-Mg-Al-based alloys remain unavailable. In this section, the relationship between the maximum amorphous strip thickness and critical cooling rate was predicted by mathematical modeling, and are given in Figure 5. If the critical cooling rates were the same for all these three alloys, the maximum amorphous thickness of these three alloys were quite close, although the maximum thickness of 60Ca-20Mg-20Al alloy was slightly greater than the other two alloys', because of its higher thermal conductivity and lower density.

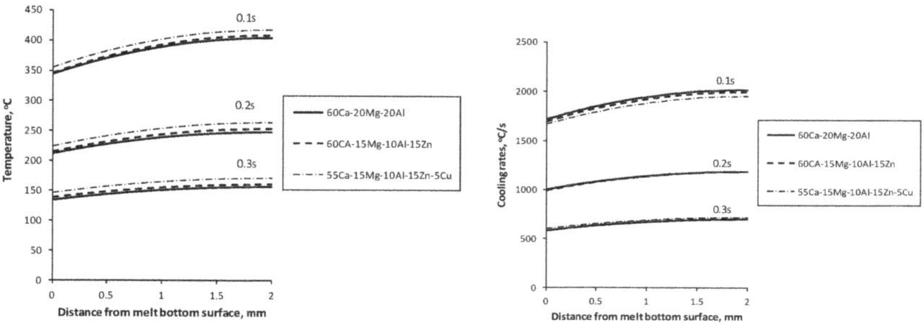

Figure 4. Predicted temperature profiles and cooling rates of the thin liquid metals, thickness=2mm.

For example, if the critical cooling rates is 200°C/s, the maximum amorphous strip thickness of these three alloys will be about ~4.86-4.96mm, compared to the fully amorphous strip of 2.70~2.80mm thick, at a critical cooling rate of 400°C/s; whereas the maximum amorphous strip thickness would decrease to 1.47~1.52mm, if the critical cooling rate is 800°C/s.

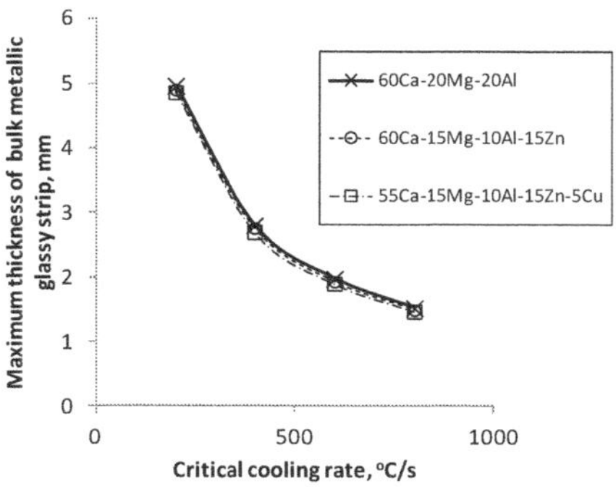

Figure 5. Relationship between maximum amorphous thickness of the three Ca-Mg-Al-based alloys and their critical cooling rates.

Therefore, the critical cooling rate has a great effect on the maximum amorphous strip thickness that can be obtained by the horizontal single belt strip casting process. Later, by casting these alloys on HSBC simulator, their maximum possible amorphous thickness would be obtained, and then the critical cooling rates could be roughly estimated in accord with the predictions in Figure 5.

Casting Experiments

Experimental Apparatus and Procedures

Figure 6 provides a schematic view of the single belt strip casting simulator designed for casting Ca-Mg-Al-based alloys via the HSBC process. The simulator mainly comprises a moving substrate, fitted with a stationary refractory-lined tundish and a compression spring system to propel the substrate laterally under a stationary tundish. The substrate was 500mm long, 110mm wide and 12.7mm thick. Before any casting experiment, the spring was compressed for storing energy, the head of the substrate was located beneath the tundish nozzle slot, and a refractory plate on the substrate blocked the tundish nozzle slot for holding the melt in the tundish. The Ca-Mg-Al-based alloys were initially melted in a closed stainless steel crucible in an electrical furnace with a protective argon atmosphere. A stainless steel stopper was used to block the crucible's bottom outlet. When the melt temperature reached about 600°C, then, in order to melt in all the alloy materials, the temperature settings of the furnace were adjusted downwards to 530°C. Once the melt temperature inside the crucible had decreased to 530~540°C, the crucible was taken out of the furnace and quickly fixed inside the tundish on the HSBC simulator; the stainless steel stopper was then lifted up to open the crucible's bottom outlet, allowing the fresh liquid metal to enter the tundish; at the same time, the compressed spring was activated so as to propel the substrate underneath the tundish, causing the liquid metal to flow out through the tundish nozzle slot and to deposit on the moving substrate, thereby forming a thin metallic strip. The substrate moved at constant speed because the friction of the roller bearings carrying the substrate was very small. Strip samples of thicknesses ranging from 0.5mm to 3mm were cast, by appropriately adjusting the casting speed and the width of the nozzle slot.

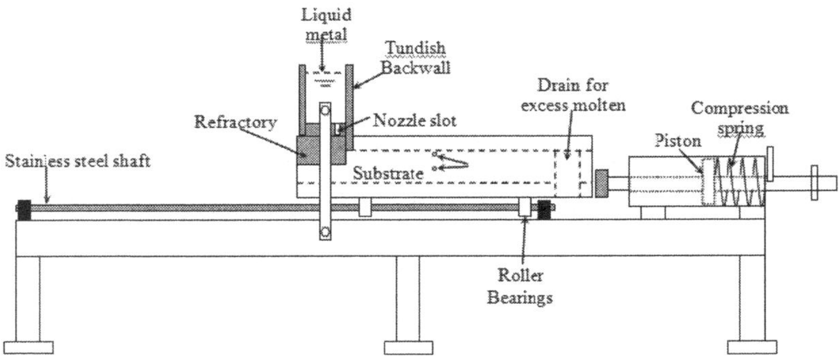

Figure 6. Schematic view of the Horizontal Single Belt strip casting simulator.

In this paper, pure copper was used as the substrate material for extracting heat from the molten Ca-Mg-Al-based alloys. The copper surface was sandblasted by aluminum oxide particles in order to obtain a uniform surface roughness of about 5 microns. A black enclosure box was placed on the simulator in order to provide for a protective gas containing argon during the casting process. Two T-type thermocouples with extremely rapid response times were embedded in the copper substrate at different vertical positions for measuring the temperature variations of the copper substrate [13]. The IHCP (Inverse Heat Conduction Problems) method was used to calculate the transient interfacial heat fluxes, using the temperature-time data acquired.

In order to enhance the interfacial heat fluxes and to form thicker amorphous strips, helium gas was injected through the holes drilled into the nozzle refractory in front of the casting meniscus, forming a helium curtain and trying to flush the air out of the meniscus region, shown in Figure7. The diameter of the holes is about 1 mm, and the holes were distributed on the nozzle refractory surface at an interval of every 3mm in the transversal direction. The interfacial heat fluxes obtained were compared with the heat fluxes when casting the melt in the air, under the same casting parameters. Helium flow rates of 3 LPM (liters per minute) and 10LPM were considered.

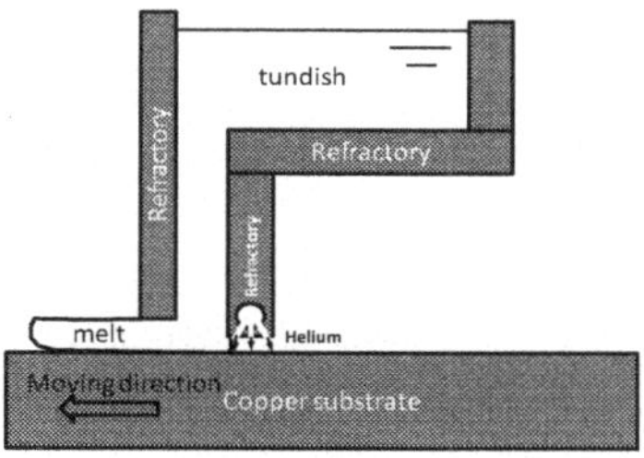

Figure 7. Nozzle configuration for helium injection at meniscus region.

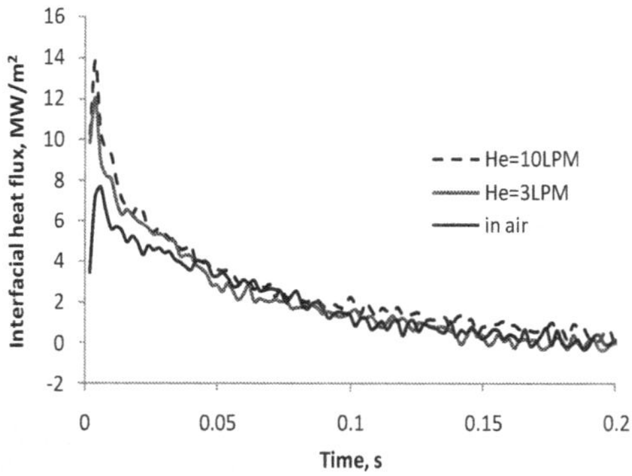

Figure 8. Interfacial heat flux curves of casting 60Ca-15Mg-10Al-15Zn alloy within/without helium.

Experimental results

Interfacial heat fluxes

The typical curves of interfacial heat fluxes obtained in casting experiments have been given in Figure 8, for the 60Ca-15Mg-10Al-15Zn alloy cast on the HSBC simulator. When the melt was cast in air, the maximum heat flux that was reached, was about 8.0MW/m^2 in 5 mille seconds; when a helium flow rate of LPM was used, the maximum heat flux went up to 12.0MW/m^2; while the helium flow rate=10LPM was used, the maximum heat flux obtained became 14.0MW/m^2. After about 0.05s, all the interfacial heat flux curves became similar because the thin metallic strips started to bend up and lost contact with the copper substrate. In the casting experiments, when the helium flow rate exceeded 10LPM, too much helium gas was blown/entrapped into the melt/substrate interface, and the melt started to float on top of helium layer. This compromised heat transfer efficiency, and simultaneously caused the nucleation of crystalline elements. Therefore, a maximum flow rate of helium of 10LPM was used in all subsequent casting experiments. Although a fivefold increase in interfacial heat fluxes was expected by replacing air with helium was anticipated, only 40% ~70% increases of the maximum value of heat fluxes was obtained in the experiments. A possible reason is that not all of the air could be flushed out of the micro-valleys in the copper surface, when the surface was given a rapid flush of helium, just prior to meeting falling metal alloy. If so, the average thermal conductivity of the interface between the melt and substrate would be greatly compromised by the entrapped air. In order to further enhance interfacial heat transfer, more effort should be made in flushing or purging the substrate surface by helium atmosphere before the substrate starts to contact the melt.

Ternary Alloy of 60Ca-20Mg-20Al

Figure 9 displays the typical sample of the 60Ca-20Mg-20Al alloy strips, cast on the HSBC simulator. The strip thickness was 0.8mm, and the strip's length and width were 500mm and 70mm respectively. Large, "eye-like" marks could be seen on the strip's bottom surface in Figure 9(b). Only about ~5-10% of the whole strip surface had a bright appearance, and was flat in the central part of the strip, without any marks. The strip samples were tested by X-Ray Diffraction (XRD), and the results are given in Figure 10. The sample taken from the central bright area was verified to have an amorphous matrix, while the samples from the marked areas displayed crystallization peaks on the XRD curves. For a strip thickness of 1.2mm, it was found that the entire strip's bottom surface was covered by these "eye-like" marks and that the strip had crystallized completely. According to the predictions in Figure 5, the 60Ca-20Mg-20Al alloy must require much higher cooling rates than 800°C/s during solidification, if it is to form an amorphous matrix. It was therefore ascertained that the HSBC process, in its present format, could not provide enough cooling to form metallic glassy strips for the 60Ca-20Mg-20Al alloy.

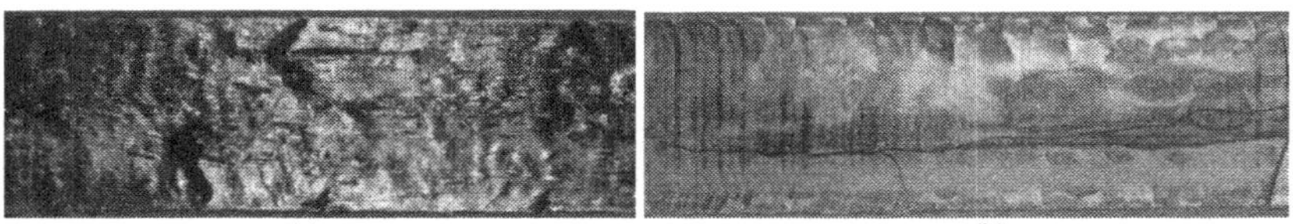

(a) (b)

Figure 9. Typical strip sample of 60Ca-20Mg-20Al alloy, thickness 0.8mm, length 500mm, width 70mm. (a) top surface; (b) bottom surface

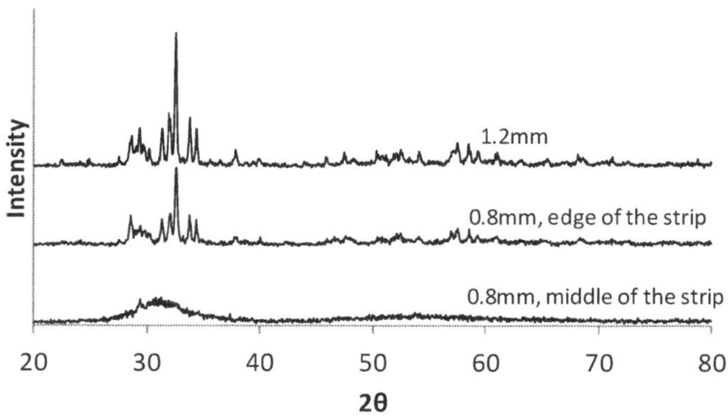

Figure 10. XRD curves of 60Ca-20Mg-20Al alloy.

Quaternary Alloy of 60Ca-15Mg-10Al-15Zn

The 60Ca-15Mg-10Al-15Zn alloy was next cast on the HSBC simulator. When alloy strips, 0.9mm and 1.1mm thick were cast, the strip's bottom surface was flat and bright. Figure11 gives a typical strip sample, 1.1 mm thick. The length of the strip was 500mm and its width was 70mm. The fracture surfaces of the cracked strips were shiny, sharp, with a glassy appearance. When the strip thickness was increased to 1.8mm, some dark areas appeared on the bottom surface, starting from the edges and moving towards the longitudinal centerline of the strip, as shown in Figure 12. With increasing strip thickness, the dark areas grew more and more, and finally occupied the whole strip surface.

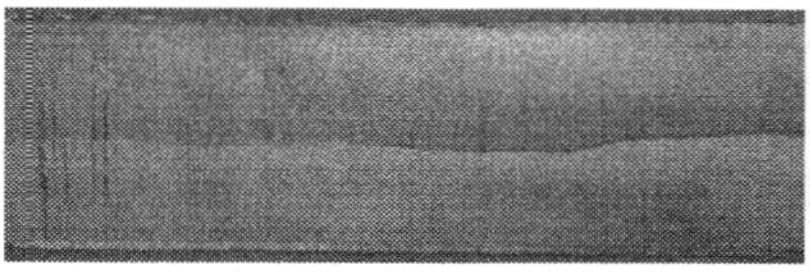

Figure 11. Photo of the strip 60Ca-15Mg-10Al-15Zn alloy, thickness 1.1mm, length 500mm, width 70mm.

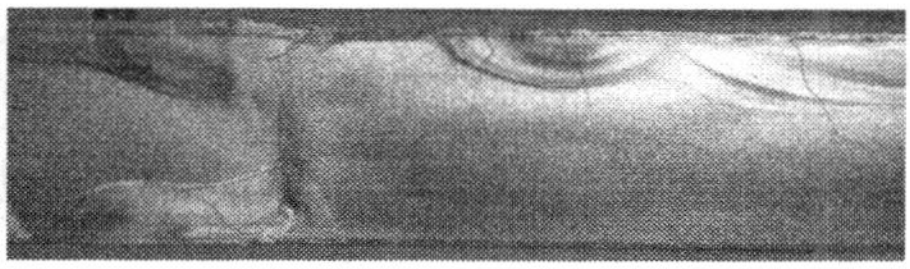

Figure 12. Photo of the strip of 60Ca-15Mg-10Al-15Zn alloy, thickness 1.8mm, length 500mm, width70mm.

The microstructures of the 60Ca-15Mg-10Al-15Zn alloy strip samples were investigated by X-ray diffraction (XRD). Figure 13 gives the XRD curves for the 60Ca-15Mg-10Al-15Zn strips, for samples of different thicknesses. When the strip thickness was 0.9mm and 1.1mm, samples were taken from both the central part of the strips and the edge part of the strips. The XRD curves verified that the nucleation and crystallization had been prevented, because of the rapid cooling by the copper substrate, and that an amorphous matrix had been formed in the strips.

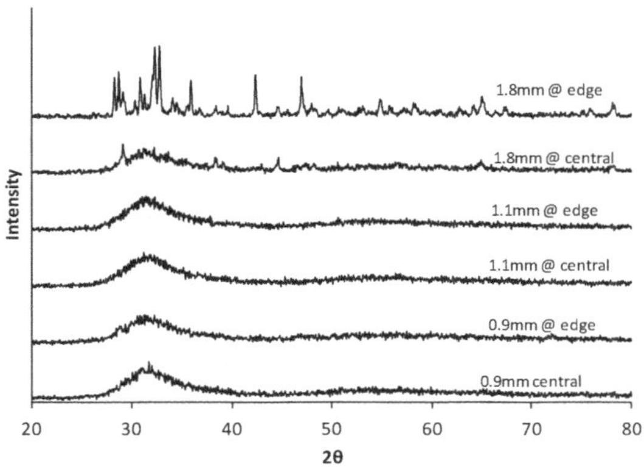

Figure 13. XRD curves of 60Ca-15Mg-10Al-15Zn alloy at different thicknesses.

Adding the element Zn into the Ca-Mg-Al-based alloy certainly improved the glass forming ability and decreased its critical cooling rate, in comparison with the results of 60Ca-20Mg-20Al alloy. When the thickness of these strips was increased to 1.8mm, samples were taken from both the dark area at the strip edge and central flat area for XRD test. In the central area of the strip it had an amorphous-based matrix, together with some peaks generated by crystals or metallic compounds that had precipitated during the rapid cooling process. However, the strip's edges in the dark area were completely crystallized. In the casting experiments, it was found that when the strip thickness was increased to more than 2.7mm, the strips formed were completely crystallized, and had become dark.

During solidification, the melt shrinks, causing the whole strip to become curved up at the edges in the transverse direction. Once the strip at the edges had been released from the copper surface, the air gap between the melt and the copper surface at the edge regions was increased. This significantly increased the thermal resistance for interfacial heat transfer, and reduced the cooling rate as a consequence, in comparison with the cooling rate in the central area of the strip. The crystallized metal at the edge changed the way light was reflected from its surface, thereby forming obvious marks on the strip's surface. This made it very easy to determine whether crystallization had occurred, or not, in a given strip.

Quinary 55Ca-15Mg-10Al-15Zn-5Cu Alloy

Adding elemental Cu into the alloy further improved the glass forming ability of the Ca-Mg-Al-based alloy. This was verified in the strip casting tests. When strip samples of 0.5mm, 1.3mm and 2.0mm thickness were cast on the HSBC simulator, all the sample surfaces were flat and bright, with no obvious signs of crystallization. A typical strip sample, 1.3mm thick, is given in Figure 14. When the strip thickness was less than 2.0mm, the microstructure of the amorphous matrix was obtained and confirmed by XRD tests, as shown in Figure 15. When the strip thickness increased to 3.0mm, crystallization marks started to appear on the edge of the strip, similar to the 60Ca-15Mg-10Al-15Zn alloy, shown in Figure 16. The XRD curve of the sample taken from the central bright area verified that it has a glassy-based matrix with some small peaks showing crystallization was about to start, while the sample taken from the dark edge of the strip displayed complete crystallization peaks (Figure 15).

Figure 14. Photo of the strip of 55Ca-15Mg-10Al-15Zn-5Cu alloy, thickness 1.3mm, length 500mm, width 70mm.

Figure 16. Photo of the strip of 55Ca-15Mg-10Al-15Zn-5Cu alloy, thickness 3.0mm, length 500mm, width 70mm.

From the experimental tests, it was found that the maximum amorphous strip thickness of 55Ca-15Mg-10Al-15Zn-5Cu alloy, cast by the HSBC process, is between 2.0mm ~ 3.0mm. This allows us to estimate that the critical cooling rate of 55Ca-15Mg-10Al-15Zn-5Cu alloy is about 400~500°C/s, based on the predictions of the mathematical model presented in Figure 5.

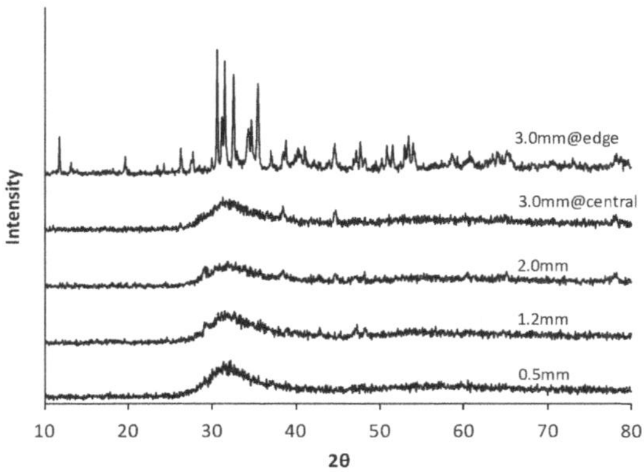

Figure 15. XRD curves of 55Ca-15Mg-10Al-15Zn-5Cu strips with different thicknesses.

Scanning Electron Microscopy (SEM)

In the casting experiments, the strips cracked or shattered shortly after initial solidification (~ 20 – 40 secs). The fracture surfaces were analyzed using scanning electron microscopy (SEM). Figure 17 gives the SEI image of a typical strip's fracture surface of 60Ca-15Mg-10Al-15Zn alloy at a magnification of 100×. The sample of the surface was scanned, and found to be comprised by a small white area. This small white area, marked by a white circle, was amplified in Figure18 to 750×, and proved to be a region of porosity. Within this porous region, many small particles of about 1μm diameter were attached to the fracture surface. Using EDS, the chemical compositions of the particles were analyzed and found to be rich in calcium and chlorine.

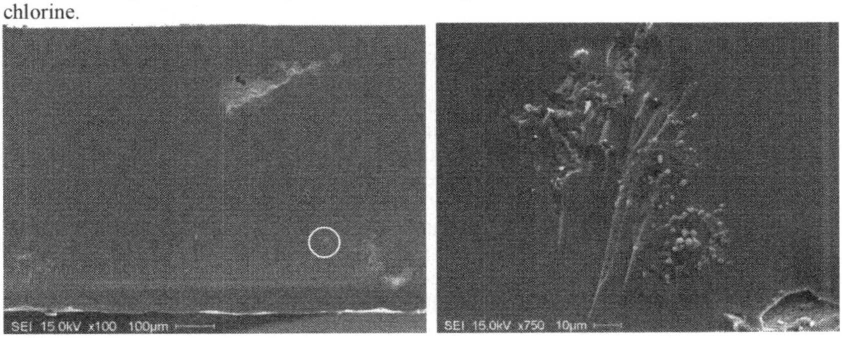

Figure 17a. SEI image of the strip fracture surface. The circled area is amplified in Figure 17b.

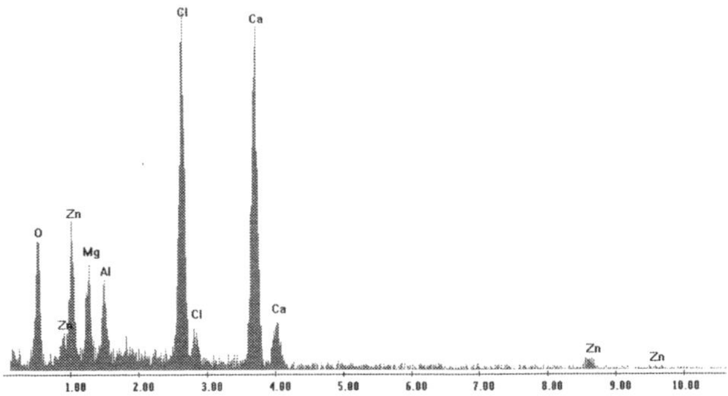

Figure 18. EDS chemical analysis of the particles in the amplified SEI image in Figure 17b.

The samples of the other two alloys were also checked on their fracture surfaces, and similar results were obtained. By analyzing the XRD curves, it was verified that the particles were $CaCl_2$. Presumably these particles will be found in the raw materials provided, and became impurities within the strip. These kinds of impurities will become the nucleation sites or weak points for thermal stresses induced during rapid cooling processes. In any future work, it will be important to remove these impurities from the melt, before any further strip casting experiments aimed at producing Ca-Mg-Al-based amorphous strips, in bulk quantities. In addition, a degassing treatment will also be necessary for eliminating the porosity in the bulk metallic glassy strip.

Conclusions

Ca-Mg-Al-based alloys of bulk metallic glassy materials were cast on a copper substrate using the Horizontal Single Belt Strip Casting (HSBC) process. It was found that the HSBC process is able to provide sufficient cooling ability to form amorphous strips for the 60CA-15Mg-10Al-15Zn and 55Ca-15Mg-10Al-15Zn-5Cu alloys. The maximum amorphous strip thickness for 60CA-15Mg-10Al-15Zn alloy obtained was 1.1mm and the maximum amorphous strip thickness for 55Ca-15Mg-10Al-15Zn-5Cu alloy was 2.0mm. For the case of 60Ca-20Mg-20Al alloy, the critical cooling rate was too high to form metallic glassy strip via the HSBC process. The possible reason for strips cracking after casting test was analyzed. The strips exhibited problems with impurities and with crystals forming at their edges. The presence of compounds, rich in Ca-Cl-Mg-Zn-Al-O, that we suppose are present in the preparation of the alloys, acted as nucleation sites for spontaneous breakages. Similarly, thermal stresses generated by the rapid cooling of the strips made them susceptible to brittle fracture and cracking. Clearly, impurities in the raw materials and any gas bubbles in the melt will need to be removed before further casting experiments are made.

References

1. Chapter 8, Treatise on Process Metallurgy, 1st Edition, Editor-in-Chief, S. Seetharaman (Elsevier Press), to be published 2014.
2. V. Keppens, Z. Zhang, O.N. Senkov, and D.B. Miracle, "Localized Einstein modes in Ca-based bulk metallic glasses", *Philosophical Magazine*, 87(2007), 503–508.
3. O. N. Senkov, D. B. Miracle, J. M. Scott, "Development and characterization of Ca–Mg–Zn–Cu bulk metallic glasses", *Intermetallics,* (2006), vol. 14, pp 1055.
4. M. L. Morrison, R. A. Buchanan, O. N. Senkov, D. B. Miracle, P. K. Liaw, "Electrochemical behavior of Ca-based bulk metallic glasses", *Metall. Mater. Trans. A*, (2006), vol. 37, pp 1239.
5. J.E.Dahlman, O.N.Senkov, J.M.Scott and D.B.Miracle, "Corrosion Behavior of a Ti-Based Bulk Metallic Glass and Its Crystalline Alloys", *Mater. Trans.* 48(2007), pp 1857-1885.
6. K. Amiya, A. Inoue, "Formation and Thermal Stability of Ca-Mg-Ag-Cu Bulk Glassy Alloys", *Mater. Trans. JIM*, (2002), vol. 43, pp 81.
7. K. Amiya, A. Inoue, "Formation and Thermal Stability of Ca-Mg-Ag-Cu Bulk Glassy Alloys", *Mater. Trans. JIM*, (2002), vol.43, pp 2578.
8. Senkov, Oleg N.; Scott, James M.; Miracle, Daniel B. Miracle. "Development of Low Density Ca-Mg-Al- Based Bulk Metallic Glasses", *Materials Transactions* , (2007), vol. 48(8), pp 2258.
9. O.N.Senkov, J.M.Scott, D.B.Miracle;"Development of Low Density Ca-Mg-Al-Based Bulk Metallic Glasses", *Materials Transactions* , (2007), vol. 48(7), pp 1610-1616.
10. E.S. Park and D.H. Kim, "Formation of Ca–Mg–Zn bulk glassy alloy by casting into cone-shaped copper mold" *J. Mater. Res*., (2004), vol. 19, pp. 685–88.
11. E.S. Park, W.T. Kim, and D.H. Kim, "Effect of Alloy Composition on the Glass Forming Ability in Ca-Mg-Zn Alloy System" *Mater. Sci. Forum*, (2005), vols. 475–479, pp. 3415–18.
12. G.X Wang and E.F. Matthys, in E.F. Matthys, W.G. Truckner (Eds). (*Proceedings of Melt Spinning, Strip casting and Slab Casting*, Warrendale, 1996), 205-236.
13. D. Li, M. Isac and R. Guthrie, (*2011 GUTHRIE SYMPOSIUM PROCEEDINGS*, Montreal, Canada, June 2011, MMPC, ISBN: 978-0-986968-0-6), pp. 445-451.

CELEBRATING THE MEGASCALE

Proceedings of the Extraction and Processing Division Symposium on Pyrometallurgy in Honor of David G.C. Robertson

Ferro-Alloys

Celebrating the Megascale: Proceedings of the Extraction and Processing Division Symposium on Pyrometallurgy in Honor of David G.C. Robertson
Edited by: Phillip J. Mackey, Eric J. Grimsey, Rodney T. Jones, and Geoffrey A. Brooks
TMS (The Minerals, Metals & Materials Society), 2014

DEVELOPMENTS IN MANGANESE FERROALLOY RESEARCH AND PRODUCTION IN THE LAST 25 YEARS

Merete Tangstad[1], Ragnar Tronstad[2]

[1]Norwegian University of Science and Technology; 7491Trondheim; Norway
[2]Elkem AS, Technology; Kristiansand, Norway

Keywords: manganese ferroalloys, thermodynamics, emissions

Abstract

Ferromanganese production in electrical furnaces is quite an old process. However, as the furnaces are getting bigger, the requirements of cost, yield, product quality, emissions and security are raised. With the increased expectations, the fundamental knowledge must follow. Within the last 25 years there has been a great effort both in the industry as well as the academic environment to push the limits of this process. This article will show some examples of improvements within fundamental data, reaction mechanisms and emissions over this time period.

Introduction

In many areas of our society there has been an extraordinary technical development the last 25 years. The most obvious area is developments in the information technology. Going from a state where only a few knew about the internet, till today's everyday life, where we cannot check the date or pay for a bus ticket without using it, is dramatic. However, gadgets and products have also had a riveting development. Some of us remember the mobile phone from the early 1990ties, and most of us agree that it was not very mobile. Many will maybe say that the cars are more or less the same, as the appearance is basically the same. Though the motor is basically the same, there is a number of improvements in the cars control system and engine, changing the performance and the emissions of a car. This is also the case of course for the metal producing industries, where the appearance in some cases looks similar, but where there are major changes in the performance and emissions.

There are however some things that have not changed. The laws of nature are the same today as they were 25 years ago, and before. For an oxide ore, we still need to remove the oxygen by a reductant or by electrolysis. We still have the same basic elements we had 25 years ago, and hence still carbon is the most used reductant. This has not changed. The fundamentals of the process are still the same and can be explained by a low temperature zone (the pre-reduction zone) and a high temperature zone (the coke bed zone). In the pre-reduction zone the raw materials enters, and water evaporates, carbonates decomposes and higher manganese oxides will decompose to MnO:

$$MnO_2 + CO = MnO + CO_2 \qquad (1)$$

Instead of a MnO_2 ore, the ore could be a Mn_2O_3 ore, or a sinter containing Mn_3O_4, where all oxide forms will be reduced with CO gas producing CO_2. If the pre-reduction happens deep in the furnace, at high temperatures, the produced CO_2 will react according to the Boudouard reaction:

$$C + CO_2 = 2\,CO \qquad (2)$$

The Boudouard reaction (reaction 2) is energy consuming and will increase the total energy consumption of the process.

In the high temperature zone the important metal producing reactions are the reduction of MnO and SiO_2 in the liquid slag to Mn and Si:

$$MnO + C = Mn + CO \qquad (3)$$
$$SiO_2 + 2C = Si + 2CO \qquad (4)$$

The iron is easily reducible compared to Mn and Si, and hence the iron will be reduced mainly before the ore melts. The product will hence be HC FeMn, a carbon saturated Mn-Fe alloy, or if higher temperature and higher SiO_2 content in the raw materials is present, a carbon saturated Mn-Si-Fe alloy, standard SiMn will be produced.

The development within manganese ferroalloy production has by and large been driven by the major alloy producers. However, on the research side, also a substantial amount of knowledge is produced in the research institutions. Before 1990, one of the main contributors to basic knowledge was National Institute of Metallurgy (now Mintek) in South Africa which did a tremendous job of establishing the fundamentals in the process. The last 25 years there has also been a considerable amount of Mn specific research at University of Missouri Rolla (USA), the University of New South Wales (Australia) as well as a joint effort by the Norwegian industry and SINTEF/NTNU in Norway. In addition the general knowledge about thermodynamic has been developed and made more easily accessible by program and database developers like Thermocalc and Factsage .

As previously mentioned, the laws of nature are still the same as they were 25 years ago. However the great improvements that have occurred during this time period, are numerous. These can be grouped in several areas. First, the knowledge of the fundamentals has increased dramatically. Today one can model, calculate, and maybe more importantly, interpret the furnace operation, based on a completely different knowledgebase. Next, the quality and the availability of control systems have increased, and hence one may control the furnaces and furnace operation today on a changed level. A third group is the HSE focus and knowledge. Both the security and the environmental focus are on very different levels, compared to the 1990ties. In the following some examples of increased knowledge and competence obtained during the last 25 years will be given.

Thermodynamic data

Both in the high temperature and the low temperature areas, the thermodynamic systems are quite complex, mainly due to the many elements present. The typical Mn-ore contains a number of elements in addition to Mn. The 6 major elements in the furnace are Mn-Fe-Si-Al-Ca-Mg, in addition to oxygen and carbon. There are also elements of lower concentration level like Ba, Ti, K, Zn and trace elements of importance like P, B, Hg, As.

In the beginning of the 1990ties thermodynamic data for 3 component systems was known. Examples of this are the phase diagrams of Muan and Osborne (Muan and Osborne,1966). An example is shown in Figure 1. However, equilibrium data between 5 component slag (MnO-SiO_2-Al_2O_3-CaO-MgO) and 4 component metal phase (Mn-Fe-Si-C_{sat}) was not known, with the consequence that it was unclear if the furnaces was tapped at equilibrium, or if kinetic parameters determined the slag and metal analyses.

The commencement of extensive research in the fundamentals in Norway started with the investigations of thermodynamic equilibrium between slag and metal. Experimentally it is not easy to obtain 4 phase equilibrium, as the attainment of slag-metal equilibrium is quite fast, and the attainment of equilibrium between slag-metal-gas and solid carbon, is much slower. Hence, the solution was achieved with a very small amount of slag and a large metal quantity, as shown in Figure 2 A), giving the advantages that the slag-metal-gas equilibrium could be easily obtained, and also that the composition of the metal would not change during the experiment. Some of the results from the equilibrium data are shown in Figure 2 B). The data is shown in pseudo-ternary diagrams where the apex's was the sum of several components, but with given ratios. The follow-up of the experimental work was later to develop databases that can be used to determine chemical equilibrium. This was done during the first decade of the 2000 millennium. Today the equilibrium data for the major components are well known and can be easily accessed.

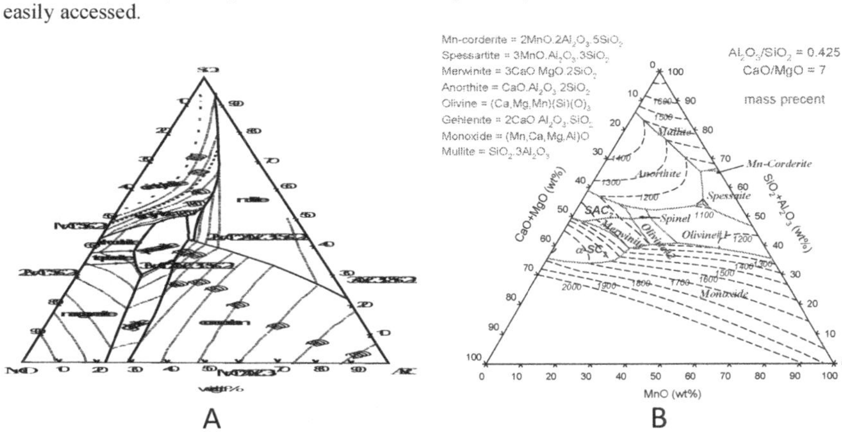

Figure 1. A) MnO-SiO$_2$-Al$_2$O$_3$ phase diagram from Muan and Osborne (1966) and B) MnO-MgO-CaO-Al$_2$O$_3$-SiO$_2$ phase diagram with Al$_2$O$_3$/SiO$_2$=0.425 and CaO/MgO=7 (Olsen et al. 2007).

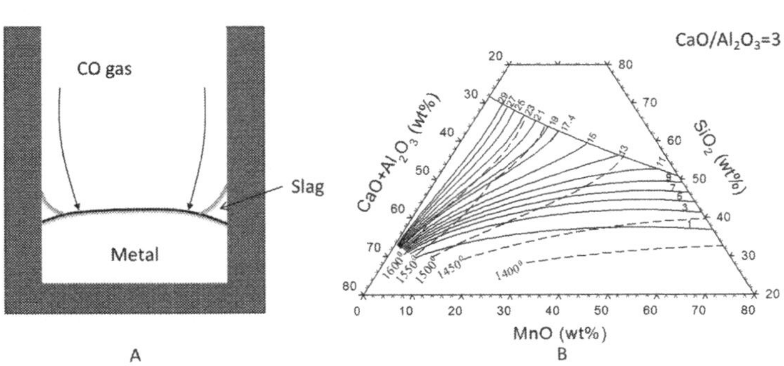

Figure 2. A) Experimental set up for equilibrium measurements and B) partial and total equilibrium lines for Mn-C$_{sat}$ and MnO-CaO-Al$_2$O$_3$-SiO$_2$ slags with CaO/Al$_2$O$_3$=3. (Ding, 1993)

Reaction mechanisms and its influence of furnace operation

Though a furnace may have a number of different reaction zones dependent on how the furnace has been operated, the furnace was previously defined into two zones; a pre-reduction zone and a coke bed zone. The general view on the reduction to Mn and Si was previously that the ore melted on top of the coke bed, and as the melted ore trickled trough the coke bed it would gradually be reduced. By a combination of industrial excavations, pilot scale experiments and lab scale experiment determining the phase conditions and the rate of the reaction, this has been found not to be the reaction mechanism.

During industrial excavations and pilot scale experiments, it has been found that almost all of the reduction takes place on the top of the coke bed , which is before it enters the coke bed. While the slag on top of the coke bed has the composition of a smelted ore, the slag composition a couple of cm's into the coke bed is close to the tapped slag. This can be seen in Figure 3. The main reason for this is explained by the viscosity of the slag as will be discussed below.

Another feature is that it was found industrially that the slag was not in equilibrium with the metal at the tapping temperature. The equilibrium temperature for the tapped slag and metal in a HC FeMn process would typically be around 1350 °C, while the tapped slag was between 1450-1500 °C. For the basic slags that are typically produced in Norway, this could be explained by the phase composition, the kinetics and the energy development in the furnace.

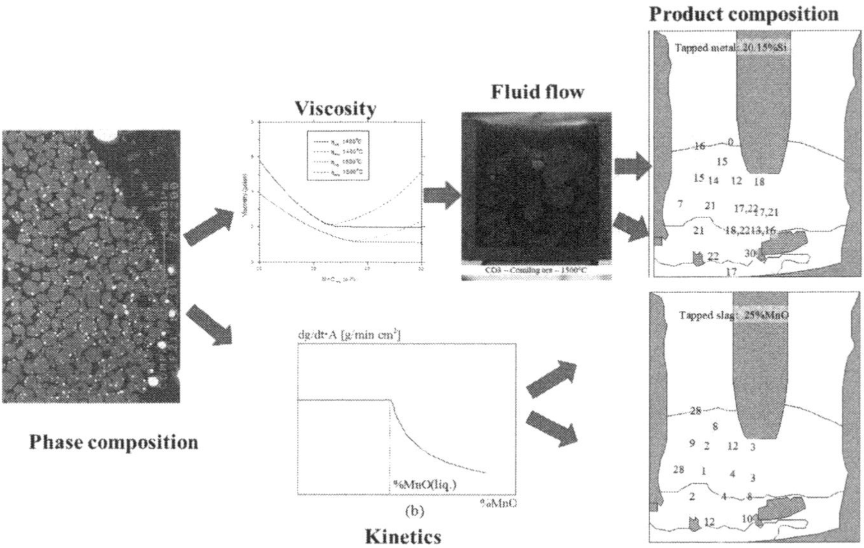

Figure 3. The product quality, given as %Si in the metal and %MnO in the slag, is given by the phase composition, and its effect on the viscosity and the reduction rate.

Phase compositions for three component systems were known before the 1990ies. During the 1990ties also measurement was done to establish the phase composition in 4 and 5 component slags. For medium Al/Si ratio slag systems, a smelted ore would contain a solid MnO phase and a liquid slag phase. This has two consequences; first the activity of MnO is high and hence that the viscosity of the smelted ore is high. Hence, the partly smelted ore has too high viscosity to flow into the coke bed. As the reduction proceeds one will at a certain point reach the liquidus composition. Here the slag viscosity is at its lowest, and the slag enters the coke bed. However, as the solid MnO phase is now dissolved, the activity of MnO decreases dramatically, and hence the reduction rate will also decrease. This leads to the fact that not much more reduction will take place in the coke bed. Hence, the slag is tapped close to the liquidus composition.

The fact that the slag is tapped close to the liquidus composition can also explain some of the features experienced during operation. The furnace may be operated dynamically from a very large coke bed to a very small coke bed, depending on the coke addition. This will change the temperatures in the process. However, the tapped slag changes typically very little. This is due to the fact that while equilibrium composition is very temperature dependent, the liquidus composition is not changing much with temperature. It also explains why a higher temperature gives a higher MnO content in the slag.

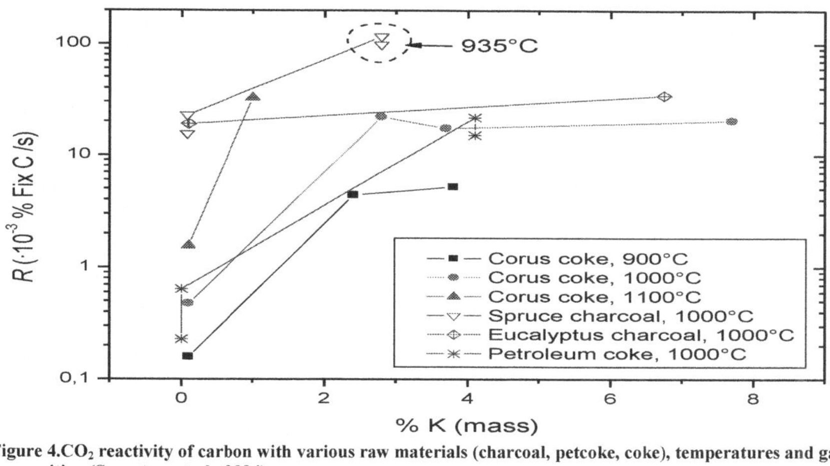

Figure 4.CO_2 reactivity of carbon with various raw materials (charcoal, petcoke, coke), temperatures and gas composition (Syvertsen et al., 2004).

Pre-reduction zone, K- and Zn- circulation

One of the major items in the pre-reduction zone is the degree of pre-reduction in the furnace, which have a large impact on the energy and carbon consumption. The degree of pre-reduction is both dependent on the reduction rate of the ore, represented by equation (1) and the following oxidation of the carbon, shown in equation (2). Reaction (2) will occur at high temperatures, and hence, if the ore is pre-reduced at low temperature, that is a high reduction rate, there will not be any CO_2 present at high temperatures and reaction (2) will not occur. The work regarding pre-reduction has then been focused along two axes, first find the reactivity of various ores, and next find the CO_2 reactivity of the various carbon sources used.

The CO_2 reactivity of the various carbon sources was to certain degree investigated, both by the traders of the carbon materials and by research institutions. However, any potassium present in the gas will affect the reduction rate of the carbon materials, and since it was established that there is at all times potassium recirculating in the furnace, the carbon reactivity will at all times be much higher than measured for CO/CO_2 gas. An example is that while charcoal has a CO_2 reactivity about 100 times higher than coke for pure CO/CO_2 gas, the reaction rate is equal when more than 3-4 % K is present in the carbon material as seen in Figure 4.

As mentioned, the CO reactivity of the ores should be high to avoid an increased carbon and energy consumption (Swamy et al, 1998). There is a continuous investigation of various ores, and their CO reactivity. The general aspect is that it seems likely that the CO reactivity follows the topochemical model and that diffusion through product layer of CO/CO_2 is rate determining. As the reactivity then will be determined by porosity and mineralogy, it seems like the CO reactivity is individual for the different ores.

The K- and Zn-potassium circulation has also been a topic of discussion, where the major issues have been how much is circulating and the effect of it. The major consequence of Zn circulation is its ability to create hard crusts, which may cause explosions. This was first documented in industrial scale by Lee and Kozak (1993), and later verified by lab scale experiments (Slizovskiy, 2012). The potassium circulation effect on furnace operation is not clear. A significant amount of work has been done do simulate the circulation of alkalis. Although a major amount of potassium is circulating, only the effect on the degree of pre-reduction has been verified, as already discussed. It may theoretically affect the lining materials; however, the major problems regarding lining materials in Mn furnaces are caused by slag attack. It is quite rare that there are large lining problems in the pre-reduction area, but liquid potassium has been tapped through drilled holes in the furnace shell at taphole levels. It has also been established that the potassium will not cause the same crust problems as Zn does. Hence, the potassium circulation and its effect on furnace operation is still a matter of debate.

HSE with a focus on emissions

The most noticeable changes during the last 25 years are the increased focus on health, security and emissions at the plants. Some examples from the Norwegian plants are the continuous use of health gear like goggles and masks everywhere in the operational areas. The diffuse emissions have also been focused on and reduced by rebuilding and installing better off gas equipment. When it comes to safety, one of the major visible changes are the trend of having no, or as few as possible, operators around the furnaces. The control room has often been moved to premises some distance from the furnace. While previously the operators controlled the furnace by seeing the top of the furnace and listening to the sound of the furnace, the operator must now control the furnace by measuring vital signs of the furnace. The vital signs are temperatures of gas, slag and metal, the composition of the gas, slag and metal and electrical parameters. The continuous development in robotics that is applied to the tapping equipment is of course also a part of the safety- and control- strategy of keeping the tappers away from potentially hazardous situations.

The focus on emissions to air, land and sea is on another level compared to the situation 25 years ago. First of all, the total amount of the individual elements has to be much lower, typically ___. In addition, there is also a governmental demand today that one should have full control of all trace elements going through the plant. This may of course be a challenge when there is some uncertainty in analyzing the trace elements in the raw materials and the products are in the same order of magnitude as the total content of trace elements itself. However, this has also resulted in extra efforts into the analyzing methods. One result of this development in Norway has been the new mercury cleaning plants added to the existing off gas systems. As new and more strict policy regarding mercury emissions were implemented by the government, the Norwegian industry had to implement state of the art equipment for Hg cleaning of furnace gas. One results of this step-change development is the step-change in levels of emissions as can be seen happening in the mid-1990ties as seen in Figure 5.

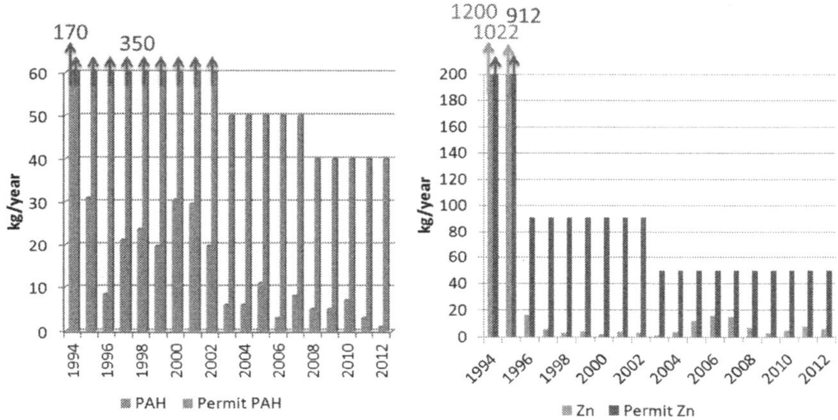

Figure 5. PAH and Zn emissions and their permit levels from 1994 to 2012 from two Norwegian Eramet plants.

Education

The cooperation between the industry and the educational institutions vary quite much between the different countries. The examples here are hence given for Norway, but the trends are the same in many countries worldwide. The education system was previously focused towards the various industries. Hence, many universities had metallurgy and metal production as educational choices. Today the education system is more general, and includes a number of various possibilities e.g. towards material technology in addition to more non-technological courses. Hence, previous 5 year of study gave you an occupation, while today 5 year of study will give you a general competence in a number of different directions.

As for the society in general, also the education level among the employees at the plants have increased. Today, the majority of operators are skilled workers often with additional competence qualifications.. Also engineers usually have a degree MSc or a PhD, while previous they were BSc and MSc. 25 years ago it was commonly expressed by plant management that "Our Company has no use for engineers with a PhD". This has definitely changed, as higher degree of competence is needed in every step of the value chain. Since 1990 14 PhD candidates has undertaken their thesis within the field of FeMn production in Norway. Of these 5 have started in the FeMn industry in Norway, some in other industries, 3 in commercial research institutions and a couple in academic institutions.

Current and future trends

There are some overall trends that has been present for 25 years, and which also will increase in the future. The focus on environmental impact will definitely continue. Maybe the most important issues are to reduce the consumption of resources like raw materials and energy. The energy consumption per ton of alloy may be reduced overall, by the use of energy recovery but also by a higher yield. Hence the trend towards more agglomeration of the raw materials will definitely continue. This will also help towards the scarcity of high quality raw materials in the future.

Incremental improvements of the present process for HC FeMn/SiMn will take place worldwide. However, one can also be so bold as to dream of new types of operation that will exploit the natural laws more optimally. In Japan there are examples of using separate prereduction units for Mn-ores and thereby dividing the furnace processes into unit operations. Though higher investments, one should believe that the possibility of controlling the various parts of the process separately will give a better control. As the time passes and even more knowledge is gained, the ultimate FeMn process may be in a different shape than the electrical smelting furnace found today.

However, going back to the process used today, many metallurgical challenges are still interesting. The security and resource optimization using more fines is an interesting issue. How much fines can a furnace take, and what kind of fines? As much of the reduction is occurring at the top of the coke bed, more information regarding different raw materials is of quite significant interest. It is among other things shown that agglomerated acid ores (sinter and pellets) melts and reduces at lower temperatures compared to their original ore. One would believe that a process that is close to 100 years in age, would be operated at the theoretical limit. As more knowledge is gained, it is, however more clear than ever that there is still a way to go to meet the theoretical limits on environment, personal security, control system and also on process details. The Norwegian University of Science and Technology will therefore continue its fundamental work on FeMn/SiMn processing in cooperation with the Norwegian industry.

References

1. Muan, A, Osborne, E.F., *Phase equilibria among oxides in steelmaking*, Addison Wesley, (1966)

2. Olsen, S, Tangstad, M, Lindstad, T: *Production of manganese ferroalloys*, Tapir forlag, 2007

3. Ding, W:*Equlibrium relations in the production of manganese alloys*, PhD Thesis 1993:15 at NTH, Trondheim, Norway (1993)

4. K. N. Swamy, K, N, Robertson, D.G.C, Calvert, P, Kozak, D, *Factors Affecting Carbon Consumption in the Production of High Carbon Ferromanganese*, INFACON 9, pp. 293-301.

5. Syvertsen, M, Lindstad, T, Grøntvedt, P, O: *Boudouard reaction in coke and charcoal*, Presentation at NorskMetallurgiskSommermøte in Oslo, Norway, 2004.

6. Lee, Y, Kozak, D: The role of zinc in the eruption of high carbon FeMn smelting furnace, Proceedings of the Electric Furnace Conference (1993)

7. Slizovskiy, D: *Agglomeration in ferromanganese submerged arc furnace with focus on K and Zn deposition,* PhD thesis no 2012:137 at NTNU, Trondheim, Norway, ISBN 978-82-471-3558-7 (2012)

Celebrating the Megascale: Proceedings of the Extraction and Processing Division Symposium on Pyrometallurgy in Honor of David G.C. Robertson
Edited by: Phillip J. Mackey, Eric J. Grimsey, Rodney T. Jones, and Geoffrey A. Brooks
TMS (The Minerals, Metals & Materials Society), 2014

DC ARC FURNACES – PAST, PRESENT, AND FUTURE

Rodney T. Jones

Mintek; 200 Malibongwe Drive, Private Bag X3015, Randburg, 2125, South Africa

Keywords: Pyrometallurgy, DC arc furnace, electric arc, smelting

Abstract

DC arc furnaces were first used industrially for the reductive smelting of chromite fines to produce ferrochromium thirty years ago. Since then, they have been used for a variety of applications, including the smelting of ilmenite to produce titania slag and pig iron, the recovery of cobalt from non-ferrous smelter slags, and the smelting of nickel laterite ores to produce ferronickel. The power of these furnaces has increased from 12 MW to 72 MW for ferrochromium, and to 80 MW for ferronickel. The largest of these furnaces requires two electrodes to carry sufficient current to generate this much power. A review is presented of various DC arc furnaces in use, along with a discussion of the likely ways in which furnace power might be increased further in the future.

Introduction and History

Early Description of an Arc by Humphry Davy

The earliest known written description of a man-made electric arc emanated from Humphry Davy's work in the early 1800s. His early experiments in 1800, only a few months after Volta's introduction of the electric battery, involved the production of sparks during the making and breaking of an electric circuit.

"I have found that this substance [well burned charcoal] possesses the same properties as metallic bodies in producing the shock and spark when made a medium of communication between the ends of the galvanic pile of Signor Volta. The spark is most vivid when the charcoal is hot." [1]

In his 1812 book, Elements of Chemical Philosophy [2], Sir Humphry Davy provides the first description and diagram (reproduced in Figure 1) of the long horizontal arch of flame that gives the arc its name.

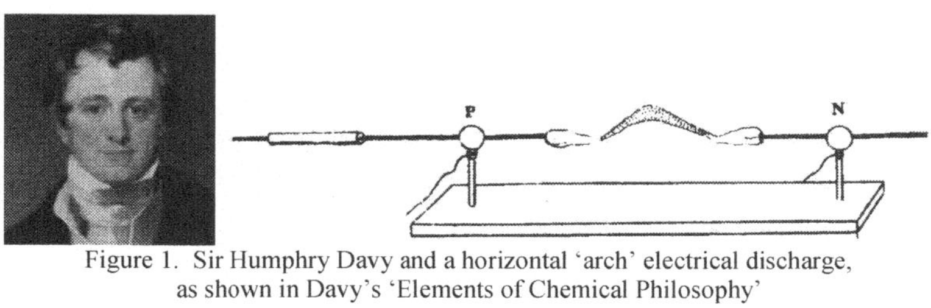

Figure 1. Sir Humphry Davy and a horizontal 'arch' electrical discharge, as shown in Davy's 'Elements of Chemical Philosophy'

"When pieces of charcoal about an inch long and one sixth of an inch in diameter, were brought near each other (within the thirtieth or fortieth part of an inch), a bright spark was produced, and more than half the volume of the charcoal became ignited to whiteness, and by withdrawing the points from each other a constant discharge took place through the heated air, in a space equal at least to four inches, producing a most brilliant ascending arch of light, broad, and conical in form in the middle. When any substance was introduced into this arch, it instantly became ignited; platina melted as readily in it as wax in the flame of a common candle; quartz, the sapphire, magnesia, lime, all entered into fusion; fragments of diamond, and points of charcoal and plumbago, rapidly disappeared, and seemed to evaporate in it, even when the connection was made in a receiver exhausted by the air pump; but there was no evidence of their having previously undergone fusion." [2]

DC Arc Furnace by Sir William Siemens

The idea behind the (direct current) DC arc furnace has been around for a very long time. Its use for the bulk melting of metals dates back at least to 1878, when Sir William Siemens used a DC arc furnace with a vertical graphite cathode, with the arc transferred to the melt in contact with a water-cooled bottom anode (shown in Figure 2) [3].

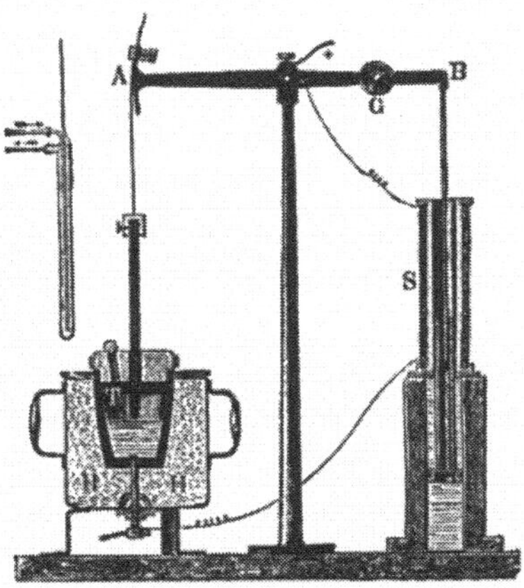

Figure 2. Sir William Siemens and the DC furnace he patented in 1878

About two decades after this, the first AC (alternating current) electric arc furnace, invented by Paul Héroult, was patented and first operated in La Praz, France, in 1900. The first AC furnace in the USA was installed in 1906 in Philadelphia. Electric furnace technology became almost entirely AC based thereafter, because of the use of AC for efficient power transmission from large central power stations (following developments by Nikola Tesla and George Westinghouse in 1887 and 1888) and the use of local step-down transformers to supply the high currents required in the furnaces.

Significant commercialisation of DC furnace technology occurred in the 1990s, when there was a huge demand for DC arc furnaces for steel scrap melting, following the development of high-power solid-state rectifiers. Around that time, about 80 DC arc furnaces were built in the northern hemisphere.

Features of DC Arc Furnaces

A DC arc furnace typically comprises a refractory-lined cylindrical steel shell, with a central graphite (cathode) electrode vertically positioned through an opening in the centre of the roof. The anode connection in the hearth of the furnace is in direct contact with the layer of molten metal that is covered by a layer of molten slag. The energy is supplied by means of an open plasma arc that is generated between the bottom tip of the cathode and the upper surface of the molten slag. At least a central portion of the slag surface is open (uncovered by feed material). Because the furnace is electrically powered, very high temperatures (> 1500°C) can be attained. The open bath allows fine feed materials to be fed into the furnace, without risk of blocking the gas emanating from the chemical reactions. (By contrast, in a blast furnace, or in a submerged-arc furnace, coarse feed materials are required in order to provide a porous bed that allows the reaction gases to percolate away and escape from the reaction zone.)

Finer ores and a requirement for very high temperature lead one to the choice of an open-arc furnace (either AC or DC). It should be noted that fine ores allow the use of a fluidized-bed reactor for pre-heating or pre-reduction, and this can provide significant energy savings when used in conjunction with a DC arc furnace.

In the case of open-arc furnaces, DC furnaces have a number of advantages over AC furnaces. One particular advantage in a circular furnace is that there is no arc repulsion in the case of the single DC arc, compared to the AC case where the arcs repel one another, flaring towards the walls, leading to hot spots on the areas of the side-walls in closest proximity to the electrodes. DC furnaces also experience lower electrode consumption. In large DC furnaces, a higher current can be carried per electrode (or smaller electrodes can be used for the same current), because of the AC 'skin effect' where current is concentrated in the outer periphery of the electrodes.

Some Past and Current Applications of DC Arc Furnaces

Early Developments in the Application of DC Arc Furnaces to Chromite Smelting

South Africa is very fortunate in having large reserves of chromite. Unfortunately a lot of the 'Transvaal chromites' are highly friable, resulting in large stockpiles of fine ore materials. This posed a particular challenge to Mintek to provide a solution to this problem. Peter Jochens of Mintek identified 'plasma furnaces' as a possible solution to the 'chromite fines' problem. Mintek and Middelburg Steel & Alloys (now part of Samancor Chrome) conducted smelting trials on Tetronics' pilot transferred-arc plasma furnaces in the UK in 1979/80. These trials showed successful metallurgy, but this type of furnace was seen as difficult to scale up to the very large furnaces that were required for the ferro-alloys industry.

Around about the same time, also in the 1970s, ASEA in Sweden developed high-power thyristor rectifiers. Sven-Einar Stenkvist investigated the conversion of AC open arc furnaces to DC, principally for steelmaking. He implemented the idea of using a graphite cathode electrode arcing onto a slag/metal bath as the anode, and devised an electrically conductive hearth and a hollow graphite electrode for finely sized iron ore smelting, where the feed material is fed through the electrode and through the heart of the plasma arc. This is shown in Figure 3.

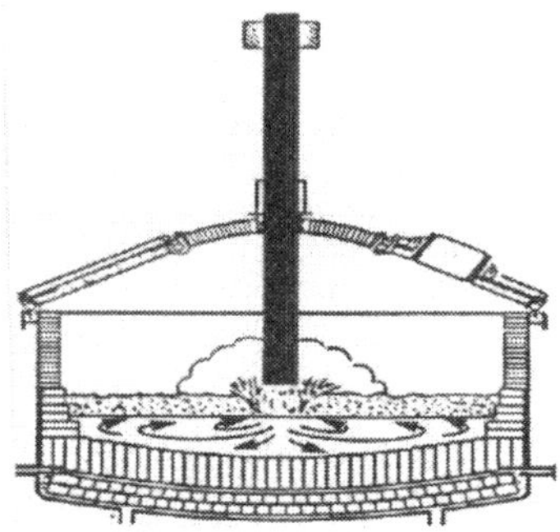

Figure 3. ASEA's DC arc furnace

Nic Barcza of Mintek recognised the synergy between the metallurgy that was proven during the Tetronics testwork, and the scale-up potential of ASEA's DC arc furnace. This is where the application of pre-baked graphite electrodes in a DC arc furnace came about. Mintek built a 1.2 MW DC arc furnace in 1983 to support this development. Since then, that concept has been applied in a number of areas.

The first industrial application of a DC arc furnace for smelting came about when Middelburg Steel & Alloys converted an existing AC furnace at Palmiet Ferrochrome (now Mogale Alloys) in Krugersdorp to a 12 MW DC arc furnace of ASEA design in 1984.

Chromite Smelting

Mintek [4] began investigating alternative smelting methods for ferro-alloy production in the mid to late 1970s. The growing ferro-alloy industry in South Africa faced the problem of how to deal with the significant quantity of fine material (< 6 mm) that was generated from the friable local chromite ores. In 1976, studies commenced on the possible use of plasma smelting (DC transferred-arc technology), motivated by its potential ability to utilise fine feed materials directly, without costly agglomeration. A further advantage was that the metallurgical and electrical parameters of the smelting process are independent, unlike in a conventional submerged-arc furnace [5].

In 1979, Middelburg Steel & Alloys, now part of Samancor Chrome, asked Mintek to take part in tests on plasma smelting at Tetronics in the United Kingdom. After that, Mintek committed itself to the investigation of plasma technology, and to the demonstration to industry of its benefits, particularly the ability to smelt fine materials. Equipment for this purpose was installed in Mintek's pilot bays, and was based initially on a power supply of 100 kVA, and later on another of 3.2 MVA. The first such ferrochromium was produced in a bench-scale DC arc furnace in 1979. The installation of the large experimental furnaces in Mintek's pyrometallurgy pilot bay, Bay 1, demanded an extension of the original building, and, in June 1982, a formal opening ceremony marked not only the completion of the building operations, but also the inauguration of the larger 3.2 MVA furnace. The 1 t/h DC arc furnace pilot plant was commissioned in 1984.

As a result of successful initial testwork, Mintek and Middelburg Steel & Alloys undertook a longer-term R&D programme to develop the technology commercially. Based on testwork at 0.3 – 0.5 MW in furnaces 1 – 2 m in diameter, the initial industrial-scale 16 MVA (12 MW) transferred-arc furnace was installed at the Krugersdorp (Palmiet Ferrochrome) plant of Middelburg Steel & Alloys at the end of 1983, then upgraded (around 1987) to 40 MVA (~33 MW) in 1988. An additional 10 MW furnace was later built on the same site (for Mogale Alloys).

Chromite is smelted together with some form of carbon to produce ferrochromium, along with some slag and some CO gas. The DC arc furnace operates with an open-arc, open-bath configuration, so there is no heaped burden through which the reaction gases would need to escape. There is simply a molten bath onto which the fine material is dropped, and it almost immediately assimilates into the molten bath, melting and dissolving into the slag phase where the reactions take place. This furnace does not require coke because there is no burden above the bath so it doesn't require the porosity that would otherwise be needed. This is a very significant advantage, in that it allows the use of inexpensive reductants, thereby avoiding the high cost and relative scarcity of coke.

The power supplied to the furnace is largely independent of the slag composition, because there is an open electric arc that allows one to adjust the amount of power going into the furnace independently of the electrical resistivity of the slag. This provides extra flexibility and an extra degree of freedom of a change in slag composition unconstrained by electrical properties, so that the chemical activities of the important metals can be changed to achieve higher chromium recovery. A typical increase in chromium recovery could be from about 85% in an AC furnace to about 95% in a DC arc furnace – a very significant increase. In addition, the DC arc furnace has lower electrode consumption. The DC-arc process allows the use of unagglomerated chromite fines and cheaper, non-coking coal. This furnace technology is regarded as one of the lowest-cost options for the production of ferrochromium.

Based on the successful operation of the DC arc furnace for chromite smelting at Krugersdorp, Middelburg Ferrochrome (now part of Samancor Chrome) built a 44 MW (62 MVA) DC arc furnace, the largest of its type, in Middelburg in 1997. This was followed, in 2009, by an additional 60 MW furnace in Middelburg, currently the largest DC arc furnace in South Africa [6]. The 44 MW furnace was subsequently upgraded to 60 MW.

Ilmenite Smelting

Ilmenite is smelted to produce titania slag (mostly for pigment production) and pig iron. Ilmenite from beach sands in South Africa is of too low a grade to be used directly for the production of pigment or synthetic rutile. Instead, it is smelted to produce a titania slag suitable for the production of TiO_2 pigments. The principal reaction involved is shown simplistically as:

$$FeO.TiO_2 + C = TiO_2 + Fe + CO \qquad (1)$$

The high electrical conductivity of titania slags and the required accurate control of the slag composition effectively rule out the use of conventional submerged-arc technology for the smelting of ilmenite. Richards Bay Minerals (RBM) uses process technology originally developed by Quebec Iron & Titanium (QIT Fer et Titane) of Sorel, Canada, employing rectangular six-in-line graphite-electrode furnaces in open-bath mode with AC open-arc operation.

An alternative to this process, based on single hollow-electrode DC-arc furnace technology, was developed by Mintek and Anglo American Corporation for the Namakwa Sands project. The problem that needed to be solved was to find alternative equipment in which to produce a very conductive slag. Because the slag is so conductive, it is not feasible to obtain sufficient resistive heating in the molten slag, and it is, therefore, necessary to use an open arc. (This is a very different reason for using a DC arc furnace from that requiring it for smelting chromite fines.)

Phase one of the testwork, in 1990, involved four 15 kg batch tests at 30 kW, and showed that the process was feasible, and that a freeze lining was required. Phase two, in 1991, involved smelting two metric tons of ilmenite, with continuous feeding at 50 kg/h, at a power level of 100 kW. This produced an on-specification slag, and a metal that was high in Ti. Phase three, also in 1991, saw the smelting of 35 tons of ilmenite, fed continuously at 300 kg/h, at power levels of 500 kW (to 1 MW), with on-grade slag and metal produced. Phase four, in 1995, processed 200 metric tons of ilmenite at 1 t/h and 1.5 MW, and primarily involved the development of the furnace start-up procedure, process control, and operator training.

Following the 0.5 MW pilot-plant testwork in a 1.8 m diameter furnace, the first 25 MW DC arc furnace at Namakwa Sands was constructed in 1994, followed by a 35 MW DC furnace in 1998. The first (25 MW) furnace began production of ilmenite slag and pig iron in June 1995, and the second furnace (35 MW) was brought on line in February 1999 [7].

Based on that success, without any further testwork being required at Mintek, two further 36 MW DC furnaces for ilmenite smelting were built for Ticor SA, near Empangeni, and were commissioned in 2003.

A further 30 MW furnace was built by Bateman and commissioned for CYMG in China in 2009.

Nickel laterite smelting

Laterites and other oxidized nickel ores constitute a very important part of worldwide nickel reserves. In the conventional production of ferronickel from these ores, much fine material is produced which cannot readily be accommodated directly in existing three-electrode or six-in-line AC furnaces. DC arc furnace technology allows ore particles less than 1 mm in size to be treated directly, thereby improving the overall recovery of nickel without the need for expensive agglomeration techniques. Because of the high moisture content of laterites, the ores should be dried and calcined before smelting. A further saving in energy consumption can be achieved by pre-reducing the ores.

Mintek has been working on the production of unrefined ferronickel from nickel-containing laterite in DC arc furnaces since 1993. In this process, lateritic material is fed, together with a carbonaceous reducing agent, to the central region of the molten bath of a cylindrical DC arc furnace. A wide compositional range of nickel laterites can be smelted in this way. The flexible operation of a DC arc furnace (especially its lower dependence on electrical properties of the slag, because of open-arc operation, in addition to the ability to run at an optimum slag temperature, due to the open-bath mode of operation) allowed for the successful treatment of ores with a SiO_2/MgO ratio between 1.2 and 3.0, as well as ores containing up to 30 per cent by mass of iron (which tends to cause slag foaming in a conventional immersed-electrode furnace). A frozen lining can be maintained between the molten bath and the refractory lining, in order to minimize refractory wear (especially at high SiO_2 contents). The smelting of nickel laterite to produce ferronickel was piloted at Mintek [8] from 1993 to 2006, using a wide variety of feed materials.

The first commercial application of the DC FeNi process was a 12 MW DC arc furnace in Orsk in the Southern Urals, Russia, commissioned in September 2011 [9].

Present

Kazchrome

Kazchrome (a ferrochromium producer in Kazakhstan) contracted with Mintek in 2007 to undertake smelting testwork. Based on this, and on other design inputs provided by Mintek in 2009, SMS Siemag has built four 72 MW DC arc furnaces in Aktobe, Kazakhstan, and commissioning was scheduled to begin in 2013. Each of these furnaces features an automated online electrode changer for the 75 cm graphite electrodes, and a suspended roof.

Koniambo Nickel

The largest application of DC arc furnace smelting technology is the Koniambo Nickel project in New Caledonia – a greenfield joint venture partnership between Société Minière du Sud Pacifique (SMSP) and Glencore Xstrata. The process is based on milled ore, fluidized beds, and two twin-electrode 80 MW DC arc furnaces, with cyclone pre-heaters. Each 71 cm graphite electrode has a separate 40 MW power supply (transformer and rectifier). The smelting testwork for this project was carried out at Mintek in the late 1990s. The Koniambo ferronickel smelter, which will employ about 800 people, aims to produce 60 kt/a of Ni in FeNi from one of the world's largest and highest-grade nickel laterite deposits. The smelter commissioning began in 2012, and the first metal from the furnace was produced in April 2013 [10].

List of DC arc furnaces used for smelting applications

Table I. DC furnace installations

Process	Year	Furnaces
Steel		> 80 DC furnaces, up to 175 MW
FeCr	1985	10MW, 30MW, 60MW, 60MW, (4 x 72MW)
TiO2	1994	25MW, 30MW, 35MW, 36MW, 36MW
Co	2001	40MW
Stainless steel dust	2004	32MW
Battery recycling	2008	2.5MW
FeNi	2011	12MW, (2 x 80MW)

Future

Very large DC arc furnaces are currently in use in the steel industry. For example, in 2007, Tokyo Steel ordered the world's largest electric arc furnace (EAF) from Danieli. This furnace has a 420 t liquid steel capacity. It is a twin-cathode DC arc furnace, with four bottom electrodes. The power supply includes 8 x 32 MVA transformers, making up a total of 256 MVA. The operating power is 175 MW. Danieli claims that it overcomes the limits of power input in the EAF, and reduces graphite electrode consumption, as compared with AC and single-cathode DC furnaces, and disturbances on the power supply network are minimized due to the double-arc effect.

Figure 4. Tokyo Steel's 175 MW twin-cathode DC arc furnace

The largest DC furnace used in the steel industry (175 MW) is substantially larger than the largest of the DC smelting furnaces (currently 80 MW). Perhaps the steel industry will lead the way for future expansions in capacity elsewhere.

Speculative Design of Very Large Furnaces

The economy of scale creates an inexorable push towards larger furnaces in the ferro-alloy industry. However, the AC type of submerged-arc furnace as is commonly used at present has certain fundamental limits, and so conversions to other forms of furnace are likely. Once this 'AC barrier' has been overcome, scale up will probably involve larger electrodes and more electrodes. This could be applied to either DC open-arc furnaces or submerged-arc furnaces. The potential for the scaling up of Söderberg electrodes will only be known once this AC barrier has been overcome, and larger furnaces have been tried in practice.

Barker and Barcza *et al.* [11,12] pointed out how DC power can overcome some of the constraints imposed on AC submerged-arc furnaces used for ferro-alloy production, namely the current-carrying capacity of the electrodes, and the interaction effect in the control of the electrodes.

The limiting current I (in kA) carried by a Söderberg electrode of diameter d (in m) is represented by equation (2).

$$I = 55\ d^{1.5} \tag{2}$$

Graphite electrodes can carry almost double the current of Söderberg electrodes of the same diameter.

A larger Söderberg electrode tends to give more trouble than a smaller one, for two principal reasons: the strength-to-weight ratio deteriorates as the electrode is scaled up, and the skin effect (the tendency of alternating current to keep to the outside of a conductor) distorts the distribution of the current flowing in the electrode. The skin effect applies only to AC, not DC. At 50 Hz, the 'skin depth' is around 35 cm for Söderberg electrodes, and about 15 cm for graphite electrodes. Large electrodes are expected to behave better in a DC furnace.

Probably the greatest challenge for the operator of a ferro-alloy furnace with Söderberg electrodes is to balance the baking-slipping rates of the electrodes, as required to maintain electrode lengths. Pre-baked and graphite electrodes do not have this limitation, because they can be slipped as much or as little as required.

Because the power factor typically decreases as the size of an AC furnace increases, the interaction effect is probably one of the main constraints on building larger AC furnaces. Electro-magnetic inductance creates reactance only with alternating current. Therefore, with DC power, there are no constraints from the reactance side. In particular, with a multi-electrode DC furnace, each electrode behaves individually - *i.e.*, there is no interaction effect.

Because the voltages on adjacent electrodes in a multi-electrode DC furnace are similar, and the electrodes do not interact with each other operationally, they can in theory be mounted closer together, should the need arise. It is also worth noting that the flares in a multi-cathode DC furnace will be directed inwards, unlike the outwardly directed flares in an AC furnace.

As the nature of power generation changes in response to environmental challenges, and more renewable energy is used, the ability to swing the furnace load is likely to bring significant economic advantages, and so might become common practice, in spite of the nuisance factors that this will incur with the operation of the plant. It would seem that larger furnaces might also be better than present furnaces at handling power dips and outages for a longer time, and DC furnaces could offer significant advantages when it comes to easy and fast swinging of the load.

Barker [11] proposed a speculative 6-electrode furnace design with the electrodes arranged in a circle. Each electrode would be fed by its own power supply module, using DC power. Barker suggests that, in the future, one could perhaps contemplate Söderberg electrodes with a diameter of 3.0 m (only about 50% larger than the largest electrodes that are presently in use), able to carry a current in excess of 250 kA. A large 6-electrode DC arc furnace could then have a total power of over 200 MW.

The questions remain as to how much current can be removed from the bottom of the furnace, and how long such connections can be kept functioning.

Conclusions

DC arc furnaces have some wonderful features that have been proven in industrial practice over the past thirty years. They are good at accommodating finely sized feed materials (because of the open bath). They do not require coke or char (no burden porosity required). DC arc furnaces can treat feed materials with a wide range of composition (because of the extra degree of freedom coming from power being supplied by an open arc); this allows choice of chemistry for metallurgical benefit. These furnaces are geometrically simple and elegant, thereby reducing uneven wear on side-walls, and leading to lower costs.

However, there are some drawbacks. Thermal efficiency is decreased by the hot off-gas leaving the furnace (unless some of this energy is recovered). The absence of a burden does not allow for the capture of volatile species.

DC arc furnaces are not a panacea for all metallurgical problems, but are very well suited to a number of reductive smelting processes where they have been applied successfully in a number of industrial contexts, and many further applications are expected.

Acknowledgements

This paper is published by permission of Mintek. Support from the National Research Foundation (NRF) of South Africa is gratefully acknowledged.

References

1. Humphry Davy, "VII. Additional Experiments on Galvanic Electricity", *Journal of Natural Philosophy, Chemistry, and the Arts (Nicholson's Journal)*, London, Vol.IV, October 1800, 326-328.

2. Sir Humphry Davy, *Elements of Chemical Philosophy: Part 1 Vol.1* (London: J. Johnson and Co., St Paul's Church Yard, 1812), p.85 and Plate III, Fig.18.
https://archive.org/details/elementschemica00davygoog

3. English patents, No. 4208 of 1878 and No. 2110 of 1879.

4. R.T. Jones and T.R. Curr, "Pyrometallurgy at Mintek", *Southern African Pyrometallurgy 2006*, Edited by R.T. Jones, SAIMM, Johannesburg, 5-8 March 2006, pp.127-150.
http://www.mintek.co.za/Pyromet/Files/2006JonesMintek.pdf

5. R.T. Jones, T.R. Curr, and N.A. Barcza, "Developments in Plasma Furnace Technology", *High-Intensity Pyrometallurgy*, The Institution of Mining and Metallurgy, 4 February 1993, London.
http://www.mintek.co.za/Pyromet/Files/PlasmaDev.pdf

6. D. Sager, D. Grant, R. Stadler, and T. Schreiter, "Low cost ferroalloy extraction in DC-arc furnace at Middelburg Ferrochrome", (Infacon 12, Helsinki, 6-9 June 2010), *SAIMM Journal*, Vol.110, No.12, December 2010, pp.717-724.
http://www.pyrometallurgy.co.za/InfaconXII/803-Sager.pdf
http://www.saimm.co.za/Journal/v110n12p717.pdf

7. M. Gous, "An overview of the Namakwa Sands ilmenite smelting operations", *Southern African Pyrometallurgy 2006*, Edited by R.T. Jones, South African Institute of Mining and Metallurgy, Johannesburg, 5-8 March 2006, pp.189-201.
http://www.pyrometallurgy.co.za/Pyro2006/Papers/189_Namakwa.pdf

8. H. Lagendijk and R.T. Jones, "Production of ferronickel from nickel laterites in a DC-arc furnace", *Nickel-Cobalt 97*, 36th Annual Conference of Metallurgists, Sudbury, Canada, 17-20 August 1997, pp. 151-162.
http://www.mintek.co.za/Pyromet/Laterite/Laterite.htm

9. C.P. Naude and M.D. Shapiro, "Implementation of the first commercial scale DC smelter for ferronickel production from low grade laterite ores—technology building blocks and lessons learned", *Journal of the Southern African Institute of Mining and Metallurgy*, vol.110, no.12, December 2010. pp.725–732.
http://www.saimm.co.za/Journal/v110n12p725.pdf

10. Mining Journal, "Xstrata's Koniambo produces first nickel", *Mining Journal*, 11 April 2013.
http://www.mining-journal.com/exploration--and--development/xstratas-koniambo-produces-first-nickel

11. I.J. Barker, "Some considerations on future developments in ferro-alloy furnaces", *Southern African Pyrometallurgy 2011*, Edited by R.T. Jones & P. den Hoed, SAIMM, Johannesburg, 6-9 March 2011, pp.1-13.
http://www.pyrometallurgy.co.za/Pyro2011/Papers/001-Barker.pdf

12. N.A. Barcza, I.J. Barker, M.S. Rennie, and P.J. Brereton-Stiles, "The application and scale-up of AC and DC smelting furnaces for ferro-alloys", *60th Electric Furnace Conference Proceedings*, Vol. 60, Warrendale, PA: Iron & Steel Society, 2002. pp. 425-437.
http://www.mintek.co.za/Pyromet/Files/2002Barcza.pdf

Celebrating the Megascale: Proceedings of the Extraction and Processing Division
Symposium on Pyrometallurgy in Honor of David G.C. Robertson
Edited by: Phillip J. Mackey, Eric J. Grimsey, Rodney T. Jones, and Geoffrey A. Brooks
TMS (The Minerals, Metals & Materials Society), 2014

RECENT DEVELOPMENTS IN FACTSAGE THERMOCHEMICAL SOFTWARE AND DATABASES

Christopher W. Bale[1], E. Bélisle[1], P. Chartrand[1], S.A. Decterov[1], G. Eriksson[2], A. Gheribi[1], K. Hack**, I.-H. Jung[1], J. Melançon[1], A.D. Pelton[1], S. Petersen[2] and C. Robelin[1]

[1]CRCT - Center for Research in Computational Thermochemistry
Dept. of Chemical Eng., École Polytechnique (Campus de l'Université de Montréal)
Box 6079, Station Downtown, Montréal, Québec, Canada, H3C 3A7
www.crct.polymtl.ca

[2]GTT-Technologies, Kaiserstrasse 100,
52134 Herzogenrath, GERMANY
www.gtt-technologies.de

Abstract

The **FactSage®** package consists of a series of information, database and calculation modules that enable one to access pure substances and solution databases and perform thermochemical equilibrium calculations. FactSage was originally founded by process pyrometallurgists and has since expanded its applications to include hydrometallurgy, electrometallurgy, corrosion, glass technology, combustion, ceramics, geology, environmental studies, etc. With the various modules one can perform a wide variety of thermochemical calculations and generate tables and graphs of complex chemical equilibria and phase diagrams for multicomponent systems.

This paper presents examples of recent and current developments in FactSage in areas of particular interest to Dr. David Robertson. These include both ferrous and non-ferrous processing, as well as solidification and refining simulation. The paper also includes reference to the Internet **Fact-Web** education package.

Introduction

FactSage® was introduced in 2001 as the fusion of two well-known software packages in the field of computational thermochemistry: **F*A*C*T / FACT-Win** and **ChemSage**. It is used worldwide at approximately 900 installations in universities, governmental and non-governmental research laboratories and industry. Articles on FactSage are published elsewhere [1, 2] and details of the software and databases are extensively covered at the website - **www.factsage.com**. For purposes of education and training, free but limited access via the Internet to some of the FactSage modules and FACT pure substances database is offered via **Fact-Web** [3].

While an understanding of chemical thermodynamics is useful in order to run the modules, it is not essential to be an expert in the field. With program usage and the assistance of extensive documentation, one can acquire a practical understanding of the principles of thermochemistry especially as these relate to complex phase equilibria. FactSage has become an invaluable educational tool for teaching and applying chemical thermodynamics. This paper outlines some of the recent developments in the databases and in the calculations that involve the modules Equilib, Fact-Optimal, Phase Diagrams and Fact-Web as well as viscosity calculations.

Equilib

The **Equilib** module calculates the conditions for multiphase, multicomponent equilibrium, with a wide variety of tabular and graphical output modes, under a large range of possible constraints through Gibbs energy minimization (Figure 1).

There are many different types of Equilib calculations, for example :

- **Equilibrium** using themochemical data from single or multiple databases
- **Scheil-Gulliver** and **Equilibrium Cooling**
- **Casting** algorithms
- **Open** calculations
- **Simulation** of processes – Macro Processing
- **Streams** and recycling
- **Fact-XML** customized output – tables, spreadsheets, graphs
- **Fact-Function-Builder** – customized functions
- **FactOptimal** – identify the optimal conditions for alloy and process design using thermodynamic and property databases

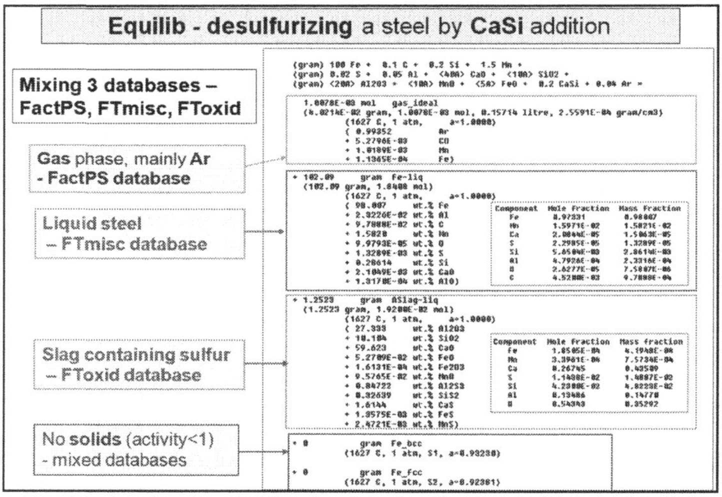

Figure 1. The results of an Equilib calculation in which steel is being desulfurized by CaSi additions where the data are taken from a combination of databases – FactPS, FTmisc and FToxid.

When more than one database is used in such an equilibrium calculation, there may be duplicate and possibly conflicting data for the same compound species. In the case of duplicate gas species this causes an error in the calculation. In FactSage an <u>exclusion strategy</u> is now automatically applied to compound species such that duplicate or conflicting species are <u>no longer selected</u>. The species exclusion priorities have been set by the database developers.

Process simulation

It is not possible to use FactSage to fully simulate the processes of an industrial flowsheet. Integrated simulations need to be carried out by software packages such as **METSIM** as taught by Dr. David Robertson [4]. Nevertheless, employing Equilib in the simulation of a unit process is possible by *Macro Processing* where

Equilib commands are stored in a file. The *Macro Processor* reads the file and executes the commands. In this manner one is able to dynamically define the input reactants (initial amounts, temperature, input streams), modify the conditions of the calculation (product T, P, species selection, heat loss) and then direct the results to a file (Excel, text). *Macro Processing* is particularly useful for large and repetitive Equilib calculations that may involve simulations which include kinetic models, input and recycled streams.

The combination of *Equilib Macro Processing*, Excel spreadsheets and a recently developed thermodynamic database for copper smelting and converting has been employed [5] to simulate the ISASMELT® copper smelting process (Figure 2). Because of good mixing in the furnace, it was possible to use a one-stage equilibrium calculation to describe the process. The following tables in Figure 2 show the mass and energy balance, as well as the distribution of elements between output streams (slag, gas and matte) and a comparison with the plant assay data [6]. Good agreement is obtained.

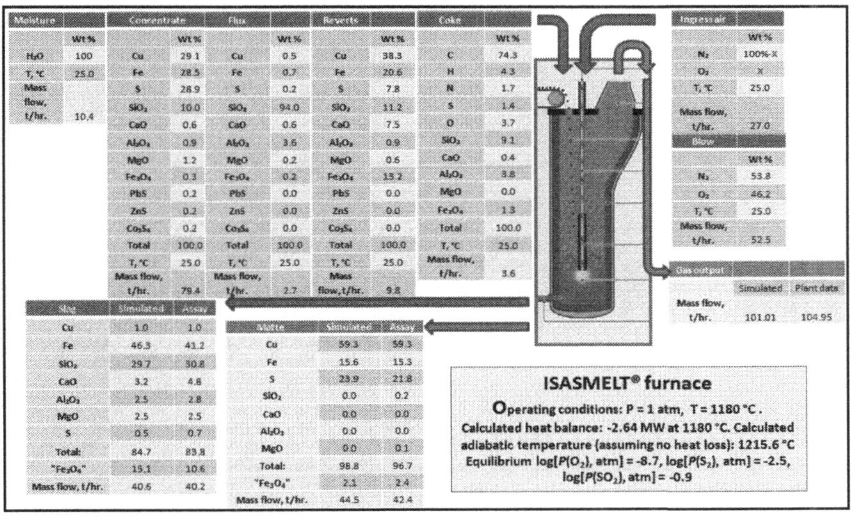

Figure 2. The results of a process simulation.

Viscosities

The viscosity of oxide melts can vary by orders of magnitude depending upon the composition and temperature and it is strongly related to the connectivity of the silica network. The connectivity of the network (which is related to the probability of Si-O-Si bridges) can be calculated from the model parameters of the FactSage thermodynamic database for molten oxides. This database has been developed over 20 years from critical evaluation and optimization of thermodynamic and phase equilibrium data.

The parameters of the viscosity model are obtained by critical evaluation and optimization of data for pure oxides and selected binary and ternary systems. The model can then predict viscosities in multicomponent systems [7, 8].

There are two viscosity databases:

- database for molten slags : Al_2O_3-B_2O_3-CaO-FeO-Fe_2O_3-K_2O-MgO-MnO-Na_2O-NiO-PbO-SiO_2-TiO_2-Ti_2O_3-ZnO-F
- database for glasses (super-cooled melts) : Al_2O_3-B_2O_3-CaO-K_2O-MgO-Na_2O-PbO-SiO_2

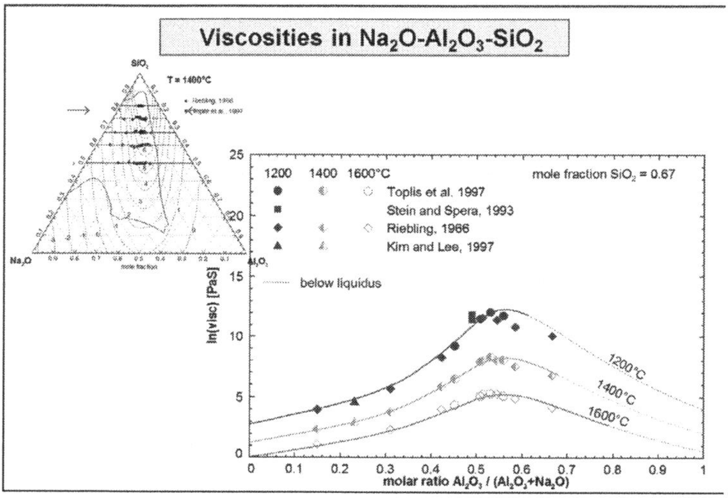

Figure 3. shows a comparison between calculated and experimental viscosities in the Na_2O-Al_2O_3-SiO_2 ternary system with the viscosities calculated using the molten slag database.

FactOptimal

Equilib can be used to screen potential systems, searching for compositions having a desired set of properties and phase constitution, under a given set of constraints. For example molten salts can be used to store solar energy. Equilib could search for a molten salt within a given composition range, with a liquidus temperature below x°C, with a desired freezing range. In the case of alloys one could also specify a maximum or minimum amount of precipitates after annealing. One could also search for optimal annealing or rolling temperatures. It is possible to perform such searches "by hand" for a multicomponent salt or alloy by carrying out thousands of calculations over a grid of compositions. But this approach is inefficient and extremely time-consuming.

FactOptimal minimizes and/or maximizes a set of functions: **{ f1(x1, x2... T, P); f2(x1, x2...T, P) }** where

- the functions f (e.g. T, P, V, heat, mass, density, conductivity, ...) are calculated by **Equilib**
- the functions may be non-smooth (e.g. liquidus)
- the estimation of derivatives may be problematic
- the evaluations of f may be time consuming
- the function calculation may fail unexpectedly at some points
- the constraints may be non-linear, non-smooth or Boolean

FactOptimal [9] identifies the optimal conditions for alloy and process design (Figure 4). It employs **Equilib** and the **Mesh Adaptive Direct Search** (MADS) algorithm [10].

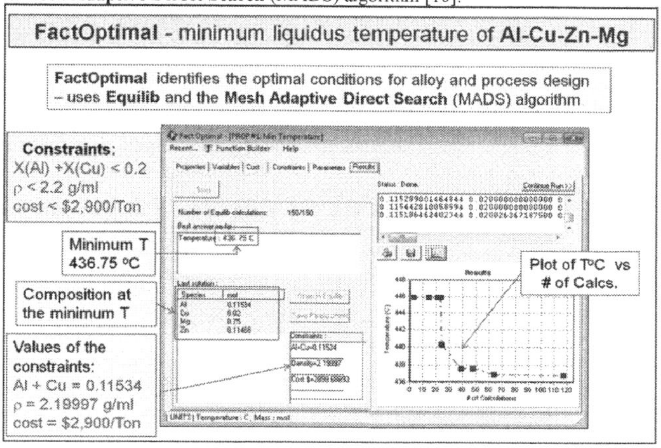

Figure 4. The results of over 100 Equilib calculations for the minimum liquidus temperature (436.75 °C) of **Al-Cu-Zn-Mg** alloys using data from the **FTlite** database. The following three constraints were applied :composition $X_{Al} + X_{Cu} < 0.2$, (2) density < 2.2 g/ml, (3) cost < $2,900/Ton.

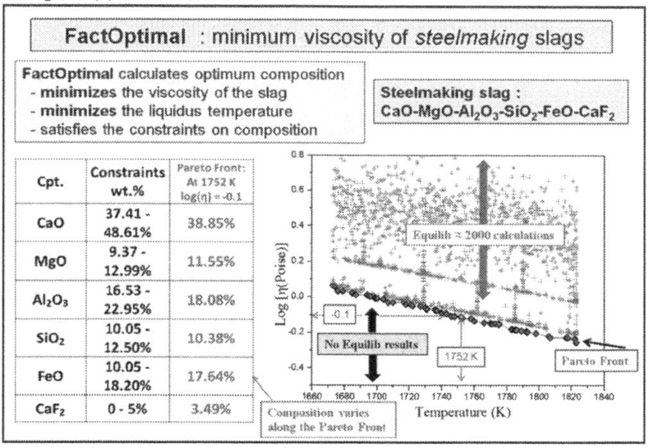

Figure 5. The calculated viscosity versus liquidus temperature points for **CaO-MgO-Al_2O_3-SiO_2-FeO-CaF_2** steelmaking slags.

Figure 5 shows calculated viscosity versus liquidus temperature points for **CaO-MgO-Al_2O_3-SiO_2-FeO-CaF_2** *steelmaking* slags where there are constraints on the range of slag compositions. The **Pareto Front** is a plot of *minimum slag viscosity* versus *minimum liquidus temperature*. For example, in Fig. 5 at the arbitrary liquidus temperature 1752 K the calculated minimum slag viscosity (poise) is $10^{-0.1}$ and the composition (wt.%) is 38.85% CaO, 11.55% MgO, 18.08% Al_2O_3, 10.38% SiO_2, 17.64% FeO, 3.49% CaF_2.

Phase Diagram

In the **Phase Diagram** module two thermodynamic properties are plotted on the X- and Y-axes, while the other properties are held constant. The properties that may be selected as axes or constants are:

> **T** (or **1/TK**) - temperature; **P** - total pressure; **V** - volume; **Comp** - Composition (mole fractions, molar ratios, weight fractions, weight ratios); **Potl** - Chemical potentials (RT ln(a_i), RT ln(P_i), log(a_i), log(P_i) where a_i and P_i = activity or partial pressure of component i); **Delta H** - Enthalpy relative to a standard state temperature; **E**, **pH** - Potential (volts) and pH axes of an aqueous Pourbaix-type diagram.

Not all combinations are permitted. For example, **T** (Y-axis) vs **Comp** (X-axis) is permitted, but not **Comp** (Y-axis) vs **T** (X-axis).

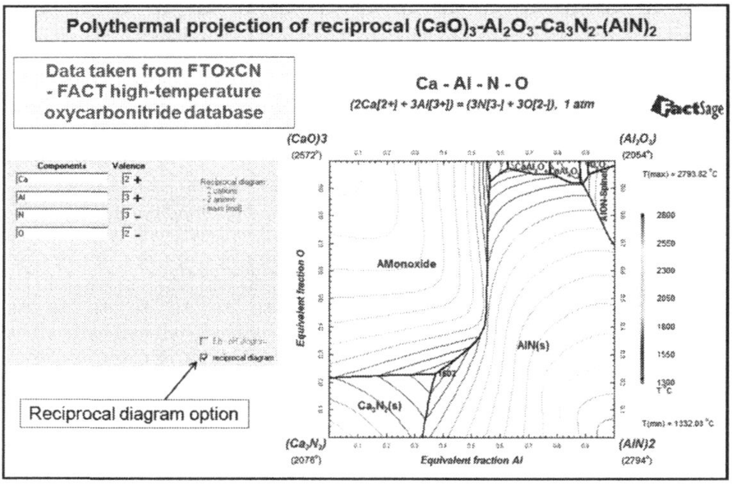

Figure 6. Results of phase diagram computations

The input and output for **reciprocal** phase diagrams has been vastly improved (Figure 6). New types of phase diagrams recently added to **Phase Diagram** are **first-melting** (Figure 7) and **paraequilibrium** phase diagrams (not shown). A detailed explanation of **first-melting** phase diagrams is given elsewhere [11]. There are hundreds of calculated phase diagrams stored in FactSage and most are posted at www.factsage.com which also includes a phase diagram search engine.

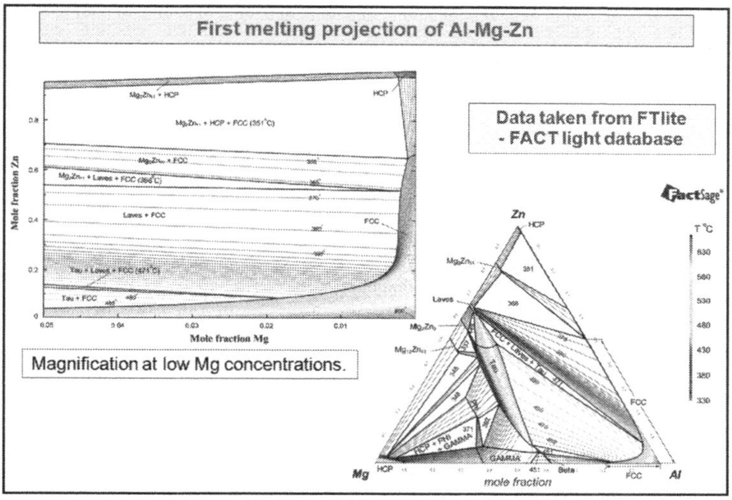

Figure 7. Example of first melting projection for Al-Mg-Zn.

Fact-Web

FactSage is a powerful tool in teaching chemical thermodynamics and applications particularly in the field of materials processing. For purposes of teaching and training **Fact-Web** [3] is offered on the Internet as a free educational package (Fig. 8) at www.factsage.com and www.crct.polymtl.ca/factweb.php. Fact-Web provides limited access to some of the FactSage modules (View Compound, Reaction, EpH, Predom, Equilib, Phase Diagram) and access to all the compounds in FactPS - the FACT pure substances database (Figure 8).

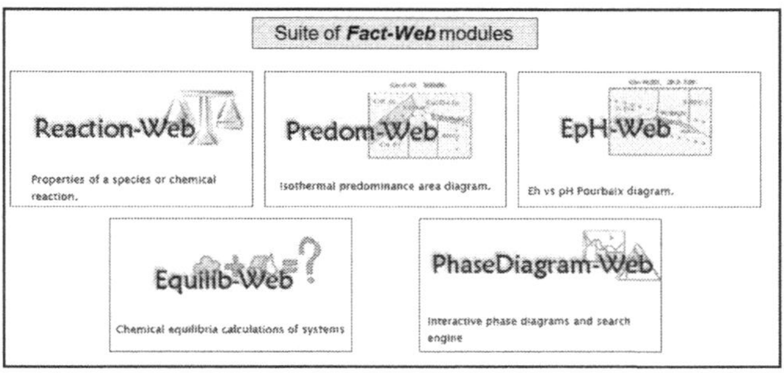

Figure 8. Suite of Fact-Web modules.

The **Reaction-Web Manual** (Figure 9) is very informative and compliments well the FactSage **Reaction Manual**.

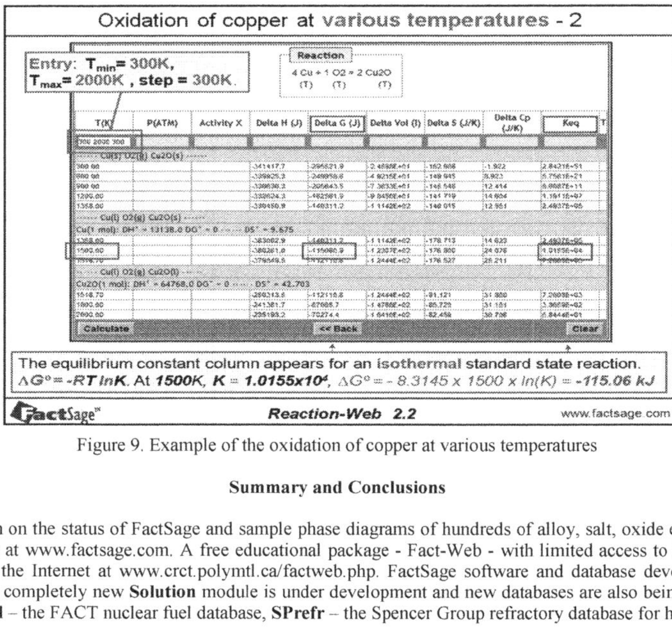

Figure 9. Example of the oxidation of copper at various temperatures

Summary and Conclusions

Information on the status of FactSage and sample phase diagrams of hundreds of alloy, salt, oxide etc. systems is available at www.factsage.com. A free educational package - Fact-Web - with limited access to FactSage is offered on the Internet at www.crct.polymtl.ca/factweb.php. FactSage software and database development is ongoing. A completely new **Solution** module is under development and new databases are also being prepared, e.g. **FTnucl** – the FACT nuclear fuel database, **SPrefr** – the Spencer Group refractory database for hard metals.

References

[1] C.W. Bale, P. Chartrand, S.A. Decterov, G. Eriksson, K. Hack, R. Ben Mahfoud, J. Melançon, A.D. Pelton and S. Petersen: *Calphad J.*, (2002), vol. 26, pp. 189-228.
[2] C.W. Bale, E. Bélisle, P. Chartrand, S. A. Decterov, G. Eriksson, K. Hack, I. H. Jung, Y. B. Kang, J. Melançon, A. D. Pelton, C. Robelin and S. Petersen, *Calphad J.*, (2009), vol. 33, pp. 295-311.
[3] C.W. Bale and E. Bélisle: *Fact-Web suite of interactive programs*, <www.crct.polymtl.ca/factweb.php>
[4] D.G.C. Robertson. *METSIM Short Course (2 days) at FactSage Workshop*, (Montreal 2008).
[5] D. Shishin, "*Development of a thermodynamic database for copper smelting and converting*" (Ph. D. thesis, Ecole Polytechnique, University of Montreal, 2013).
[6] R.L. Player, "Copper Isasmelt-process investigations, "*Howard Worner International Sympos. on Injection in Pyrometallurgy*, (Eds M. Nilmani and T.Lehner, 1996, The TMS, Warrendale, PA, USA), pp. 439-446.
[7] W.-Y. Kim, A.D. Pelton and S.A. Decterov, *Int. J. Mat. Res.*, (2012), vol. 103, pp. 313-328.
[8] E. Brosh, A.D. Pelton and S.A. Decterov, *Int. J. Mat. Res.*, (2012), vol. 103, pp. 494-501 and pp. 537-550.
[9] A.E. Gheribi, C. Audet, S. Le Digabel, E. Bélisle, C.W. Bale, A.D. Pelton, *Calphad J.*,(2012), vol 36, pp. 136-143.
[10] C. Audet and S. Le Digabel, *Pacific Journal of Optimization*, (2012), vol. 8(1), pp. 103-119.
[11] G. Eriksson, C.W. Bale and A.D. Pelton, *J. Chem. Therm.*, (2013), (in press) .

Celebrating the Megascale: Proceedings of the Extraction and Processing Division Symposium on Pyrometallurgy in Honor of David G.C. Robertson
Edited by: Phillip J. Mackey, Eric J. Grimsey, Rodney T. Jones, and Geoffrey A. Brooks
TMS (The Minerals, Metals & Materials Society), 2014

REDUCTION OF AGGLOMERATED MANGANESE ORES IN A 150 kW PILOT SCALE FURNACE

Merete Tangstad[1], Thomas Brynjulfsen[1], Eli Ringdalen[2]

[1]Norwegian University of Science and Technology; 7491Trondheim; Norway
[2]SINTEF, Trondheim, Norway

Keywords: manganese ferroalloys, manganese sinter, pilot scale experiments

Abstract

Manganese ore fines cannot be added to the submerged arc furnace directly as they will prevent even gas flow through the burden. Low gas permeability in the burden will lower the degree of pre-reduction of ore subsequently increase the carbon and energy consumption of the process. In order to utilize manganese ore fines in the furnace, they are typically agglomerated into sinter. In this work, the melting and reduction properties of Gabonese sinter and Gabonese ore have been investigated in pilot scale experiments. Two experiments were conducted, one with 100% sinter as manganese source and one with 50/50 sinter and ore. Lime was used to achieve a charge basicity of 0.72 in both experiments. The composition of slag samples from the excavation of the furnace was established using EPMA(WDS), and the composition of the tapped slag and metal was found using XRF. It was found that while the coke-bed size would determine the tapped slag composition, the melting behavior of the Mn source would determine the mixing with lime.

Introduction

In the FeMn furnace process, a carbon-saturated manganese-iron alloy with typically 76-78% Mn is produced. The furnace may be divided into two main zones. The low temperature region is typically called the pre-reduction zone, where the higher manganese oxides are reduced. As the temperature is increased, close to the coke bed, the ore will be reduced from a two phase mixture of solid MnO and a liquid slag[1-7], according to the following reaction: MnO(l)+C=Mn(l)+CO. The solid MnO phase will be dissolved as the reduction continues, and for many slag basicities the tapped slag is closer to the liquidus composition rather than the equilibrium composition[1].

Previously, it was believed that the reduction of MnO to Mn occurred as the slag trickled through the coke bed. However, pilot scale experiments and industrial excavations[8-11] in the previous year have shown that the major part of the reduction occurs on top of the coke bed. As the slag contains a solid MnO phase, the viscosity will be too high for it to enter the coke bed. When the slag reaches its liquidus composition, the slag viscosity is at its lowest, and the slag flows into the coke bed. Hence, a major part of the reduction occurs at the top of the coke bed. Two pilot scale experiments were performed to investigate the reduction behavior at the top of the coke bed and compare it to the tapped slag analyses. Another important goal with the experiments was to study if two materials with similar chemical composition, e.g. an ore and a sinter from the same ore, would affect the operation. Especially interesting is that the sinter is known to melt and be reduced at lower temperatures than the ore due to low melting silicate phases in the sinter[12]. It has been discussed in previous work how the melting behavior of the raw materials would affect the furnace operation[13-15], and these experiments will highlight this phenomenon.

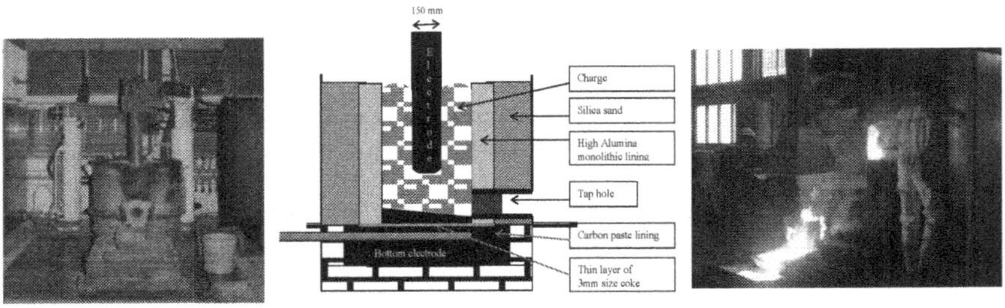

Figure 1. Pictures of pilot scale furnace: from left, cold furnace, sketch of furnace, furnace during tapping.

Equipment and procedures

The two HC FeMn experiments were performed in the 440 kVA pilot furnace at SINTEF/NTNU with the electrical characteristics shown in Table 1. Pictures of the furnace are shown in Figure 1. The experiments were run in AC with a top and bottom electrode. The lining is designed to account for an excavation technique whereby it is possible to make casted cross sections of the furnace. The inner lining of this furnace consists of high-alumina monolithic lining.

Table 1 Electrical characteristics of pilot furnace

	Voltage, U	Current, I
AC	10 - 215	50 -5700
DC	10 – 300	100 -8000

Table 2. Chemical analyses of sinter, ore, lime and coke (wt %).

	Gabonese sinter	Gabonese ore	Limestone	Coke
MnO_2	19.9	72.3		
MnO	55.5	3.5		
Fe_2O_3		4.69	0.16	0.91
FeO	5.0			
SiO_2	9.2	6.1	1.0	3.7
Al_2O_3	6.1	6.72	0.3	2.5
CaO	0.18	0.01	54.0	0.4
MgO	0.14	0.01	1.0	0.20
CO_2			47.5	
Fix C				90.1

Table 3. Charge composition of the 2 experiments (kg).

	Sinter experiment	Sinter/ore experiment
Gabonese ore	-	10.7
Gabonese sinter	21.6	10.7
Limestone	4.4	4.1
Coke	4.0	4.4

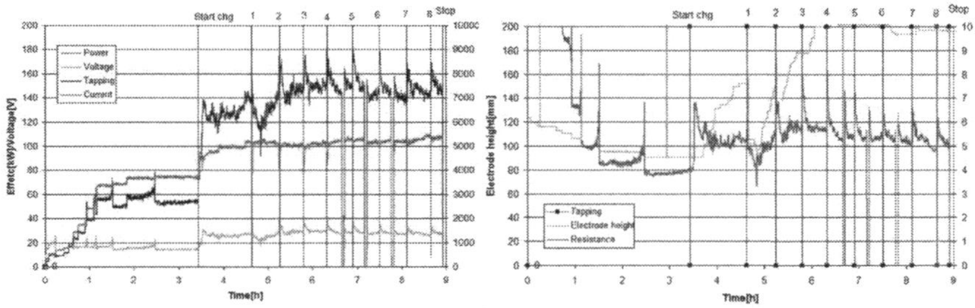

Figure 2. Measured electrical parameters of load, voltage, current, resistance and electrode height and tapping for the sinter experiment.

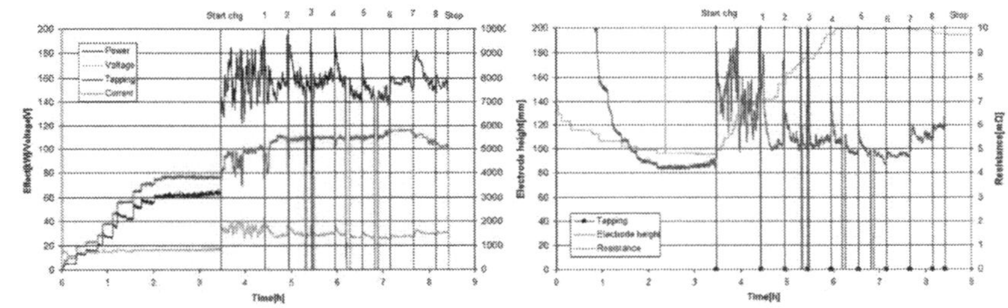

Figure 3. Measured electrical parameters of load, voltage, current, resistance and electrode height and tapping for the ore/sinter experiment.

A starting coke bed of 10 kg was added into the furnace initially. This coke bed was preheated using 150 kWh of energy. The raw materials were then charged into the furnace. The analyses of the raw materials and the charge composition are shown in Table 2 and Table 3. Based on previous experiments it was assumed that the off-gas was mainly CO gas. The aim for the operation was to fix the electrode tip position 20 cm above the furnace bottom and to control the voltage to achieve a stable load of 150 kW. The electrode tip was positioned at the top of the coke bed at the start of the experiments, and gradually lifted up to 20 cm above the furnace bottom at a rate that made it possible to maintain the furnace load at 150 kW. The first tapping was done after an additional 150 kWh of energy input, and the following tappings each after an additional 80 kWh of energy input. During the experiment the electrical parameters, current and voltage were logged automatically as a part of the furnace control system. Time for charging and tapping were registered manually. These numbers are shown in Figure 2 and Figure 3. It can be seen that the furnace resistances for the two experiments are quite similar even though the coke bed sizes are quite different, as will be discussed later in the paper.

The furnace was excavated using a technique developed by B. Heiland at NTNU (now Glencore). The furnace is shut off 40 kWh after the last tapping. It is then cooled for approximately 72 hours before epoxy is poured into the furnace. After another 48 hours the epoxy has hardened and the electrode is cut. The furnace is taken away in one piece and a cross sectional slab through the center of the furnace is sawed out for further investigations. The distribution of the different zones in the furnace can then be studied.

Tapped slag and alloy from the experiment were separated and weighed. Representative samples of slag and alloy from each tapping were taken out and analysed by Molab. In some of the tappings slag and alloy was intermixed, and the exact amounts were difficult to determine.

Results and discussion

The cross-sections from the furnace are shown in Figure 4. The interior of the furnace is divided into several zones. In the charge zone, the raw materials are still solid. As the raw materials descend in the furnace, they are heated and the ores started to melt into a solid MnO phase and a liquid slag. This zone is called the partly molten charge. At even higher temperature the coke bed will be present. A cavity below the electrode was also found, and is believed to be due to low horizontal mass flow. In the lowest part of the coke bed, the reduced slag fills up the void between the coke particles. Finally, the metal will be collected at the bottom.

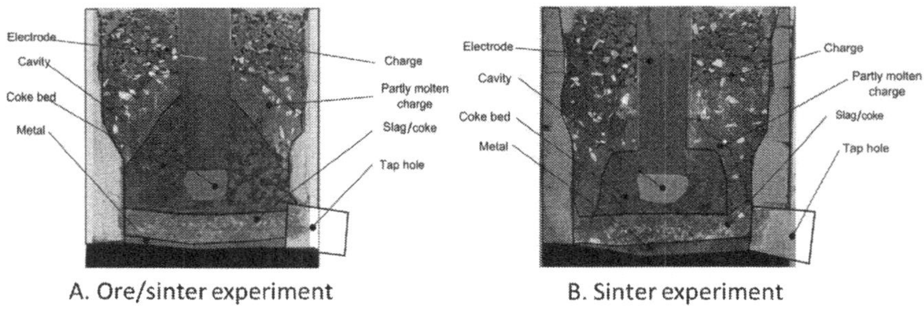

Figure 4. Furnace cross-sections showing the interior zones of the furnace.

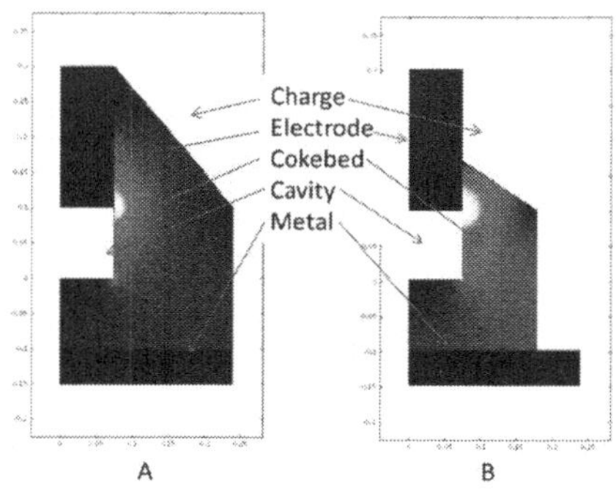

Figure 5. Calculated specific effect (kW/m^3) based on the geometry of the coke beds and a total effect of 150 kW in A) the ore/sinter experiment and B) the sinter experiment.

The first visual impression from the cross-sections is that the quantity of coke is much larger in the ore/sinter experiments. This experiment was over-coked, indicated by the fact that the off-gas contained more CO_2 than expected. This can be explained by the easily-reduced MnO_2 content of the Gabonese ore. The consequence of this over-coking is that the coke bed will be larger and growing. It is understood that a larger coke bed will give a lower temperature at a fixed load. This can be explained by a lower kW developed per unit volume, as shown in Figure 5.

The tapped slag analyses are shown in Figure 6. The Mn wt % values in the metal for the last 5 taps are all about 80%. The basicity is quite similar in the two experiments, except for the higher alumina content in the ore/sinter experiments. The higher alumina content is due to dissolution of the lining. The dissolution of the lining is more extensive in the ore/sinter experiments as the coke bed is wider and is in direct contact with the lining. In the sinter experiment, the coke bed is narrower and does not come into contact with the lining. In addition to the difference in alumina content, the most important result is the lower MnO content in the slag in the ore/sinter experiment. This is again explained by the difference in temperature. Since the temperature is lower in the ore/sinter experiment, the liquidus MnO content will also be lower, as the MnO content is believed to be determined by the liquidus composition. The temperature in the high-temperature zone was measured during the operation, and though somewhat uncertain, the temperatures measured in the sinter experiments were higher at 5 of 6 positions[12] (the remaining position was above the coke bed).

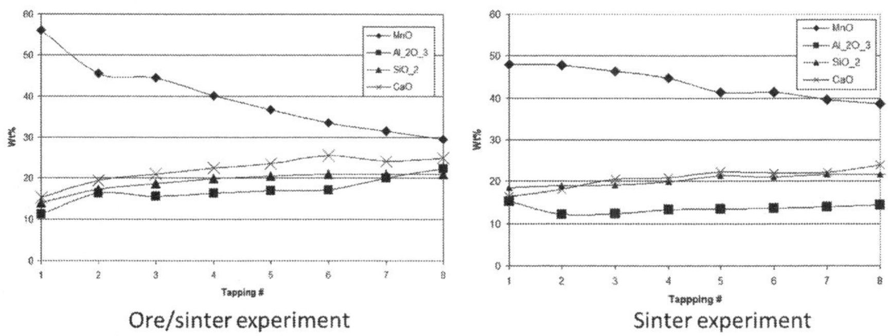

Figure 6. Chemical analyses of tapped slag during the two experiments.

Samples were taken from the top of the coke bed and studied at the EPMA. The sample positions for the two experiments are shown in Figure 7. All the samples contained a structure that originated from a solid MnO phase and a liquid slag, as shown in the right-hand image in Figure 7. The light grey particles are MnO particles, occasionally with some MgO. The white dots are metallic phases, and the mixture of darker grey phases is believed to be the liquid slag at high temperatures. Based on the analyses of the individual phases and the areas of the individual phases, the composition of the liquid slag can be calculated. The MnO content in the liquid phase is shown as a function of both $CaO+K_2O$ content and basicity, defined as $B=(CaO+MgO+K_2O)/(SiO_2+Al_2O_3)$. As the MgO content in these raw materials is quite low, they would be expected to give the same trend. The tapping analyses are also shown – they show the same basicity but have somewhat lower MnO content in the case of the ore/sinter experiment due to the lower temperature, as was discussed previously. Another interesting point is that the samples on top of the coke bed dissolved the lime in the ore/sinter experiment. The MnO content in the liquid phase is therefore very similar to the tapped composition. In the sinter experiment, hardly any lime was dissolved into the liquid phase, as the basicity was generally below 0.2. This means that the reduction occured before the lime was dissolved in the sinter experiment. This is believed to be due to the fact that the sinter starts to melt (and is reduced) at about 100 degrees lower temperature than the ore[12].

Figure 7. Position of samples for EPMA analyses shown in left and middle figure. At right, the typical structure of the partly molten ore found in all samples. (ore/sinter sample, 2400x magnification).

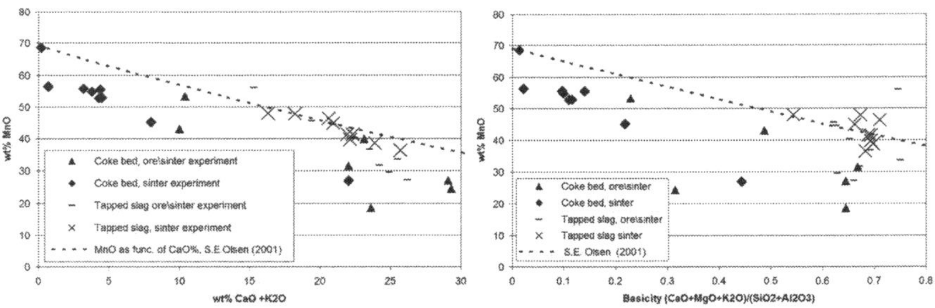

Figure 8. MnO content in tapped slag and liquid phase in the partly molten slag on top of the coke bed.

Table 4. Comparison of liquid phase on top of the coke bed and tapped slag (MS=ore/sinter, S=sinter).

Sample	MnO	SiO_2	Al_2O_3	CaO
MS2-MS4	32.8	26.8	15.8	24.6
Tapped slag	32.8	20.8	19.1	24.5
S2-S6 + Limestone + reduction	40.0	20.8	15.9	20.4
Tapped slag	40.3	21.4	13.8	22.6

Table 4 and Figure 9 summarize the reduction mechanism in the case of a low-melting Mn sinter and a high-melting Mn ore. As the ore melts into a liquid phase and a solid MnO phase at high temperature, it dissolves the lime. The reduction will therefore proceed until it reaches its liquidus composition on top of the coke bed. If a low-melting sinter is used, the reduction will proceed at a lower temperature, and the lime will not be dissolved until after the reduction step has occurred, at which point the potential for dissolution is much higher. As the slag dissolves the lime, a new liquidus will be obtained and additional reduction will occur. However, based on the liquidus composition, about 85% of the reduction occurs before the lime is dissolved.

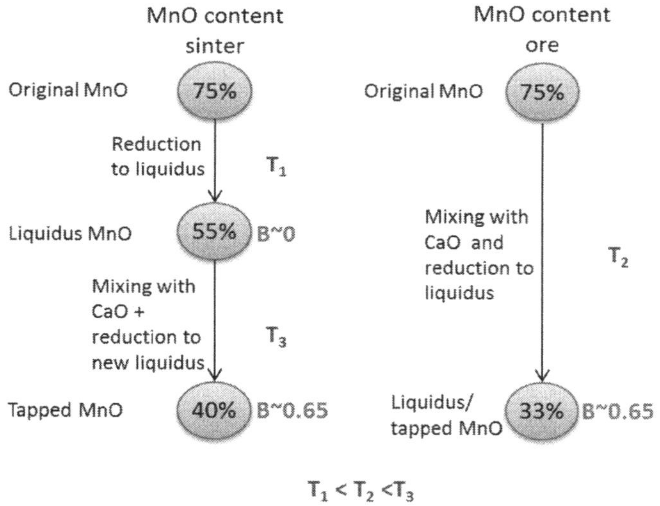

Figure 9. Summary of reduction mechanism with a low-melting Mn source (sinter) versus a high-melting Mn source (ore).

Conclusion

Two high-carbon ferromanganese experiments were performed, and HC FeMn with about 80% Mn was produced. The experiments were performed at 150 kW, and in the final part of the experiment, the height between the electrode tip position and bottom lining was 20 cm. The same basicity was used in both experiments and was controlled by the addition of lime. The Mn source was in one case sinter, which was found to melt and reduce at lower temperature. In the second case a mixture of sinter and ore was used, which melted and reduced at higher temperatures. Due to over-coking in the ore experiment, the coke bed was larger and cooler than in the ore/sinter experiment.

The tapping composition (that is, the MnO content in the slag) was mostly determined by the temperature in the coke bed. The larger coke bed in the ore experiment gave a lower temperature and hence a lower MnO fraction in the slag, as the composition is believed to be mainly determined by the liquidus composition.

The mixing mechanism of the flux was determined by the melting temperature of the ore. The sinter, which melted at lower temperatures, did not dissolve the lime until it was reduced to its liquidus. Since it melts at higher temperatures, reduction of the ore and dissolution of the lime occurred simultaneously.

References

(1) S.E. Olsen, M. Tangstad, and T. Lindstad. *Production of manganese ferroalloys.*Tapir academic press, Trondheim, 2007

(2) V. Olsø, M. Tangstad, and S. Olsen. *Reduction kinetics of MnO saturated slags.* Infacon VIII, 1998.

(3) O. Ostrovski, S.E. Olsen, M. Tangstad, and M. Yastrebo_. Kinetic modeling of MnO reduction from manganese ore. Canadian Metallurgical Quarterly, Vol 41, 2002.

(4) T.A. Skjervheim and S.E. Olsen. The rate and mechanism for reduction of manganese oxide from silicate slags. Conference proceedings Infacon 7, 1995.

(5) T. Shimoo, S. Ando, and H. Kimura. Rates of MnO reduction from silicate slags with solid carbon. J.Japan Inst. metals Vol. 48 No.9, 1984.

(6) G. Roghani, E. Jak, and P. Hayes. *Phase equilibrium studies in the MnO-Al2O3-SiO2 system.* Metallurgical and materials transactions B, 2002.

(7) SlagAtlas. Slag atlas 2nd edition, page 116. Verlag Stahleisen GmbH, 1995.

(8) B. Monsen, M. Tangstad, I. Solheim, M. Syvertsen, R. Ishak, and H. Midtgaard: *Charcoal for manganese alloy production.* Infacon XI conference proceeding, 2007.

(9) P.A. Eidem. *Electrical resistivity of coke bed.* PhD-Thesis , Norwegian university of science and technology, 2008.

(10) E. Ringdalen and M. Tangstad. *Study of SiMn production in pilot scale experiments-comparing Carajas sinter and Assmang ore.* Infacon XIII conference proceedings, volume 1, pp 195-206, 2013.

(11) Olsen, S.E, Tangstad, M, *Silicomanganese production – Process Understanding,* Infacon X, 2004, Cape Town, South Africa

(12) T.Brynjulfsen; *Reduction of manganese ore agglomerates,* PhD thesis, NTNU, Trondheim, Norway, to be published fall 2013.

(13) D. Slizovskiy, M. Tangstad, and S. Wasbo. *Melting temperatures of ore forferromanganese production.* To be published 2013.

(14) S. Gaal, D. Lou, S. Wasbo, B. Ravary, and M. Tangstad. *Melting phenomena in ferromanganese production.* Infacon XI conference proceedings, 2007.

(15) E. Ringdalen, S. Gaal, M. Tangstad, and O. Ostrovski. *Ore melting and reduction in silicomanganese production.* Metallurgical and materials transactions 41B, 2010.

Celebrating the Megascale: Proceedings of the Extraction and Processing Division
Symposium on Pyrometallurgy in Honor of David G.C. Robertson
Edited by: Phillip J. Mackey, Eric J. Grimsey, Rodney T. Jones, and Geoffrey A. Brooks
TMS (The Minerals, Metals & Materials Society), 2014

ARC DETECTION IN DC ARC FURNACES

Quinn G Reynolds[1], Christopher J Hockaday[1], Dominic T Jordan[1], Ian J Barker[2]

[1]Mintek; Private Bag X3015; Randburg 2125; South Africa
[2]Private Consultant; c/o Mintek; Private Bag X3015; Randburg 2125; South Africa

Keywords: Furnace, Plasma Arc, Control, Measurement, Modelling

Abstract

A direct-current (DC) plasma arc furnace is a type of electric furnace used for metallurgical operations. The successful operation of DC furnaces depends to a large extent on gaining a fundamental understanding of the arc phenomenon itself, and ensuring its presence in the furnace at all times. A method for detection of the presence of the arc in a DC circuit is presented, along with discussion of why this may be of value for certain modes of furnace operation such as brush-arc. The theoretical development of the method is presented along with supporting experimental work conducted on large-scale pilot-plant facilities.

Introduction

Direct-current (DC) plasma arc furnaces have seen widespread usage in the steel industry as a scrap remelting unit. More recently, process applications for DC furnaces in the smelting of primary ores have been gaining ground in industry. These applications include the smelting of chromite to produce ferrochromium, the smelting of ilmenite to produce titania slag and pig iron, and the smelting of nickel laterite to produce ferronickel [1].

A typical DC furnace consists of a cylindrical shell topped with a conical roof, through which the ore and chemical reagents needed for the process are fed. This vessel is lined with refractory material and may also include additional cooling systems such as forced-convection copper cooling elements. The vessel typically contains an open molten bath consisting of several different phases, frequently a light slag phase above a denser metal phase. One or more electrodes made of pre-baked graphite serve as cathodes, and enter through the center of the roof. The base of the furnace is designed with an electrically conductive anode in direct contact with the molten bath. The plasma arc exists between the tip of the electrode and the surface of the bath, and completes the circuit between the graphite electrode and the anode connection. The electrode and anode are connected to a DC power supply, which generally consists of a transformer unit, a rectifier unit, and one or more chokes. The transformer adjusts alternating-current power from the grid to the appropriate voltages and currents, and the rectifier then converts the alternating current into DC. The chokes provide inductive capacity in order to damp any large spikes in voltage resulting from fluctuations in the furnace operation.

The arc is the primary heating and stirring element in the DC furnace, converting electrical energy into thermal energy and transferring it into the molten bath. It is a high-temperature, high-velocity jet consisting of plasma – material in the fourth state of matter. The thermal plasma in the arc is formed by the dissociation of the gases in the furnace freeboard region when they are heated to very high temperatures (above 5 000 K) and break down into a mixture of positive ions and negative electrons. Due to the presence of free electrons, the arc plasma conducts electricity effectively.

Since the arc and slag bath are connected in series in the DC electric circuit through the furnace, part of the total operating voltage (and hence power dissipation) is split across each. The proportion in which the voltage is divided between the two is determined by the physical properties of the arc plasma and slag, as well as the geometry of the furnace [2]. In particular, for processes involving slags of high electrical resistivity (for example, in ferronickel production where slags have a moderate to high SiO_2 content), the voltage drop across the slag layer is likely to be significant. This can result in either a high total voltage requirement for the furnace, or a limited voltage available for the arc with the majority of the voltage drop occurring across the slag. Very high voltage specifications can add considerable expense to the design of the DC power supply. The preference may therefore be to operate such a furnace with a reduced arc length, thereby limiting the total voltage while retaining the process advantages [1] of operating with an arc present. This brush-arc mode of operation has been applied in AC submerged arc furnace operations [3], and is likely to see application on DC plants in the future.

In addition to the expansion of DC applications into the smelting of high-resistivity materials, there is growing interest in the viability of retrofitting three-electrode circular AC furnaces to operate using DC power. Due to the cost and complexity involved in redesigning the hearth to include a bottom anode, a potentially attractive means of accomplishing the switch to DC is to use a dual-electrode design [4], with two separate graphite electrodes serving as anode and cathode respectively. In such a design, there are two arcs present inside the furnace vessel, and current passes through them in series via the slag layer. This can result in a significantly higher voltage requirement (approximately double) at a given current than a traditional single-electrode DC furnace, and again the preference may be to operate such furnaces with short arc lengths, in brush-arc mode [4].

Since many of the advantages of DC furnace operation are related to the presence of an open arc in the furnace, a means of determining when this is the case would be of some value. This is especially true for processes requiring brush-arc operation, in which it may be very difficult for operators to discern between open-arc (with the electrode tip a small distance above the slag) and resistance-heating (with the electrode tip in direct contact with the slag) conditions without visual inspection of the interior of the furnace during operation. In addition to being a safety hazard, this may be difficult or impossible depending on the design of the vessel.

Theory of arc behaviour

The DC plasma arc is a complex coupled system resulting from strong interactions between fluid flow, thermal, and electromagnetic fields inside the furnace vessel [5]. Due to the large velocities and driving forces involved, the plasma arc exhibits dynamic behaviour and changes shape rapidly on very short time scales, typically of the order of milliseconds or less [5]. Signals of interest to arc measurement studies are therefore expected to occur in the kHz and higher frequency range.

The arc as a magneto-hydrodynamic (MHD) system

Arcs may be described using three key physical aspects: a momentum transfer model, a heat transfer model, and an electromagnetic field model. Together these form an MHD system. A number of approximations are applied to the governing equations in order to simplify the mathematical description in the case of arcs consisting of thermal plasma. The MHD equations commonly used to model plasma arcs are presented below – more detail on the derivation of these equations and the approximations involved can be found in other publications [5].

The Navier-Stokes (1) and continuity (2) equations govern fluid flow for Newtonian fluids.

$$\rho \frac{\partial \mathbf{v}}{\partial t} + (\rho \mathbf{v} \cdot \nabla)\mathbf{v} + \nabla P = \nabla \cdot (\mu \nabla \mathbf{v}) + \mathbf{j} \times \mathbf{B} \tag{1}$$

$$\frac{\partial \rho}{\partial t} + \nabla \cdot (\rho \mathbf{v}) = 0 \tag{2}$$

Here, $\mathbf{v}$ is the velocity vector field, ρ is the plasma fluid density, P is the pressure, and μ is the dynamic viscosity. $\mathbf{B}$ and $\mathbf{j}$ are the magnetic field intensity and the current density vector fields, which together produce the Lorentz force source term in (1) – this acts on the plasma to compress and accelerate it, resulting in the formation of an arc jet.

The temperature field is described by the energy conservation equation (3).

$$\rho C_P \frac{\partial T}{\partial t} + (\rho C_P \mathbf{v} \cdot \nabla)T = \nabla \cdot (\kappa \nabla T) + \frac{\mathbf{j} \cdot \mathbf{j}}{\sigma} - Q_R \tag{3}$$

Here, T is the plasma temperature, C_P is the heat capacity, and κ is the thermal conductivity. σ is the electrical conductivity, and Q_R is the volumetric rate of energy loss from the plasma by radiation. Q_R and σ are both strong functions of temperature. The last two terms on the right-hand side of (3) constitute the energy source term for the plasma – energy is generated by the passage of electric current (Ohmic heating), and lost by thermal radiation.

The electromagnetic fields are given by the electrostatic and magnetostatic approximations to Maxwell's equations (4)-(6).

$$\mathbf{j} = -\sigma \nabla \phi \tag{4}$$

$$\nabla \cdot \mathbf{j} = 0 \tag{5}$$

$$\nabla \times \mathbf{B} = \mu_0 \mathbf{j} \tag{6}$$

Here, ϕ is the electric potential (voltage) field, and μ_0 is the permeability of free space, a physical constant.

This coupled set of equations governs $\mathbf{v}$, T, $\mathbf{j}$, and $\mathbf{B}$ as they evolve in time to produce a self-sustained plasma arc.

<u>The arc as an inertial system</u>

It is reasonable to assume that the spatial shape of the arc jet, as determined by its temperature distribution, at any given time is approximately proportional to its electrical resistance (and therefore voltage at a fixed current). This is due to the fact that the temperature distribution strongly influences the electric potential distribution, via the temperature dependence of σ in (4). Also, since the velocities in the arc are generally very high, convective energy transport is dominant in (3) and regions of high velocity tend to correlate with regions of high temperature as energy from the hot arc root near the electrode tip is carried down the jet into the body of the arc.

In order to gain an understanding of the dynamic behaviour of arc voltage, it is therefore of interest to examine how the momentum transport equation, (1), affects the arc jet's shape. Much of the high-velocity flow and rapid dynamic motion of arcs originates in regions where the $\mathbf{j} \times \mathbf{B}$ Lorentz force term in (1) is dominant. (1) may therefore be roughly approximated as:

$$\rho \frac{\partial \mathbf{v}}{\partial t} \approx \mathbf{j} \times \mathbf{B} \equiv \mathbf{L_F} \tag{7}$$

Defining **v** as the first derivative in time of a displacement vector **r**, and simplifying and rearranging (7), gives:

$$\frac{\partial^2 \mathbf{r}}{\partial t^2} \approx \frac{\mathbf{L_F}}{\rho} \tag{8}$$

Here, $\mathbf{L_F}$ is the Lorentz force term, and **r** indicates how the shape of the arc changes in space with velocity field **v**. **r** therefore directly affects the voltage and related external measurable parameters of the arc, and may be taken as a proxy for them in the analysis that follows.

(8) is seen to describe a simple inertial response system, in which a measured dependent variable (**r**) is affected by an independent control variable ($\mathbf{L_F}$) via the second derivative in time. It is interesting to consider the frequency response of such systems – if the Fourier transform of (8) is taken, we obtain:

$$-(2\pi\omega)^2\mathcal{F}(\mathbf{r}) = \frac{\mathcal{F}(\mathbf{L_F})}{\rho} \tag{9}$$

Considering the magnitude of the terms on either side of the equation, rearranging, and taking the logarithm then gives:

$$\log_{10}|\mathcal{F}(\mathbf{r})| = \log_{10}\left[\frac{|\mathcal{F}(\mathbf{L_F})|}{4\pi^2\rho}\right] - 2\log_{10}\omega \tag{10}$$

If a white noise perturbation signal is assumed for $\mathbf{L_F}$, consistent with random noise on the current input to the arc (which determines both **j** and **B**), the first term on the right-hand side of (10) becomes constant. Plotting the magnitude of the Fourier transform of **r** against frequency ω on a log-log graph would then give a straight line with a gradient of -2, or a fall-off of two decades per decade, as illustrated in Figure 1.

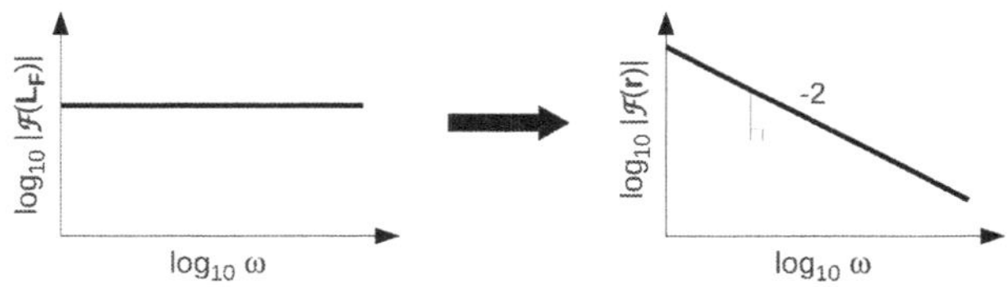

Figure 1: Signal analysis of a simplified inertial response system.

As **r** directly affects voltage and other measurable parameters of the arc in the present model, it is reasonable to propose that arcs may be identified from voltage signals measured on DC furnaces by performing a Fourier transform operation on the data and seeking parts of the spectrum that exhibit a two decades per decade falloff.

Computational modelling of the arc

A high resolution, three-dimensional computational model of a DC plasma arc in air was developed using numerical solutions of (1)-(6) [5]. The model calculates both the spatial and temporal development of the arc from specified initial and boundary conditions. The plasma gas used in the model was air, which determines the physical properties and temperature dependence of electrical conductivity and radiation energy loss. Other parameters used in the model are shown in Table I.

The model uses a box-shaped domain, with the upper surface adjacent to the tip of the electrode (which forms the cathode), and the bottom surface adjacent to the molten bath (which forms the anode). Constant temperatures and zero velocities are assumed at the electrode, bath and other boundary surfaces. An arc attachment zone is defined at the centre of the electrode tip, on which the current density is specified as $j_k = 3.5$ kA/cm^2 [6]. This zone forms the root of the arc, which is established in the space between the electrode and bath.

Table I: Simulation parameters used in computational model of the arc.

Parameter	*Value*	*Parameter*	*Value*
Region dimensions	0.1 x 0.1 x 0.05 m	Arc current	500 A
Electrode diameter	0.05 m	Simulation time	10 ms
Arc root diameter	0.00426 m	Mesh resolution	384 x 384 x 192
FFT sample size	25 000	FFT time step size	0.2 μs

Operating under perfect DC conditions, the plasma arc model exhibits a large degree of high-frequency chaotic behaviour resulting from various fluid-dynamic and electromagnetic instabilities characteristic of plasma systems. This can be seen in plots of the temperature field at various times shown in Figure 2.

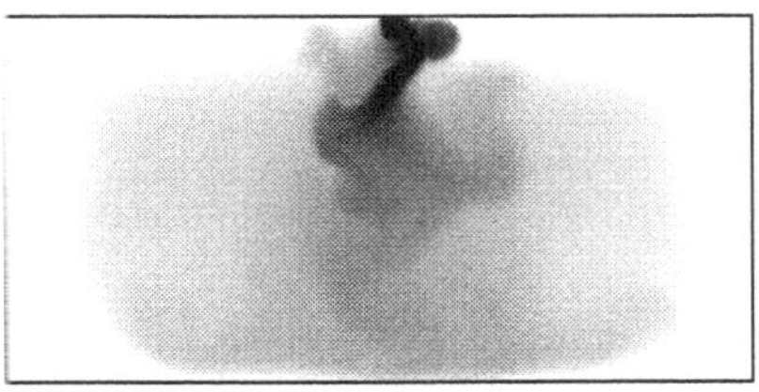

(a) 2.962 ms

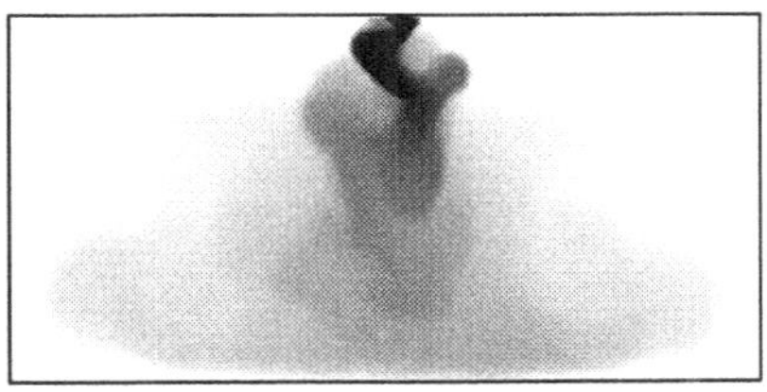

(b) 4.871 ms

Figure 2: X-Z projections of temperature field from plasma arc model at various times, scale 5 000 K (white) to 12 000 K (black).

The variability of the arc shape manifests in the calculated voltage (the peak value of the electric potential field at the arc root surface on the electrode) as shown in Figure 3(a). Taking a discrete Fourier transform of the voltage data for the last 5 ms using the fast Fourier transform (FFT) algorithm [7] gives the result shown in Figure 3(b) – the gradient of the falloff in the frequency range between 5 kHz and 100 kHz is seen to be in excess of 2 decades per decade, indicating the strong presence of high-frequency inertial components in the model arc behaviour.

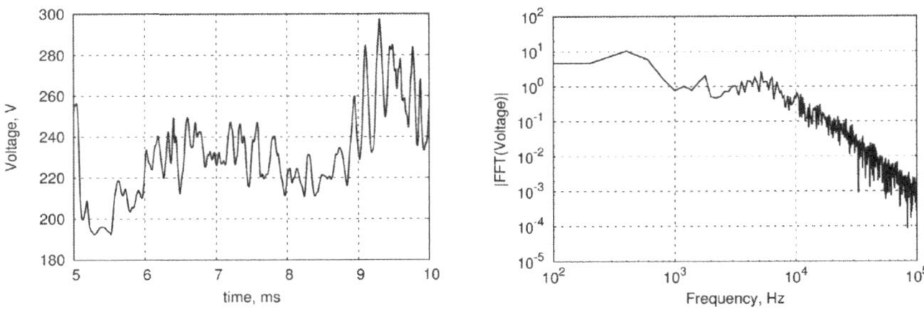

(a) (b)

Figure 3: Voltage data generated from computational arc model,
(a) time series, (b) Fourier transform magnitude spectrum.

Experimental equipment

A prototype DC arc detection instrument was constructed and used for measurement of the electrical variables on various pilot plant furnaces and test facilities at Mintek. High sampling rates (up to 30 kHz) were used in order to accurately measure fluctuations induced by the arc in the electrical circuit. Some photographs of the prototype unit are shown in Figure 4.

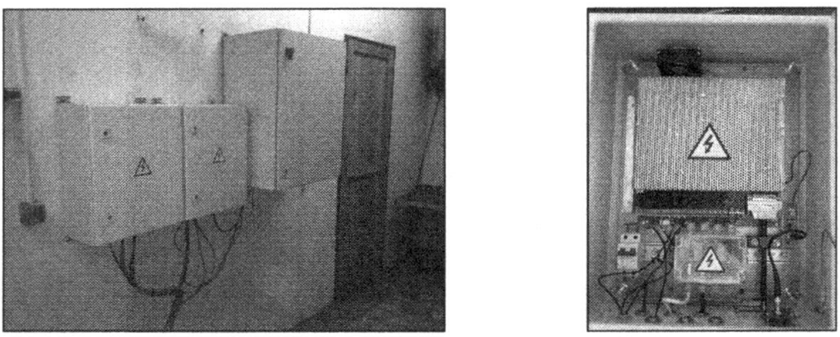

(a) (b)

Figure 4: Photographs of arc detection instrument prototype.

In Figure 4(a), the complete installation is shown. The two boxes at left contain signal conditioning hardware and connect directly to the busbars exiting the DC furnace power supply units located in the room below. The larger box at right contains a desktop computer with a Data Translation data acquisition card capable of logging measurements at the required rate of 30 kHz. The computer also includes customised software for the purpose of computing and analysing the frequency spectrum of the signal using fast Fourier transform (FFT) methods.

Figure 4(b) shows detail of the voltage signal conditioning hardware. At top is an Arcmon [8,9] signal conditioning and isolating board, which prepares the raw voltage signal for input to the data acquisition system. The smaller box at bottom contains surge arrestor circuitry. This is designed to limit the peak voltage that is passed through to the sensitive Arcmon circuit board from the voltage measurement input. The open-circuit voltage for rectifiers with large inductances in series can rise significantly higher than the normal operating voltages for short periods of time; the surge arrestor hardware safely steps down the voltage before it enters the Arcmon board, and redirects any extreme spikes to ground.

Testing of the arc detection prototype was accomplished using two pilot-plant furnace facilities at Mintek. For the first series of tests, a 3.2 MVA power supply using an IGBT (Insulated-Gate Bipolar Transistor) high-frequency switching DC rectifier was used to generate arcs in the air gap between a 200 mm pre-baked graphite electrode and a flat graphite block connected to the anode. No furnace shell or roof was present for these tests, enabling the arc to be observed directly. An Olympus iSpeed 3 high-speed video camera with a Tokina 80-200 mm f/4 lens was used to photograph the arc during these tests. The Olympus camera and the arc detection prototype were triggered simultaneously using a signal splitter, enabling the electrical and visual measurements to be synchronized and compared.

For the second series of tests, a 5.6 MVA power supply using a 12-pulse thyristor DC rectifier was used to power a test furnace 4.25 m in diameter smelting an iron and nickel-bearing raw material. The position of the electrode relative to the slag surface was determined by visual inspection, and tests using the arc detection prototype were conducted with the electrode tip positioned both above and below the slag-gas interface.

Test results and discussion

IGBT power supply – arc in air

For this test, the prototype arc detection unit was connected to the 3.2 MVA IGBT-type DC power supply. The facility was configured with a single 200 mm diameter graphite electrode mounted above a 500 mm x 500 mm x 500 mm graphite block connected to the anode. The arc was struck and operated in the air gap between the electrode and the graphite block. The presence of an arc was confirmed visually and photographically for each of the tests. The arc detection prototype's sample rate was set at 15 kHz. Sample sets of 2 seconds in length were obtained, portions of which were used for the FFT analysis in each test.

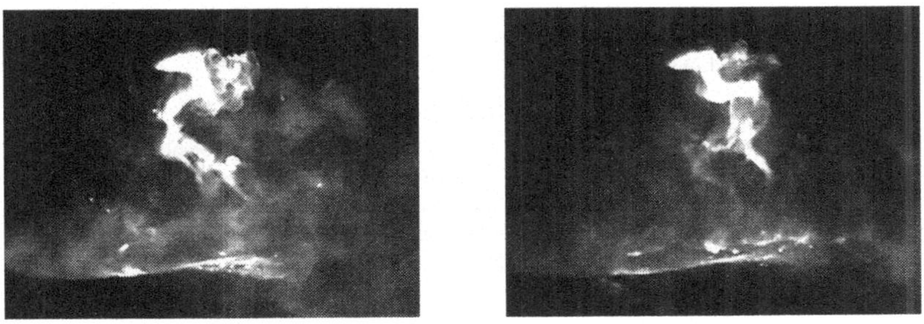

(a) -0.710 s (b) -0.552 s

Figure 5: Images from high-speed video of arc jet, shutter speed 5 μs. Times relative to trigger point.

Test 1 was conducted at 3 kA current and an arc length of approximately 15 cm. Some images of the turbulent arc jet are shown in Figure 5.

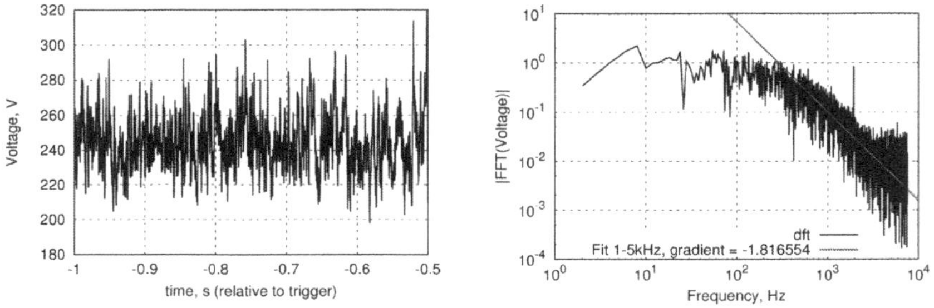

(a) (b)

Figure 6: Test 1 results, (a) time series of measured arc voltage, (b) the Fourier transform magnitude spectrum.

The measured voltage data for the test are shown in Figure 6. Figure 6(a) shows the time series, while Figure 6(b) shows the Fourier spectrum together with the fitted gradient in the power-law falloff region between 1 and 5 kHz.

An additional test was conducted at 1.5 kA current and 35 cm arc length. The results from both tests are shown in Table II.

Table II: Results of IGBT tests, arc in air.

Test ID	*Arc present?*	*Current*	*Arc length*	*Falloff, decade/decade*
Test 1	Yes	3.0 kA	15 cm	-1.82
Test 2	Yes	1.5 kA	35 cm	-1.90

Both tests at different currents and arc lengths yielded a falloff gradient of nearly -2 decades per decade in the high-frequency region between 1 and 5 kHz. This behaviour correlates well with the arc theory and model examined earlier.

Thyristor power supply – arc in furnace conditions

A second series of tests was conducted on Mintek's Bay 2 smelter pilot-plant during normal operation. This facility uses a 5.6 MVA Thyristor-type DC power supply to operate a DC arc furnace. 200 mm prebaked graphite electrodes are used, and the furnace anode features a multiple-pin design. The process operates with both a slag and metal pool, ensuring optimal conditions for testing the arc detection principle.

During these tests, the hydraulic electrode hoist was set to manual control. An electrode touch-down (locating the tip of the electrode at the surface of the molten slag) was performed giving a hoist zero position. A visual inspection was then performed to confirm that the tip of the electrode was touching the surface of the slag bath. With this reference point, arc length or electrode immersion depth could be calculated using the position of the hoist at any given time. The arc detection prototype's sample rate was set at 30 kHz. Measurements of furnace voltage were obtained at between 4 and 7 kA current, using arc lengths and electrode immersions up to 15 cm. Up to five sample sets, each 2 seconds long, were obtained during each test. The full sample length was used for FFT analysis.

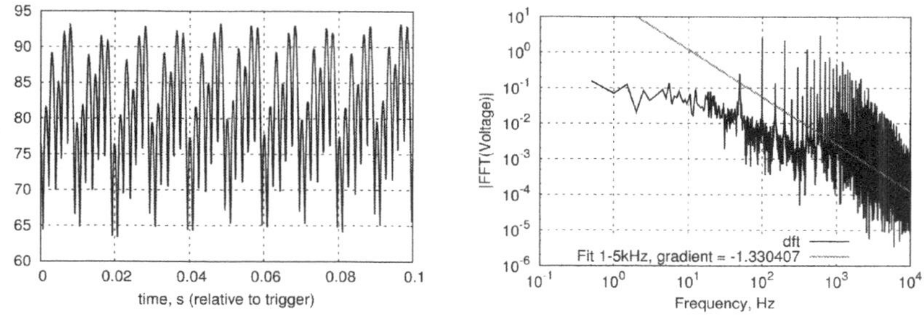

(a) (b)

Figure 7: Test results at 4 kA current and 11 cm electrode immersion, (a) a section of the voltage time series, (b) the Fourier transform magnitude spectrum.

The time series and Fourier spectra for example tests with the electrode immersed in the slag (no arc present) and the electrode above the slag (arc present) are shown in Figures 7 and 8 respectively.

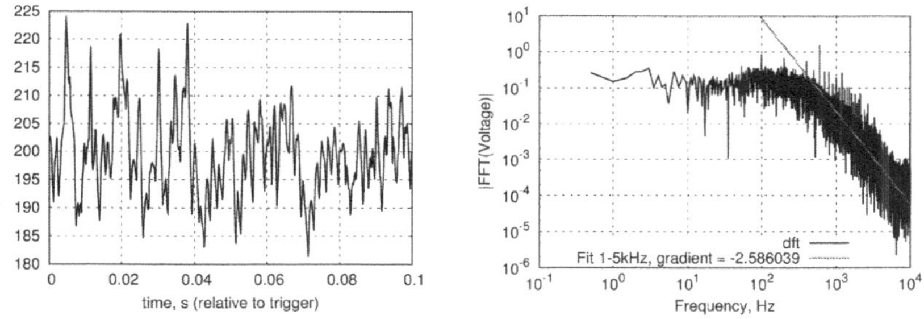

(a) (b)

Figure 8: Test results at 4 kA current and 10 cm arc length, (a) a section of the voltage time series, (b) the Fourier transform magnitude spectrum.

Results from all tests on the 5.6 MVA facility are collected in Table III. Negative arc lengths indicate immersion of the electrode in the slag, with the furnace operating in resistance-heating mode with no arc present.

For all test cases in which the furnace was operated with the electrode submerged in the slag to some degree, the falloff in the 1 – 5 kHz range exhibited a shallower slope. As soon as an arc was present in the circuit the shape of the Fourier spectrum changed noticeably, with a sharp knee point arising in the data around 1 kHz. This is as a result of a significant increase in the high-frequency falloff values, with all open-arc tests under furnace conditions yielding a steeper slope; this is consistent with the presence of a dynamic arc exhibiting an inertial response as discussed earlier.

Table III: Results of Thyristor tests, arc in furnace conditions.

Test ID	*Arc present?*	*Current*	*Arc length*	*Falloff mean, decade/decade*	*Falloff variance s^2, $decade^2/decade^2$*
Test 1	Yes	4.0 kA	10 cm	-2.62	0.00229
Test 2	No	4.0 kA	-6 cm	-1.41	0.00300
Test 3	No	4.0 kA	-11 cm	-1.31	0.01415
Test 4	No	4.0 kA	-16 cm	-1.41	0.00301
Test 5	Yes	4.0 kA	10 cm	-2.51	0.03154
Test 6	Yes	6.8 kA	10 cm	-2.15	0.00121

Conclusions

Development and application of a prototype instrument for the problem of measuring the presence of a plasma arc in a DC furnace has been largely successful.

Study of the theory of plasma arc dynamics suggested a practical measurement technique based on Fourier analysis that may be used to discern between open-arc and submerged operating conditions.

The arc detection prototype was successfully operated on various pilot-scale DC power supply and DC furnace facilities. Data gathered during these tests supported the theory of the arc as an inertial response system, with a characteristic Fourier spectrum falloff of approximately -2 decades per decade or lower observed in the voltage signal at high frequencies when an arc was present in the circuit.

It would therefore appear that the decade per decade falloff in the high-frequency 1 – 5 kHz range is a suitable indicator for the presence of an arc, at least at pilot-plant scale. The identified process for implementing an arc detection device based on this technique is as follows:

- Measure the DC voltage signal between the furnace anode and cathode connections at high sample rates, in excess of 15 – 30 kHz
- Perform FFT analysis on sections of the measured signal
- If certain threshold criteria are met, issue a signal indicating that the furnace is operating with an open arc

Such an instrument may be usefully applied as a control or operator-guidance tool on DC plasma arc furnaces used for any of a variety of applications, especially those in which the presence of an arc is necessary but difficult to verify.

Acknowledgements

This paper is published by permission of Mintek. Access to high-performance computing facilities at the Centre for High Performance Computing in Cape Town, South Africa was invaluable for performing model arc calculations.

References

1. R.T. Jones and T.R. Curr. "Pyrometallurgy at Mintek", *Proceedings of Southern African Pyrometallurgy 2006* , ed. R.T. Jones (Johannesburg, South Africa: Southern African Institute of Mining and Metallurgy, 2006), 127-150.

2. Q.G. Reynolds and R.T. Jones, "Semi-empirical modelling of the electrical behaviour of DC-arc smelting furnaces", *Journal of the South African Institute of Mining and Metallurgy*, 7 (2004), 345-351.

3. C. Walker, K. Sutherland, B. Marshall, J.M.A. Geldenhuis, J. Els, N. Voermann, and B. Wasmund. "Conversion to Partially Open Bath Smelting on Highveld Ironmaking Furnaces", *Proceedings of the European Metallurgical Conference 2007*, ed. GDMB Gesellschaft der Metallurgen und Bergleute e.V. (Clausthal-Zellerfeld, Germany: GDMB Gesellschaft der Metallurgen und Bergleute e.V., 2007) 895-914.

4. Q.G. Reynolds, "The dual-electrode DC arc furnace – modelling brush arc conditions", *Journal of the Southern African Institute of Mining and Metallurgy*, 7 (2012), 605-611.

5. Q.G. Reynolds, R.T. Jones and B.D. Reddy. "Mathematical and computational modelling of the dynamic behaviour of direct-current plasma arcs", *Proceedings of the 12th International Ferroalloys Congress (INFACON XII)*, ed. A. Vartiainen (Espoo, Finland: Outotec Oyj, 2010), 789-801.

6. B. Bowman. "Properties of arcs in DC furnaces", *Proceedings of the 52nd Electric Furnace Conference*, ed. Iron and Steel Society (Nashville TN, USA: Iron and Steel Society, 1994), 111-120.

7. FFTW, http://www.fftw.org.

8. I.J. Barker, M.S. Rennie, C.J. Hockaday and P.J. Brereton-Stiles. "Modes of electrical conduction in industrial silicon-type furnaces", *Proceedings of Silicon for the Chemical Industry VIII*, ed. H.A. Oye, H. Brekken, T. Foosnes and L. Nygaard (Trondheim, Norway: Norwegian University of Science and Technology (NTNU), 2006), 79-90.

9. I.J. Barker, M.S. Rennie, C.J. Hockaday and P.J. Brereton-Stiles. "Measurement and Control of Arcing in a Submerged-Arc Furnace", *Proceedings of the 11th International Ferroalloys Congress (INFACON XI)*, ed. R.K. Das and T.S. Sundaresan (Mumbai, India: The Indian Ferro Alloys Producers Association, 2007), 685-694.

Celebrating the Megascale: Proceedings of the Extraction and Processing Division Symposium on Pyrometallurgy in Honor of David G.C. Robertson
Edited by: Phillip J. Mackey, Eric J. Grimsey, Rodney T. Jones, and Geoffrey A. Brooks
TMS (The Minerals, Metals & Materials Society), 2014

AN ELECTROMAGNETICALLY STIRRED SLURRY MODEL FOR THE SMELTING ZONE OF A FERROALLOY FURNACE

Ben Bowman

Consultant, Westlake, Ohio

benbowman4@gmail.com

Keywords: smelting region model, slurry, electromagnetic stirring, CO stirring

Abstract

For certain ferroalloy furnaces, such as ferrochrome or ferromanganese, there is still doubt about the mechanisms operating under the electrode tips. In this paper a model is built that does not call upon an arc. The medium is best described as an electrically conducting slurry, a mixture of liquids, solids and gas. There are two sources of propulsion, the Lorentz force within the conducting part and buoyancy generated by CO bubbles, creating two opposed poloidal circulation loops, downward below the electrode axis and upward further out.

Introduction

Although the electrically-powered ferroalloy furnace has been with us for a good hundred years it is still not clear, for certain alloys, what happens under and around the electrode tips. Access to this interesting region is extremely difficult. In a few cases an 'autopsy' has been performed on a furnace after it has been switched off and allowed to cool [1, 2]. But as some of the authors themselves have remarked, what is found, weeks later, is not necessarily indicative of its on-power operation.

It is known that for certain alloys, such as ferrosilicon or ferronickel, an arc runs between the electrode tips and the metal bath inside a cavity of a diameter larger than the tip [3, 4]. However for ferrochrome and ferromanganese such a cavity has not been detected, nor is it suspected. There seem to be several schools of thought – either pure electroslag conduction or some combination involving an arc, such as an arc on top of slag or arcing to a bed of coke.

By concentrating on models and theory of slag conductivity several authors [e.g. 5] have favored implicitly an electroslag type operation. In particular, Yonemochi [6] stated categorically that there was no arc in ferromanganese furnaces. The performance data he presented of an Italian furnace, particularly the low electrode consumption, seems to support his case (see below). On the contrary, current harmonics measured by Barker and colleagues [7] seem to have convincingly demonstrated intermittent arcing in a ferrochrome furnace, but far less than in a ferrosilicon furnace.

A comparison with steelmaking furnaces is pertinent here. Historically referred to as an 'open arc' operation, the steelmaking furnace for the last 30 years or so nevertheless has operated, for a significant fraction of its heat, with *submerged* arcs, in this case submerged in foaming slag. Indeed some furnaces operating with continuously charged directly reduced iron (DRI) run with submerged arcs for 100 % of the time. They could also be classed as *reduction* furnaces, in the sense that the remaining iron oxide (about 10%) is eliminated by reaction with the carbide contained in the charge material, the CO evolved sustaining the foam. Such foaming slags in steelmaking contain some 50 to 70 % CO bubbles, yielding a corresponding low conductivity allowing only a minor current fraction. There is little doubt that an arc is present, spanning the electrode - metal gap.

Electrode consumption in ferroalloy and steelmaking furnaces

The 'safe' slipping rate for 1.55 m diameter Søderberg electrodes in a FeCr furnace corresponds to a specific consumption of 6.22 10^{-3} kg/kA^{2}h [8]. The actual consumption is lower. In a recent publication on consumption in steelmaking furnaces Potey et al. determined values between 6 and 9 10^{-3} kg/kA^{2}h for furnaces charging directly-reduced iron [9]. This range refers to consumption at the electrode tips; there is a further 30 to 40 % consumption along the sides. It is therefore surprising that the on-furnace baked Søderberg system can achieve a lower consumption rate than graphite electrodes at the same current.

Furnaces charging directly-reduced iron are characterized by lower specific consumption and lower current harmonics compared to the more common scrap melters. An explanation has been given by Bowman [10] based on the existence of a broader arc with a lower temperature attachment at the electrodes for the DRI furnaces. It seems likely that a similar explanation– an even lower current density attachment to the electrode – could also explain the Søderberg data, supporting the idea that no arc is involved.

The model presented here

An attempt is made here to create a model of the reduction zone in a ferroalloy furnace without recourse to an arc. The medium is best described as a slurry – a mixture of solids (ore, coke and quartz), liquids (slag and metal) and gas (CO bubbles). Slurry technology is widely used to transport a solids/liquid mixture through pipelines. Obviously a pump is required. In the case of a ferroalloy furnace the propulsion is provided by two mechanisms – electromagnetic in the main current carrying region, and buoyancy generated by CO bubbles.

The geometry of the reduction zone

A relatively recent 'dig-out' on a SiMn furnace has been described by Olsen and Tangstad [2]. The furnace was a 16 MW, 6.7 m diameter unit using 1.4 m diameter Søderberg electrodes. They determined that the electrode tips were respectively 60, 110 and 50 cm above the metal bath. They believed that a separation of 60 cm would have been suitable. The (cold) material between the electrodes and metal was described as a 'cokebed with slag'. The model described below is based on this geometry.

Development of the model

A 3-phase furnace is assumed but only one phase is treated; current flowing directly between the electrodes through the slurry is taken to be a small fraction of the total. Accordingly an axially symmetric system is treated, with coordinates of z in the downwards vertical direction and r as radius. Under the electrode tip the temperature is higher than in the bulk of the furnace. Because of the exponential increase of slag conductivity with temperature via the Arrhenius factor this higher temperature tends to 'attract' current towards the electrode axis, diminishing the direct flow between electrodes.

Initially the DC current from an ellipsoidal tip to a lower metal bath has been calculated. First, the conductivity of a typical slag at 1800 K was assumed, allowing voltage isopotentials to be calculated. Keeping the same isopotentials, a Gaussian shaped temperature hump, of 200 K amplitude, raises the axial temperature into the 2000 K range and 'attracts' current towards the axis. The result at this stage is a new current distribution, more concentrated around the axis, with streamlines normal to the isopotentials.

With an ellipsoidal tip radius of 53 cm, depth 30 cm and a separation of 62 cm the current in the example given is 110 kA at a voltage of 105 V, as illustrated in figure 1. Such a tip radius would apply to a tapered electrode of about 1.5 m diameter.

Based on measurements by Ossin for slags typical of FeCr production [11] a conductivity of 2.25 S/m was used at 1800 K. With an activation energy of 200 kJ/mol the conductivity increases at 2000 K by a factor of 3.8. At these conductivities the skin depth at 50 Hz is respectively 100 and 50 m, much larger than the dimensions of the current domain. The AC effect is therefore ignored.

The model is aimed at a ferrochrome operation since there has been generally much more data available for this alloy, especially from South Africa. For example, recently Pan [12] has presented useful details of several such furnaces, all charging chromite, in lump form, coke and quartz. The conductivity and viscosity of the slag has been estimated from the data of Ossin [11].

Slurry properties: effect of particles

Particles within a liquid influence the overall properties, for example, the relevant properties of electrical/thermal conductivity and viscosity. If the particles are more conductive then the result is an increase in bulk conductivity. This is the case here with the coke, which has conductivity some 10^2 to 10^3 times higher than the slag. On the contrary the CO bubbles have effectively zero conductivity so act in the opposite sense. There is little doubt that there is a significant volume of CO in the slurry. The carbon content however is difficult to estimate; it would depend on the lifetime and details of the reduction mechanism. Nevertheless judging by 'dig-outs' the carbon content is also significant.

The effect is such that 10% by volume of carbon would increase the conductivity by about 30 %. In comparison to the strong temperature effect on conductivity the overall C/CO effect has therefore been ignored.

Stirring produced by the electromagnetic Lorentz force

Current diverges from the electrode tip to the metal bath, both surfaces being treated as isopotentials. As a result the Lorentz force acts to push the slurry downwards along the electrode axis. Calculations by three previous groups of authors serve as a guideline of what we might expect here. Jardy et al. [13] and Kelkar et al. [14] have determined motion in slag baths, at, respectively, 2.6 kA and 6.4 kA, both AC. Trenty et al. [15] treated the motion in a bath of liquid metal due to an arc on the top (2 kA DC). In all three cases two loops of flow were seen, an inner due to the electromagnetic force, an outer due thermal buoyancy forces. Flow velocities were of the order of cm/s, with a maximum of about 26 cm/s in Kelkar's case.

Calculation of current streamlines and isopotentials

Current streamlines from an ellipsoidal electrode tip leave the surface normally and follow hyperbolic paths [16]. Isopotentials form ellipsoids of larger size further from the tip. In order to force current entry to a horizontal metal bath (an isopotential) a fictitious 'mirror ellipsoidal current source' is placed inverted and equidistant below the bath surface, receiving the same total current. In this way the combination of upper and lower currents forces the net current streamlines to cross the horizontal bath vertically, necessary to fulfill the isopotential condition. Such a calculation was made initially for an isothermal condition, that is, uniform conductivity.

To be more realistic a higher temperature region is assumed under the electrodes. A maximum of 2000 K is based on a Gibbs free energy plot against temperature for the reaction of carbon with chromia, which shows that silicon can be reduced from silica above about 1900 K [17]. Accordingly a Gaussian shaped temperature hump of 200 K amplitude was added to a background of 1800 K. To estimate the resulting current streamline coordinates the same isopotentials as found in the isothermal case were used while the 'width' of the Gaussian hump was adjusted to maintain current streamlines normal to them.

Over the 62 cm gap the Gaussian 'width' (standard deviation) varies from 14 cm at the tip to 21 cm at the bath; these are pure guesses. The resulting current streamlines and isopotentials are shown in figure 1.

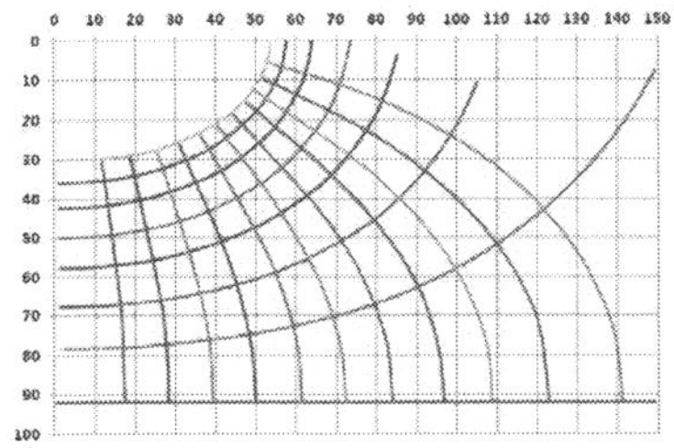

Figure 1: Current streamlines (10 to 110 kA) and isopotentials (0 to 105 V). Scales in cm.

From the distribution of current the Lorentz force, the vector product of current density and magnetic flux density, with units of Pa/m, can be calculated. There are both inward radial and downward components to this force. An integration over radius around the isopotentials results in an electromagnetic compression at the axis, with a maximum at the electrode tip. Pressure at the tip axis has a value of 1587 Pa, falling to 531 Pa at the bath. It is this pressure gradient that drives the slurry down below the tip.

Flow velocities down the electrode axis.

An estimate the velocity flow field at mid gap is made using the analogy with an efflux (nozzle) jet, an analogy that has proved useful for arcs [18], where the plasma is similarly propelled by electromagnetic forces. Mass continuity allows further details of the flow near the electrode tip and over the metal bath to be made.

A measure of the turbulent viscosity is required, for which the commonly used relation between turbulent kinetic energy, k, and turbulent power dissipation, ε, [19] provides a measure of turbulent viscosity, μ_t:

$$\mu_t = 0.09\, k^2/\varepsilon \tag{1}$$

Both these parameters are available for jets, the kinetic energy from Bridges and Wernet [20] and the energy dissipation from Friehe et al. [21]. Jet data is normally correlated to nozzle parameters, such as diameter and fluid exit velocity. Anticipating a velocity below 1 m/s, based on the work of previous authors, a nozzle diameter of 10 cm with an exit velocity of 2 m/s has been chosen. At distances between 1.5 and 2.5 m from this fictitious nozzle the axial velocity falls into the 0.5 to 1.5 m/s range, covering the range expected here. In this way a value for the turbulent viscosity some 45 times higher than the molecular viscosity was calculated. This is not an unusually high ratio; Kelkar et al. found a range of 20 to 150 in their model of the ESR process [14]. In this way a turbulent viscosity of 5 Pa.s was determined from a molecular level of 0.11 Pa.s for the slag at 2000 K.

Particles in the slurry (carbon, ore and metal) have the effect of increasing the effective viscosity, due to an increase in the proportion of small scale eddies in which viscous dissipation occurs. The increase depends on the solids volume fraction. For micron-sized polyethylene particles in water Tanaka et al. found a doubling of viscosity at 30 % solids by volume [22]. However transporting pebbles of 8 to 16 mm diameter, also with water, Ming et al. found an increase by a factor between 3.5 for 7-8 % volume and 6.3 for 10-12 % by volume [23] over the turbulent viscosity of water at flow rates around 1 m/s. This latter data is more likely to apply to our slurry, so we may expect an increase by a factor between 4 and 10.

The axial pressure gradient, *gradP(z,r)*, known from the integration of the Lorentz force, can now be used to drive the flow against the viscous drag. The balance reads:

$$gradP = \mu_t \nabla^2 v \tag{2}$$

where $v(z,r)$ is the velocity. At the mid-gap, using cylindrical coordinates for simplicity, a double integration over radius at constant z yields the relation:

$$v(z,r) = v_0 + \frac{r^2}{4\mu_t}\frac{dP}{dz} \tag{3}$$

with v_0 the velocity on the axis. With the turbulent viscosity around 30 to 40 Pa.s, a value for v_0 in the range 60 to 70 cm/s causes the velocity to drop to zero at a radius of 25 to 30 cm. The volume flow up to the zero crossing for these conditions is around 100 +/- 20 litre/s.

The vertical jet impacts the metal bath and spreads horizontally. Velocity distribution has been estimated from the work of Bradshaw and Love [24].

Stirring generated by CO bubbles

The high generation rate of CO in FeCr furnaces – about 6.5 m^3/s at a pressure of 1.5 bar and a temperature of 1900 K (calculated from Pan's data) – can be accommodated at moderate volume fractions (for example, 12 %) only if it is generated over a large volume, such as over a diameter of 10 m and depth of 1m. If it were mostly generated just below the electrode the volume fraction would be so high as to severely restrict current flow, perhaps triggering an arc. Thus reduction probably occurs over most, or all, of the volume of the slag/slurry bath.

Bubbles injected by nozzles are commonly used for stirring liquids. According to Iguchi et al. [25], away from the nozzle, it is the buoyancy due to the bubbles that propels the fluid. CO bubbles diminish the bulk density of the slurry by an amount depending on the volume fraction, which, in turn, depends on the rate of evolution and the residence time. It is likely that in the higher temperature region near the electrode axis the rate of reduction and therefore CO evolution is higher so bulk density is lower. As a result there is a pressure gradient at the metal bath level, directed *towards* the axis. The equation describing the flow is the same equation 2, but now we are interested in the circulation within the slurry outside the conducting region. The depth is taken to be about 1 m.

To estimate the circulation due to the CO bubble buoyancy we need to remember that the electrode is not at the centre of the furnace; it has been placed 2 m from the furnace centre. A mean CO transit time of 1.42 s has been estimated from bubble 'slip velocity' (relative to fluid) of 25 cm/s [26] and an average fluid up velocity of 10 cm/s. A Gaussian distribution of fraction bubble volume (voidance) varying from 12.5 % at the furnace axis to 10.5 % at the walls fits the required generation rate of 6.5 m^3/s. From these figures the outer loop slurry circulation is of the order of 1000 litre/s or ~3 tonne/s, about ten times higher than that under the electrode.

As this inward flow meets the outward flow from the impact of the 'jet' with the metal bath a combined 'upwelling' must occur taking both loops towards the top layer of the slurry, where the charge pile is resting. A volume flow balance allows the position of this boundary to be estimated. In this case the boundary is estimated to be around a radius of 52 cm. A summary of the velocity estimates is shown in figure 2.

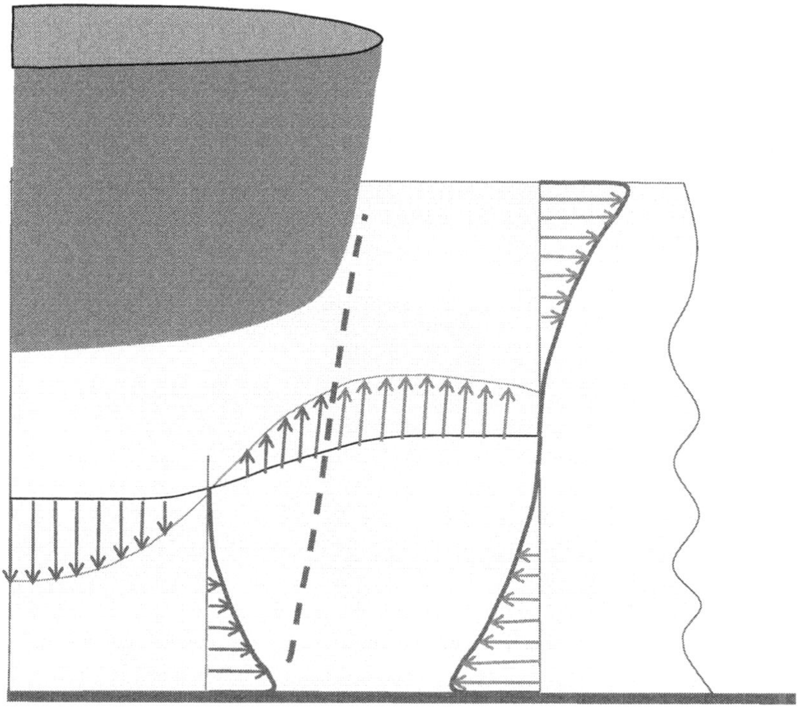

Figure 2: Estimated velocity distributions within the slurry.

The position of the boundary between the loops, shown by the dashed line, is based on a volume flow balance. On the right, 0.8m from the electrode axis, the horizontal velocity distribution has been calculated from equation 2 by integrating over z, at constant radius, yielding the volume flow estimate, around 1 m^3/s. On the left the volume flow estimated at the mid gap is about 100 litre/s. These two numbers allow the position of the boundary to be estimated, and consequently an estimation of the flow streamlines.

Ore dissolution

It seems likely that the initial mechanism leading to reduction in the slurry is the dissolution of the chromite in the slag. The rate of dissolution according to measurements by Demir and Eric [27] depends on temperature and relative particle-to-slag velocity. As an example, at a relative velocity of 10 cm/s the lifetime for a lump of 50 mm equivalent diameter would be about 68 s at 1800 K and 23 s at 2000 K. Now the volume of the inner loop is of the order of 510 litre, and since the flow rate is about 100 litre/s that means the circulation time is ~ 5 s. Thus ore lumps either make several turns of the inner loop or become transferred to the outer loop even before dissolution is completed.

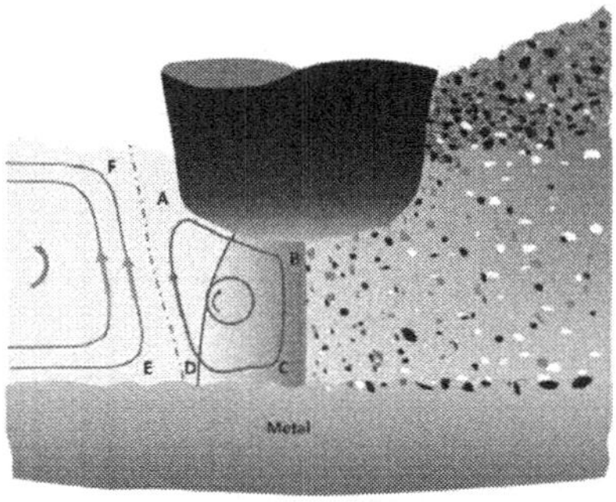

Figure 3: sketch of the proposed model: carbon in black, ore light gray, metal dark gray and white bubbles

Summary

Figure 3 shows a sketch of the model - slurry and heaped charge on the right, circulation loops on the left. The isotherm at 1820 K is shown on the left reaching to the centre of the inner loop. At the axis temperature is 2000 K. Charge material enters predominately around the electrode periphery at A. It is drawn along the tip towards B from which it is pumped down with the slag towards C. C is a stagnation point for this jet flow. The flow spreads towards D. In the region of D and E stirring due to the CO bubbles takes over, lifting the slurry towards F. A boundary between the two loops occurs, shown by the dashed line. Further out in radius the CO driven loop is closed by a downward turn next to the freeze lining of the furnace walls.

Conclusions

Although electromagnetic forces at currents over 100 kA generally generate considerable stirring in conductive fluids (e.g. liquid metals and plasmas, where the thrust can reach of the order of 1000 N) the effect of the high rate of generation of CO in the slurry seems to be, remarkably, equally important, if not predominant. In particular, the volumetric distribution of the bubbles plays a dominant role in driving the slurry circulation in most of the slurry bath.

There are many assumptions involved in this model; certainly many of the values quoted are probably in significant error. However the overall impression is that the double loop circulation system, propelled as explained, can offer a credible qualitative explanation of the phenomena that occurs under the electrodes.

References

1. N. Barcza, A. Kousaris, J.B. See, and W.A. Gericke, "The 'dig-out' of a 75 MVA high carbon ferromanganese electric smelting furnace", Electric Furnace Conference, ISS, 1979, p 19-33.

2. S.E. Olsen and M. Tangstad, "Silicomanganese production – process understanding", INFACON X, Cape Town, 2004.

3. Y. Otani, M. Saito, K. Usui, and N. Chino, "The inner structure in submerged arc furnace", Proc. 6th Int. Congress on Electroheat, Brighton, 1968, p 1-7.

4. A. Schei, J.K. Tuset, and H. Tveit, *Production of high silicon alloys*, Tapir Forlag, 1998.

5. J.H. Downing and L. Urban, "Electrical conduction in submerged arc furnaces", J. of Metals, Mar, 1966, p337-344.

6. J.Yonemochi, "Innovations and improvements in metallurgical processes", Electric Furnace Conference Proceedings, ISS, 1976, pp.73-84.

7. I.J. Barker, M.S. Rennie, C.J. Hockaday, and P.J. Brereton-Stiles, "Measurement and control of arcing in a submerged arc furnace", INFACON XI, 2007.

8. C.R. Ray, P.K. Sahoo and S.S. Pao, "Electrode management-investigation into soft breaks at 48 MVA FeCr furnace", INFACON XI, 2007.

9. D. Potey, B. Bowman, and S. Alameddine, "An update on electrode consumption", 8th European electric steelmaking conference, Birmingham, May, 2005.

10. B. Bowman, "Broad arcs in steel and ferroalloy furnaces", 10th European electric steelmaking conference, Graz, Sept, 2012.

11. D.I. Ossin, D.D. Howat, and P.R. Jochens, "Liquidus temperature, viscosities, and electrical conductivities of lime-containing slags produced during the smelting of HCFeCr and FeCr-Si alloys", Electric Furnace Proc., ISS, 1971.

12. X. Pan, "Effect of South African chrome ores on ferrochrome production", ICMMMC, Johannesburg, Apr., 2013.

13. A. Jardy, D. Ablitzer, and S. Jorget, "Modelling of slag behaviour in a non-ferrous smelting electric furnace", Reinhardt Schumann Int. Symp. Innovation technologies and reactor design in extractive metallurgy, Colorado Springs, Nov., 1966, TMS-AIME, p 419-431.

14. K.M. Kelkar, S. Patankar, and A. Mitchell, "Computational modeling of the electroslag remelting (ESR) process used for production of ingots of high-performance alloys", Int. Symp. on liquid metal processing and casting, Santa Fe, Sept., 2005.

15. L. Trenty, A. Bouvier, C. Delalondre, O. Simonin, and J.B. Guillot, "Numerical modeling of metallic bath heating with a transferred arc", Int. Symp. Plasma Chemistry, Vol 4, 1995, 2327-32.

16. R. Holm, *Electric contacts*, Springer-Verlag, Berlin, 4th Edn., 1967.

17. P.C. Pistorius, "Reduction selection in ferro-alloy production: the case for the importance of dissolution in the metal", J. S.A. Inst. Mining & Metall., Jan/Feb 2002, p 33-36.

18. B. Bowman and K. Krüger, *Arc furnace physics*, Stahleisen communications, 2009

19. J. Bredberg, "On two-equation eddy-viscosity models", Chalmers University of Technology, Dept. of thermo and fluid mechanics, Internal rept. 01/8, 2001.

20. J. Bridges and M.P. Wernet, "Establishing consensus turbulence statistics for hot subsonic jets", AIAA/Aeroacoustics conf., Stockholm, June 2010, p 1-41.

21 C.A. Friehe, C.W. van Atta, and C.H. Gibson, "Jet turbulence: dissipation rate measurements and correlations", in AGARD Turbulent Shear Flows, CP-93, 18-1-18-7, 1971.

22. A. Tanaka, M. Takahasi, and H. Ohkubo, "Measurement of apparent viscosity of liquid-solid mixtures", Int. Symp. "Next generation of air conditioning and refrigeration technology", Feb. 2010, Tokyo.

23. G. Ming, L. Ruixiang, N. Fusheng, and X. Liqun, "Hydraulic transport of coarse gravel – a laboratory investigation into flow resistance", Global dredging congress, May, 2007, p 703-710.

24. P. Bradshaw and E. Love, "The normal impingement of a circular jet on a flat plate", Aerodynamics Division, Nat. Phys. Lab., R & M 3205, 1961.

25. M. Iguchi, Z. Morita, H. Tokunaga, H. Tatemichi, Y. Sakamoto, and S. Tagaki, "Estimation of bubble rising velocities in bottom blowing bubbling jet", ISIJ Int., V 34, No. 4, 1994, p 322-9.

26. R. Clifft, J.R. Grace and M.E.Weber, *Bubbles, drops and particles*, Academic Press, 1978.

27. O. Demir and R.H. Eric, "Rate and mechanism of reduction-dissolution of chromite in liquid slags", High temperature materials and processes, V 32, Issue 3, p 215-263, 2012.

Celebrating the Megascale: Proceedings of the Extraction and Processing Division Symposium on Pyrometallurgy in Honor of David G.C. Robertson
Edited by: Phillip J. Mackey, Eric J. Grimsey, Rodney T. Jones, and Geoffrey A. Brooks
TMS (The Minerals, Metals & Materials Society), 2014

ROLES OF Mn CARBIDES IN CARBOTHERMIC PROCESSES FOR Mn ALLOYS

Byeong Deok Lee, Hyun Kyu Shin, and Young Eun Lee

Dongbu Metal Company
Donghae, Gangwon-do, 240-130, Republic of Korea

Keywords: Mn carbides. smelting of Mn oxides, Mn alloys

Abstract

The reduction of Mn oxides at pre-smelting steps is an integral part of smelting processes for Mn alloys, and an understanding of its reaction mechanism is expected to contribute to the improvement of processing efficiency for Mn alloys smelting. The phase diagram shows that five Mn carbides are operative in the Mn-C system. Although they are expected to participate in the reduction of Mn oxides, no specific information is available yet to clarify their roles in pre-smelting carbothermic reaction of Mn oxides.

Experiments with composite pellets prepared with Mn oxides and graphite powders show that Mn_3O_4 is reduced initially to MnO at about 1130°C and, subsequently, to Mn_7C_3 carbide as temperature increases further. At 1331°C, Mn_7C_3 carbide dissociates to Mn alloy and carbon. The degree of reducing reaction increases with increasing carbon balance in composite pellets and increasing temperatures.

Introduction

Most Mn alloys for steelmaking processes are produced by carbothermic reduction of Mn ores in submerged arc furnaces. The smelting processes are designed, developed, and practiced from the consideration of reactions between Mn oxides and carbon, but such activities usually do not consider the participation of Mn carbides in the smelting reactions. This approach is justified because Mn carbides are not stable at smelting temperatures.

The information of phase diagrams shows that the Mn-C system is characterized by five Mn carbides ($Mn_{23}C_6$, Mn_3C, Mn_5C_2, Mn_7C_3, and ε carbide) and that they are stable at temperatures less than 1331°C[1,2]. Because of their stability, Mn carbides may play important roles in carbothermic reducing reaction of Mn oxides at temperatures lower than 1331°C. Such reactions are characteristic phenomena taken place in a preheating zone of submerged arc furnaces. As it is an integral part of the smelting process, the processing efficiency at the preheating zone affects the performance of a whole smelting operation for Mn alloys. Therefore, a good understanding the role of Mn carbides is expected to improve the performance of pre-smelting carbothermic reducing reactions in submerged arc furnaces.

Experiments

The experiments are performed to examine the carbothermic reducing reactions of Mn oxides in the range of pre-smelting temperatures for Mn alloys smelting. As the reactions will take place in a solid state, specimen is designed to attain maximum interfaces and intimate contacts between the reactants. In order to facilitate this objective, the fine sized raw materials are prepared as composite pellets by pressing thoroughly mixed powders into cylindrical sample.

Raw materials

Industrial grade Mn oxide dust is collected as fine powder from the refining operation of MC and LC FeMn alloy. As Mn oxides is the major ingredient of it, it is chosen as the raw materials for Mn oxide in this study. Table 1 lists its chemical analysis.

Table 1. Chemical composition of Mn oxide dust

Mn oxide dust	Mn	Al_2O_3	CaO	SiO_2	MgO	C	S	P	Fe
wt%	66.83	1.16	1.12	0.85	0.09	0.13	0.010	0.047	3.76

An XRD analysis shows Mn to exist as Mn_3O_4 in Mn oxide dust. The analyzed Mn content, 66.83 wt%, in Table 1 is assumed to be for Mn as Mn_3O_4, and the content of Mn_3O_4 in Mn oxide dust is taken to be 92.76 wt%. This qualifies the Mn oxide dust as a high grade Mn raw material. The particle size of it is found to be in the range of 0.1 to 3.4 μm.

Graphite powder (a high purity grade) as supplied by Sigma-Aldrich is taken as a reducing agent. Its particle size is found to be 2.8 to 7.7 μm, which is comparable to that of Mn oxide dust.

Mix order

The required carbon to reduce Mn oxide powder is calculated by the stoichiometric relationship given by the equation (1)

$$Mn_3O_4 + 4C = 3Mn + 4CO \qquad (1)$$

In order to examine the effect of carbon balance, mix orders for composite pellets are prepared in two carbon ratios; 100 and 130% required carbon as determined by the equation (1).

Composite pellet

Weighted Mn oxide dust and graphite powders are thoroughly mixed without binder, and the mixture is pressed into a pellet of 21 mm diameter and 16 mm height (H). It is stored in a desiccator until its use.

Experimental procedure

Prior to experiments, the composite pellet is placed in an alumina crucible, and the set of pellet and alumina crucible is weighed together. The sample-alumina crucible assembly is positioned in a uniform temperature zone of horizontal resistance furnace. After closing both ends of furnace tube, Ar gas is introduced at the flow rate of 1 l/min. After a sufficient time to purge it, the furnace is heated at the rate of 10°C/min until temperature reaches 500°C. Thereafter, the furnace temperature is raised at the rate of 2 to 3°C/min up to reaction temperature. A preliminary run shows that the weight loss of sample reaches to a steady state at 20 min. In order to assure a sufficient reaction time, all runs are made with reaction time of 1 hour. After reaction for 1 hour at the reaction temperature, the sample-alumina assembly is cooled in the furnace under a continuing Ar gas flow. At room temperature, the sample-alumina assembly is withdrawn from the furnace and weighted to determine the weight loss. After a visual examination, samples are collected for chemical and XRD analyses.

Results

Visual examination of reacted pellets

All reacted pellets show a greyish appearance. They are shrunken but maintained the cylindrical shape and integrity of pellets at temperatures less than about 1300°C. However, these physical features of pellets are destroyed once temperatures reach at about 1300°C or higher by the formation of a liquid phase. The liquid phase is collected as buttons of Mn alloy in the bottom of crucible, and some non-metallic substances are observed to float on top of it.

Degree of Reduction

It is assumed that the carbothermic reducing reaction of Mn oxides proceeds to form CO, and the observed weight loss of pellets is due to the loss as CO. In order to evaluate the performance of the reducing reaction, the degree of reduction is defined by the ratio of oxygen lost as CO to oxygen in raw pellet.

$$\text{Degree of Reduction (\%)} = [O_{CO} / O_i] \times 100 \qquad (2)$$

where O_{CO} is oxygen lost as CO calculated from the observed weight loss and O_i oxygen in raw pellet.

Table 2 lists the observed degree of reduction with the mix order of 100 and 130% required carbon.

Table 2. Degree of reduction with 100 and 130% required C in mix order

	temperature, °C	raw pellet, wt(gr)	weight lost, wt(gr)	deg red. (%)
100%required C	1150	17.95	2.32	34.33
	1200	17.98	3.18	47.10
	1250	17.96	4.59	67.96
	1300	17.98	5.97	88.28
	1350	18.00	6.20	91.67
	1375	17.96	6.00	88.85
	1400	18.00	6.20	91.67
	1450	17.97	6.44	95.47
130%required C	1150	8.04	2.03	55.83
	1200	8.01	2.46	65.46
	1250	8.00	2.78	76.46
	1300	18.16	5.83	96.18
	1350	17.93	6.23	102.89
	1375	18.05	6.28	103.74

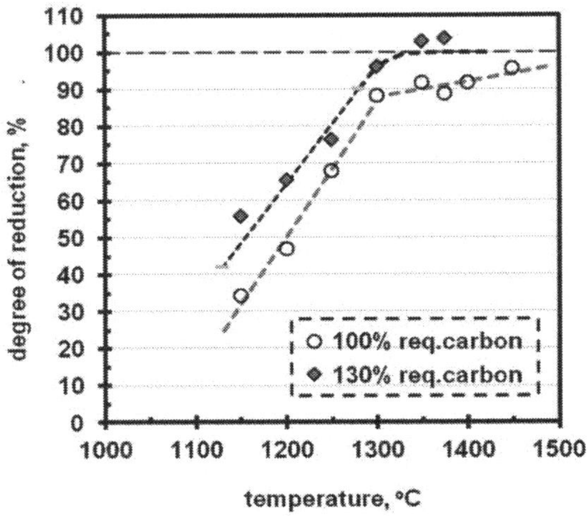

Figure 1. Degree of reduction with mix order of 100 and 130% required carbon

Figure 1 shows the variation of degree of reduction with reaction temperatures. The degree of reduction for the mix order with 100% required C is shown to increase with temperature as it increases from 1150 to 1300°C. It reaches to 88.28% at 1300°C. At temperatures higher than about 1300°C, however, the degree of reduction increases at significantly reduced rates and does not appear to attain a complete reduction.

The pellets with increased carbon balance at 130% required carbon show similar performance of 100% required carbon in mix order but with the enhanced degree of reduction at a given temperature. It attains about 50% degree of reduction at 1150°C and continues to increase to 96.8% at 1300°C. Ultimately, a complete reduction is obtained at about 1350°C.

Analysis of Mn alloy buttons

Buttons of Mn alloy are analyzed for Mn and C, and the results are summarized in Table 3.

Table 3. Chemical analysis of Mn alloy buttons

		chemical analysis	
	temperature, °C	wt% Mn	wt% C
100% required C	1350	88.26	5.14
	1375	88.66	4.42
	1400	88.77	4.87
	1450	88.31	4.23
		avg. 88.5	avg. 4.67
130% required C	1300	75.97	10.57
	1350	77.14	10.27
	1375	78.99	8.86
		avg. 77.37	avg. 9.90

Carbon content in Mn alloy button is shown to be higher with higher carbon balance in composite pellets. Table 3 shows that the carbon content in Mn alloy is 4.67wt%C on average with mix order of 100% required carbon but 9.90wt%C with that of 130% required carbon. The phase diagram of Mn-C system shows that C content in Mn melt at saturation with graphite is in the range of 7.3 to 7.7 wt%C at temperatures of 1300 to 1450°C [1,2]. When it compares with the analyzed carbon content in Mn alloy buttons, it is apparent that the mix order with 100% required carbon is short of carbon for saturation with carbon, i.e. a under coked condition, and that the mix order of 130% required carbon provides sufficient carbon not only to reduce Mn oxides but also to make Mn alloy melt saturated with carbon.

XRD analysis

Figure 2 summarizes the result of XRD analysis with the mix order of 100% required carbon, and Figure 3 with that of 130% required carbon.

With the mix order of 100% required carbon, samples are observed to contain phases of MnO, Mn_7C_3, and graphite at temperatures lower than about 1300°C (Figure 2a). At higher temperatures than 1300°C, samples contain phases of Mn, MnO, and $Mn_{23}C_6$ (Figure 2b). It is noteworthy of attention to the fact that no Mn_3O_4 phase is detected in these samples. This is due to the fact that Mn_3O_4 is already reduced to MnO at 1130°C and that MnO, not Mn_3O_4, participates in the reducing reaction of Mn oxides at temperatures higher than that. Another noticeable feature in Figure 2b is the absence of graphite in samples at temperatures higher than 1300°C. Combining it with the observed low carbon analysis in Mn alloy button (4.67wt%C), it is evident that the carbon balance in composite pellets with mix order of 100% required carbon does not provide sufficient enough carbon for Mn alloys to be in saturation with carbon. It is judged that $Mn_{23}C_6$ carbide observed in Mn alloy button is not originated from the carbothermic reduction processes but from the precipitation in Mn alloys melt during its cooling.

As shown in Figure 3, Mn alloys with mix order of 130% required carbon are observed to contain Mn_5C_2, Mn_7C_3, and graphite but no MnO. Also, the carbon content in Mn alloys button is reported to be 9.9 wt%C on average. It is evident that the mix order of 130% required carbon does provide sufficient carbon balance to complete the reduction of MnO and for Mn alloys to be in saturation with carbon.

Discussion

Reduction of Mn_3O_4 to MnO

The information of reduction as shown Figure 1 suggests that Mn_3O_4 (Hausmannite) in composite pellet is reduced to MnO (Manganosite) at about 1130°C without formation of Mn carbides. This reaction is described by the equation (3).

$$Mn_3O_4 + C = 3\ MnO + CO \qquad (3)$$

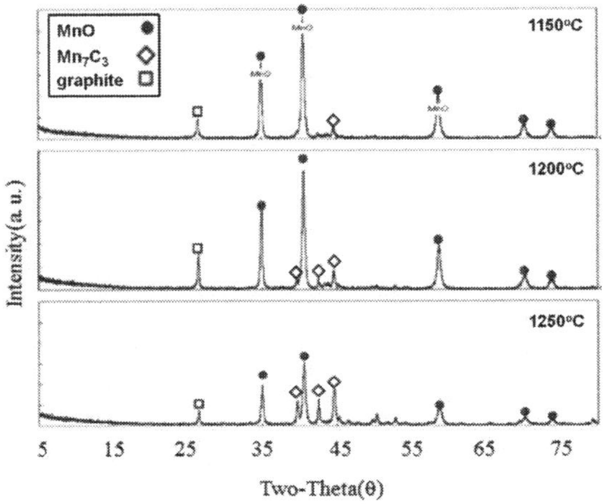

(a) 1150, 1200, and 1250°C

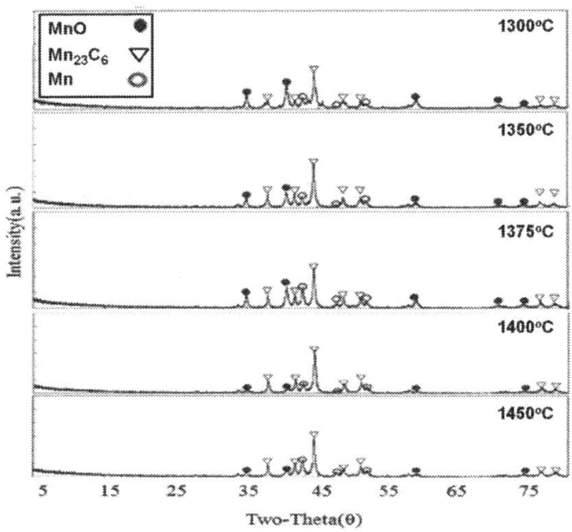

(b) 1300, 1350, 1375, 1400, and 1450°C

Figure 2. XRD analysis with mix order of 100% required carbon

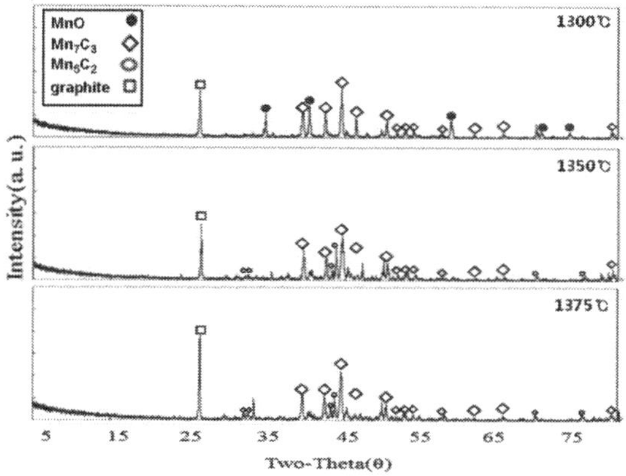

Figure 3. XRD analysis with mix order of 130% required carbon at 1300, 1350, and 1375°C

Formation of Mn_7C_3 from carbothermic reduction of Mn oxide

As temperature increases beyond 1130°C, MnO in the composite pellets is reduced to Mn_7C_3 carbide. This takes place in a solid state, and the reacted pellets have maintained the mechanical integrity although its volume experiences some degree of shrinkage due to the mass loss of C and O. The observed $Mn_{23}C_6$ and Mn_5C_2 are not formed by the carbothermic reduction reaction of MnO but originated from Mn alloys by precipitation during cooling.

The phase diagram of Mn-C system shows that $Mn_{23}C_6$, Mn_3C, Mn_5C_2, and ε carbide are stable only in a limited range of temperatures but only Mn_7C_3 carbide is stable at all temperatures up to 1331°C. Apparently, only Mn_7C_3 carbide is favorable energetically to be produced from carbothermic reduction of Mn oxides. Based on the experimental evidence, the carbothermic reducing reaction of MnO is described by the equation (4).

$$7\ MnO + 10\ C = Mn_7C_3 + 7\ CO \tag{4}$$

Mn_7C_3 carbide is an integral component in carbothermic reduction of Mn oxide, and the calculation of mix order for composite pellets should account for the formation of Mn_7C_3 carbide. In this respect, the equation (1) is not adequate to correctly address the required carbon balance for carbothermic reduction of Mn oxides.

Production of Mn alloys

At 1331°C and higher, Mn_7C_3 carbide dissociates into Mn alloy and graphite. This dissociation at 1331°C yields 99% Mn alloy melt which contains 7.57%C and 1% graphite. Any excess MnO reacts with carbon. This reaction appears to proceed at slow rates.

Conclusion

Reduction of Mn oxides at pre-smelting step is an integral part of smelting process for Mn alloys and takes place carbothermically in a solid state condition. Because of low prevailing temperatures in pre-smelting zone, Mn carbides are expected to form by carbothermic reducing reaction of Mn oxides.

Experiment with composite pellets of Mn oxide dust and graphite powders shows that Mn_3O_4 is reduced to MnO at about 1130°C and, subsequently, MnO to Mn_7C_3 as temperature increases further. At 1310°C, Mn_7C_3 dissociates into Mn alloy melt and carbon. This Mn alloy melt is in saturation with carbon.

Acknowledgment

The authors thank the management of Dongbu Metal Company for the permission to publish this article.

References

1. H. Okamoto, eds., 2000 Desk Handbook Phase Diagrams for Binary Alloys, ASM International, Materials Park, OH

2. Y. Yamada, et al., "Metal carbide-carbon peritectic systems as high-temperature fixed points in thermometry", Metrologia, 43 (2006), L23-L27

CELEBRATING THE MEGASCALE

Proceedings of the Extraction and Processing Division Symposium on Pyrometallurgy in Honor of David G.C. Robertson

Non-Ferrous Smelting, Converting, and Refining

Celebrating the Megascale: Proceedings of the Extraction and Processing Division Symposium on Pyrometallurgy in Honor of David G.C. Robertson
Edited by: Phillip J. Mackey, Eric J. Grimsey, Rodney T. Jones, and Geoffrey A. Brooks
TMS (The Minerals, Metals & Materials Society), 2014

REDOUBLING PLATINUM GROUP METAL SMELTING INTENSITY – OPERATIONAL CHALLENGES AND SOLUTIONS

Rodney Hundermark[1], Lloyd Nelson[1], Bertus de Villiers[1], July Ndlovu[1], Diale Mokwena[1], Phillimon Mukumbe[1], Bart Pieterse[1], Whitey Seyanund[1], Paul van Manen[1]

[1]Anglo American Platinum, 55 Marshall Street, Johannesburg, South Africa, 2001

Keywords: PGM smelting, electric furnace

Abstract

The 2003 redoubling in power to 68MW on the Polokwane Smelter furnace represented a significant intensification in platinum group metal (PGM) smelting. Combined with onerous 'green' PGM concentrate smelting requirements, this yielded conditions unusually corrosive to copper coolers and refractories. This presented unexpected operational and design challenges to reliable crucible operation and maintenance.

Combined with specific operational control intervention, development of protective coatings has led to the life of water-cooled copper components improving from 9 to 40 months. The furnace matte endwall was extended in 2010 to address accelerated wear of refractories and the potential risk for contact of copper components by superheated matte. An 18 month planned endwall rebuild cycle has been demonstrated (versus catastrophic failure within 9 months).

Finally, benefits including lower energy consumption, improved metal recoveries and higher productivity resulting from operational and in-house design developments will be described, that justify "Celebrating the Mega-scale" in PGM smelting.

Introduction

Anglo American Platinum's Polokwane Smelter, located in the Limpopo Province in South Africa, was commissioned in March 2003. Polokwane Smelter operates the world's largest Platinum Group Metal (PGM) smelting furnace, which is rated at 68MW. Prior to 1991, the highest power ratings on PGM smelting furnaces were 7.5 to 18MW. In response to greater demand for PGMs during the 1990s, several furnaces were doubled in power rating to 34-40MW. Thus, the 68MW power rating on the Polokwane furnace was a near redoubling of the original 18MW furnaces (in a similar furnace hearth area), and resulted in more intense operating conditions. Various operational challenges have arisen over the last ten years and the associated solutions will be discussed.

PGM smelting is somewhat of a misnomer as the furnaces are required to smelt nickel-copper concentrates, which contain only trace amounts of PGMs. The geology of the PGM deposits in Southern Africa results in low concentrations of base metal sulfides being present in furnace feeds, with considerable amounts of gangue (primarily MgO and SiO_2). Chromite is an important mineral in the Upper Group 2 (UG2) reef, also containing PGMs, which adds further complications to PGM smelting by increasing the Cr_2O_3 input. The concentrates are not readily amenable to roasting or autogenous flash smelting given the low base metal and sulfur content, and thus electric furnace smelting is the standard treatment route for PGM concentrates. The high MgO (>20%) and Cr_2O_3 (>1.5 to 5%) contents in the furnace slag result in high slag liquidus temperatures (1450 to >1650°C depending on the Cr_2O_3 content). The consequence is an unfortunate disconnect between the nickel-copper matte liquidus temperatures of approximately 900°C, and slag operating temperatures ranging from 1500 to 1700°C. The outcome is common matte operating temperatures of 1400 to 1500°C, with matte superheats of 500-600°C and this matte must be contained and safely tapped from the furnaces [1].

Operational Challenges

Copper corrosion

Before the introduction of the Polokwane furnace, the standard slag zone sidewall designs in the PGM furnaces were either refractories with natural air convection or, on furnaces operating up to 38MW, shallow water-cooled copper plate coolers interleaved between the refractories. The higher smelting intensity in the Polokwane furnace, original hearth power density of 244 kW/m^2, necessitated deep water-cooled copper coolers in the slag zone (Figure 1). Unfortunately, the Polokwane copper coolers provided a good surface for condensation of the labile sulfur, any volatile chlorine bearing species, and moisture contained in the unroasted "green" concentrate feed. The corrosion rate of the copper coolers in contact with the sulfur was unexpectedly rapid, and in the worst case, the coolers failed within 9 months. In addition to the sulfur, there is suspected to be contribution from chloride species in the furnace feed resulting in chloride-accelerated sulfidation of the copper [1].

Sidewall cooling type	Natural Air Cooling (e.g. Waterval pre-1994 and Mortimer pre-2008)	Plate Coolers at Waterval and Mortimer	Waffle Coolers at Polokwane (from 2003)	Split waffle coolers at Polokwane (from 2005) with extended matte endwall (from 2010)
Typical PGM furnace power level(s)	7.5 to 18 MW	34 to 38 MW	68 MW	68 MW
Approximate maximum slag zone heat flux	5 kW/m^2	50 kW/m^2	500 kW/m^2	500 kW/m^2
Hearth power density at maximum power	91 kW/m^2	187 kW/m^2	244 kW/m^2	229 kW/m^2
Distance from E1 to matte endwall	4.0m	4.0m	4.8m	7.5m
Matte end volumetric power density*	68 kW/m^3	124 kW/m^3	176 kW/m^3	125 kW/m^3

* Maximum energy input at Electrode 1 (E1) divided by slag volume up to matte endwall

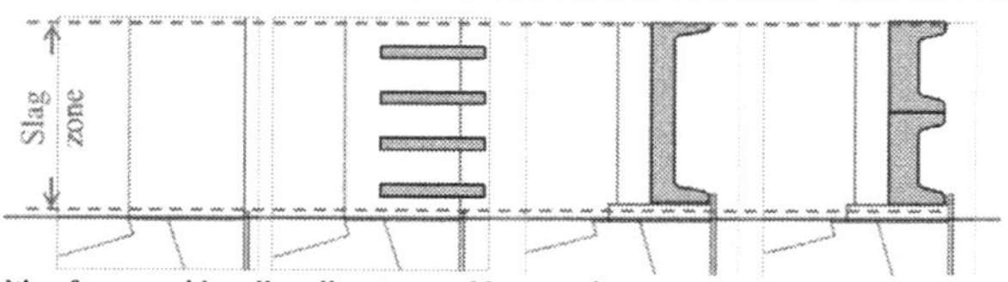

Figure 1: Intensification of PGM smelting furnace, sidewall cooling type and key metrics

Over several years and iterations, coatings were developed to protect the copper from corrosion, and coated coolers have now been proven to last beyond 40 months. Another site development, which arose out of the copper cooler corrosion problem, is an automated water leak detection system which routinely checks for leaks on each of the water-cooled circuits on the furnace by means of a brief static pressure test [2]. The system is run every 4 to 12h and is able to detect very small water leaks (<1L/h), and has proven invaluable since introduction in 2006.

Matte endwall wear

The three matte tapholes on the Polokwane furnace are located on the northern endwall. Adjacent to the matte tapblocks are three rows of deep-cooled copper flanker coolers (FC) and mag-chrome refractories, approximately 300mm long, are positioned at the hot face of the flanker coolers and the matte tapblocks. The bricks protect the copper coolers from exposure to matte, while relying on the cooling from the copper blocks. The slag zone waffle coolers are placed above the matte tapblocks (Figure 2).

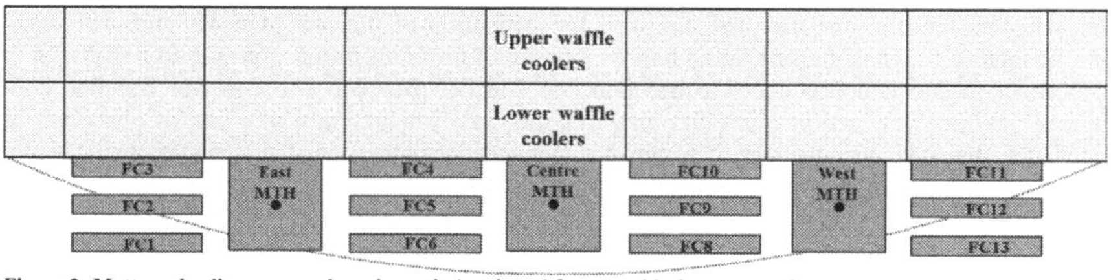

Figure 2: Matte endwall copper cooler schematic (as viewed from outside furnace looking south)

In 2005, the matte endwall was inspected after 9 months of operation and the residual brick thickness in front of the matte tapblocks and flanker coolers was found to be ~30mm in the thinnest areas (Figure 3a). During 2008, the furnace was operated at the maximum power input of 68MW for 4 continuous months until, in Nov 2008, a catastrophic failure of the centre matte tapblock occurred, and the furnace drained substantially in 45 minutes. The residual brick thickness measured on the matte endwall elsewhere was approximately 70mm. Despite having extremely comprehensive instrumentation on the entire furnace, the concerning issue highlighted by the failure was that the instruments provided no more than 90s warning of the impending problem (Figure 3b). This again emphasises the risks inherent with operating a furnace containing highly superheated matte.

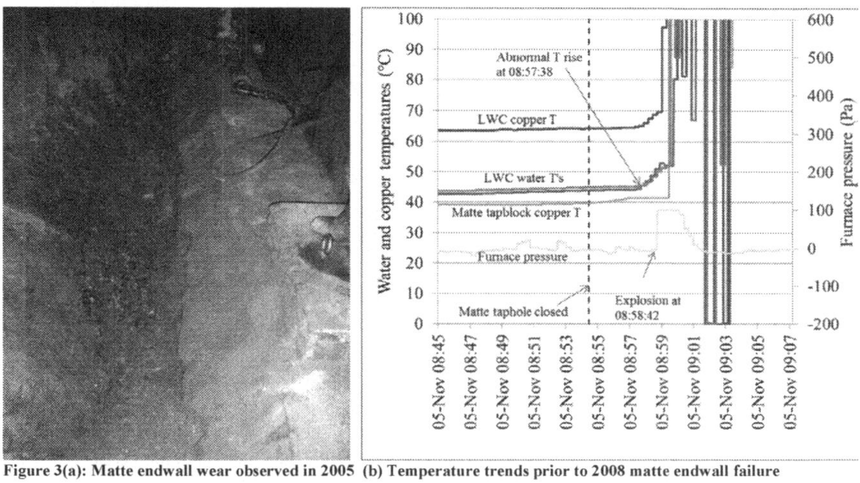

Figure 3(a): Matte endwall wear observed in 2005 (b) Temperature trends prior to 2008 matte endwall failure

The hypothesis for the failure mechanism of the centre matte tapblock was as follows:

- The residual brick thickness on the matte endwall was very thin in front of the block
- The matte taphole was closed approximately 4 minutes before the failure and it is suspected that reaction between the taphole clay and the matte created sufficient turbulence for the residual brick to be washed away
- With no refractory protecting the copper tapblock, superheated matte melted the copper and cooling water pipe resulting in an explosion which dislodged the block from the wall.

Following the failure, options were assessed for mitigation of the risk of matte endwall wear. Two actions were implemented immediately, namely limiting the furnace power to 55MW, and limiting the matte temperature to a maximum of 1480°C (beyond which power setpoint was reduced). While these actions partially mitigated the rate of matte endwall wear (as observed in Jan 2010), the reduced power limit somewhat negated the objective of process intensification, and it was clear that more drastic measures were required.

Solutions

The option to extend the matte wall originated from observations at the other Anglo American Platinum furnaces (at Waterval and Mortimer Smelters) where matte endwall refractory conditions, after similar durations of operation, were considerably better than at Polokwane. The furnace geometries, maximum power and corresponding matte endwall volumetric power densities were calculated (Figure 1). A 2.7m extension of the Polokwane furnace matte endwall would achieve reduction in endwall volumetric power density, comparable to existing furnaces, and was also a practical distance from a constructability perspective.

In January 2010, a full furnace rebuild was undertaken and the matte endwall was moved 2.7m further away from the closest electrode (Figure 4). New feed and sounding ports were added to ensure that the new area of the furnace could be adequately fed, and the levels monitored.

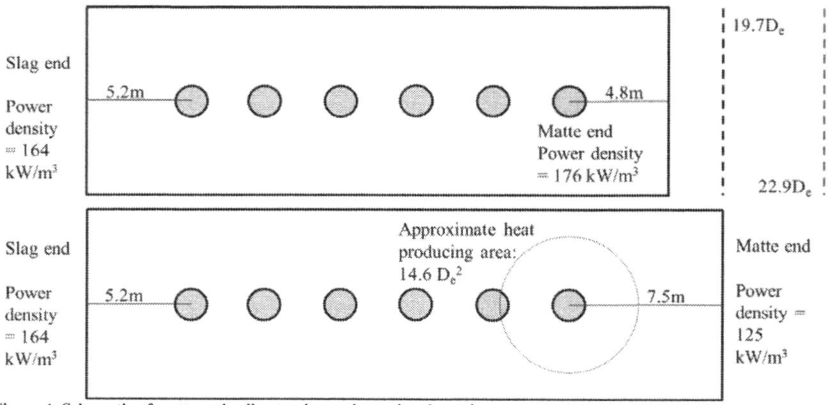

Figure 4: Schematic of matte endwall extension and associated metrics

It is of some interest to consider the original dimensions of the Polokwane furnace, particularly the length. Furnace dimensions are often empirically ratioed to the electrode diameter (D_e), with typical length to D_e ratio values of 19.7 to 22.9 for six-in-line electrode rectangular furnaces [3]. Given the Polokwane electrode diameter of 1.6m, a furnace length of 31.5m to 36.3m is calculated from the ratios, whereas the original length (28.6m) to D_e ratio was 17.9. With extension of the matte endwall, the ratio was increased to 19.6. The approximate heat producing area ($14.6D_e^2$) is also plotted [4]. Of further interest is that the original matte endwall location was closer to its nearest electrode, at a distance of 4.8m, than the slag endwall was to its nearest electrode at 5.2m (Figure 4). This may be beneficial for tapping metal in furnaces smelting high melting point ferroalloys, close to their liquidus temperatures, but with excess matte superheat this is less of a concern in PGM smelting.

Owing to the lack of proven condition-based monitoring techniques for the refractory thickness on the hotface of the matte endwall, a conservative inspection program was established to determine the efficacy of the matte endwall extension. Inspections were carried out on an annual basis for the first two years (Jan 2011 and Jan 2012) and then after 18 months (Jun 2013). With matte temperatures controlled below 1480°C, the positive impact of the endwall extension was a marked improvement in the condition of the matte endwall as observed during the inspections. After the second annual inspection confirmed that much of the hotface brick was still at original thickness (maximum wear of 120mm measured), the maximum power setpoint on the furnace was nominally raised from 55 to 60MW for the 18 month campaign. After the 18 months operation, the matte endwall was actually found in better condition (40mm maximum wear) despite operation up to 60MW. Almost the full brick thickness was still intact on the endwall and the brick imprints in the frozen matte were visible after demolition of the bricks (Figure 5).

Figure 5: Matte endwall demolition after 18 months of operation, from which endwall brick thickness can be inferred

Possible explanations for the reduced wear following the matte endwall extension are:

- Heat fluxes were reduced on the matte endwall due to decreased convective heat transfer at a greater distance from the nearest Electrode 1 energy source.
- Operational control on matte and slag temperatures, through improved matching feed to power, and (occasionally) reducing power to limit process temperature excursions.
- During upset conditions, the additional furnace volume between Electrode 1 and the matte endwall provided some thermal buffer to direct endwall contact.
- Improved control was implemented on clay volume used for matte taphole closure, with reductions in clay leading to less turbulent conditions, and also greater drill depths.

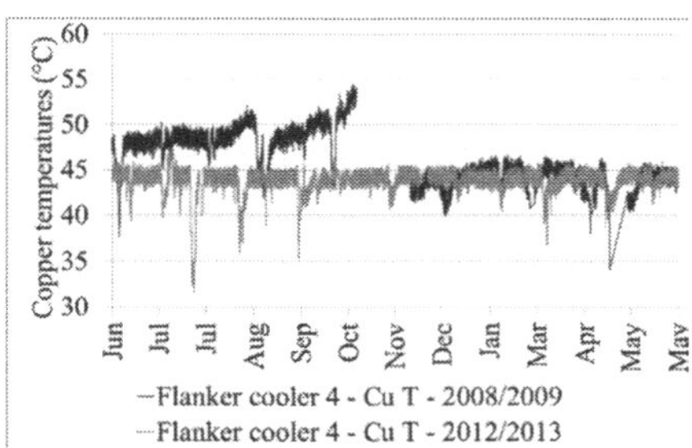

Figure 6: Flanker cooler 4 copper temperatures

With the exception of the brief period of high power (>65MW) operation between Jun and Oct 2008, heat fluxes and flanker cooler copper temperatures (Figure 6) were not noticeably different between the original and extended matte endwall. Intuitively it was expected that the increased distance from the energy source at Electrode 1 to the extended matte endwall would reduce the driving force for energy transfer.

Thus consideration was given to the rate equation for convection between the slag bath and the furnace sidewall:

$$q/A = h(T_{bulk\ slag} - T_{liquidus}) \qquad (1)$$

where q is energy (W), h is the convective heat transfer coefficient ($W/m^2.K$), A is the area (m^2), $T_{bulk\ slag}$ is the bulk slag bath temperature (K) and $T_{liquidus}$ is the slag liquidus temperature (K). Given that the measured mean heat fluxes on the matte endwall after the extension did not change substantially suggests that there were no major changes in the convective heat transfer coefficient or slag superheat (Table 1).

Table 1: Convective heat transfer coefficient estimation for periods before and after matte endwall extension

		May 2009: pre-endwall extension	July 2010: post-endwall extension
Average power input	MW	52.6	50.3
Average matte endwall heat flux	kW/m^2	30.6	33.2
Average slag operating temperature	°C	1602	1586
Liquidus temperature (FactSage calculation)	°C	1491	1466
Slag superheat	°C	111	120
Calculated convective heat transfer coefficient	$W/m^2.K$	274	278

In laboratory measurements and modelling of heat transfer at furnace sidewalls, the mechanism is often treated as natural convection and convective heat transfer coefficients of up to $500W/m^2.K$ are estimated [5]. There is little to suggest, from the data, that the situation at the Polokwane matte endwall is decidedly different. The improved condition of the matte endwall refractory after extending the endwall does not, therefore, appear convincingly attributable to any reduction in the matte endwall heat flux.

Three distinct periods of matte tapping temperatures are discernible (Figure 7a). Elevated temperatures (often exceeding 1500°C) associated with high power operation (>65MW) between mid to end 2008. A period of low matte tapping temperatures (<1400°C) associated with feed of unusually low chromium oxide content (< 1.5 % Cr_2O_3) and high matte fall. Since the matte endwall extension, and coupled with more active power down if 1480°C exceeded, a more controlled matte tapping temperature of 1420-1480°C has been maintained.

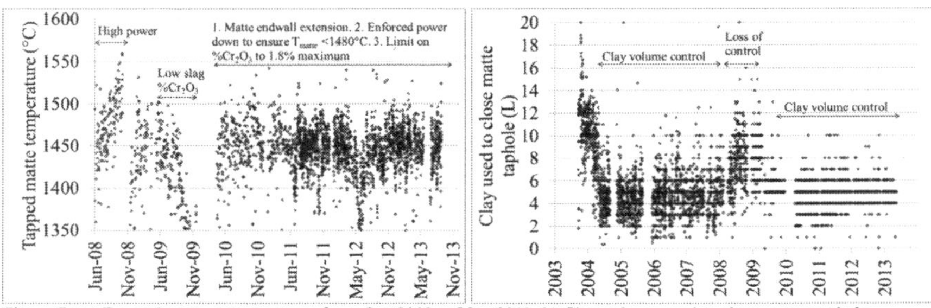

Figure 7: (a) Tapped matte temperatures from Jun 2008 to May 2013 (b) Clay volume used to close matte taphole

Another possible contribution to matte endwall wear is localized events related to matte taphole closure conditions, where the clay injected into the taphole reacts violently with the matte and results in highly turbulent conditions. It has been demonstrated that consistent usage of the minimum clay requirement to close a taphole results in deeper drill depths and better overall taphole life [6]. Unfortunately, these lessons were unlearnt at Polokwane during 2008 and better clay control was only established again after April 2009 (Figure 7b). Sustained improved control has likely contributed to improved matte endwall condition.

Intensification

The performance of the Polokwane furnace has achieved the objectives of redoubling PGM smelting intensity, namely: better energy efficiency, higher metal recoveries and greater productivity.

Energy efficiency

As previously described for increased power input on the same furnace [7], heat losses are proportionally lower (Figure 8a). Some of the observed scatter is explained by changing concentrates and a concentrate specific energy requirement for smelting, coupled with associated differences in process matte, slag and offgas temperatures. None the less, the expected trend in lower specific energy consumption with increased power is undeniable.

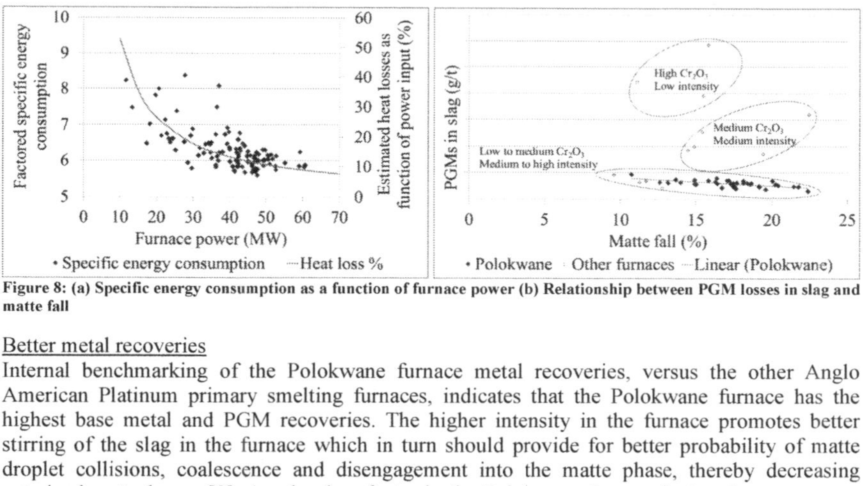

Figure 8: (a) Specific energy consumption as a function of furnace power (b) Relationship between PGM losses in slag and matte fall

Better metal recoveries

Internal benchmarking of the Polokwane furnace metal recoveries, versus the other Anglo American Platinum primary smelting furnaces, indicates that the Polokwane furnace has the highest base metal and PGM recoveries. The higher intensity in the furnace promotes better stirring of the slag in the furnace which in turn should provide for better probability of matte droplet collisions, coalescence and disengagement into the matte phase, thereby decreasing entrained matte losses [8]. Another key factor in the Polokwane furnace is that the concentrate feed into the furnace generally has higher matte fall, and lower Cr_2O_3. Combined with better stirring, greater matte fall allows for greater probability of entrained matte droplet collision and formation of larger droplets which settle out more quickly into the matte phase, while the lower slag Cr_2O_3 content promotes reduced slag viscosity, further enhancing matte droplet disengagement as observed (Figure 8b). Of course, the generally lower PGM grade of the matte produced at Polokwane, due to the specific concentrates treated, also helps to promote a lower loss of PGMs to slag. This is supported by the general relationship between PGM losses in slag (indicative of entrained matte) and the matte fall (Figure 8b).

Greater productivity

A key measure of productivity is the concentrate throughput achieved per permanent employee. At Polokwane, over 60% higher productivity is achieved. Another benefit of a large, intense, single-furnace operation is that all of the management, operational and maintenance focus is on one unit. The consequent attention to detail provides for better understanding and faster resolution of any issues which arise. Also, both short and long term planning is more effective.

Conclusions

Over the last ten years, many challenges have been overcome in operating and maintaining the Polokwane Smelter furnace. The copper corrosion problems have been resolved proactively by means of various coatings, and supported by an automated water leak detection system. A sustainable matte endwall campaign life appears to have been achieved with extension of the matte endwall, with greatly reduced risk of catastrophic failure. The benefits of a mega-scale furnace are being realized in terms of greater productivity, higher metal recoveries and better energy efficiencies.

Acknowledgements

The authors thank Anglo American Platinum for permission to publish this paper. In addition, the efforts of past and current employees at Polokwane Smelter are recognized for their contributions to getting the furnace to where it is today.

References

1. A. Shaw et al., Challenges and Solutions in PGM Furnace Operation: High Matte Temperature and Copper Cooler Corrosion, *Journal of the Southern African Institute of Mining and Metallurgy*, 113 (2013), 251-261.
2. T. de Vries, Automatic pressure testing of furnace cooling water circuits at Anglo Platinum, *Hatch Furnace Technology Symposium*, Johannesburg, 2010.
3. R.H. Eric and A.A. Hejja, Dimensioning, Scale-up and Operating Considerations for Six Electrode Electric Furnaces, Part II: Design and Scale-up Considerations for Furnaces Treating PGM-Containing Copper-Nickel Concentrates, *EPD Congress*, ed G.W. Warren,1995, 239-257.
4. M.W. Kennedy, Electric Slag Furnace Dimensioning, *International Smelting Technology Symposium: Incorporating the 6th Advances in Sulfide Smelting Symposium,* ed J. P. Downey et al., John Wiley & Sons, Inc., USA, 2012.
5. D. G. C. Robertson and S. Kang, "Model Studies of Heat Transfer and Flow in Slag-Cleaning Furnaces", *Fluid Flow Phenomena in Metals Processing*, ed N. El-Kaddah, The Metallurgical Society, Warrendale, PA, 1999, 157-168.
6. J. Ndlovu et al., "Operational Readiness" – A Value Proposition, *Nickel and Cobalt 2005 – Challenges in Extraction and Production*, COM2005, Calgary, 2005, 389-404, paper and presentation.
7. P.K. van Manen, Furnace energy efficiency at Polokwane Smelter, Third International Platinum Conference 'Platinum in Transformation', *The Southern African Institute of Mining and Metallurgy*, 2008, 191-196.
8. L.R. Nelson, et al., Role of Technical Innovation on Production Delivery at the Polokwane Smelter, *Nickel and Cobalt 2005 – Challenges in Extraction and Production*, COM2005, Calgary, 2005, 91-116.

Celebrating the Megascale: Proceedings of the Extraction and Processing Division Symposium on Pyrometallurgy in Honor of David G.C. Robertson
Edited by: Phillip J. Mackey, Eric J. Grimsey, Rodney T. Jones, and Geoffrey A. Brooks
TMS (The Minerals, Metals & Materials Society), 2014

PYROMETALLURGICAL PROCESSING TECHNOLOGIES FOR TREATING HIGH ARSENIC COPPER CONCENTRATES

Patrick R. Taylor and Teuku A. R. Putra

Kroll Institute for Extractive Metallurgy
George S. Ansell Department of Metallurgical and Materials Engineering
Colorado School of Mines
Golden, Colorado 80401 USA

Keywords: Pyrometallurgy, Copper Concentrates, Arsenic

Abstract

Various pyrometallurgical methods for treating copper concentrates that contain appreciable arsenic are reviewed. Various aspects of the engineering of these treatment methods are discussed. The methods discussed include: complete oxidation with arsenic fixation; selective volatilization of arsenic; acid baking, soda ash roasting, and others methods. Methods for the ultimate disposal, or marketing, of the arsenic are discussed.

Introduction

Conventional methods to extract copper from sulfide ores using pyrometallurgy have been used for decades and about 75% of primary copper is recovered using these methods [1]. Even though conventional processing is highly efficient, the depletion of clean copper sulfide ore reserves requires new strategies to extract the copper with certain levels of impurities. Group 15 elements (As, Sb and Bi) are difficult to handle in conventional copper smelting. Environmental regulations have resulted in avoidance of high-arsenic copper sulfide concentrates and led to increased smelter penalties and lower acceptable amounts of arsenic in the concentrates treated (2). An example of limits and penalties from the US is presented in Table I.

Hydrometallurgy

After more than 50 years of using hydrometallurgy to recover copper from low grade ores, certain problems and issues are still a concern, especially for the treatment of high arsenic bearing ores or concentrates. Various hydrometallurgical methods have been proposed for treating high arsenic copper ores and concentrates. This topic is not covered in this review.

Arsenic

Arsenic in copper presents several problems. In the final product, cracks appear during the forming process and reduce the electrical conductivity of copper. The maximum allowable arsenic concentration in copper for electrical applications is set at 1 ppm [4]. A main objective for the control of arsenic in mineral processing and extractive metallurgy operations is to avoid its emissions in either the gaseous and aqueous phases. This is addressed through arsenic removal and converting it to a disposable, environmentally acceptable stable compound [5]. There is a small market for arsenic currently. The El Indio operations is one example of this, the operations separated and recovered arsenic as the commodity As_2O_3 until 2002, at which time operations were closed due a combination of factors, including the lack of a market for this arsenic product [6]. Table II presents USGS [7] data on world production and reserves of arsenic trioxide.

Table I. Limits and Penalties at a US Copper Smelter [3]

Element	Symbol	Penalty Formula
Alumina	Al_2O_3	$0.50 ea 0.1% > 5%
Iron	Fe	> 15% = increased treatment charge for more flux needed
Arsenic	As	$0.50/lb > 1% (20 lb) OR $2/dt ea 0.1% > 0.1% Max 0.2%
Barium	Ba	0.5 to 1% limit
Beryllium	Be	< 10 ppm limit
Bismuth	Bi	($1.10 to $7.50) /dt ea 0.1% > (0.1% to 0.4%) Max 0.4 %
Cyanide	CN	< 10 ppm
Cadmium	Cd	($2.20 to $7.50) /dt ea 0.1% > (0.05% to 0.2%) Max 0.4 %
Chloride	Cl	BAD PLAYER, DO NOT WANT ANY
Cobalt	Co	0.5% limit
Chromium	Cr	$0.50 dt ea 0.1% > 3% no hex chrome, 5% max on tri v Cr
Fluoride	F	$5 dt ea 0.1% > 0.2% 0.5% max
Mercury	Hg	($1.85 to $ 2) /dt ea 10 ppm > 10 ppm
Magnesium Ox	MgO	Normally 10% limit
Manganese	Mn	2.0% limit
Sodium	Na	5.0% limit
Nickel	Ni	$2 dt ea 0.1% > 2%
Phosphorus	P	3.0% limit
Lead	Pb	$1 dt ea 0.1& > 1% OR $1/lb >0.5% (more severe)
Antimony	Sb	BAD PLAYER, DO NOT WANT ANY
Selenium	Se	0.1% limit
Tin	Sn	($1.10 to $3) dt ea 0.1% > (0.2 to 3%) Max 3 %
Tellurium	Te	0.01% limit
Thallium	Tl	0.01% limit
Zinc	Zn	$0.50 dt ea 0.1% > 3% 4.0% limit
Moisture	H_2O	$2.50 wt ea 1% >

dt is dry ton, wt is wet ton

Process Options

Both pyrometallurgical and hydrometallurgical processes have been evaluated for the extraction of copper from enargite concentrates. Most research has focused on hydrometallurgical methods, as they are generally thought to be more environmentally benign and less capital intensive.

Pyrometallurgy

High arsenic content represents a problem in direct smelting of enargite concentrates and makes the value of the concentrates greatly diminished. For those reasons, the arsenic must be minimized before smelting to obviate the need to collect and treat volatile arsenic compounds and to decrease slag and metal contamination [8]. During the pyrometallurgical processing of primary and secondary materials, arsenic represents a risk and most countries require its removal from the processing residues. Controlled oxidation roasting is one of the options for pyrometallurgical pre-treatment to remove the arsenic from enargite concentrates by volatilization. The final product of the arsenic must be in a stable disposable form. Numerous studies on the use of roasting and smelting have been published for enargite concentrate treatment [9, 10, 11]. Several very good review papers have been published on this topic [12, 13, 14]. Several of the operating and potential processes will be discussed in this paper.

Table II. USGS Data on Arsenic Trioxide (metric tonnes) [7]

Source	Production of Arsenic Trioxide 2011	Production of Arsenic Trioxide 2011	Reserves
United States	-	-	
Belgium	1,000	1,000	
Chile	10,000	10,000	
China	25,000	25,000	
Morocco	8,000	6,000	
Russia	1,500	1,500	
Other countries	300	300	
World total (rounded)	45,800	44,000	World Reserves are thought to be about 20 times annual world production

Thermal Decomposition of Enargite

Enargite can be decomposed through roasting methods. The mechanisms for decomposition of enargite are known. Padilla [15, 16, 17]) showed that enargite decomposition starts at about 525°C, but maximum arsenic removal occurred between 700°C and 750°C. Enargite starts to thermally decompose at 525°C according to reaction (1):

$$2\ Cu_3AsS_4\ (s) \rightarrow 3\ Cu_2S\ (s) + As_2S_3\ (g) + S_2\ (g) \qquad (1)$$

The reaction is very endothermic with an enthalpy of +327.4 kJ/mol.

Enargite started to decompose at about 550°C in a nitrogen atmosphere by reactions (2) and (3).

$$(Cu_3AsS_4)_4\ (s) \rightarrow Cu_{12}As_4S_{13}\ (s) + 3/2\ S_2\ (g) \qquad (2)$$

$$Cu_{12}As_4S_{13}\ (s) \rightarrow 6Cu_2S_{1+x}\ (s) + As_4S_4\ (g) + (1.5\text{-}3x)\ S_2\ (g) \qquad (3)$$

Enargite transforms to tennantite and digenite followed by the formation of chalcocite as the final decomposition product in the temperature range 575°C - 900°C. The arsenic was removed completely as arsenic sulfide species in neutral atmospheres at temperatures below 727°C.

Controlled Oxidative Roasting of Enargite

In controlled oxidative atmospheres, enargite decomposes to chalcocite, sulfur dioxide and arsenic trioxide according to Yoshimura [18] by reaction (4).

$$4\ Cu_3AsS_4\ (s) + 13\ O_2\ (g) \rightarrow 6\ Cu_2S\ (s) + 10\ SO_2\ (g) + 2\ As_2O_3\ (g) \qquad (4)$$

Another possible reaction (5) is [19]:

$$2\ Cu_3AsS_4\ (s) + 2\ O_2\ (g) \rightarrow 3\ Cu_2S\ (s) + 2\ SO_2\ (g) + As_2S_3\ (g) \quad (5)$$

Comparison between enargite roasting in neutral and oxidative atmospheres revealed that the thermal decomposition in oxidative atmospheres is much faster than in neutral atmospheres. Temperature and oxygen potential play an important role in the rate of oxidative roasting of enargite in the temperature range 600°C – 900°C.

Oxidative Roasting of Enargite

The intermediate cuprous oxide in reaction (6) is followed by the formation of cupric oxide at extended reaction times by reaction (7).

$$4\ Cu_3AsS_4\ (s) + 22\ O_2\ (g) \rightarrow 6\ Cu_2O\ (s) + 16\ SO_2\ (g) + As_4O_6\ (g) \quad (6)$$

$$4\ Cu_3AsS_4\ (s) + 25\ O_2\ (g) \rightarrow 12\ CuO\ (s) + 16\ SO_2\ (g) + As_4O_6\ (g) \quad (7)$$

Enargite oxidative roasting showed a transformation to an intermediate Cu_2O and then final CuO reaction product with the complete removal of arsenic in the temperature range 500°C - 900°C.

Lime Roasting

Lime roasting was investigated by Terry [20, 21]. The arsenic and sulfur can be captured during roasting, where As_2O_3 and SO_2 gases react with CaO. Rapid oxidation of As_2S_3 gas upon contacting with CaO was estimated to be the mechanism of arsenic fixation according to reaction (8).

$$As_2S_3\ (g) + 3\ CaO\ (s) \rightarrow 1/2\ As_4O_6\ (g) + 3\ CaS\ (s) \quad (8)$$

The reaction of As_4O_6 gas with CaO results in the formation of calcium arsenite as shown in reaction (9). This reaction happened in the absence of oxygen.

$$3\ CaO\ (s) + \tfrac{1}{2}\ As_4O_6\ (g) \rightarrow Ca_3(AsO_3)_2\ (s) \quad (9)$$

Liu [22] performed research regarding lime roasting on a refractory gold concentrate. The SO_2 and As_2O_3 gases were fixed based on reactions (10) and (11).

$$2\ CaO\ (s) + 2\ SO_2\ (g) + O_2\ (g) \rightarrow 2\ CaSO_4\ (s) \quad (10)$$

$$3\ CaO\ (s) + As_2O_3\ (g) + O_2\ (g) \rightarrow Ca_3(AsO_4)_2\ (s) \quad (11)$$

The free energy estimation for the reactions showed $Ca_3(AsO_4)_2$ is thermodynamically more stable than $CaSO_4$. Based on this, theoretically, CaO will preferentially react with As_2O_3 gas rather than SO_2. The application of hydrated lime can help in arsenic fixation. Taylor [23, 24] conducted test work on lime roasting of refractory precious metal ores. This study used hydrated lime to react with arsenic and sulfur under typical roasting conditions. The results showed more than 99.9% of the arsenic and 98% of the sulfur were retained in the calcine when roasting.

The problem with lime roasting for industrial application is that the arsenic reacted with lime stays in the solid calcine and is very difficult to remove. One possible approach would be to do this reaction in another reactor where the arsenic off gas is contacted with lime at elevated temperature. The question of the acceptability of this product for disposal would have to be addressed.

Soda Ash Roasting

Taylor [25] studied soda ash roasting of As-Sb precious metal concentrates. The research focused on treating complex sulfide concentrate i.e. stibnite and arsenopyrite with soda ash to recover antimony and precious metals while changing arsenic and sulfur to sodium arsenate and sodium sulfate. The result showed the solid residue from roasting maintained nearly all the sulfur, arsenic, and antimony. The arsenic was removed by leaching using heated caustic solution. The arsenic can then be precipitated using a mixture of calcium hydroxide and calcium chloride or ferric salts.

Putra and Taylor evaluated soda ash roasting on two enargite concentrate samples [26]. They looked at partial and complete oxidation. Conditions were found to reduce the arsenic content of the concentrate, as a water soluble compound, while maintaining much of the sulfur in the roast calcine. Controlled oxidation, soda ash roasting may be a method to eliminate the need for extensive arsenic collection and conversion in the off gas and acid manufacturing. The major obstacle is the cost of the soda ash for arsenic capture and ferric sulfate for arsenic precipitation. Research should be performed to evaluate reagent regeneration using methods such as ferric sulfate production for hematite ores and salt splitting technology for acid and base regeneration using methods such as electro-dialysis.

El Indio [6]

The El Indio operation in Chile processed enargite concentrates (23 wt.% Cu, 15 wt.% Fe, 35 wt.% S, 10 wt.% As and 0.8 wt.% Sb). The concentrates were treated in a 14-hearth 6.5 m diameter Nicholas-Herreschoff roaster in the temperature range of 500°C - 700°C with an oxygen content of less than 0.5%. This made the process essentially a neutral roasting. The El Indio roaster flow sheet is shown in Figure 1.

The residence time of the feed was about 3 hours. The outlet gases were captured in two cyclones, where the collected dust was fed back to the roaster and the un-oxidized gases (sulfur and arsenic trisulfide) were oxidized in a combustion chamber with excess air through an exothermic reaction at a temperature of 750 ± 25°C. Then the oxidized gases were cooled down to 350°C - 400°C in a heat exchanger and directed to an electrostatic precipitator. The dust, which contained high precious metal values, was mixed with flotation concentrates. The dust free gases were further quenched down to 120 ± 5°C with cold air in which a major part of the arsenic was condensed and removed from the gas phase as arsenic trioxide (white powder). The final gas stream was released to the atmosphere.

Sulfation Roast

B. Chambers [27] evaluated sulfation roasting of enargite. Calcines were produced at two temperatures, 500 and 750 C. He indicated that while high copper extraction and arsenic fixation rates could be achieved, the sulfation roasting reactions are highly exothermic and significant cooling water would need to be added. The proposed flow sheet from Chambers in shown in Figure 2.

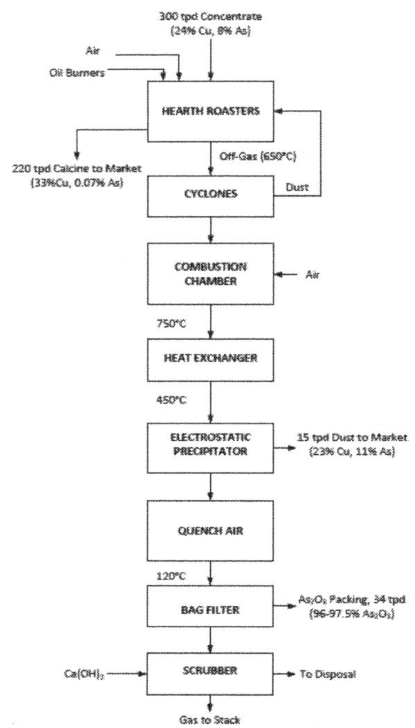

Figure 1. Schematic diagram of the El Indio flow sheet [6]

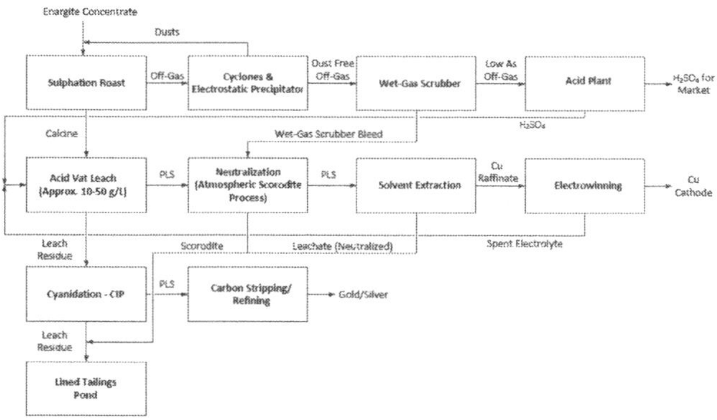

Figure 2. Sulfation Roast Proposed Flow Sheet [27].

Acid Bake Process

Safarzadeh et al. [28] evaluated acid bake and leach of enargite. Preliminary experiments showed that the acid-bake leach process could be adapted and used for the treatment of enargite concentrates. Low-temperature (100-400 °C) sulfuric acid baking of an enargite concentrate followed by water-leaching was evaluated. Results indicate the thermal transformation of enargite to copper sulfate during sulfuric acid baking, with most of the arsenic being released from the concentrate into the gas phase for baking at 400 °C. The condition for maximum copper extraction (~100%) from the concentrate was found to be 400 °C, sulfuric acid amount of 0.5 mL/g of concentrate, and 2 h baking time. Part of the arsenic was captured in the form of arsenopyrite (FeAsS). XRD analysis proved the presence of some arsenic in the form of claudetite (As_2O_3) in the baked concentrate.

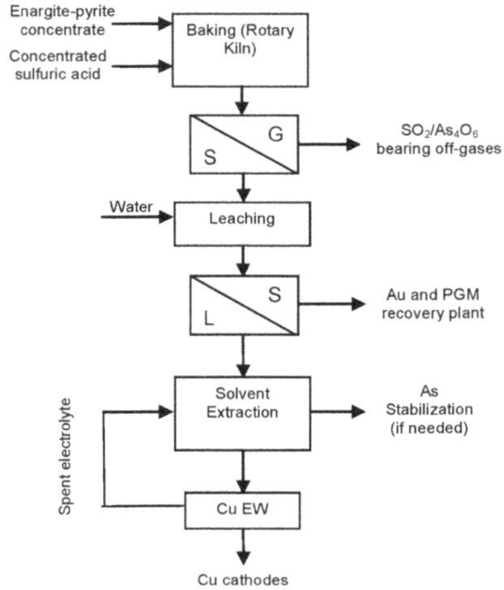

Figure 3 – Proposed acid bake flow sheet [28]

CMI NESA Process

Oudenne [29] reviewed thirty years of effort by CMI NESA in the treatment of arsenical and antimonical bearing concentrates and other complex metallurgical byproducts. The importance of preliminary mineralogical characterization is emphasized as a preliminary step for the success of the roasting. Stage selective roasting appears may be an attractive and efficient route for the treatment of complex concentrates. They found that a reducing atmosphere produces the greatest arsenic removal while treating copper bearing materials containing tennantite, enargite, chalcopyrite and pyrite. In the selective roasting process, the complex sulfide materials are heated in reducing or starved air conditions. Depending on the specific characteristics of the concentrates, such as particle size, different controllable variables appear to influence the volatilization of Cd, As, Sb, Bi and Hg during roasting. These variables include ultimate temperature, heating rate, gaseous atmosphere and presence of sulfur.

Selective roasting is carried out under the conditions explain above to facilitate impurities removal i.e. adequate ultimate temperature, low and smooth heating rate and reducing atmosphere with optimized CO/CO_2 ratio. A roasting furnace operated under a single given atmosphere such as in a fluidized bed roaster, makes it difficult to obtain appropriate roasting conditions for complex sulfide materials. Since the volatile sulfides are unaffected by a reducing atmosphere, it is thus possible to effect an elimination of them from nonvolatile minerals. Moreover, in a multiple stage reactor, the separate control of temperature and atmosphere is possible for each hearth of the reactor and equally important above and within the material bed.

Enargite selective roasting. The purpose of the pretreatment by selective roasting of enargite concentrate is the removal of a maximum of the contained arsenic and antimony to obtain a saleable product, while retaining a high sulfur level for smelting. The copper flotation concentrate contains enargite (Cu_3AsS_4), covellite (CuS), chalcopyrite ($CuFeS_2$), pyrite (FeS_2), elemental sulfur and quartz (SiO_2) as major minerals. The average size of the flotation concentrate is in the range of 40 microns.

In a reducing atmosphere, the main sulfides minerals will decompose according to reactions (12), (13) and (14).

$$2\,Cu_3AsS_4 = Cu_2S + 4\,CuS + As_2S_3\,(g) \quad (12)$$

$$2\,FeS_2 = 2\,FeS + S_2\,(g) \quad (13)$$

$$8\,FeAsS = 4\,FeAs + 4\,FeS + As_4S_4\,(g) \quad (`14)$$

These decompositions proceed as the concentrate is progressively heated in a controlled atmosphere. Moreover as the elimination of volatile sulfides occurs, one observes that the sintering temperature of the material being roasted is increasing. The oxygen-free atmosphere in and just above the reacting bed is protected because oxygen is consumed by reacting with rising volatile sulfides and labile sulfur arising from the lower hearths. Part of the heat used in the multistage reactor is generated by controlling air flow into the upper hearths taking advantage of the exothermic reactions with the rising gases:

$$S_2(g) + 2\,O_2(g) = 2\,SO_2\,(g) \quad (15)$$

$$As_4S_4(g) + 7\,O_2\,(g) = As_2O_3(g) + 4\,SO_2(g) \quad (16)$$

In the bottom hearths, external fuel is burnt in order to control hearth temperatures and gas atmosphere. The presence of pyrite in the concentrate is favorable because it helps the volatilization of the arsenic sulfide.

Table III Data from the selective Roasting of an enargite concentrate [29]

% Weight	%As	%Sb	%Cu	%S tot	%Fe	g/t Au	g/t Ag
Concentrate							
Pilot Plant	10.8	0.4	28.1	33.8	13.5	32	395
Industrial operation	8.5	0.6	25.6	30.1	13.0	159	561
Selective calcine							
Pilot Plant	0.08	0.2	24.6	24.6	19.6	43	608
Industrial operation	0.09	0.11	34.7	20.9	19.8	194	789

Ministro Hales

The treatment of the Ministro Hales deposit in Chile by Codelco [30,31] is expected to start in 2013 with a yearly throughput of 550,000 tonnes of copper concentrate and 250,000 tonnes of sulfuric acid production. They will utilize controlled oxidation (or neutral roast) to remove arsenic as a vapor compound. Arsenic dust will be collected and converted to a disposable form or stockpiled. Outotec in Frankfurt, Germany operated a 25 kg per hour pilot plant that uses the roasting method to treat copper concentrate [31].

Smelters

The ASARCO smelter at El Paso, Texas allowed a maximum of 0.2% As in the copper concentrate in its feed [32]. Enargite concentrates are smelted directly by a few copper smelters but most smelters limit their arsenic input for environmental reasons and only process clean copper concentrates, containing less than about 0.5% As [6]. The Lepanto roaster at Lepanto Consolidated Mining Company, Philippines treated about 180 tpd of enargite concentrates containing 11% As, 31% Cu, 15% Fe, 34% S and produced calcine containing 0.3% As, 43% Cu, 23% Fe, and 20% S. The Boliden roaster at Boliden Oy, Finland, with a capacity of up to 800 tpd concentrate, treated concentrates containing 1-2% As and typically produced a calcine containing about 0.1% As [6]. Dundee Precious Metals Inc. in Chelopec, Bulgaria transports enargite concentrates (175,000 t/a) to the NCS smelter in Tsumeb, Namibia. This smelter can treat high arsenic and lead bearing copper concentrate. Not many smelters in the world can do that. Copper blister and arsenic trioxide (As_2O_3) were produced from this smelter [33].

Smelter Arsenic

As_4O_6 or As_2S_3 vapors in the gas phase may be present in some streams when extracting the copper. Electrostatic precipitators or wet gas scrubbers are used to collect As_2O_3 dusts. Arsenic trioxide dust cannot be stockpiled because of the long term problems associated with the high solubility in water and toxicity. Two example flow sheets for the treatment of arsenic containing dusts and weak acid streams from El Teniente are presented in Figures 4 and 5 [34].

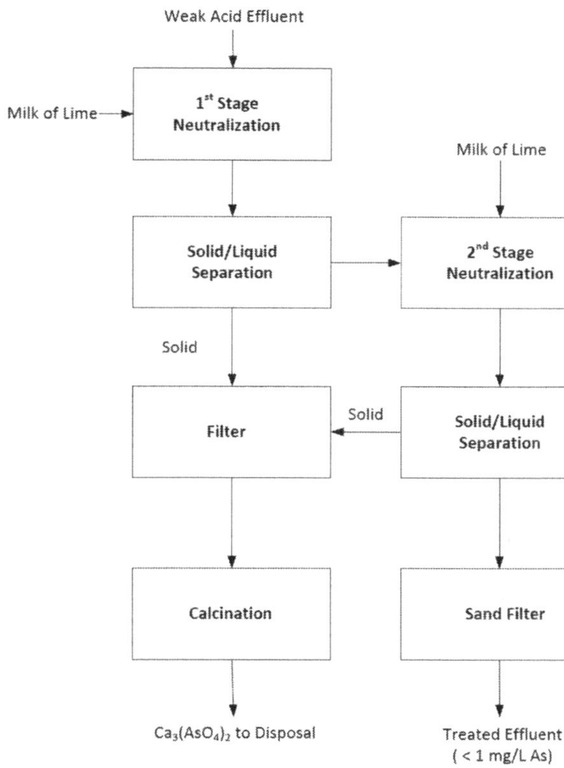

Figure 4. El Teniente plant for treatment of effluents from sulfuric acid plants [34]

Ferrihydrate Precipitation

Co-precipitation removal of arsenic from solutions containing both dissolved iron and arsenic species by ferrihydrite precipitation is practiced [35]. Post-precipitation is the removal of arsenic from solutions by the addition of a solution of aqueous ferric bearing species at a pH where ferrihydrite forms (usually at pH 7 to 8). Adsorption is removal of arsenic from solutions by exposing arsenic bearing solutions to a previously precipitated ferrihydrite solid. The precipitation depends on time, temperature, pH, Fe/As mole ratio, agitation rate, arsenic valence state, associated ion, etc.

Ferric Precipitation

The precipitation of dissolved arsenic with ferric ions has been declared by The United States Environmental Protection Agency (EPA) as "The Best Demonstrated Available Technology"

(BDAT) [36]. Several factors that have influences on the stability of any disposed arsenic compound are characteristics of the disposal location, crystallinity and size distribution of the compound, and the presence of oxygen or any complexing agents.

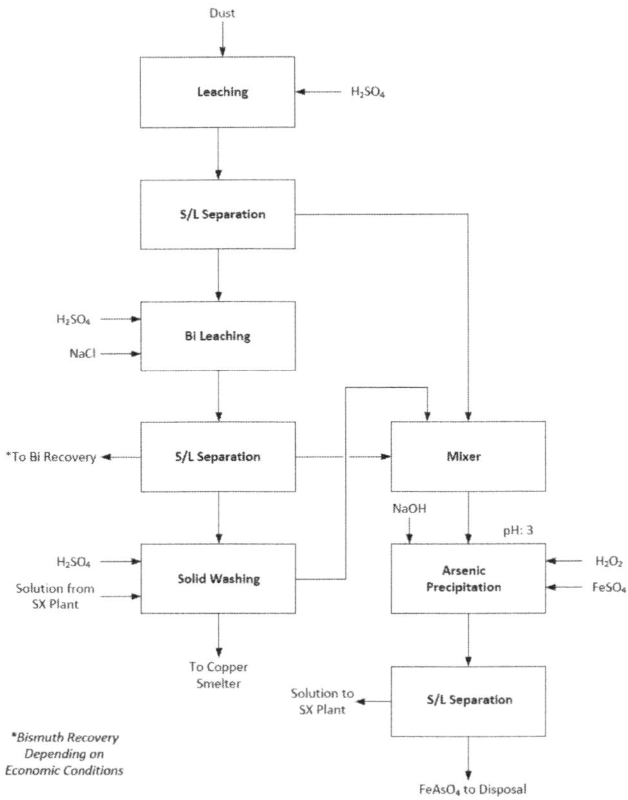

Figure 5. El Teniente plant for treatment of smelter dusts [34]

Conclusions

Smelters generally accept copper concentrates with low arsenic levels (less than 0.2% to 0.5%) due to environmental and product purity concerns. Treating the waste stream from high arsenic copper concentrates requires significant investment. The separation and collection of arsenic as arsenic trioxide is problematic, primarily due to its toxicity and lack of market value.

Pyrometallurgical processing has several concerns: strict environmental regulations on arsenic release, complexity of gas/dust capture, separation facilities and the stabilization of the final arsenic compound. Recent application in the industry to treat high arsenic concentrate (enargite) has occurred at Barrick's El Indio smelter in Chile and CENA in Belgium. Both employed reductive roasting-smelting process for the treatment of enargite concentrates.

Despite the perceived shortcomings for roasting to selectively volatilize arsenic, the process could an option under specific conditions, provided that arsenic precipitation is possible. Roasting is the only process that has been operated commercially for enargite concentrates; therefore the level of project risk is lower, i.e. based on past commercial experience. Another advantage of roasting is it is less sensitive to feed mineralogy than hydrometallurgical options and arsenic disposal as a compact crystalline compound with low solubility would appear possible.

References

1. W.G. Davenport, M. King, M. Schlesinger, A.K. Biswas, *Extractive metallurgy of copper 4th edition* (New York: Pergamon Press, 2002), 1-2.
2. J.G. Peacey, M.Z. Gupta and K.J.R. Ford "Review of process options to treat enargite concentrates." *Proceedings of Copper 2010, Vol. 3, Pyrometallurgy II* (Hamburg, GDMB, 2010), 1035-1050.
3. Anonymous, personal communication, 2013.
4. H. Fabian, H.W. Richardson, F Habashi, and R. Besold, "Chapter 8, Copper." *Handbook of Extractive Metallurgy*, (F. Habashi, Ed.), Wiley-VCH, (1997) pp. 491-579.
5. B.S. Terry and M. and Sánchez, "Environmentally friendly processing of copper minerals bearing arsenic and/or sulfur by pyrometallurgical route, *Proceedings of Copper 95 International Conference*, vol. II, (A. Casali, G.S. Dobby, C. Molina and W.J. Thoburn, Eds.), Santiago, Chile, (1995), pp. 337-345.
6. E.H. Smith and E. Paredes, " How St. Joe Gold's El Indio mine has become a major producer of high quality crude arsenic trioxide, *Arsenic Metallurgy: Fundamentals and Applications*, (R.G. Reddy, J.L. Hendrix, and P.B. Queneau, Eds.), Warrendale, PA: The Metallurgical Society, (1988), pp. 145-160.
7. U.S. Geological Survey, *Mineral Commodity Summaries*, January (2013), Arsenic, pp. 20-21.
8. R. Padilla, Y. Fan, M. Sanchez, and I. Wilkomirsky, I., 1997, "Arsenic volatilization from enargite concentrate." *EPD Congress 1997*, (B. Mishra, Ed.), Warrendale, PA: The Minerals, Metals and Materials Society, (1997) pp. 73-83.
9. R. McElroy, T. Lipiec, and M. Tomlinson, M., 2008, "Roasting – the neglected option." *Hydrometallurgy 2008*, Proceedings of the SME Sixth International Symposium, (2008) pp. 425-430.
10. N.L. Piret, and A.E. Melin, "An engineering approach to the arsenic problem in the extraction of non-ferrous metals", *Productivity and Technology in the Metallurgical Industries*, (M. Koch and J.C. Taylor, Eds.), TMS, Warrendale, PA, USA, (1989), pp. 735-814
11. K. Baxter, H. Scriba, and I. Vega, "Treatment of high-arsenic copper-gold concentrates- An options review", *Copper 2010*, Hamburg, Germany, vol. 5, (2010), pp. 1783-1802.
12. G. Lindkvist and A. Holmström, "Roasting of complex concentrates with high arsenic content", *Advances in Sulfide Smelting, Technology and Practice*, (H.Y. Sohn, D.B. George, and A.D. Zunkel, Eds.), Warrendale, PA: The Metallurgical Society, vol. II, (1983), pp. 451-472.

13. M.S. Safarzadeh, M.S., Moats, M.S.and J.D. Miller, "Recent trends in the processing of enargite concentrates." *Mineral Processing and Extractive Metallurgy Review*, DOI: 10.1080/08827508.2012.723651. (2012a).
14. M.S. Safarzadeh, M.S., Moats and J.D. Miller, J.D., 2012b, "An update to recent trends in the processing of enargite concentrates." *Mineral Processing and Extractive Metallurgy Review*: DOI: 10.1080/08827508.2012.725683 (2012b)
15. R. Padilla, Y. Fan, and I. Wilkomirsky, I., Thermal decomposition of enargite." *EPD Congress 1999*, (B. Mishra, Ed.), Warrendale, PA: The Minerals, Metals and Materials Society, (1999) pp. 341-351.
16. R. Padilla, Y. Fan, Y. and I. Wilkomirsky, "Decomposition of enargite in nitrogen atmosphere." *Canadian Metallurgical Quarterly*, 40, (2001) pp. 335-342.
17. R. Padilla, A. Aracena and M.C. Ruiz, "Reaction mechanism and kinetics of enargite oxidation at roasting temperatures." *Metallurgical and Materials Transactions B*, 43B, (2012) pp. 1119-1126.
18. Yoshimura, Z., 1962, "The fundamental investigation of desarsenifying roasting of copper concentrate and its industrial practice." Journal of the Mining and Metallurgical Institute of Japan, 78, pp. 447-453
19. A.C. Secco, G.A. Riveros and A.A. Luraschi, "Thermal decomposition of enargite and phase relations in the system copper-arsenic-sulfur",*Copper 87*, Pyrometallurgy of Copper, (C. Diaz, C. Landolt, and A. Luraschi, Eds.), Santiago, Chile: Universidad de Chile, vol. 4, (1988), pp. 225-238.
20. A. Ulloa, M. Sánchez and B.S. Terry, "Lime roasting of enargite" *IMM Pyrometallurgy 95*, Cambridge, UK, (1995), pp. 275-283.
21. B.S. Terry, M.A. Sánchez and A.G. Ulloa, "Calcium oxide as a reagent for the capture of arsenic emissions during the roasting of enargite," *Extraction and Processing for the Treatment and Minimization of Wastes*, (J.P. Hager, B.J. Hansen, W.P. Imrie, J.F. Pusateri, and V. Ramachandran, Eds.), Warrendale, PA: TMS, (1994), pp. 201-215.
22. J. Liu, R. Chi, R., Z. Zeng, J. Liang and Z. Xu, Z.,"Selective arsenic-fixing roast of refractory gold concentrate." *Metallurgical and Materials Transactions B*, 31B, (2000) pp. 1163 -1168.
23. P.R. Taylor, Z.B. Yin and R.W. Bartlett, R.W., 1990, "Lime roasting of pyrite." *Gold 90*, D. Hausen, Ed., SME, Littleton, CO (1990), pp. 411-419.
24. P.R. Taylor, Z.B. Yin, S.L. Bell and R.W. Bartlett, R.W., 1991, "Lime roasting of refractory precious metal ores." *EPD Congress 91*, D.R. Gaskell, Ed., TMS, Warrendale, PA, (1991) pp. 725-742.
25. P. R. Taylor and Z.B. Yin, Soda Ash Roasting of Sb-As-Precious Metal Concentrates," Mineral and Metallurgical Processing, (Nov. 1993) pp. 163-169.
26. T.A.R. Putra "Arsenic Removal from Enargite with Sodium Carbonate Using Complete and partial Oxidation," Ph.D. Thesis, Colorado School of Mines, Golden, CO (2013).
27. B.T. Chambers, "An investigation into the sulfation roasting of enargite concentrates." MSc. Thesis, Queen's University, Canada (2012).
28. M. S. Safarzadeh, M. S. Moats, J. D. Miller (2012). "Acid bake-leach process for the treatment of enargite concentrates", *Hydrometallurgy*, **119–120**, 30–39.

29. P. D. Oudenne , "The Stage- Selective Roasting, An Efficient way for the Treatment of Arsenical and Antimonical Bearing Concentrates and Other Complex Metallurgical By Products", *Sohn International Symposium,* Edited by F. Kongoli and R.G. Reddy TMS, Warrendale PA (2006) pp. 519-527.
30. R. McElroy, T. Lipiec and M. Tomlinson, M., "Roasting – the neglected option." *Hydrometallurgy 2008*, Proceedings of the SME Sixth International Symposium,(2008) pp. 425-430.
31. G. Ondrey, "A new roasting process slated for commercialization." *Chemical Engineering*, March (2012) pp. 9.
32. Malcolm Pirnie, Inc., 2010, "Review of ASARCO El Paso smelting processes." Texas Custodial Trust, El Paso, TX.
33. Dundee Precious Metals Inc. website, 2012, http://www.dundeeprecious.com/files/annual_report/122195%20DUNDEE%20AR%2 Complete_final_v001_h3aq6j.pdf.
34. A. Valenzuela, E. Balladares and D. M. Cordero, " Arsenic management in the Metallurgical Industry: The Chilean Experience", *Sohn International Symposium* Vol 9, Edited by F. Kongoli and R.G. Reddy, TMS, (2006) pp. 408-422
35. C.G. Anderson and L.G. Twidwell, "The control of iron and arsenic in the treatment of gold bearing enargite concentrates", *Iron ControlTechnologies*, (J.E. Dutrizac and P.A. Riveros, Eds.), Montreal, Canada, (2006), pp. 943-956.
36. P.A. Riveros, J.E. Dutrizac.and P. Spencer, P., 2001, "Arsenic disposal practices in the metallurgical industry." *Canadian Metallurgical Quarterly*, 40, (2001) pp. 395-420.

Celebrating the Megascale: Proceedings of the Extraction and Processing Division
Symposium on Pyrometallurgy in Honor of David G.C. Robertson
Edited by: Phillip J. Mackey, Eric J. Grimsey, Rodney T. Jones, and Geoffrey A. Brooks
TMS (The Minerals, Metals & Materials Society), 2014

Arsenic and Antimony Capacities in Ni-Cu Mattes and Slags

Ramana G. Reddy[1*] and Jonkion M. Font[2]

[1]ACIPCO Endowed Professor,
[1]Department of Metallurgical and Materials Engineering,
The University of Alabama, Tuscaloosa, Al 35487-0202, USA.
[2]Formerly, Research Fellow,
Department of Metallurgical and Materials Engineering,
The University of Alabama, Tuscaloosa, Al 35487-0202, USA.
[2]At present, Senior Engineer and Head of the Process Engineering,
Chuquicamata smelter, Chuquicamata Division, CODELCO-Chile

Abstract

A thermodynamic model for *a priori* predictions of arsenic and antimony impurity capacity in Ni-Cu mattes and slags was developed based on the Reddy-Blander (RB) capacity model. The As and Sb capacities were calculated *a-priori* considering the RB capacity model in Ni-Cu mattes and FeO-$FeO_{1.5}$-MgO-$CuO_{0.5}$-NiO-SiO_2 slag at 1573 K and 0.1 to 1 atm partial pressure of SO_2. The results showed an excellent agreement between the model calculated and experimental data of arsenic and antimony capacities between the mattes and slag. The impurity capacity model developed here, can be extended for prediction of impurity capacities in multi-component base metal slags and base metal mattes. Such *a priori* predictions of impurity capacities can lead to develop or improve the efficiency of impurity removal from the base metal smelting, converting and refining processes.

Keywords: Arsenic, Antimony, Impurity Capacity, Ni-Cu mattes, Slags

Introduction

The arsenic and antimony are the two major impurities in nickel and copper ores that are difficult to remove during the smelting and refining processes. Elimination of these unwanted impurities in the matte smelting process step is responsible for the high cost of the refining process and the detrimental effect on the environment. Behavior of impurities between mattes and slags by considering thermodynamic and operational approaches was studied [1-13]. The high oxidative smelting/refining conditions are employed to enhance their removal from the metal/matte into the slag and gas phases. The information on the *a priori* prediction of these impurity behavior in the slag is very important for controlling the matte smelting processes. However, such available information is limited to specific process conditions such as temperature, P_{SO2} and composition of matte and slag. The Reddy-Blander (RB) model is a thermodynamic model which was successfully applied for *a priori* prediction of sulfur in metal and slag systems [14-17] and impurities capacities (i.e. As, Sb and Bi) in several non-ferrous metals/mattes and slags systems [11-13, 17].

[1*] Corresponding Author, E-mail: rreddy@eng.ua.edu

The Reddy-Blander (RB) model, *a priori*, is based on a simple solution model and on knowledge of the chemical and solution properties of impurities and oxides. It was developed to calculate the impurities capacities (C_R). It is basically a measure of the ability of an oxide system (slag) to hold an impurity. This *a priori* model is very useful in predicting the sulfide capacities of various types of slags since it utilizes all the chemical and solution properties of sulfides and oxides. Sulfide capacities of several binary silicate [14-16, 18], aluminate [19], titanate [20], multi-component silicate [19, 21, 22, 1] and some industrial slags [23, 24] were predicted using the RB model. The RB model was also applied for sulfate [22, 1], arsenate [11, 12] and antimonite [13] capacities in slags, molten metals and mattes, and the results were found in good agreement with the experimental data. However, very limited information on the arsenic and antimony capacities in the Ni-Cu mattes and slag systems is available.

In the present study, an arsenic and antimony capacity model was developed using the RB model and results were compared with the experimental data. The activity of impurity in matte, activity of metal oxide, temperature, partial pressure of SO_2 and compositions of mattes and slags were considered in developing *a priori* thermodynamic model for capacities of arsenic and antimony between the slag and the Ni-Cu mattes. The results showed very good agreement between the model calculated and experimental data for arsenic and antimony capacities.

Theoretical Considerations

Impurity Capacity

The prediction of impurity capacity of impurity R (R=As and Sb) in the slag is essential for evaluating the smelting processes of non-ferrous metals like copper and nickel. However, there is very limited availability of experimental or thermodynamic data for impurities in Ni-Cu matte and slag systems. Therefore, an expression was developed for the correlation between the thermodynamic impurity data and the impurity capacity.

The impurity species RO_4^{3-} (for arsenic AsO_4^{3-} and antimony SbO_4^{3-}) are considered as stable species at high temperatures. The fundamental concepts of the Reddy-Blander model was used in developing the impurity capacity model [1, 11, 13]. The impurity capacity becomes,

$$C_{RO_4^{3-}} = \frac{\left(\text{wt pct } RO_4^{3-}\right)}{a_R\, p_{O_2}^{5/4}} \tag{1}$$

Where, a_R is the activity of impurity (R = As, Sb), and P_{O2} is the partial pressure of oxygen. The impurity equilibrium reaction with MO is given as [1, 11, 13],

$$\frac{3}{2}\text{MO (l)} + \text{R (l)} + \frac{5}{4}\text{O}_2\text{ (g)} = \text{M}_{3/2}\text{RO}_4\text{ (l)} \tag{2}$$

Where, M (= Fe, Mg, Cu, Ni, etc.) is the impurity compound forming element and K_M is the equilibrium constant for the reaction equation (2). By combining the equilibrium constant, K_M, of reaction (2) with equation (1), the impurity capacity is derived for measurable quantities given as,

$$C_{RO_4^{3-}} = \left(\text{wt pct } RO_4^{3-}\right)\frac{K_M\, a_{MO}^{3/2}}{a_{M_{3/2}RO_4}} \tag{3}$$

Where, $a_{M_{3/2}RO_4}$ is the activity of the super cooled liquid $M_{3/2}RO_4$.

For basic melts in the $MO\text{-}SiO_2\text{-}FeO_{1.5}$ system $(0 \le (X_{SiO_2} + X_{FeO_{1.5}} + ...) \le 0.33)$, the impurity capacity is given as [1, 11, 13-15],

$$C_{RO_4^{3-}} = \frac{100\, K_M\, a_{MO}^{3/2}\, W_{RO_4^{3-}}\, [1 - 2\,(X_{SiO_2} + X_{FeO_{1.5}} + ...)]}{\gamma_{M_{3/2}RO_4}\, \overline{W}} \tag{4}$$

Here, $\gamma_{M_{3/2}RO_4}$ is the Henrian activity coefficient of $M_{3/2}RO_4$, $W_{RO_4^{3-}}$ is the molecular weight of RO_4^{3-}, $\overline{W}$ $(= W_{MO}X_{MO} + W_{SiO_2}X_{SiO_2} + W_{FeO_{1.5}}X_{FeO_{1.5}} + ...)$ is the average molecular weight of the solution.

For acidic melts in the $MO\text{-}SiO_2\text{-}FeO_{1.5}$ system $(0.33 \le (X_{SiO_2} + X_{FeO_{1.5}} + ...) < 1)$, the impurity capacity is given as [1, 11, 13-15],

$$C_{RO_4^{3-}} = \frac{100\, K_M\, a_{MO}^{3/2}\, W_{RO_4^{3-}}\, (X_{SiO_2} + X_{FeO_{1.5}} + ...)}{\overline{W}}\, e^{\frac{5}{4}\left(\frac{1}{m} - 1\right)} \tag{5}$$

Where, m is the average polymer chain length expressed as [14, 15],

$$\frac{1}{m} = (1 - a_{MO})\left[\frac{1}{(X_{SiO_2} + X_{FeO_{1.5}} + ...)} - 2\right] \tag{6}$$

The impurity capacity for the $FeO\text{-}MgO\text{-}SiO_2\text{-}FeO_{1.5}$ multi-component system at a constant $(SiO_2\text{+}FeO_{1.5})$ mole fraction is given as [1, 19, 21],

$$\log C_{RO_4^{3-},(Fe,Mg)O} = N_{FeO} \log C_{RO_4^{3-},FeO} + N_{MgO} \log C_{RO_4^{3-},MgO} + \ldots \tag{7}$$

The N_{FeO} and N_{MgO} are the electrical equivalent cationic fractions ($N_{FeO} = \frac{X_{FeO}}{X_{FeO} + X_{MgO}}$ and $N_{MgO} = \frac{X_{MgO}}{X_{FeO} + X_{MgO}}$), $C_{RO_4^{3-},FeO}$ and $C_{RO_4^{3-},MgO}$ are the impurity capacities in the $FeO\text{-}SiO_2$ and $MgO\text{-}SiO_2$ binary systems at constant $(SiO_2 + FeO_{1.5} + \ldots)$ mole fraction, and $C_{RO_4^{3-},(Fe,Mg)O}$ is the impurity capacity of the $FeO\text{-}MgO\text{-}SiO_2\text{-}FeO_{1.5}$ system. Hence, for the entire composition range $(0 \le X_{SiO_2} + X_{FeO_{1.5}} + ... < 1)$ and given temperature, the impurity capacity of binary and multi-components slag systems can be predicted by using equations (4) to (7).

For the $MO\text{-}SiO_2$ binary system, equilibrium constants for reaction (2) for arsenic and antimony at 1573 K are presented in Table 1 [11, 12, 25, 26]. The $\Delta G°$ for the reaction involving Fe, the values of $\Delta H°$ and $\Delta S°$ for As(l) and $Fe_{3/2}AsO_4$(s) were extrapolated from

1200 K and 811 K, respectively [11, 12]. The ΔG° for the liquid $Fe_{3/2}AsO_4$(l) was estimated from the experimental arsenic distribution data at 1573 K. The a_{FeO} in FeO-SiO_2 binary system was calculated at 1573 K. The $\Delta G°$ for the reaction of $Ca_{3/2}AsO_4$(s) and $Cu_{3/2}AsO_4$ in equation (2) are taken from the available data [11], and the $\Delta G°$ for Mg, the values of $\Delta H°$ and $\Delta S°$ for As (l) and $Mg_{3/2}AsO_4$(s) [25] were extrapolated from 1200 K and 1225 K, respectively. Due to the lack of availability of liquids data, the solid $Ca_{3/2}AsO_4$, $Mg_{3/2}AsO_4$, $Cu_{3/2}AsO_4$, $Ca_{3/2}SbO_4$, $Mg_{3/2}SbO_4$, $Ni_{3/2}SbO_4$ and $Cu_{3/2}SbO_4$ data were used in calculating the Gibbs energy of the reactions (2). It is expected that the use of liquids data will increase in some extent the arsenic and antimony capacities of these systems.

Table 1: Equilibrium constants for arsenic and antimony forming reactions at 1573K

	Log K_M (Reaction 2)	
M(1)	Arsenic	Antimony
Fe(2)	11.0	10.4
Ca	10.3	-
Mg	10.2	9.6
Cu	-4.7	-6.0
Ni(2)	3.4	3.1

(1) Solid state (2) Estimated for liquid

Results and Discussion

The experimental phase equilibrium studies for arsenic and antimony between Ni-Cu mattes and slags at 1573 K were reported [6-9]. The impurity experimental data were reported in the form of distribution coefficients in the slags. An expression was derived between the distribution coefficient and impurity capacity of slags in equilibrium with mattes, and from the equation, experimental data are derived. Development of such expression is given below.

Expression between Distribution Coefficient (L_R) and Impurity Capacity ($C_{RO_4^{3-}}$)

The impurity distribution coefficient between slag and matte phases is defined as [11, 13],

$$L_R = \frac{(\text{wt pct R in slag})}{\{\text{wt pct R in matte}\}} \tag{8}$$

The weight pct of R in matte is expressed as,

$$\{\text{weight pct of R in matte}\} = \frac{a_{R(l)}\, W_R \{n_T\}}{\gamma_R} \tag{9}$$

Where $a_{R(l)}$ is activity of impurity in matte, γ_R is activity coefficient of impurity in matte and W_R is the molecular weight of R, and $\{n_T\}$ is the total number of moles of matte phase. Combining equations (8), (9) and (1), the relationship between the impurity capacity and the L_R is obtained as,

$$C_{RO_4^{3-}} = \frac{L_R \, W_{RO_4} \{n_T\}}{\gamma_R \, p_{O_2}^{5/4}} \quad (10)$$

The experimental impurity capacities were derived using equation (10) for each of the experimental L_R at various wt pct Ni-Cu in matte and 1573 K.

Comparison of RB Model Data with Experimental Data

For evaluation of the effect of P_{SO2} or P_{O2} on the impurity capacities, these were calculated for each of the slag composition derived from the reported composition of the experimental data for the Ni+Cu matte in equilibrium with the FeO-$FeO_{1.5}$-MgO-NiO-$CuO_{0.5}$-SiO_2 slag at 1573 K and P_{SO2} of 0.1atm, 0.5 atm and 1atm [6]. The *a priori* RB model calculated impurity capacity and the experimental impurity capacity data are shown in Figures 1 and 2. The experimental impurity capacity calculations were performed by considering the reported experimental data of γ_R, $\{n_T\}$ for the Ni+Cu matte, P_{O2}, a_{MO}, and $L_R^{s/m}$ [6-9, 11, 13] in equation (10). As shown in the figures, good agreement was observed between the *a priori* RB model calculated impurity capacity data and experimental reported data for these slag systems (with high acidic components, $X_{SiO_2} + X_{FeO_{1.5}} + \dots$). However, for basic slag systems very limited data were available. Hence, impurity capacity in basic slag melts was calculated using the RB model for slag systems with $X_{SiO_2} = 0$. It is also important to note that these good agreements may be due to the accurate determination of composition of the slag. These include the dissolved quantities of NiO or $CuO_{0.5}$ that affect the composition of X_{MO} in the slag, and the a_{MO} for the *a priori* RB model impurity capacity calculations using equations (4) to (10). This is especially true for high matte grades where the dissolutions of Ni and Cu into the slag are considerably high. Moreover, the small differences between the RB models calculated data and the experimental data may be mainly due to the uncertainties of the experimental $L_R^{s/m}$ values.

Arsenic Capacity

In Figure 1, the *a priori* RB model arsenic capacity calculations data show good agreement with the experimental data. These good agreements for given matte grade and P_{SO2} are ascribable mainly to the oxidation of FeS ($FeS+1.5O_2=FeO+SO_2$) in the matte which affects the P_{O2} and the a_{FeO} values in equation (2). The deviations found at low P_{SO2} in Figures 1 (a) may be due to uncertainties in the experimental L_{As}, and γ_{As} values [6-9, 11, 13] used in equation (10). In addition, it may be also due to the extrapolation in calculating the Gibbs energy using the solid $Ca_{3/2}AsO_4$, $Mg_{3/2}AsO_4$ and $Cu_{3/2}AsO_4$ data [25, 26]. It is expected that the use of liquids data will increase to some extent the arsenic capacities of these systems.

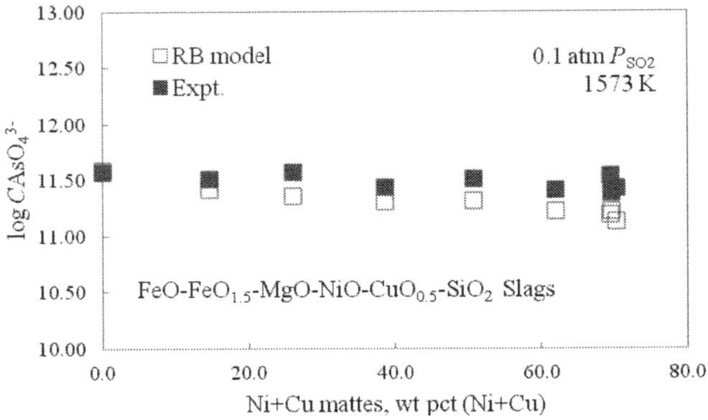
RB model
Expt.
0.1 atm P_{SO2}
1573 K
FeO-FeO$_{1.5}$-MgO-NiO-CuO$_{0.5}$-SiO$_2$ Slags
log $CAsO_4^{3-}$
Ni+Cu mattes, wt pct (Ni+Cu)
13.00
12.50
12.00
11.50
11.00
10.50
10.00
0.0
20.0
40.0
60.0
80.0

(a)

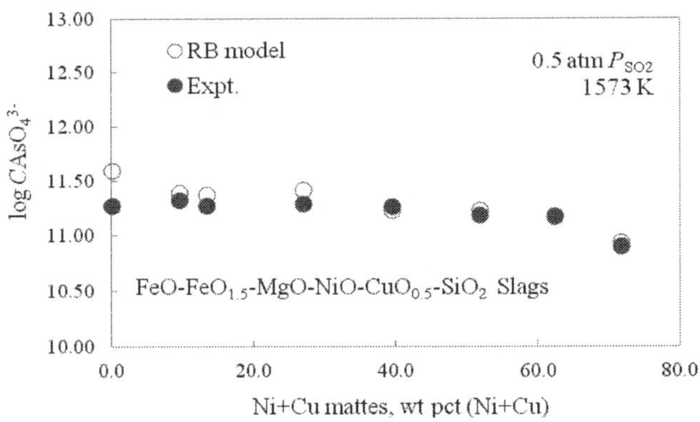
RB model
Expt.
0.5 atm P_{SO2}
1573 K
FeO-FeO$_{1.5}$-MgO-NiO-CuO$_{0.5}$-SiO$_2$ Slags
log $CAsO_4^{3-}$
Ni+Cu mattes, wt pct (Ni+Cu)
13.00
12.50
12.00
11.50
11.00
10.50
10.00
0.0
20.0
40.0
60.0
80.0

(b)

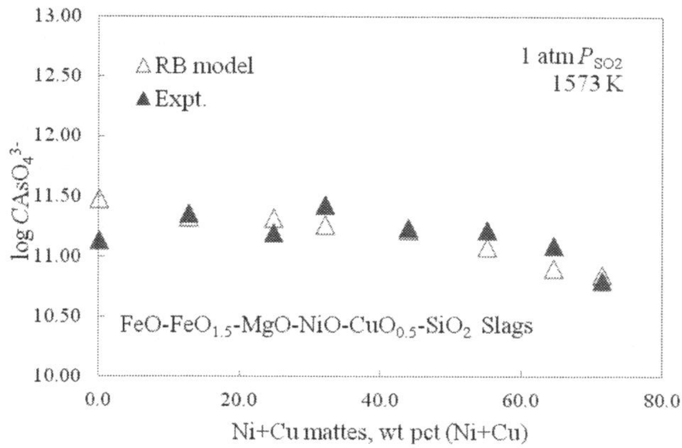

(c)

Figure 1. Arsenic capacity for the $FeO-FeO_{1.5}-MgO-NiO-CuO_{0.5}-SiO_2$ slags and Ni+Cu mattes at 1573 K and P_{SO2} of (a) 0.1atm, (b) 0.5 atm, and (c) 1 atm.

Antimony Capacity

The antimony capacities were calculated for each of the slag composition derived from the reported composition of the experimental data for the Ni+Cu matte in equilibrium with the $FeO-FeO_{1.5}-MgO-NiO-CuO_{0.5}-SiO_2$ slag at 1573 K and P_{SO2} of 0.1atm, 0.5 atm and 1 atm [6]. The calculated RB model antimony capacity and the experimental antimony capacity data are shown in Figure 2. The experimental antimony capacity calculations were done by considering the reported experimental data of γ_{Sb}, $\{n_T\}$ for the Ni+Cu matte, P_{O2}, a_{MO}, and $L_{Sb}^{s/m}$ [6-9, 11, 13] in equation (10). In Figure 2, the *a priori* RB model antimony capacity calculated data show good agreement with the experimental data. The deviations found at low P_{SO2} and at high matte grades in Figure 2 (a) are may be due to the uncertainties of the experimental L_{Sb}, and the γ_{Sb} values used in equation (10). In addition, it may be also due to the extrapolation in calculating the Gibbs energy data using the solid $Ca_{3/2}SbO_4$, $Mg_{3/2}SbO_4$, $Ni_{3/2}SbO_4$ and $Cu_{3/2}SbO_4$ data [25, 26]. It is expected that the use of liquids data will increase in some extent the antimony capacities of these systems.

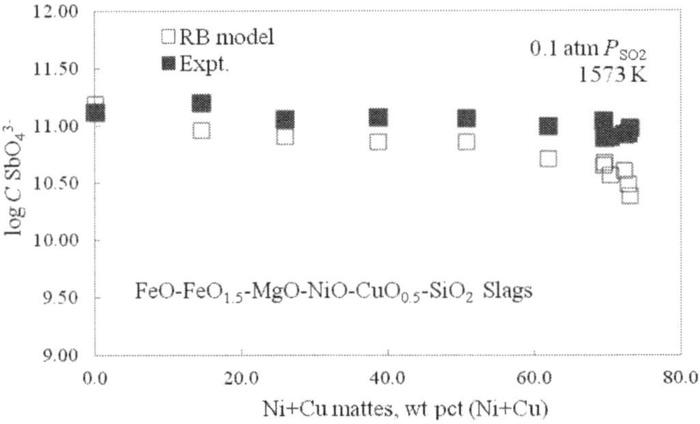

RB model
Expt.
0.1 atm P_{SO2}
1573 K
$\log C\ SbO_4^{3-}$
FeO-FeO$_{1.5}$-MgO-NiO-CuO$_{0.5}$-SiO$_2$ Slags
Ni+Cu mattes, wt pct (Ni+Cu)

(a)

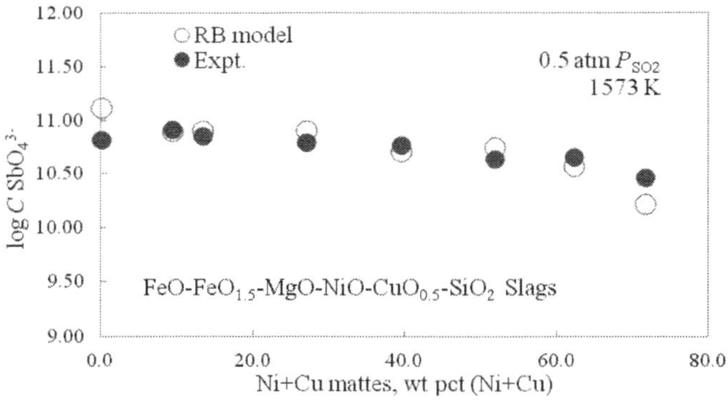

RB model
Expt.
0.5 atm P_{SO2}
1573 K
$\log C\ SbO_4^{3-}$
FeO-FeO$_{1.5}$-MgO-NiO-CuO$_{0.5}$-SiO$_2$ Slags
Ni+Cu mattes, wt pct (Ni+Cu)

(b)

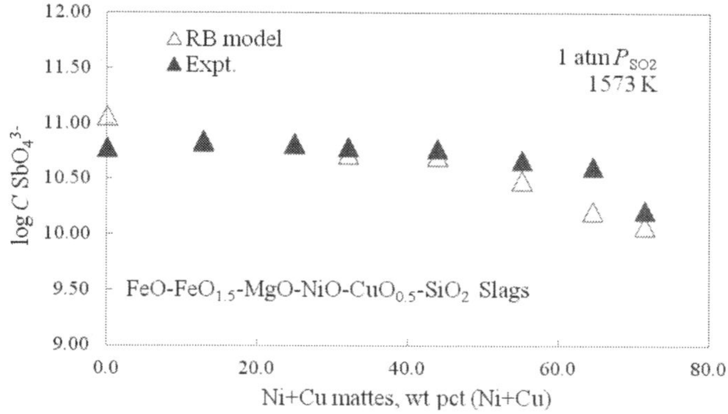

(c)

Figure 2. Antimony capacity for the FeO-$FeO_{1.5}$-MgO-NiO-$CuO_{0.5}$-SiO_2 slags and Ni+Cu mattes at 1573 K and P_{SO2} of (a) 0.1atm, (b) 0.5 atm, and (c) 1 atm.

Conclusions

The thermodynamic Reddy-Blander model was extended for *a priori* predictions of arsenic and antimonate capacities between the slags and the Ni-Cu mattes. The derived RB model was evaluated for basic and acidic slags, mattes, P_{SO2} (0.1, 0.5 and 1atm) and temperature (1573 K). The *a priori* predicted RB model impurity capacity data was compared with the available experimental data. An excellent agreement between the calculated RB model and the reported experimental data for both arsenic and antimony capacities in the FeO-$FeO_{1.5}$-MgO-NiO-$CuO_{0.5}$-SiO_2 slags and Ni+Cu mattes was observed. The reliable thermodynamic data for arsenic and antimony components is essential for the accurate prediction of capacities and distribution ratio. The use of this model for the *a priori* prediction of other impurity capacities, such as, Bi, Se and Te in the non-ferrous smelting slags and mattes is very feasible. Such predictions will be very useful in understanding the behavior of impurities, and therefore the development of new improvements for impurity control in the non-ferrous smelting and refining processes can be achieved.

Acknowledgments

The authors are pleased to acknowledge the financial support for this research from American Cast Iron Pipe Company (ACIPCO), and Pyrometallurgy Funds and The University of Alabama.

References

1. R. G. Reddy, "Emerging Technologies in Extraction and Processing of Metals," *Metallurgical and Materials Transactions B,* 34, (2003), 137-152.

2. R. G. Reddy. "Recovery of Pollution Causing Elements from Copper Slags," Minor Elements 2000: Processing and Environmental Aspects of As, Sb, Se and Te, C. Young (editor), (SME, Littleton, CO, USA, 2000), 239-249.

3. M. Nagamori and P. J. Mackey, *Metall. Trans. B*, 9B, (1978) 255-265.

4. P. C. Chaubal and H. Y. Sohn, *Metall. Trans. B*, 17B, (1986), 51-60.

5. T. Rosenqvist, *Principles of Extractive Metallurgy,* (McGraw-Hill, N.Y, 1983) 324-342.

6. J.M. Font, M. Hino and K. Itagaki, "Minor Elements Distribution between Iron-Silicate Base Slag and Ni_3S_2-FeS Matte under High Partial Pressures of SO_2," *Mater. Trans. JIM,* 39, (1998), 834-840.

7. J.M. Font, M. Hino and K. Itagaki, "Thermodynamic Evaluation of Distribution Behavior of VA Elements in Nickel Matte Smelting," *Metal. Rev. of the MMIJ,* 15 (2), (1998), 202-220.

8. Kimio Itagaki and Akira Yazawa. "Thermodynamic Evaluation of Distribution Behavior of Arsenic, Antimony and Bismuth in Copper Smelting," Advances in Sulfides Smelting Symposium, H. Y. Sohn, D.B. George and A. D. Zunkel (editor). (TMS, Warrendale, PA, 1983), pp. 119-142.

9. G. Roghani, Y. Takeda and K. Itagaki, "Phase Equilibrium and Minor Element Distribution between FeO_X-SiO_2-MgO-Based Slag and Cu_2S-FeS Matte at 1573 K under High Partial Pressures of SO_2," *Metall. Trans. B,* 31, (2000), 705-712.

10. J.M. Font, M. Hino and K. Itagaki, "Phase Equilibrium and Minor-Element Distribution between Ni_3S_2-FeS Matte and Calcium Ferrite slag under High Partial Pressures of SO_2," *Metall. Trans. B,* 31, (2000) 1231-1239.

11. R. G. Reddy and J. M. Font, "Arsenic Capacities of Copper Smelting Slags", *Metall. Mat. Trans. B*, 34B, (2003), 565-571.

12. J. Font and R. G. Reddy, "Modeling of Impurity Distribution between Mattes and Slags", Vol. IV- Pyrometallurgy of Copper, C. Diaz, C. Landolt and T. Utigard, (editors), (Copper 2003, Chile, 2003), 301-313.

13. J. M. Font and R.G. Reddy, "Modeling of Antimonate Capacity in copper and Nickel Smelting Slags," *Trans. Inst. Min. Metall. C,* 114, (2005), 160-64.

14. R. G. Reddy and M. Blander, "Modeling of Sulfide Capacities of Silicate Melts," Metall. and Mat. Trans.B, 18, (1987), 591-596.

15. R. G. Reddy and M. Blander, "Sulfide Capacities of MnO-SiO_2 Slags," Metall. Mat. Trans. B, 20, (1989), 137-140.

16. R. G. Reddy and W. Zhao, 1995, "Sulfide Capacities of Na_2O-SiO_2 Melts," Metall. Mat. Trans. B, 32, (1995), 925-928.

17. R. G. Reddy, "Impurity Capacities in Metallurgical Slags", Yazawa International Symposium, Materials Processing Fundamentals and New Technologies, F. Kongoli, K. Itagaki, C. Yamauchi and H.Y. Sohn, (editors), (TMS, Warandale, PA, USA, 2003), 1, 25-48.

18. B. Derin, O. Yucel and R. G. Reddy, "Sulfide Capacity Modeling of FeO_x-MO-SiO_2

(MO=CaO, MnO, MgO) Melts", *Minerals & Metallurgical Processing*, 28 (1), (2010), 33-36.

19. A. Yahya and R. G. Reddy, "Sulfide Capacities of CaO-MgO-AlO1.5, MgO-MnO-AlO1.5 and CaO-MgO-MnO-AlO1.5 Slags", *Trans. Inst. Min. Metall. C*, 120 (1), (2011), 45-48.
20. B. Derin, O. Yucel, and R. G. Reddy, "Modeling of Sulfide Capacities of Binary Titanate Slags," EPD Congress, (TMS, Warrandale. PA, USA, 2004), 155-160.
21. B. Chen, R. G. Reddy and M. Blander, "Sulfide Capacities of CaO-FeO-SiO2 Slags," 3rd International Conference on Molten Slags and Fluxes, *The Institute of Metals*, (London, UK, 1998), 270-272.22.
22. A. D. Pelton, G. Eriksson and A. Romero-Serrano, "Calculation of Sulfide Capacities of Multi-component Slags," *Metall. Trans. B*, 24, (1993), 817-825.
23. B. Derin, O. Yucel and R. G. Reddy, "Predicting of sulfide capacities of industrial lead smelting slags". F. Kongoli and R. G. Reddy (editors), Sohn International Symposium; Advanced Processing of Metals and Materials Volume 1: Thermo and Physicochemical Principles: Non-Ferrous High-Temperature Processing, (TMS Warrendale, PA, USA, 2006), 1, 237-244.
24. B. Derin and R. G. Reddy, "Sulfur and Oxygen Partial Pressure Ratios Prediction in Copper Flash Smelting Plants using Reddy-Blander Model", F. Kongoli, K. Itagaki, C. Yamauchi and H.Y. Sohn, (editors), Yazawa International Symposium on Metallurgical and Materials Processing, (TMS, Warandale, PA, USA, 2003), 1, 625-632.
25. HSC Chemistry software, A. Roine, ver. 4.1, *Outokumpu Research Oy*, Pori, Finland.
26. FactSageTM 5.0 software, *Thermfact Ltd. (Montreal) and GTT-Technologies (Aachen)*, 2001.

Celebrating the Megascale: Proceedings of the Extraction and Processing Division Symposium on Pyrometallurgy in Honor of David G.C. Robertson
Edited by: Phillip J. Mackey, Eric J. Grimsey, Rodney T. Jones, and Geoffrey A. Brooks
TMS (The Minerals, Metals & Materials Society), 2014

QUARTZ-CRISTOBALITE TRANSFORMATION AND ITS EFFECT ON REACTIONS IN Si PRODUCTION: INITIAL STUDIES

Eli Ringdalen[1], Dian Adisty[2], Leiv Kolbeinsen[3]

[1] Sintef Materials and Chemistry Pb 4760 Sluppen, N 7465- Trondheim, Norway
[2]Det Norske Veritas as, Veritasveien 1,1363 Høvik, Norway
[3]NTNU, Trondheim, Norway

Keywords: silicon, quartz, cristobalite, reaction rates

Abstract

In Si and FeSi production, the main Si source is SiO_2, in the form of quartz. Reactions with SiO_2 generate SiO-gas that further reacts with SiC to Si. During heating, quartz will transform to other SiO_2 modifications with cristobalite as the stable high temperature phase. Transformation to cristobalite has been investigated and shown to be a slow process where the rate varies between the investigated quartz types.

The effect of quartz versus cristobalite as SiO_2 source has been studied for reaction (1) $2SiO_2 + SiC = 3SiO + CO$, by heating of pellets with SiO_2 + SiC in Ar-gas. The reaction was faster with cristobalite than with quartz. Initial experiments indicates that this is the case only in Ar and not in CO-atmosphere The higher specific surface area of cristobalite is believed to contribute to the higher rate. The industrial consequences of the observed difference between quartz types are discussed.

Introduction

Metallurgical silicon (Si) and ferrosilicon (FeSi) are produced in submerged arc furnaces (SAF), where the two main raw materials are quartz and reductants such as coke, coal, charcoal and woodchips. For FeSi production iron sources are added. Energy, around 11-13MWh/ton alloy for Si production, is another important input factor. The reduction takes place by the overall reaction (1)

$$SiO_2 + C = Si + CO \qquad \Delta H^o{}_{2000^oC} = 687 \text{ kJ/mol} \qquad (1)$$

In industrial furnaces, the reactions take place in several steps as described by Schei et al [1]. In the upper part of the furnace, descending raw materials react with ascending SiO and CO gas by reactions (2) and (3). The ascending gases react according to reaction (4) and (5) as a result of the decreasing temperature.

$$2C(s) + SiO(g) = SiC(s) + CO(g) \qquad \Delta H^o{}_{1800^oC} = -78 \text{ kJ/mol} \qquad (2)$$

$$SiO_2(s) + C(s) = SiO(g) + CO(g) \qquad \Delta H^o{}_{1800^oC} = 667 \text{ kJ/mol} \qquad (3)$$

$$2SiO(g) = SiO_2(s,l) + Si(l) \qquad \Delta H^o{}_{1800^oC} = -606 \text{ kJ/mol} \qquad (4)$$

$$3SiO(g) + CO(g) = 2SiO_2(s,l) + SiC(s) \qquad \Delta H^o{}_{1500^oC} = -1380 \text{ kJ/mol} \qquad (5)$$

In the lower part of the furnace, where the temperatures reach more than 2000°C, Si is produced by reaction (6). SiO gas required for this reaction and for SiC production in the upper part of the furnace is produced by reactions (7) and (8). Dominant reactions and their stability depend strongly of the partial pressures of SiO and CO as shown in Figure 1.

$$SiC\,(s) + SiO\,(g) = 2\,Si(l) + CO\,(g) \qquad \Delta H^{o}_{2000^{o}C} = 167\ kJ/mol \qquad (6)$$

$$SiO_2(l) + Si(l) = 2SiO\,(g) \qquad \Delta H^{o}_{2000^{o}C} = 599\ kJ/mol \qquad (7)$$

$$2SiO_2(s,l) + SiC\,(s) = 3SiO\,(g) + CO\,(g) \qquad \Delta H^{o}_{2000^{o}C} = 1364\ kJ/mol \qquad (8)$$

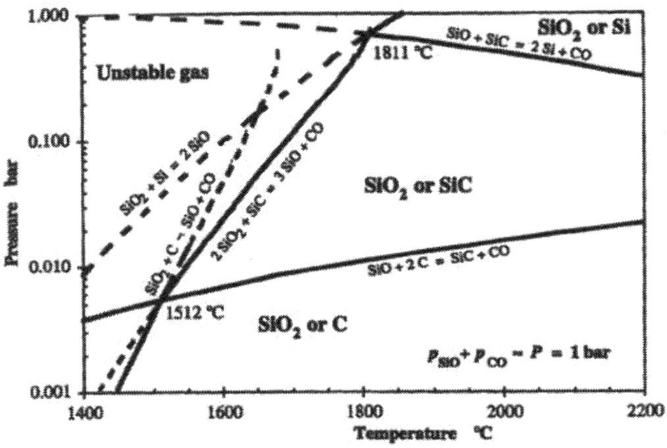

Figure 1. Partial pressure of SiO (g) in equilibrium with SiO_2, SiC and C as a function of temperature. Total pressure of SiO and CO gas are 1 atm [1].

Reaction (3) is the only of the reaction involving quartz that take place in the upper part of the furnace. As seen in Figure 1, this will only occur at very low SiO partial pressures. Production of Si takes place through SiO by reaction (6). All quartz must therefore be converted to SiO by reactions (7) and (8). These reactions takes place at high temperature and the high temperature properties of quartz are therefore important. Reactions (7) and (8) are both strongly endothermic, but reaction (8) requires nearly three times as much energy as reaction (7). Energy distribution in the furnace will therefore depend on which of these reactions that are dominating.

The silicon producing reaction, (6), requires a SiO partial pressure above 0.67 to run at 1811°C, slightly decreasing with increasing temperature. The SiO producing reactions, e.g. the reactions with quartz provides SiO for this reaction and should therefore have a reaction rate that is balanced with the rate for the Si producing reaction. SiO produced higher up in the furnace, at a temperature lower than 1811°C will not be used for Si-production by reaction (6), but at temperatures above 1511°C react with carbon to produce SiC by reaction (2). A surplus of SiO may react to SiO_2 and Si or SiC by reactions (3) and (4). These are highly exothermic and will transport energy to the upper part of the furnace and give higher temperature at the charge top. SiO that does not react by these reactions will go to the off-gas and decrease Si-yield.

The reaction rate for the SiO producing reactions at various temperatures will therefore affect Si-yield, temperature distribution in the furnace and energy consumption and production in different zones.

The rate of reaction (3) between silica and carbon was found by Wiik [2] to be nearly twice as fast when cristobalite was used as the silica source instead of quartz. The difference was attributed to a higher specific surface area for cristobalite versus quartz.

Quartz Properties

When quartz is heated, its structure will change to different SiO_2 polymorphs as shown in the phase diagram in Figure 2[3]. When heated in air at 1 atm in air α-quartz is transformed to β quartz at 573°C. This transformation is fast and will be reversed during cooling. According to the phase diagram, tridymite will be formed at 873°C. It is not agreed if this phase change really will take place. At 1450°C cristobalite is formed. The α-quartz – β-cristobalite transformation goes through an amorphous transition phase. [4]. Silica is expected to me more reactive when it has high percentage of amorphous silica during the transition

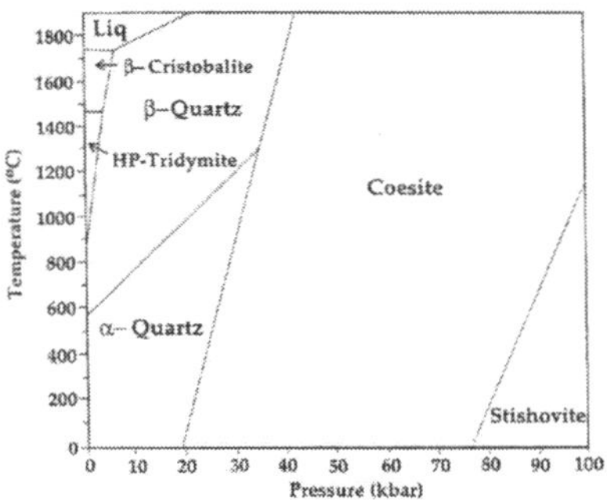

Figure 2. SiO_2 phase diagram.

When α-quartz is transformed to β-quartz, the volume will increase by 0.26 – 0.45 %. Further transformation from β-quartz to cristobalite will increase the volume by 17.4 %. This large increase in volume will give a lower density and a higher specific surface area. Cracking as a result of the volume change will contribute to the higher surface area. Wiik [2] measured the original surface area of 0.025 m^2/g for quartz had increased to 0.50 m^2/g after the material had been heat treated and transformed to cristobalite. The α-quartz – β-cristobalite transformation is slow and in Wiik's experiment the quartz was heat treated at 1400°C for 5 days in order to achieve complete transformation to cristobalite.

All of the above studies were performed in air. During Si-production in SAF, the quartz will be heated in a CO/SiO atmosphere. The effect on phase transformations of such an atmosphere is not known.

SiO producing reactions

The rates of the two SiO-producing reactions involving silica, (7) and (8), have been investigated by Andersen [5] and by Bao et al. [6], in DTA-TGA equipment from 1550°C to 1800°C with quartz as Si-source. Both reactions are slow at 1550°C. At 1550 to 1730°C, reaction (7) with Si is around 1.5 times faster than reaction (8) with SiC. The rate for the reaction with SiC increases most with temperature and is the dominating reaction at 1800°C. SiO production from SiO_2 and SiC had a reaction rate of 0.1 mg/min at 1550°C while at 1730 °C it was 1.3 mg/min and at 1830 °C it was 4.1 mg/min. A conversion rate of 0.8 was reached after 56 minutes at 1820°C and after 112 minutes at 1730°C for this reaction.

Experimental

Two different types of investigations were performed:

1. Changes in quartz during heating
2. Effect of cristobalite versus quartz on SiO production from SiC

Parts of this work are reported by Andersen [5] and Adisty [6,7].

Changes in Quartz during Heating

Phase transformations Most of the experiments and the comparison of different quartz types were done by heat treatment in a Super kanthal rapid heating furnace. Around 50 gram of quartz in the size fraction 1-3 mm was used in each experiment. The furnace was heated as fast as possible to target temperature and held at that temperature for a predetermined time. Heating from 25°C to 1500°C took between 21 and 47 minutes. Samples Qz 2 and Qz 9 had the longest heating time, while all the other samples had lower times. The tests conducted for 1 hour at 1500°C were all heated 21-22 minutes.

The surface area of the samples was measured before and after heating by nitrogen adsorption. 1-3 mm material was used for the measurements. The amounts of different phases in samples after heat treatment were measured by quantitative XRD. A Bruker D8 advance with the software DIFFRAC^{PLUS}TOPAS was used. The quantity of amorphous silica was determined by the method described by Sørensen and Vedeler [8].

Softening temperature and volume expansion Melting and softening temperatures were investigated in a wetting furnace. The apparatus essentially consisted of a horizontal graphite heater surrounded by graphite radiation shields, located in a water-cooled vacuum chamber. The chamber was fitted with windows to allow a digital video camera to record the shape of the sample every second. The maximum temperature of the furnace was 2400°C, and it was controlled by a type C thermocouple. A pyrometer, functioning from 900 to 2400°C, was used as an extra control of the temperature. The linear dimensions of the images were measured directly from the image of the droplet using shape analysis software.

The test samples were drilled from lump quartz samples. They were formed as cylinders with a diameter of 4 mm and ground the two sides to flat surface. The quartz sample were placed on an ISO88 graphite substrate and heated in CO atmosphere first to 900°C in about 3 min, then with a rate of 50°C/min to 1500°C, and 10°C/min further to final melting.

Images were recorded and analysed afterwards to determine the:

- Softening - when the shape of the sample has lost the sharp edges;
- Final melting - when the droplet is round, with no artifacts;
- Volume expansion

Softening temperatures and volume expansion in 4 different quartz types had earlier been studied by Andersen [5] with the same procedure and same apparatus as applied here. In his work softening in both CO and Ar were investigated and the amount of different phases after heating to 1500°C and 1700°C in CO and Ar measured. Amorphous phases were not included in the measurements and they are therefore only indicative.

Effect of Cristobalite Versus Quartz on Rate of SiO Production from SiC

The reaction rate for reaction (6) between SiO_2 and SiC has been investigated in two different tube furnaces. One series was carried out in a graphite resistance thermogravimetry tube furnace. A graphite crucible, 35 mm in diameter and 70 mm height filled with around 30 -40 grams of material was used. It was covered with a lid and hung in the thermo balance that made it possible to record the weight continuously during the experiment. The furnace was evacuated and flushed with Argon twice before the furnace was heated up. A gas flow of 1l/min of either Ar or CO was blown through the furnace during the experiment. To enhance gas flow out of the crucible, there were holes in the lid. To avoid contamination of the furnace chamber, the lid was covered with graphite wool that captured SiO-gas. The setup is shown in Figure 3. The furnace was heated up 15°C/min up to target temperature and the kept at this temperature for the predetermined time. The raw materials were SiO_2 and SiC in the size 1-3 mm. They were mixed in the same molar ratio as equation (8): 2:1, e.g. 35 gram SiO_2 and 8.34 gram SiC.

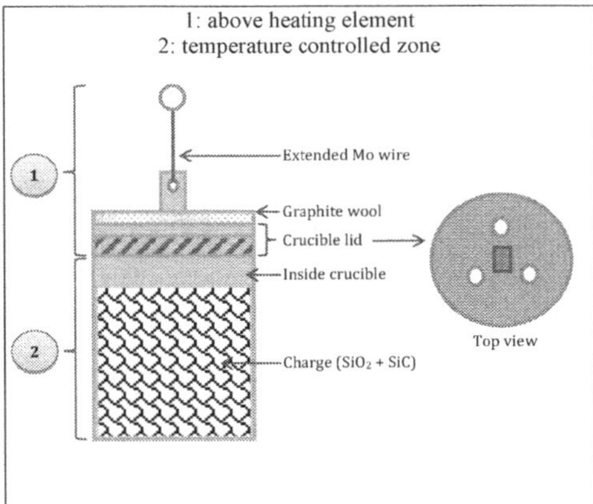

Figure 3. Sketch of crucible for resistance furnace [7].

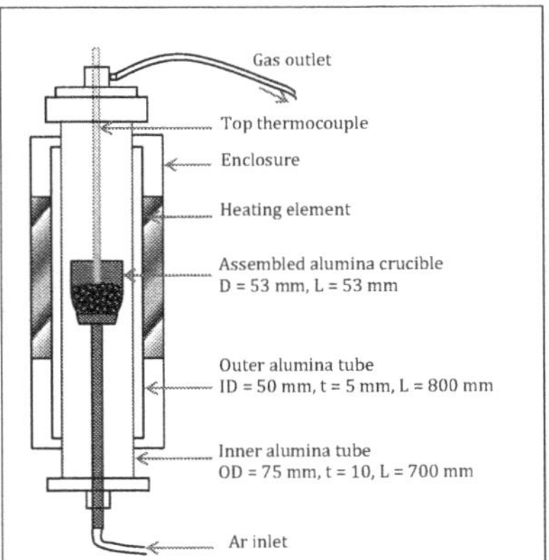

Figure 4. Sketch of vertical alumina tube furnace[9].

The other series of experiments were run in an alumina vertical tube furnace shown in Figure 4. All parts in contact with raw materials and gases were made of alumina. Composition of the outlet gas was measured with a mass spectrometer (MS).

A sample of 10 gram was used in each experiment and a gas flow of 1 l/min of Ar, CO or H_2 was flushed through the furnace during the experiment. The furnace was heated at 22°C/min to 400 °C and further with a rate of 10°C/min up to target temperature. Quartz and SiC used in the experiments were crushed to around 20 μm and pelletized to 1-2 mm pellets with water as binder. They were hardened at 1200°C for 15 minutes. A molar ratio SiO_2/SiC of 1:3 was used- e.g. an excess of silica compared to reaction (8).

Type and amount of different phases were determined by quantitative XRD in both raw materials and products. Their O and C content were analysed by LECO at Elkem Research. Degree of conversion was calculated based on changes in oxygen content from LECO analysis. Measurements of %CO in the off-gas gave additional information about the reaction. Specific surface areas were also measured. Some of the samples were studied by Electron Probe Micro Analyzer (EPMA). This work was a part of MSc thesis of Adisty [9].

Materials

Quartz from different sources for industrial Si-production has been used in the experiments. They are all relatively pure with 97.6 to 99.8 %SiO_2. Quartz Q0 with 99.3% SiO_2 is the quartz type used for studies of reaction rates and is also the type where phase transformations have been studied in most detail. Available information about the quartz varies.

In the experiments with SiC, commercial SiC from Washington Mills was used. Type of SiC is expected to affect the reactions. Studies of this were not a part of the current investigations and this parameter is therefore kept constant.

Cristobalite was found to have a higher specific surface area than quartz. In the experiments where a mixture of SiO_2 grains and SiC were used, quartz had a specific surface area of 0.06 m^2/g while cristobalite had 0.28 m^2/g. In the experiments with pellets, the pellets with quartz had a specific surface area of 1.35 m^2/g while pellets with cristobalite had 0.81 m^2/g.

Results

Changes in Quartz during Heating

Results from the earlier experiments by Andersen [5] that also included some results from Aasly [3] are shown in Table I. The measured values are different from expected from the phase diagram. Softening temperatures are higher than 1750°C, volume expansion is for several of the samples larger than 17.4 % and the quartz had not been converted to cristobalite at 1450°C. This shows that kinetics is of importance for changes in quartzes during heating. The differences between the investigated quartz samples show that the kinetics of the changes varies with quartz type dependent properties. The investigation also indicated faster cristobalite formation in Ar than in CO, and that volume expansion increased with increasing amount of cristobalite. Since the quantification of cristobalite has a lot of uncertainty these correlations are only indicative.

Table I .Softening Temperatures, Volume Expansion and Cristobalite Formation in quartz [5]

Sample	Gas Atmosphere	Softening Temperature, °C	% Volume Expansion 1700 °C	% Cristobalite 1500 °C	% Cristobalite 1700 °C
C	CO	1790	42	1	25
D	CO	1800	15	0	2
E	CO	1750	39	10	68
G	CO	1810	21	2	1
C	Ar	1760	59		33-71
D	Ar		15		4
E	Ar	1750	55		
G	Ar	>1800	15		17

In the current investigation, 12 different quartz types have been investigated. Some important data are listed in Table II. The heating for 6 hours of samples Qz 2 and Qz 9 at different temperatures are presented in Figure 5 and this showed that the phase transformation started at 1400°C, a lower temperature than expected from the phase diagram in Figure 2. The quartz had not completely transformed to quartz even after 6 hours at 1600°C. The kinetics are very slow and the difference between the two quartz types shows that it is affected by quartz properties. Specific surface area increases when quartz is transformed to cristobalite. For Qz 2, it was 8 times higher than the original value after 2 hours at 1500°C.

Table II. Important data from quartz investigations

Sample	% SiO_2	Softening temp in CO	Complete melting °C	Max volume expansion %	Specific surface area m^2/g original	% Christo-balite, 1500 °C,1h	% Amorphous silica, 1500 °C,1h	% Christo-balite, 1500° C, 2h	Specific surface area m2/g, 1500 °C,1 h	Specific surface area m2/g, 1500 °C, 2 h	Specific surface area m2/g 1600 °C, 6 h
Q0	99.3				0.06			51		0.28	
Q1	99.81	1755	1900	28		58	20		0.47		
Q2		1785	1900	23	0.06	61	13	81	0.6	0.49	0.37
Q3		1790	1900	37		8	17		0.14	0.14	
Q4		1735	1880	20		6	19		0.13		
Q5		1800	1920	15		12	4		0.1		
Q6		1600	1880	18		11	28		0.12		
Q7	98.73	1800	1900	35		40	8		0.11		
Q8	98.39	1755	1830	10		8	5		0.09		
Q9	97.63	1775	1800	14	0.07	19	8	29	0.11	0.09	0.25
Q10	99.97	1765	1910	32		10	7		0.09		
Q11		1770	1900	15		1	0		0.011		

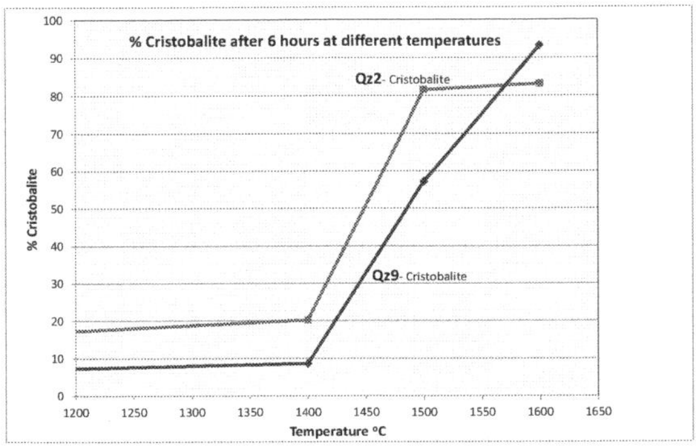

Figure 5. The % cristobalite in the sample after 6 hours heating in air at different temperatures

The effect of time was investigated by heating to 1500°C at different holding times and the results are presented in Figure 6. For samples Qz 2, Qz 3 and Qz 9, the amount of amorphous silica in the heat treated samples were measured while only percentage of cristobalite was measured [6] in sample Qz0 which is used in further experiments. The relatively low amounts of amorphous silica measured here are expected to have large uncertainty. The transformation rate differs between the different quartz with Qz 2 as the fastest and Qz 3 as the slowest. Qz 0 that has been used in reaction tests is between these.

The amount of amorphous silica increased quickly at the start of the transformation and then slowed down and then decreased. From Qz 2 and Qz 9, peak amounts of amorphous silica seem to be at around 40% cristobalite. Qz 3 does not reach this value for cristobalite formation. The ratio between cristobalite and amorphous silica also varied between the samples and this was especially pronounced for Qz 3 that had a higher content of amorphous silica than cristobalite.

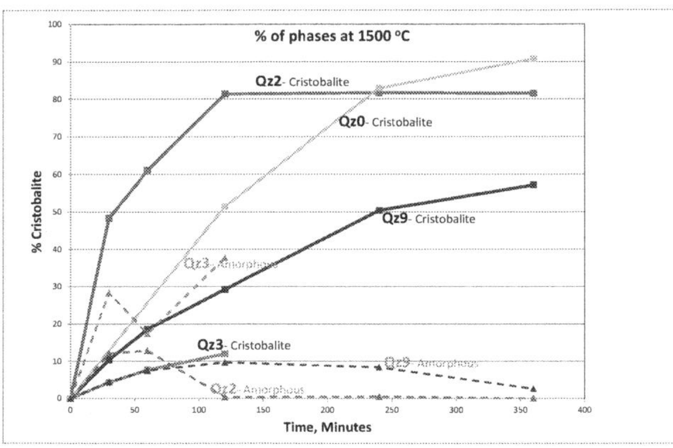

Figure 6. Effect of time on formation of cristobalite and amorphous silica when quartz is heated in air at 1500°C.

The amounts of cristobalite and amorphous silica after 1 hours heating at 1500°C in all the investigated samples are shown in Figure 7. There are large differences between the samples with Qz 2 as the sample with highest cristobalite content. Sample Qz 3, Qz 4 and Qz 6 had higher content of amorphous silica than of cristobalite while the ratio was opposite for the other quartz type. Only one sample of each quartz type has been investigated so the reproducibility is not known.

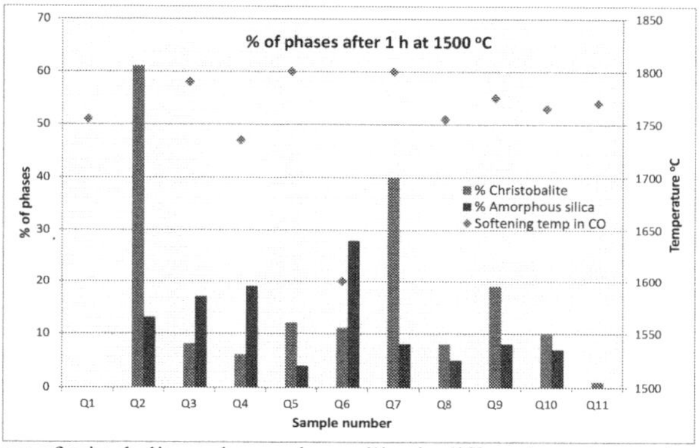

Figure 7. Amount of cristobalite and amorphous silica in different quartz samples after heating 1 hour at 1500°C. Softening temperatures for the different samples are also included.

Effect of Cristobalite versus Quartz on Rate of SiO Production from SiC

The first series of experiments run in a graphite resistance furnace are presented in Figure 8. The results show that at 1700°C in Ar atmosphere, reaction (8) will proceed faster when SiO_2 is in the form of cristobalite compared to the quartz. This is not the case in CO where the SiO_2 polymorph does not affect reaction rate.

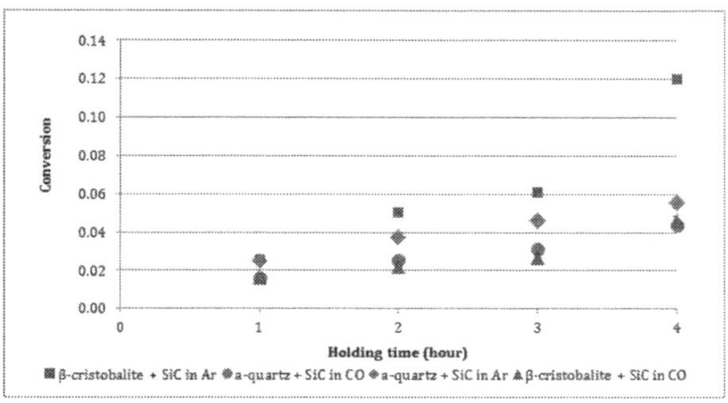

Figure 8. Conversion rate for reaction between SiO_2 and SiC at 1700°C in different gas atmospheres [7]

In the experiments run with pellets in a gas flow in alumina crucible, described in Figure 3, the reaction rates were, as shown in Figure 9, much faster. And after 2 hours the quartz was nearly completely reacted. At 1500°C, quartz had reacted slightly less than cristobalite while no significant differences were observed at higher temperatures.

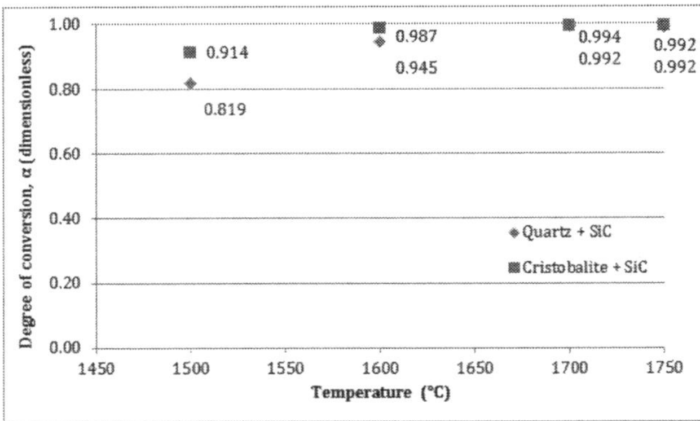

Figure 9. Conversion rate for SiO_2 + SiC reaction after 2 hours in Ar atmosphere [9].

Discussion

These experiments are the initial study of a work to determine how quartz and quartz properties might affect silicon and ferrosilicon production in sub merged arc furnaces.

The experiments and the results raise different discussion points:

1. How reliable are the results?
2. Does the results give any guidance about how further experiments should be designed and evaluated?
3. Do the results give any information about reaction mechanisms and reaction rates and is this relevant for industrial Si/FeSi production?

This initial investigation was designed to determine the most important parameters. Reproducibility and variance of the experiments are therefore not investigated and cannot be quantified. The main uncertainties for cristobalite formation are related to quantitative determination of amorphous silica in the samples, variation in heating time to reach target temperature, and variance in measured target temperature. For softening and volume expansion where small samples are used, representativeness of the samples will be the main uncertainty. In the investigation of reaction rates, sampling and characterisation of the final products have been a main challenge. Since SiO gas condense in the furnace and other equipment, determination of weights and weight losses create several uncertainties. For the first series of reaction rates in a crucible with lid, the gas atmosphere is nearly stagnant and gas composition is not known. The experiments have indicate interesting effects that will be further investigated in controlled experiments designed based on achieved results.

Heat treatment of quartz showed as expected and similar to the results of Aasly [3] that the quartz-cristobalite phase transformation is slow. Cristobalite is formed already at 1400°C, a lower temperature than expected from the phase diagram. It is not completed even after 6 hours at 1600°C. The rate varies considerably between different quartz types and both the effect of time and of temperature varies. No tridymite was formed. The differences in rate for cristobalite formation could not be correlated to other parameters as chemical composition or softening temperature. Cristobalite formation will increase the specific surface area for materials by as much as 6-8 times. This will in reactions governed by kinetics, contribute to a higher reaction rate.

Reactions involving SiO_2 are often studied at temperatures above 1500 °C. When quartz is heated to these reaction temperatures, it will be partly transformed to cristobalite. The extent depends on heating rate and quartz type. Reaction rates for quartz will therefore be difficult to study and results from investigation with different quartz types cannot be compared directly.

Transformation from quartz to cristobalite goes through a transition phase with amorphous silica [2, 3]. The amount of amorphous silica increases fastest at the beginning of the transformation until it stabilises at a given level. The mechanism behind this has not been further investigated. Maximum amount of amorphous phase and ratio between cristobalite and amorphous silica vary between the investigated quartzes. Wiik [2] proposed that a high amount of amorphous silica will increase reduction rates. The varying amounts of amorphous silica is therefore together with the differences in amount of cristobalite and in specific surface area expected to give different reaction rate depending on quartz type.

The rate for reaction (8) between SiC and SiO_2 was found to be higher with cristobalite than with quartz as the silica source for experiments in Ar, while there was no difference in CO. A similar difference is reported by Wiik [2]. More data are needed to understand the reaction mechanism and the effect of different parameters.

In industrial Si production, SiO is produced from reactions with SiO_2 and either SiC or Si. The relative importance of the reactions is not known and this depends on their relative reaction rates at different temperatures. Specific surface areas together with the amounts of cristobalite and amorphous silica are expected to affect reaction rates. Since these parameters vary with quartz type, choice of quartz can affect performance in industrial furnaces. The importance of quartz type will depend on retention time for the quartz in different temperature zones. The effect on SiO_2 properties of the CO and SiO gas atmosphere that is found in industrial furnaces has not been studied in this first step of the investigations.

Conclusions

The rate of quartz – cristobalite transformation and formation of amorphous silica vary with quartz type. This is expected to be of importance for reaction rate for SiO formation and quartz type may thus affect performance of furnaces for Si-production.

In the investigations of reactions with SiO_2 above 1400°C, cristobalite will be formed during heating to reaction temperature. The amount varies with quartz type and investigations with different quartz types or with different heating rates, can therefore not be compared

Acknowledgements

These investigations are funded by the Norwegian Research council (RCN) through the ReSiNa project and by RCN and Elkem through the ESiP project. They are both acknowledged.

References

1. A Schei A., J K Tuset J.K., and H.Tveit, "Production of High Silicon Alloys", *Tapir forlag,* (Trondheim, Norway, 1998), ISBN: 82-519-1317-9.

2. Kjell Wiik, "Kinetics of reactions between silica and carbon" (PhD thesis, NTNU Trondheim, Norway 1990), IUK report 1990:60; ISBN: 82-7119-170-5.

3. Kurt Aasly, "Properties and behavior of quartz for the silicon process" (PhD thesis, NTNU Trondheim ,Norway, 2008).

4. Chaklader A C D,. and Roberts A L, "Transformation of quartz to cristobalite", *Journal of the American Ceramic Society,* (1961), 44, 35-4.

5. Vegar Andersen, "Investigation of thermal properties of quartz for the silicon industry under reducing atmosphere", (NTNU Project work in TMT 4500, Trondheim, Norway, 2009).

6. Dian Adisty, " Influence of Holding Time on Solid-Solid Reaction between quartz + SiC and Cristobalite + SiC", (NTNU Report, Trondheim, Norway, 2012).

7. Dian Adisty, "Kinetics study of silicon production from different polymorphs of quartz", (NTNU Project work, TMT 5500, Trondheim, Norway 2012).

8. Bjørn Eske Sørensen and Sebastian Vedeler, "Preliminary study to quantify amorphous silica with X-Ray Diffraction", (NTNU report 228.11.11, Trondheim, Norway 2011).

9. Dian Adisty, "Kinetic Study of SiO_2 + SiC Reaction in Silicon Production", (MSc thesis, NTNU, Trondheim, Norway, 2013).

Celebrating the Megascale: Proceedings of the Extraction and Processing Division Symposium on Pyrometallurgy in Honor of David G.C. Robertson
Edited by: Phillip J. Mackey, Eric J. Grimsey, Rodney T. Jones, and Geoffrey A. Brooks
TMS (The Minerals, Metals & Materials Society), 2014

MODIFICATIONS TO A SMELTER TO ACCOMMODATE RECYCLED MATERIALS

Greg Roset[1], Dayle Flynn[1] and Katie Schumacher[1]

[1]Stillwater Mining Company
1321 Discovery Drive; Billings, MT, 59102, USA
E Mail: KSchumacher@stillwatermining.com

Keywords: Recycle, smelter, PGMs, Stillwater

Abstract

The Columbus Processing Complex of the Stillwater Mining Company began operations at Columbus, Montana in 1990 based on electric furnace smelting treating a precious metals concentrate from the nearby Stillwater mine and mill. Since that time, the smelter plant has been expanded in stages in order to treat recycled materials in addition to the concentrate. This paper describes a number of changes made at the plant to effectively handle recycled materials.

Introduction

The Columbus Processing Centre was commissioned in 1990 to treat precious metal concentrate from the Stillwater mine and mill [1]. The Stillwater Mining Company has been processing recycled materials in its Columbus Processing Center since 1997. To increase capacity, efficiency and hygiene in the smelter at the Columbus Processing Center, a recycling plant was constructed, an automated recycle laboratory was designed and installed, feed systems were modified and a slag cleaning furnace was also installed. This paper describes these facilities.

Sampling Plant

Originally, ceramic automotive catalytic converters were crushed, bagged and fed with concentrate into the concentrate drier. This was a time consuming, labor intensive endeavor that allowed for approximately 15 tons of catalytic converters to be treated per day. To alleviate this bottleneck and allow for more recycled materials to be processed, a recycling plant was designed with construction completed in August 2010. A simplified flow diagram of the plant for receiving and processing ceramic automotive catalytic converters is shown in Figure 1. The recycling plant includes a dedicated unloading dock and warehouse storage facilities for recycled materials. A sampling plant simultaneously crushes and samples the ceramic auto catalyst which simplifies the process and saves time. At the smelter, the material is fed into a storage bin to be automatically blended with the concentrate at a preset ratio prior to smelting.

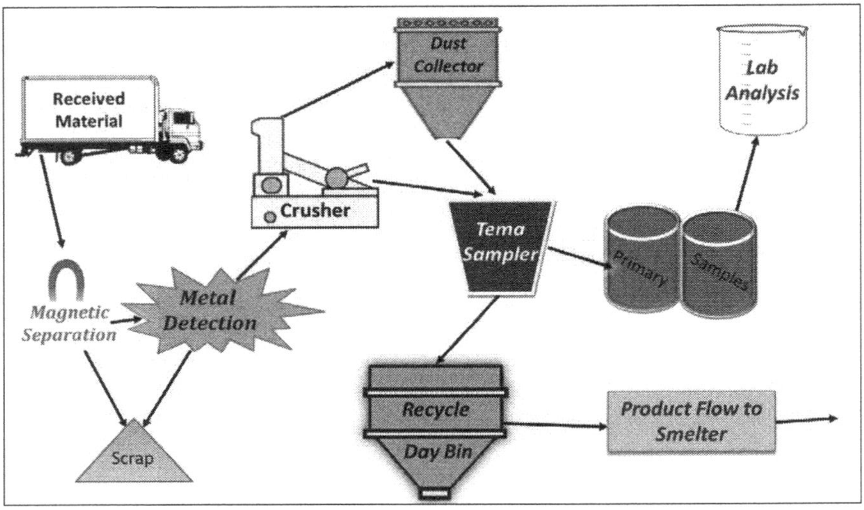

Figure 1. Process flow diagram of the ceramic catalytic converter circuit.

Laboratory Changes

In addition to the recycling plant, an Automated Recycle Laboratory (referred to as the ARL) was designed and implemented in March, 2011. The layout of the ARL facility is shown in Figure 2. A robot unit (Figure 3) pulverizes the sample, creates a pellet, x-rays the pellet and provides an exchange quality assay. After the implementation of the recycle plant and ARL, assays can typically be provided within 7 days of receiving the material. This efficiency provides suppliers with the fastest exchange quality assays in the market, which reduces risks caused by lengthy traditional assay times.

Along with the ARL, a dedicated team of chemists and specialists was assembled to develop improved methods for assaying new recycle materials. The concentrate coming from the mill has a very consistent composition which allows it to be assayed much more easily than the widely variable recycled materials. Approximately 70% of the received recycled materials can be assayed by the ARL, while the remaining materials must be analyzed by fire assay methods. New incoming materials often require new techniques in order to obtain an accurate assay. With a dedicated team of researchers, Stillwater Mining is able to quickly and accurately assay a wide variety of materials.

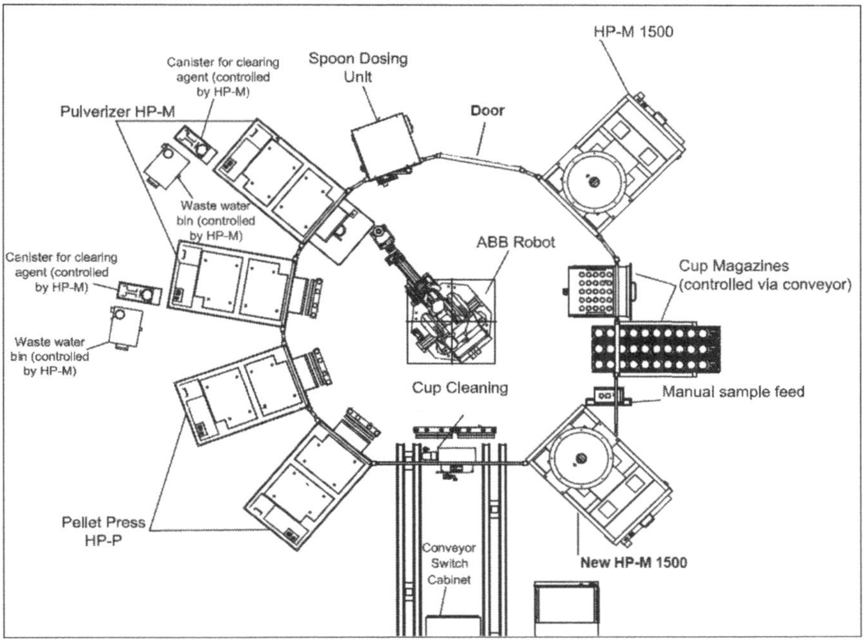

Figure 2. Layout of the ARL facility.

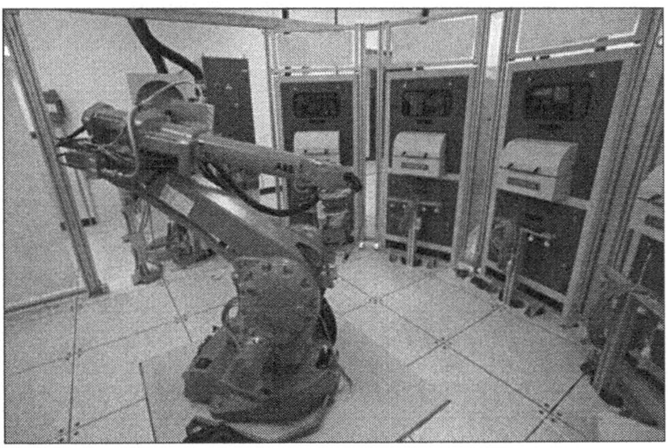

Figure 3. Photograph of the ARL robot at the Columbus Processing Center.

Feeding High Carbon Materials

Spent petroleum catalyst dust is a fine particulate material that is high in carbon (>20% C) and PGMs. In the past, drums of dust were emptied by smelter operators into 907 kg (2,000 lb) super sacks. The super sacks were then dumped into the furnace every few days. When this material was introduced to the furnace, the iron content of the matte increased through reduction transfer from the iron in the slag while the slag temperature showed a sharp increase. The operators feeding the spent petroleum catalyst would become covered in dust. A more hygienic and efficient system of feeding fine, high carbon materials to the furnace was sought. A drum dumper and tubular conveying system were designed and installed (Figure 4). The new system confined the fine dust to within the system without exposing smelter operators. With a slower but constant feed rate of the high carbon material, this also prevented the temperature spikes in the furnace, thus allowing smoother operating conditions. Following the success of the new tubular conveyor feed system, a similar but larger capacity system was added to allow materials of slightly lower carbon content (between 10 to 20 % C) to be handled.

Figure 4. Conveying system for moderately high carbon materials (~15-20% C).

Slag Cleaning

Because recycled materials have a higher concentration of PGMs than the concentrate coming from the mill, the slag can show an increase in PGM levels in slag, increasing PGM losses. Stillwater Mining Company has seen a steady increase in recycled materials (with exception of the market crash in 2009), Figure 5. As the amount of recycled material being fed to the furnace increased, so did the losses in the slag.

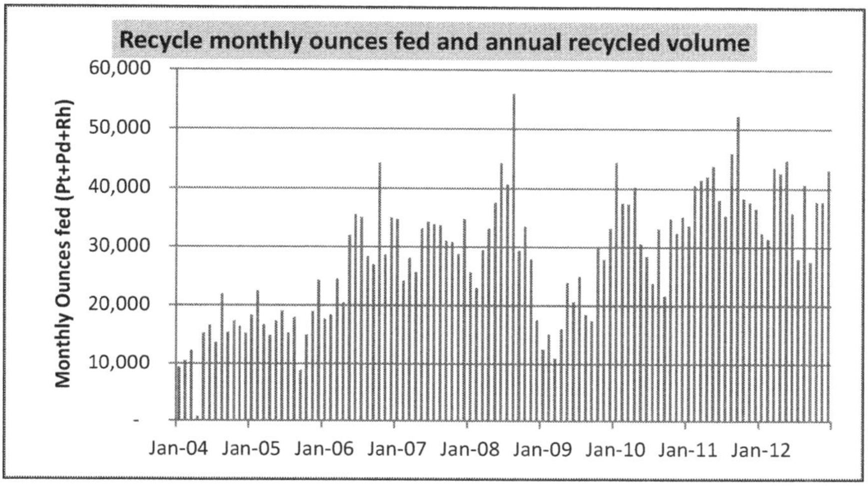

Figure 5. Ounces of PGMs fed from recycled materials by month, 2004 to 2012.

To help reduce the amount of PGMs being lost via the slag, a slag cleaning furnace was installed. Molten slag is transferred from the primary furnace, down a launder and into the slag cleaning furnace. A reductant is added to the slag cleaning furnace to facilitate the collection of PGMs. With the addition of the slag cleaning furnace, the final slag has a greatly reduced level of PGMs.

Conclusions

A recycling plant was constructed, an automated recycle laboratory was created, feed systems were modified and a slag cleaning furnace was installed in the Stillwater Mining Company smelter. This has allowed an increase capacity, better efficiency and improved hygiene conditions.

Reference

1. C.J. Hodges, T.J. McNeill, L.M. Millard and G.K. Roset, Electric Furnace Operation at the Stillwater Mining Company's Precious Metals Smelter, *Proceedings of the International Symposium on Non-ferrous Metals, Trace Elements, Furnace Practices and Energy Efficiency*, (Eds. R. Bergman, B.D. Chamberlain, P. Godbehere, L. Davis, P. Mahant, A. Vahed, K. Molnar and P. Barr (The Metallurgical Society of CIM, 1992), 279-287.

Celebrating the Megascale: Proceedings of the Extraction and Processing Division
Symposium on Pyrometallurgy in Honor of David G.C. Robertson
Edited by: Phillip J. Mackey, Eric J. Grimsey, Rodney T. Jones, and Geoffrey A. Brooks
TMS (The Minerals, Metals & Materials Society), 2014

Removal of Pb from Molten Copper by Fe_tO-SiO_2 (-CaO, Al_2O_3) Slag Treatment in Mitsubishi Process

Soo Sang Park[1], Joo Hyun Park[2,*]

[1]Metals and Materials Research Department, LS-Nikko Copper, Ulsan 70, Korea
[2]Department of Materials Engineering, Hanyang University, Ansan 426-791, Korea
*email: basicity@hanyang.ac.kr

Keywords: Pb distribution ratio, Copper smelting slag, Fe^{3+}/Fe^{2+} ratio, Mitsubishi process

Abstract

The distribution behavior of Pb between molten copper and Fe_tO-SiO_2 (-CaO, Al_2O_3) slags was investigated at 1473 K (1200 °C) and $p(O_2)=10^{-10}$ atm in view of the reaction mechanism of Pb dissolution into the slag. The distribution ratio of Pb (L_{Pb}) decreases with increasing CaO content (~6 mass pct) irrespective of Fe/SiO_2 ratio (1.4~1.7). However, the addition of alumina into a slag with Fe/SiO_2=1.5 linearly decreases the L_{Pb}, whereas a minimum value is observed at about 4 mass pct Al_2O_3 at Fe/SiO_2=1.3. The log L_{Pb} continuously decreases with increasing Fe/SiO_2 ratio, and the addition of Al_2O_3 (5 to 15 mass pct) into the silica-saturated iron silicate slag (Fe/SiO_2 < 1.0) yields the highest Pb distribution ratio. The log L_{Pb} linearly decreases by increasing the log (Fe^{3+}/Fe^{2+}) value. The Pb distribution ratio increases and the excess free energy of PbO decreases with increasing Cu_2O content in the slag. However, from the viewpoint of copper loss into the slag, the silica-saturated system containing small amounts of alumina is strongly recommended to stabilize PbO in the slag phase at a low Cu_2O content.

Introduction

It is very important in copper smelting and refining processes to remove several minor elements such as Pb, Bi, Sb, As, and Te [1-2]. Specifically, Pb is a troublesome impurity which increases the amount of slime in the electro-refining process. Also, from the viewpoint of copper smelting and refining processes, Pb is a detrimental element that is always present in sulfide concentrates and must be eliminated. Therefore, the content of Pb in molten copper should be lowered within an appropriate concentration range during plant operation.

Comparing Outokumpu flash smelting process and Mitsubishi continuous smelting process (Figure 1), there is a difference in the content of Pb in anode Cu as shown in Figure 2, which probably originates from a difference in slag chemistry in between both processes (The concentrate was nominally the same in each case).

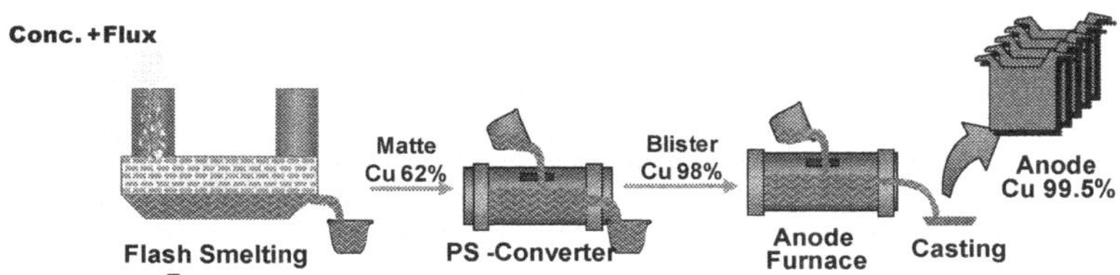

(a)

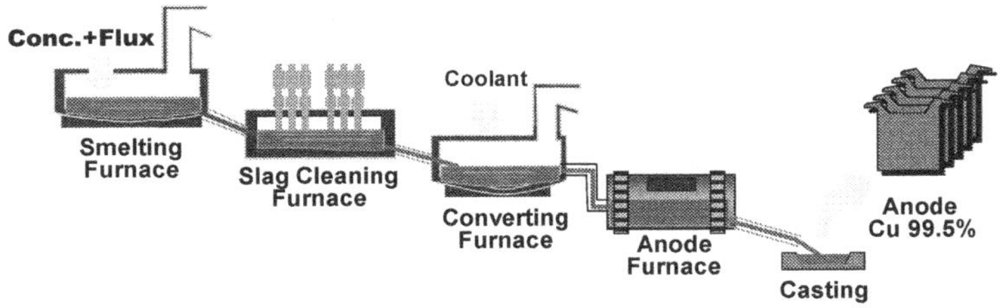

(b)

Figure 1. (a) Outokumpu flash smelting process and (b) Mitsubishi continuous smelting process.

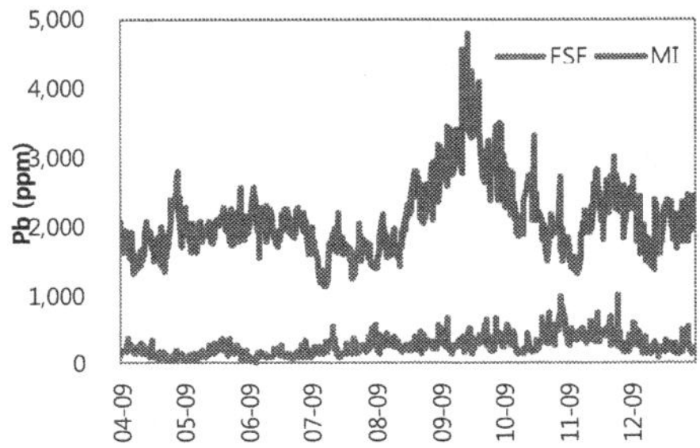

Figure 2. Residual Pb content in Cu anode in Outokumpu flash smelting process (FSF) and Mitsubishi continuous smelting process (MI).

Therefore, we need to increase the Pb removal ratio in the Mitsubishi continuous smelting process based on the fundamental knowledge for the slag chemistry of Pb removal.

Kim and Sohn examined the effects of CaO, Al_2O_3, and MgO additions on the distribution behavior of Pb between an iron silicate slag and liquid copper under a CO-CO_2 atmosphere $p(O_2)=10^{-12} \sim 10^{-6}$ atm at 1523 K (1250 °C) and found that the distribution ratio of Pb and the activity coefficient of PbO were not significantly affected by the additives [3]. The governing factors affecting Pb solubility in iron silicate slag were proposed as the oxygen potential and the ratio of Fe_tO/SiO_2 (or SiO_2 content).

Matsuzaki *et al.* [4] calculated the activity coefficient of PbO (γ_{PbO}) with the regular solution model and plotted the iso-γ_{PbO} contours for the CaO-SiO_2-Al_2O_3 ternary slag at 1623 K (1350 °C). The value of γ_{PbO} decreased as CaO was replaced with SiO_2 at any fixed concentration of Al_2O_3, indicating the basic character of PbO in the slag. A similar tendency of the effect of CaO/SiO_2 ratio on the γ_{PbO} was also reported by Kudo et al. [5] using the Fe_tO-CaO-SiO_2 system in equilibrium with metallic iron at 1573 K (1300 °C). In that study [5[, the changes in temperature from 1473 to 1573 K (1200 to 1300 °C) did not have a significant effect on γ_{PbO} in the slag. Decterov and Pelton computed the distribution behavior of Pb among matte, slag, and liquid copper at 1523 K (1250 °C) and $p(O_2)=10^{-12} \sim 10^{-5}$ atm under conditions from silica saturation to magnetite saturation [6]. The distribution ratio of Pb decreased with increasing Fe/SiO_2 ratio from silica to magnetite saturation. The small addition of CaO (3 mass pct) also decreased the Pb distribution ratio.

Even though the effectiveness of slag treatment for the removal of Pb from the Cu melt has been investigated as mentioned above, many studies focused on the equilibrium distribution behavior of Pb in the Cu/matte/slag system during the smelting and converting processes. Therefore, in the present study, the distribution behavior of Pb between molten copper and Fe_tO-SiO_2 (-CaO, Al_2O_3) slags was investigated from the view of the reaction mechanism of Pb dissolution into the slag.

Experimental

The experiments were carried out using a super-kanthal electric furnace with a $MoSi_2$ heating element. An alloy sample with the composition of Cu-1 mass pct Pb was preliminarily prepared from pure (99.99 pct purity) Cu and Pb. The slag samples were prepared by mixing reagent-grade FeO, SiO_2, Al_2O_3, and CaO calcined from $CaCO_3$ at 1273 K (1000 °C) for 10 hours. The equilibration time was predetermined to be 6 hours at 1473 K (1200 °C) [7].

The metal (8 g) and slag (5 g) were loaded in a fused magnesia (99.9 pct purity) crucible (OD: 18 mm, ID: 14 mm, HT: 50 mm) placed in an alumina holder (OD: 50 mm, ID: 45 mm, HT: 70 mm) and attached to a molybdenum wire. The oxygen partial pressure of the system, $p(O_2)=10^{-10}$ atm, was controlled by the following reaction [8],

$$CO(g)+\frac{1}{2}O_2(g)=CO_2(g),\quad \Delta G^{o}=-280700+85.2\,T\ (J/mol) \qquad [1]$$

Under the same experimental conditions as mentioned above, additional experiments were carried out for the equilibration between the silica-saturated Fe_tO-SiO_2-Al_2O_3 slags (5 g) and copper melt (8 g) in a fused quartz crucible (OD; 20 mm, ID; 16 mm, HT; 40 mm) in order to investigate the effect of Al_2O_3 addition on Pb distribution ratio between molten copper and silica saturated slag.

The composition of metal and slag samples was determined using inductively coupled plasma spectrometry. Furthermore, the ferric (Fe^{3+}) to ferrous (Fe^{2+}) ratio in the slags was determined by titration using a potassium dichromate.

Results and Discussion

Influence of Additives (CaO, Al_2O_3) and Fe/SiO_2 Ratio on the Distribution Ratio of Pb

The distribution ratio of Pb (L_{Pb}) between molten copper and Fe_tO-SiO_2 (-CaO, Al_2O_3) slags at 1473 K (1200 °C) and $p(O_2)=10^{-10}$ atm is shown in Figure 3 as a function of the CaO and Al_2O_3 content, respectively, at different Fe/SiO_2 ratios. The distribution ratio of Pb between metal and slag phases in this study is defined as follows:

$$L_{Pb}=\frac{(\text{mass pct Pb})_{slag}}{[\text{mass pct Pb}]_{metal}} \qquad [2]$$

The value of L_{Pb} decreases with increasing CaO content irrespective of Fe/SiO_2 ratio. This tendency is qualitatively similar to the results in the literature even at different temperatures and oxygen potentials [3-6], indicating that CaO unambiguously behaves as a basic component (Eq. [4]) in the iron silicate slag based on the following ionic dissolution mechanism.

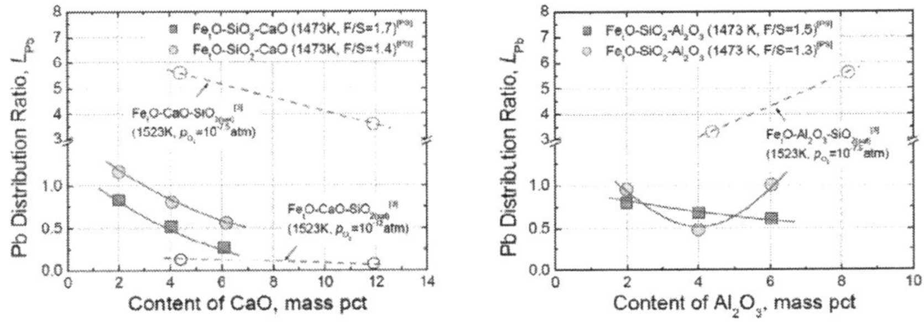

Figure 3. Influence of CaO and Al_2O_3 on the distribution ratio of lead between molten copper and Fe_tO-SiO_2-CaO slag at 1473 K (1200 °C) and $p(O_2)$= 10^{-10} atm. [PS]: present study.

$$[Pb]+\frac{1}{2}O_2(g)=(Pb^{2+})+(O^{2-}) \quad [3]$$

$$(CaO) = (Ca^{2+}) + (O^{2-}) \quad [4]$$

The addition of alumina in the iron silicate melts with the Fe/SiO_2=1.5 linearly decreases the Pb distribution ratio, whereas a minimum value is observed at about 4 mass pct Al_2O_3 with Fe/SiO_2=1.3. Therefore, alumina is considered as a basic component in the former system (Fe/SiO_2=1.5) and in the composition of Al_2O_3 content less than 4 mass pct in the latter system (Fe/SiO_2=1.3) based on Eq. [5]. However, the alumina behaves as an acidic component when Al_2O_3 content greater than 4 mass pct in the latter system (Fe/SiO_2=1.3) based on Eq. [6].[9]

$$(Al_2O_3) = 2(Al^{3+}) + 3(O^{2-}) \quad [5]$$

$$(Al_2O_3) + 5(O^{2-}) = 2(AlO_4)^{5-} \quad [6]$$

The value of L_{Pb} between molten copper and Fe_tO-SiO_2 (-CaO, Al_2O_3) slags is shown in Figure 4 as a function of Fe/SiO_2 ratio. The log L_{Pb} continuously decreases with increasing Fe/SiO_2 ratio and this tendency is in good accordance with the literature data [1,6]. Furthermore, the addition of Al_2O_3 (5 to 15 mass pct) into the silica saturated iron silicate slag (Fe/SiO_2 < 1.0) shows the highest Pb distribution ratio.

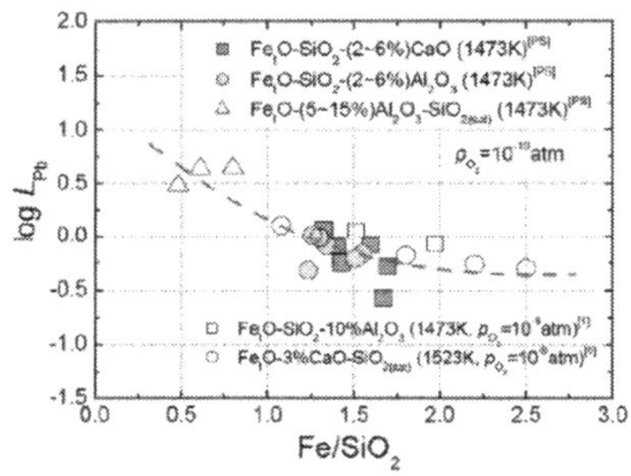

Figure 4. Influence of Fe/SiO_2 ratio on log L_{Pb} in the Fe_tO-SiO_2 (-CaO, Al_2O_3) slags at 1473 K (1200 °C) and $p(O_2)$= 10^{-10} atm. [PS]: present study.

Influence of Fe^{3+}/Fe^{2+} Ratio on the Distribution Ratio of Lead

Generally, the ferric ion, Fe^{3+} is suggested to be predominant rather than ferrous ion, Fe^{2+} by addition of basic oxide such as CaO, which produces the free O^{2-} ions. Therefore, the equilibrium between Fe^{2+} and Fe^{3+} ions is expressed as follows [9].

$$(Fe^{2+}) + \frac{1}{4}O_2(g) + \frac{3}{2}(O^{2-}) = (FeO_2^-) \qquad [7]$$

By combining Eqs. [3] and [7], the Pb↔Fe exchange reaction can be deduced as follows.

$$\frac{3}{2}[Pb] + (Fe^{2+}) + O_2(g) = \frac{3}{2}(Pb^{2+}) + (FeO_2^-) \qquad [8]$$

From the equilibrium constant of Eq. [8] ($K_{[8]}$), the relationship between Fe^{3+}/Fe^{2+} ratio and Pb distribution ratio is deduced as follows.

$$\log L_{Pb} = -\frac{2}{3}\log\left(\frac{pct\,Fe^{3+}}{pct\,Fe^{2+}}\right) - \frac{2}{3}\log\left(\frac{f_{Pb^{2+}}^{1.5}\cdot f_{FeO_2^-}}{f_{Fe^{2+}}}\right) + \log f_{Pb} + \frac{2}{3}\log p_{O_2} + \log K_{[8]} \qquad [9]$$

Where, f_i and p_{O_2} are the activity coefficient of species i and the oxygen partial pressure, respectively. Thus, log L_{Pb} is expected to be inversely proportional to log (Fe^{3+}/Fe^{2+}) with the slope of -2/3 at a given temperature and oxygen potential assuming that the second activity coefficient term on the right hand side in Eq. [9] would not be significantly affected by slag composition. Figure 5 shows a relationship between the log L_{Pb} and log (Fe^{3+}/Fe^{2+}), wherein the linear relationship between them with the slope of about -0.44 is obtained from a linear regression analysis.

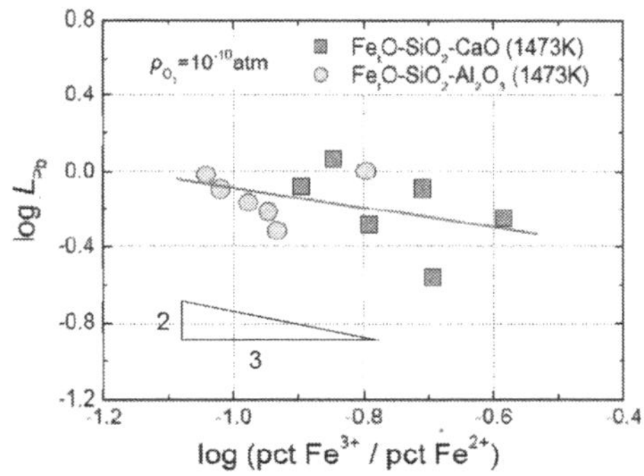

Figure 5. Relationship between log L_{Pb} and log (Fe^{3+}/Fe^{2+}) in the Fe_tO-SiO_2 (-CaO, Al_2O_3) slags at 1473 K (1200 °C) and $p(O_2)$= 10^{-10} atm.

The slope of the line is slightly lower than the expected value (-0.67) in Eq. [9], which probably originates from the attraction between Pb^{2+} and ferric complex ions with increasing Fe^{3+}/Fe^{2+} ratio. This is thermodynamically consistent to the fact that there are several stable lead ferrite compounds such as $Pb_2Fe_2O_5$ (2:1, δ-phase), $PbFe_4O_7$ (1:2, γ-phase), and $PbFe_{12}O_{19}$ (1:6, β-phase) [10].

Influence of Cu_2O on the Distribution Ratio of Lead and Stability of PbO in Slag Phase

The effect of Cu_2O on the Pb distribution ratio between molten copper and iron silicate slags at 1473 K (1200 °C) is shown in Figure 6.

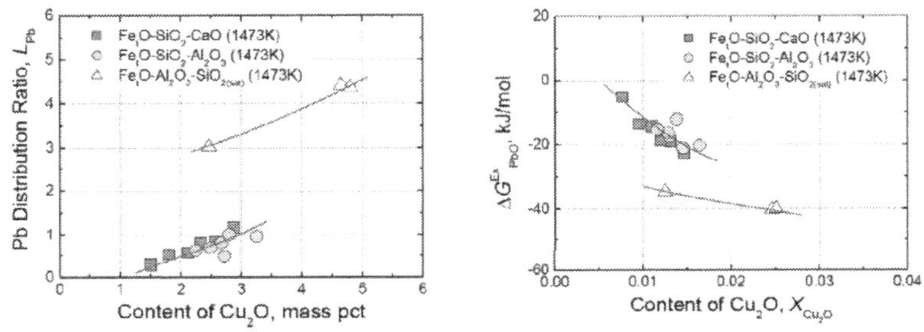

Figure 6. Influence of Cu_2O content on the distribution of lead between molten copper and Fe_tO-SiO_2 (-CaO, Al_2O_3) slags at 1473 K (1200 °C) and $p(O_2)= 10^{-10}$ atm (Left) and Excess free energy of PbO in the Fe_tO-SiO_2 (-CaO, Al_2O_3) slags at 1473 K (1200 °C). (Right).

The Pb distribution ratio increases with increasing content of Cu_2O in the slag. It was known from the previous studies that Pb and Cu are present in the slag as PbO and Cu_2O in molecular form, respectively [4], and thus the following equilibrium reaction is given at slag/metal interface.

$$[Pb] + (Cu_2O) = (PbO) + 2Cu(l) \quad [10]$$

$$K_{[10]} = \frac{a_{PbO}}{a_{Pb} \cdot a_{Cu_2O}} = \frac{\gamma_{PbO} \cdot X_{PbO}}{\gamma_{Pb} \cdot X_{Pb} \cdot \gamma_{Cu_2O} \cdot X_{Cu_2O}} \quad [11]$$

Where, $K_{[10]}$, a_i, γ_i and X_i are the equilibrium constant of Eq. [10], the activity, the activity coefficient and the mole fraction of component i, respectively.

On the other hand, the activity coefficient (γ_{PbO}) and thus an excess free energy of PbO (ΔG^{EX}_{PbO}) can be estimated from the following equations [11].

$$[Pb]_{Cu} + \frac{1}{2}O_2(g) = PbO(l), \quad \Delta G^o = -217270 + 78.8\,T \text{ (J/mol)} \quad [12]$$

$$\gamma_{PbO} = \frac{K_{[12]} \cdot \gamma^o_{Pb} \cdot X_{Pb} \cdot \sqrt{p_{O_2}}}{X_{PbO}} \quad [13]$$

$$\Delta G_{\text{PbO}}^{\text{Ex}} = -RT \ln \gamma_{\text{PbO}} \qquad [14]$$

where γ_{Pb}^{o} is the Henrian activity coefficient of Pb in molten copper, which is $\gamma_{Pb}^{o} = 5.27$ at 1473 K (1200 °C) [11]. The excess free energy of PbO calculated from Eqs. [13] and [14] is plotted against the mole fraction of Cu_2O in Fig. 4. The excess free energy of PbO decreases with increasing Cu_2O content, indicating that Cu_2O is more acidic than PbO in the copper smelting iron silicate slags. However, from the viewpoint of copper loss into the slag, the silica-saturated system containing small amounts of alumina is strongly recommended to stabilize PbO in the slag phase at low Cu_2O content.

Conclusions

The distribution behavior of Pb between molten copper and Fe_tO-SiO_2 (-CaO, Al_2O_3) slags was investigated at 1473 K (1200 °C) and $p(O_2)=10^{-10}$ atm to observe the reaction mechanism of Pb dissolution into the slag. The experimental results are summarized below.

1) The value of L_{Pb} decreases with increasing CaO content (~6 mass pct) irrespective of Fe/SiO_2 ratio (1.4~1.7). However, the addition of alumina in the slag with $Fe/SiO_2=1.5$ linearly decreases the L_{Pb}, and there is a minimum value observed at about 4 mass pct Al_2O_3 at $Fe/SiO_2=1.3$.
2) The log L_{Pb} continuously decreases with increasing Fe/SiO_2 ratio, and the addition of Al_2O_3 (5 to 15 mass pct) into the silica-saturated iron silicate slag ($Fe/SiO_2 < 1.0$) shows the highest Pb distribution ratio.
3) The log L_{Pb} and log (Fe^{3+}/Fe^{2+}) values show a linear relationship with a slope of about -0.44, which was slightly lower than the expected value of -2/3. This probably originates from the attraction between Pb^{2+} and ferric complex ions with increasing Fe^{3+}/Fe^{2+} ratio.
4) The Pb distribution ratio increases and the excess free energy of PbO decreases with increasing Cu_2O content in the slag, indicating that Cu_2O is relatively more acidic than PbO in the copper smelting iron silicate slags.

References

1. M. Nagamori, P.J. Mackey and P. Tarassoff, "Copper Solubility in FeO-Fe_2O_3-SiO_2-Al_2O_3 Slag and Distribution Equilibria of Pb, Bi, Sb and As Between Slag and Metallic Copper", *Metall. Trans, B* 6B (1975) 295-301.
2. A.K. Kyllo and G.G. Richards, "Kinetic Modeling of Minor Element Behavior on copper converting", *Metall. Mater. Trans. B* 29B (1998) 261-68.
3. H. G. Kim and H.Y. Sohn, "Effects of CaO, Al_2O_3, and MgO Additions on the Copper Solubility, Ferric/Ferrous Ratio, and Minor-Element Behavior of Iron-Silicate Slags", *Metall. Mater. Trans. B* 29B (1998) 583-90.
4. K. Matsuzaki, T. Ishikawa, T. Tsukada and K. Ito, "Distribution Equilibria of Pb and Cu between CaO-SiO_2-Al_2O_3 Melts and Liquid Copper", *Metall. Mater. Trans. B* 31B (2000) 1261-66.
5. M. Kudo, E. Jak, P. Hayes, K. Yamaguchi and Y. Takeda, "Lead Solubility in FeO_x-CaO-SiO_2 Slags at Iron Saturation", *Metall. Mater. Trans. B* 31B (2000) 15-24.

6. S.A. Decterov and A.D. Pelton, "Thermodynamic modeling of Lead distribution among Matte, Slag, and Liquid Copper", *Metall. Mater. Trans. B* 30B (1999) 1033-44.
7. J.H. Heo, S.S. Park and J.H Park, "Effect of Slag Composition on the Distribution Behavior of Pb between Fe_tO-SiO_2 (-CaO, Al_2O_3) Slag and Molten Copper", *Metall. Mater. Trans. B* 43B (2012) 1098-1105.
8. E.T. Turkdogan, *Physical Chemistry of High Temperature Technology*, Academic Press, New York, NY, (1980) 1-24.
9. N. Sano, Thermodynamics of Slags (Chapter 2), in *Advanced Physical Chemistry for Process Metallurgy*, ed. by N. Sano, W.K. Lu, P.V. Riboud and M. Maeda, Academic Press, San Diego, CA, (1997) 46-51.
10. A.J. Mountvala and S.F. Ravitz, "Phase Relations and Structures in the System PbO-Fe_2O_3", *J. Am. Ceram. Soc.* 45 (1962) 285-88.
11. G.K. Sigworth and J.F. Elliot, "The thermodynamics of dilute liquid copper alloys", *Can. Metall. Q.* 13 (1974) 455-61.
12. S. Ban-ya, "Mathematical Expression of Slag-Metal Reactions in Steelmaking Process by Quadratic Formalism Based on the Regular Solution Model", *ISIJ Int.* 33(1993) 2-11.

Celebrating the Megascale: Proceedings of the Extraction and Processing Division Symposium on Pyrometallurgy in Honor of David G.C. Robertson
Edited by: Phillip J. Mackey, Eric J. Grimsey, Rodney T. Jones, and Geoffrey A. Brooks
TMS (The Minerals, Metals & Materials Society), 2014

SIMULATION OF THE GAS FLOW IN A PEIRCE-SMITH CONVERTER

Wagner Moulin Silva[1], Bruno Augusto Batista Ribeiro[1], Felipe Terra Elias[1]

[1]Magnesita Refratários S.A.; Praça Louis Ensch 240, Contagem, MG, 32210-902, Brazil

Keywords: Numerical Simulation, Peirce-Smith Converter, Refractory, Gas Injection

Abstract

In the pyrometallurgical production of copper from its sulfides, converting in the Peirce Smith (P-S) Converter is still the most widespread method to obtain white metal and blister copper from the matte. Despite the fact of being a century-old process, it is still possible to achieve improvements in the process, by the use of modern tools, one of the most valuable being numerical calculation and computer-aided simulation. The present study presents three cases in which the angle for the gas injection and the number of working tuyeres are changed during the slag blow phase of the converting process. It was noticed that a decrease of the number of tuyeres, followed by an increase in the injection gas velocity, results in improved reaction volume and less stress to the refractory lining. However, an increase in the amount of copper lost to the slag as oxide is likely to occur.

Introduction

In the pyrometallurgical production of copper from its sulfides, converting is a key process for the refining of matte into blister copper. More than 90% of all converters in use worldwide are Peirce-Smith converters. A key parameter for the efficient and economic converting in this horizontal vessel is the flow pattern of gas through the tuyeres and inside the metal. In order to evaluate the flow during the injection of gas into the molten matte and/or metal, water models have been traditionally in use, and are still an important tool to visualize flow patterns in molten media [1,2]. However, the limitations of this methodology impair a full overview of the real phenomena that take place during the production of metals through pyrometallurgical treatment.

The improvement of computational power and of numerical tools, especially through ever higher mathematical sophistication of standard softwares, such as ANSYS™ CFX, ANSYS™ FLUENT and ABAQUS™ FEA, reduces the working time for experts to elaborate useful fluid-dynamic models that can predict metallurgical behavior by using real properties of the fluids in real-sized geometries, with low cost and high accuracy. Thus, an increasing number of technical papers in the literature present numerical models for various different metallurgical processes and vessels, among which references [2-5] can be cited.

The present work analyses numerically the behavior of the gas injected in a P-S converter during the slag blow stage of the process. The major aim of the work was to evaluate the impact on the treatment time and infer the effect on the refractory lining of two changes in the process: (i) change in the total number of open tuyeres from 42 to 30 (maintaining the total gas flow as a function of time), and (ii) changing the inclination of the tuyeres during the injection of air from 7° (as relative to the floor level) to 0° (effectively parallel to the floor surface).

Methodology

The software ANSYS™ 14.0 CFX was used to calculate numerically the cases described in this paper. The P-S converter was drawn in a 1:1 scale with AUTOCAD™ 2012 and SolidWorks™ 2012 CAD tools. The dimensions of the converter were entered into the CFX software, and a mesh was generated. For the present case, the geometry was considered as a cylindrical barrel, since the mouth of the converter does not take part in the fluid-dynamic processes studied inside the melt. This barrel-like geometry can be easily modeled with a hexahedral mesh, with good detail and resolution.

Some of the boundary conditions considered are hereby given: (i) metal = copper matte; (ii) gas = air; and (iii) matte temperature = 1,250 °C. The turbulence model adopted was the standard K-Epsilon. One advantage of this model for the present work is that it provides the turbulent kinetic energy in one of its two basic equations. Refining of the mesh near the walls of the converted was according to a variable Y+ model supplied by the CFD software. The use of a variable Y+ parameter (a non dimensional measurement for the distance of a given point to the wall) is necessary to account for the changes in turbulence according to the variation of the Reynolds number, thus providing a better definition close to the tuyeres, the critical part of the model. Other parameters, such as air pressure in the bustle pipe, diameter of the tuyeres, flow rate and velocity of gas at each tuyere are proprietary data and not disclosed in this paper. No slag layer was considered at this stage of the work.

Numerical Simulation Results

First Case: 7° Angle and 42 Open Tuyeres

Figure 1 presents four representative slices of the converter during the injection of air, whereas Figure 2 depicts a transverse cut of a section containing a tuyere/tuyere brick, for better visualization. As can be noticed, the volume of matte reacted with gas is small, compared to the bulk, and matte is projected to the top of the converter. Thus, despite the high air velocities involved in the process, penetration depth is poor, and a significant amount of the energy contained in the blown air is spent to project matte droplets out of the bath region. "Dead zones", i.e. zones with low or no stirring, are present close to the end walls and in the bulk of the matte bath. This behavior of the injected air was already described by previous authors [5].

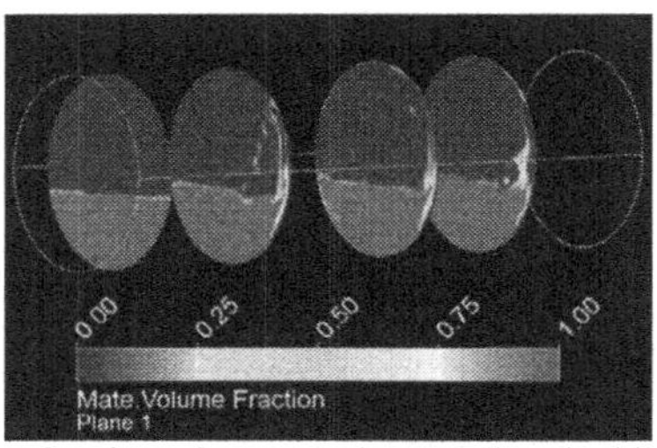

Figure 1. Volume fraction of matte and gas during the blow stage in the Converter. Red color indicates matte and blue color indicates air. Lighter colors are representative of regions with mixture of gas and matte.

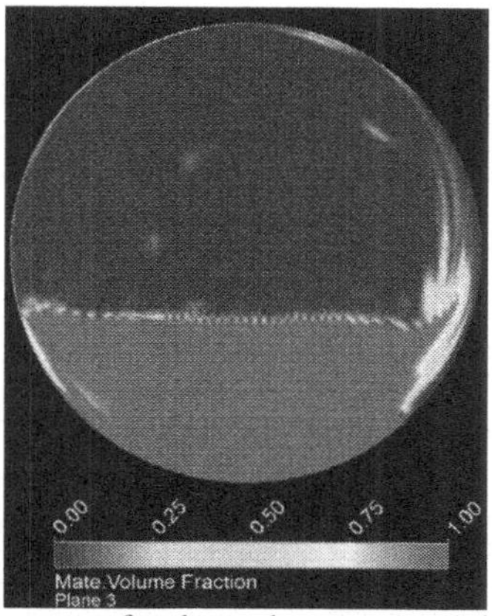

Figure 2. Transverse cut of a plane where a tuyere brick is located (on the right hand side). It can be noticed that, due to the movement of the matte, a zone with possible reaction between gas and matte appears in the opposite side to the tuyere.

The preferred way to qualitatively evaluate how effectively the momentum of the gas is transferred to the matte is by plotting the relative kinetic energy inside the bath and on the surrounding atmosphere and evaluating the turbulence inside the fluids, as shown in Figure 3. The relative kinetic energy is normalized to the mass of matte and gas, and is used for a better comparison of the three cases, in order to rule out local effects due to large differences in mass and velocity.

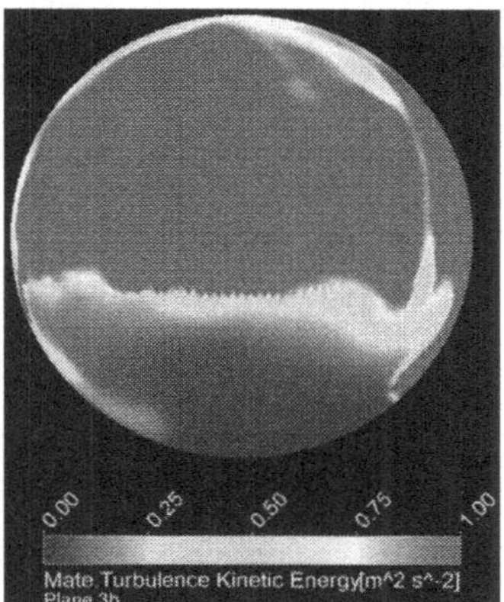

Figure 3. Relative kinetic energy in $m^2.s^{-2}$ as a measurement of turbulence in the same cut as presented in Figure 2. Refer to text for explanation.

It is possible to see that, due to the very low residence time of the gas inside the molten matte, turbulence in the atmosphere of the converter is roughly the same as that observed in the exit of the tuyere. The air flows onto the surface of the refractory lined vessel and generates turbulence mostly on the surface of the bath and on the opposite side. This kind of movement on the surface of the bath is probably related to the wave oscillation behavior described elsewhere for Teniente Converters [2].

Second Case: 7° Angle and 30 Open Tuyeres

For this study, six tuyeres in the center of the barrel were considered plugged, as well as the three tuyeres closest to each of the end walls were also considered being closed. Figures 4a and 4b present the volume fraction of matte and the kinetic energy in the same cut considered for Figures 2 and 3, respectively.

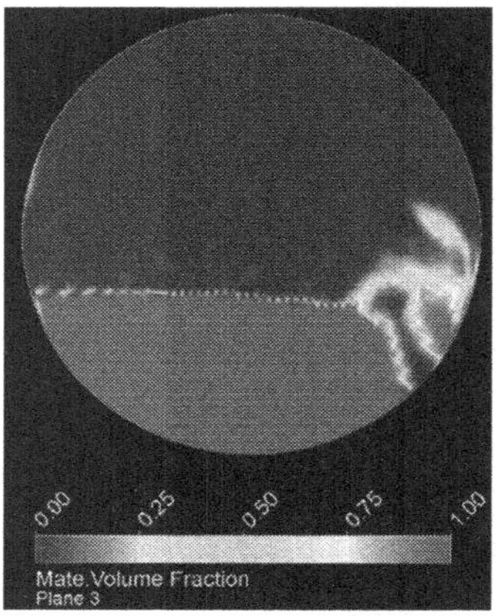

(a)

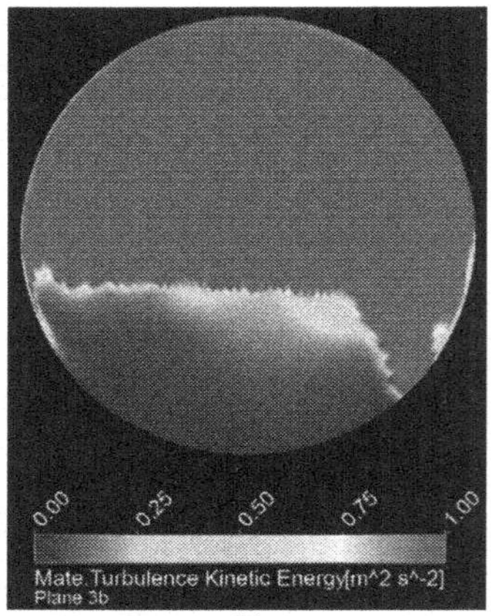

(b)

Figure 4. Second case: (a) Transversal cut of a plane containing a tuyere brick, presenting the volumetric fraction of matte and air during the injection of gas stage; (b) Kinetic energy in $m^2.s^{-2}$ as a measurement of turbulence.

One can notice that, with the closure of 12 tuyeres, the increase in the pressure of the blow drives the gas towards the melt, thus reducing the splash of matte and driving the gas flow out of the surface of the refractory lining. The turbulence of the melt is also reduced, especially in the region opposite to the blow. The kinetic energy is more effectively transferred to the melt in the region where the tuyeres are located in the present case than in the previous one, but a loss close to the end walls is observed, as would be expected.

Third Case: 0° Angle and 30 Open Tuyeres

For these calculations, the same array of open tuyeres was set as in the previous case, but the air was considered to blow parallel to the floor. Figures 5a and 5b present the volume fraction of matte and the kinetic energy, respectively, in the same cut as considered for Figure 4.

As a consequence of the neutral inclination towards the metal (at the 7° case, the air is injected downwards), a smaller volume of interaction area is perceived in this case, despite the higher gas velocity observed, as compared to the first case. The air flow is deviated from the plane depicted, thus no exit for the gas is observed in Figure 5. Nonetheless, it can be observed less matte splash and also a much smaller turbulence of the melt, as indicated by the lower volume with positive kinetic energy.

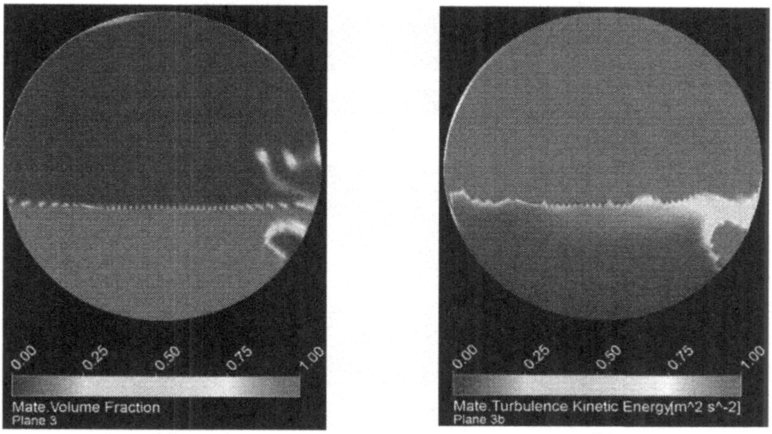

(a) (b)

Figure 5. Third case: (a) Transversal cut of a plane containing a tuyere brick, presenting the volumetric fraction of matte and air during the injection of gas stage; (b) Kinetic energy in $m^2.s^{-2}$ as a measurement of turbulence.

Comparison Between the Three Cases – a Study of Kinetic Energy

The numerical comparison of the three cases presented was best represented by means of the study of the kinetic energy inside the melt, since it generates necessary turbulence to improve the converting reactions in the matte. Through our calculations, a threshold of 0.1 $m^2.s^{-2}$ was considered as a boundary between regions of low and high kinetic energy in the melt. The larger the volume with a thus defined high kinetic energy, the better is the efficiency of the gas injection process. Figures 6 to 8 present the low energy surfaces calculated for each of the three cases. The calculated volume of low kinetic energy was:

- First case: 18.4 m^3
- Second case: 18.1 m^3
- Third case: 21.7 m^3

The Figures plotted show a much flatter surface of turbulence for the case where air is blown at an angle of 0°, while the reduction of the number of tuyeres in the 7° case lead to a deeper penetration of the process gas. Nevertheless the mixture close to the end walls is less effective, and results in roughly an equal volume as obtained in the original case.

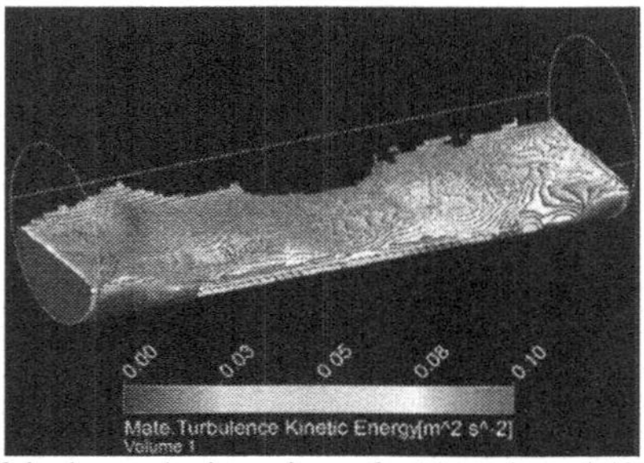

Figure 6. Plot of the low velocity volume for 7° angle and 42 open tuyeres case.

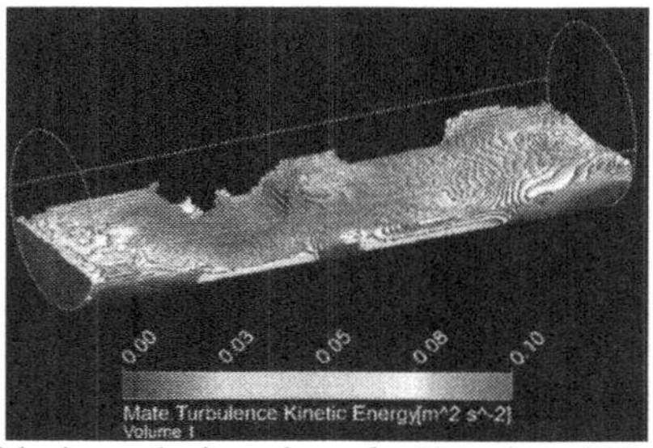

Figure 7. Plot of the low velocity volume for 7° angle and 30 open tuyeres case.

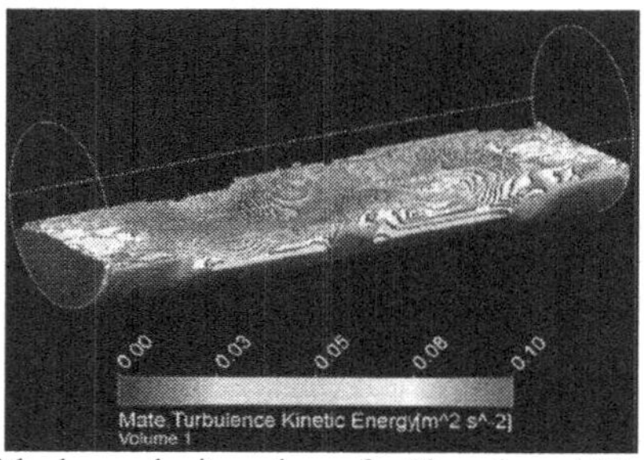

Figure 8. Plot of the low velocity volume for 0° angle and 30 open tuyeres case.

Discussion of the Results and Final Considerations

From the point of view of matte converting, the volume of melt effectively treated during the blowing of gas is more dependent on the injection angle than on the velocity of the gas, considering a same flow rate as in the bulk of the melted matte. The closure of some tuyeres improves the depth of penetration of the gas, but this is counter-effected by an increase in volume of dead areas, since the tendency is that the injected gas is rapidly driven upwards by the difference in density. Moreover, the high viscosity of the matte reduces the transfer of energy between the areas frontal and lateral to the tuyere blast.

However, the increase in the volume of the calculated dead zones may pose a real problem to the quality of the white metal produced, as well as to the potential loss of copper to the slag as cuprous oxide. If the change of mass between the dead zones and the turbulent area in the tuyere front volume is ineffective, the process gas will start to oxidize the copper in the regions depleted in iron sulfide close to the tuyeres, while the regions far from them will still present volumes of iron sulfide in the matte above acceptable levels. By our results, it is not expected a significant mass transfer due to the bath turbulence. Nonetheless, the movement of the converter during matte and scrap charge, as well as slag skimming should proportionate enough mixing to minimize this issue.

The poorest scenario is with the tuyeres at 0°, at which a longer blow time would be needed to achieve the same degree of conversion. Even though the lower turbulence of the melt can have a positive impact on the entrapment of matte in the slag, as well as improve the rate of settling of the entrapped droplets, the residence time of the gas and the reaction area would be reduced. Hence, the total amount of iron sulfide oxidized by the process gas would be diminished in a given time.

As for the expected refractory life, the major cause of wear is the very high erosion and spalling that occur as a result of the high turbulence close to the areas where gas is injected. As the simulations show, this high turbulence at the refractory surface occurs as a result of insufficient air velocity to drive the blast towards the matte, instead of allowing it to flow over the lining. This flow is not only harmful for the integrity of the refractory, but also reduces the interaction volume between the gas and the melt. The process gas that flows onto the refractory may also oxidize entrained metal, thus causing the oxidation and spalling by the increase in volume during the formation of cuprous oxide. Thus, the case in which 30 tuyeres are used for blowing air instead of 42, it is less demanding for the refractory life, and an extended service life for the lining is expected. Moreover, the reduction in the splash of matte on the aerial areas of the converter should also reduce refractory wear and skull formation.

Trials are ongoing at the copper smelter, and results are not yet consolidated. The reduction in the number of tuyeres is one part of a full process review to increase the refractory life and predictability of campaign of the converters (change in the charging practice and reduction in the total metal load are two other important measures). Thus, the impact of this measure by itself could not be properly evaluated. At the time this paper went to press, life of the refractory lining in the tuyere and surroundings increased from a range between 20,000 to 33,000 metric tons to a narrower gap of 39,600 to 45,000 metric tons. Not only is campaign life improved, but also the variability is reduced.

The presented model is incomplete, since no slag is considered inside the converter. The next steps for the work should be the incorporation of a slag layer on the top of the matte, and to verify its effect on the calculated patterns. No draft force was considered either, and this should bring changes to the observed turbulence in the aerial areas, and may eventually decrease the reaction in the regions opposite to the tuyeres, favoring higher gas velocities over larger number of tuyeres. After the improvement of this part of the process, simulation of the metal blow stage is planned, at which differences may be noticed due to the different properties of the blister copper.

Conclusions

Despite the simplicity of the model, the main objective of the work, which was evaluating the impact of such changes on the refractory life, was achieved. The most important observations are:

- An improvement in the efficiency of the conversion of matte to white metal is expected by the reduction of the number of working tuyeres and, thus, an increase in the velocity of the injected gas;
- The volume of dead zones in the melted matte is increased by the reduction of the number of active tuyeres, mostly close to the end walls. If all the blocked tuyeres were positioned in the center of the barrel, this effect would probably be minimized, but a dead zone might have appeared in the area with no active blowing;
- The increase in the gas velocity during the blowing phase will probably result in increased refractory lining life;
- Higher injection velocities also lead to fewer splashes and to reduction in the formation of skull in the converter.

Acknowledgement

The authors would like to express their thankfulness to Magnesita Refratários S.A., for allowing the publication of this paper.

References

1. C. López et al., *Experimental Estimation of the Residence Time Distribution in a P-S Converter* (Proceedings of the Copper 2010 Conference, Clausthal-Zellerfeld, Germany, GDMB, 2010), Vol. 2, 919-930.

2. M. Rosales et al., *A Fluid-Dynamic Review of the Teniente Converter* (Proceedings of the Cu 2010 Conference, Clausthal-Zellerfeld, Germany, GDMB, 2010), Vol.3, 1095-1113.

3. Chen Zhuo et al., *Simulation Study of Intensified Flash Smelting Process* (Copper 2010, Proceedings of the Copper 2010 Conference, Clausthal-Zellerfeld, Germany, GDMB, 2010), Vol. 3, 1313-1323.

4. Zhou Jun et al., *Numerical Simulation of Fluid Flow and Melt Temperature in Settler* (Proceedings of the Copper 2010 Conference, Clausthal-Zellerfeld, Germany: GDMB, 2010), Vol. 2, 823-830.

5. C. López et al., *Numerical Simulation of Air Blowing into a Copper Matte in a P-S Converter Using a Convergent / Divergent Nozzle* (Proceedings of the Copper 2010 Conference, Clausthal-Zellerfeld, Germany: GDMB, 2010), Vol. 2, 931-942.

Celebrating the Megascale: Proceedings of the Extraction and Processing Division
Symposium on Pyrometallurgy in Honor of David G.C. Robertson
Edited by: Phillip J. Mackey, Eric J. Grimsey, Rodney T. Jones, and Geoffrey A. Brooks
TMS (The Minerals, Metals & Materials Society), 2014

FROM PHASE EQUILIBRIUM AND THERMODYNAMIC MODELING TO FREEZE LININGS – THE DEVELOPMENT OF TECHNIQUES FOR THE ANALYSIS OF COMPLEX SLAG SYSTEMS

Ata Fallah Mehrjardi, Peter Hayes, Evgueni Jak

PYROSEARCH, School of Chemical Engineering, The University of Queensland,
Brisbane, Queensland, 4072, Australia

Keywords: Phase equilibria, Freeze lining

Abstract

Modern analytical tools, such as electron probe X-ray microanalysis (EPMA), are used in experimental characterization of phase equilibria and microstructures of complex slags. Integrated thermodynamic computer packages, such as FactSage, are used to provide more accurate descriptions of complex slag systems. These advanced methodologies have been applied to characterize dynamic steady state freeze linings in slag systems using submerged cold finger probes at controlled laboratory conditions, establishing the effects of bath chemistry, temperature, heat extraction rate and bulk fluid flow.

It has been found that stationary freeze lining deposit interface temperatures at steady state conditions can be lower than the bulk slag liquidus temperature, and that stable operation below the liquidus is possible. A conceptual framework has been developed to explain the phenomenon and the range of interface temperatures that can be obtained in dynamic steady state conditions. These findings have important implications for the design of freeze linings and possible improvements to high-temperature metallurgical operations.

Introduction

The control of slag chemistry is critical to metallurgical processes influencing a wide range of properties including liquidus temperatures, proportion of solids present, and the partitioning of minor elements between phases and process streams. These properties in turn determine key process operating parameters, such as, tapping temperatures, slag fluxing requirements, slag volume and residence times. Developing accurate methods of determining and predicting the chemical and physico-chemical properties of slags has been an ongoing quest for metallurgical researchers and engineers. Whilst most low order systems have now been well-characterized the need for accurate data on multi-component systems of interest to industrial operations continues. The availability of the electron probe X-ray microanalysis in particular has enabled new approaches to analyze complex systems, which has greatly extended research capabilities. Not only have these new approaches lead to more accurate characterization of phase equilibria but they have also been extended to the examination of kinetic phenomena. An issue of particular importance to metallurgical practice is the emerging role of freeze linings to furnace integrity [1, 2].

A freeze lining is a solidified layer of slag bath, as material formed on the cooled wall of a furnace. This regenerative layer then protects the refractory of the furnace from mechanical, chemical and thermal attack [3-5]. This paper demonstrates how the experimental and analytical techniques developed for phase equilibria studies can provide new insights into the behavior of freeze lining systems.

Experimental Procedures

Phase Equilibria Study

The experimental approach developed by PYROSEARCH to determine the phase equilibria in slag systems (see Figure 1) involves high temperature equilibration in controlled gas atmospheres, rapid quenching and direct measurement of equilibrium phases with electron probe X-ray microanalysis (EPMA) [1, 2]. Pure oxide and metal powers are mixed, pelletized, equilibrated in crucibles or on substrates at carefully controlled temperature and gas atmosphere, then quenched, and sections of the samples are prepared. The compositions of phases are measured directly by EPMA. An important advantage of the technique is that the compositions of phases are measured directly after the experiment, so that changes in bulk compositions of samples taking place during high temperature experiments do not affect the accuracy of the final results. The approach overcomes most of the limitations associated with conventional phase equilibria studies. The phase equilibria for many high temperature systems previously difficult or impossible for investigation have now been accurately characterized. In particular, this approach allowed the uncertainties due to interaction of sample with crucible, vaporization, and metal / slag / gas equilibration reactions to be eliminated. Importantly, this technique enabled the use of substrates made of the primary phase so that no crucibles are used – complex systems such as Fe- , Cu- and S- containing slag/metal/matte systems can now be accurately investigated [6, 7]. Many uncertainties, however, can still exist. In particular, uncertainties related to the difficulty to quench, and difficulty to achieve equilibria. This aspect is paid particular attention at PYROSEARCH. The equilibration experiment is considered as a kinetic process to achieve the equilibrium, and the closer to the equilibrium – the lower "driving force". A four-point approach has been introduced to ensure and confirm the achievement of equilibria: 1. equilibration time, 2. homogeneity of phases, 3. achievement of equilibrium from different starting compositions, and 4. systematic analysis of reactions specific to the particular sample. The latter may include dissolution, equilibration, vaporization, gas /slag /alloy /matte reactions and others. The capability of measuring specific local (within 2-10 micron) and long-distance (10-500 micron) compositional profiles using EPMA provides important information that is used as a foundation to identify possible incomplete reactions and therefore to introduce corrections into experimental procedure (e.g. in the Sn-Fe-O-Si system [8])

Thermodynamic modeling undertaken using computer package FactSage with the quasi-chemical model for the liquid slag phase is closely integrated with the parallel experimental research. Experiments are planned to provide specific data for thermodynamic model development as well as for pseudo-ternary liquidus diagrams that can be used directly by process operators. Thermodynamic assessments are used to identify priorities and plan experiments, and to interpret results. Experimental and modeling studies are combined into an integrated research program contributing to and enhancing outcomes of each and of the overall program [2].

Freeze-Lining Experiments

In the experimental study of the freeze-lining deposit [3, 4], the cold finger technique was used. "Moving" slag contained in an alumina or magnesia crucible was rotated with a given speed (see Figure 2). Approximately 3.5 kg of synthetic or industrial slag is used. The finger is cooled by passing compressed air through two concentric tubes. A series of thin thermocouples are installed at different distances from the cold probe surface, and the temperatures are recorded continuously during the experiments. The probe is submerged into the slag, the bath temperature is maintained constant, the quenched samples of the slag bath are taken during experiment, and after a steady-state condition is achieved, the probe is removed fast from the bath, quenched into water (that takes approximately 10 seconds); cross sections of the deposit are prepared, and microstructure and composition of phases formed across the deposit, as well as compositional profiles, are analyzed using Scanning Electron Microscope (SEM) and EPMA. The morphology and compositions of the phases are used to identify the conditions, the temperature and the reactions taking place in the different layers of the deposits, in the slag near the interface and in the slag bath away from the freeze lining. The interpretation of the results obtained from characterization of the freeze-lining deposits is supported by the parallel phase equilibrium studies and thermodynamic modeling using FactSage.

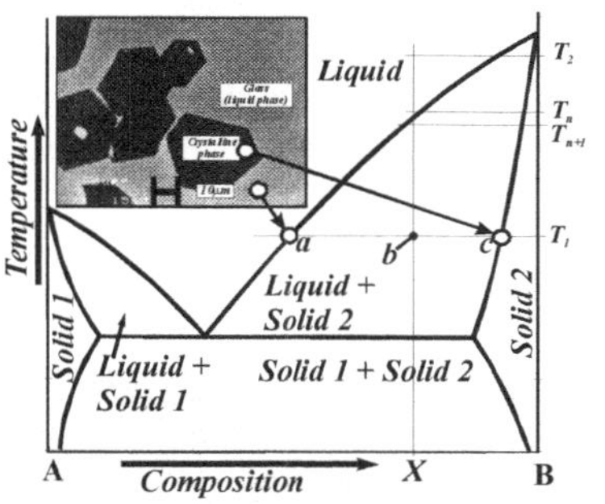

Figure 1. Hypothetical diagram to illustrate experimental technique for determination of phase equilibria [1].

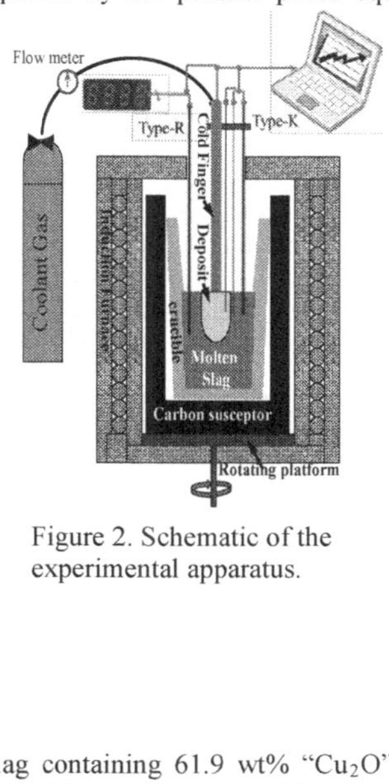

Figure 2. Schematic of the experimental apparatus.

Results

<u>Phase Equilibria Study</u>

Equilibrium experiments were conducted on bulk slag containing 61.9 wt% "Cu_2O", 15.6% "FeO", 15.0% SiO_2, 7.5% Al_2O_3 in equilibrium with metallic copper to provide accurate basis for further characterization of the dynamic freeze lining deposits. The liquidus temperature was measured to be 1140 °C with delafossite ($CuFeO_2$) to be the primary phase. The cuprite phase (Cu_2O) was observed to be present in addition to the delafossite at 1100 °C (see Figure 3). The eutectic temperature was found to be between 1050 °C and 1000 °C. The final phase assemblage at the temperatures below the solidus consists of delafossite, cuprite, tridymite and copper metal.

The temperature of the interface between stationary steady state freeze-lining deposit, T_f, is introduced as the key parameter that reflect the actual conditions of phases at the interface, and this helps to describe the other freeze lining properties, such as, the deposit thickness, heat transfer and other. All previous descriptions and modeling approaches found in open literature [9]assumed $T_f = T_{liquidus}$. Recent experimental results [3-5] have conclusively proven that in many cases of the slag systems $T_f < T_{liquidus}$ or T_f is even close to $T_{solidus}$. For example, Figure 4 presents microstructure of the deposit of the slag having $T_{liquidus}$ = 1140 °C and delafossite as the primary phase. The figure shows the detached crystals of cuprite in addition to the detached primary delafossite crystals present in the liquid layer adjacent to the stationary freeze lining interface (note that cuprite starts to precipitate at temperatures significantly below the liquidus of 1140 °C). Direct measurements using series of the thin thermocouples installed at different distances in this deposit gave the interface temperature T_f = 1113 °C, which is below $T_{liquidus}$ = 1140 °C. Precipitation of the detached cuprite on the detached delafossite (see Figure 4.c) supports the interface temperature measurement and conclusively proves that T_f can be below $T_{liquidus}$ at steady state conditions.

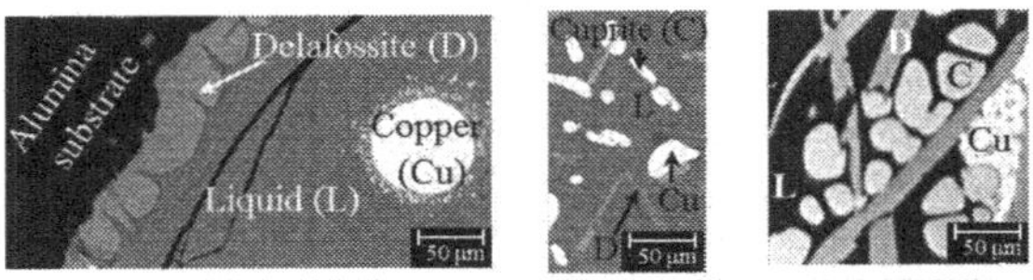

Figure 3. Phase assemblages formed on equilibration of slag with metallic copper at different temperatures for a given bulk bath composition. Note that C, cuprite; Cu, metallic copper; D, delafossite; and L, liquid [3]. Note the liquid matrix surrounding delafossite in 1100 °C is quenched poorly since the cooling rate is insufficient to prevent crystallization.

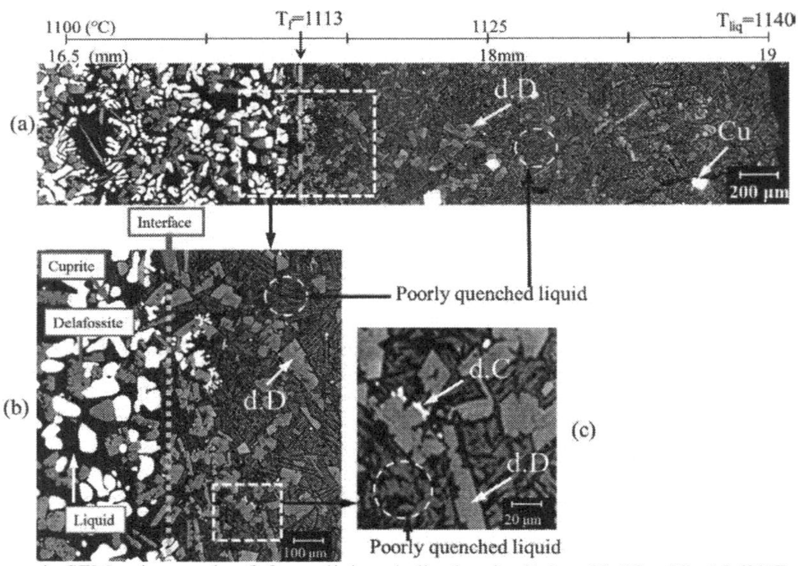

Figure 4. SEM micrograph of freeze-lining, bulk slag is 61.9 wt% "Cu_2O", 15.6%"FeO", 15.0%SiO_2, 7.5%Al_2O_3 at bath temperature: 1165 °C, $T_{liquidus}$=1140 °C, cooling air flow rate: 100 l/min, speed of rotation: 20 RPM and immersion time 9 h. Detached delafossite (d.D), detached cuprite (d.C) and metallic copper (Cu). Note the liquid matrix surrounding the detached delafossite is marked as "poorly quenched liquid" since the cooling rate is insufficient to prevent crystallization during the quench.

A new conceptual framework of the processes involved in the formation of the freeze lining has been developed to explain this phenomenon, that at steady state $T_f < T_{liquidus}$ (see Figure 5). It is argued that as the result of the natural and forced convection, fresh portions or liquid cells move from the bulk bath towards the stationary deposit with the temperature decreasing from the bath temperature, T_{bath} towards T_f. As the liquid passes into the zone where $T < T_{liquidus}$, nucleation and crystal growth on homogeneous nuclei and on pre-existing detached solids entrained in the liquid takes place; the composition of the remaining liquid in the cell changes, the local equilibrium between the detached crystals and the surrounding liquid is maintained through short range diffusion within this cell during this passage across the "subliquidus layer", and the liquid arrives at the deposit interface with no thermodynamic driving force to add further solid crystals onto the deposit or to cement new detached crystals to the stationary deposit interface. The turbulent eddies also transport cells containing the detached precipitated crystals away from the deposit interface back into the bulk liquid where the temperature is higher and liquid is close to the bulk slag composition, and the crystals re-dissolve. The steady-state is attained at condition at which there is a balance between crystallization on the detached crystals and their dissolution processes. The extent of crystallization of the melt on the detached crystals in the subliquidus liquid bath layer is the key factor determining the steady state interface temperature of the freeze lining, T_f.

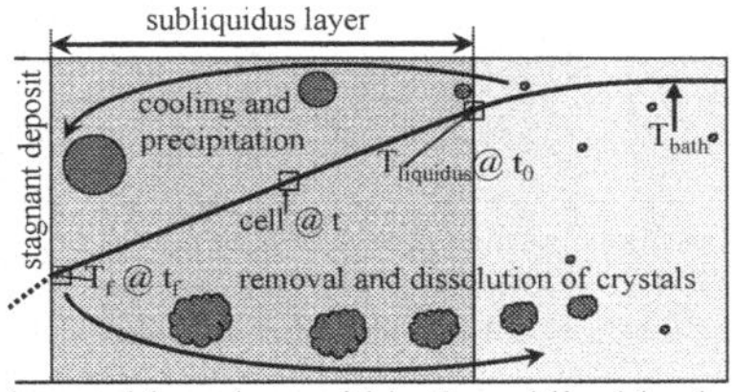

Figure 5. Proposed mechanisms taking place within the subliquidus layer at steady-state conditions at the deposit interface.

Further Experimental Evidence of the Proposed Mechanism

The proposed mechanism (schematically shown in Figure 5) assumes that at steady state conditions primary phase delafossite crystals precipitating and moving in the subliquidus layer do not become attached to the deposit surface but are removed by the force of the bath agitation into the bulk liquid where they redissolve. The process of solids precipitation in the subliquidus layer, transportation of the newly-formed crystals into the bulk liquid bath and further dissolution is said to be repeated continuously at steady-state conditions. The evidence to support the proposed mechanism is the appearance of the primary phase delafossite crystals (up to 50μm in diameter) in the bulk liquid bath samples after extended time experiments. These crystals were observed in the sample at steady-state condition taken 50 mm away from the deposit interface only at the end of the freeze-lining experiment (see Figure 6) but were not present before the experiment. Further analysis of the liquid and the characteristics of these entrained (remelting or dissolving) delafossite crystals and the liquid close to the crystals provide further proof of the proposed mechanism as follows;

Morphology of Delafossite Crystals in the Bath: Previous studies have shown that delafossite crystals in the deposit formed from the molten Cu-Fe-Si-Al-O slags are faceted, forming solids with plate-like microstructure. The entrained delafossite crystals (10-100 μm in diameter) in the bulk liquid bath samples are, however, irregular and rounded in shape; this morphology indicates that dissolution/remelting of the solid particles has occurred (see Figure 6). The shape of the entrained delafossite appears independent of the speed of rotation of the bath.

<u>Composition of delafossite in the bath and in the subliquidus layer:</u> The compositions of the entrained delafossite crystals obtained from bath samples taken at the end of experiments for various speeds of bath rotation were measured using electron probe X-ray microanalysis. These analyses are consistent with the measured compositions of delafossite found in the near deposit interface as well as in equilibrium experiments at 1100 and 1140 °C. Again, note that no solid phase was observed in any bath samples taken before conducting the experiments (see Figure 6).

<u>Liquid composition surrounding entrained delafossite crystals in the bath for various RPM:</u> The compositions of the liquid as a function of distance from the entrained delafossite solid phase were measured. Despite the difficulties in obtaining areas of microcrystalline material for EPMA analysis it can be seen there is an unmistakable change in composition surrounding each of the particles demonstrating that these are dissolving into the surrounding melts. The compositions of the liquid phase (Al_2O_3, Cu_2O, Fe_2O_3 and SiO_2) approach that of the bulk bath at distances of approximately 30 μm from delafossite/liquid interface.

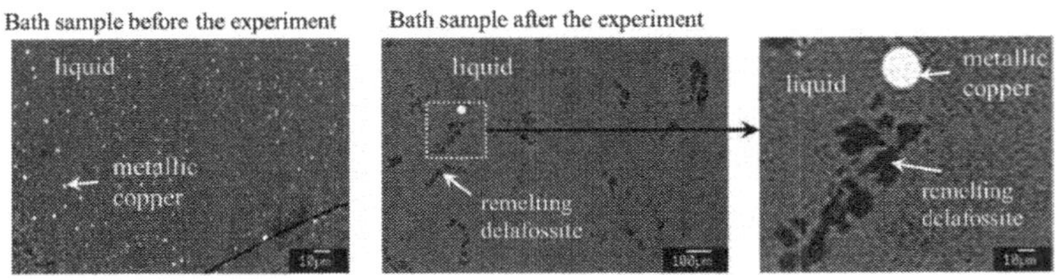

Figure 6. Backscattered micrographs of bath samples taken before and after deposit experiments interface for different speeds of rotation/ immersion times 20 RPM/9h. Note the liquid matrix surrounding the detached delafossite is quenched poorly since the cooling rate is insufficient to prevent crystallization during the quench.

<u>Proposed Conceptual Framework</u>

The following approach is suggested to provide a basis for future quantification of the processes based on the developed conceptual framework. The mean interface temperature T_f corresponds to X_f - the mean composition of the local portions of liquid in the vicinity of the interface in contact with the crystals of the stationary deposit (see Figure 7 where example is given for a binary A-B system with eutectic and no solid solution of B in A). A cell of fluid material with composition X_{bath} moving to the freeze lining interface (see Figure 5) reaches $T_{liquidus}$ at time $t_0 = 0$, then crystallization starts and continues as the cell passes through the subliquidus layer where temperatures decrease from $T_{liquidus}$ to T_f. The mean time for the cell to arrive at the interface is defined as t_f. The rate of deposition of A on the detached crystals within the cell at any temperature and composition is given by R(t) [moles s^{-1}]. The total quantity of material deposited on the detached crystals over the time interval t_f, $\Delta M_{A,f}$, is then given by $\Delta M_{A,f} = \int_{t_0=0}^{t_f} R(t)dt$ [moles]. The concentration of B in the liquid X_f in the partially crystallized cell is then given by $X_f = M_{B0}/(M_{B0} + M_{A0} - \Delta M_{A,f}) = M_{B0}/(M_{B0} + M_{A0} - \int_{t_0=0}^{t_f} R(t)dt)$, where M_{B0} and M_{A0} are the initial number of moles of A in B in a given cell volume of liquid. At the dynamic steady state, the particular mean interface temperature, T_f is determined by the mean liquid composition X_f, which in turn is determined by the extent of crystallization on the detached crystals described by the $\int_{t_0=0}^{t_f} R(t)dt$ term. The $\int_{t_0=0}^{t_f} R(t)dt$ term is determined by the kinetics of both i) crystallization (R(t)) and ii) convection (t_f) phenomena and is proposed to be taken as a key term in determining T_f for a given chemical system (see Figure 7). If the mean crystallization rate on the detached crystals is introduced as $R_f = \left(\int_{t_0=0}^{t_f} R(t)dt\right) / t_f$, then this term may be expressed as ($R_f t_f$).

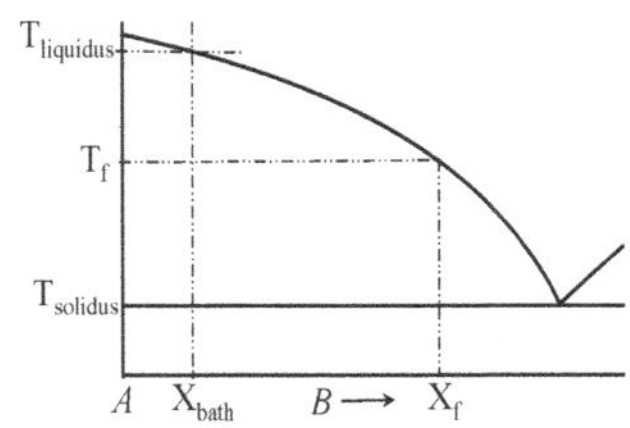

Figure 7. Hypothetical binary phase diagram (mol fractions).

There are clearly different ways in which the mean crystallization rate R_f and time t_f terms can be described as a function of key system characteristics, such as, liquid properties (e.g. viscosity, thermal and mass diffusivity, nucleation and crystallization kinetics, liquidus shape of phase equilibrium and dissolution kinetics), fluid flow characteristics and mass transfer rates close to and at the stationary deposit interface, influence the stability, thickness, interface temperature and the heat flow through the freeze lining deposits. However, the above analysis indicates a potential approach to quantify the interface temperature of the deposit/bath. Qualitatively, for example, (a) increasing the time taken to cross the sub-liquidus layer will lead to increased extent of crystallization and lower interface temperature, T_f. or (b) increasing the rate of nucleation and crystallization of solids on detached crystals during passage across the sub-liquidus layer will also lead to the increased extent of crystallization and thus lower interface temperature, T_f. In the limit, if equilibrium crystallization takes place across the layer, the extent of solids formation and the interface temperature T_f is described by the solidus, $T_{solidus}$, in the particular chemical system. If no crystallization on the detached crystals occurs before the material arrives at the deposit interface then the interface temperature T_f is $T_{liquidus}$. In the slag systems that have been investigated in the recent studies, the steady state interface temperature T_f was found to always be between these limits, i.e. $T_{solidus} < T_f < T_{liquidus}$. Experimental observations [3-5] supported the proposed mechanism with $T_f = f(R_f{*}t_f)$ formalism.

Further experiments proved qualitatively the proposed mechanism. A series of experiments with all the same conditions and a variable rotational speed of crucible [5] demonstrated that increased agitation (and therefore decreased t_f) results in the increased T_f (that is explained by the decreased ($R_f t_f$)value).

Experiments undertaken for the copper-containing slag system with low viscosity and for a cryolite salt system with highest diffusive mass transfer rates resulted in the interface temperature to be close/equal to the liquidus temperature of the liquid bulk bath [4]. This is explained to be a result of the increase of the mass transfer rates to the deposit interface and corresponding relative decrease of the crystallization on the detached crystals. These experiments provide qualitative proof of the proposed framework for description of the phenomena.

Conclusions

Advanced analytical tools and new experimental techniques have been applied to the characterization of microstructures formed in freeze linings. The studies have demonstrated that at steady state the slag/deposit interface temperature, T_f can be (and in many cases is) below $T_{liquidus}$ – this is an significant finding since it is very different from the previous descriptions of these systems found in literature. The finding has important practical implications since it suggests that the processes can be operated at temperatures below the liquidus. It also opens the way to new strategies about the selection of slag chemistry and the potential process benefits that flow from slag chemistry control. Further studies are now under way to establish the relative effects of key variables in these freeze lining systems.

Acknowledgements

The authors would like to thank Australian Research Council Linkage program, Rio Tinto Kennecott Utah Copper, Corp., Xstrata Technology, Xstrata Copper, BHP Billiton Olympic Dam Operation, and Outotec Oy for financial support.

References

1. E. Jak, H.G. Lee, and P.C. Hayes, "Improved methodologies for the determination of high temperature phase equilibria," *Korean IMM. J*, 1 (1995), 1-8.
2. E. Jak, "Integrated experimental and thermodynamic modelling research methodology for metallurgical slags with examples in the copper production field," *Intl. Conf. on Molten Slags and Fluxes*, Beijing, China, 2012, Paper W077.
3. A. Fallah-Mehrjardi, P.C. Hayes, and E. Jak, "Investigation of freeze linings in copper containing clag cystems: Part I. preliminary experiments," *Metall. Mater. Trans. B.*, 44B (2013), 534-48.
4. A. Fallah-Mehrjardi, P.C. Hayes, and E. Jak, "Investigation of freeze linings in copper containing clag cystems: Part II. mechanism of the deposit stabilisation," *Metall. Mater. Trans.B.*, 44B (2013), 549-60.
5. A. Fallah-Mehrjardi, P.C. Hayes, and E. Jak, "Investigation of freeze-lining in copper containing-slag systems: Part III. high temperature experimental investigation of the effect of bath agitation," *Metall. Mater. Trans. B*, (2013, DOI: 10.1007/s11663-013-9925-6.).
6. A. Ilyushechkin, P.C. Hayes, and E. Jak, "Liquidus temperatures in calcium ferrite slags in equilibrium with molten copper," *Metall. Mater. Trans. B*, 35B (2004), 203-15.
7. S. Nikolic, P.C. Hayes, and E. Jak, "Experimental Techniques for Investigating Calcium Ferrite Slags at Metallic Copper Saturation and Application to the Systems "Cu_2O"-"Fe_2O_3" and "Cu_2O"-CaO at Metallic Copper Saturation," *Metall. Mater. Trans. B*, 40 (2009), 892-99.
8. X. Xu, P.C. Hayes, and E.Jak, "Experimental study of phase equilibria in the "SnO"-CaO-SiO_2 system in equilibrium with tin metal," *Int. J. Mat. Res.*, 104 (2013), 235-43.
9. K. Verscheure, M. Van Camp, B. Blanpain, P. Wollants, P. Hayes, and E. Jak, "Continuous fuming of zinc-bearing residues: Part II. The submerged-plasma zinc-fuming process," *Metall. Mater. Trans. B.*, 38B (2007), 21-33.

Celebrating the Megascale: Proceedings of the Extraction and Processing Division
Symposium on Pyrometallurgy in Honor of David G.C. Robertson
Edited by: Phillip J. Mackey, Eric J. Grimsey, Rodney T. Jones, and Geoffrey A. Brooks
TMS (The Minerals, Metals & Materials Society), 2014

MODELLING SIMULATION AND COMPARISON OF REFRACTORY CORROSION AT RHI'S TECHNOLOGY CENTER

Dean Gregurek[1], Angelika Ressler[1], Anna Franzkowiak[2], Alfred Spanring[2]

[1]RHI AG, TC Leoben, Magnesitstrasse 2, Leoben, A-8700, Austria
[2]RHI AG, Wienerbergstrasse 9, Vienna, A-1100, Austria

Keywords: Corrosion tests, RHI, refractories, magnesia-chromite bricks

Abstract

In order to determine the most suitable refractory products and improve the lining lifetime for the diverse furnaces used in the nonferrous metal industry, corrosion tests are performed at RHI's Technology Center. The practical facilities include the cup test, induction furnace, rotary kiln, and drip slag test described in this paper, which enable a comprehensive understanding of the chemo-thermal brick wear on a pilot scale. The corrosion trials are performed with actual slags generated during operations at a customer's plant. To determine the highest influencing wear parameter, every single test is combined with a detailed mineralogical investigation and thermochemical calculations performed using FactSage. Based on the results, tailored refractory solutions for the nonferrous metal industry can be provided in combination with trials conducted at the customer's site.

Introduction

The complexity of metallurgical processes, frequent changes in the process - conditions and routes, in combination with the diverse metal processing furnaces [1, 2], require a precise understanding of each individual system for optimized refractory lining solutions to be developed.

Based on the experience acquired through postmortem studies carried out on bricks from nonferrous furnaces [3, 4] together with practical knowledge from laboratory-scale trials, appropriate refractory product selection for specific nonferrous metal applications can be made. In addition, to achieve this goal test procedures have been developed and installed at the RHI Technology Center in Leoben, Austria, to simulate the various demanding service conditions [5]. For example static and dynamic corrosion test methods and equipment such as the cup test, induction furnace, rotary kiln, drip slag test, together with microscopic analyses and thermochemical calculations, enable a comprehensive understanding of brick wear following such pilot scale trials.

Experimental Procedure

The main corrosion testing facilities located in the Pilot Plant at RHI's Technology Center Leoben are described below.

Cup Test

A static cup test is performed to provide an initial impression of the brick behaviour when in contact with a specific slag. The crucible, 80 x 80 x 70 mm, is cut from the brick grade to be tested. A cylindrical hole, 40 mm in diameter and 45 mm high, is drilled into this test sample (see Figure 1). The crucible is filled with 80–100 grams of customer-provided process slag, sealed with a lid made of the same refractory material, and heated in a laboratory electric furnace. The test can be performed up to a temperature of 1700°C in an oxidizing atmosphere, usually for 6–8 hours. Additionally, the crucible can be covered with coke fines to generate a reducing atmosphere during the test. After testing, a mineralogical investigation is carried out to evaluate depth of slag penetration in the brick microstructure.

Figure 1. Cup sample before testing (left side) and after testing with residual slag in the cup bottom (right side). Sample dimensions: 80 x 80 x 70 mm. The inner cylindrical hole diameter is 40 mm.

Induction Furnace and Rotary Kiln

To extend the results obtained from the static cup test, dynamic testing in an induction furnace and/or rotary kiln can be performed. The refractory wear during these tests is principally dependent on the chemical composition of the customer's slag, the chemical composition of the refractory, testing temperature, and trial duration. In order to generate maximal refractory wear in a relatively short testing time, the furnace operating temperature is usually higher than that occurring during the actual customer process.

The induction furnace can be lined with up to 16 bricks (4–5 different brick brands). The furnace diameter is 250 mm and the temperature can range between 1500–1750°C. Up to 60 kg of metal and 1.5–3 kg of slag (per cycle) are used during the test. The process slag is changed every 0.5–1.5 hours to achieve a constant, highly corrosive medium in contact with the installed refractory material. Testing is usually performed for 6–8 hours.

Figure 2 is an example of a magnesia-chromite brick after a test performed in the induction furnace with a fayalitic-type slag at 1650°C. Due to Marangoni convection, the highest macroscopically visible wear is always at the interface between the refractory, metal, and slag.

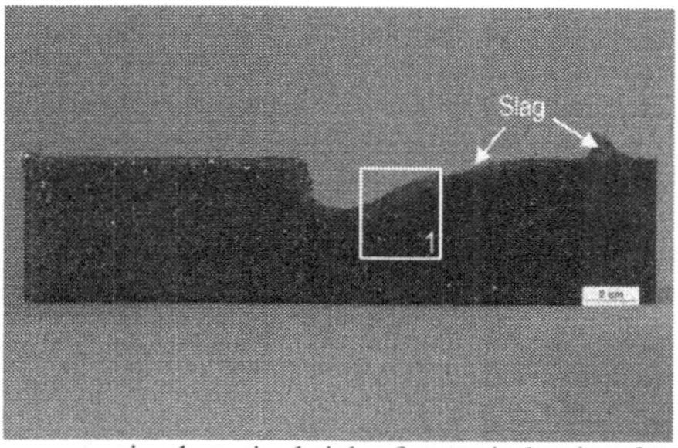

Figure 2. Example of a magnesia-chromite brick after an induction furnace test carried out at 1650°C. The immediate brick hot face is coated in slag. Sample for mineralogical investigation (1).

The Technology Center Leoben is equipped with two rotary kilns. The larger rotary kiln (300 mm in diameter with a furnace length of 1350 mm) can be lined with up to 64 bricks (four rings; 15 brick brands). The furnace is heated up to a testing temperature of 1800°C using a propane-oxygen gas mixture.

Two different inner diameters (i.e., 165 mm and 92 mm) are possible when testing in the small rotary kiln. Up to 18 bricks (6 different brick brands) can be installed and tested simultaneously with the rotary kiln inner diameter of 165 mm. The furnace is heated with a propane-oxygen gas mixture. The furnace rotation can be adjusted between 3–20 rpm. Trials are usually performed within a temperature range of 1500–1750°C for 20–24 cycles using 1.5–3 kg of slag per cycle. Up to six different brick brands can be installed and tested simultaneously in the rotary kiln when it is setup with an inner diameter of 92 mm.

Figure 3 is an example of a magnesia-chromite brick after a rotary kiln test performed using a fayalitic-type slag at 1600°C.

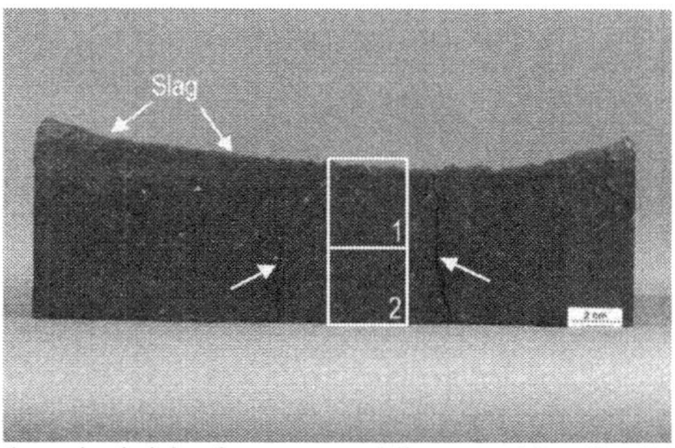

Figure 3. Magnesia-chromite brick after a rotary kiln test with a fayalithic-type slag at 1600 °C. Slag coating on the immediate brick hot face. Cracks formed when the kiln cooled down (arrows). Samples for mineralogical investigation (1 and 2).

<u>Drip Slag Test</u>

The principle of the drip slag test method, according to ASTM C-768, is that molten slag is dripped onto the test specimen housed in a heated furnace [6]. Slag in the form of a slag rod is fed with an automatic feeding mechanism through a water-cooled tube into the furnace where it melts and drops onto the brick (see Figure 4, left side). The initial impingement has an erosive effect. The molten slag then diffuses into and reacts with the refractory as it flows, forming a groove. The operating temperature is up to approximately 1480°C and the testing time is usually up to 6 hours.

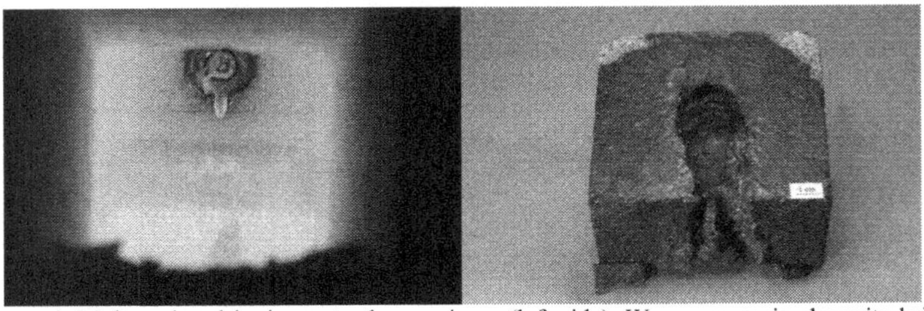

Figure 4. Molten slag dripping onto the specimen (left side). Worn magnesia-chromite brick after the drip slag test (right side).

$$\rho = \frac{m}{v} \rightarrow v = \frac{m}{\rho} \rightarrow v\,[cm^3] = \frac{m\,[g] \cdot 1000}{1644\,[g/dm^3]} \qquad (1)$$

$\rho = Packing\ density\,[g/dm^3]$

$m = Bulk\ of\ tabular\ alumina\,[g]$

$v = Wear\ volume\,[cm^3]$

Evaluation of the wear rate (see Figure 4, right side) is performed by calculating the wear volume at the end of the test and comparing it to tabular alumina (grain size of 0.3–0.6 mm, and a packing density of 1644 g/dm^3) exposed to the same conditions (see Equation 1).

Analytical Procedures

Slag Characterization

To produce the optimal conditions for corrosion testing, initially a complete chemical and mineralogical characterization is carried out on slag samples received from the customer; this is followed by thermochemical analysis using FactSage. An example of a slag chemical analysis is shown in Table I. Additionally, the slag melting points are determined using a heating microscope, according to DIN 51730.

Table I. Chemical analysis of a customer slag (wt.%)*

MgO	Al_2O_3	SiO_2	CaO	Cr_2O_3	Fe_2O_3	Na_2O	SO_3	NiO	CuO	ZnO	SnO_2	PbO
2	3	28	0.8	0.3	43	0.2	1	0.7	5	6	1	7

* Semi-quantitative X-Ray fluorescence analysis (data as reported without calibrating standards)

Macroscopic Evaluation

The corrosion trial results, in terms of wear area and wear depth, following the induction furnace and rotary kiln tests are evaluated from the final brick cross-sectional profile. The macroscopic appearance (cross sections) also provides a first impression regarding the brick wear behaviour during corrosion testing.

Chemical Analysis, Mineralogical Investigation, and Thermochemical Calculations

The chemical analyses are carried out using X-ray fluorescence analysis (Bruker S8 TIGER), after dissolution of the sample in $Li_2B_4O_7$ (according to DIN 51001), as well as on the original sample. Additionally, mineralogical investigations are performed on polished sections using a reflected light microscope, X-ray diffraction (Bruker D8 ADVANCE), and a scanning electron microscope (JEOL JSM-6460) combined with an energy- and wave-dispersive X-ray analyser.

The results of a microstructural investigation performed after testing a magnesia-chromite brick in the rotary kiln with fayalitic-type slag are shown in Figure 5. The brick microstructure at the immediate brick hot face was coated with slag and a reaction zone was clearly visible. Below the reaction zone, an infiltrated and corroded brick microstructure was observed (see Figure 5). The infiltration depth was measured microscopically enabling a detailed comparison between the different brick brands examined. In the infiltrated brick microstructure, corrosion of the magnesia component had taken place. The main reaction product was Mg-silicate of a forsteritic type (Mg_2SiO_4) [7]. Due to diffusion phenomena, the chemical composition of the chromite was also altered, having become mainly enriched with Fe-oxide at the rims. This was also observed with the primary and secondary chromite precipitations.

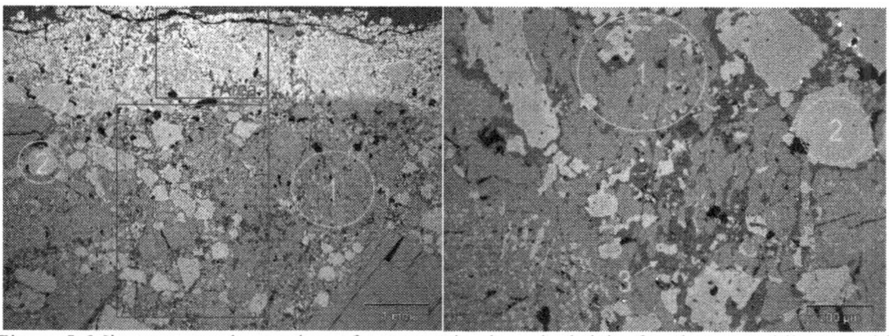

Figure 5. Microstructural overview of a magnesia-chromite brick after the rotary kiln test with fayalitic-type slag. Immediate brick hot face (left side): Reaction zone (area 1), infiltrated and corroded brick microstructure (area 2), corroded magnesia component (1), chromite (2). Approx. 2 mm from the hot face (right side): Corroded magnesia component (1), chromite (2), forsterite (Mg_2SiO_4) (3).

In addition to the mineralogical investigations, the interactions between different combinations of refractory materials and slag are described thermochemically with regard to potential reaction products formed at the interface and the solubility of refractory components in the slag. The calculations were performed using the FactSage 6.3 software, within the same temperature range used during the physical testing.

FactSage is a Gibbs energy minimization tool and calculates the concentrations of chemical species when specified elements or compounds react, or partially react, to reach a state of chemical equilibrium. Users have access to databases containing thermodynamic data for thousands of compounds as well as evaluated and optimized databases for hundreds of solutions including metals, liquid and solid oxide solutions, mattes, molten and solid salt solutions, and aqueous solutions. The FactSage software automatically accesses these databases. The evaluated databases for oxides, slags, and mattes were developed by optimizing data from the literature using advanced modelling techniques [8–9].

The reaction products include pure substances (e.g., liquid and solid), ideal solutions (e.g., gaseous, liquid, solid, and aqueous), and non-ideal solutions (e.g., real gases, slags, molten salts, mattes, ceramics, alloys, and diluted solutions) from the FactPS and FToxide databases. The range of compositions, temperature, and total pressure are defined. The equilibrium phases and phase compositions are determined as a function of the refractory/slag ratio.

Conclusion

The combination of different test facilities and testing methods, such as the cup test, induction furnace, rotary kiln, and drip slag test enable the best possible understanding of chemothermal brick wear on a pilot scale. Corrosion trials that are performed with customer-provided slags can be run at temperatures up to 1750–1800°C. In order to generate maximal refractory wear in a relatively short testing time, the operating temperature of the furnace is usually higher than that at the customer site. A dynamic slag testing method such as the drip slag test produces very relevant results, also at lower testing temperatures, especially with regard to determining the relative refractory resistance to molten slags. In addition to the aforementioned test facilities, RHI is currently installing a new high-frequency induction furnace for dynamic corrosion testing.

In order to determine the highest influencing wear parameter, every corrosion test is combined with comprehensive chemical analyses, mineralogical investigations, and thermochemical calculations performed using FactSage.

Based on the research results and specific process operating knowledge, product development is performed together with the Marketing Department so that appropriate brick lining solutions can be recommended for any metallurgical furnace. This approach, conducted in collaboration with RHI customers including Outotec, Xstrata, Boliden, and Stillwater Mining Company, has resulted in improved lining lifetimes and generated both technical and economic benefits.

References

1. W.G. Davenport et al., *Extractive Metallurgy of Copper* (Kidlington, Oxford OX5 1GB, UK: Elsevier Science Ltd., 2002), 432.

2. F.K. Crundwell et al., *Extractive Metallurgy of Nickel, Cobalt and Platinum-Group Metals* (Kidlington, Oxford OX5 1GB, UK, Elsevier Science Ltd., 2011), 610.

3. H. Barthel, "Wear of Chrome Magnesite Bricks in Copper Smelting Furnaces," *Interceram*, 30 (1981), 250–255.

4. D. Gregurek et al., "Refractory Wear Mechanisms in the Non-Ferrous Metal Industry – Testing and Modelling Results," *JOM*, 65 (11) (2013).

5. D. Gregurek and C. Majcenovic, "Wear Mechanism of Basic Brick Linings in the Non Ferrous Metals Industry – Case Studies From Copper Smelting Furnaces," *RHI Bulletin*, 1 (2003), 17–21.

6. J.C. McMullen, "A New Slag Testing Furnace," *Bulletin Am. Ceramic. Soc.*, 19 (11) (1940), 439–442.

7. W.A. Deer, R.A. Howie, and J. Zussman, *An Introduction to the Rock-Forming Minerals*, (Essex CM20 2 JE, UK, Pearson Education Ltd., 1992), 3–15.

8. C.W. Bale, et al., "FactSage Thermochemical Software and Databases," *Calphad*, 26 (2) (2002), 189–228.

9. G. Eriksson and K. Hack, "ChemSage – A Computer Program for the Calculation of Complex Chemical Equilibrium," *Metallurgical and Materials Transactions B*, 21B (1990), 1013–1023.

CELEBRATING THE MEGASCALE

Proceedings of the Extraction and Processing Division Symposium on Pyrometallurgy in Honor of David G.C. Robertson

Iron and Steel Production

Celebrating the Megascale: Proceedings of the Extraction and Processing Division Symposium on Pyrometallurgy in Honor of David G.C. Robertson
Edited by: Phillip J. Mackey, Eric J. Grimsey, Rodney T. Jones, and Geoffrey A. Brooks
TMS (The Minerals, Metals & Materials Society), 2014

SUSTAINABILITY IN IRONMAKING: THE RISE OF DIRECT REDUCTION

Thomas P. Battle[1]

[1]Research and Development Technology Center
Midrex Technologies
Pineville, North Carolina 28134

Keywords: Ironmaking, Direct Reduction, DRI, HBI

Abstract

Modern-day direct reduction of iron first developed as a small-scale, low capital and operating cost alternative to the blast furnace. Since commercialization of continuous DR technology in the late 1960s, the market for the products of direct reduction has grown to more than 74 million tonnes in 2012. The initial advantages of DR plants over BF facilities have grown over the years, for several reasons, including the increased size of DR modules, lower energy and emissions (particularly CO_2), along with the flexibility to use a number of different reductants. The development of DR technology over the past forty years will be emphasized in this presentation, including recent developments that allow for even more direct sustainability comparisons with the iron blast furnace – and even combine the two technologies for improved synergies. Also briefly discussed will be the possibilities of using direct reduction for non-ferrous ores.

Introduction

When continuous shaft-furnace based direct reduction of iron was commercialized in the late 1960s, it was a small-scale, low capital and operating cost alternative to the blast furnace, a perfect fit to supply the nascent market of electric-arc furnace based minimills with a source of consistent, high-grade metallic iron units to supplement or replace scrap. Since that time the market for direct reduced iron products such as Cold Direct Reduced Iron (CDRI), Hot Briquetted Iron (HBI), and Hot Direct Reduced Iron (HDRI) has grown to more than 74 million tonnes in 2012, with expectations that this number will increase considerably in upcoming years. The initial advantages of DR plants over integrated mills for certain scenarios have grown as the size of the plants and their flexibility has increased, and the advantages of lower energy and emissions (particularly CO_2) become more and more relevant. Now the advantages of DR have become even more obvious, as the capacity of individual DR modules begins to approach that of smaller blast furnaces, and technology such as MXCOL® allow the MIDREX® Direct Reduction Process to use syngases derived from coal or coke oven gas as a supplement or alternative to natural gas to produce DRI in a shaft furnace. Most recently, options have begun to appear directly linking integrated mills with direct reduction.

Early history of ironmaking

It's quite possible that the first discovery of metallic iron and its uses in antiquity was a result of meteorite falls, thousands of years ago [1]. After this, ancient man may have been better prepared to see the potential value in the spongy mass that was left in the residue of fires burned with charcoal in the presence of iron-bearing rock. These early engineers were able to take this spongy mass and beat it into a valuable product, which became known as wrought iron. This development was so important that it gave rise to the name given by archaeologists to this period of history, the Iron Age [2]. Since the product was made in basically one high-temperature step, it later became defined as the product of *direct* reduction.

Over time, reaction temperatures were elevated, until carbon began dissolving in the iron. Eventually local temperatures exceeded the melting point of the iron-carbon alloy. The result was the melting of the iron and separation of slag. Although the purity of the iron was now increased by separation of slag, it was saturated with carbon and thus not suitable for many applications. Impurity removal at elevated temperature to make wrought iron was another expensive step beyond the original process, thus *indirect* reduction.

The ultimate evolution of indirect reduction was the blast furnace, which was developed more than 700 years ago. It has evolved in size and capabilities, leading to the behemoths found today around the world. 'The advent of the blast furnace moved ironmaking from a craft into an efficient technology well suited to the industrial age.' [2]. To this day, there is no more efficient way to make iron.

Although 'indirect reduction' long ago eclipsed 'direct reduction' as the main technology to make iron, research did not cease on direct reduction methods. As a result of the great savings to be achieved with economy of scale, eventually the larger blast furnaces were 'not well suited for small, local needs' [2]. Because of this large scale, there are very high capital costs for integrated mills. Also, the coke ovens of that time were a significant environmental issue. Different sources list many of the 100 or more direct-reduction processes developed in the last 100 or so years; many never made it beyond concept; some as far as pilot plant; only a few commercial [1,2]. But only two DR technologies have achieved long-term success – coal-based rotary kilns, which represent about 20% of the current market (mainly in India), and the subject of this paper, gas-based direct reduction in a shaft furnace.

Development of gas-based direct reduction

The pioneer in commercial operation of gas-based direct reduction was HyL in Mexico. Due to the lack of affordable scrap to feed their electric furnaces, they spent several years exploring different concepts for making their own iron. They commercialized what became known as the HyL I Process in Monterrey, Mexico in 1957. This was a batch process involving four different reactors, with a capacity of 75,000 tonnes per year, and was their mainstay until the continuous HyL III was commercialized in 1980. HyL I was slow to be commercialized outside Mexico, so the next commercial developments were made by the Surface Combustion Division of Midland-Ross Corporation in Toledo, Ohio, several years later.

Surface Combustion Corporation (SCC) was formed in 1915 to exploit new understandings in how combustion occurs, and how to best make it happen in commercial installations. They soon became known for their gas generators and combustion systems [3,4]. They became directly involved in the iron ore business when they partnered with four mining and steel companies to develop a shaft furnace for induration of a new product in the Mesabi Range, iron oxide pellets made from beneficiated taconite ore. This development was successful; 41 furnaces were sold in late 1940s and early 1950s [4]. In the early 1960s, SCC took a turn at direct reduction by developing a coal-based technology using a rotary-hearth furnace, HeatFast. As they worked through developmental issues with that technology, SCC's head of research, Donald Beggs, thought of how much easier it would be to simply use reformed natural gas as the reductant. Combining SCC's experiences in processing pellets through a shaft furnace, and the idea of near-stoichiometric reforming of natural gas, what became known as the MIDREX Process was developed.

The basic patent of Beggs and Scarlett, granted in 1973, can be seen in Figure 1 [5]. Hot reducing gas is fed into the side of the shaft furnace, and passes countercurrent to the iron oxide pellets or lump fed from the top. The spent top gas still has thermal value and unreacted CO and H_2. The off-gas is passed through a scrubber to remove water and solids, then split to form a purge stream and process gas. The process gas is mixed with fresh gas, and the mixture fed to the reformer, where the CO_2 and H_2O from the top gas reforms the fresh gas to more CO and H_2. The product of the reformer is then sent to the shaft furnace, to reduce more oxide.

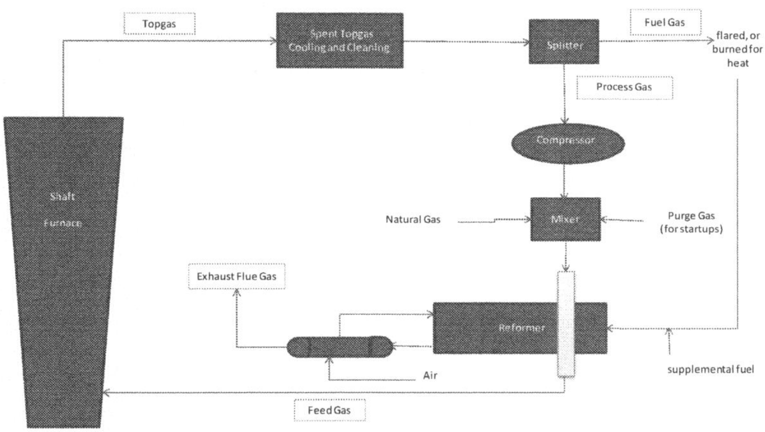

Figure 1. Base MIDREX Process patent

The MIDREX Process pilot plant was operated in Toledo, Ohio, in 1967, with the first commercial modules beginning operation at Oregon Steel in 1969. The Oregon Steel DR plant consisted of a pelletizer and twin 3.7 meter diameter shaft furnaces, each with a capacity of 150,000 tpy. During operation of this plant, key issues such as carbon deposition, DRI degradation, and clustering were examined and overcome. Also during this time, the business was spun off as a new division of Midland-Ross, Midrex [4].

In the normal course of events, after the process had been proven in at the Oregon site Midrex would have scrambled to find a client willing to build another, perhaps larger, plant. But they were fortunate enough to commercialize the technology much more quickly, due to one visionary – the German industrialist, Willy Korf. At a 1968 meeting [4]:

> Korf told his hosts that he wanted a direct reduction plant designed for him immediately for Georgetown Steel in South Carolina, where he was building a mini-mill, and for Hamburg, Germany, where he planned to develop a steel complex. The SCD [Surface Combustion Division] people were reluctant, because the Oregon Steel direct reduction plant was just completing construction and would not be operating for several weeks. Korf persisted, and wrote a check

on the spot for SCD to begin design work. Both plants were rated at 400,000 tons per year, included 4.9 m diameter furnaces, and started up in 1971.

(Korf liked the technology and bought the company in 1974, moving it to North Carolina. In 1983, after Korf Industries applied for bankruptcy, Kobe Steel acquired Midrex.)

The other factor in the early success of the MIDREX Process was its commercialization at the same time that mini-mills were evolving. Steel mini-mills had achieved great success by building small electric furnaces making lower-grade steels out of local scrap, generally for a local or regional market. It didn't take long for the US mini-mills to take over the American market for rebar and other long products. Increasing sales further required penetration of the flat products market. However, even the best scrap contains too many residual elements to be successfully utilized at a 100% feed to make this higher-value product. Mini-mills needed 'Alternate Iron Units' (AIU) – but not necessarily large amounts, as might come as pig iron from a captive blast furnace. Enter DRI from shaft furnaces: minimills needed AIU at the same time DR plants needed customers. The market for gas-based shaft furnaces took off in the mid-1970s, and hasn't slowed down for significant periods of time since. Figure 2 shows the evolution in annual production for all gas-based products (for details on these products, see below), which was 74 million tonnes in 2012 [6].

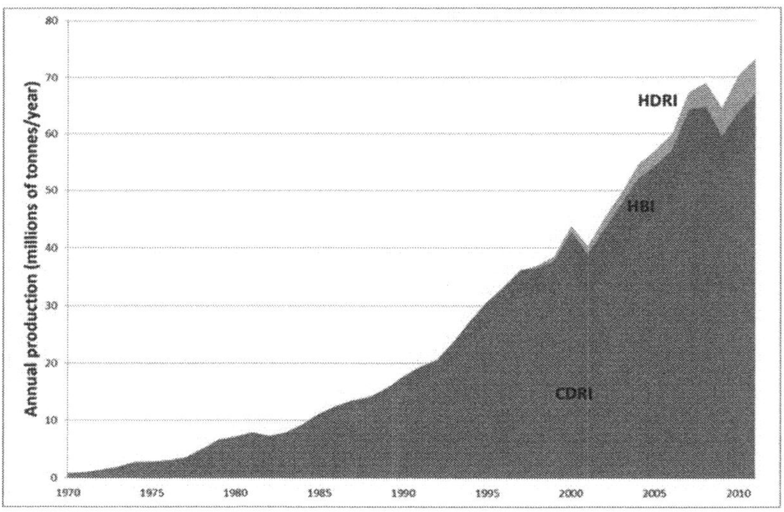

Figure 2. Historical production of Direct Reduced Iron

Sustainability

Not much has yet been said about sustainability. This is because sustainability, though a concept that has been around for at least two hundred years, had not really taken hold in the steel industry in these times. When the MIDREX Process was being sold to clients, what was stressed was the fact that the shaft furnace process was economical at small scales, was significantly lower in capital and operating costs than a blast furnace, and was much simpler – for example, no coke ovens or sinter plants were required. When sustainability began to be something worth talking about, direct reduction had a lot to say. First, of course, by using natural gas as the reductant instead of carbon in the form of coke, gas-based DR automatically has the advantage of using a far cleaner fuel - besides lower emissions per tonne steel, there are fewer emissions, period, because the mills are much smaller than an integrated mill, and can easily start and stop as conditions demand.

Actually quantifying the environmental benefits of DR versus the BF is not simple, as the two unit operations cannot be compared directly, as the products are different, and the raw materials can have separate footprints. The logical common basis is per tonne of liquid steel, which involves an analysis of the particular agglomeration, ironmaking, and steelmaking steps. Special care must be taken so that the assumptions are the same; this is not always achieved in literature comparisons [7, 8]. Figures 3 and 4 below are taken from a recent publication [8] and compare different scenarios for energy use (MJ/tonne of steel) and CO_2 production (kg CO_2/tonne of steel), two commonly-used measures of sustainability; in all cases, pellets were used as the ironmaking process feed. In both cases, using the combination of DR with an EAF was better than BF-BOF, particularly in the case of CO_2 production. The numbers for energy are much closer than they used to be, since the average blast furnace is vastly more efficient than it was even 10-20 years ago [9].

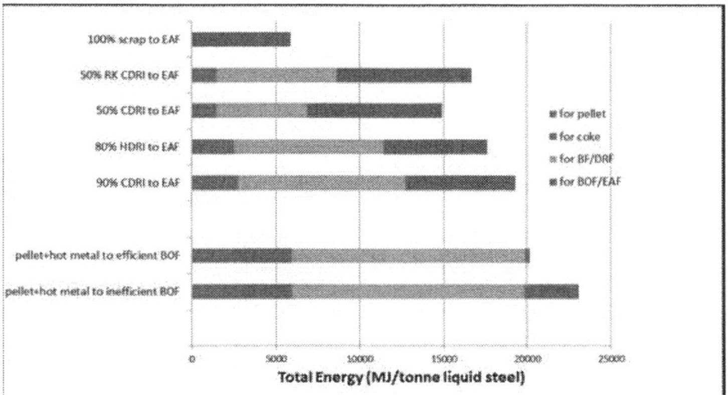

Figure 3. Energy Consumption for typical ironmaking-steelmaking scenarios

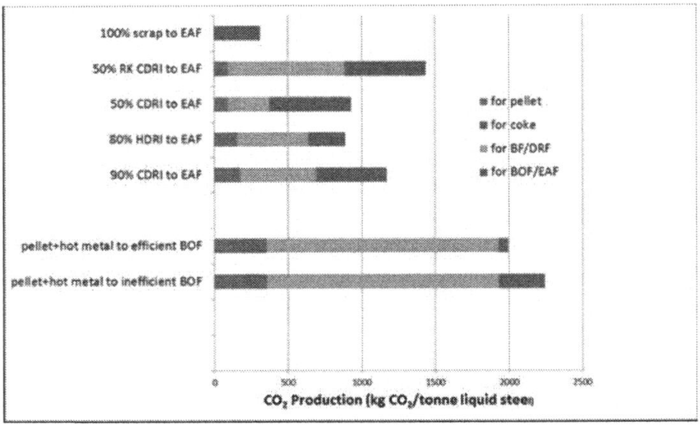

Figure 4. CO_2 production for typical ironmaking-steelmaking scenarios

The modern DR plant

There have been many modifications to the earliest DR modules to improve their operability and meet customer needs. This has included changes to plant size, energy efficiency, products, and reductants.

As mentioned before, the earliest plants had capacities of a few hundred thousand tonnes of DRI products a year; now the largest operating plant is 1.85 mtpy, and modules with official capacities of 2.0 to 2.5 million tpy have been announced. In terms of energy efficiency, Figure 5 shows how average natural gas consumption has decreased over the last forty years [4]. The six improvements given in the figure are listed below:

1: 480°C combustion air preheat
2: *in situ* reforming
3: 540°C combustion air preheat and 400°C process gas preheat
4: 650°C combustion air preheat and 540°C process gas preheat
5: Top gas fuel preheat
6: Preheat of natural gas used for transition zone and enrichment

The original product of the DR process technologies was Cold DRI (CDRI), which would then be shipped to the steel mill, whether next door or across the world. However, it was early realized that CDRI is not completely stable; it is prone to rapid oxidization, particularly in the presence of water, and, even worse, salt water. After several incidents involving piles of DRI catching fire, strict guidelines were set in place for shipping DRI. Considerable research was conducted to passivate DRI; more successful was the compaction of hot DRI through rollers to make Hot Briquetted Iron (HBI). HBI is much more dense and stable, and can be easily and safely stored and shipped anywhere. Most recently, for steel mills in close proximity to a DR plant, the DRI can be transferred hot by one of several methods, arriving in the EAF at a temperature of several hundred degrees, thus Hot DRI (HDRI). Figure 2 showed the total amount of DRI produced over the years, divided into CDRI, HBI, and HDRI. The increasing popularity of HBI and HDRI is evident.

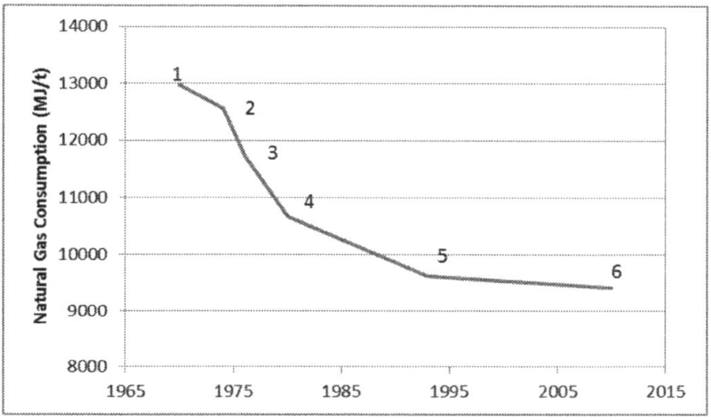

Figure 5. Improvements in energy efficiency in MIDREX Process

The reducing gas used in the initial plants was natural gas, which is mainly methane but varies widely in quality in different locations. There has been considerable interest, particularly in recent years, with using alternate sources of the CO and H_2 used in direct reduction. The world's first application of coal gasification to produce DRI in a MIDREX Plant began operation in 1999 at ArcelorMittal Steel South Africa (formerly Saldanha Steel). This is now known as the COREX®/MXCOL® Process. Midrex has multiple projects in the construction phase in India that will use syngas. These include: JSPL Angul I, which will be the world's First DR Plant with direct coal gasification. The plant is currently under construction and will use indigenous iron ore and coal. JSW Projects Ltd., a group company of JSW Steel Ltd., is building another COREX/MXCOL Plant featuring multiple discharge options. This new facility is being erected at JSW Steel Ltd., located in Karnataka, India. Lastly, JSW Dolvi is modifying their existing MIDREX Plant to utilize coke oven gas (COG) to supplement its natural gas supply for production. The plant is located at the JSW-Dolvi Works (formerly JSW Ispat, Ltd.), in Maharashtra, India.

Direct Reduction – Indirect Reduction Synergies

For the last 40-50 years, marketing and sales personnel from DR suppliers have worked to convince clients that ironmaking by direct reduction was a better choice for them than indirect reduction via the blast furnace. Figure 2 shows the significant growth in DRI production over the last forty years. How does that compare to total ironmaking production? Figure 6 shows the comparison [11], where overall production has increased over that time period from approaching

500 million tpy to almost 1100 mpty. The DR gain in this diagram is visible, but only just (about 7% of the total). Perhaps a better choice than trying to 'beat them' is to 'join them', which is occurring as a result of two initiatives – use of integrated mill waste off-gases as partial or complete sources of reducing gases for DR, and use of metallic iron created in a DR plant as part of the blast furnace feed.

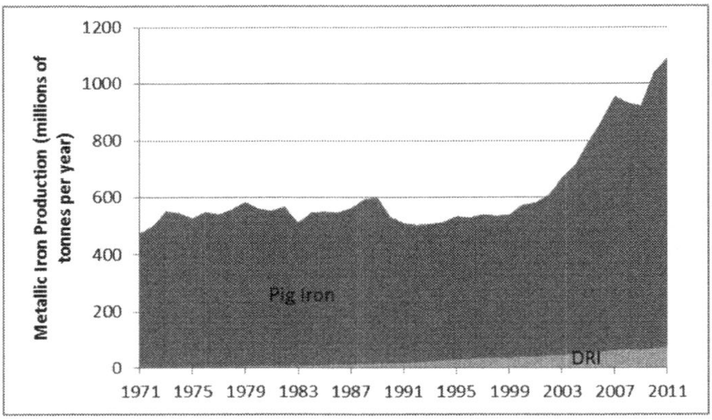

Figure 6. World Metallic Iron Production

It has been known for many years that Coke Oven Gas, and to a lesser extent Blast Furnace Gas and Basic Oxygen Furnace Gas, contain reducing components, and might be more efficiently utilized in reducing a metal oxide rather than being burned as fuel, and certainly better than being flared. While COG does have considerable reducing power, it also contains tars and higher hydrocarbons that would cause problems in the standard reformer. In the past year, Midrex and Praxair have piloted a new partial oxidation system, the Thermal Reactor System™ (TRS™), to remove the undesired species from simulated COG. Partial use of COG in one Indian steel mill/DR furnace has already been announced, and when TRS is proven out, a DR plant may become an integral part, ironically, of improving the emissions footprint of an integrated steel mill.

For fifty years, researchers have looked at the impact of adding a small amount of iron metallics to a blast furnace [12]. Research has shown that adding 10% metallic (scrap, DRI, or HBI) to the feed mix will reduce coke consumption by 7%, and increase productivity by 8% [12, 13]. This has been utilized by a number of mills to temporarily increase production when they are hot-metal short, or more permanently (AK Steel being the most prominent example) to shut down an old blast furnace while still making the hot metal needed in the overall facility [14].

However, this approach has been hampered by the high relative cost of HBI compared to Blast Furnace grade pellets. An approach being explored to reduce this expense is to make the HBI not from DR-grade oxide pellets, but less expensive BF grade – the only reason standard HBI is currently used in BFs is that it's the only grade available. Laboratory and pilot-plant experiments have shown it is possible to make 'Blast Furnace Briquetted Iron' (BFBI) from BF-grade pellets [15]. This is not surprising because DR-grade pellets are used in DR processes mainly because of the needs of the client electric arc furnace, not the shaft furnace itself. Commercial trials are planned in the near future.

In this situation the economic question comes down to the best use of a given amount of BF-grade pellets – produce hot metal, BFBI, or some fraction of each. Essentially this comes down to coke use to make metallic iron, against natural gas use. Figure 7 graphically shows the economics in this situation, where negative numbers (in red) indicate conditions where coke use in a standard BF is cost-effective [15]. The positive numbers (representing a lower operating cost for the BFBI option), particularly those in green, favor utilizing BFBI from a natural gas DR plant. Even without a CO_2 tax, using at least some fraction of BFBI in a blend is attractive – and this BFBI could have been created using waste gases from the same facility.

The first major DR project to provide HBI or BFBI to a blast furnace on a regular basis will be undertaken by the Voestalpine Group. Voestalpine is in the process of building a 2.0 million tonne per year MIDREX HBI plant in Texas, that will ship at least 1.0 mtpy to their steelplants in Austria [16].

		Price of Gas per mmBTU												
	$ 0.78	$ 3.00	$ 3.25	$ 3.50	$ 3.75	$ 4.00	$ 4.25	$ 4.50	$ 4.75	$ 5.00	$ 5.25	$ 5.50	$ 5.75	$ 6.00
Price of Coke per tonne	$ 120	$ (21.41)	$ (23.74)	$ (26.07)	$ (28.40)	$ (30.73)	$ (33.06)	$ (35.40)	$ (37.73)	$ (40.06)	$ (42.39)	$ (44.72)	$ (47.05)	$ (49.39)
	$ 130	$ (17.81)	$ (20.14)	$ (22.47)	$ (24.80)	$ (27.13)	$ (29.46)	$ (31.80)	$ (34.13)	$ (36.46)	$ (38.79)	$ (41.12)	$ (43.45)	$ (45.79)
	$ 140	$ (14.21)	$ (16.54)	$ (18.87)	$ (21.20)	$ (23.53)	$ (25.86)	$ (28.20)	$ (30.53)	$ (32.86)	$ (35.19)	$ (37.52)	$ (39.85)	$ (42.19)
	$ 150	$ (10.61)	$ (12.94)	$ (15.27)	$ (17.60)	$ (19.93)	$ (22.26)	$ (24.60)	$ (26.93)	$ (29.26)	$ (31.59)	$ (33.92)	$ (36.25)	$ (38.59)
	$ 160	$ (7.01)	$ (9.34)	$ (11.67)	$ (14.00)	$ (16.33)	$ (18.66)	$ (21.00)	$ (23.33)	$ (25.66)	$ (27.99)	$ (30.32)	$ (32.65)	$ (34.99)
	$ 170	$ (3.41)	$ (5.74)	$ (8.07)	$ (10.40)	$ (12.73)	$ (15.06)	$ (17.40)	$ (19.73)	$ (22.06)	$ (24.39)	$ (26.72)	$ (29.05)	$ (31.39)
	$ 180	$ 0.19	$ (2.14)	$ (4.47)	$ (6.80)	$ (9.13)	$ (11.46)	$ (13.80)	$ (16.13)	$ (18.46)	$ (20.79)	$ (23.12)	$ (25.45)	$ (27.79)
	$ 190	$ 3.79	$ 1.46	$ (0.87)	$ (3.20)	$ (5.53)	$ (7.86)	$ (10.20)	$ (12.53)	$ (14.86)	$ (17.19)	$ (19.52)	$ (21.85)	$ (24.19)
	$ 200	$ 7.39	$ 5.06	$ 2.73	$ 0.40	$ (1.93)	$ (4.26)	$ (6.60)	$ (8.93)	$ (11.26)	$ (13.59)	$ (15.92)	$ (18.25)	$ (20.59)
	$ 210	$ 11.00	$ 8.66	$ 6.33	$ 4.00	$ 1.67	$ (0.66)	$ (3.00)	$ (5.33)	$ (7.66)	$ (9.99)	$ (12.32)	$ (14.65)	$ (16.99)
	$ 220	$ 14.60	$ 12.26	$ 9.93	$ 7.60	$ 5.27	$ 2.94	$ 0.60	$ (1.73)	$ (4.06)	$ (6.39)	$ (8.72)	$ (11.05)	$ (13.39)
	$ 230	$ 18.20	$ 15.86	$ 13.53	$ 11.20	$ 8.87	$ 6.54	$ 4.20	$ 1.87	$ (0.46)	$ (2.79)	$ (5.12)	$ (7.45)	$ (9.79)
	$ 240	$ 21.80	$ 19.46	$ 17.13	$ 14.80	$ 12.47	$ 10.14	$ 7.80	$ 5.47	$ 3.14	$ 0.81	$ (1.52)	$ (3.85)	$ (6.19)
	$ 250	$ 25.40	$ 23.06	$ 20.73	$ 18.40	$ 16.07	$ 13.74	$ 11.41	$ 9.07	$ 6.74	$ 4.41	$ 2.08	$ (0.25)	$ (2.59)
	$ 260	$ 29.00	$ 26.66	$ 24.33	$ 22.00	$ 19.67	$ 17.34	$ 15.01	$ 12.67	$ 10.34	$ 8.01	$ 5.68	$ 3.35	$ 1.01
	$ 270	$ 32.60	$ 30.26	$ 27.93	$ 25.60	$ 23.27	$ 20.94	$ 18.61	$ 16.27	$ 13.94	$ 11.61	$ 9.28	$ 6.95	$ 4.61
	$ 280	$ 36.20	$ 33.86	$ 31.53	$ 29.20	$ 26.87	$ 24.54	$ 22.21	$ 19.87	$ 17.54	$ 15.21	$ 12.88	$ 10.55	$ 8.21
	$ 290	$ 39.80	$ 37.46	$ 35.13	$ 32.80	$ 30.47	$ 28.14	$ 25.81	$ 23.47	$ 21.14	$ 18.81	$ 16.48	$ 14.15	$ 11.82
	$ 300	$ 43.40	$ 41.06	$ 38.73	$ 36.40	$ 34.07	$ 31.74	$ 29.41	$ 27.07	$ 24.74	$ 22.41	$ 20.08	$ 17.75	$ 15.42
	$ 310	$ 47.00	$ 44.66	$ 42.33	$ 40.00	$ 37.67	$ 35.34	$ 33.01	$ 30.67	$ 28.34	$ 26.01	$ 23.68	$ 21.35	$ 19.02
	$ 320	$ 50.60	$ 48.26	$ 45.93	$ 43.60	$ 41.27	$ 38.94	$ 36.61	$ 34.27	$ 31.94	$ 29.61	$ 27.28	$ 24.95	$ 22.62
	$ 330	$ 54.20	$ 51.86	$ 49.53	$ 47.20	$ 44.87	$ 42.54	$ 40.21	$ 37.87	$ 35.54	$ 33.21	$ 30.88	$ 28.55	$ 26.22
	$ 340	$ 57.80	$ 55.46	$ 53.13	$ 50.80	$ 48.47	$ 46.14	$ 43.81	$ 41.47	$ 39.14	$ 36.81	$ 34.48	$ 32.15	$ 29.82
	$ 350	$ 61.40	$ 59.06	$ 56.73	$ 54.40	$ 52.07	$ 49.74	$ 47.41	$ 45.07	$ 42.74	$ 40.41	$ 38.08	$ 35.75	$ 33.42
	$ 360	$ 65.00	$ 62.66	$ 60.33	$ 58.00	$ 55.67	$ 53.34	$ 51.01	$ 48.67	$ 46.34	$ 44.01	$ 41.68	$ 39.35	$ 37.02
	$ 370	$ 68.60	$ 66.26	$ 63.93	$ 61.60	$ 59.27	$ 56.94	$ 54.61	$ 52.27	$ 49.94	$ 47.61	$ 45.28	$ 42.95	$ 40.62
	$ 380	$ 72.20	$ 69.86	$ 67.53	$ 65.20	$ 62.87	$ 60.54	$ 58.21	$ 55.87	$ 53.54	$ 51.21	$ 48.88	$ 46.55	$ 44.22
	$ 390	$ 75.80	$ 73.46	$ 71.13	$ 68.80	$ 66.47	$ 64.14	$ 61.81	$ 59.47	$ 57.14	$ 54.81	$ 52.48	$ 50.15	$ 47.82
	$ 400	$ 79.40	$ 77.06	$ 74.73	$ 72.40	$ 70.07	$ 67.74	$ 65.41	$ 63.07	$ 60.74	$ 58.41	$ 56.08	$ 53.75	$ 51.42

Figure 7. Economics of BF pellet use in integrated mills

The Future

It is quite likely that direct reduction plants will continue to be built well into the future, and current ones will continue to operate. In the short term, there is great interest in building DR plants to utilize the abundant shale gas in the United States, with one large project under construction, another announced, and still more companies interested. In conditions where gas-based shaft furnaces aren't cost effective, there are other options where the reduction source is ultimately coal. This includes MXCOL (as discussed above), and the direct utilization of coal in rotary kilns and rotary hearth furnaces. Rotary Kilns are used extensively in India, but are much smaller single units than gas-based furnaces. RHFs have been used in several locations to process pelletized or briquetted waste oxides (for example, the FASTMET® Process in Japan), and virgin iron ore (Mesabi Nugget using the ITmk3® technology in Minnesota).

Many of these technologies have the capability to be harnessed for non-ferrous applications as well, particularly those that can be scaled down to the sizes typically needed for reducing pure non-ferrous oxide (shaft furnace), or those that can handle large amounts of a low-grade ore and make valuable co-products (RHF). Midrex alone in recent years has successfully processed nickel laterites, ilmenite and titano-magnetites (in the case of ilmenite, using an ITmk3-like process to make pig iron and a high-TiO_2 slag in one simple reactor [17]), and creation of a ferrochrome product from chromite ore. This could open the way to the use of combination ores to create all types of ferroalloys. The future looks bright for these technologies.

Summary and Conclusions

The current version of direct reduction of iron ore now has a distinguished history going back more than fifty years. It is an integral part of the overall steel industry, because of its flexibility and sustainability. The industry will continue to grow because of its original attributes of low cost, low capital, and flexible operation, as well as developments over the years in plant capacity, new products, and ability to use a range of different reductants. Applications will only increase as synergies with integrated mills become more apparent.

References

1. William T. Lanford, Jr., et al., ed., *The Making, Shaping and Treating of Steel*, 10th edition, (Pittsburgh, Herbick & Held, 1985).

2. John T. Kopfle. "The History of Direct Reduction," *Direct Reduced Iron: Technology and Economics of Production and Use*, ed. J. Feinman and D. R. MacRae, (Warrendale, PA: ISS-AIME, 1999).

3. Wade N. Harris. *Midland-Ross Corporation: A Story of Diversification*, (Newcomen Society in North America, 1964).

4. John T. Kopfle. *A Better Mousetrap: The History of Midrex Technologies*. Direct from Midrex, 1st Quarter, 2004, pp 3, 2nd Quarter, 2004, pp 6, 3rd Quarter, 2004, pp 11, and 4th Quarter, 2004, pp 6, Midrex Technologies. http://www.midrex.com/.

5. Donald Beggs and John C. Scarlett. *Method of and Apparatus for Reducing Iron Oxide to Metallic Iron*. US Patent 3748120, granted July 24, 1973.

6. 2012 World Reduction Statistics. (Charlotte, NC, Midrex, 2012). http://www.midrex.com/.

7. T. P. Battle and J. M. McClelland: "A New Model for Energy and Carbon Dioxide Calculations: Model Development, Assumptions, and Applications to EAF Steelmaking," *AISTech 2011*, pp 943.

8. Seshadri Seetharaman, ed., *Treatise on Process Metallurgy*. Elsevier, to be published, 2014.

9. H. B. Luengen and P. Schmoele. "Measures to reduce CO_2 and other emissions in the steel industry in Germany and Europe." *METEC InSteelCon*, 2010.

10. G. Metius and J. McClelland, "Syngas Options for Coal-Based Direct Reduction" (Presentation at AISTech 2012, Atlanta, Georgia).

11. USGS/USBM Minerals Yearbooks. http://www.usgs.gov/

12. E. J. Ostrowski, R. L. Stephenson, D. A. Kercsmar, "Use of DRI in Ironmaking", *Direct Reduced Iron: Technology and Economics of Production and Use*, ed. J. Feinman and D. R. MacRae, (Warrendale, PA: ISS-AIME, 1999), pp 147-157.

13. Robert Hunter, "Massive CO_2 Savings by Use of HBI Made with Natural Gas: Blast Furnace Economics", *Direct from Midrex*, 2Q2010, pp 5-8.

14. D. A. Kercsmar, et al., "Sustained Production in excess of 9 tons per day 1100 Cu Ft WV at Middletown No. 3 Blast Furnace", *Electric Furnace Conference*, 1994.

15. Thomas Battle, Urvashi Srivastava, and Mark Boedecker, "Use of hot briquetted iron made from low-grade iron ore pellets in a blast furnace", *AISTech 2013*, Pittsburgh, Pennsylvania

16. *voestalpine signs contract with Siemens and Midrex for DR Plant.* http://www.midrex.com/

17. J. M. Shekiro, et al., "TiRON – A New Process for the Production of High-Grade Feedstock for TiO_2 Pigment Manufacture", *8th International Heavy Minerals Conference*, Perth, Western Australia, 2011, pp 283.

Celebrating the Megascale: Proceedings of the Extraction and Processing Division Symposium on Pyrometallurgy in Honor of David G.C. Robertson
Edited by: Phillip J. Mackey, Eric J. Grimsey, Rodney T. Jones, and Geoffrey A. Brooks
TMS (The Minerals, Metals & Materials Society), 2014

KINETICS OF REACTION IMPORTANT IN OXYGEN STEELMAKING

Kenneth S. Coley[1], Elaine Chen[2], and Michael Pomeroy[1]

[1]McMaster Steel Research Centre, Department of Materials Science and Engineering, McMaster University, 1280 Main St West, Hamilton, Ontario, L8S 4L7, Canada

[2]Arcelor-Mittal, Global R&D, East Chicago I 3001 East Columbus Drive, East Chicago, IN 46312-2939, USA

Keywords: Steelmaking, BOF Modelling, Kinetics of Droplet Decarburization

Abstract

Recent work on modeling of BOF steelmaking is reviewed, highlighting the critical aspects of each approach. It is concluded that the most successful models should be based on a deep understanding of the mechanisms and kinetics of the critical reactions. The importance of the decarburization mechanism is discussed with particular reference to its role in droplet swelling or bloating which has a profound influence on the droplet residence time in the slag. Conditions which cause bloating are discussed and the rate determining step is proposed to be primarily nucleation of CO bubbles inside the metal droplet with some influence from growth by reaction at the bubble/metal interface. The discrepancy in the super-saturation ratio required for classical nucleation is discussed and an approach using a surface tension modifying parameter is illustrated. Finally, the role of CO nucleation in controlling the driving force for dephosphorization is discussed.

Introduction

It is a great pleasure and an honour for the authors and in particular the presenting author(KSC) to present this paper at a symposium in honour of Dr David Robertson. Dr Robertson is a leader in modeling of metallurgical processes and his work on coupled reaction modeling[1, 2] is an inspiration to those of us who wish to see fundamental kinetics applied to the "big picture". The appeal of this model is that it is elegant in its simplicity whilst remaining true to the fundamental kinetics. Its power is probably greater today than when it was first proposed and it forms the basis for ongoing work in the McMaster Steel Research Centre on ladle metallurgy and inclusion formation[3] Dr Robertson has also shown a long standing interest in the mechanistic details of metallurgical reactions and on many occasions in recent years this presenter has enjoyed a stimulating discussion with him on the apparent impossibility of high rates of carbon monoxide nucleation inside iron droplets although we have taken quite different approaches in "explaining the impossible"[4-6]. The Robertson Symposium is the ideal venue to present this paper as it discusses the use of detailed mechanistic and kinetic studies in developing big picture models, and in particular it addresses the carbon monoxide nucleation discrepancy in relation to such a model for oxygen steelmaking.

Modeling of Oxygen Steelmaking

In recent years, modeling of metallurgical processes has started to move from an activity which offers insight into the process to something that may offer prediction and even control. Kitamura[7] has suggested that now is the time for kinetics to be applied in achieving quantitative predictions, in the way thermodynamics has been for several decades. This is "music to the ears" of practitioners in the field of kinetics but requires the development of models which combine many different kinetic equations in describing the process in its entirety. It also, as will be illustrated in the current paper, may require reevaluation of the kinetic data including a detailed understanding of the mechanisms and the way in which different reactions are likely to interact. The current paper will review recent work on the modeling of oxygen steelmaking to illustrate the power of an approach which combines detailed kinetic studies with big picture modeling. In doing this it will highlight the importance of droplet swelling (the bloated droplet model) and discuss in detail the kinetics of swelling and the carbon monoxide nucleation contradiction. Finally there will be discussion of the way in which the Bloated Droplet affects dephosphorization.

Kattenbelt and Roffel[8] developed a model to describe the response of oxygen steelmaking to step changes in process variables such as oxygen blow rate, lance height and oxide additions. In defining the parameters to be considered in an empirical relation for decarburization rate, these workers discussed the mechanisms involved in some detail including work on droplet generation, size distribution and residence time. However, they did not employ any of the fundamental relations directly. These workers showed excellent agreement between their predictions and measured data from BOF steelmaking.

A more recent study by Li et al[9] employed a similar approach to Kattenbelt and Roffel but in this case modeled the process with respect to time rather than response to step changes in operating parameters. Li et al used three generalized rate equations; one for each stage of decarburization. These equations were modified to account for bath mixing which was described as a function of lance height. Li et al determined rate constants by fitting plant data from 67 heats and further verifying their model in comparison to 30 additional heats. In common with Kattenbelt and Roffel the equations employed by Li et al did not take account of detailed mechanisms. However, their data showed remarkable agreement with plant data. Sukala et al(10) took a very different approach to modeling the BOF. Based on a review of work on the important reactions, these workers suggested that for the main part of the blow, the process could be modeled by assuming equilibrium partition of supplied oxygen. However, to get good predictions of observed features these workers employed "efficiency factors". These factors were related to the kinetics of the individual reactions which strongly suggests an element of kinetic control.

Arguably the most successful attempt at modeling the BOF is that of Dogan et al(11-13) who built their model based on a fundamental understanding of the mechanisms involved. These workers developed models for each of the key components of the process and combined these in an overall model which they tested against plant data. The model was tested against the plant data of Ciccuti et al(14, 15) and showed excellent agreement through most of the blow but under-predicted the extent of decarburization towards the end of the blow. The most likely explanation for this is an underestimate of the importance of the hot-spot as this would be expected to dominate towards the end of the blow. The success of the Dogan model is that it recognizes that several components of refining must be considered:

1/ The formation of slag
2/ The generation of droplets and their ejection from the bath into the slag
3/ Droplets have a finite residence time in the slag during which time they can be refined
4/ Rate of individual refining reactions with droplets
6/ Subsequent return of droplets to the bath resulting in an incremental change in bath composition
7/ Decarburization of the bath by direct reaction with the impinging oxygen jet. It should be noted that this is a viable mechanism for decarburization but not for dephosphorization.

Dogan et al developed individual models to describe each of these components and combined them in a global model of the process. It should be noted that any fitting of constants was conducted during the creation of the individual models and no further adjustment was made on combination. Given this fact, the agreement achieved by Dogan et al is indeed remarkable.
It is the belief of the authors, that future attempts to model the BOF should build on the structure proposed by Dogan et al. This work shows the power of combining lab-scale kinetic studies and plant trial data with process scale modeling. It also has the benefit that individual researchers can apply their specific skills to further developing one aspect of the approach as appropriate. This in fact, has been the approach taken in our laboratory at McMaster, where we have been working on various aspects of 2, 3 and 4 over the last several years(16-19)

Generation of Droplets

Li and Harris[20] developed an equation, based on the Kelvin-Helmhotz criterion that predicts the onset of splashing from top gas injection:

$$\frac{\rho_G u_G^2}{2\sqrt{\sigma g \rho_L}} = 1.0 \tag{1}$$

where where r_g and r_L denote the density of gas and liquid, respectively, and g is gravitational acceleration, σ is surface tension and u_G denotes the critical gas velocity, related to the centre line velocity of the jet, u_i, by [20]

$$u_G = \eta u_i \tag{2}$$

where η is a constant. Plant data for oxygen steelmaking by Deo and Boom[21] gives the value of η as 0.447, which agrees closely with values collected by Li and Harris for a wide range of liquids.[20] Subagyo et al [17] demonstrated that the term on the left hand side of Equation (1) is useful in characterizing metal droplet generation through gas impingement and termed the resulting dimensionless parameter the "blowing number", N_B :

$$N_B = \frac{\rho_G u_G^2}{2\sqrt{\sigma g \rho_L}} \tag{3}$$

Whilst noting that He and Standish[22] had employed a similar number, to analyse cold modeling data Subagyo et al favoured the Blowing Number because of its fundamental basis on the Kelvin-Helmholtz criterion. Figure 1 shows data taken from He and Standish and Subagyo et al[17] for droplet generation rate per unit volume of blown gas plotted against the blowing number. The solid line on this figure is represented by Equation (5).

$$\frac{R_B}{F_G} = \frac{(N_B)^{3.2}}{\left[2.6x10^6 + 2.0x10^{-4}(N_B)^{12}\right]^{0.2}} \tag{4}$$

Although the high temperature data of Subagyo et al is more scattered the two sets of data are in remarkable agreement when plotted as a function of blowing number. Recent work by Alam et al[23] suggests that η is not in fact a constant but is a function of lance height and that the blowing number correlation breaks down when the jet is in the penetrating mode.

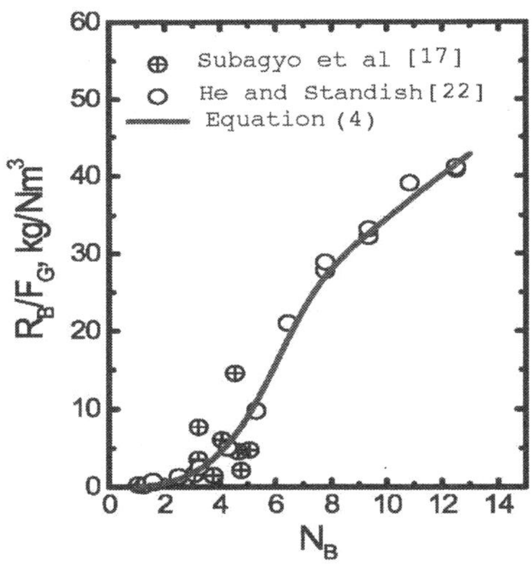

Figure 1. Effect of Blowing Number on the Generation Rate of Metal Droplets

Droplet Size Distribution

Having determined the rate of droplet production one must also determine the size distribution which can then be used in predictions of residence time and, when combined with data on kinetics of the important reactions, allow the prediction of overall refining rates. Subagyo et al(18) also investigated droplet size and found, in agreement with previous workers Koria and Lange(24) that size distribution followed the Roisin-Rammler-Sperling distribution function. In addition Subagyo et al found that the parameters for the distribution function could be described as a function of the blowing number. These authors were not able to offer any theoretical basis for this relationship.

Residence Time

If one knows the size and residence time in the slag of an individual droplet, this may be combined with kinetic data to predict the extent of refining for that droplet prior to its re-entry into the metal bath. This approach can be summed over all droplets produced over time to calculate the contribution of droplets to the overall refining process. Subagyo et al attempted to predict residence time of different droplets by, determining their velocity on ejection from the bath and using that velocity in a ballistic calculation of droplet trajectory[16]. These workers found that predicted residence times were extremely short, less than 0.25s which is not at all consistent with estimates based on refining rates. In subsequent work, Brooks and coworkers[19] identified the source of this discrepancy. Based on work by Molloseau and Fruehan[25], which showed iron carbon droplets swelling due to CO evolution on contact with oxidizing slag, these workers were able to predict density changes in the droplet as a function of time. When these density changes were included in the ballistic calculations, the predicted residence times were much more consistent with observed refining rates. This work resulted in the Bloated Droplet Model, which became a subsequent component of the overall model of BOF steelmaking developed by Dogan et al[11, 12]. It is worth noting, in Dogan's work the model was not terribly sensitive to droplet size and in fact assuming a uniform size resulted in fairly reasonable predictions of decarburization.

Kinetics and Mechanism of Selected Reactions in BOF Steelmaking

Kinetics of Carbon-monoxide Gas Formation in Iron Carbon Droplets

In light of its importance in determining residence time in the slag, and in possible mechanisms for of important refining reactions, work in the authors' laboratory over the last few years has focused on micro-kinetics of droplet reactions in particular those related to swelling[4, 33-35]. In this paper we will focus primarily on the reaction of carbon to form CO as this is the critical reaction in droplet swelling and has been the subject of much discussion between the presenting author and Dr Robertson.

Since the early work of Mulholland et al[26], our picture of carbon bearing metal droplets reacting with slag has been of a droplet surrounded by a halo of CO/CO2, the reaction proceeds via a number of steps: 1/ mass transport of oxide in the sag, 2/ slag/gas reaction; mass transport in the halo, 3/reaction at the gas/metal interface and 4/ mass transport in the metal.

Min and Fruehan[27] conducted an analysis of each of the possible rate determining steps and concluded that for high sulphur iron, the reaction was controlled by the metal/ gas reaction because of the retarding effect of sulphur. For low sulphur metal the reaction was controlled by a combination of mass transport in the slag and in the halo. This finding is contrary to the accepted mechanism for reaction with solid carbon, however, these workers calculated that the slag/gas reaction and transport in the metal should be much faster than the other steps. A model was developed by Murthy et al.[28, 29] to explore the possibility of gas halo transport control. These workers claimed that it was feasible for transport in the halo to control the reaction, but the suggestion by Murthy et al that increased conductivity of the slag would enhance the reaction, implies that mass transport in the slag is at least partly rate controlling.

For CO formation to cause droplet swelling the gas must nucleate inside the droplet rather than form a gas halo or bubble swarm on the surface. Gare and Hazeldean[32] defined a number of different "reaction regimes" based on phenomenological observations. These workers, discussed a regime in which the halo was somewhat unstable and gas nucleation inside the droplet occurred. Gaye and Riboud [31] made similar observations but also observed swelling of the droplet at lower carbon contents; these workers termed this phenomenon emulsification. Molloseau and Fruehan also suggested that droplets became emulsified at FeO contents in the slag greater than 10%, however, similar experiments by the current authors suggest that the droplets do not become emulsified but swell to a size where they are no longer sufficiently dense to show on the x-ray image. When the reaction slows, droplets can be observed to reappear on the image as they shrink. Molloseau and Fruehan [25] acknowledged that it was difficult to distinguish between emulsification and boiling but attributed the fast reaction at high oxygen potentials to emulsification. These workers proposed that because oxygen is not consumed sufficiently rapidly at the surface of the droplet, it diffuses into the droplet where CO can nucleate causing boiling. They also proposed that once "emulsification/boiling" started that the reaction may be controlled by mass transport in the slag.

Chen and Coley [4, 32] demonstrated that CO formation in the droplet was directly proportional to droplet size as is shown in Figure 2. This observation suggests the rate-determining step to be nucleation of CO inside the droplet. Control by mass transport in the slag/droplet or chemical reaction at the surface would all show a dependency on droplet surface area rather than mass or volume. Chen also showed that the rate of CO evolution was strongly affected by sulphur dissolved in the metal. Figure 3 shows that at lower levels of sulphur the rate increased with increasing sulphur going through a maximum at about 0.01wt% S. The low sulphur behaviour is consistent with rate control by nucleation which would show an increase in rate at lower surface tensions. The behaviour at higher sulphur levels suggests an element of control by interfacial chemical reaction. A maximum in rate was also found in other work by the current authors, over a wide range of oxygen potential [33, 34]. Molloseau and Fruehan[25] found the same behaviour at 1713K.

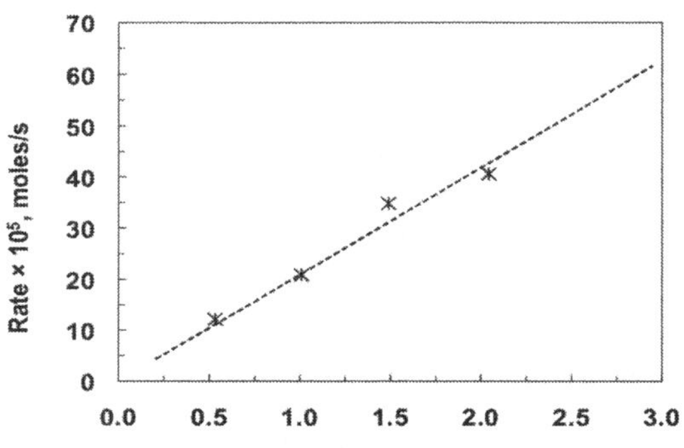

Figure 2. Dependence of CO evolution on droplet size at T=1773K,
Droplet composition; 2.64%C, 0.022%S,
Slag composition; CaO/SiO2=1.2, MgO=12%, FeO=20%,

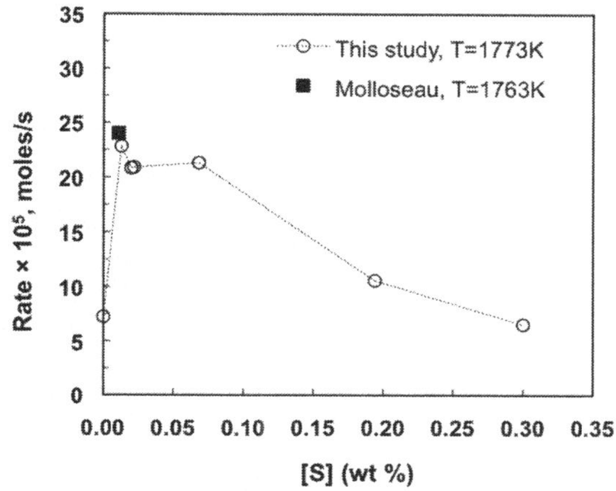

Figure 3. Dependence of CO evolution rate on S content.

Chen and Coley[4] found that for the conditions employed in their work, classical nucleation theory could not justify the observed rates. This is consistent with the work of El Kaddah and Robertson[6] who showed that CO nucleated in iron carbon alloys at levels of supersaturation considerably lower than those predicted by classical nucleation theory. These workers evaluated several possible sources of this nucleation discrepancy but This discrepancy is well known for solute species forming gas bubbles in liquids[36-37], in fact one author [37] suggests that the discrepancy between theory and practice might be the largest observed in science, but a convincing explanation remains elusive.

Bower et al[36] proposed formation of a metastable blob phase prior to bubble formation. The interfacial energy between the proposed blob is considerably less than for a bubble with a sharp interface, hence nucleation can proceed a much lower supersaturation levels. These workers suggested that the blob could take one of two forms; either as a zone within the liquid which has a higher concentration of the gas forming solute, or as something similar to a bubble with a diffuse interface. Theoretical analysis showed that either model would allow nucleation under conditions consistent with experimental observation however, the diffuse interface model involved a more physically likely number of molecules. Lubetkin[37] reviewed data from a wide rage of systems dismissing most previously proposed theories for the nucleation discrepancy. Lubetkin proposed a mechanism involving preferential adsorption of the gas species under the influence of a high saturation pressure. This mechanism resulted in a modification of classical nucleation theory where surface tension is replaced with a linear function of the super-saturation ratio. When employed on the CO2 water system this relation agreed well with experiment when the modified surface tension was between one ninth and one eighteenth of the bulk value.

Chen and Coley[4] employed a similar method to Lubetkin but the surface tension modification took a simpler form. Chen and Coley proposed that the rate of nucleation could be represented as follows:

$$J_s = N_0 \exp(\frac{-\Delta H}{kT})[\frac{3(\psi\sigma_0)}{\pi m}]^{\frac{1}{2}} \exp[\frac{-16\pi(\psi\sigma_0)^3}{3kT[P_{ve} - P_L]^2}] \quad (5)$$

Where: N_0: is the number concentration of CO embryos in the liquid; σ is surface tension at liquid gas interface; ΔH is heat of formation of one CO molecule; m is the mass of one molecule; T is temperature, in Kelvin, k is Boltzman constant; P_{ve} is the pressure in the vapor bubble at equilibrium; P_L is the liquid pressure and ψ is the surface tension modifying parameter.

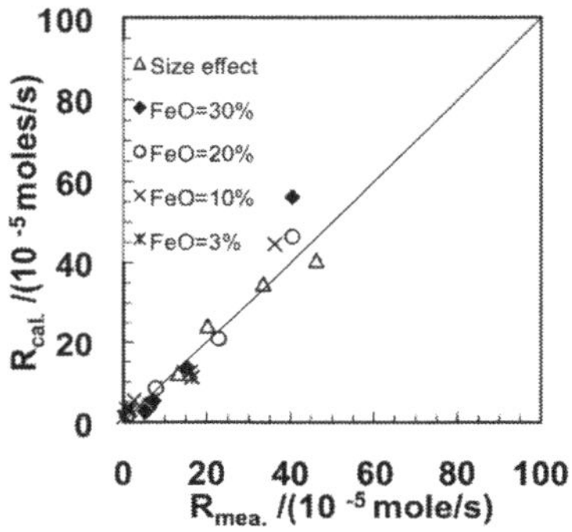

Figure 4. Comparison of Calculated Versus Measured Rates; Equation 5, ψ, Chen and Coley[4]

Figure 4 shows the agreement between experimental data and nucleation rates calculated using Equation 5. The data presented in this figure represent experiments conducted at four temperatures between 1673 and 1873 K there is apparently no appreciable effect of temperatutre on the surface tension modifying parameter. The values of the surface tension modifying parameters are given in Table 1. We see that these values are typically very small although those for %FeO= 20 and 30 are consistent with the modification found necessary by Lubetkin for the CO2 /water system. The FeO content of the slag appears to make a significant difference however on inspection we see that for slags with 20 and 30% FeO there is not much difference in ψ. In addition, the slags at 3 and 10% FeO did not cause droplet swelling hence ψ for these experiments may reflect the lower nucleation barrier for CO formed at the slag metal interface.

Internal nucleation of CO as a rate determining step is consistent with the observation that the rate is proportional to droplet mass. In fact, the authors can think of no other rate determining step that would lead to this result. However, the data shown in Figure 3, which have been verified by several other workers [25, 33, 34] with almost exact agreement on the position of the maximum, suggests that some element of control by interfacial reaction. Chen[32] proposed that this effect was caused by sulphur poisoning oxyen transfer from the slag thereby reducing the driving force for nucleation. No direct evidence was offered for this mechanism. On the other hand the rate of formation of CO is the product of the nucleation rate in bubbles per second and the volume of these bubbles. If each bubble has a finite lifetime prior to escaping from the droplet, it may grow during that time. Assuming sulphur poisons the bubble/metal interface, it can be expected that higher sulphur levels will result in smaller bubbles. Increasing sulphur concentration from extremely low levels will be expected to increase nucleation rates thereby increasing the rate of gas generation. Subsequent increases in sulphur concentration might be expected to slow bubble growth rate thereby decreasing the final bubble size and overall gas generation rate. The authors currently have no direct evidence for this mechanism but offer it as a possible explanation to rationalize all observations. Work is currently ongoing in the McMaster Steel Research Centre to explore this mechanism further.

Table 1.Surface Tension Modifying Parameter
As a Function of %FeO

%FeO	ψ
3	0.0019
10	0.0053
20	0.023
30	0.029

Assuming the rate to be controlled by nucleation and growth as discussed in the foregoing paragraph, one might expect some error in the values of ψ. Whilst this is true the authors do not expect this parameter to be greatly affected. Pomeroy[38] found that in modeling the onset of nucleation, where growth is not an issue, the value of ψ which gave the best fit to the experimental data was very close to that proposed by Chen[4, 32]and presented in Table 1.

Prediction of the On-set of Nucleation

Experimental observation by Chen and Coley[4, 32] show that all droplets go through an incubation period prior to the onset of swelling. For swelling to play a role in determining the droplet residence time in the slag, it is important that swelling initiates within the residence time of a dense droplet i.e. before that droplet falls back into the metal bath. In an attempt to predict the incubation time for CO nucleation, Pomeroy[35, 38] developed a model based on the relative rates of oxygen supply to the droplet, diffusion of oxygen and carbon within the droplet and the rate of CO nucleation. Pomeroy expanded on the framework of a model previously developed by Sun[39] and combined this with the nucleation model developed by Chen and Coley[4]. Pomeroy then validated the model with experimental data obtained using a combination of x-ray fluoroscopy and constant volume pressure increase. Results of Pomeroy's model are shown in Figure 5 where the solid lines representing contours of equal residence time are plotted as a function of initial carbon concentration in the droplet and %FeO in the slag. Points representing experimentally determined residence times are included on the diagram. Comparison of the experimental points with the model suggests that the model under-predicts the incubation time by between 0.15 and 0.28 seconds. The reason for this discrepancy can be found in the criteria for the onset of nucleation adopted in the model and experiment. Pomeroy's model assumes that once the threshold for nucleation of a single bubble is crossed, gas evolution ensures a well-mixed droplet and bubbles nucleate uniformly thereafter i.e. the incubation time is the time to nucleation of the first bubble. However, Pomeroy also recognized that it was not possible experimentally, to identify the first nucleated bubble and therefore defined nucleation to have started at the point where the droplet started to float in the slag. Clearly, many bubbles must nucleate to initiate floatation leading to a significant difference between the time at which the model identifies nucleation to have begun and that at which it can be observed. If one considers this difference, the agreement between prediction and model is quite remarkable.

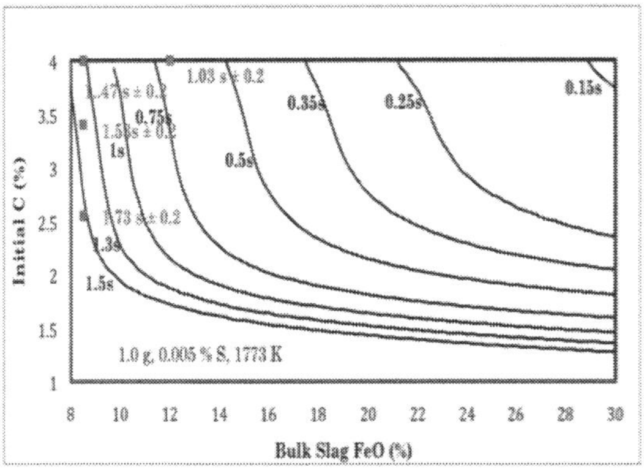

Figure 5. Predicted Incubation Times for CO Bubble Nucleation

The consequences of Figure 5 should be examined in the context of droplet residence time. Brooks et al[19] have shown convincingly that swelling is necessary to ensure sufficient residence times for droplets to participate in refining. However, swelling will only occur in the BOF if bubbles nucleate prior to the dense droplet settling back into the slag. Subagyo et al [16] and Brooks et al[19] showed that the residence time of dense droplets will be around 0.25 s. Examination of Figure 5 shows that this is very close to the predicted incubation time for high carbon drops and for lower carbon contents the incubation time increases quite sharply. This result appears to be at odds with published data on droplets sampled from real BOFs [14, 15] which shows that droplets are always at a level of refining significantly in advance of the bath. For this to be so, the residence time must be greater than that predicted for dense droplets. Again, the answer to this discrepancy may lie in one of the underlying assumptions in Pomeroy's model; that the initial oxygen content of the droplet is zero. In a real BOF the droplets are generated under the oxygen lance, which will offer ample opportunity for oxygen buildup. Thus, droplets in a real BOF will start with a significant oxygen content, likely lowering the incubation time to less than the residence time of dense droplets.

Rate of Dephosphorisation

Dephosphorization in the BOF is much more dependent on droplet formation than decarburization. Decarburization can occur in the "hot-spot" whereas dephosphorization requires a slag. There have been numerous studies on both the kinetics and thermodynamics of dephosphorisation under a variety of conditions, however some specific issues must be addressed before existing studies can be applied to droplets. It is well established that one of the critical issues in dephosphorisation is the oxygen potential at the slag metal interface. However, the dynamic nature of this parameter renders its precise value somewhat elusive. Both Monaghan et al[40] and Wei et al[41, 42] showed that the oxygen potential depends very much on the level of stirring; high levels tend to favour control by carbon whereas low levels favour control by FeO in the slag. Coley et al[43] suggested that this may be related to the ease with which CO nucleates at the slag metal interface which is in keeping with similar data on desulphurization[44]. One might expect that in the case of bloated droplets, which depend on high rates of CO nucleation, the driving force for dephosphorization will be relatively low. However, preliminary research in the authors' laboratory[45] suggests that this is not the case. It can only be assumed that internal nucleation of bubbles implies sluggish transport of carbon to the droplet surface which might favour higher interfacial oxygen potentials. Work is on going in the McMaster Steel Research Centre to devlop a full understanding of the dephosphorization of bloated droplets.

Conclusions

1/ A review of recent modeling work on BOF steelmaking suggests that we are close to achieving a quantitatively predictive model of the process

2/ Such a model will be most effective if it combines detailed kinetic studies of important reactions with in a global model of the process, verified with plant data.

3/ In oxygen steelmaking the carbon oxidation reaction plays a key role in controlling droplet residence time in the slag and therefore in controlling refining rates.

4/ The rate of droplet swelling in oxygen steelmaking is primarily controlled by the rate of nucleation of CO gas in side the droplet.

5/ Classical nucleation theory can be used to predict CO nucleation ratesin liquid iron if the nucleation discrepancy is accommodated by using a surface tension modifying parameter. The precise source of the nucleation discrepancy, the exact form of the surface tension modifying parameter and its relationship to other physical properties of the system have yet to be determined.

6/ Preliminary data for dephosphorization of bloated droplets suggests a complicated picture where subtle effects of sulphur in the metal and CO nucleation rate have important effects on the balance between oxygen supply and consumption.

References

1. S. Ohguchi, D.G.C. Robertson, B. Deo, P. Grieveson, and J.H.E Jeffes, "Simultaneous Dephosphorization And Desulphurization Of Molten Pig Iron," *Ironmaking and Steelmaking*, v 11, (1984) 202-213.

2. D.G.C. Robertson, B. Deo, and S.Ohguchi, "Multicomponent Mixed-Transport-Control Theory For Kinetics Of Coupled Slag/Metal And Slag/Metal/Gas Reactions: Application To Desulphurization Of Molten Iron," *Ironmaking and Steelmaking*, 11 (1984) 41-55.

3. K.J. Graham, and G.A. Irons, "Toward Integrated Ladle Metallurgy Control,"*Iron and Steel Technology*, 6 (1) (2009), 164-173.

4. E. Chen, and K. S.Coley, "Kinetic Study of Droplet Swelling in BOF Steelmaking," *Ironmaking & Steelmaking*, 37 (7) (2010), 541-545.

5. N. El-Kaddah and D. Robertson, "The Nucleation of CO Bubbles in Molten Iron Carbon Drops Reacting with Oxidizing Gases," *Metallurgical Transactions B* 19 (6) (1988), 831-837.

6. N. El-kaddah, and D. Robertson, "Homogeneous Nucleation of Carbonmonoxide in Iron Drops" *J. Colloid and Interface Science*, 60 (1977) 349- 360.

7. S Kitamura, "Importance of Kinetic Models in the Analysis of Steelmaking Reactions", *Steel Research int.* 81 (9) (2010) 766-771

8. C. Kattenbelt, and B. Roffel, "Dynamic Modeling of the Main Blow in Basic Oxygen Steelmaking Using Measured Step Responses," *Metall and Mater Trans B*, 39 (2008) 764-769

9. G. Li, B. Wang, Q. Liu, X. Tian, R. Zhu, L. Hu, and G. Cheng, "A process Model for BOF Process Based on Bath Mixing Degree," *International Journal of Minerals, Metallurgy and Materials*, 17 (6) (2010), 715-722

10. A. K. Shukla, B. Deo, S. Millman, B. Snoeijer, A. Overbosch, and A. Kapilashrami, "An Insight into the Mechanism and Kinetics of Reactions In BOF Steelmaking: Theory vs Practice," *Steel Research Int.* 81 (2010) 940-948.

11. N. Dogan, G. A. Brooks, and M. A. Rhamdhani, "Comprehensive Model of Oxygen Steelmaking Part 1: Model Development and Validation," *ISIJ International*, 51 (7) (2011), 1086-1092.

12. N. Dogan, G. A. Brooks, and M. A. Rhamdhani, "Comprehensive Model of Oxygen Steelmaking Part 2: Application of Bloated Droplet Theory for Decarburization in Emulsion Zone," *ISIJ International*, 51 (7) (2011), 1093-1101.

13. N. Dogan, G. A. Brooks, and M. A. Rhamdhani, "Comprehensive Model of Oxygen Steelmaking Part 3: Decarburization in Impact Zone," *ISIJ International*, 51 (7) (2011), 1102-1109.

14. C. Cicutti, M.Valdez, T. Perez, R. Donayo, and J. Petroni, "Analysis of Slag Foaming During the Operation of an Industrial Converter," *J. Latin Am. Appl. Res.*, 32 (2002) 237.

15. C. Cicutti, M. Valdez, T. Pérez, J. Petroni, A Gómez, R. Donayo, L. Ferro, "Study Of Slag-Metal Reactions In An Ld-Lbe Converter," 6th International Conference on Molten Slags, Fluxes and Salts, Stockholm-Helsinki, 2000,

16. Subagyo, G.A. Brooks, and K. Coley: "Residence Time of Metal Droplets in Slag-Metal - Gas Emulsions through Top Gas Blowing," *Canadian Metallurgical Quarterly*, 44 (1) (2005), 119-129.

17. Subagyo, G.A. Brooks, K.S. Coley, and G.A. Irons, "Generation of Droplets in Slag-Metal Emulsions Through Top Gas Blowing," *ISIJ Int.*, 43 (7) (2003), 983-989.

18. Subagyo, G.A. Brooks, and K. Coley, "Interfacial Area in Top Blown Oxygen Steelmaking", Steelmaking Conference Proceedings, ISS, Warrendale Pa., 85 (2002), 749-762.

19. G. A. Brooks, Y. Pan, Subagyo and K. S. Coley "Modeling of Trajectory and Residence Time of Metal Droplets in Slag-Metal-Gas Emulsions in Oxygen Steelmaking," *Metall and Mater Trans B*, 36B (2005), 525-535.

20. R. Li, and R. Harris, "Interaction Of Gas Jets With Model Process Liquids Proceedings Of Pyrometallurgy" 95, I.M.M, London, 1995, 235.

21. B. Deo, and R. Boom, : Fundamentals of Steelmaking Metallurgy , Prentice Hall,

22. Q.L. He, and N. Standish, "A Model Study Of Droplet Generation In The BOF Steelmaking," *ISIJ International*, 30 (1990), 305-309.

23. M. Alam, J. Nasr, G. Brooks, and A. Fontana, "A Computational Fluid Dynamics Model of Shrouded Supersonic Jet Impingement on a Water Surface," *ISIJ Int*, 52(2012),1026-1035.

24. S.C. Koria, and K.W. Lange, "A New Approach To Investigate The Drop Size Distribution In Basic Oxygen Steelmaking," *Metall. Trans. B*, 15B, (1984) 109-116

25. C.L. Molloseau and R.J. Fruehan, "The Reaction Behavior Of Fe-C-S Droplets In CaO-SiO2-MgO-FeO Slags," *Metall. Trans. B*, 33B (2002), 335-344.

26. E.W. Mulholland, G.S.F. Hazeldean, and M.W. Davies, "Visualization Of Slag-Metal Reactions By X-Ray Fluoroscopy: Decarburization In Basic Oxygen Steelmaking," *Journal of The Iron and Steel Institute*, 211(1973), 632-639.

27. D.J. Min, and R.J. Fruehan, "Rate Of Reduction Of FeO In Slag By Fe-C Drops," *Metall. Trans. B*, 23B (1992), 29-37.

28. G.G. Krishna Murthy, Y. Sawada, and J.F. Elliott, "Reduction of FeO Dissolved in CaO-SiO2-Al2O3 Slags by Fe-C Droplets," *Ironmaking and Steelmaking*, 20 (3) (1993)179-190.

29. G.G.Krishna Murthy, A.Hasham, and U.B.Pal, "Reduction Rates of FeO Dissolved in CaO-SiO2-Al2O3 Slags by Fe-C Droplets," *Ironmaking and Steelmaking*, 20 (3) (1993) 191-200

30. T. Gare and G.S.F. Hazeldean, "Basic Oxygen Steelmaking: Decarburization of Binary Fe-C Droplets and Ternary Fe-C-X Droplets in Ferruginous Slags," *Ironmaking and Steelmaking*, (4) (1981), 169-181.

31. H. Gaye and P.V. Riboud, "Oxidation Kinetics of Iron Alloy Drops in Oxidizing Slags," *Metallurgical Transactions B*, 8B, (9) (1977), 409-415.

32. E. Chen, "Kinetic Study of Droplet Swelling in BOF Steelmaking" PhD Thesis, McMaster University 2011

33. M. Pomeroy, G. Brown, and K.S. Coley, "Mechanism of Enhanced Cr2O3 Reduction by Liquid Fe-C in Steelmaking Slags," AISTech - Iron and Steel Technology Conference Proceedings, 2011, 1267-1276.

34. K. S. Coley and G. Brown, "Kinetics of BOF Steelmaking," International conference on the Technology of Ironmaking and Steelmaking, Jamshedphur, India, Dec 16-18, 2013

35. M. Pomeroy, E. Chen, K. S. Coley and G. Brown, "Kinetic Study of Droplet Swelling in BOF Steelmaking" Proceedings of the European Oxygen Steelmaking Conference, Stockholm, Sweden September 2011.

36. P. G.Bowers, K. Bar-Eli and R. M. Noyes . "Unstable Supersaturated Solutions of Gases and Nucleation Theory" *Journal of Chemical Society, Faraday Transactions*, 92(1996), 2843-2849.

37. S.D.Lubetkin . "Why Is It Much Easier to Nucleate Gas Bubbles Than Theory Predicts?" *Langmuir*, 19 (2003), 2575-2587

38. M. D. Pomeroy, Decarburization Kinetics of Fe-C-S Droplets in Oxygen Steelmaking Slags," MASc Thesis McMaster University 2011

39. [35] H. Sun, "Reaction Rates And Swelling Phenomenon Of Fe-C Droplets In FeO Bearing Slag," *ISIJ Int.*, 46 (2006), 1560-1569.

40. B.J. Monaghan, R.J. Pomfret, and K.S. Coley, "The Kinetics of Dephosphorization of Carbon-Saturated Iron Using an Oxidizing Slag", *Metall and Mater Trans B*, 29B (1998), 111-118.

41. P. Wei, M. Sano, M. Hirasawa, and K. Mori, "Kinetics Of Phosphorus Transfer Between Iron-Oxide Containing Slag And Molten Iron Of High-Carbon Concentration Under Ar-O2 Atmosphere," *ISIJ.Int.*, 33 (1993), 479-87.

42. P. Wei, M. Ohya, M. Sano, and K. Mori, "Estimation Of Slag Metal Interfacial Oxygen Potential In Phosphorus Reaction Between Fe(T)O Containing Slag And Molten Iron Of High-Carbon Concentration" *ISIJ. Int.*, 33 (1993), 847-54.

43. K. S. Coley, M Pomeroy, G. Brown and E. Chen, "Kinetics of CO Nucleation During Slag Metal Reactions" Guthrie Honorary Symposium, June 6-9 2011, Montreal, Canada

44. N. W. Jones, "Desulphurisation Kinetics of Fe - C - S Alloys by CaO ˜ Al2O3 Based Slags at 1550 °C to 1650 °C," PhD Thesis University of Strathclyde, 1996

45. G. Brown and B Jamieson, McMaster University Unpublished Work

Celebrating the Megascale: Proceedings of the Extraction and Processing Division
Symposium on Pyrometallurgy in Honor of David G.C. Robertson
Edited by: Phillip J. Mackey, Eric J. Grimsey, Rodney T. Jones, and Geoffrey A. Brooks
TMS (The Minerals, Metals & Materials Society), 2014

CURRENT STATUS AND FUTURE DIRECTION OF LOW-EMISSION INTEGRATED STEELMAKING PROCESS

S Jahanshahi[1], A Deev[1], N Haque[1], L Lu[1], J G Mathieson[1], T E Norgate[1], Y Pan[1], P Ridgeway[2], H Rogers[3], M A Somerville[1], D Xie[1], P Zulli[3]

[1]CSIRO Minerals Down Under Flagship, Clayton, Victoria, Australia

[2]Arrium, Newcastle, NSW Australia

[3]BlueScope Steel, Port Kembla, NSW, Australia

Keywords: GHG emission, steelmaking, biomass, charcoal, waste heat recovery, dry slag granulation, techno-economics, life cycle assessment.

Corresponding author: sharif.jahanshahi@csiro.au

Abstract

In 2006 the Australian steel industry and CSIRO initiated an R&D program to reduce the industry's net greenhouse emission by at least 50%. Given that most of the CO_2 emissions in steel production occur during the reduction of iron ore to hot metal through use of coal and coke, a key focus of this program has been to substitute these with renewable carbon (charcoal) sourced from sustainable sources such as plantations of biomass species. Another key component of the program has been to recover the waste heat from molten slags and produce a by-product that could be substituted for Portland cement.

This paper provides an overview of the low-emission Integrated Steelmaking Process, progress made over the past seven years and the program's future direction which includes proposed demonstrations of the technologies developed including large scale piloting and full scale plant trials.

Introduction

The global steel industry produced about 1.5 Gt of crude steel in 2012 with an emission of about 2.5 Gt of $CO_{2\text{-}e}$. This level of GHG emission is significant and represents about 5-6% of the total global emission. While the steel industry has reduced its energy consumption and hence GHG emission by about 50% over the past decades, it is conscious of high the volume of CO_2 emissions and has been actively developing a number of step change technologies to reduce its net emission by at least another 50%. About 10 years ago, the World Steel Association initiated an R&D program called CO_2 Breakthrough Program which focused on step change reduction in their emissions. This collaborative program has benefited from contributions by many steel companies, engineering firms and R&D institutes in Europe, Japan, Korea, North America, Australia, etc. The Australian CO_2 Breakthrough Program was initiated by BlueScope Steel in 2006, through collaboration with OneSteel/Arrium and CSIRO. This program aims to reduce net GHG emission by the steel industry through use of renewable carbon sources as fuels and reductants, as well as the recovery and utilization of waste heat from molten slags. The key research questions behind this program have been:

- Can deep cuts into CO_2 emissions be made using current process equipment?
- Can charcoal provide a low-capital path to a low-CO_2 future?
- Can waste heat recovery from molten slag provide a further cut in CO_2 emission?

The combination of (1) sustainable supply of biomass and its processing to produce charcoal, bio-oil and other renewable energy; (2) utilization of charcoal as partial replacement for coal and coke products in iron and steelmaking; (3) recovery of high grade waste heat from molten slag and (4) conversion of molten blast furnace slag into a Portland cement substitute has been defined as the low-emission Integrated Steelmaking Process (ISP). In Figure 1, the conceptual "flowsheet" for the ISP is presented.

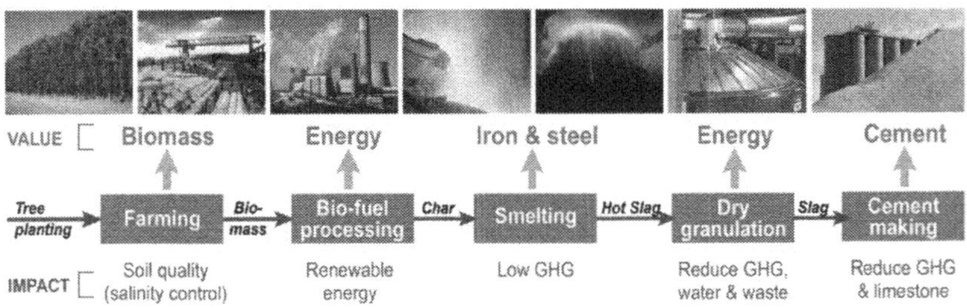

Figure 1: The conceptual flowsheet across the value chain for the low-emission Integrated Steelmaking Process (ISP).

The present paper provides a summary of the key findings from a series of investigations covering the full value chain, i.e. from the sustainable supply of biomass to production of charcoal, iron and steel as well as dry granulation of slag to produce feedstock for the cement industry and energy recovered from waste heat. The paper also provides an outline of the future direction of the ISP development program.

Supply of biomass from sustainable sources

One of the questions that we were faced with was "do we have sufficient supply of biomass resources in the short and medium terms to support the steel industry?" To address this and related questions a survey of existing resources and residues available from plantations in some regions of Australia had to be carried out. Thus the quantities of biomass available from some regions of Australia (northern, central, southern New South Wales, western Victoria, and southern South Australia) were estimated by Haque et al.[1]. These regions were selected due to their proximity to existing steel plants at Port Kembla, Sydney, Geelong and Whyalla. Our estimate was made based on information gathered through site visits, meetings with the various State forestry agencies, wood processing industries and where possible from the open literature.

The annual forestry and wood processing residues were estimated on an oven-dry basis to be 3.4 Mt and 2.3 Mt, respectively. The non-forestry residues were estimated to be around 1.2 Mt/year. Non-forestry residues included biomass from grain crops such as wheat and maize, olive pomace, grape marc, almond waste and cut flowers waste. A significant part of non-forestry waste was from sugarcane bagasse, in-field cane crop residues, woody weeds (Camphor Laurel) and macadamia nut processing waste in Northern New South Wales. Assuming 30% charcoal yield of dry residues and 50% of total dry residues being available immediately, this equates to 3.7 Mt/year of dry biomass that can potentially generate about 1.14 Mt of charcoal per year.

These results indicate that currently the biomass residues from forestry and agricultural industries are sufficient to satisfy the short-to-medium term needs of the steel industry. In the longer term, some dedicated plantation of short cycle biomass species should be put in place to ensure a secure supply of biomass material from sustainable sources.

<u>Life cycle assessment and techno-economic evaluation of biomass and charcoal supply for steel industry</u>

The second question that arises is whether the biomass/charcoal option is environmentally and economically sound. To answer this question we have carried out some life cycle assessments (LCA) covering both environmental and economic dimensions. An initial social LCA on the biomass/charcoal scenario was recently completed by Weldegioris and Franks [2].

LCA methodology was used to estimate the greenhouse gas footprint of charcoal production from Mallee eucalypt biomass, and is described in detail elsewhere by Norgate et al [3]. The results of this LCA, showing the contributions of the various stages to the greenhouse gas footprint of charcoal production, are shown in Figure 2.

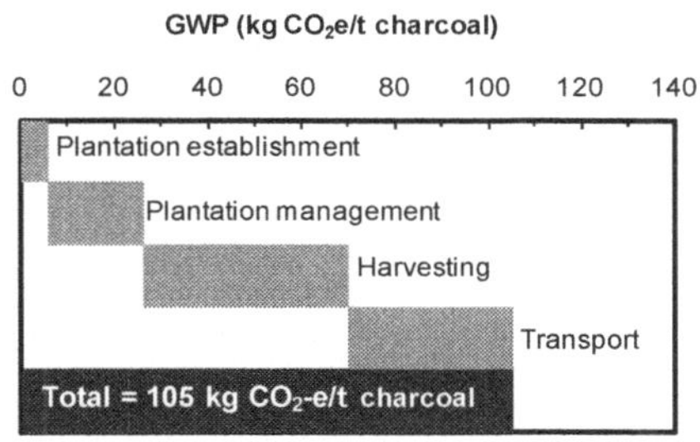

Figure 2: Stage contributions to Global Warming Potential (GWP) or greenhouse gas footprint of charcoal production (no by-product credits).

The greenhouse gas footprint for charcoal production was 105 kg CO_2-e/t charcoal, which reflects the fossil fuel used in its production (i.e. plantation management including fertilizer, harvesting and transport). Since the charcoal plant, which includes drying and pyrolysis of biomass, is fuelled entirely by combustion of biomass by-products, any emissions associated with these stages are not fossil fuel-based.

Greenhouse gas credits for the bio-oil and combustible bio-gas by-products produced during charcoal production (i.e. electricity is generated using the gas which replaces electricity generated from black coal, and bio-oil replaces diesel) can significantly reduce the greenhouse gas impact of charcoal production. It is worth noting that the credits will normally outweigh the 105 kg/t charcoal, meaning overall GWP is expected to be negative [3]. However, it should be appreciated that the magnitude of the by-product credits depends on the by-product yields in the charcoal retort, which in turn are dependent on a number of factors, in particular, the nature of the pyrolysis process (fast or slow) and the biomass feed composition. Slow pyrolysis tends to maximize the charcoal yield, with less gas and bio-oil by-products produced, thus by-product credits are less with slow pyrolysis.

A techno-economic evaluation study was carried out on a hypothetical biochar plant with an annual capacity of 100 kt of charcoal. For this exercise a process flow sheet was developed and key items of equipment were identified and costs were collected from various sources [4,5]. The required raw material feed and final charcoal product output rates were estimated based on the flowsheet and product yields obtained from small scale slow pyrolysis of biomass at CSIRO (~0.3 t charcoal and 0.3 t of condensate per t of oven-dry biomass, with about 20% of the condensate being bio-oil).

The evaluation was based on the scenario that:

- The hypothetical charcoal plant is located at or near a sawmill and charcoal is transported to a steel plant that is about 300 km away.
- Raw biomass is collected, chipped, and dried close to the biomass source and transported 100 km to the sawmill and charcoal plant.
- The bio-oil product has a value of A$88/t (August 2013 price of crude oil in Australian dollar used and the energy content of bio-oil is 50% of crude oil. It is to be noted that if diesel price is used, the value of bio-oil will almost be double).
- The value-in-use of softwood charcoal as injectant into a blast furnace is about $89/t higher than a standard high volatile PCI coal [6] with estimated price of A$135/t [7] in 2013.
- A carbon credit of A$23/t CO_2-e is applied to the charcoal and by-products such as bio-oil.

The net cost of charcoal to a steel plant was estimated to be A$169/t, which increases with decreasing carbon price/tax and decreases as the price of PCI coal increases. It should be noted that we have not valued the excess combustible gas from pyrolysis process as quantities are uncertain.

It is worth mentioning that 45% of the total operating cost of A$446/t at the charcoal plant was for supply of raw materials and 16% for transport. Maintenance, interest on capital, labour and utilities accounted for the remaining 39% of the cost.

The estimated total capital cost of the charcoal plant was about A$1100/t of charcoal per year. A sensitivity analysis showed that for the same percentage change in base case value, changes in the raw material cost had the greatest effect on the operating cost followed by plant capital-related costs (maintenance and interest/depreciation) and raw material transport, with the labour cost having the least effect.

It should be emphasized that the making of charcoal and supplying it to the steel plant is financially viable if all the byproduct credits and benefits are taken into account.

Development of an energy and resource efficient pyrolysis process

The optimum substitution of charcoal for coal-based materials used for iron and steel making imposes a number of requirements such as acceptable price, density, contents of volatile matter and ash, and others (see Table 1). Furthermore, even at moderate levels of replacement, millions of tonnes of charcoal are needed annually for a large integrated steel plant. Existing charcoal making technologies cannot satisfy requirements for the quality of the product, its net price and the high production volumes [8]. Hence, the development of a new technology is needed. The ability to convert smaller-sized wood (such as wood chips or wood pellets) to charcoal is to be an important feature of a new technology as it will permit the utilization of lower-cost materials (wood waste/residues) and densified wood pellets as a feedstock. Charcoal made from wood pellets has up to three times higher density than that of conventional charcoal, which is essential for some of its applications (Table 1). An important contribution to the economic efficiency of charcoal making can be made if a new technology permits the extraction of maximum value from its by-products, namely; pyrolysis condensate (also called bio-oil) and pyrolysis gas (syngas).

The main barrier in the way to processing large volumes of smaller-sized wood in a pyrolysis reactor is associated with the combination of low gas permeability of such material with its low thermal conductivity. The former makes heating through flow of hot gas impracticable and the latter diminishes the efficiency of external heating of the reactor. These difficulties can be overcome if the principle of autogenous pyrolysis is used. That is the arrangement of the process can be such that the material is heated to the maximum pyrolysis temperature spontaneously by the heat of internal exothermic reactions, which occur within the material even without any air supply to the reactor [9]. Based on this principle a pilot scale plant was developed at CSIRO. The flow diagram given in Figure 2 shows its main units. The reactor is a well insulated vertical shaft with a wood charging mechanism at the top and a charcoal discharging valve at the bottom. The simplicity of the mechanical design of the reactor, with no moving parts in its hot zone, ensures the minimization of its installation and maintenance costs. Additionally, cooling of charcoal in the bottom section of the reactor can be accelerated by circulating a non-combustible gas through it, which also permits the recovery of heat from charcoal [10]. The flow of the cooling gas is confined to the bottom section of the reactor and the gas does not permeate the material in the rest of the reactor. Other main units are the dryer, the screw conveyors to transfer wood chips or pellets to the top of the reactor, condensers and the afterburner.

The advantages of this technology are:

- High energy efficiency, as no high-grade heat is needed to bring wood to the maximum pyrolysis temperature under steady-state (low-grade heat is still needed for drying the material);
- High-yield of charcoal from wood, as the conditions in the reactor are favourable for that [8];
- High value of pyrolysis by-products, since;
    - pyrolysis gas is not diluted by any inert gases or combustion products; and
    - the condensate is not diluted by free moisture from wood (the material is supplied to the reactor bone-dry)
- No heat-transfer limitation to scaling up the reactor to large sizes, as no supply of external heat to the material is needed; production of 100,000 tonnes of charcoal per year by a single reactor unit may be achievable.

The pyrolysis pilot plant (internal diameter of reactor ~ 600 mm, maximum depth of material bed in the reactor ~ 2700 mm, nominal productivity of 800 t of charcoal per year, if operated continuously) is currently being commissioned and it has been demonstrated that in a batch mode operation the autogenous heating of the material can bring its temperature to above 450 °C. It is expected that a maximum temperature of 520 °C will be reached in continuous operation of the reactor.

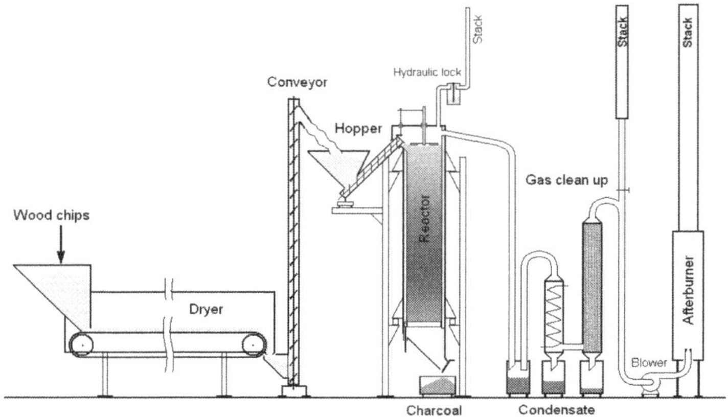

Figure 3: Process flow diagram of the pilot plant based on the principle of autogenous pyrolysis of wood.

<u>Concept of designer chars for various applications in iron and steel making</u>

Raw biomass is not suitable for iron and steelmaking applications because of its high moisture content, low carbon content and specific energy. Since pyrolysis is necessary to produce chars that are chemically similar to coal/coke, it is appealing to produce optimized products that are either superior to traditional fuels/reductants or that minimize any weaknesses in physical properties.

The substitution of charcoal for coal/coke in iron and steel making operations requires the preparation of charcoal with specific properties. Mathieson et al [6, 11-13] summarized the optimum quality criteria for charcoal substitution and showed that different applications require different charcoal properties, e.g. charcoal for sintering fuel should be of low volatile content (<3 %), high density, low reactivity and in the 0.3-3 mm size range. In comparison, charcoal as a blast furnace injectant should be higher in volatile matter (10-20 %), low in ash and low in alkalis. Recent research conducted at CSIRO as part of the Australian steel industry CO_2 Breakthrough Program has focused on controlling the pyrolysis process to produce charcoal with the desired properties. This concept has been called "designer charcoal".

Somerville et al [14] showed that the volatile content of charcoal can be controlled through the pyrolysis temperature (shown in Figure 4a). Charcoal density can also be controlled through pyrolysis temperature as well through the choice of biomass feedstock, i.e. dense wood (Figure 4b). The use of dense biomass fuel (DBF) pellets as a pyrolysis feedstock has been very effective at increasing the bulk density and particle density of charcoal. Densification of charcoal during pyrolysis and after pyrolysis have also been investigated. Similarly, alternative methods for controlling the ash content of the charcoal have been studied. These studies have shown selection of biomass species as well as different fractions of the plants could be used to produce charcoals with different ash contents and compositions. Our study also suggests that soil chemistry affects the ash composition and content.

Charcoal with low levels of moisture and volatile matter has been prepared for successful recarburisation plant trials at the Sydney Steel Mill [14]. Dense charcoal has also been prepared for pilot scale iron ore sintering trials [15].

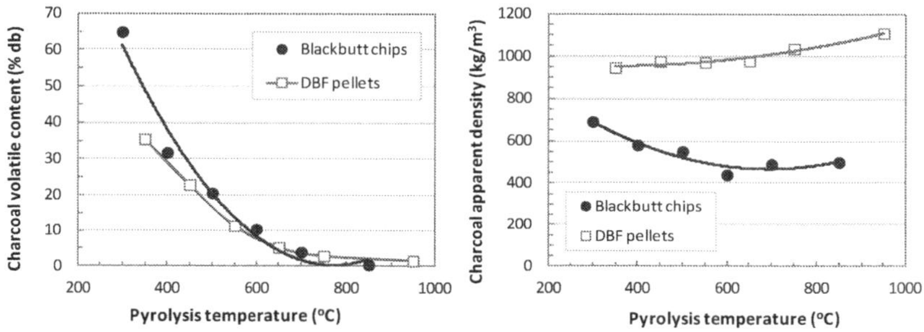

Figure 4: The effects of pyrolysis temperature on the (a) volatile content (% dry-basis) and (b) apparent density (particle density) of charcoal made from wood chips and dense biomass fuel (DBF) feedstocks.

Life cycle assessment of charcoal substitution for coal/coke

Life cycle assessments of the integrated, mini-mill and direct smelting steelmaking routes were carried out with charcoal (88.3% C, 31.1 MJ/kg, dry basis) substituted for coke, coal and carbon in various applications for each route by Norgate et al. [16]. For the integrated route, charcoal was assumed to replace coke (92% C, dry basis) and coal (75% C, dry basis) in the blast furnace on an equivalent fixed carbon basis and pulverised coal injectant (30 MJ/kg) on an equivalent energy basis. In the direct smelting route, charcoal was assumed to replace coal (81.3% C, dry basis) in the bath smelting reactor on an equivalent fixed carbon basis over a range of substitution rates up to 100%, while for the mini-mill/EAF route, charcoal was assumed to fully replace charge, injectant and recarburiser carbon (85% C) in the EAF on an equivalent fixed carbon basis. The results of these LCAs in terms of the total potential reduction in greenhouse gas emissions from charcoal use compared to the coal/coke base case values for the three steelmaking routes are listed in Table 2.

Table 2: LCA based total potential reduction in greenhouse gas emissions resulting from substitution of coal/coke with charcoal for different steelmaking routes.

Steelmaking route	Total potential Reduction without by-product credits (t CO_2-e/t steel)	Total potential reduction with by-product credits (t CO_2-e/t steel)
Integrated BF-BOF	0.69 – 1.21	0.91 – 1.61
EAF mini-mill	0.028 – 0.056	0.037 – 0.075
Direct smelting	0.34 – 1.70	0.45 – 2.15

The physical significance of the results for the integrated route (sintering) in terms of greenhouse gas emissions is shown in Figure 5, where the various components contributing to the greenhouse gas footprint of the integrated route are shown, along with the reductions achieved from the use of charcoal in this process for the case without any by-product credits. The results shown in Figure 5 mean that without any by-product credits, the greenhouse gas footprint for the integrated route (sintering) is reduced from 2.17 t CO_2e/t steel to between 0.96 t CO_2e/t steel and 1.48 t CO_2e/t steel over the range of charcoal substitution rates considered.

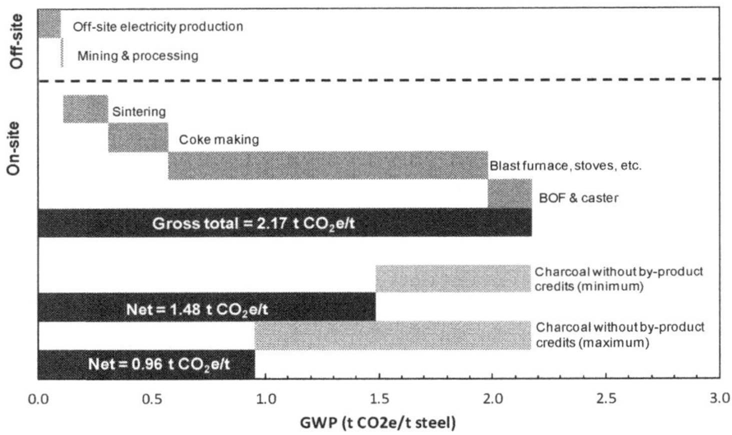

Figure 5: Global warming potential footprint of the integrated steelmaking route (sintering-BF-BOF) showing reduction with use of charcoal.

Value of Using Charcoal as a Substitute for Coal and Coke

Because of low ash, sulfur and phosphorus levels, charcoal can generally be considered to be chemically superior to coal-based fuels/reductants. This means that optimized charcoal products (see Table 1) should lead to fuel and other savings in ironmaking and steelmaking processes, as well as the advantages of using renewable carbon to combat net CO_2 emissions. These advantages mean that designer charcoal products may have greater value than traditional fuels/reductants in several of the applications. For example, Mathieson et al. [6,12] estimated that BF coke rate could be reduced by 23 – 30 kg/t-HM by using charcoal as the tuyere injectant and 4.5 – 9 kg/t-HM by charging 5 – 10% charcoal/ore composites. Our industrial trial [11] also indicated that charcoal was potentially a more efficient liquid steel recarburiser.

Mathieson et al. [6] illustrated the value-in-use (VIU) of various charcoal types compared with coal types for BF tuyere injection at 140 kg/t-HM. This is shown in Figure 6, along with the net CO_2 emissions involved. It will be observed that the three charcoal samples had calculated values 50% greater than the reference high-volatile bituminous coal, reflecting predicted lower BF coke rates and improved hot metal composition (Si, S and P).

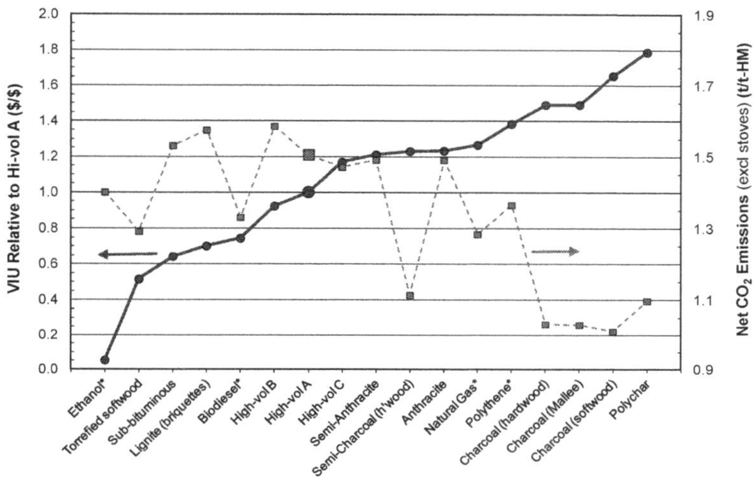

Figure 6: Calculated VIUs for BF Injectants and Consequent Net Carbon Dioxide Emissions from the Blast Furnace. (Order of Injectants: Increasing VIU. * indicates injection at 60 kg/t-HM, rather than 140 kg/t-HM)

Current status and future directions of designer charcoal applications

Table 1 provides a summary of the charcoal applications proposed for ironmaking and steelmaking processes. Key quality parameters that optimize each application are given [13] as well as estimated net CO_2 emissions savings based on simple substitutions [6] (gate-to-gate). Additional savings of around 50% in net emissions are indicated by cradle-to-gate life-cycle assessments [3, 16] because it appears possible to produce charcoal with negative global warming potential if pyrolysis co-products (bio-oil and combustible gases) are captured and utilized.

The aggregate CO_2 emissions savings shown in Table 1 are calculated for Australian conditions and are 0.7 – 1.3 t-CO_2/t-crude steel (31 to 57%) for the integrated BF-BOF steelmaking route, but only 0.04 – 0.06 t-CO_2/t-crude steel for the EAF route (8 – 12%), because of the dominance of CO_2 emitted for electricity generation in that sector. It will be observed that BF charcoal injection is the largest of the applications and has the potential to decrease net CO_2 emissions by up to 25% in the integrated route. This application is already practiced 100 – 200 kg/t-HM in some Brazilian mini-BFs (up to 700 m^3 inner volume). The R&D phase has been completed for several of the applications making them ready for industrial trials or commercial operations.

Table 1: Charcoal Applications Proposed for Ironmaking and Steelmaking [6, 12]

Application	Key Parameters	Net CO_2 Emissions Savings (t-CO_2/ t-crude steel)	Current Status	Next Step
Sintering solid fuel	Low VM: <3% High density*: >700 kg/m^3 Size: 0.3 - 3 mm	0.12 – 0.32 (5 – 15%)	Pilot scale testing [15]	Industrial trials
Cokemaking blend component	Low to mid VM: <10% High density*: >700 kg/m^3 Size: <1 mm Low alkalis	0.02 – 0.11 (1 – 5%)	Bench-scale R&D [17]	Pilot oven trials
BF tuyere injectant	Higher VM: 10 - 20% Low ash: <5% Low alkalis	0.41 – 0.55 (19 – 25%)	Theoretical analysis [6] & combustion testing [18]; Mini-BF commercial operations	Trial on large BF
BF nut coke replacement	Low to mid VM: <7% Higher density Size: 20 - 25 mm	0.08 – 0.16 (3 – 7%)	Mini-BF commercial operations [19]	Trial on large BF
Carbon/ore composites	Low VM: <5% Size: 80% passing 75 μm	0.06 – 0.12 (3 – 5%)	Bench-scale R&D [20]	Industrial trials
Steel recarburiser	Low VM: <3% Low moisture: <2% High density*: >500 kg/m^3	0.001 – 004 (0.3 – 1.0%)	Industrial trial [14]	Commercial operations
EAF charge carbon	Low to mid VM: <7% Size: 20 - 30 mm Low alkalis	0.02 – 0.04 (4 - 8%)‡	Proposed	Industrial trial
EAF foaming agent/ inject carbon	Low to mid VM: 2 - 7% Moisture: 1 - 7% Size: 0.5 - 5 mm Low alkalis	0.02 (4%)‡	Industrial trial [21]	Industrial trial with optimized charcoal

† Applications not requiring very low moisture levels require relatively dry charcoal, say <12% moisture.
* This is particle (not bulk) density, *e.g.* made from DBF pellets.
‡ Refers to savings in an EAF plant

Waste heat and value recovery from BF slag through dry granulation

Molten slags are high volume by-products from pyrometallurgical industries, with ironmaking typically producing 250-300 kilograms of slag for every tonne of hot metal produced, amounting to some hundreds of millions of tonnes of blast furnace slag each year worldwide (e.g. 230-270 million tonnes in 2010). Blast furnace slag is commonly granulated using water to produce a glassy product that can be used as a valued feed in cement production or to substitute for Portland cement. However, wet granulation consumes a large amount of fresh water (approximately one tonne of evaporative loss of water for one tonne of molten slag processed), can generate H_2S and/or acid mist causing air pollution, and more importantly, does not allow for recovery of a large amount of high grade waste heat in the slag (about 1.8 GJ of heat for every tonne of molten blast furnace slag cooled from a discharge temperature of 1500 to 100 °C). To overcome these shortcomings, dry granulation of slag has attracted extensive research efforts over the last 30-40 years to develop an alternative and more sustainable process than the conventional wet slag process.

In the dry process molten slag was first broken up into small droplets by several mechanical means such as air blast, rotary drum(s) and spinning disc/cup. Slag droplets were quenched and solidified using air to recover high grade heat released from the slag in a fluidized or moving bed. A number of dry granulation methods have been proposed and some trialled at pilot and demonstration plant scales in the 1980s, but none of them have been successfully commercialized. Nevertheless, there is a renewed interest in dry slag granulation since early 2000 due to increasing pressure and drive by the industry for more sustainable slag processing.

CSIRO started working on dry slag granulation (DSG) via a spinning disc in 2002. Following an extensive investigation, major breakthroughs have been made to resolve some common issues associated with slag atomization at a spinning disc, in particular the formation of slag "wool". A novel disc design has been developed to produce fine granulates without the formation of slag wool. A cyclonic air flow is employed to quench the hot granules to produce a highly glassy product. These breakthroughs make it possible to significantly reduce the droplet flying distance required, contributing to a compact reactor design that can efficiently recover heat. An integrated dry granulation and heat recovery process has since been developed and the concept has been proved through laboratory pilot trials at scales up to 5 t/h using a 3 m diameter pilot plant at CSIRO's Clayton laboratory as shown in Figure 7. The granulated products were highly glassy and suitable for cement manufacture [22] and the heated air stream reached 300°C during a relatively the short tapping time of about 1-2 min (which was limited by the amount of molten slag available at the laboratory).

(a) (b)

Figure 7: (a) the 3 m diameter DSG pilot plant (5 tonne/h) and (b) a still image from a high speed video (1000 frame per second) of disc atomisation (slag 1450°C, 40 kg/min, 800 rpm)

Techno-economic and environmental assessment of dry slag granulation

A techno-economic and environmental assessment was carried out [23,24] to estimate the capital and operating costs (reported in 2012 Australian dollars A$) of a DSG plant to handle 300,000 t/y of slag (or 1 Mt/y steel) and to estimate its potential greenhouse gas benefits. The capital cost was estimated at A$9.1M. However, as CSIRO's DSG process is still at the pilot plant scale, this estimate is an order-of-magnitude cost only, as further development work on the process, particularly the heat recovery stage, is required before more definitive capital costs can be estimated. This estimate excludes any capital costs associated with utilization of the recovered waste heat. The operating cost of the dry slag granulation process was estimated to be A$4.80/t slag, with the major contributor being the labour component. Further details of the DSG process economics are given by Norgate et al. [23, 24]. The estimated capital and operating costs for the slag dry granulation process are compared with the corresponding costs for the conventional wet granulation process (estimated by the authors based on various data sources) in Figure 8. The results in this figure indicate that both the capital and operating cost of the dry granulation process are roughly half of those for wet granulation, bearing in mind the preliminary nature of these costs.

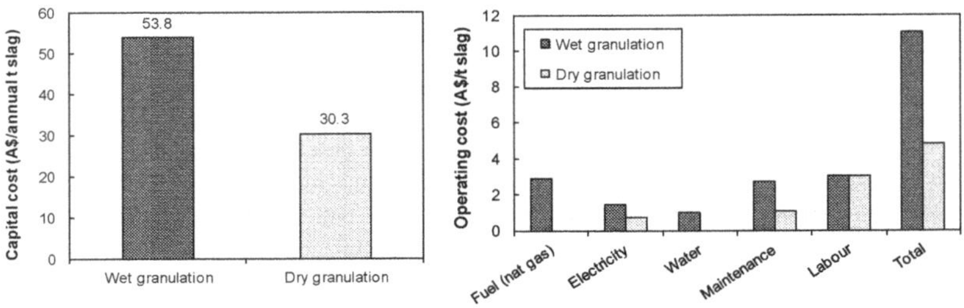

Figure 8: Comparison of capital and operating costs for dry and wet granulation.

Dry granulation also offers possible process credits which have the potential to improve the overall process economics compared to wet granulation and simple air-cooling of slag, as well as offering potential environmental benefits in the form of reduced greenhouse gas emissions. Compared to wet granulation these potential credits are the elimination of water for granulation and energy for slag drying, utilization of recovered waste heat, and avoided greenhouse gas (GHG) emissions and any associated carbon cost. The potential credits compared to air-cooling include the revenue derived from selling the granulated blast furnace slag to Portland cement producers for use as a cement clinker substitute, and the avoided GHG emissions and carbon costs associated with this substitution. The latter credit does not apply when comparing to wet granulation as both wet and dry granulated slags can be used in this way. Compared to wet granulation the process credits amounted to A$15.7/t slag, while compared to air cooling the credits were A$59.8/t slag. The combined avoided GHG emissions for dry granulation over wet granulation were 87 kg CO_2e/t slag, and 867 kg CO_2e/t slag compared to air cooling of slag. Calculation of these process credits are given in detail elsewhere [23, 24]. The Net Present Value (NPV) of the dry slag granulation plant without any process credits included was A$33 M. With all potential process credits included, the NPV increased to A$56 M compared to wet granulation and A$126 M compared to air-cooling. Detailed calculation of these process credits and benefits have been provided elsewhere [23, 24]

Future direction of integrated heat recovery and DSG process

The work carried out to date has led to development of know-how that has addressed some of the issues that prevented further development and commercialization of the spinning disc based process in the past. We believe that our recent progress in understanding the fundamentals of the DSG process and the advanced computational fluid dynamic model developed and validated against our pilot plant results at CSIRO place us in an excellent position to take on the next stage in the scale up of the process and its demonstration under continuous operation at a steel plant with access to tonnage amounts of molten slag. We expect the commercialization of the process to follow within a few years after the demonstration phase. Given the growing interest in this technology by the non-ferrous industry, one could see that dry granulation of base metals slags and mattes is the logical extension of this application in future years.

Concluding remarks

For the integrated BF-BOF route, the use of renewable biomass-derived chars in ironmaking and steelmaking has the potential to decrease net CO_2 emissions by 32-58% directly, and more under full life-cycle assessment. The potential in the EAF route is less, at around 10-15%, because its energy principally comes from the electricity grid and therefore depends on emissions in that sector. Through combining the carbon credits for waste heat recovery from molten slag and production of "green cement" from granulated blast furnace slag where viable, then the net reduction in CO_2 emissions for integrated BF-BOF route increases by 10% to about 42 to 68%.

The ISP represents an integrated, innovative and potentially very effective program to decrease the net CO_2 emissions from the steel industry using current process equipments. ISP is based on 7 years of productive collaboration.

The World Steel Association has recognised these innovative technologies, which do not depend on carbon capture and storage, as the best options for short and medium term CO_2 reduction within the steel industry and, when proven, they will be widely applicable throughout the world.

References

1. N. Haque, M. Somerville, S. Jahanshahi, J.G. Mathieson and P. Ridgeway, "Survey of sustainable biomass resources for the iron and steel industry", *2nd Annual Conference, Centre for Sustainable Resource Processing (CSRP '08)*, Brisbane, November 2008.
2. F.S. Weldegiorgis and D.M. Franks, "Social dimensions of energy supply alternatives in steelmaking: Comparison of biomass and coal production scenarios in Australia", *Journal of Cleaner Production*, in press. http://www.csrm.uq.edu.au/publications/246-the-social-dimensons-of-charcoal-use-in-steelmaking
3. T. Norgate, N. Haque, M. Somerville and S. Jahanshahi, "Biomass as a source of renewable carbon for iron and steelmaking", *ISIJ International*, 52/8, 2012a, 1472-1481
4. W. Emrich: Handbook of charcoal making: *The traditional and industrial method*. Series E: Energy from Biomass, Volume 7, publication arranged by Commission of the European Communities. D. Reidel Publishing Company, Dordrecht, The Netherlands, 1985.
5. C. Stucley, "Integrated Tree Processing of Mallee Eucalypts", RRIDC Publication No 01/160, 2001.
6. J.G. Mathieson, H. Rogers, M. Somerville, P. Ridgeway, and S. Jahanshahi, "Use of Biomass in the Iron and Steel Industry – an Australian Perspective", *EECR-METEC InSteelCon*, Dusseldorf, Paper 252, CD, June 2011.
7. S. Wang, Private communication, Port Kembla, September 2013.
8. A. Deev, S. Jahanshahi, "Development of a pyrolysis technology to produce large quantities of charcoal for the iron and steel industry", *6th International Congress on the Science and Technology of Ironmaking*, Rio de Janeiro, October 2012, 1132-1142.
9. O.F. Stafford, "Destructive Distillation Process", US patent 1,380,262 (1921).
10. A.V. Deev, "Apparatus and process for continuous carbonisation of wood chips or wastes and other charring organic materials for charcoal production", International Patent Application WO 2012174587, A1 (2012).
11. J. G. Mathieson, H. Rogers, M. A. Somerville, S. Jahanshahi and P. Ridgeway, "Potential for the use of biomass in the iron and steel industry", *in Chemeca 2011*, paper 252, September 2011.

12. J.G. Mathieson, T. Norgate, S. Jahanshahi, M.A. Somerville, N. Haque, A. Deev, P. Ridgeway, and P. Zulli, "The Potential for Charcoal to Reduce Net Greenhouse Emissions from the Australian Steel Industry", *6th International Congress on the Science and Technology of Ironmaking*, Rio de Janeiro, October 2012, 602-1613.
13. M.A. Somerville, J.G. Mathieson and P. Ridgeway, "Overcoming problems of using charcoal as a substitute for coal and coke in iron and steelmaking operations, in *6th International Congress on the Science and Technology of Iron making*, Rio de Janeiro, October 2012, Brazil
14. M.A. Somerville, M. Davies, J.G. Mathieson, P. Ridgeway, and S. Jahanshahi, "Addition of Renewable Carbon to Liquid Steel: Plant Trials at OneSteel Sydney Steel Mill", *CHEMECA 2011*, Sydney, Paper 223, CD, September 2011.
15. L. Lu, M. Adam, M. Somerville, S. Hapugoda, S Jahanshahi, and J.G. Mathieson, "Iron Ore Sintering with Charcoal", *6th International Congress on the Science and Technology of Ironmaking*, Rio de Janeiro, October 2012, 1121-1131.
16. T. Norgate, N. Haque, M. Somerville, and S. Jahanshahi, "Biomass as a Source of Renewable Carbon for Iron and Steelmaking", *ISIJ International*, 52 (2012) 1472-1481.
17. K.W. Ng, L. Giroux, T. MacPhee, and T. Todoschuk, "Incorporation of Charcoal in Coking Coal Blend - A Study of the Effects on Carbonization Conditions and Coke Quality", *AISTech 2012: Proceedings of the iron & steel technology conference*, Atlanta, Georgia, USA, 7-10 May, 2012, 225-236.
18. J.G. Mathieson, H. Rogers, M. Somerville, and S. Jahanshahi, "Reducing CO_2 Emissions Using Charcoal as a Blast Furnace Tuyere Injectant", *ISIJ International*, 52 (2012) 1489-1496.
19. L.J. Gonclaves, V.M. de Oliveira, and F.R. da Silva, "Consumption of Small Charcoal in Blast Furnace 2 of Aperam South America", *6th International Congress on the Science and Technology of Ironmaking*, Rio de Janeiro, October 2012, 1765-1776.
20. T. Usui, H. Konishi, K. Ichikawa, and H. Ono, "Preparation and Reduction Behaviour of Carbon Composite Pellets Using Semi-Charcoal", *6th International Congress on the Science and Technology of Ironmaking*, Rio de Janeiro, October 2012, 698-709
21. L. Wibberley, J. Nunn, P. Scaife, A. Zivlas, L. Andrews, B. Coomber, R. Dickson, D. Gardner, and A. Cowie, "Large Scale use of Forest Biomass for Iron and Steelmaking", NSW Government Sustainable Energy Research Development Fund (SERDF) Report by BHP Minerals Technology and NSW State Forests, 2001.
22. S. Jahanshahi, D. Xie, Y. Pan, P. Ridgeway, J. Mathieson, "Dry slag granulation with integrated heat recovery" In: *1st Int. Conf on Energy Efficiency and CO_2 Reduction in the Steel Industry (EECR Steel 2011)* - 27 June - 1 July 2011; Dusseldorf, Germany. The Steel Institute VDEh; 2011. Session 13, 7 pages.
23. T. Norgate, D. Xie and S. Jahanshahi, "Technical and economic evaluation of slag dry granulation", *AISTech2012 Iron & Steel Technology Conference and Exposition*, May 2012b, Atlanta, USA, 2012.
24. T. Norgate, D. Xie and S. Jahanshahi, "Economic and environmental evaluation of slag dry granulation", *SCANMET IV, 4th International Conference on Process Development in Iron and Steelmaking*, Lulea, Sweden, June 2012c, 507-516. 2012.

Celebrating the Megascale: Proceedings of the Extraction and Processing Division
Symposium on Pyrometallurgy in Honor of David G.C. Robertson
Edited by: Phillip J. Mackey, Eric J. Grimsey, Rodney T. Jones, and Geoffrey A. Brooks
TMS (The Minerals, Metals & Materials Society), 2014

ANALYSIS OF STEELMAKING REACTIONS BY COUPLED REACTION MODEL

Shin-ya Kitamura

Institute of Multidisciplinary Research for Advanced Materials, Tohoku University
2-1-1 Katahira, Aoba ku, Sendai, Miyagi 980-8577, Japan

Keywords: Steelmaking, Kinetics, Decarburization, Dephosphorization, Inclusion

Abstract

Kinetic modeling is very important to understand steelmaking reactions. The coupled reaction model proposed by Prof. D.G.C Robertson is the best suited program, and therefore, the author has applied it to the various steelmaking reactions, a few examples of which are discussed. Hot metal dephosphorization occurs under non-equilibrium conditions as it takes place between the slag, which has a high oxygen potential, and the hot metal, which has a low oxygen potential. A process simulation model based on the coupled reaction model was constructed. This model considers the solid and liquid slag, liquid metal phases, and the reaction between the solid and liquid slag, besides the reaction between the liquid slag and liquid metal. Using this model, the ruling factor to increase the reaction efficiency of dephosphorization has been verified. In the secondary refining process, the values of inclusion composition calculated by the thermodynamic model, under the assumption of equilibrium with the metal composition, were found to be different from those observed in practice. A kinetic model, incorporating the reactions between slag, metal, inclusion, and refractory has been postulated, and the composition change of inclusion during treatment has been analyzed.

1. Introduction

Steelmaking reactions have been traditionally analyzed based on thermodynamics. Software packages, used to simulate reactions based on equilibrium conditions, too have advanced much. The information obtained through these packages is useful to analyze the steelmaking process in a few cases. However, in industrial operations, steelmaking reactions such as decarburization, dephosphorization, desulfurization, nitrogen removal, and control of inclusion composition in molten steel are not completed to equilibrium. In decarburization, the actual relation between carbon and oxygen contents is different from that calculated under conditions of equilibrium. In hot metal dephosphorization, the oxygen activity calculated using the metal composition and that calculated using the slag composition is widely different. In the vacuum degassing process, the actual contents of carbon, nitrogen, oxygen are higher than their contents calculated on the basis of thermodynamics. In addition, once the entire ladle-treatment system reaches equilibrium, the composition of the oxide phase is determined by its equilibrium relation with molten steel. However, the observed compositions of the slag and inclusions are both different from that obtained through equilibrium-based calculations, and various inclusions of various compositions are observed. Therefore, kinetic models are very important in understanding the steelmaking reactions. In addition, during the steelmaking process, many reactions occur simultaneously, with each reaction affecting the kinetic behaviors of others. Figure 1 shows an example of the reactions that take place during the hot metal dephosphorization process. To analyze these reactions, the coupled reaction model[1)] developed by Prof. D.G.C Robertson is the the best suited program, and therefore, the author has applied it to the various steelmaking reactions. A few specific examples of the application of the kinetic model to analyze the steelmaking process are discussed in this paper.

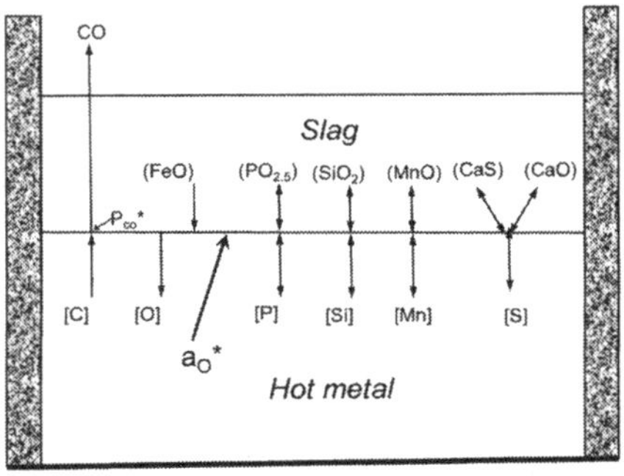

Figure 1. Various reactions that occur simultaneously with the dephosphorization reaction during the industrial steelmaking process.

2. Kinetic Model for Hot-Metal Dephosphorization

Figure 2 shows the change in the oxygen potential of the metal and slag phases with slag basicity [2]. The oxygen potential of the hot metal is very low as it is controlled by the carbon content in the hot metal. On the other hand, the oxygen potential of the slag is high as it is controlled by the FeO content in the slag. This result indicates that the dephosphorization occurs under non-equilibrium conditions, and that the oxygen potential at the interface of the hot metal and the slag plays a critical role in increasing the reaction rate.

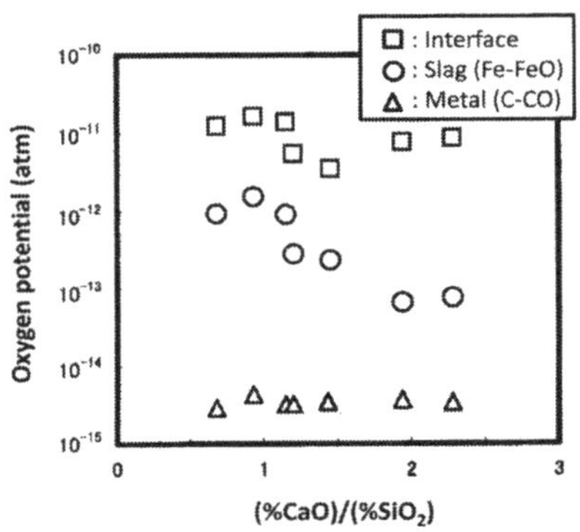

Figure 2. Differences in the oxygen potentials of the metal phase, the slag phases, and their interface under various conditions of slag basicity.

Because of the large difference in the oxygen potentials of both phases, the validity of the process analysis based on conditions of thermodynamic equilibrium is very limited. Therefore, the analysis needs to be based on reaction kinetics; and the model that does so is the coupled reaction model, which, thus, makes it the most important model. In this model, the reactions are

described using the double film theory assuming equilibrium conditions at the interface. Although its various parameters need to be defined empirically, this model can calculate the composition change of the various components in metal and slag phases simultaneously.
In this model, the oxidation reaction of element "M" is described by equation (1), the molar flux density is described by equation (2), and the equilibrium relation at the interface is given by equation (3).

$$[M] + n[O] = (MO_n) \qquad (1)$$

$$J_M = (k_m \cdot \rho_m/(100M_M))\{[\%M]^b - [\%M]^*\} = (k_s \cdot \rho_s/(100M_{Mon}))\{(\%MO_n)_L^* - (\%MO_n)_L^b\} \quad (2)$$

where, J_M represents the molar flux density of element M (mol/(m$^2 \cdot$s)); k_m and k_s represent the mass transfer coefficients in the boundary layer of the liquid metal and slag phases, respectively (m/s); ρ_m and ρ_s are the densities of the liquid metal and slag, respectively (kg/m^3); M_M and M_{Mon} represent the atomic weight of M and molecular weight of MO_n, respectively.

$$E_M = (\%MO_n)_L^*/\{[\%M]^* \cdot ao^{*n}\} = 100 \cdot C \cdot N_{MOn} \cdot f_M \cdot K_M/(\rho_s \cdot \gamma_{MOn}) \qquad (3)$$

where E_M is the equilibrium distribution ratio of element M; ao is the oxygen activity; C is the total number of moles in the liquid slag phase; f is the activity coefficient; K_M is the equilibrium constant; γ_{MOn} is the activity coefficient of MO_n; and the superscripts b and * indicate the bulk and interface concentrations, respectively, while the subscript L indicates liquid slag. The composition of each component in the liquid slag phase was used instead of the average over the liquid and solid slag phases.
Decarburization by the reaction between the slag and metal is expressed using the mass transfer rate in the metal phase, and the CO gas formation rate, as shown in equation (4).

$$J_C = (k_m \cdot \rho_m/(100M_C))\{[\%C]^b - [\%C]^*\} = G_{CO}(P_{CO}^* - 1) \quad (4)$$

where, P_{CO}^* denotes the partial pressure of CO at the interface, and G_{CO} denotes the formation rate of CO gas bubbles (mol/(m$^2 \cdot$s)).
From these equations and from the condition for electric neutrality shown in equation (5), the concentration change in all the elements in the metal and slag phases can be calculated.

$$\sum J_M = 0 \cdots\cdots (5)$$

Generally, in the case of hot metal dephosphorization, phosphorus, sulfur, silicon, manganese, iron, and oxygen are considered element "M." However, it is possible to consider other elements such as vanadium, niobium, or magnesium.
Initially, this model was applied to simulate the simultaneous reactions of dephosphorization and desulfurization in the hot-metal pretreatment process using flux injection technology. It has been confirmed by this model that dephosphorization by the highly basic slag with low oxygen activity at the interface induces the desulfurization reaction.
Based on this model, the authors have proposed the process simulation code that considers the reactions of the injected flux (transitory reaction), the top slag (permanent reaction), and the top-blown oxygen gas (Figure 3 (a))[3]. Various parameters of the coupled reaction model were evaluated by fundamental experiments. Using this process simulation code, the various treatment conditions were simulated, as shown in Figure 3 (b). In this figure, the dephosphorization

efficiencies of oxygen and CaO (η_{O2}, η_{CaO}) are shown as functions of the ratio of the unit consumption of lime to the unit consumption of total oxygen (CaO/O). This code can be used as a basic model for dynamic and static control systems in industrial operations.

To carry out the flux injection process using a torpedo car, the addition of CaF_2 is necessary. However, the leaching of fluorine is restricted by the environmental regulation and therefore, utilization of slag becomes difficult. To address this limitation, hot-metal dephosphorization was carried out using BOF recently. In this process, foaming slag of low basicity and a high FeO content can be used. The slag of hot-metal dephosphorization using a converter is characterized by a solid and liquid and a solid solution of di-calcium silicate ($2CaO\cdot SiO_2$) and tri-calcium phosphate ($3CaO\cdot P_2O_5$) precipitated in the liquid phase. The presence of the solid solution phase in the slag, with respect to its effects on the reaction efficiency, is an important factor that requires attention. In 2005, the ISIJ (Iron & Steel Institute of Japan) had launched a research project—"New Refining Process by using Multi-phase Fluxes" chaired by Prof. Tsukihashi—that focused on using slag with coexisting solid and liquid phases to increase the efficiency of dephosphorization refining[4]. Subsequently, another project—"Process Simulation for Dephosphorization of Pig Iron by Multi-Phases" chaired by Prof. Ito—was initiated to create a process simulation model of the dephosphorization process considering the characteristics of the slag comprising coexisting solid and liquid phases[5]. A basic conceptual illustration of this model is shown in Figure 4.

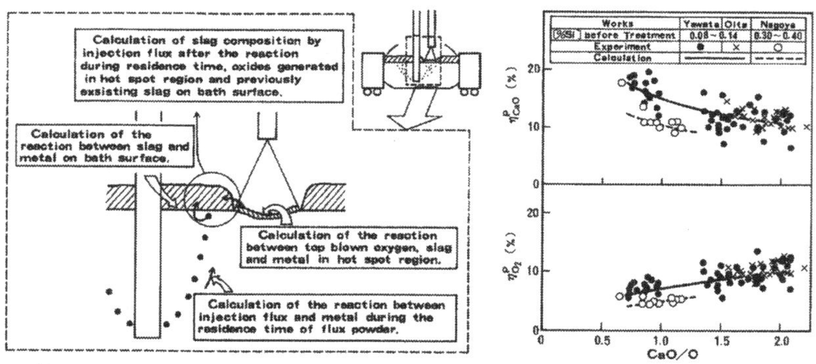

(a) (b)

Figure 3. (a) Schematic of the hot-metal dephosphorization model[3], (b) and comparison between the calculation results with the results from industrial operation.

This simulation model considers the following:

1) Chemical reaction kinetics between the liquid phase of slag and the liquid metal.

2) Precipitation and dissolution of the solid phase in the slag, and the distribution of phosphorus between the liquid phase and the solid solution of $2CaO\cdot SiO_2$ and $3CaO\cdot P_2O_5$ phase.

3) Flux dissolution rate into the liquid phase of slag, and the melting rate of scrap into molten metal.

4) Formation of FeO, direct decarburization, and post-combustion of top-blown oxygen.

5) Heat balance calculation.

The coupled reaction model was used to calculate the chemical reaction and the empirical equation to calculate the flux dissolution rate. The distribution ratio of phosphorus between the liquid and solid solution phases in the slag is calculated based on the activity of P_2O_5 in both phases. The regular solution model and the evaluation method proposed previously[6] are used to calculate the activity of P_2O_5 in liquid phase and in solid solution, respectively. The mass transfer coefficient and CO gas generation rate by slag and metal reactions have been estimated using equations formulated as functions of stirring intensity, size of the furnace, and slag composition as part of the ISIJ project[7].

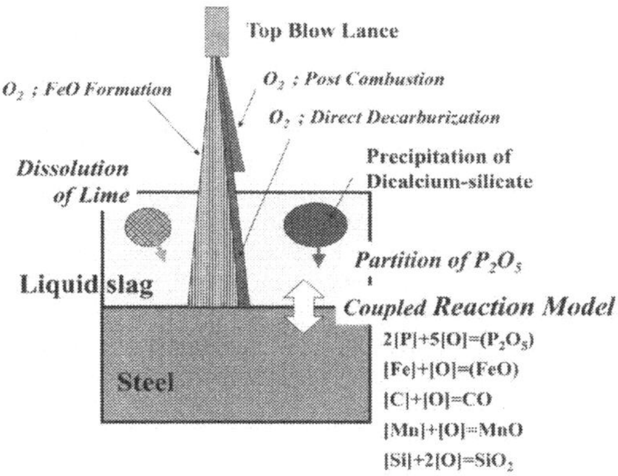

Figure 4. Schematic of the simulation model for hot-metal dephosphorization process[6].

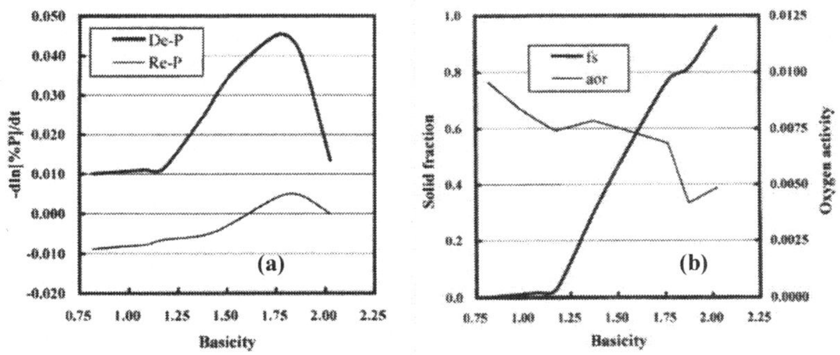

Figure 5. Optimum basicity for the dephosphorization (De-P) and phosphorus recovery (Re-P) (a) and the changes in solid fraction of slag and oxygen activity at interface (b) calculated by the model[8].

Using this model, factors such as the optimum conditions for change in slag composition during treatment, the influence of lime particle size, the effect of stirring, can be determined. Figure 6[8] shows examples of the calculation results. Figure 5 (a) shows the optimum basicity required for dephosphorization (De-P) and phosphorus recovery (Re-P). Figure 5(b) shows the changes in the solid fraction of slag and the oxygen activity at the interface.

A CD-ROM of this simulation model program was distributed to the member companies of ISIJ to investigate the optimum operating conditions for their respective steelmaking shops. The validity of this simulation model has already been well documented for the dephosphorization[6),8)] and the decarburization[9)] processes of BOF.

3. Composition Change of Inclusions during Ladle Treatment

Given that the quality requirements of steel have become stricter, it has become necessary to control the composition of inclusions in steel. Therefore, many studies have been conducted on controlling the composition; ladle treatment plays an important role in achieving this objective. In the past, the reactions that occur during ladle treatment were investigated under the assumption of an equilibrium state by many researchers. On the assumption that the entire ladle treatment system reached equilibrium, the compositions of the oxide phases were determined using their equilibrium relations with molten steel. However, the compositions of the slag and inclusions measured in actual ladle treatment are different to those calculated. Moreover, various inclusions of various compositions are observed in actual conditions. Recently, an automated system to evaluate steel cleanliness was developed, which involved the use of a computer-controlled scanning electron microscope. The size and composition distributions of inclusions have since been easily obtained using this system. Certain results have indicated that the variations in inclusion compositions are very large; therefore, these variations cannot be ignored. These facts indicate that the chemical reaction in the ladle does not reach equilibrium, and that variation in the reaction rate of each inclusion causes non-uniformity in the composition of that inclusion. In addition, it is well known that $MgO \cdot Al_2O_3$-type inclusions are formed even when Mg is not added during treatment. Many researchers have noted that Mg is formed in steel due to the reduction of MgO in the slag by Al, which is added as a deoxidizer, based on thermodynamic considerations. Graham *et al*[10)] applied the coupled reaction model to the ladle treatment to calculate the composition change of slag and metal phases, and demonstrated good agreement experimental results.

The authors[11,12)] have expanded this model to calculate the composition change of inclusions by considering the reactions between molten steel, slag, inclusions, and refractory, as shown in Figure 6.

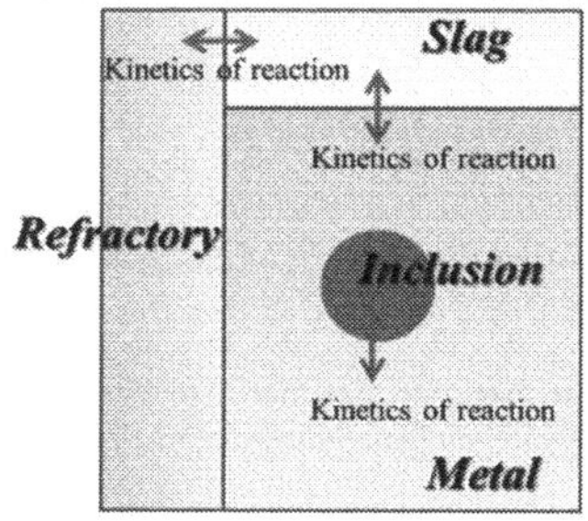

Figure 6. Illustration of the model that simulates the reaction kinetics in the ladle[11)].

In the proposed model, the coupled reaction model was applied to calculate the reaction between the molten metal and the inclusions in liquid phase, in addition to the reaction between the metal and slag. The thermodynamic calculation software, FactSage6.2, was connected to the developed program, which was coded in Visual C++, through ChemApp to calculate the activity of each component in the slag phase at each step of the calculation. The MgO dissolution rate into slag from the refractory was also taken into account. The added deoxidizing elements, such as Al and Fe-Si, etc. form inclusions as products of the deoxidization process in accordance with the equilibrium relations and mass balance. A fraction of slag is entrapped in the molten metal and forms inclusions. The coagulation and floatation of these inclusions are also taken into account. Figure 7 (a) shows the change in Mg and Ca content as calculated by this model on being applied to the ladle treatment process. Although these elements were not added, their contents gradually increased owing to the reaction between the metal and the slag after the second addition of Al at 18 min. Figure 7 (b) shows the changes in the contents of deoxidation products with time. As the content of Mg increases, the composition of the deoxidation product changes from Al_2O_3 to $MgO{\cdot}Al_2O_3$.

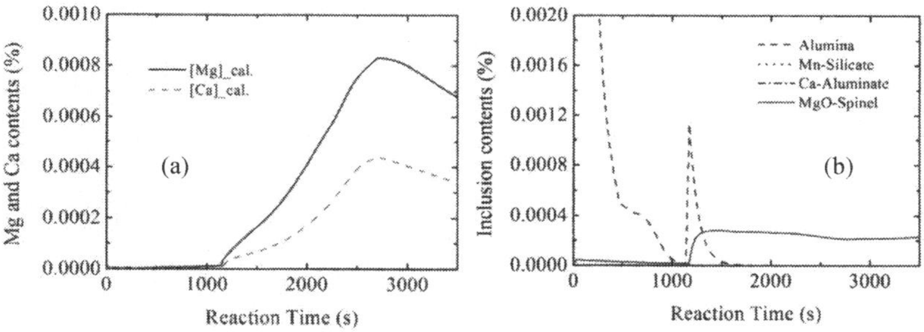

Figure 7. (a) Calculated changes in Mg and Ca contents in metal phase and (b) changes in deoxidation product content with time[11].

These results indicate the importance of kinetic modeling in analyzing inclusion compositions. However, currently it is impossible to simulate the variations in the inclusion compositions, and further study is required.

4. Conclusions

A few examples of the application of the kinetic model, which is based on the coupled reaction model, to analyze the steelmaking process are discussed. Hot-metal dephosphorization occurs under non-equilibrium conditions as it takes place between slag with a high oxygen potential and hot metal with a low oxygen potential. Process simulation model that considers solid slag, liquid slag, liquid metal phases, and the reaction between the solid and liquid slag, besides the reaction between the liquid slag and liquid metal, was constructed by an ISIJ research group under the project named "Process Simulation for Dephosphorization of Pig Iron by Multi-Phases." Through the model developed here, the ruling factor to increase the reaction efficiency of dephosphorization was verified. In the secondary refining process, the observed content and composition of inclusions is different from those calculated assuming their equilibrium with metal content. The kinetic model, considering the reaction between slag, metal, inclusion, and refractory, was established, and the composition change of inclusions during treatment was analyzed.

References

1. S. Ohguchi, D.G.C. Robertson, B. Deo, P. Grieveson and J.H.E. Jeffes, "Simultaneous dephosphorization and desulphurization of molten pig iron", *Ironmaking and Steelmaking*, **11** (1984), 202-213.
2. K. Miyamoto, K. Naito, I. Kitagawa and M. Matsuo, "Estimation of oxygen potential at slag/metal interface and effect of initial slag condition on hot metal dephosphorization", *Tetsu-to-Hagané*, **95**(2009), 199-206.
3. S.Kitamura, T.Kitamura, E.Aida, R.Sakomura, T.Kan eko and T.Nuibe, "Development of analysis and control method for hot metal dephosphorization process by computer simulation", *ISIJ Int.* **31**(1991), 1329-1335.
4. F. Tsukihashi, "Innovative development of refining processes in steelmaking by multi phase fluxes", *Tetsu-to-Hagané*, 95(2009), 187.
5. K. Ito and M. Mori, "Development of simulation program for hot metal dephosphorization processes", *CAMP-ISIJ*, 25(2012), 188-190.
6. F. Pahlevani, S. Kitamura, H. Shibata and N. Maruoka, "Distribution of P_2O_5 between solid solution of $2CaO \cdot SiO_2$-$3CaO \cdot P_2O_5$ and liquid phase", *ISIJ Int.*, 50(2010), 822-829.
7. S. Kitamura, K. Ito, F. Pahlevani and M. Mori, "Development of simulation model for hot metal dephosphorization process", *Tetsu-to-Hagané*, 100(2014), to be published.
8. S.Kitamura, K. Miyamoto, H. Shibata, N. Maruoka and M. Matsuo, "Analysis of dephosphorization reaction using simulation model of hot metal dephosphorization by multiphase slag", *ISIJ Inter.*, 49(2009), 1333-1339
9. S. Kitamura, H. Shibata and N. Maruoka, " Simulation model of dephosphorization by liquid and solid coexisting slag", *Proceedings of the 3rd International Conference of Process Development in Iron and Steelmaking, SCANMET III*, MEFOS, Sweden (2008), vol.1, 283-294.
10. K.J. Graham and G.A.Irons, "Integrated control in the ladle metallurgy furnace", *Proceedings of the 3rd International Conference of Process Development in Iron and Steelmaking, SCANMET III*, MEFOS, Sweden (2008), vol.1, 385–396.
11. A. Harada, G. Miyano, N. Maruoka, H. Shibata and S. Kitamura, "Composition change of inclusions during ladle treatment by the reaction with slag and refractory", *Proceedings of 8th International Conference on Clean Steel*, Budapest, Hungary, (2012), CD-ROM
12. A. Harada, N. Maruoka, H. Shibata and S. Kitamura, "A Kinetic Model to Predict the Compositions of Metal, Slag and Inclusions during Ladle Refining. Part 1. The Basic Concept and the Application to the Industrial Operation", *ISIJ Int.*, to be published

Celebrating the Megascale: Proceedings of the Extraction and Processing Division Symposium on Pyrometallurgy in Honor of David G.C. Robertson
Edited by: Phillip J. Mackey, Eric J. Grimsey, Rodney T. Jones, and Geoffrey A. Brooks
TMS (The Minerals, Metals & Materials Society), 2014

COLD MODELLING OF SPLASHING PHENOMENA IN OXYGEN STEELMAKING

Shabnam Sabah[1], Geoffrey Brooks[1]

[1]Swinbune University of Technology;
John Street; Hawthorn, VIC 3122, Australia

Keywords: steelmaking, splashing, penetrating, cavity oscillation

Abstract

In oxygen steelmaking, splashing and droplet formation play a key role in the kinetics of the process. Though splashing has been studied by previous investigators, there is only limited understanding of how different cavity modes affect splashing. Therefore, a cold model study (at various lance heights and gas flow rates) has been carried out focusing on the issue of cavity modes and how it affects splashing phenomenon. Analysis using Fast Fourier Transform (FFT) was carried out on cavity oscillation that showed that frequency and amplitude of oscillation was highest in penetrating mode. Measurements of splashing over a range of conditions showed splashing greatly reduced when cavities went from splashing to penetrating mode. The results were validated and compared with plant data and previous model investigations.

Introduction

In oxygen steelmaking, the metal droplets are created due to impact of supersonic jets. This phenomenon is referred to as "splashing". In top blown oxygen steelmaking, there is significant evidence to show that many of the key reactions occur in the emulsion, which in turn is directly related to the quantity of splash generated from the injection and subsequent behavior of the droplets in the emulsion. Several studies [1, 2] have identified the formation of waves inside the cavity. But it is not yet clearly understood how waves inside the cavity affect splashing. Molloy [3] termed various cavity shapes as dimpling mode, splashing mode and penetrating mode. In dimpling mode, there is a slight depression but no droplet formation. In splashing mode, there is a shallow depression as well as large outwardly directed splash. In the case of penetrating mode, there is a decrease in outward splash with a much deeper cavity. Lee et al. [2] predicted that cavity modes can be distinguished by the frequency of oscillation of cavity, rather than visual observation. In summary, there is limited understanding of the influence of different cavity modes on droplet generation in oxygen steelmaking.

In estimating the amounts of droplets present in the emulsion, various cold models, hot models, and plant trials have been carried out in the past. In most of the cases, samples were collected from one place of the emulsion, and droplets found at that sampling place were taken as being representative of the total emulsion. A survey of the literature shows that there is a large variation in the reported amount of droplets present in the emulsion. A recent investigation on improving phosphorus refining (IMPHOS) [4, 5], commissioned by the EU, used a 6 t pilot-plant BOS converter. An automated sampling lance was designed and used to take samples from seven places at a time, every 2 minutes from the start of the oxygen injection. The percentage of metal circulation rate (% tap weight/second) in the emulsion phase (in three heats) varied between 7 and 13%.

There are insufficient studies that show quantitatively how splashing behaviour is affected by different cavity modes. To predict droplet generation in steelmaking, Subagyo et al. [6] proposed a new dimensional number, called the "Blowing number" (N_B) which is a ratio of inertia force to buoyancy and surface tension forces. It is defined as follows

$$N_B = (\rho_g * U_g^2/2) / (\sigma * g * \rho_l)^{0.5} \quad (1)$$

Then Subagyo et al. [6] proposed an empirical formula relating metal droplet generation rate (R_B) per unit volume of blown gas (F_g) and N_B using their measured data of high temperature modelling and cold modelling data of He et al. [7, 8]. The formula [6] is given as follows,

$$R_B / F_g = N_B^{3.2} / [2.6*10^6 + 2*10^{-4}*N_B^{12})^{0.2} \quad (2)$$

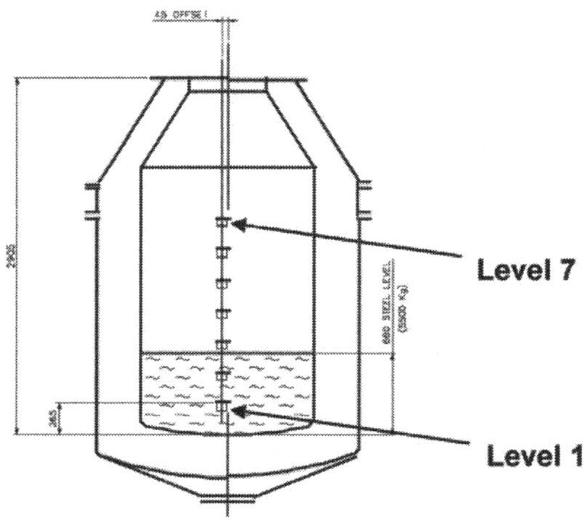

Figure 1. Schematic of IMPHOS investigation [4]

This relation suggests that an increase in N_B always results in an increase in the generation of droplets. Alam et al. [9] found that at high N_B and penetrating mode, splashing was lower than that of low N_B in splashing mode. It showed that the effect of cavity mode on Blowing number theory is not understood and needs to be investigated.

The aim of the present work is to investigate cavity oscillation using Fast Fourier Transform (FFT) analysis, and to identify if it is possible to differentiate among various cavity modes quantitatively. In addition, a comprehensive splashing study has been carried out to investigate the issue of splashing and Blowing number in different cavity modes and find how the sampling position can affect the estimation of droplets present in the emulsion. Water at 25°C has a similar kinematic viscosity to molten steel at 1600°C. The work of Sharma et al. [10] showed that the motion in the liquid bath is influenced by the impact force of the jet, not by the presence of the slag and carbon content of the liquid steel. Therefore, in the present work, the water-air system has been used to represent the case of oxygen injected on the surface of molten iron.

Experimental Description

For FFT analysis, the model used in the present investigation (stated in Table I) is a 1/10th scaled model of the converter stated in the work of Luomala et al. [11]. Compressed air and water were used to simulate oxygen and liquid steel respectively. No slag phase was considered. A transparent cylindrical rig was used as the vessel of the experiment. A flow meter (range: 0.01 m^3/min to 0.1 m^3/min) was used to control the flow rate of compressed air. A high-speed camera named MotionPro Y (model Y4L) was used to capture photographs of the cavity and the splashing of droplets. Pictures of the cavity were taken at a rate of 30 frames/ second for duration of 4 minutes. In order to observe the droplet generation, a video of 2 seconds was taken at 2000 frames/second rate.

Table I. Experimental Setup Description

Dimensions	BOF [11]	1/10th model
Nozzle diameter, d (m)	0.029	0.003
Vessel diameter, D (m)	4.30	0.420
Vessel height, l (m)	7.20	0.500
Bath height, H (m)	1.50	0.15,0.20
Number of nozzles, N	5	1
Lance height, L (m)	1.50-2.00	0.05,0.08,0.10,0.12,0.15,0.16,0.17
Nozzle angle, θ (°)	15	0

For the study of splashing, Heat S1845 of the IMPHOS investigation [4, 5] was taken as the target heat number to compare the results with the present work. Table II shows the details of the experimental set up and IMPHOS investigation. In order to cover a wide range of experimental conditions, splashing studies have been carried out at various lance heights and gas flow rates in addition to the IMPHOS conditions. In both the splashing and the FFT study, the geometric and dynamic similarities were maintained as much as possible. The blowing number similarity criteria has been given preference over other dynamic similarity criteria, as suggested by Subagyo et al. [6] as it incorporates the Kelvin-Helmholtz instability criteria into droplet generation condition.

Table II. Comparison of IMPHOS Work and Present Model

Dimensions	IMPHOS BOF	1/5th model
Nozzle diameter, d (m)	0.0155	0.003
Vessel diameter, D (m)	1.340	0.420
Vessel height, l (m)	2.905	0.500
Bath height, H (m)	0.660	0.200
Number of nozzles, N	1	1
Lance height, L (m)	1.70, 1.60, 1.50 (Heat S1845)	0.170,0.160,0.150 (compared with S1845), 0.120,0.100,0.080,0.050 (Additional)
Nozzle angle, θ (°)	0	0

To collect droplets in different sampling positions, a stainless steel structure was put above the cylindrical bath. Also, a stainless steel strip was perpendicularly attached to the structure. The cylindrical sample pots, with a height of 0.014 m and diameter of 0.013 m, were attached with the strip. The sampling positions of the pots were scaled down 1/5th from the IMPHOS investigation. For clarity of identification, sample pots of IMPHOS pilot investigation will be termed as P1, P2, P3, P4, P5, P6, and P7 (as shown in Figure 1) whereas sample pots of the present model will be labeled as M1, M2, M3, M4, M5, M6 and M7. There were no M1 and M2 pots in this study. The sampling position of M3 was equivalent P3; position of M4 was equivalent to P4 and so on. The sampling rod was kept in various radial positions (0.033 m, 0.060 m, 0.090 m, 0.120 m, 0.150 m, and 0.180 m radially away from the lance) and vertical positions (0.024 m, 0.074 m, 0.124 m, 0.174 m, and 0.224 m above the bath surface). At a specific radial position, droplets were collected in the vertically positioned five sample pots (M3 to M7) for duration of 3 seconds to 150 seconds, depending on the droplet generation rate. Each set of data was taken two times and an average was determined. After each sampling process, the weights of the sample pots were measured in a digital weight scale. Subtracting the weight of empty pots from the measured weight, the collected droplet weight was calculated.

Results and Discussion

FFT Analysis

At each operating condition, 30 images of the cavity were acquired over duration of 1 second. Using ImageJ1.45s software, the depth of cavity was measured from the 30 images of each operating condition. Thus, a time series of cavity depth in each operating condition was produced. In order to find the frequency and amplitude of the cavity oscillation, the Fast Fourier Transform (FFT) was applied on the time series in Origin Pro 8.5 software. At a particular operating condition, FFT results showed a number of frequencies rather than "one" dominant frequency. The total range of the frequencies was between 3 to 9 Hz which is compatible with previous studies [2, 12]. As there was no "one" dominant frequency, an average of three frequencies which had the highest amplitudes were calculated and termed as the frequency of cavity oscillation at that operating condition. Frequency and amplitude of cavity depth oscillation in each cavity mode was calculated and presented in Figure 2 and Figure 3. Frequencies and amplitudes of dimpling, splashing and penetrating modes were derived from 7, 13, and 19 various experimental conditions respectively. How various cavity modes were identified will be discussed in the later sections of the paper.

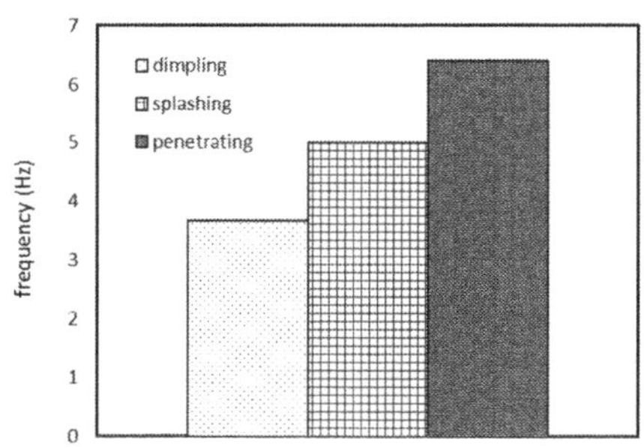

Figure 2. Frequency of oscillation in different cavity modes

Figure 2 shows that the frequency of cavity depth oscillation in dimpling mode changed from 3.67 Hz to 5 Hz as the cavity turned into splashing mode. As the cavity mode became penetrating, the frequency was found to be the highest, which was 6.40 Hz. In the case of amplitude of oscillation, Figure 3 shows that amplitude of cavity oscillation gradually increased as the cavity mode shifted from dimpling to splashing, and splashing to penetrating. The amplitude of oscillation was found highest in penetrating mode.

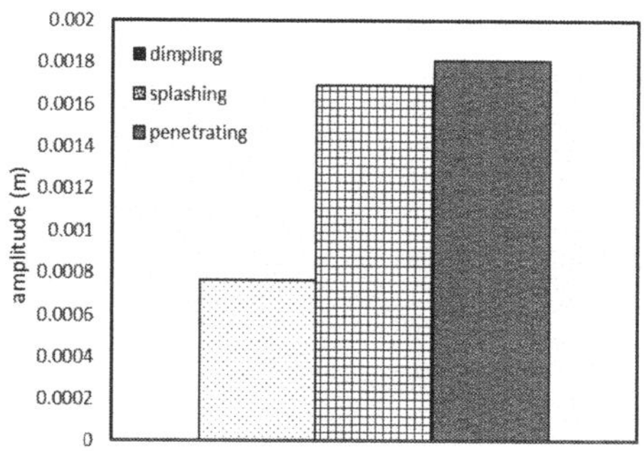

Figure 3. Amplitude of oscillation in different cavity modes

The energy of a wave is dependent on the frequency and amplitude of the wave. If the amplitude of a wave becomes double, or if the frequency of the wave is doubled, it will have a quadrupling effect on the energy transported by that wave. From the FFT results, frequency and amplitude of cavity oscillation was found highest in penetrating mode. Therefore, from an energy balance perspective, it can be concluded that more energy is used in oscillating the cavity in penetrating mode compared to splashing and dimpling modes [13].

Comparison with Previous Works

When calculating the droplet generation rate at each operating condition, the bath above the water surface was divided into six cylinders, as there were six radial sampling positions (i.e., 0.033 m, 0.060 m, 0.090 m, 0.120 m, 0.150 m, and 0.180 m away from the lance). At each radial sampling position, an average of droplets collected in various vertically placed sample pots was taken as being representative of the droplet generation rate in that cylinder. The droplet generation rate, R_B was calculated using the following equation[4].

$$R_B = (\Sigma W_{sp}) * V_b / (t_s * V_{sp} * N_{sp}) \quad (3)$$

Here, ΣW_{sp}= weight of droplets collected in pots, V_b= volume of each cylinder above the bath, t_s=sampling time, V_{sp}=volume of a sample pot, N_{sp}=no of sample pots. Droplet generation rate at each cylinder was calculated using equation 3. Then an average of the rates of all cylinders was taken as the droplet generation rate (R_B) at that operating condition.

High-speed imaging of the cavity (as shown in Figure 4) showed the generation of "splash sheets". Splash sheets were found to form by the reflected gas velocity and fell into the sample pots closer to the bath (i.e., M3 and M4). For this reason, amounts of droplets collected in the positions close to the bath (i.e., M3 and M4) were much greater than that of other pots (i.e., M5, M6, and M7). Therefore, it was reasonable that the droplet generation rate was to be calculated excluding the sample pots closer to the bath surface (i.e., M3 and M4). As sampling positions of P3 and P4 of IMPHOS are in equivalent positions as M3 and M4, droplets collected in both P3 and P4 should be excluded from the R_B calculation in IMPHOS work. Millman et al. [4] excluded only P3 when calculating droplet generation rate.

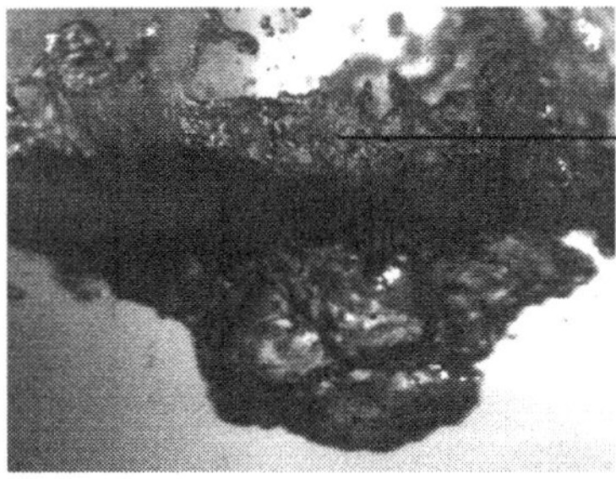

Figure 4. Splash sheets formation

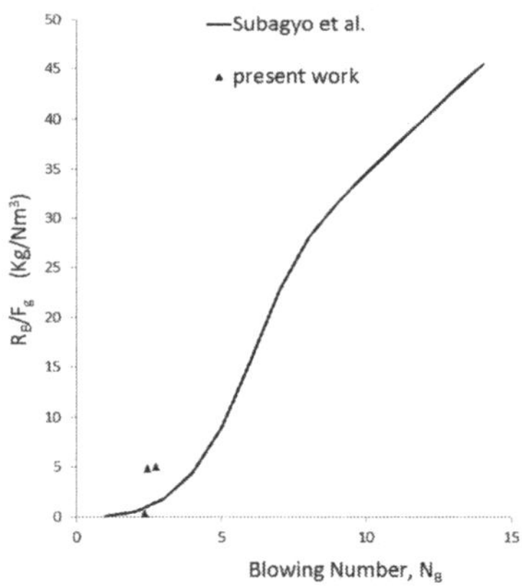

Figure 5. Present work in comparison with Subagyo et al.'s [6] investigation

Figure 5 shows the results of the present splashing study scaled down from IMPHOS, and equation 2 proposed by Subagyo et al. [6]. Fg was converted from operating conditions to pressure 101325 Pa and temperature 273 K condition to be consistent with the work of Subagyo et al. [6]. Figure 5 shows that the outcomes of the present work are in good agreement with the investigations of Subagyo et al. [6]. R_B/F_g in heat S1845 of IMPHOS was found in the range of 180 to 360 kg/Nm^3. Present analysis showed that sampling too close to the bath can give overestimation in the droplet amount. Therefore, it implied that the value of R_B/Fg was overestimated in the IMPHOS investigation. Due to the formation of splash sheets, droplets collected in P4 also needs to be excluded from R_B calculation in IMPHOS.

Splashing Study at Different Blowing Numbers

Figure 6 shows how the droplet generation rate per volumetric flow rate of blown gas (R_B/F_g) was affected by cavity modes and Blowing number (N_B). At a constant lance height, with an increase of N_B, droplet generation rate per volume flow rate of blown gas increased in general. At higher lance height of 0.170 m, 0.160 m, and 0.150 m, R_B/F_g was greater than that of lower lance heights. As lance height was lowered to 0.120 m, there was drastic reduction in R_B/F_g. From lance height 0.120 m to 0.050 m, the values of R_B/F_g were found to be lowest. So, the cavity modes at lance height 0.15, 0.16, 0.17 were titled as splashing modes whereas the modes of the cavity for the rest were termed as penetrating modes.

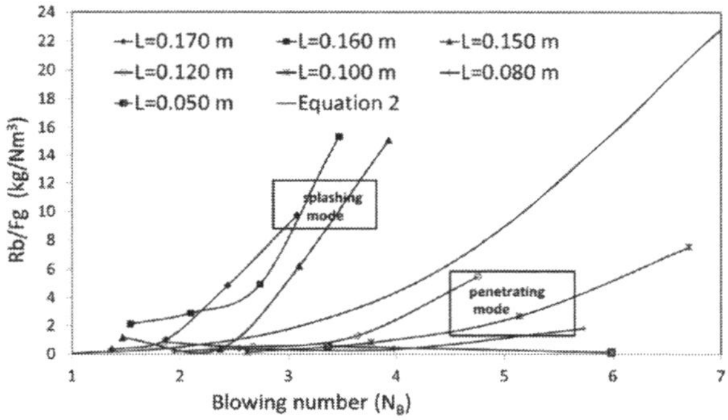

Figure 6. Droplet generation rate per volumetric flow rate of blown gas (R_B/F_g) in various N_B

Figure 6 shows quite clearly how splashing rate was affected by the change in the cavity modes. At a constant N_B of 3, R_B / F_g was found to be dependent on the cavity mode. If the cavity was in splashing mode, the R_B/F_g was found to be in the range of 5 kg/Nm^3 to 9 kg/Nm^3 (approx.) whereas R_B/Fg became 0.4 kg/Nm^3 (approx.) when the cavity was in penetrating mode at the same Blowing number. It indicated that N_B along with cavity mode is needed in estimating droplets amount[14]. Present findings are consistent with the results of Alam et al. [9]. The results of the present work were also compared with equation 2 given by Subagyo et al. [6]. The equation was developed for a fixed lance height. Present results consisted of various lance heights. Also, there were unavoidable experimental errors associated with the present experiments. Therefore, variations could be expected between the results of present work and Subagyo et al. [6]. From the present results, it was found that when $L/d \geq 50$, the cavity was in splashing mode and at $L/d \leq 40$, the cavity was in penetrating mode [14]. This criteria needs further investigation over a wide range of conditions before we can appreciate its applicablity to steelmaking conditions.

Conclusions

From the present cold modelling work, the following can be concluded:

1. FFT analysis of cavity oscillation showed a clear difference among frequency and amplitude of oscillation in various cavity modes. In penetrating mode, frequency and amplitude of cavity oscillation was found to be highest.

2. Droplets found in sample pots varied both across the radius of the bath and along the vertical distance from the bath surface. Largest variations in the droplet weight were found in sample pots M3 and M4 (which were closest to the bath).

3. Splash sheets formation was identified from high-speed images that caused higher droplet count in the sample pots closest to the bath.

4. The rate of droplet generation rate reduced greatly as cavity mode changed from splashing to penetrating at a constant Blowing number.

This cold model study of splashing and image analysis identified two important aspects of cavity modes. Firstly, FFT analysis showed how cavity modes were noticeably distinct in frequency and amplitude of cavity oscillation. Secondly, the splashing study quantified how splashing was affected by the changing of cavity modes. Also, formation of splash sheets indicated the effect of sampling positions on the estimation of droplets present in the emulsion that is quite important for plant trials or hot model experiments to avoid overestimation or underestimation in the amounts of droplets in the emulsion.

References

1. K.D. Peaslee and D.G.C. Robertson, "Model studies of splash, waves, and recirculating flows within steelmaking furnaces," *Steelmaking Conference Proceedings*. Warrendale, PA: Iron & Steel Society 77, (1994), 713-722.
2. M. Lee, V. Whitney, and N. Molloy, "Jet-liquid interaction in a steelmaking electric arc furnace," *Scandinavian Journal of Metallurgy*, 30 (5), (2008), 330–336.
3. N. Molloy, "Impinging jet flow in a two-phase system: the basic flow pattern," *Journal of the Iron and Steel Institute*, 216 (1970), 943–950.
4. M. Millman, A. Overbosch, A. Kapilashrami, D. Malmberg, and M. Brämming, "Study of refining performance in BOS converter," *Ironmaking & Steelmaking*, 38 (7), (2011), 499-509.
5. M.S. Millman, A. Kapilashrami, M. Bramming, and D. Malmberg, *Imphos: improving phosphorus refining*, 2011, Publications Office of the European Union: Luxembourg.
6. Subagyo, G.A. Brooks, K.S. Coley, and G.A. Irons, "Generation of droplets in slag-metal emulsions through top gas blowing," *ISIJ International*, 43 (7), (2003), 983-989.
7. Q.L. He and N. Standish, "A Model Study of Droplet Generation in the BOF Steelmaking," *ISIJ International*, 30 (4), (1990), 305-309.
8. N. Standish and Q.L. He, "Drop generation due to an impinging jet and the effect of bottom blowing in the steelmaking vessel," *ISIJ International*, 29 (6), (1989), 455-461.
9. M. Alam, G. Irons, G. Brooks, A. Fontana, and J. Naser, "Inclined Jetting and Splashing in Electric Arc Furnace Steelmaking," *ISIJ International*, 51 (9), (2011), 1439-1477.
10. S. Sharma, J. Hlinka, D. Kern, "The Bath Circulation, Jet Penetration and High-Temp Reaction Zone in BOF Steelmaking," *Iron and Steel Maker* 4(7), (1977) 7-18
11. M.J. Luomala, T.M.J. Fabritius, E.O. Virtanen, T.P. Siivola, and J.J. Harkki, "Splashing and spitting behaviour in the combined blown steelmaking converter," *ISIJ International*, 42 (9), (2002), 944-949.
12. K.D. Peaslee and D.G.C. Robertson, "Fluid dynamics of inclined jetting on a slag/metal bath," *EPD Congress*. San Francisco, California, USA: TMS, (1994), 1129-1145.
13. S. Sabah and G. Brooks, submitted to ISIJ International 2013.
14. S. Sabah and G. Brooks, submitted to ISIJ International 2013.

LEAN OPERATIONS STRATEGY TO COMBAT UNCERTAINTIES IN TEMPERTURE AT BOF END POINT, TAPPING, DEOXIDATION, ALLOY ADDITION AND THERMAL HISTORY

Ishani Shukla[1], G Rajesh[2], Pratik Agarwal[3], Ajay Kumar Shukla[4], Deepu Philip[1],

[1]Industrial and Management Engineering, Indian Institute of Technology Kanpur, 208016, India
[2]Steelmelting Shop, Visakhapatnam Steel Plant, Visakhapatnam, 531001, India
[3]Materials and Metallurgical Engineering Department, Indian Institute of Technology, Kanpur, 208016, India
[4]Department of Metallurgical and Materials Engineering, Indian Institute of Technology, Madras, Chennai, 600036, India
Emails: Ishani.shukla@gmail.com, shukla@iitm.ac.in, dhhilip@iitk.ac.in

Keywords: BOF end point, ladle, finite difference, model, heat transfer, decision support system.

Abstract

Steel plants generally employ static BOF end point control models to arrive at a given temperature and chemistry at the end of blow. These end point models have now been supplemented with chaos control models to steer the blowing process in the right direction while the blow is in progress. Merely arriving at the correct end point temperature is however not adequate because in the subsequent stages as well the temperature variations can be large and unpredictable. The present paper deals with an integrated model to take into account the effect of all parameters affecting temperature and composition from tapping to the start of casting to minimize the use of LF or aluminum heating, and also minimize grade mixing. The application of lean operations strategy is explained in which the previous "Push System' is changed to a "Pull System", minimizing the use of ladle furnace or aluminum heating during steelmaking.

1.0 Introduction

Strategies are required for attacking a problem in an industrial environment. In this study, the problem of designing a lean decision support system (DSC) is addressed for improving the management aspects of scheduling in a large scale integrated manufacturing plant. A case study of a steelmelt shop of a 3.4 million ton steel plant at Visakhpatnam (VSP) is undertaken.

The literature is replete with scheduling models for steelmelt shop (comprising of steelmaking units as well as continuous casting machines) [1,2]. Though various claims have been made in literature, the practical implementation and continuous use of the scheduling models faces many problems because the published models focus more on the rigors of flow shop-job shop scheduling rather than the crucial and dual requirement of "correct temperature and composition at correct time". The scheduling of steelmelt shop, because of unexpected delays and breakdowns, is so complex that it is still considered as an unsolvable NP hard problem. To get a working model simplifying assumptions have to be made for a given plant situation. The approach used in the present work is heuristic and suites a steel plant in which breakdowns and delays are not infrequent.

The steel melt shop chosen for present study comprises of three 140 ton basic oxygen furnaces (BOFs), three argon rinsing stations, one intermediate heating and up temperature (IRUT), one ladle furnace ((LF) and six billet casters. A simplified ideal flow situation for one converter operation is shown through the Wenn diagram in Fig. 1 and disrupted flow situation is shown in Fig. 2. Since no temperature increase can be done at ARS, the scheduling has to be such that the time spent in the intersection areas i1, i2, i3 (marked in Fig. 1 and Fig. 2) is minimum required. Further, the planning has to be such that the additional units IRUT and LF are used for a minimum number of heats, implying that the liquid steel from BOF can be directly sent (as in Fig. 1) to the casters with only argon rinsing and alloying (limited amount) at ARS.

In the initial stages of work, in-depth discussions were held at corporate level as well as with the experts within the plant to identify the main factors affecting scheduling: delivery time, prediction of temperature at a given time (based on ladle history), alloy recovery, and position of ladles on the track waiting for clearance to go to caster. The study of past data showed that the use IRUT unit varies from 30- 60 % of the heats; large dependence on IRUT implies reduced production rate and in some cases reduced quality as well. The conditions which necessitated the use of IRUT or were documented.

From a deep study, it was found that the problems of heat loss from the ladle, unpredictability of tapping temperature and alloy recovery, heat losses during taping, and unexpected delays, needed attention. In critical cases, even one degree Celsius drop in temperature of molten steel can have a significant impact on the quality and castability. In monetary terms saving one minute can lead to a saving of more than $4500. A small variation in alloy recovery can make 140tons of steel (one batch of steel produced in the furnace) off grade (rejected heat) with respect to carbon and phosphorus and hence lead to a loss of nearly $10000.
Predictive models were developed to help determine, in advance, the possible temperature and composition of steel for different operating conditions. The proposed decision support system, based on these model, thus caters to advising the exact time of start of blow in converter, followed by prediction of expected converter tapping duration (based on tap hole life), transfer time to argon rinsing station (ARS), and transfer time to turret at the caster, and finally the expected time and temperature at the start of casting.

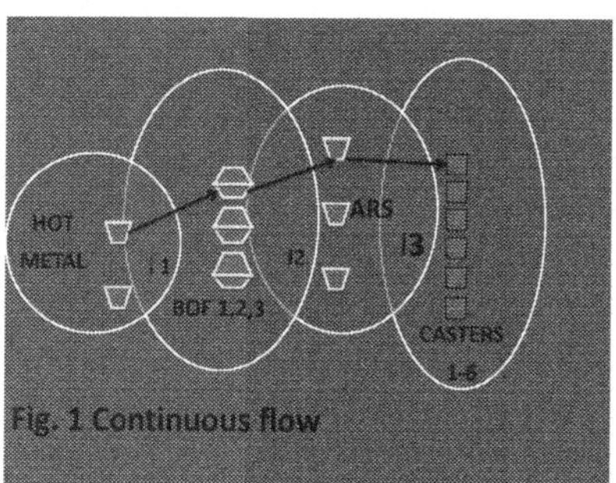

Fig. 1 Continuous flow

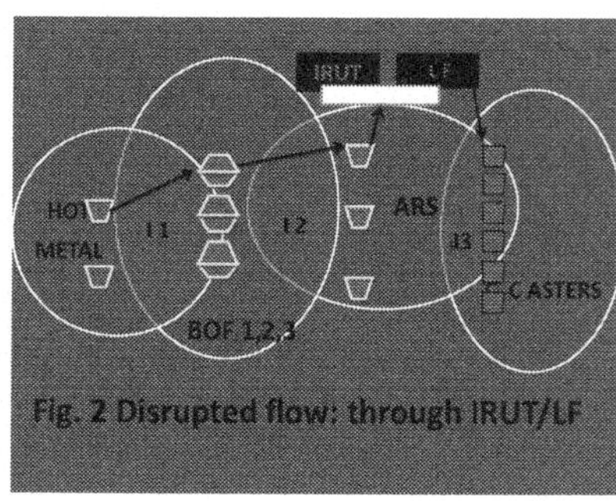

Fig. 2 Disrupted flow: through IRUT/LF

The individual models which are an integral part of the decision support system include the following

(1) Hot metal model to calculate the temperature, mass and time of delivery of hot metal to BOFs
(2) Converter taping and alloying model: based on tapping time, tap hole diameter, and amount and composition of alloys added during tapping.
(3) Liquid steel transfer model: to calculate temperature loss during transfer by Steel Transfer Cars (STC) from Basic Oxygen Furnace (BOF) to Argon Rinsing Station (ARS).
(4) ARS stirring model: to calculate temperature and composition of steel prediction after stirring and alloy addition at ARS
(5) Liquid steel transfer model to predict temperature drop from ARS to the turret stage
(6) Ladle return model: the thermal condition of ladle after return from caster to BOF

In support of the above models 1-6, the time and motion studies were also carried out to determine the accuracy levels of data recording through hand held terminals and also automatically recorded data (through activation of photo switches Level II system). Discrepancies in recorded time data to the tune of ±10-15 seconds (time lag or over estimate) were identified. Each second counts for temperature. The main focus is on the actual 'flow' status of predefined operations at the time of taking decision.

The above models 1-6 are integrated in the overall 'Tapping Advice Model' which advices the aim tapping temperature in order to obtain the desired temperature, at a given time, on arrival at ARS, and subsequently on arrival at turret. In contrast with other scheduling models published in literature, in the strategy adopted in this work, the shop floor in charge positions himself at the middle station ARS, overlooking the progress of casting at different casters and then decides instantaneously which particular converter should start blowing, aiming for the desired tapping temperature dictated by the situation at that time. A safe limit for variation (small temperature cushion) from desired temperature, with or without reblow is built into the decision support system so as to absorb unexpected delays. The new decision control system thus monitors

- The time at which the hot metal is supposed to reach BOF and thereafter the blow in BOF has to start
- The type of ladle placed at BOF for tapping (lining life and ladle return time from caster)
- The time the operators need to start tapping
- Amount of alloy to be added during tapping after consulting grade chart
- Making necessary back calculations of desired tapping temperature in case of delay
- Making quick decisions on alloy additions at ARS to achieve a specific grade

Since the decisions about desired tapping temperature are taken at ARS and not at the caster, the present decision support system is not a 100% "pull system", as practiced in the case of lean management, but the following positive aspects of Lean management are well incorporated in the procedures followed:

- 'Just in time, *kanban*' to ensure delivery at a required temperature, time and composition' at ARS and finally at the caster (CC).
- 'Autonomation or *Jikoda*': A ladle is not sent to caster directly if it does not meet the time, composition and temperature criteria; it has to be routed either through IRUT or

LF. The planning and management of operations is however such that the additional units (LF and IRUT) are needed in the minimum number of cases.

- Minimization of wastage: the problem of 'Return Heat' (wasted heat) is also reduced; 'return heat' means the liquid steel not being cast fully in the caster and the ladle with liquid steel is returned back to BOF for refining.
- Flow variability *Heijunka*: The converter, casters IRUT and LF operate only when needed. Not all units need to be operated; out of six casters usually five are in operation.

In view of above, the name given in the present work is 'Lean Decision Support System'.

2. Model simplification, validation, and actual operation

To start with, detailed heat transfer models based on finite difference approach were developed for heat loss calculations from BOF and ladle (both full and empty). These models calculate detailed temperature profiles in refractory for different thicknesses of refractory, waiting time, etc. The calculations are time intensive and cannot be used directly in actual production situation for quick assessment changes in metal temperature. Since the main deciding parameters (tapping temperature, tapping duration, ladle life, ladle return time from caster, liquid steel weigh, carry over slag, transportation time from BOF to caster) vary in a defined small ranges only, the following equation was fitted to the results obtained from finite difference models:

$$\mathbf{T = Tinitial - \alpha_1 . Time\_Hold - \alpha_2 . (Time\_Hold)^{\beta} - \alpha_3 . Life\_fraction - \alpha_4 . (Waiting\_Time\ empty\ ladle) + \alpha_5} \qquad (1)$$

'Tinitial' is temperature of steel in ladle after tapping and alloying. 'Time_Hold is ladle waiting time (minutes) full of steel, Life_fraction is actual lining life of ladle divided by maximum ladle lining life, Waiting_Time (minutes) of empty ladle is ladle return time from caster back to BOF. The value of 'Tinitial' is obtained from BOF waiting model (after the heat has been made in BOF)and tapping plus alloying model combine.

The standard deviation of various parameters under actual shop floor conditions is listed in Table 1. A rough production schedule for the day is drawn up by the production planning department on the basis of these parameters. But once the production starts, parameters slide or change depending upon delays and breakdowns and immediate adjustments are made in the schedule by the Lean Decision Support System. The responsibility of decision lies with the shift in-charge based on the estimates provided by Lean Decision Support System.

Table 1: Time and temperature at various steps of steelmaking

	Hot metal Temp (C)	Time Hot metal Charge to blow Start (min)	Time Blow in BOF (min)	Time Reblow in bof (min)	Turn down temp (C)	Time Tap start to finish (min)	Time Tap finish to ARS (min)	Time ARS to caster (min)	Casting Time (min

Standard deviation	32.0	0.90	3.35	155.6	29.2	1.6	9.4	12.2	18.3
Average	1365.8	2	16.6	310.8	1704.8	5.5	7.6	18.8	84.1
Maximum	1446	9	85	999	1770	19	19	83	138
Minimum	1250	2.8		100	1600				

Thus the decision process is heuristic, evolves with time in spite of breakdown and delays, and that is why it works. The effort to optimize the situation at each step based on predictions of the decision support system. The implementation of above strategy has increased hit rate of achieving target temperature and composition on delivery of liquid steel at ARS and caster.

The maximum uncertainty occurs at tapping stage because of the continuous change in tap hole diameter and the corresponding error in heat loss calculation while the metal is tapped from converter into the ladle. The effect of tap hole life on tap duration is shown in Fig. 4, on the basis of actual plant data. The effect of tap duraion on loss of temperature (calcualed) is shown in Fig. 5

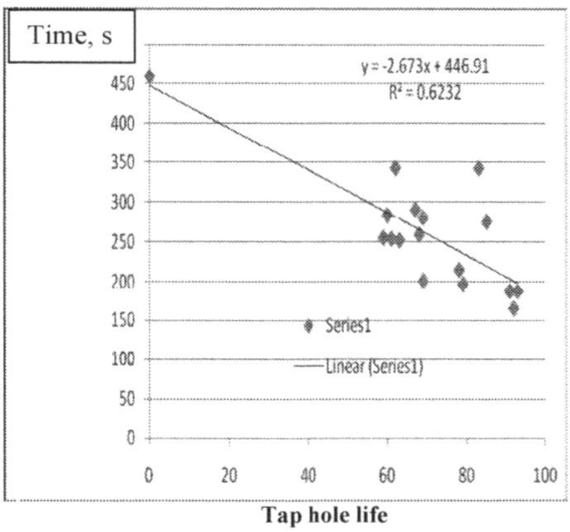

Fig. 3 Effect of tap hole life on tap duration

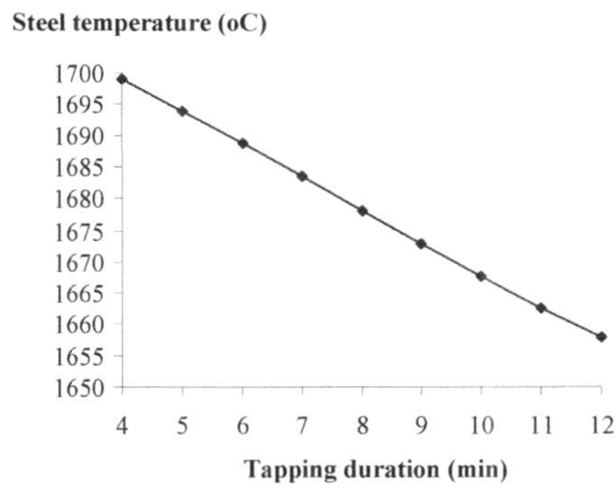

Fig 4. Effect of tap duraion on liquid steel temperature

The next highest possible source of error is the difference between the aim tapping temperature and temperature actually obtained at first turn down (FTD) in BOF. In order to reduce this error a Level II chaos control system has been introduced in the BOFs [3] In case of re-blow the error is less than ±10 C. The third source of error is ladle return time and heat lost from ladle when the ladle becomes empty after the casting is over. It takes finite but variable time (ladle return time) for each ladle to return to BOF. A ladle monitoring and tracking system has been put into operation and all information about the ladle is made available before the blowing is over. Both ladle life and ladle return time affect significantly the drop in steel temperature and hence it is important to take this in to account.

The suitable values of $\alpha_1 - \alpha_5$ and β in Eq. 1 were first estimated by tuning to the results of finite difference models. It was observed that the overall correlation coefficient was more than 0.93 (Fig. 5). Having established the appropriateness of equation, the values of $\alpha_1 - \alpha_5$ and β were then re-tuned on actual operational data. This equation thus obtained was used in the decision support system. The predicted and actual values of temperature at ARS for actual heats in the plant are shown in Fig. 6 and agree within ±20 C.

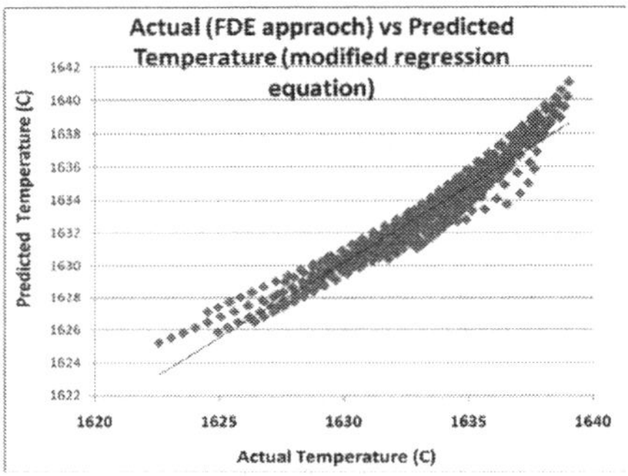

Fig. 5 Comparison of FDE predicted and regression equation prediced values

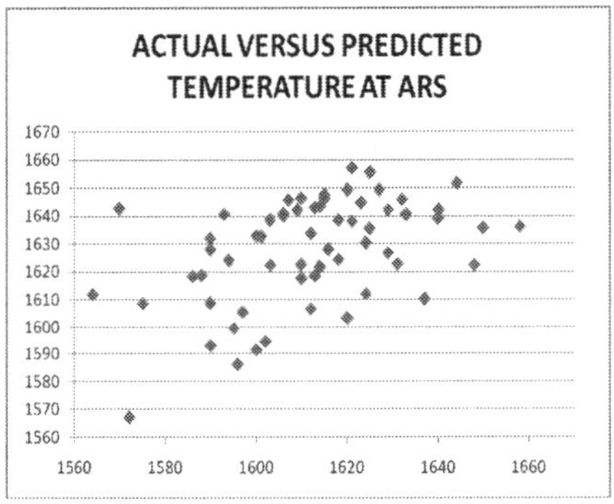

Fig. 6 Actual versus predicted temperature at ARS

The decision support system uses Eq. 1 in combination with tapping model and alloy additions model and is effective in prediction of temperature on arrival at ARS and this helps to ensure the prediction of metal temperature at turret as well by using the steel holding model before the start of casting. In case the temperature on arrival at ARS is below the minimum expected value, the particular heat is sent to IRUT or LF (almost in 30% of the cases). A slight adjustment in casting speed can be made to compensate for error in temperature at turret. The alloy recovery model helps to predict the chemical composition of metal but still in a large number of heats the prediction of carbon and phosphorus (due to reversion) is inaccurate and this leads to grade diversion and/or grade mixing in more than 30% of heats. Effort is being made to reduce this error by taking bath samples at first turndown and thereafter regulating the variability in the chemical composition of ferro-alloys added for composition adjustment during tapping and at also ARS.

Conclusions

A decision support system based on heat transfer models and shop logistics has been made. It is practically possible to predict the temperature on arrival at argon rising station within ±20 C. The model execution is guided by Lean principle and uses a nearly 'pull' system because the decision place is mid point of ladle transfer (at argon rinsing station) for tactical reasons of better control.

In the opinion of the authors, the scheduling issues of steel melting shop, due to its very nature of unexpected breakdowns and/or delays. cannot be solved through mathematical models of job-shop or flow-shop scheduling methods or by using optimization techniques such as linear programming, evolutionary methods (GA etc), and its variations. The Lean based approach aide with a decision support systems appears to be better and simpler alternative.

References

[1] L.Tang, J Liu, A.Rong, Z.Yang; A mathematical programming model for scheduling steelmaking-continuous casting production, European J. of Operations Research, 2000, Vol. 120, p. 423-435

[2] Sun Liang-Liang, Wang Xiu-ying; Application of mix optimization scheduling approach for steelmaking continuous casting process based on actual steelmaking industry J. of Iron and Steel Research International, 2013, 20(10), p. 01-09

[3] Brahma Deo, Aart Overbosch, Bert Snoeijer Debasis Das, Karumanchi Srinivas: Technical paper 2712, Trans IIM, published on-line July 4, 2013, to be assigned to October 2013

Celebrating the Megascale: Proceedings of the Extraction and Processing Division Symposium on Pyrometallurgy in Honor of David G.C. Robertson
Edited by: Phillip J. Mackey, Eric J. Grimsey, Rodney T. Jones, and Geoffrey A. Brooks
TMS (The Minerals, Metals & Materials Society), 2014

LADLE METALLURGY KINETICS: INCLUSION-INCLUSION REACTIONS

P. Chris Pistorius

Center for Iron and Steelmaking Research, Department of Materials Science and Engineering, Carnegie Mellon University, 5000 Forbes Avenue, Pittsburgh PA 15213, USA

Keywords: Mass transfer control, sulfide inclusions, steel, ladle metallurgy

Abstract

An example is presented to illustrate the joint effect of local reaction equilibria and mass transfer limitations, for reactions during ladle refining of steel. The example relies on some of the kinetic principles that David Robertson has employed to quantify many metallurgical processes. In calcium treatment of alumina inclusions in aluminum-killed steels, solid CaS forms as an intermediate reaction product. During subsequent reaction, CaS disappears and calcium aluminate forms; at the same time, aluminum and sulfur dissolve in the steel. Kinetic analysis shows that the rate of this reaction is not limited by mass transfer of dissolved aluminum and sulfur away from the reacting inclusions. The reaction rate is likely limited by transport of dissolved calcium. This example also illustrates the use of FactSage macros for kinetic modeling.

Introduction

Calcium treatment is often applied in the production of aluminum-killed steel; calcium is injected into liquid steel to convert alumina inclusions to molten or partially molten calcium aluminates [1]. Conversion of alumina to molten calcium aluminates avoids clogging of submerged-entry nozzles during continuous casting.

Recent work has confirmed that, for aluminum-killed steels containing more than approximately 10 parts per million (ppm) by mass of dissolved sulfur, the main product that forms first during calcium treatment is solid calcium sulfide [2]. As indicated in Figure 1, CaS is generally observed to be associated with alumina [2], suggesting that calcium sulfide nucleates on alumina inclusions. The calcium sulfide subsequently reacts with alumina inclusions to yield calcium aluminate inclusions, according to the following reaction (see also Figure 1):

$$3CaS + Al_2O_3 \rightarrow 3CaO + 3[S] + 2[Al] \quad (1)$$

In this reaction, solid CaS reacts with solid Al_2O_3 to form CaO; the CaO is present as various calcium aluminates. At a typical temperature of 1550°C, the expected sequence of calcium aluminate reaction products, as reaction (1) proceeds, is $CaO.6Al_2O_3$ (CA_6), then $CaO.2Al_2O_3$ (CA_2), $CaO.Al_2O_3$ (CA), and finally liquid $CaO\text{-}Al_2O_3$.

As reaction (1) indicates, for each reaction step the equilibrium product $[\%S]^3[\%Al]^2$ would have a constant value (assuming constant values of the activity coefficients of dissolved S and Al). As the reaction proceeds, and different oxide phases (with different CaO and Al_2O_3 activities) equilibrate, the equilibrium product $[\%S]^3[\%Al]^2$ changes (being smaller for later reaction steps, involving increased CaO activity, and decreased Al_2O_3 activity).

Sampling of liquid steel (both laboratory melts and plant heats) indicated that unreacted CaS typically persists in liquid steel for several minutes [2]. The kinetics of reaction (1) is important; persistence of CaS may lead to clogging of submerged-entry nozzles, because of incomplete conversion of alumina to molten aluminates, and because CaS itself can cause clogging.

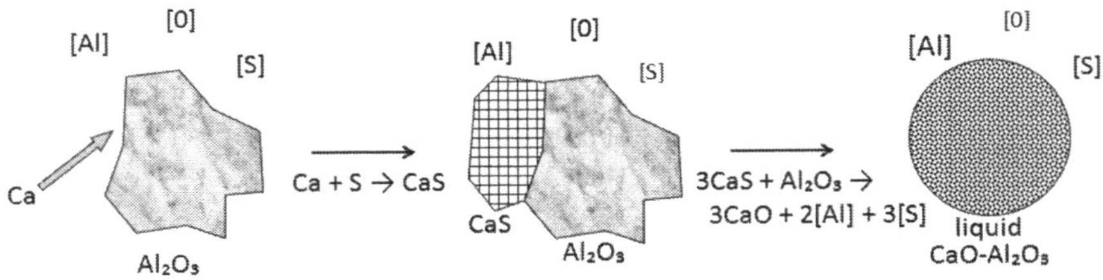

Figure 1. Schematic illustration of the reaction steps upon calcium treatment of sulfur-containing aluminum-killed steel. Solid calcium sulfide forms as the initial reaction production, and modified inclusions form by reaction of CaS with Al_2O_3.

As written, Reaction (1) is between two solids. The true reaction mechanism is expected to involve dissolved calcium in the liquid steel (surrounding the inclusions) as intermediate. The solution thermodynamics of calcium in liquid steel is not known accurately (published values for concentrations of dissolved calcium equilibrated with CaO and dissolved oxygen differ by several orders of magnitude) [3]. However, the dissolved calcium concentration is expected to be 1 ppm or less [4]. A low concentration of dissolved calcium would limit the rate of the CaS-Al_2O_3 reaction.

Another potentially rate-determining step is transport of Al and S (which form in reaction (1)) away from the inclusions into the bulk liquid steel. One indirect way to test whether the CaS-Al_2O_3 reaction (by way of dissolved Ca as intermediate) could be rate-determining is to calculate the expected rate of inclusion conversion, if transport of Al and S were to control the reaction rate. This paper reports the result of such a calculation.

Kinetic model

Equations

The main features of the kinetic model are illustrated in Figure 2. The inclusions are assumed to be mixtures of CaS and Al_2O_3 before reaction, changing to mixtures of CaS and calcium aluminates as the reaction proceeds. The inclusions are typically 0.5-4 μm in diameter. For such small inclusions, mass transfer between the bulk steel and the inclusion is dominated by diffusion under effectively stagnant conditions, for which the Sherwood number is 2 (assuming the inclusions to be approximately spherical); that is, the mass transfer coefficient between bulk steel and the inclusion surface is given by:

$$m = D/r \tag{2}$$

where m is the mass transfer coefficient, D the diffusivity in liquid steel (of Al and S in this case), and r is the radius of the inclusion.

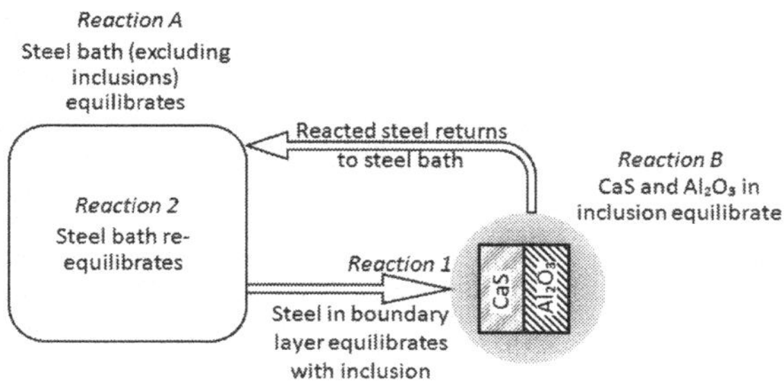

Figure 2. Flow diagram illustrating the principle of the kinetic model, for local equilibrium at the inclusion, and limited mass transfer between the steel around the inclusion and the bulk steel. Reactions A and B occur once to set up the initial conditions, and Reactions 1 and 2 take place repeatedly, once each time step.

It was assumed that reaction (1) is limited by mass transfer in the steel. In the model, mass transfer was quantified by transporting a finite volume of steel to each inclusion in each time step, allowing this volume of steel to equilibrate with the inclusion, then mixing the reacted steel back into the bulk steel, and repeating. Since it was assumed that reaction (1) reaches local equilibrium between the inclusion and the reacting steel volume, the solid-solid reaction (mediated by dissolved calcium) was assumed not to be rate-determining.

For each time increment Δt, the mass of steel which reacts with one inclusion is then given by:

$$\Delta M_{\text{steel}} = mA\rho_{\text{steel}}\Delta t = 4\pi\rho_{\text{steel}} rD\Delta t \tag{3}$$

If the total mass of inclusions in a given steel volume is $M_{\text{inclusion}}$, the total number of inclusions is

$$n_{\text{inclusion}} = M_{\text{inclusion}}/[(4/3)\pi r^3\rho_{\text{inclusion}}]. \tag{4}$$

The mass of steel reacting relative to the mass of the inclusions is then given by:

$$\Delta M_{\text{steel}}\, n_{\text{inclusion}} / M_{\text{inclusion}} = 3(\rho_{\text{steel}}/\rho_{\text{inclusion}})D\Delta t/r^2 \tag{5}$$

In this formulation, it is implicitly assumed that all the inclusions have approximately the same size and composition, and react in parallel.

From equation (5), the time required for the mass of steel reacted to be equal to the inclusion mass is given by:

$$\tau = r^2\rho_{\text{inclusion}}/(3D\rho_{\text{steel}}) \tag{6}$$

The numerical value of this characteristic time is small. For a typical values of r=1 μm, D=9×10^{-9} m^2/s [5], $\rho_{\text{inclusion}}$ = 2500 kg/m^3 and ρ_{steel} = 7000 kg/m^3, τ = 13 μs. However, because the concentrations of dissolved Al and S in equilibrium with the inclusion are small, the total time required to complete inclusion modification would be many multiples of the characteristic time. The kinetic model was used to estimate the required number of multiples.

FactSage macro model

The FactSage thermodynamic package [6,7] includes the option of using a macro to perform repetitive equilibrium calculations. Such a macro can model a mass transfer controlled reaction, since mass transfer control implies local equilibrium in the reaction volume. Quite complex processes can be modeled in this way; a recent example is a model of RH degassing [8].

The basic structure of the macro used here is illustrated in Figure 2: the macro calculations involve repeated equilibration of a small fraction of the steel mass (equation 3) with each inclusion, each time followed by mixing the reacted steel with the bulk volume of unreacted steel.

FactSage 6.4 was used for the calculations, employing the "Fe-liq" model (from the FTmisc database) for liquid steel, "ASlag" (from FToxid) for liquid slag, and also considering the pure solids CaS, Al_2O_3, CA_6, CA_2, CA and C_3A_2 as possible reaction products. The assumed conditions are listed in Table I.

Table I. Assumed conditions for model calculations

Temperature: 1550°C; pressure: 1 atm
Inclusion density: 2500 kg/m^3; steel density: 7000 kg/m^3
Initial steel composition: 0.03% Al, 50 ppm S, 4 ppm O, balance Fe
Initial inclusion composition: CaS and Al_2O_3, in mass ratio 0.7:1
Initial inclusion concentration in steel, expressed as total bound oxygen content: 20 ppm

Results and Discussion

A typical simulation result, for the conditions listed in Table I, is shown in Figure 3. The time is given as multiples of the characteristic time τ (defined in Equation (6)).

The transition of the phases in the oxide inclusions shows the expected sequence, namely alumina→ CA_6 → CA_2 → CA → liquid; the alumina inclusions are modified to calcium aluminates by reaction with CaS.

The time required for conversion of alumina inclusions to mostly liquid oxide inclusions depends on the assumed conditions, but is typically around 2000τ. In absolute terms, this is a short time – around 26 ms for inclusions with a typical radius of 1 μm. The implication is that reaction of CaS with Al_2O_3 is not limited by Al and S mass transfer in the liquid steel (since the reaction actually takes several minutes); this conclusion is consistent with the suggested limiting role of mass transfer of dissolved calcium between CaS and the oxide inclusions.

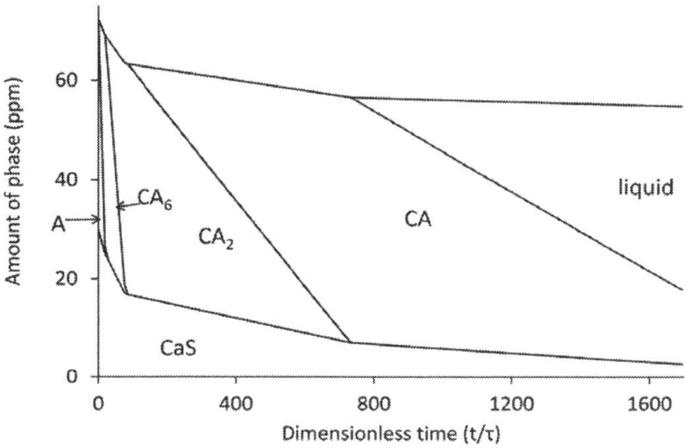

Figure 3. Calculated change in the amounts of the inclusion phases, for mass transfer controlled reaction of steel (composition as given in Table I) with inclusions initially consisting of Al_2O_3 and CaS.

As reaction (1) shows, consumption of solid CaS gives a direct measure of the extent of the inclusion modification reaction. For each step of inclusion reaction, the calculated conversion rate (as indicated by the rate of removal of CaS) is constant; see Figure 3, which shows that the CaS mass decreases linearly during each reaction step. (Here, each step is defined by the inclusion phases present: Al_2O_3-CA_6-CaS is one step, followed by the CA_6-CA_2-CaS step, and so forth.)

The reaction rate is constant for each step because it is assumed that modification is controlled by the rate of transport of dissolved Al and S away from the inclusion, with local equilibrium between dissolved Al and S and the inclusion phases. The concentrations of Al and S in local equilibrium with partially modified inclusions (i) remain constant while the same oxide phases are equilibrated with CaS and dissolved Al and S, and (ii) are high compared with the bulk concentrations (for the earlier reaction steps, involving solid calcium aluminates). As an example, one pair of equilibrium concentrations of Al and S in steel in contact with Al_2O_3, CA_6 and CaS at 1550°C is 0.16% Al and 0.28% S – much higher than their bulk concentrations. Because the concentrations of Al and S in the reaction volume around the inclusion are much higher than the (slowly changing) bulk concentrations, there are nearly constant concentration differences (of Al and S) between the reaction volume and bulk steel. The nearly constant concentration differences result in the observed constant reaction rate, for each reaction step.

The reaction rate is lower for each subsequent reaction step, because of lower equilibrium concentrations of Al and S. As an example, for the four-phase equilibrium steel-CaS-CA-liquid slag at 1550°C, an equilibrium combination is 0.035% Al and 0.0083%S (much lower than the values given above for steel-CaS-alumina-CA_6 equilibrium).

Conclusions

Reaction of CaS with alumina in calcium-treated steel, under control of Al and S mass transfer to the liquid steel surrounding the inclusions, can be readily modeled using a FactSage macro.

The results show that the CaS-alumina reaction would be very fast if it were controlled by Al and S mass transfer – the reaction would be complete in tens of milliseconds for typical conditions, compared with the observed reaction times of several minutes. Instead of rate control by Al and S mass transfer, the reaction must be limited by another reaction step, which is expected to be mass transfer of dissolved calcium (at low concentrations) between CaS and oxide inclusions.

Acknowledgments

This work is supported by the industrial members of the Center for Iron and Steelmaking Research; special thanks go to Dr Eugene Pretorius for helpful information on FactSage macro calculations.

References

1. G.M. Faulring, J.W. Farrell and D.C. Hilty, "Steel flow through nozzles: influence of calcium," *I&SM*, 7 (2) (1980), 14-20.
2. N. Verma, P.C. Pistorius, R.J. Fruehan, M. Potter, M. Lind and S.R. Story, "Transient inclusion evolution during modification of alumina inclusions by calcium in liquid steel: Part II. Results and discussion," *Metallurgical and Materials Transactions B*, 42 (2011), 720-729.
3. M. Hino, K. Ito, *Thermodynamic data for steelmaking* (Sendai, Japan: Tohuku University Press, 2010), 16-17.
4. W. Tiekink, R. Boertje, R. Boom, R. Kooter, and B. Deo, "Aspects of CaFe cored wire injection into steel," *ISSTech 2003 Conf. Proc*, Iron and Steel Society, Warrendale, PA, 2003, 157–164.
5. R.I.L. Guthrie, *Engineering in Process Metallurgy*, second edition (Oxford, United Kingdom: Clarendon Press, 1991), 64.
6. C.W. Bale, P. Chartrand, S.A. Decterov, G. Eriksson, K. Hack, R. Ben Mahfoud, J. Melançon, A.D. Pelton and S. Petersen, "FactSage Thermochemical Software and Databases." *Calphad*, 26 (2002), 189-228.
7. C. W. Bale, E. Bélisle, P. Chartrand, S. A. Decterov, G. Eriksson, K. Hack, I.-H. Jung, Y.-B. Kang, J. Melançon, A. D. Pelton, C. Robelin and S. Petersen, "FactSage Thermochemical Software and Databases – Recent Developments," *Calphad*, 33 (2009), 295-311.
8. M.-A. van Ende, Y.-M. Kim, M.-K. Cho, J. Choi and I.-H. Jung, "A kinetic model for the Ruhrstahl Heraeus (RH) degassing process," *Metallurgical and Materials Transactions B*, 42 (2011), 477-489.

Celebrating the Megascale: Proceedings of the Extraction and Processing Division Symposium on Pyrometallurgy in Honor of David G.C. Robertson
Edited by: Phillip J. Mackey, Eric J. Grimsey, Rodney T. Jones, and Geoffrey A. Brooks
TMS (The Minerals, Metals & Materials Society), 2014

VALORIZATION OF ELECTRICAL ARC FURNACE OXIDIZING SLAG

Joonho Lee[1], Sung Beom An[1], Minsoo Shin[1], Ki Joo Sim[1]

[1]Department of Materials Science and Engineering, Korea University
145 Anam-ro, Seongbuk-gu, Seoul, 136-713, Republic of Korea

Keywords: Electrical arc furnace, Oxidizing slag, Recovery of iron.

Abstract

Valorization of the electrical arc furnace oxidizing slag by coke was carried out by the reduction of iron oxide in the slag. Since the reduction of iron oxide by solid carbon is an endothermic reaction, supply of heat energy is required. In addition, slag has a low thermal conductivity; traditional fuel injection is not suitable for this reaction. Microwave can be used to supply heat energy to the system. Commercial oxidizing slag was reduced by coke under the microwave irradiation (1.7 kW, 2.45 GHz). The maximum recovery ratio of iron was 87%. In addition, slag composition modification was attempted to use the remaining slag for a mixture in cement clinker. Free-CaO in the slag was successfully suppressed, and the use of the remaining slag as a potential mixture was confirmed.

Introduction

Growth of slag generation is a social issue with the growth of steel industry in Republic of Korea. Steel production in Korea in 2012 was 69.1 million tons, which is continuously growing. Among them, approximately 38% (26 million tons) are produced in electrical arc furnace. With the growth of steelmaking industry, slag generation has been increased as well. In 2012, slag generation was 24.4 million tons; 13.7 million tons from the blast furnaces, 7.1 million tons from the converters, and 3.7 million tons from the electrical arc furnaces. (**Fig. 1**)

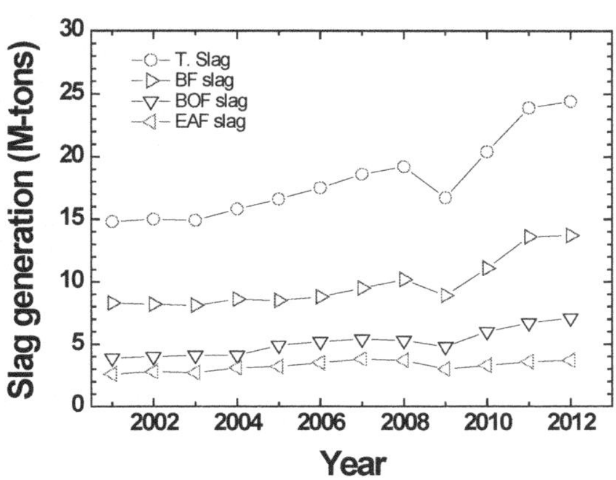

Figure 1. Domestic slag generation in Republic of Korea [1,2].

Most of the blast furnace slag is utilized as a high value-added raw material in cement clinker (about 75%). Some of the blast furnace slag is also used in road construction, backfill, and fertilizer, but the portions in these areas are very limited. On the other hand, the steelmaking slag is utilized as road construction (35%) and backfill (41%) [1, 2].

Table I shows the chemical composition of the various slags from individual processes. The basic oxygen steelmaking (BOF) slag and the electrical arc furnace (EAF) oxidizing slag contain high content of iron, which prevents the application of them as high-value added materials. Consequently, the reduction of iron oxide in the steelmaking slag is highly anticipated.

Table I. General chemical composition of the various slags [1, 2].

<table>
<tr><th colspan="2" rowspan="2">Classification</th><th colspan="4">Composition (wt%)</th><th rowspan="2">Applications</th></tr>
<tr><th>CaO</th><th>SiO_2</th><th>T.Fe</th><th>Al_2O_3</th></tr>
<tr><td rowspan="2">Blast furnace slag</td><td>Granulated</td><td rowspan="2">35-45</td><td rowspan="2">30-40</td><td rowspan="2"><1</td><td rowspan="2">10-20</td><td>Cement</td></tr>
<tr><td>Aircooled</td><td rowspan="4">Lower-value added material</td></tr>
<tr><td rowspan="3">Steelmaking slag</td><td>BOF</td><td>35-45</td><td>10-20</td><td>20-30</td><td><5</td></tr>
<tr><td>EAF oxidizing</td><td>20-30</td><td>10-20</td><td>20-30</td><td><10</td></tr>
<tr><td>EAF reducing</td><td>50-60</td><td>10-20</td><td><1</td><td>10-20</td></tr>
</table>

On the other hand, the reduction of iron oxide in the oxidizing slag by solid carbon is a strong endothermic reaction given by Eq. (1).

$$FeO(l) + C(s) = Fe(l) + CO(g) \qquad (\Delta H^{o} = 141.7 \text{ kJ [3]}) \qquad (1)$$

At a fixed temperature, heat energy required for the reduction of iron oxide in the commercial oxidizing slag can be calculated using the FactSage program [4]. Enthalpy required for the reduction of iron oxide with α part of carbon (weight part) was calculated at 1450°C and shown in **Fig. 2**. It is found that the heat supply is required during the reduction process continuously. The weight fraction changes of slag and reduced iron in this process are shown in **Fig. 3**. In the initial stage of the reduction, carbon content was very low and solid iron was produced. As the reaction proceeded, carbon content in the reduced iron increased, and finally form liquid Fe-C alloy.

In order for the reduction to proceed successfully, heat supply and mixing of the reducing agent carbon in the slag are essential. However, due to high viscosity and low thermal conductivity of slag, the conventional fuel injection technology cannot be applied [5, 6].

Recently, the microwave processing has been considered as an alternative method to supply heat energy in pyrometallurgical process. Once raw materials in powder are mixed with reducing agent carbon, microwave may penetrate through the raw materials (mostly oxides) and selectively heat the carbon. Due to the excellence in rapid and selective heating mechanism, many researchers used this technique to reduce metallic components from fly ash, sludge, mill scale and electrical arc furnace dust [7-12].

In this paper, our recent works on the reduction of the BOF slag and EAF oxidizing slag by carbonaceous materials under the microwave irradiation are reported [13-14].

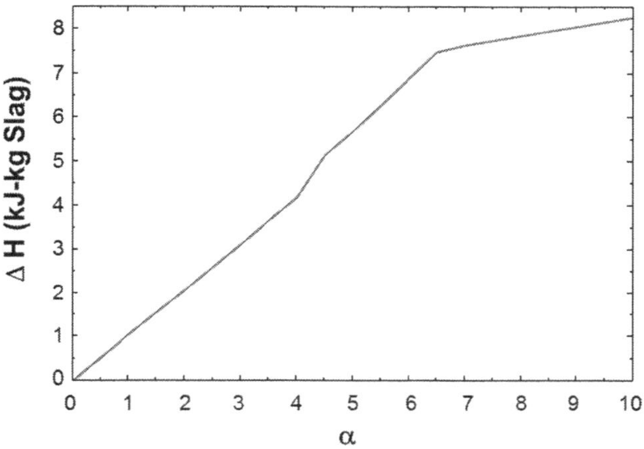

Figure 2. Energy consumption for the reduction of oxidizing slag by solid carbon.

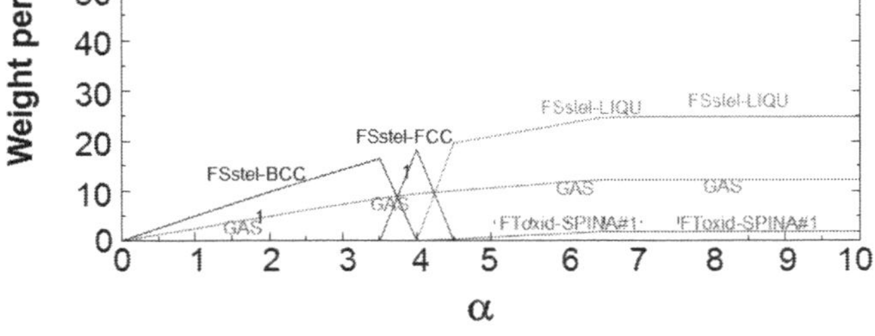

Figure 3. Reduction of the oxidizing slag by solid carbon.

Microwave Heating Fundamentals

Microwaves are electromagnetic waves with frequencies ranging from 300 MHz to 300 GHz (the wavelength is in the range of 1 mm to 1 m). The microwave may supply energy on the valence electrons, resulting in electron fluctuation [15]. This electron fluctuation is transferred to lattice vibrations, yielding heat energy. Consequently, the energy absorption in a material is expressed by Eq. (2) [15].

$$E = \omega \varepsilon_0 \varepsilon''_{eff} E^2_{rms} + \omega \mu_0 \mu''_{eff} H^2_{rms} \tag{2}$$

where ω is the angular frequency, ε_0 is the permittivity of free space (8.854×10^{-12} F/m), ε''_{eff} is the effective relative dielectric loss factor, E_{rms} is the root mean square of the electric field, μ_0 is the permeability of free space ($2\pi \times 10^{-7}$ H/m), μ''_{eff} is the effective relative magnetic loss factor, and H_{rms} is the root mean square of the magnetic field.

It is difficult to distinguish the effect of the electric field from that of the magnetic field, because they mutually affect with each other. As a result, "thermal runaway" behavior is frequently observed in the heating process of ceramic materials [16]. For carbon-oxide mixtures, it is considered that the heating is dominated by the selective heating of carbon [11-14]. Consequently, the particle size of carbonaceous material strongly affects the heating rate. **Fig. 4** shows the heating rate of coke-alumina powder mixtures of various sizes (1.7 kW, 2.45 GHz) [14]. As the coke particle size decreases, the heating rate and the maximum temperature increase. Coke particles smaller than 75 μm show a similar heating behavior of fine graphite particles. The maximum heating rate with coke in this study was 550 K/min [14].

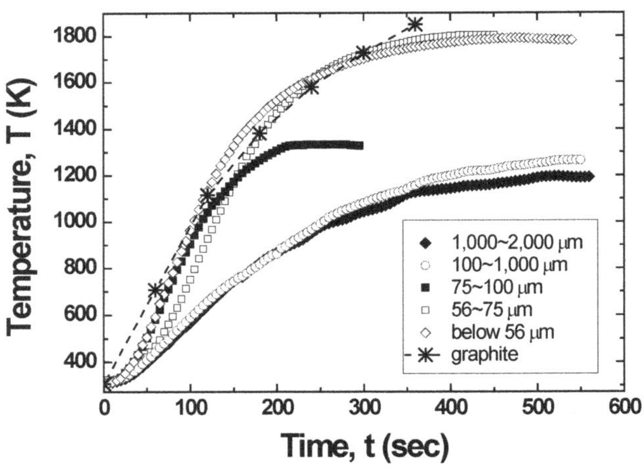

Figure 4. Heating behavior of coke-alumina powder mixtures [14].

Recovery of Metallic Iron from Slag

Recovery of metallic iron from BOF slag was studied by Kim and Lee, and Kang et al. [11,13]. Fig. 5 shows a typical cross-section of the sample after 15 min microwave processing. Reduced metal was clearly separated from the remaining slag [11]. **Fig. 6** shows the total Fe content in the BOF slag after microwave processing [11]. With increasing the processing time, the total Fe content decreases, and shows a minimum value at 600 s, then it increases again. In the microwave processing with carbon, the optimization of the processing time is important, because the reduced metal can be re-oxidized after full-consumption of carbon [11-14].

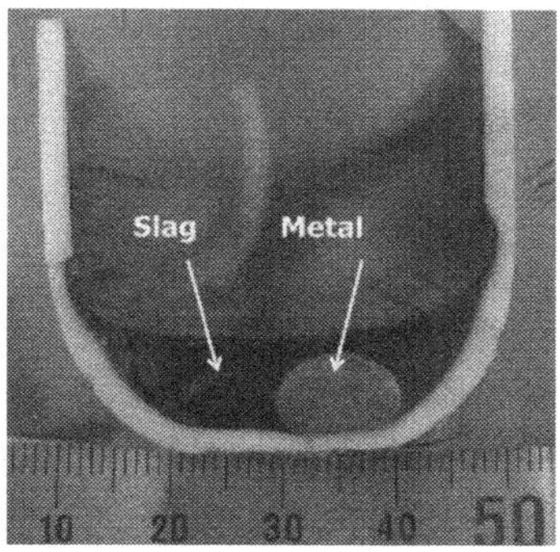

Figure 5. Typical cross-section of the sample after 15 min microwave heating experiments [11].

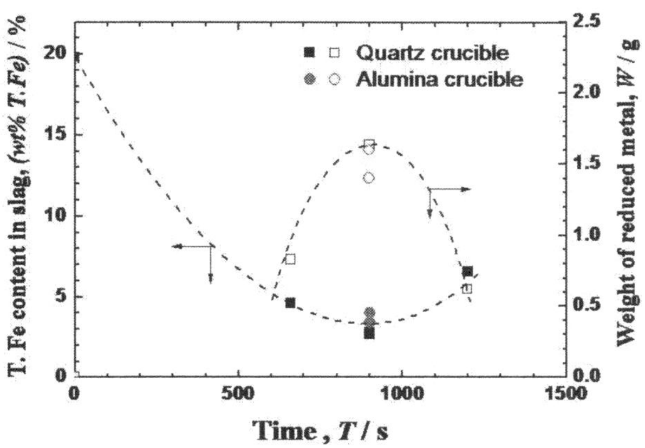

Figure 6. Total Fe content in BOF Slag and the amount of reduced iron by the microwave processing as a function of time [11].

Recovery of metallic iron from EAF oxidizing slag was studied by An et al. [14]. Since the EAF slag contains large amount of free-CaO, it was separated according to the particles size with screens in advance. Free-CaO was separated as find particles, and the remaining slag granules were in the size range of 10-40 mm in diameter, showing a relatively low content of CaO. These granules were used in the microwave processing by controlling the amount of CaO added.

Fig. 7 shows the recovery ratio of Fe from the EAF oxidizing slag as a function of the microwave processing time [14]. When the weight ratio for slag/CaO/coke was 67/26/7, the maximum recovery rate of Fe (87%) was achieved at 25 min. At a fixed microwave processing time, the recovery ratio of Fe depended on the amount of coke (**Fig. 8**).

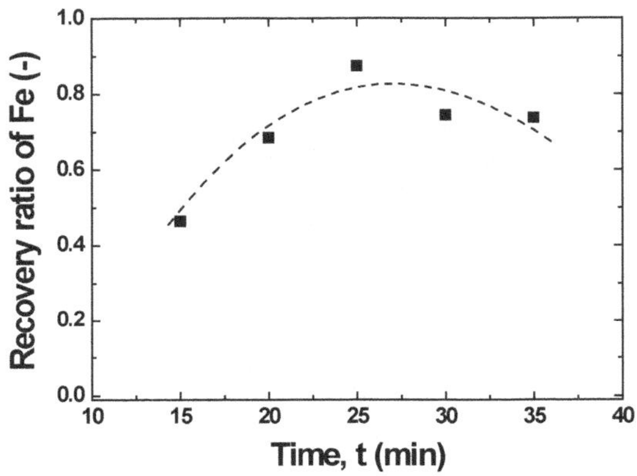

Figure 7. The recovery ratio of Fe from EAF oxidizing slag with respect to the microwave irradiation time [14].

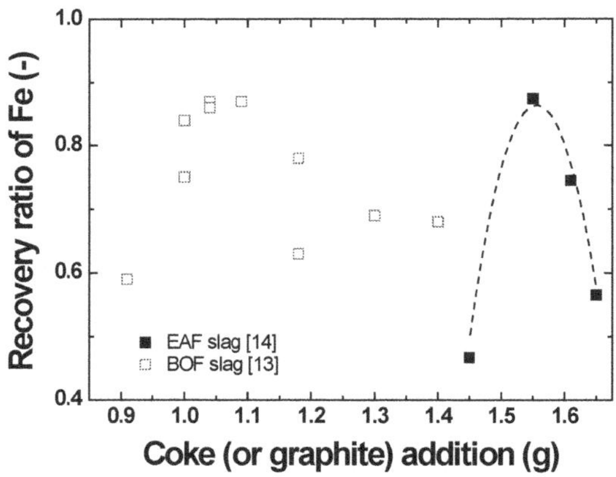

Figure 8. Recovery ratio of Fe from BOF and EAF oxidizing slag (15 g) with respect to the amount of carbonaceous material added [13, 14].

Fig. 9 shows an XRD analysis result of a typical remaining slag. Free-CaO in the slag was successfully suppressed, and the use of the remaining slag as a potential mixture was confirmed.

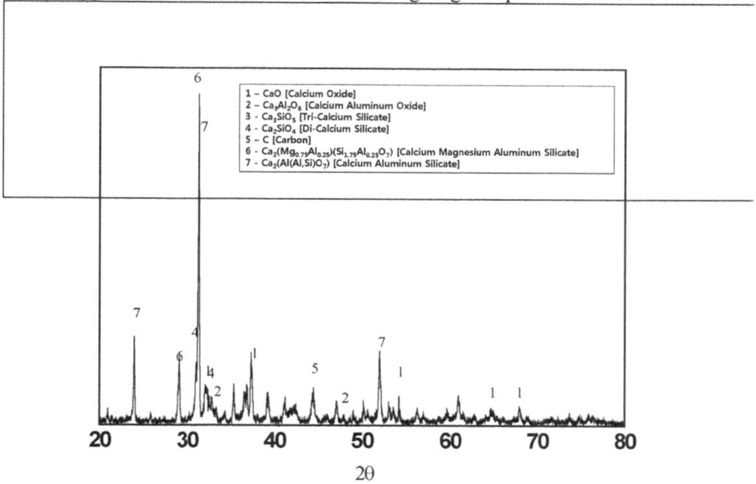

Fig. 9 XRD analysis result of the remaining slag of the EAF oxidizing slag after the microwave processing [14].

Concluding Remarks

Valorization of EAF oxidizing slag is a challenge for building sustainable society. Due to high viscosity and low thermal conductivity, the microwave heating on the carbon-slag mixture has a potential for practical applications to reduce metallic Fe. In order to increase the value of EAF oxidizing slag, the slag composition should be modified to SiO_2-rich or CaO-rich region (**Fig. 10**). Based on the present experimental results, the use of the remaining slag as a potential mixture in cement clinker was confirmed.

Acknowledgements

This work was supported by the Industrial Strategy Technology Development (No. 10035258, Development of green process for production of eco-steel slag) through a grant provided by the Ministry of Knowledge Economy, Korea. We appreciate kind supply of EAF slag from Hyundai Steels. Thank Prof. David Robertson for reviewing this paper with his kind comments.

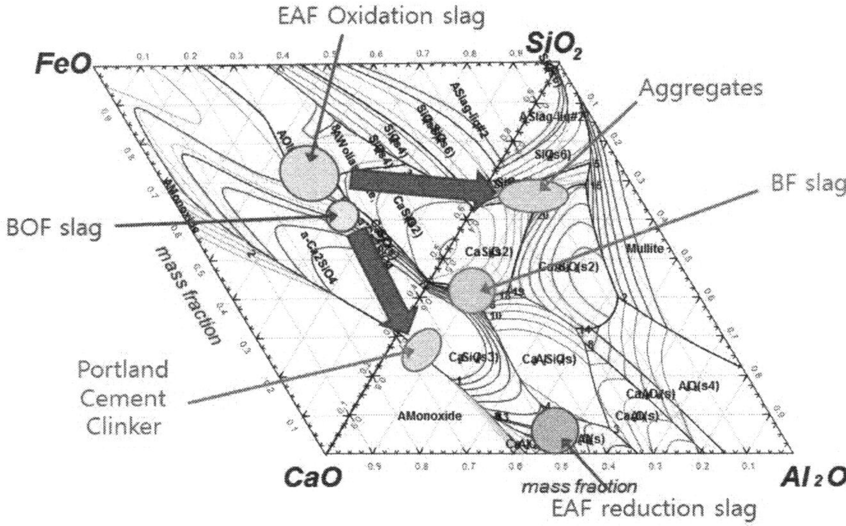

Figure 10. Perspective of valorization of metallurgical slag for high value-added materials.

References

1. J. Lee et al., "Valorization of steelmaking slag", *Euromat*, Montpellier, France, 2011.

2. I. Sohn et al., "Development of an ECO slag processing technology for iron recovery and value-added products in steelmaking", *The 7th Euroslag conference,* Ijmuiden, The Netherlands, 2013.

3. E.T. Turkdogan, *Physical Chemistry of High Temperature Technology*, (New York: Academic Press, 1980) 7, 11.

4. C. W. Bale et al., "FactSage Thermochemical Software and Databases – Recent Developments", *Calphad,* 33(2) (2009), 295-311.

5. L. Zhang and S. Jahanshahi, "Review and modeling of viscosity of silicate melts: Part II. Viscosity of melts containing iron oxide in the CaO-MgO-MnO-FeO-Fe_2O_3-SiO_2 system", *Metall. Mater. Trans. B*, 29 (1998), 187-195.

6. Y. Kang et al., "Thermal conductivity of the molten CaO-SiO_2-FeO_x system", *Metall. Mater. Trans. B*, 43(6) (2012) 1420-1426.

7. K. Morita et al., "Immobilization of ash by microwave heating", *Proceeding of the 11th Incineration Conference*, Albuquerque, Univ. California Irvine, Irvine, 1992, 507-514.

8. K. Morita et al., "Resurrection of the iron and phosphorus resource in steel-making slag", *J. Mater. Cycles Waste Manag.*, 4 (2002), 93-101.

9. S. Cho and J. Lee, "Metal recovery from stainless mill scale by microwave heating", *Met. Mater. Int.*, 14(2) (2008), 193-196.

10. X. Sun et al., "Petroleum coke particle size effects on the treatment of EAF dust through microwave heating", *J. Min. Mater. Charact. Eng.*, 8(4) (2009) 249-259.

11. T. Kim and J. Lee, "Recovery of Fe and P from CaO-SiO_2-Fe_tO-P_2O_5 slag by microwave treatment", *Mater. Trans.*, 52(12) (2011), 2233-2238.

12. E. Kim et al., "Reduction of electric arc furnace dusts by graphite powders under microwave irradiation", *Ironmaking Steelmaking*, 39(1) (2012) 45-50.

13. Y.B. Kang, T. Kim and J. Lee, "Valorisation of steelmaking slag by microwave treatment", Ironmaking Steelmaking, 39(7) (2012), 498-503.

14. S.B. An et al., "Reduction of electrical arc furnace oxidizing slag by coke under microwave irradiation", *Met. Mater. Int.*, 2014, in press.

15. M. Gupta and E. Wong, *Microwave heating, In: Microwaves and Metals*, (Singapore: John & Wiley & Sons, 2007) 43-64.

16. J. Lee and T. Kim, "Application of microwave heating to recover metallic elements from industrial waste, *Advances in induction and microwave heating of mineral and organic materials*, Intech, Rijeka, Croatia, 2011, 303-312.

CELEBRATING THE MEGASCALE

Proceedings of the Extraction and Processing Division Symposium on Pyrometallurgy in Honor of David G.C. Robertson

Process Modeling and Simulation

Celebrating the Megascale: Proceedings of the Extraction and Processing Division Symposium on Pyrometallurgy in Honor of David G.C. Robertson
Edited by: Phillip J. Mackey, Eric J. Grimsey, Rodney T. Jones, and Geoffrey A. Brooks
TMS (The Minerals, Metals & Materials Society), 2014

COMPUTATIONAL MODELLING OF METALLURGICAL PROCESSES: ACHIEVEMENTS AND CHALLENGES

Mark Cross[1], Diane McBride[1] and Nick Croft[1]

[1] College of Engineering, Swansea University, Swansea SA2 8PP, UK

Keywords: Computational modeling, high performance computing, metallurgical processes

Abstract

Extractive metallurgical processes rate amongst the most complex from the perspective of computational modeling. They typically involve multi-phase and multi-component fluid flow in very complex geometries, heat transfer driven by a number of interacting phenomena, solid-liquid-gaseous phase change and mass transfer together with complex thermodynamics and associated chemical reactions. Beyond the model building itself, key challenges have always involved being able to identify the phenomena present together with interactions to characterize processes and experimental laboratory and plant data to parameterize the arising models – it is in this milieu, which require considerable process understanding and subtlety of thought, that David Robertson has made his contributions.

We will review the achievements, over the last couple of decades, in computational modeling of extractive metallurgical processes and address some of the remaining challenges to enable simulation based process design, as we move through the 21st century.

Introduction

The first author here came across the work of David Robertson in 1981 [1] when he published a paper describing the CFD modeling of the behavior of gas when injected into a liquid bath, a phenomenon of enduring interest in metallurgical processing. This paper illustrates the key challenges of computational modeling in metallurgical processing – the physical phenomena are complex and essentially multi-phase, representing a significant challenge in CFD modeling, where even a simple two dimensional geometry, with what today is regarded as a very crude mesh (37*22), demanded considerable CPU time on one of the leading edge (CDC 6400) computers of the time, and finally the effort involved in verifying the model against experimental data. That is to say – computational modeling of metallurgical processing has always been regarded as challenging from the perspectives of the physics and chemistry involved, requiring the formulation of the mathematical models, the numerical solution procedures to generate model solutions, significant compute effort to facilitate such solutions, and the considerable challenge of validating the model against experimental data in any meaningful manner. In this paper Robertson joined the early pioneers of serious computational modeling – Julian Szekely, Rocky Sohn, Keith Brimacombe, Jim Evans and Rod Guthrie amongst others, who were amongst the first to try and capture the full spectrum of important phenomena in the analysis of real metallurgical processes. They were involved in what can only be classed as heroic computing – nothing was straightforward, in every problem something relating to modeling software technology had to be developed and results were never guaranteed.

Computational modeling emerged as a discipline in the late 1960s and was initially dominated by the analysis of the heat transfer aspects of processes, whereby either the fluid flow was easy to specify, (such as in iron ore pellet induration) or where the convection issues were considerably simplified (such as in continuous or DC casting). Whilst such models have proved their utility in enabling a more rational understanding of processes to be gathered and many years later as a useful basis for process control models, as CFD and solid mechanics technology emerged, researchers became more adventurous in trying to capture the combined thermo-fluid or thermo-structural behavior of processes. This was facilitated by numerical procedures and eventually commercial software that enabled this coupling reasonably well.

During the 1950 and 60s a substantial amount of work had been done on reaction chemistry, and much of the basic process chemistry had been delineated for most of the major metallurgical processes. Of course a good deal of these phenomena involved solid particles of 'processed' ore which were reduced and melted as they were processed. A great deal of work on the understanding and modeling of single particle phenomena was published during this period by Wadsworth, Philbrook, Nateson, Schulz amongst and then summarized by Szekely, Sohn and Evans in the book on gas-solid reactions [2]. Much of this work was later embedded in more comprehensive process models, capturing all the major phenomena including the thermodynamics for a variety of metallurgical operations. It is these sub-models, solved time and time again at each element or cell of the mesh that substantially increased solution times, adding to the overall nonlinearity of the model and making convergence even more difficult to achieve.

Much of the initial work was in two dimensions and in simplified geometries because of three key limitations:

a) In the thermo-fluid context, the solver technologies were based on structured finite volume procedures and this limited the complexity of geometries that could be captured in a practical sense
b) In the thermo-structural context, although the solver technologies used finite element procedures and so unstructured meshes, the mesh generators themselves were very limited in capability
c) The final limitation, of course, was the computational capabilities available at the time – even if you were able to generate a million element mesh, it could not be stored on the associated disk or run in the processor CPU.

The impact of Computer Technology

In the mid 1970s Young et al [3] developed a computational model of iron ore pellet induration systems. These operations consisted of a moving grate (or possibly a combined grate and kiln) with a packed bed of pellets, initially green (so at room temperature and containing around 10% moisture) which was subject to a heating or cooling gas stream across the bed as illustrated in Figure 1(a) below. Through this mechanism the bed was heated, dried, fired and cooled, whilst the gas streams entered the system at the final stage of the pellet processing, where they gathered heat from the cooling bed, were further heated from burners above the bed and finally cooled again before being exhausted. This highly recuperative process, see Figure 1(b) showing the gas streams, was modeled as a packed bed with heat transfer and a range of chemical reactions (evaporation, oxidation, limestone decomposition, etc) with shrinking core sub-models. The overall mesh was about 2000 cells/elements and to capture the effects of the coupling through the on-gas temperatures into each zone, it was solved about 10 times as a transient problem. The flow of the pellet bed was static on the grate which itself moved along, whilst the gas flowed tranversely across the bed either from the top or the bottom, and the flow could be specified within the bed. Outside of the bed it was assumed to be well mixed.

When this model was first assembled and run it took some 30 hours to reach a stable solution on what was then a latest generation DEC PDP11/45. Figure 2 shows how this runtime decreased over the following 20 years as the numerical methods were radically improved and the software re-written and migrated to PC hardware, which improved dramatically according to Moore's Law. This software now runs in a couple of seconds on current PCs. The story with regard to memory and data space is similar – the original system had 16KB of processor memory and each component of the software was loaded separately, run and results printed to a disk, whilst the code was overwritten by the following software component in the sequence.

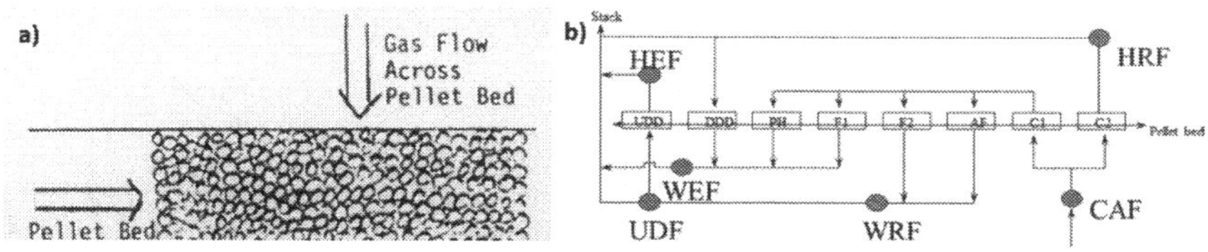

Figure 1 A brief illustration of the iron ore pellet induration model a) within zone structure and b) overall operational structure

It is worth noting that until the 1980s there were no reliable mesh generators, geometry definition was a significant challenge (no link to the CAD, and often no CAD anyway), visualization of the results was always a challenge and extremely time consuming. It is really from the 1990s onwards that accessible tools for geometry capture, mesh generation and visualization emerged to support the process modeler in using his core modeling software, which by now enabled much more nonlinearity and embedding of sub-phenomena (such as chemically reacting solid particles).

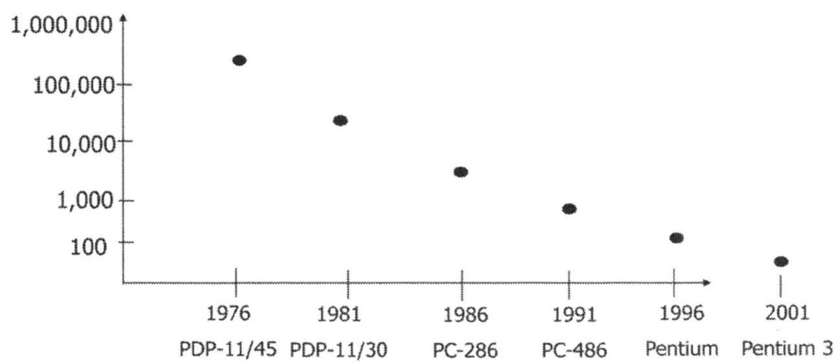

Figure 2 the improvement in the solution time for the induration model from 1976 to 2001

As Moore's law became ever more challenging to achieve with increases in single processor speed, then computer manufacturers moved towards parallel cluster systems – which were essentially a network of conventional PC processors, connected by cable with a fast bandwidth and also low latency (the time it takes to start passing data between processors). These systems, which for computational modeling needed to be high bandwidth – low latency, required the programmer to explicitly map his application onto the parallel cluster system. Most CFD codes were amenable to this through mesh partitioning, but this was not so readily the case for thermal structure codes. However, in recent years multi-core processors have emerged whereby essentially a number of CPU's on a single chip can share significant memory, and operate on a shared memory database, and this has facilitated parallel computing across the board – at least for modest scalability. Hence, it is now common to see models involving complex geometries, complex interacting physics and very fine meshes, computing significant numbers of transient phenomena in a manner hardly imagined 20 and certainly 30 years ago.

Centrifugal Casting of Ti-Al Blades – A Contemporary Challenge

One significant challenge that the authors have engaged in over the last decade is on the simulation of developments of a novel centrifugal casting process for the difficult to cast Ti-Al for aerospace turbine blades. This work was performed collaboratively with Prof Nick Green and his team at the University of Birmingham sponsored by Rolls Royce, and much of it is summarized in the following [4-10]. The CFD model developed here describes the very complex and discriminating geometries of the blades and their running and gating systems, it captures the very complex bubble and droplet behavior in a rapidly rotating environment. The model therefore has to capture the complex free surface behavior simultaneously with the cooling and solidification of the metal as it travels through the running and gating system into the mould. The flow behavior is extremely turbulent and conventional free surface capturing algorithms typical of typical commercial CFD codes were not adequate to capture the physics in anything like a discriminating enough manner. Hence, a significant effort went into the refinement of VOF based procedures to be both robust and discriminating enough for this application, for arbitrarily complex meshes in the rotating context. In this sense the work is an inheritor from where this paper started – the capture of bubble behavior in a complex environment.

The impact of the numerical improvements is shown in comparison with some experiments of a rotating vessel into which colored water is injected [7]. Figure 3 shows the bubbles of gas trapped within the pour and fluid volume for a number of different inlet conditions. Capturing the dynamics of the liquid pour in the model inlet boundary condition is critical in order to capture the gas entrainment and to accurately predict the flow dynamics. The volume of trapped gas, taking the shape of a torus, may have a volume five orders of magnitude greater than a typical sub-millimetre bubble, therefore significantly increasing the total volume of the gas/liquid mix. Figure 3b) and 3c) show the liquid entering the cylinder through a funnel. Figure 3d) employs a constant boundary inlet condition and Figure 3e) an oscillating inlet diameter with fluctuating flow conditions. The variable inlet condition was derived by tracking the discrete fluctuations in the experimental pour. The fluctuating flow condition allows the volume of gas, which is trapped beneath a layer of liquid, to escape better representing reality, where disturbances in the fluid surface naturally allow the gas to escape. Figure 3e shows that by introducing the variable inlet condition, the large re-circulating volume is broken up into smaller bubbles, which can escape the bulk volume, better representing reality (Figure 3a)

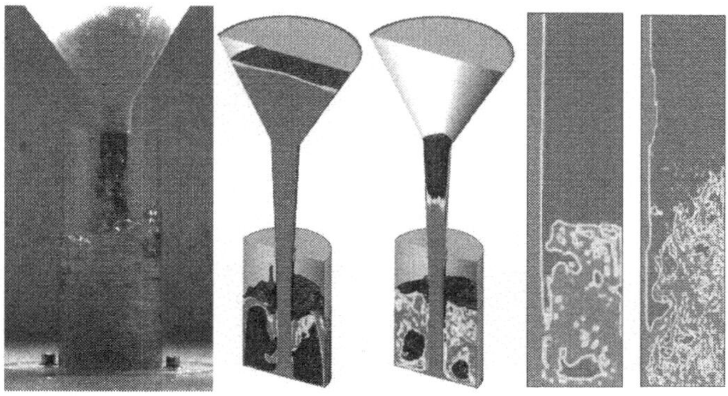

Figure 3: a) Experimental pour b,c) Funnel inlet d) Constant inlet e) variable inlet [7]

This instability, added to the variable inlet condition, creates a pour far closer to reality than the simple mass flow inlet. Introducing instabilities to the flow in the form of additional gas might be expected to increase the gas entrainment in any subsequent running system, however, similarly to the cylinder case, the ability to break up idealised entrainment events in the flow allow the resulting filling to show greater fidelity in terms of free surface capture, and reduced bubble formation. Figure 4 shows the effect of introducing the offset inlet, note in particular the lack of smeared liquid in the runner in Figure 4a compared to 4b.

Figure 4: Test geometry filling 1s a) Offset inlet b) Surface inlet [7]

These enhancements were then included in the full casting and solidification model and then used in the analysis of full scale case studies [10]. Figure 5 shows a snapshot of some simulation results when the cast is part way through the fill – where it is clearly shown how far the metal has reached, what the extent of the solidification is and the extent to which there are still micro-

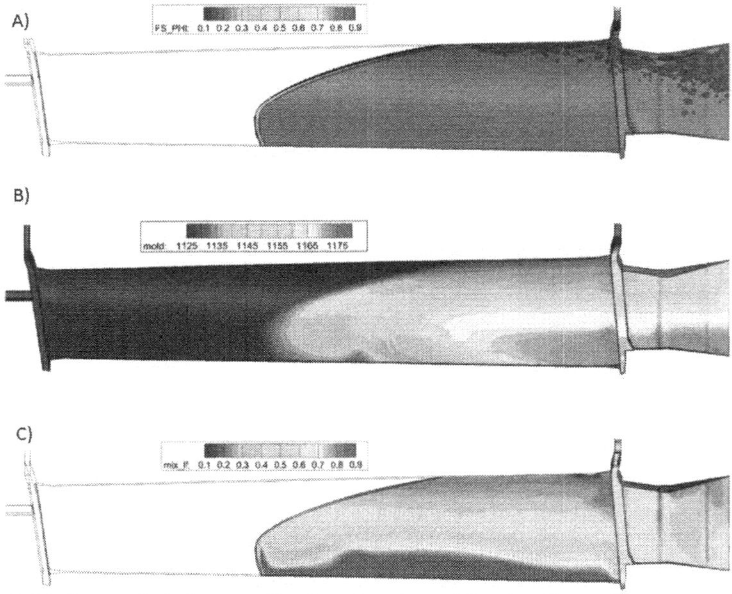

Figure 5: Blade growth during centrifugal casting a) metal fraction, b) metal temperature and c) solid fraction [10]

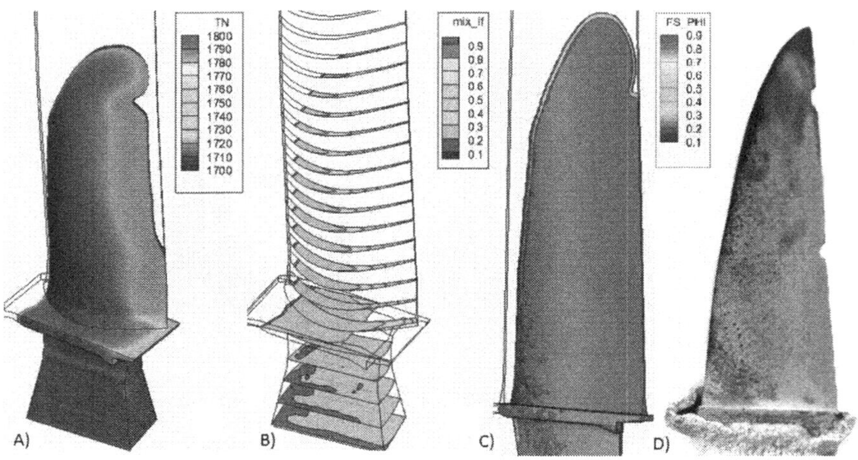

Figure 6 A similar cast with a reduced superheat and comparison with the equivalent experimental cast – a) metal temperature, b) solid fraction through the cast, c) fraction solid and d) the experimental measurement [10]

bubbles entering into the cast. The question is, of course, to what extent these remain and compromise the quality of the final product when it has fully solidified?

The turbine blade is cast at reduced superheat and the temperature of the metal and resulting solid fraction is shown in Figure 6a) and 6b) respectively. The fluidity of the metal at the cooling interface is reduced and the momentum of the moving metal front is restricted, thus causing the solidifying metal to fold over on itself introducing a defect into the cast. This defect can clearly be seen in the experimental casts, Figure 6d). The sharp resolution, with minimal smearing of the metal-gas interface, is clearly seen by the iso-surfaces plotted at metal fraction of 0.1 and 0.9. Both in the experimental cast and simulation results, small bubbles can be seen in the root of the blade which may result in defects occurring during the solidification.

These simulations required very complex meshes which typically have of the order of a million elements involving a mix of tetrahedral, wedge and hexahedral elements, requiring very small time steps, in the order of 10^{-4} to 10^{-6} to satisfy the smallest CFL number in the mesh The simulation typically takes about 36 hours on a PC cluster employing 16 processors on a Intel x5560 2.93 MHz Nehalem processor with 10q Myrinet switch configuration, dual quad core with 8G memory per node and high performance switches to minimize the latency.

The point of this example is that it shows how the challenge has moved from Robertson's 1981 paper where considerable effort was expended to capture in two dimensions the detailed flow physics of one bubble in a liquid bath, whereas here a whole array of phenomena are captured in fully three dimensional scenarios with complex geometries involving bubble swarms in a free surface flow setting subject to body forces and simultaneous cooling and solidifying.

Conclusions

The objective of this paper has been to provide a perspective on the way in which computational modeling of metallurgical processes has advanced over the last 40 years – coincidentally much of the length of the career in whose name this conference is organized and of which David Robertson has played his part. Although much has been achieved and facilitated through the broader contributions of those involved in the computational science venture, there is still much to do. Processes do not understand that they should be either thermo-fluid or thermo-structural – they are genuinely multi-physics and it is important for the metallurgical industry that this gap is bridged by the simulation community. Current simulation software technologies also do not readily cope with mixed particulate and continuum systems except in limited cases. However, there are now very strong discrete particulate modeling systems available and it would seem vital to ensure these are integrated with continuum software in such a way as to enable running on high performance clusters to enable increasingly large and more detailed analyses.

Acknowledgements

It is a pleasure to acknowledge the contribution of a number of former colleagues and collaborators to this work, especially Prof Nick Green, Dr Nick Humphries, Dr Dimitri Shevchenko and Dr Paul Withey.

References

1. M Nilmani, T T Maxwell, D G C Robertson and D B Spalding, Prediction of initial motion of gas bubbles in liquids, *Appl Math Modelling*, **5**, 24-28 (1981)

2. J Szekely, J W Evans and H Y Sohn, *Gas-Solid Reactions*, pub Academic Press (1976)
3. R W Young, M Cross and R D Gibson, A Dynamic Model of the Grate Kiln Cooler Firing Process of Iron Ore Pellets, *Ironmaking and Steelmaking*, **6**, 1-13 (1979)
4. D McBride, N Humphreys, T N Croft, N R Green and M Cross, Complex free surface flows for mould filling using centrifugal casting, *Modeling of Casting, Welding and Advanced Solidification Processes – X11 (eds. S L Cockcroft and D M Maijer), pub TMS*, pp. 77-84 (2009)
5. D Shevchenko, D McBride, N Humphreys, N Croft, P Withey, N Green, M Cross, Centrifugal casting of complex geometries: computational modelling and validation experiments, *Modeling of Casting, Welding and Advanced Solidification Processes – X11 (eds. S L Cockcroft and D M Maijer), pub TMS*, pp. 459-466 (2009)
6. D McBride, T N Croft, D Shevchenko, N Humphreys, P Withey, N Green and M Cross, Centrifugal casting of complex geometries: computational modelling and validation experiments, *Jim Evans Honorary Symposium* (Eds B Q Li et al), Pub TMS, pp 139-146 (2010)
7. N. J. Humphreys, D. McBride, D. M. Shevchenko, T.N. Croft, P. Withey, N. R. Green and M. Cross, Modelling Centrifugal Casting: The Challenges And Validation, International Symposium in Liquid Metals Processing and Casting, Nancy France, September 2011, *LMPC Proceedings, LMPC 2011, Proceedings of the 2011 International Symposium on Liquid Metal Processing and Casting*, SF2M/CNRS, Paris, France, pp 297-304 (2011)
8. N. J. Humphreys, D. McBride, D. M. Shevchenko, T.N. Croft, P. Withey, N. R. Green and M. Cross, Modelling Complex Free Surface Flows And Solidification In Conventional Casting Processes To Aid Centrifugal Casting Models, *McWASP 2012 Conference Proceedings*, July 2012
9. D. McBride, N. J. Humphreys, T. N. Croft, N. R. Green and M. Cross, Complex Free Surface Flows In Centrifugal Casting: Computational Modelling And Validation Experiments, *Computers and Fluids*, **82**, 63-72 (2013)
10. N. J. Humphreys, D. McBride, D. M. Shevchenko, T.N. Croft, P. Withey, N. R. Green and M. Cross, Modelling and Validation: Casting Of Al And TiAl Alloys in Gravity and Centrifugal Casting Processes, *Applied Math Modelling*, **37**, 7633-7643 (2013)

Celebrating the Megascale: Proceedings of the Extraction and Processing Division Symposium on Pyrometallurgy in Honor of David G.C. Robertson
Edited by: Phillip J. Mackey, Eric J. Grimsey, Rodney T. Jones, and Geoffrey A. Brooks
TMS (The Minerals, Metals & Materials Society), 2014

METALLURGICAL PLANT OPTIMIZATION THROUGH THE USE OF FLOWSHEET SIMULATION MODELLING

Mark William Kennedy
Proval Partners SA, 1004 Lausanne, Switzerland
Communicating author: m.kennedy@provalp.com

Keywords: flowsheet, model, Metsim

Abstract

Modern metallurgical plants typically have complex flowsheets and operate on a continuous basis. Real time interactions within such processes can be complex and the impacts of streams such as recycles on process efficiency and stability can be highly unexpected prior to actual operation. Current desktop computing power, combined with state-of-the-art flowsheet simulation software like Metsim, allow for thorough analysis of designs to explore the interaction between operating rate, heat and mass balances and in particular the potential negative impact of recycles. Using plant information systems, it is possible to combine real plant data with simple steady state models, using dynamic data exchange links to allow for near real time de-bottlenecking of operations. Accurate analytical results can also be combined with detailed unit operations models to allow for feed-forward model-based-control. This paper will explore some examples of the application of Metsim to real world engineering and plant operational issues.

Introduction

Modern metallurgical plants are increasingly complex; often incorporating closely coupled 'high intensity' unit operations, interacting with complicated pollution control systems. It is difficult for the designers of such systems to predict all of the interactions, particularly when combined with single or multiple recycle streams. When such a system is also associated with multiple, complex or impure feed stocks, the result during operations can be a constant hunt for 'steady state', reduced operating rate, energy efficiency, product quality and/or recovery.

Flowsheet modeling is a powerful tool to improve metallurgical plant performance at different project stages, including: (i) research and development, (ii) design, and (iii) plant start-up/ramp-up/operation and continuous improvement. The focus of this paper will be on: When, Why, and What can be usefully modeled at different project stages. How to model is specific to the chosen software and will therefore be discussed only to a limited extent.

Many different flowsheet simulators are commercially available. Some simulators are of a more 'general' nature and intended for application to many types of processes, e.g.: Aspen HYSYS or Aspen Plus [1], CHEMCAD [2], Cycad process [3], HSC Sim [4, 5], IDEAS [6, 7], Metsim [8], Pro/II [9], SysCAD [10], etc., while others are more specific, e.g. for mining/mineral processing: JKSimMet, JKSimfloat, USIMPAC, AggFlow, etc. Metsim is used as an example in this article, due to the direct experience of the author and this is not intended to imply that Metsim is the best or only choice for any particular application.

Various project stages are listed in Table I and related to the possible applications of flowsheet models and the quality of data that is typically available (given on a scale of 1-10) during each project stage. The scale of data quality is explained in Table II.

Table I. Some Applications of Flowsheet Modeling at Different Project Stages, Related to Available Data 'Quality'

Project Stage	Idea	Lab Scale	Pilot Plant	Scoping Study	Demonstration Plant	Pre-feasibility Study	Basic Engineering	Detailled Engineering	Commissioning	Ramp-up	Early Operation	Continuous Improvement
Quality of Data	1	2	3	4	5	5	6	7	7	8	9	10
Applications of Flowsheet Models												
Technical feasibility of ideas	X	X	X	X	X	X	X	X	X	X	X	X
Economic analysis of concepts	X	X	X	X	X	X	X	X				X
Equipment sizing			X	X	X	X	X	X				X
Trade-off studies	X			X		X	X					X
Process control strategy testing			X	X	X	X	X	X		X	X	X
Emergency scenario planning			X		X		X	X				
Commissioning scenarios					X		X	X				
Ramp-up check					X		X	X				
Feed forward control			X		X					X	X	X
Process improvements			X		X					X	X	X

Table II. Description of Data 'Quality', Types and Relation to Project Status

Quality of Data	Description of Data Types and Project Status
1	Guesses (hypotheses), theory, literature, order of magnitude 'models' or estimates.
2	Basic calculations, conceptual models, beaker scale measurements.
3	Advanced calculations, semi-valid mathematical models, real data at signficant 'scale'.
4	Conceptual engineering and flowsheet design, selection of final technology and scale of unit operations.
5	Real operating data of the conceptual flowsheet at a semi-industrial scale, validated mathematical models and equipment concepts, observations of operating stability and reliability, initial estimates of equipment life and maintenance costs.
6	Industrial benchmarks, engineering calculations, basic: (i) flowsheet specification, (ii) materials of construction and (iii) design criteria, initial equipment specification and vendor selection.
7	Detailled engineering calculations, material specifications, equipment designs and final vendor packages. Execution plan including: Construction, Commissioning, Ramp-up, Emergency Scenarios, e.g. power and equipment failure. How to bring plant to a safe state under all scenarios?
8	Initial operating data, first modelling of real plant data, observation of cause and effect, observations of process instability/recycle feed back loops.
9	Approach to the real bottlenecks (undersized unit operations), rapid process changes to improve stability, operability and reliability.
10	Significant historical plant data, and error-minimized mass-energy balances. Status: Debottlenecking, identification of cause and effect, search for solutions to real world problems with a corresponding return to guesses, theory and literature.

Note that incremental changes in data quality are listed in the table. Guesses, theory, and literature values will continue to be used until replaced by high quality engineering calculations, validated models or real process information from benchmarks or pilot/demo/industrial operation, e.g. overall plant operating reliability.

Modeling During Research and Development

During research and development 'unknowns' often outnumber 'knowns'. When so little information is available one may reasonably ask, "How can flowsheet modeling be usefully applied at such an early project stage?" The answer can be found by exploring two additional questions:

1. "What process option will theoretically give the best results for the project?"
2. "What data must the research program find or develop to realize the 'best' process option?"

The answers to these two questions can be usefully evaluated using flowsheet simulations at very early stages.

The way forward in a research program is often unclear at early stages, particularly when beginning with a 'blank page' on a green field project and using new technology. Many options are possible and little data exists to allow different options to be compared accurately. Flowsheet simulation can be usefully applied to explore the various processes or combinations of processes to determine which are possible (but not necessarily economically feasible) by making reasonable assumptions and applying data from standard thermo-chemical databases. Of those which are possible, basic flowsheet modeling can give 'best-case' operating cost (Op-ex) results using theoretically optimal process performance. Processes or options which give a negative or insufficient cash flow even under ideal conditions, can then be quickly eliminated, and the total number of cases reduced to a reasonable number for further consideration in the next project stage.

In order to determine the scope of a development program for new technology flowsheets, an efficient modeling approach is to 'scale-up' using the flowsheet simulator at a very early stage. A flowsheet simulator can be used to 'design' the next project scale equipment in order to determine: "What data must the research program find or develop to realize the preferred process option?" If you are at the lab scale, begin to design your pilot scale, if at the pilot scale begin to design your demonstration scale, etc.

A simple example of a single unit operation (electric smelting furnace) is shown in Figure 1. High accuracy modeling of such a unit operation would require the following information:

- Feed rate of each component (e.g. reductant, flux and ore), including uncontrolled streams like "infiltration air" and electrode usage.
- Temperature and composition of each stream totaling 100% including: mineralogy, free moisture and size distributions for solid streams (the latter two being somewhat related to dust in the off-gas).
- Complete thermodynamic data for the solid, gas, slag, metal, aqueous, and organic phases.
- All of the chemical reactions, their extents and correlation with temperature (kinetics and/or equilibria).
- Temperatures of all output streams.
- Physical properties such as density.
- Efficiencies of the phase separations between metal, slag, solid and gas.
- Heat transfer coefficients between slag and charge, slag and wall, gas and wall/roof, etc.
- Dimensions of the equipment to determine overall thermal efficiency and required power input.

As the simulation is built, many of the key data listed above will be found to be missing, e.g. the composition (totaling 100%!), mineralogy and size distribution of the natural solid raw materials, i.e. reductant, flux and ore. Most of the process data required for modeling may be also be lacking, e.g. the amount of infiltration air into the freeboard, fraction of dust carry over, heat transfer coefficients to the charge and walls, phase temperatures, efficiency of phase separations, extents of reactions, etc.

Assumptions must be used in place of actual process data and 'dummy' chemical data such as 'unknown' species (with some thermodynamic data typical of the average feed composition) added to feed streams to allow them to total 100%. Quantification of each unknown and assumption then becomes an objective/deliverable of the development program. Several project/engineering stages will likely be required to replace all guesses or assumptions with real data. Once a flowsheet model can be created with all of the key design parameters being 'known' rather than assumed, the project is clearly ready to pass through the final Stage-Gate and proceed to execution.

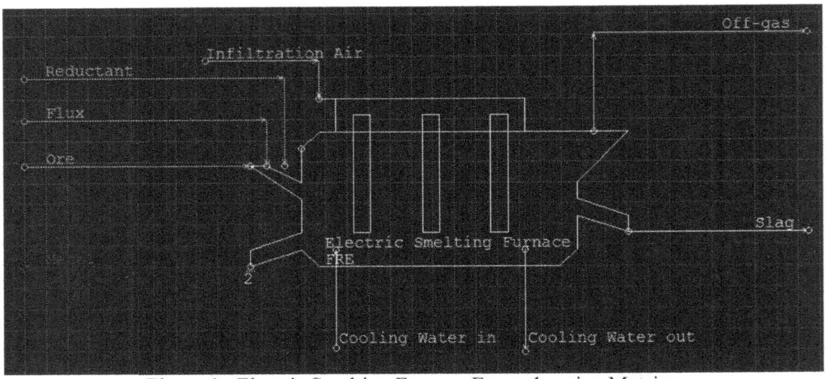

Figure 1. Electric Smelting Furnace Example using Metsim.

Modeling During the Design Stage

The design stage of most large projects usually consists of several sub-stages associated with increasing levels of engineering completion and greater precision in the operating cost (Op-ex) and capital cost (Cap-ex) estimates [20]:

- Scoping studies, including Trade-offs,
- Prefeasibility,
- Basic Engineering, including the Bankable feasibility and finishing with the Budget authorization,
- Detailed Engineering, with the associated Control and Final estimates.

<u>Scoping Studies</u>

Flowsheet modeling has a role to play at every engineering stage. At early stages such as during Scoping studies, flowsheets are created and modified with 'gross' changes. Due to the early stage of engineering, typical or assumed data often continue to be used even for key design parameters.

Important project issues are evaluated during the scoping stage, such as:

1. Economics of using different raw materials or energy sources.
2. Economics of producing process reagents (like acid) vs. purchasing.
3. Impact of plant scale on overall project return.
4. Integration of energy recovery on total consumption, project cost and reliability.
5. Open vs. closed water balances.

Many (perhaps hundreds) of mass-energy balances are typically calculated using the flowsheet simulator exploring the above factors. Sensitivity analyses are performed examining the impact of different parameters, such as raw material purity vs. production rate and quality. Entire sections or process unit operations are added and removed from the flowsheet model as decisions are made whether reagents like acid should be bought or produced on-site. At this stage of a project, it will become apparent to those working with spreadsheet models that this approach is not sufficiently flexible to allow for major flowsheet changes without considerable risk of programming errors and/or delays in delivering accurate results.

Prefeasibility and Feasibility

After many trade-off studies and one or more scoping studies, many of the key flowsheet decisions have been determined and one concept (now much more concrete) moves forward to prefeasibility. During prefeasibility and bankable feasibility, most of the assumed data used in the flowsheet simulator are replaced with validated values from sources such as demonstration scale test results, geological model or pilot mining results, industrial benchmarks, design studies and so on. The uncertainties in the calculations are thus reduced in proportion to the required accuracy of the estimates, e.g. typically -5 to +15% at Budget authorization [20].

During prefeasibility it is important to begin to explore the role of recycles and controls on process stability. As preliminary functional descriptions and Process and Instrumentation Diagrams (P&ID's) are developed, it becomes possible to program realistic controllers into flowsheet simulators like Metsim. Each recycle stream and Feed Back Controller (FBC) creates a 'loop' in the mathematical model and will also do so in the real plant. Cascaded controllers and multiple recycles create nested loops, and may result in mathematical instability in Metsim due to its Sequential Modular solution algorithm. In the Sequential Modular algorithm, each unit operation is solved in sequence, and the system is 'looped' until overall convergence is achieved. While this algorithm can result in mathematical instability, it does tend to: (i) behave in a highly analogous fashion to real plant systems and (ii) lend itself very well to dynamic simulation using discrete time steps. Control strategies which fail to converge rapidly in steady-state simulations using Metsim, will also fail to rapidly achieve steady-state in the real world; therefore, rather counter-intuitively, Metsim steady-state simulations can be used to determine the dynamic suitability of real-world control strategies on a qualitative basis.

The use of Feed Forward Control (FFC) in Metsim allows for the use of fixed ratios, e.g. fuel-to-air on a burner and avoids the creation of feedback loops. FFCs can therefore be used to increase real world process stability, as well as model computational efficiency. FFC can also be used to control recipes, e.g. slag basicity and thereby avoid nesting the entire flowsheet in a loop. In the real world this has implications for analysis of raw materials, blending and plant procedures.

Basic and Detailed Engineering

During basic and detailed engineering, a flowsheet simulator can be used as a 'warehouse' of process design criteria. No calculation anywhere within the project should be allowed to use data which is not compatible with the data used in the process flowsheet simulation. Ideally a single set of: (i) minimum, (ii) nominal and (iii) maximum values should define all valid process conditions for use both in flowsheet simulations and off-line calculations. Metsim FBC controllers allow the definition of minimum and maximum controlled variable values, as well as set point targets.

Some specific activities for which a flowsheet simulator can be used during basic and detailed engineering have been listed in Table I and will be examined in further detail in this section.

Emergency Scenarios are sometimes not thoroughly explored during engineering of metallurgical processes. Emergency scenarios include, but are not limited to failure of:

- Power,
- Major mechanical equipment,
- Total unit operations, and the
- Control system.

Using Metsim, controllers can be manually frozen at the last value (hold-last-position), set to maximum (fail-open), or set to minimum (fail-closed) in order to simulate the appropriate failure states. Heat or other energy additions can be interrupted. Cooling water can set to 'failure' values to simulate the use of back up systems or stopped altogether to simulate worst case scenarios. Such testing can explore the control and safety system interaction with the process equipment using simple steady-state simulations in order to determine if the plant achieves a safe state under each failure scenario.

Using a flowsheet simulator like Metsim, unit operations can be 'turned off' to simulate total failure and the rest of the flow sheet operated independently to observe the consequences. Observation of the mathematical instabilities created can allow an experienced Metsim modeler an initial indication of the dynamic response, even using only steady state simulations. Changes can subsequently be made to the flowsheet to ensure that the plant can always achieve a safe state, for example by installing emergency venting or flares for dangerous process gases or 'pits' for the molten contents of smelting furnaces.

Commissioning should be considered early in a project design. A plant should be divided up into sub-'Systems' that allow it to be commissioned and started-up in realistically sized portions. A plant 'System' can be defined as the equipment capable of relatively independent operation for significant periods; typically located between a sizable source and sink. Temporary vents, bleeds, back-up heating and cooling systems may be necessary in order to allow for completely independent commissioning. Plant 'Sections' as they are defined in Metsim can be operated individually. This can be used to determine if they are capable of independent operation and to establish what temporary auxiliary systems or product sinks might be required, e.g. diesel supply for nominally coal fired equipment prior to coal pulverizing plant commissioning, or a coal fired/water spray cooled 'flare' to allow for commissioning of precision coal pulverizing equipment prior to process plant start-up.

Reliability modeling of 'Systems' should be performed to ensure that individual systems each contain sufficient internal redundancy to achieve their required reliabilities. Systems should be separated by temporary sinks/supplies (e.g. surge bins, drains, vents and tanks) of sufficient size to avoid downtime in one system unduly influencing the availability of the next. Flowsheet simulators can be used to determine the likely sizes of such isolation equipment and their most effective location in the flowsheet. This represents an attractive objective for a dynamic flowsheet simulation.

Stability of operation should be studied in conjunction with the definition and modeling of 'Systems', and the determination of commissioning and reliability requirements. The size of system 'separators' or surges, may need to be increased in flowsheets containing large numbers of recycle loops in order to ensure that individual systems do not interact too closely. Strong interaction caused by recycles result in unstable simulations and poor operation in the real world. Recycles may create positive feed back loops leading to catastrophic failures in the worst case.

Overflows, vents, or auxiliary 'make-up' supplies may need to be added to the flowsheet to stabilize the volume, mass or quality of recycle streams. The impact of such unit operations, (SUBs as they are called in Metsim), on process stability can be observed easily during steady state simulations, i.e. by faster mathematical convergence.

Ramp-up is sometimes completely forgotten, even during detailed engineering. A typical error is the specification of valves which are either too large or of the wrong type to control at the low range often required during 'ramp-up'. Flowsheet simulators can be used to solve for a number of mass-energy balances from low, e.g. 10%, to high, e.g. 125% of the nominal design throughputs. Having performed such calculations, it may then be observed that some unit operations dramatically change requirements at low throughput. For example, high temperature metal vapor condensers require heating at low and cooling at high throughputs, in order to maintain a liquid discharge at all times.

Start-up (after the initial plant start-up) is a special type of 'ramp-up' encountered after every maintenance stop. Equipment designed for smooth ramp-up, will subsequently start-up well during later operating phases of the project.

Plant Operation

Flowsheet models developed early in a project cycle can 'live' and 'grow', and ultimately be used during production stages. Steady-state models developed during engineering can be converted to dynamic (at least using Metsim) and utilized during operator training. Using Dynamic Data Exchange (DDE), a Metsim model can simulate the process responses to changes in operator set points, chemical compositions or equipment conditions, e.g. heat loss. If provided with suitable interfacing, such a model can sit behind a Human Machine Interface (HMI) and act like the real process during operator training sessions. Often training models are not started early enough during engineering to be available when required. A continuously evolving model is much more likely to have reached the required stage of development prior to the commencement of operator training.

Near real time data from Plant Information Management Systems (PIMS), Laboratory Information Management Systems (LIMS), data warehouses or process historians like PI, can also be interfaced with flowsheet simulation software. This enables the flowsheet simulation to provide forewarning to operators of the consequences of changing feed grades on product quality or of the future impact of operator set-point changes on the process heat balance, before the changes have fully occurred. The model can act as a sort of 'crystal ball' on the process. This allows the plant operators to take active steps to avoid economically unfavorable or unsafe future 'steady-states' from occurring. With correct programming of steady-state simulators, even large changes occurring in complex flowsheets can be solved in seconds or at most a few minutes, using standard portables or desktop computers.

Flowsheet models can be used for feed forward recipe setting. Such models can help set feed blends, which simultaneously achieve multiple constraints or objectives, such as economically optimal product quality and recovery.

Advanced Modeling Techniques

In Metsim complex and unique unit operations can be modeled either for steady-state or dynamically using custom APL (A Programming Language) code. Functions can be defined by the user and executed within standard unit operations such as the electric smelting furnace shown in Figure 1, or in custom unit operations built from multiple simple 'blocks' of mixers (MIX), splitters (phase-SPP, stream-SPS, component-SPC), stream distributors (SUB) and free energy minimizers (FEM). Complex chemistry in many separate phases can be modeled using either the built in database, external databases such as FactSage, or manually entered data. The author has often used information from the HSC database to expand the already considerable data found in Metsim or used pre-defined look-up tables to relate complex FactSage equilibria to Metsim operating conditions. The use of pre-calculated look-up tables avoids the need to execute external software and greatly reduces convergence times.

Examples of advanced Metsim models used in academic research can be found in the literature [11-19]. These often represent a 'special' class of early stage modeling, where the focus is on technology rather than project development. The application of dynamic Metsim modeling in such academic work is particularly noteworthy [12, 14, 19] and potentially of industrial interest to those wishing to extend the scope of traditional steady-state models used during most large scale metallurgical plant engineering in the last 20 years.

Conclusions

Flowsheet simulators have a place in nearly any process research, engineering or technology development project.

Effective application at each project stage can:

- Streamline development,
- Eliminate wasted effort/improve engineering productivity,
- Increase plant operability, reliability and safety, and
- Ultimately help to ensure a financially successful project outcome.

References

[1] *www.aspentech.com/fullproductlisting/*, accessed on-line, 15 September, 2013.

[2] *www.chemstations.com/*, accessed on-line, 15 September, 2013.

[3] *www.cycadprocess.com*, accessed on-line, 15 September, 2013.

[4] R. Coleman and P. Lamberg, "Flotation Circuit Optimisation Using Modelling and Simulation Software," *Outotec Australia's Quarterly e-newsletter,* August, (2010), 1-4.

[5] *www.outotec.com/en/products--services/hsc-chemistry/calculation-modules/sim--process-simulation/*, accessed on-line, 15 September, 2013.

[6] K. Nikkhah, "Role of Simulation Software in Design and Operation of Metallurgical Plants: A Case Study," (Paper presented at SME Annual Meeting, Denver, Colorado, February, 2001), 1-11.

[7] *www.andritz.com/products-and-services/pf-detail.htm?productid=18154*, accessed on-line, 15 September, 2013.

[8] *www.metsim.com/*, accessed on-line, 15 September, 2013.

[9] *iom.invensys.com/en/pages/simsci_processengsuite_proii.aspx*, accessed on-line, 15 September, 2013.

[10] *www.syscad.net/*, accessed on-line, 15 September, 2013.

[11] H. Li, A.E. Morris, and D.G. Robertson, "Thermodynamic Model for MnO-Containing Slags and Gas-Slag-Metal Equilibrium in Ferromanganese Smelting," *Metallurgical and Materials Transactions B*, vol. 29, (1998), 1181-1191.

[12] J. Peter, K.D. Peaslee, and D.G. Robertson, "Simulations of a New Continuous Steelmaking Process," AISTech 2006: Iron & Steel Technology Conference, vol. II, 445-469.

[13] J. Peter, K.D. Peaslee, D.G. Robertson, L. Zhang, and B.G. Thomas, "Introduction of a Novel, Scrap-Based, Fully Continuous Steelmaking Process," *Urbana*, vol. 51, (2005), 61801, 1-12.

[14] S. Lekakh, D.G. Robertson, C. Rawlins, V. Richards, and K.D. Peaslee, "Investigation of a Two-Stage Aqueous Reactor Design for Carbon Dioxide Sequestration Using Steelmaking Slag," *Metallurgical and Materials Transactions B*, vol. 39, (2008), 484-492.

[15] P. Chamveha, K. Chaichana, A. Chuachuensuk, S. Authayanun, and A. Arpornwichanop, "Performance Analysis of a Smelting Reactor for Copper Production Process," *Industrial & Engineering Chemistry Research*, vol. 48, (2008), 1120-1125.

[16] P. a. C. Hernández, M. a. E. Taboada, T. A. Graber, and H. c. R. Galleguillos, "Crystallization of Hydrated Ferric Arsenate. Process Design Using Metsim," *Industrial & Engineering Chemistry Research*, vol. 48, (2009), 10522-10531.

[17] P.K.A. Raghavendra, "Recovery of Materials from Recycling of Spent Furnace Linings," (Masters thesis, Missouri University of Science and Technology, 2008), 1-82.

[18] H. Kimura, "Material and Energy Flow Simulation and Economic Analysis for a Novel Suspension Ironmaking Techonology," (Masters thesis, University of Utah, 2010), 1-126.

[19] J. Peter, K.D. Peaslee, D.G. Robertson, and B.G. Thomas, "Experimental Study of Kinetic Processes During the Steel Treatment at Two LMF's," *Urbana*, vol. 51, (2005), 61801, 1-15.

[20] M.W. Kennedy, C. Harris, and A. MacRae, "Risk Weighted Cash Flow, a Communication Tool for Engineers and Financial Professionals on New Technology Projects," *Journal of the Candian Institute of Mining and Metallurgy*, vol. 3, no. 4, (2012), 193-202.

Celebrating the Megascale: Proceedings of the Extraction and Processing Division
Symposium on Pyrometallurgy in Honor of David G.C. Robertson
Edited by: Phillip J. Mackey, Eric J. Grimsey, Rodney T. Jones, and Geoffrey A. Brooks
TMS (The Minerals, Metals & Materials Society), 2014

CHEMSHEET AS A SIMULATION PLATFORM FOR PYROMETALLURGICAL PROCESSES

Karri Penttilä[1], Justin Salminen[1] , Nagendra Tripathi[2] and Pertti Koukkari[1]

[1]VTT – Technical Research Centre of Finland, P.O. Box 1000, 02044 VTT Finland
[2]Koniambo Nickel SAS, RT1, site de Vavouto – 98833 Voh
BP 679 – 98860 Koné, New Caledonia

Keywords: Multiphase thermodynamic calculation, simulation of smelting process, online model

Abstract

ChemSheet is a thermodynamic multi-phase multi-component simulation software, which is used as an Add-in in Microsoft Excel. In ChemSheet, the unique Constrained Gibbs free energy method can be used to include dynamic constraints and reaction rates of kinetically slow reactions, yet retaining full consistency of the multiphase thermodynamic model. With appropriate data, ChemSheet models can be used to simulate reactors and processes in all fields of thermochemistry. The presentation will cover off-line modeling of Cu-flash smelters and advanced thermochemical simulation coupled with on-line process control of Cu-Ni smelting. The presentation will describe an off-line model of Cu-smelter based on critically assessed properties of the Al-Ca-Cu-Fe-O-S-Si –system (slag, matte and liquid metal) by using the quasichemical model. A four-stage reactor model (shaft, settler, uptake and bath) is used for optimizing process parameters and feed particle distribution. As a second example, an advanced thermochemical model of a Ni-Cu sulphide smelting plant will be given. The on-line model covers the operation of treating Ni-Cu-S concentrate via roasters, electric furnace and converters, producing a high grade Bessemer matte product for further refining. The model integrates the thermochemistry of the roasters and electric furnace, and predicts important process parameters such as degree of sulphur elimination in the fluid-bed roasters, matte grade, iron metallization, slag losses and the iron to silica ratio in the electric furnace slag. Both models can be used to assist process engineers and operators in calculating the addition rates of coke, flux and air for different feed scenarios.

Introduction

The common expectation within industrial companies is that commercial software vendors will supply them with user-friendly software packages that require little or no programming skills. A similar feature is encountered within universities and research institutes, in particular when highly specialized, modular software, compatible with existing 'researcher code' is required. In multiphase thermochemistry, the 'research programs' themselves may be commercialized and appear with their highly specialized databanks. The thermochemical data will then include the necessary interaction models for the complex phase mixtures, which appear as application specific for e.g. different high-temperature operations in metallurgy. To ensure the usage of such data in process simulation software, various interfaces have been developed [1, 2, 3, 4].

ChemSheet [5], when introduced in 1999, was one of the software pioneers serving the need of coupling rigorous thermochemical software with more common, easy-to-use interfaces. ChemSheet provides an MS-Excel add-in coupling to the well-known ChemApp Gibbs energy solving program [3]. Also with ChemSheet the unique Constrained Gibbs free energy method [6] can be used to include dynamic constraints and reaction rates of kinetically slow reactions, yet retaining full consistency of the multiphase thermodynamic model. Thus, with ChemSheet multi-phase multi-component thermochemical problems can be worked through the familiar spreadsheet, which is in everyday use by engineers and scientists everywhere in the world. When developing applications, writing macro-code around the basic interface is helpful but for those who do not like programming, the custom made ChemSheet simulation model may appear as just another spreadsheet file.

During its 15-year history, ChemSheet has been adapted by the industrial user, particularly for such purposes where its easy inclusion of Excel features such as special functions, presentation graphics and use of customized spreadsheet interfaces have been beneficial. Figure 1 depicts the Excel spreadsheet connection to ChemApp in ChemSheet.

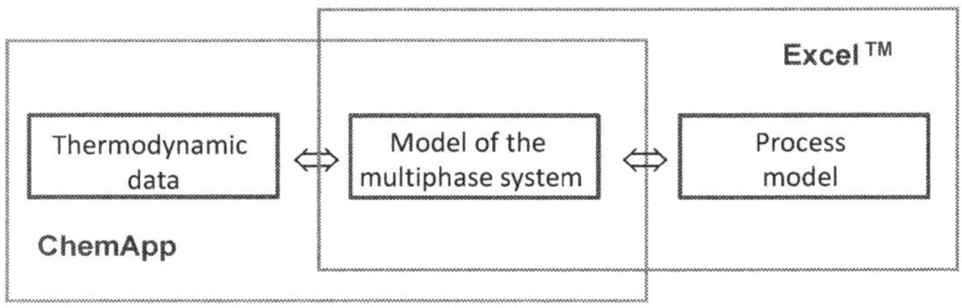

Figure 1. ChemSheet coupling of Gibbsian thermodynamics and spreadsheet calculations

ChemSheet model of the flash smelter

The copper ore (commonly chalcopyrite, $CuFeS2$) is crushed and ground before it is concentrated to between 20 and 40% copper in a flotation process. The next major step in production uses pyrometallurgical processing to convert the copper concentrate to 99% pure copper suitable for electrochemical refining. These high temperature processes first smelt the concentrate in a furnace (flash smelting), then oxidize and reduce the molten products to progressively remove remaining sulfur, iron, silicon and oxygen to leave behind relatively pure copper. The feed to flash smelting is known as copper concentrate. It contains copper and iron sulfides, silica and limestone ground to small particles (<100mm). Flash smelting is highly exothermic process. It requires almost no external heat. Concentrate and enriched air are fed to the shaft of the furnace where the iron oxides and the sulfides are converted to liquid slag. Copper sulfides are first partly oxidized and then converted to liquid called matte. At the bottom of the furnace the slag and the matte are collected in a bath and separated from each other by gravity (lighter slag is floating on top of the heavier matte). The gas from furnace consists mainly of nitrogen and sulfur oxide.

The flash smelting process is simulated with a ChemSheet model that contains four calculation steps (see Figure 2). The compositions of the feed flows are shown in Figure 3. The first step calculates the average flow temperature and the composition in the reactor shaft by using an Gibbs energy minimization method and target calculation. The second step calculates the flow temperature and the composition in the settler in a similar manner. Before entering to the settler the solid and the liquid phases are separated (using given separation efficiency) from the flow as they are assumed to go to the bath at the bottom of the settler. The third step calculates the flow temperature and the composition in the uptake shaft. Again the remaining solid and liquid phases are separated (as gas temperature decreases it is possible that also condensation takes place) from gas. The fourth step calculates the bath temperature and the composition. Flows to bath consists of separated solids and liquids from the reactor shaft and the settler.

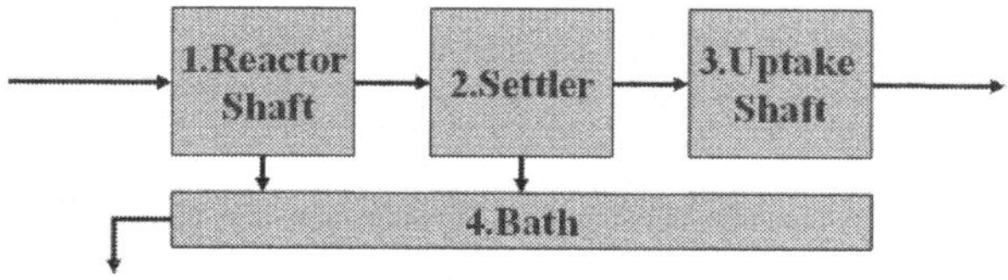

Figure 2. Four steps flash smelter ChemSheet model.

Thermodynamic data-file for copper sulfide flash smelting ChemSheet model was created with FactSage [7]. The thermodynamic properties of the slag, matte, and liquid copper phases in the Al-Ca-Cu-Fe-O-S-Si–system have been critically assessed and optimized over the ranges of compositions of importance to copper smelting/converting based on thermodynamic and phase equilibria information available in the literature and using the modified quasi-chemical model [8]. The thermodynamic system consists of ten elements (the above mentioned seven and C-H-N), 22 gas species, three liquid mixtures (nine species in slag phase, four species in liquid/blister phase and three species in matte phase) and 27 pure condensed phases.

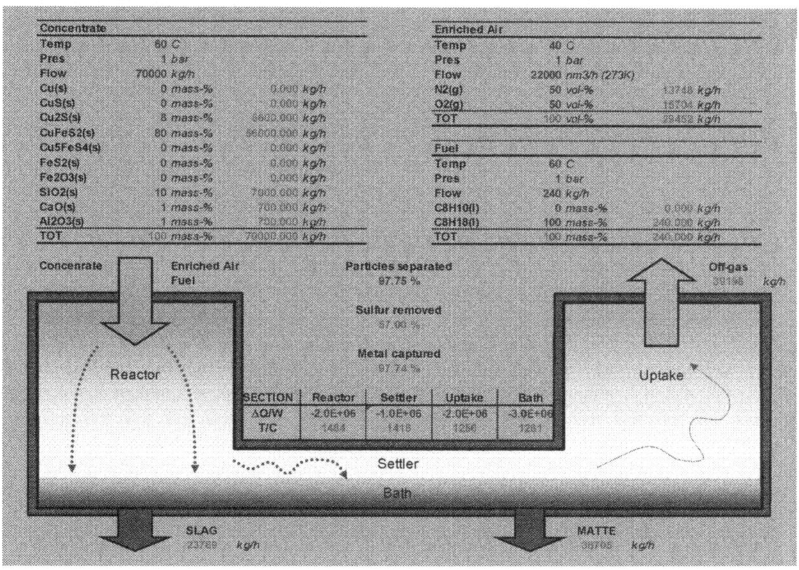

Figure 3. Excel user interface for the flash smelter model (with typical inputs).

Figure 3 shows the user interface of the four steps ChemSheet model for the flash smelter. Excel cells F28:H28 contain the estimated heat losses through the wall in each section and cells F29:H29 contain the calculated average temperatures. Table 1 shows the calculated elementary compositions of the feed and the product flows. Table 2 shows the distribution of elements between the product flows. The off-gas contains 2.25 mass-% of the concentrate feed as dust.

Table 1. Elementary compositions of the feed and product flows (as mass-%).

Composition	Cu	Fe	Ca	S	Si	Al	O	N	C	H
Concentrate	34.09 %	24.34 %	0.71 %	29.57 %	4.67 %	0.53 %	6.08 %	0.00 %	0.00 %	0.00 %
Air+Fuel	0.00 %	0.00 %	0.00 %	0.00 %	0.00 %	0.00 %	52.89 %	46.30 %	0.68 %	0.13 %
Matte	63.36 %	13.06 %	0.00 %	23.58 %	0.00 %	0.00 %	0.00 %	0.00 %	0.00 %	0.00 %
Slag	0.28 %	49.86 %	2.06 %	1.03 %	13.44 %	1.52 %	31.80 %	0.00 %	0.00 %	0.00 %
Off-gas	1.38 %	0.98 %	0.03 %	30.10 %	0.19 %	0.02 %	31.62 %	35.07 %	0.52 %	0.10 %

Table 2. Distribution of elements between the product flows.

Distribution	Cu	Fe	Ca	S	Si	Al	O	N	C	H
Matte	97.46 %	28.14 %	0.00 %	41.81 %	0.00 %	0.00 %	0.00 %	0.00 %	0.00 %	0.00 %
Slag	0.28 %	69.61 %	97.75 %	1.18 %	97.75 %	97.75 %	37.90 %	0.00 %	0.00 %	0.00 %
Off-gas	2.26 %	2.25 %	2.25 %	57.00 %	2.25 %	2.25 %	62.10 %	100.00 %	100.00 %	100.00 %

The flash smelter model was further developed to simulate the combustion of a single copper sulfide ore particle. The particle and gas (enriched air) were introduced at the top of the vertical reactor shaft. The superficial gas velocity was solved from ideal gas law and the reactor cross-sectional area and the particle velocity from its momentum equation that includes gravity, drag force and wall friction terms. Neglecting the temperature and composition gradients inside the particle, the model calculates the average temperature and the composition of the particle and the gas flow as function of the axial coordinate in the shaft. Normally the heat balance of flow is given as [9]:

$$\frac{d(\dot{m}\, c_p T)}{dx} = \dot{Q}_{rad} + \dot{Q}_{conv} + \dot{Q}_{reac} \tag{1}$$

where $\dot{m}$ is the mass flow, c_p the heat capacity and T the temperature of the flow, $\dot{Q}_{rad}$ and $\dot{Q}_{conv}$ the radiation (in radial direction) and convection heat transfer rates between the flow and its surroundings, and the last term is the heat of reactions in the flow. This time it is more convenient to select enthalpy flow as the variable instead of the temperature. Enthalpy is function of temperature, pressure and composition (mass flow rates of species) of the flow. Any changes in the enthalpy of the flow changes its temperature and/or its composition. ChemSheet contains a calculation mode that automatically solves the temperature and the composition of the flow when the enthalpy change is known (extensive property target calculation). Now equation (1) can be given for the particle and the gas flow as:

$$\frac{d\dot{H}_p}{dx} = Q_{rad} + \dot{Q}_{conv} - \dot{H}_{pg} \tag{2}$$

$$\frac{d\dot{H}_g}{dx} = \dot{Q}_{rad} + \dot{Q}_{conv} + \dot{H}_{pg} \tag{3}$$

where $\dot{H}_p$ and $\dot{H}_g$ are the enthalpy flow of the particle and the gas in axial direction, respectively, and $\dot{H}_{pg}$ is the net enthalpy flow between the particle and the gas (including the vaporized gas species). Reactor heat term is no longer needed as it is included in the enthalpy of flow term. The temperatures of the particle and the gas can now be solved from their enthalpies.

Mass transfer rate of gaseous species between the particle and the gas is given as:

$$\frac{dn_i}{dx} = k_i A_p \left(C_{i,p} - C_{i,g}\right) \tag{4}$$

where $\dot{n}_i$ is the flow rate of gas species between the particle and the gas, k_i is the mass transfer coefficient, A_p is the total surface area of the particle and C_i is the concentration of the gas species i in the particle and in the gas, respectively. The surface area of the particle is given as:

$$A_p = \frac{6\,\dot{W}_p}{\rho_p d_p |v_p|} \tag{5}$$

where $\dot{W}_p$ is the mass flow rate, ρ_p is the density, and d_p is the diameter, and v_p is the velocity of the particle, respectively. The surface area of the particle is also needed when calculating the radiation and convection heat flows between the particle, the gas and the reactor wall.

The system of independent variables consist of velocity, enthalpy and mass flow rates of the components of the particle, and enthalpy and mass flow rates of the components of the gas. The system was solved with an adaptive Runge-Kutta algorithm. The model was implemented in Excel with VBA macros and all the thermodynamic calculations were done with ChemSheet. Normally ChemSheet is operated manually via toolbar commands and dialog controls but for more advance simulations ChemSheet can also be called programmatically from a VBA macro.

Figure 4 shows the calculated particle and gas temperature profiles and the composition of the gas flow (as mass flow rates of main gas species) with average particle diameter 25 μm. The temperature of the reactor wall was set as constant (1800K). The maximum gas temperature in the reactor shaft reaches almost 2400K. Decreasing the particle size increases the combustion rate and the maximum temperature. This also increases the dusting and loss of metals.

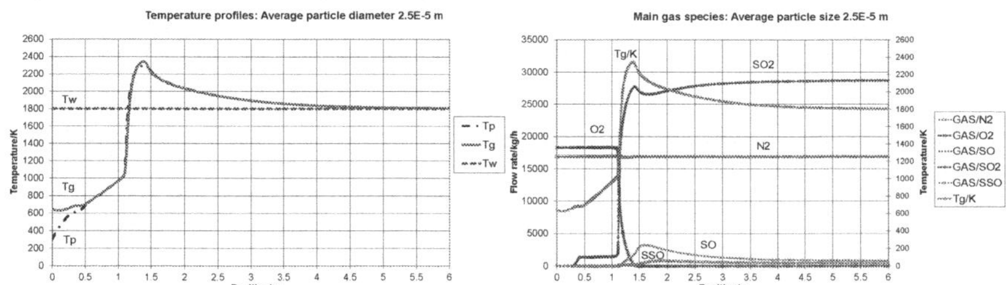

Figure 4. The calculated particle and gas temperature profiles and the composition of the main species in the gas with average particle diameter 25 μm.

On-line model for the XSTRATA Nickel Sudbury Smelter

As a pioneering application in the field of process digitalization, a predictive online thermo-chemical model was developed for Glencore-Xstrata Nickel's Sudbury smelter. To combine the practicality of a spreadsheet with Gibbs energy solving multiphase routine (FactSage), the ChemSheet platform was used. The model is connected to the PI server to communicate with the plant operation in real time via PI DataLink™. The PI server receives and archives plant measurements, while PI DataLink™ retrieves this data as input to the online model, see Figure 5 below.

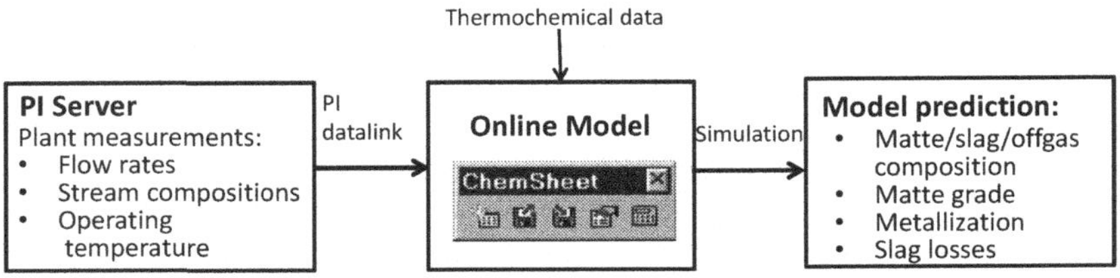

Figure 5. ChemSheet based online model operated in MS-Excel

For effective modeling, the industrial unit operations were again divided into thermochemical modules (see Figure 6). For the roasting operation, the sulphur elimination and sulphate formation steps were separately considered in two different modules. For the sulphur elimination stage, slurried concentrate, flux, revert, custom feed and air were used as the input streams. For the sulphate formation step in the off-gas handling system, dust and air infiltration were given as input streams. Calcine then exits the roasters at 760 °C, but there is a heat loss occurring during calcine transport via drag link conveyors to the electric furnace. Based upon plant measurements, the calcine enters the electric furnace at ca. 500 °C. To account for the heat loss, a separate module was implemented which represents the calcine cooling step during the conveyor transfer in the calculation.

The electric furnace model, respectively, consists of three separate zones – the freeboard, the calcine bank and the smelting zone. To properly account for the heat and mass balance model of the electric furnace, these modules were connected in a countercurrent fashion. The calcine is fed into the electric furnace from the top while reducing gases move up through the calcine layer into the freeboard. This calculating approach allows each zone to coexist with its neighbor under different operating conditions yet still remain thermochemically connected. For example, in the current model of the electric furnace the freeboard is set at a temperature of 500 °C, while the smelting zone is at 1300 °C. The model also takes into account three important process parameters – coke efficiency, degree of calcine oxidation in the freeboard and degree of reduction of the calcine bank by reducing gases from the smelting zone.

The various modules of the model are illustrated in Figure 6. For further details of the on-line model structure and its implementation background see Tripathi and Jabbar [10].

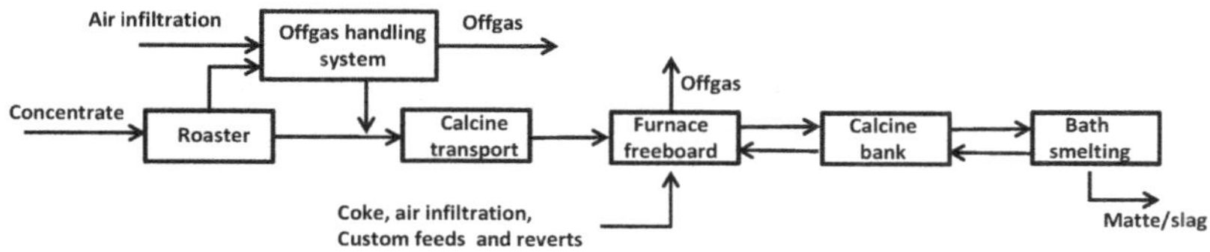

Figure 6. Modular design of the online model with co- and countercurrent connections

Online plant data, such as flow rates, compositions and operating conditions, are dynamically entered into the model. For a given feed scenario, the model converges in ca. 20 s, producing final results for calcine, matte, slag and off-gas compositions. The model could then be employed to predict the plant operating parameters in terms of the historical plant data; for example the degree of sulphur elimination in the roasters, furnace matte grade (see Figure 7) and degree of metallization as well as the composition ratio of Fe and SiO2 [%Fe/%SiO2] in the furnace slag.

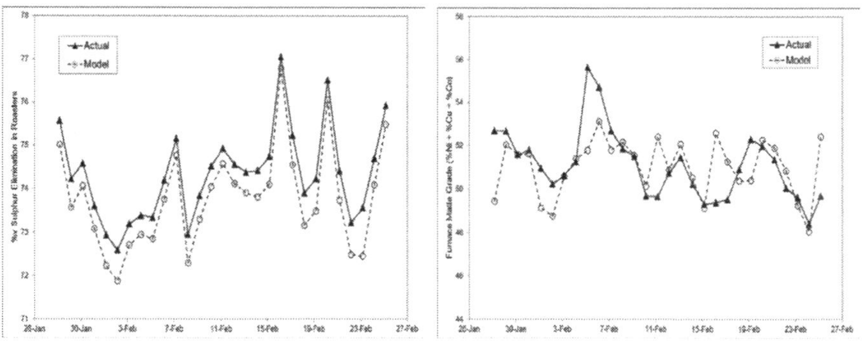

Figure 7. Comparison between the predicted values of sulphur elimination in the roasters, and furnace matte grade with the historical plant data from year 2012. The model predictions follow the plant fluctuations, mainly caused by variation in the feed grade/quality.

The interface of the online model is relatively simple and was found to be easy-to-use. The application of this model is helpful for the plant engineers in calculating the required flow rates of flux, coke and air in achieving the target furnace matte and slag compositions. The combined processing time for roasters and electric furnace is about 35 hours, therefore the model is particularly helpful in addressing plant fluctuations well in advance. The online model is also expected to help in stabilizing the metallurgical performance of the batch converter aisle operation, and is intended for eventual use by plant operations supervision.

Conclusion

Modern simulation programs and advanced thermochemical data are increasingly used for process digitalization and on-line control in metallurgical industry. The advantage of using thermodynamic models for the simulation of metallurgical processes is their inherent coupling of chemical and enthalpy changes as well as the recovery of calculation data in terms of well-defined, measurable quantities, which allows for checking the model reliability directly with plant historical trends. ChemSheet with its easily customized and application oriented spreadsheet interface provides a practical engineering approach to process modeling with multiphase thermodynamics. The spreadsheet technique allows for development of application specific traits by using Excel features as well as flexible connections with other simulation software. With continuing digitalization in the industry, robust thermochemical modeling tools will have an important role in both supporting plant operations and in supporting their long term development objectives.

References

1. Robertson, D. G. C., "The Computation of the kinetics of the reactions between multiple phases". TMS EPD Congress 1995, G. W. Warren, Editor, Warrendale, PA, (1995) 347-361.

2. Andersson, J.-O., Helander, T., Höglund, L., Shi, P., Sundman, B., "THERMO-CALC & DICTRA, Computational Tools For Materials Science",Calphad, 26 (2002), 273-312.

3. Petersen, S. , Hack, K., "The thermochemistry library ChemApp and its applications", Int. J. Mat. Res., 98 (2007), 935-945.

4. Petersen, S., Hack, K., Monheim, P., Pickart, U., "SimuSage – The component library for rapid process simulation and its applications", Int. J. Mat. Res., 98 (2007), 946 -953.

5. Eriksson, G., Hack, K., Koukkari, P. and Penttilä, K., "Combined Thermodynamic and Kinetic Models for Processing of Materials", in Taylor, P.R. (ed.), TMS EPD Congress 2001, P. R. Taylor, Editor, Warrendale, PA, (2001), 303-314.

6. Koukkari, P., Pajarre, R., "Calculation of Constrained Equilibria by Gibbs Energy Minimization", Calphad, 30 (2006), 18-26.

7. Bale, C. W., Chartrand, P., Decterov, S. A., Eriksson, G., Hack, K., Mahfoud, R. B., Melansson, J., Pelton, A. D. and Petersen, S., "FactSage Thermochemical Software and Databases", Calphad, 26 (2002), 189-228.

8. Degterov, C., Pelton, A. D., "A thermodynamic database for copper smelting and converting", Metallurgical and Materials Transactions B, Volume 30B (1999), 661-669.

9. Shook, A. A. , Richards, G. G. and Brimacombe, J. K., "Mathematical Model of Chalcocite Particle Combustion", Metallurgical And Materials Transactions B, 26B (1995), 719-729.

10. Tripathi, N., Jabbar, R., "Advanced online thermochemical model for metallurgical control at Xstrata Nickel's Sudbury smelter", Mineral Processing and Extractive Metallurgy, 122 (2013), 71-78.

Celebrating the Megascale: Proceedings of the Extraction and Processing Division Symposium on Pyrometallurgy in Honor of David G.C. Robertson
Edited by: Phillip J. Mackey, Eric J. Grimsey, Rodney T. Jones, and Geoffrey A. Brooks
TMS (The Minerals, Metals & Materials Society), 2014

A COMPUTATIONAL FLUID DYNAMIC MODEL FOR A NOVEL FLASH IRONMAKING PROCESS

Silvia E. Perez-Fontes, Hong Yong Sohn, and Miguel Olivas-Martinez

Department of Metallurgical Engineering, University of Utah;
135 S. 1460 E., Rm 412; Salt Lake City, UT 84112-0114, USA

Keywords: Ironmaking, Computational Fluid Dynamics, Iron Concentrate Flash Reduction

Abstract

A computational fluid dynamic model for a novel flash ironmaking process based on the direct gaseous reduction of iron oxide concentrates is presented. The model solves the three-dimensional governing equations including both gas-phase and gas-solid reaction kinetics. The turbulence-chemistry interaction in the gas-phase is modeled by the eddy dissipation concept incorporating chemical kinetics. The particle cloud model is used to track the particle phase in a Lagrangian framework. A nucleation and growth kinetics rate expression is adopted to calculate the reduction rate of magnetite concentrate particles. Benchmark experiments reported in the literature for a nonreacting swirling gas jet and a nonpremixed hydrogen jet flame were simulated for validation. The model predictions showed good agreement with measurements in terms of gas velocity, gas temperature and species concentrations. The relevance of the computational model for the analysis of a bench reactor operation and the design of an industrial-pilot plant is discussed.

Introduction

A novel flash ironmaking technology is being developed at the University of Utah. This technology is based on the direct gaseous reduction of iron oxide concentrates in a flash process and has the objective of significantly reducing energy consumption and environmental emissions, especially CO_2 emissions, compared with blast furnace ironmaking. These benefits arise largely from the elimination of cokemaking and pelletization/sintering steps. The proposed process uses gaseous reducing agents such as natural gas, hydrogen, or syngas.

In this process, iron ore concentrate particles are injected into a flash furnace, and are reduced in flight by hot reducing gases produced by the partial combustion of a fuel/reductant gas. The operating temperature ranges from 1650 to 1850 K.

In this work, a computational fluid dynamic model for a novel flash ironmaking process operated using hydrogen as a fuel/reductant is presented. Particular model features include a simplified chemical reaction mechanism to describe the partial combustion of H_2 with O_2; the particle cloud model to represent the motion and dispersion of concentrate particles in a Lagrangian framework; and a nucleation-and-growth kinetics rate expression to compute the reduction rate of magnetite concentrate particles. Computed results of the simulation of benchmark experiments reported in the literature for a nonpremixed H_2 jet flame are presented for model validation. The design of an industrial pilot flash reactor is discussed.

Model Formulation

The generic transport equation, in steady state, can be written in the following form:

$$\nabla \cdot \rho_g \vec{u} \phi - \nabla \cdot \Gamma_\phi \nabla \phi = S^\phi \quad (1)$$

where ϕ denotes the generic gas-phase dependent variable, ρ_g is the gas density, Γ_ϕ represents the effective transport coefficient, and S^ϕ is the source term. The component equations were solved using the standard k-ε turbulence model, combined with a discrete-ordinates method [1] to solve the radiative transfer equation.

In this work, a reaction mechanism for H_2-O_2 was adopted together with the eddy dissipation concept (EDC) approach [1, 2]. In the EDC approach, finite reaction rates are computed in control volumes, the dimensions of which are defined by the length of the small scales of turbulence. The gas-phase reaction mechanism (Table I) adopted in this work was proposed by Eklund and coworkers [3]. It consists of seven chemical reactions involving six species: H_2, O_2, H_2O, OH, H, and O. The chemical reactions were represented by the general form:

$$a\mathrm{A(g)} + b\mathrm{B(g)} = c\mathrm{C(g)} + d\mathrm{D(g)} \quad (2)$$

for which the molar reaction rate of the jth reaction was defined by

$$R_j = k_{f,j} C_\mathrm{A}^a C_\mathrm{B}^b - k_{r,j} C_\mathrm{C}^c C_\mathrm{D}^d \quad (3)$$

The forward and reverse rate constants in Equation (3) were computed from

$$k_{f,j} = A_j T^{\beta_j} \exp\left(\frac{-E_{a,j}}{RT_g}\right) \quad (4)$$

$$k_{r,j} = \frac{k_{f,j}}{K_{c,j}} \quad (5)$$

where C_i is the molar concentration of the ith species, $K_{c,j}$ is the concentration equilibrium constant, $E_{a,j}$ is the activation energy (cal mol^{-1}) of the jth reaction, T_g is gas temperature (K), and R is the ideal gas constant (1.987 cal mol^{-1} K^{-1}).

Table I. Gas-Phase Reaction Mechanism for H_2-O_2 Combustion

	Reaction	A	β	E_a
1	$H_2 + O_2 = OH + OH$	0.170×10^{14}	0.0	48151
2	$H + O_2 = OH + O$	0.142×10^{15}	0.0	16401
3	$OH + H_2 = H_2O + H$	0.316×10^{08}	1.8	3030
4	$O + H_2 = OH + H$	0.207×10^{15}	0.0	13750
5	$OH + OH = H_2O + O$	0.550×10^{14}	0.0	7000
6	$H + OH = H_2O + M$	0.221×10^{23}	-2.0	0
7	$H + H = H_2 + M$	0.653×10^{18}	-1.0	0

Particle-Phase Equations

The modeling of the iron concentrate particles was carried out from a Lagrangian viewpoint. The equation of motion for the concentrate particles in a Lagrangian framework is given by

$$\frac{d\vec{v}}{dt} = \frac{3}{4} C_D \frac{\rho_g}{\rho_p} \frac{1}{d_p} |\vec{u} - \vec{v}| (\vec{u} - \vec{v}) + \frac{(\rho_p - \rho_g)}{\rho_p} \vec{g} \quad (6)$$

where $\vec{u}$ is the gas velocity vector, $\vec{v}$ is the particle velocity vector, ρ_p is the particle density, d_p is the particle diameter, C_D is the drag coefficient, and $\vec{g}$ is the gravity vector.

The equation of energy for a particle moving along its trajectory is

$$m_p c_p \frac{dT_p}{dt} = h_s A_p (T_g - T_p) + Q_{rp} + H_r \quad (7)$$

where m_p is the particle mass, c_p is particle heat capacity, T_p is the particle temperature, h_s is particle convective heat transfer coefficient, A_p is the particle surface area, Q_{rp} is the net radiative heat-transfer rate to the particle, and H_r is the net rate of heat production by chemical reaction. The heat transfer correlation needed to calculate the convective heat transfer coefficient h_s can be found in the literature [4].

Reduction Kinetics of Magnetite Concentrate Particles

In this work, the reduction of magnetite concentrate particles to metallic iron by hydrogen is simplified by the following [5]:

$$Fe_3O_4(s) + 4H_2(g) = 3Fe(s) + 4H_2O(g) \qquad \Delta G^\circ_{1673\,K} = -13.6\ \text{KJ mol}^{-1} \tag{8}$$

The rate expression developed in this laboratory [5] was incorporated into the model. In its differential form, the rate expression is given by

$$\frac{dX}{dt} = 2k \cdot \left[p_{H_2}{}^{1/2} - \left(\frac{p_{H_2O}}{K} \right)^{1/2} \right] \cdot d_p{}^{s} \cdot (1-X) \cdot [-\ln(1-X)]^{1/2} \tag{9}$$

where X represents fractional conversion; k is the reaction rate constant (atm$^{-1/2}$ s^{-1} μm^{-s}); p_{H_2} and p_{H_2O} are the partial pressures (atm) of hydrogen and water vapor, respectively; K is the equilibrium constant for the equilibrium-limited reduction of wüstite to metallic iron [$FeO(s) + H_2(g) = Fe(s) + H_2O(g)$]; d_p is the particle diameter (μm); s is a size dependence factor; and t is time.

Sohn and coworkers [5, 6] defined the fractional reduction or 'reduction degree' as follows

$$X = \frac{n^\circ_O - n_O}{n^\circ_O} \tag{10}$$

where n°_O and n_O are the number of moles of oxygen bonded to iron in the concentrate particles before and after reduction, respectively. Therefore, the molar rate of oxygen consumption in a concentrate particle is given by

$$R_O = \frac{dn_O}{dt} = -2 \cdot n^\circ_O \cdot k \cdot \left[p_{H_2}{}^{1/2} - \left(\frac{p_{H_2O}}{K} \right)^{1/2} \right] \cdot d_p{}^{s} \cdot (1-X) \cdot [-\ln(1-X)]^{1/2} \tag{11}$$

Results and Discussion

Gas-Phase Validation

Benchmark experiments reported in the literature for a nonpremixed hydrogen jet flame [7-9] were simulated for model validation in this work. Figure 1 shows a schematic representation of the simulated geometry of the experimental hydrogen jet flame. Along the symmetry axis and the far-field boundary in the radial direction, the gradient of all dependent variables was set equal to zero, whereas the remaining boundary conditions were specified according to Table II.

Figure 2 shows predicted and experimental radial gas temperature profiles at various axial positions. The computer model yields a satisfactory prediction of the flame temperature.

Figure 3 presents the comparison between experimental and computed radial mass fraction for H_2, O_2, H_2O and OH close to the burner tip (z/L = 1/8). Overall, the computer model showed good agreement with the trends observed in the experimental measurements.

Table II. Boundary Conditions for the Simulation of the Nonpremixed Hydrogen Jet Flame [7-9]

Parameter	Value
Hydrogen Inlet	
Radius (R_{fuel}), mm	1.875
Axial velocity, m s^{-1}	296
Turbulent intensity, %	5
Hydraulic diameter, m	0.00375
Temperature, K	295
Mass Fraction, kg kg^{-1}	1.0
Air Coflow	
Axial velocity, m s^{-1}	1.0
Turbulent intensity, %	5
Hydraulic diameter, m	0.295
Outlet	
Turbulent intensity, %	5
Hydraulic diameter, m	0.3
Visible flame length (L), mm	675

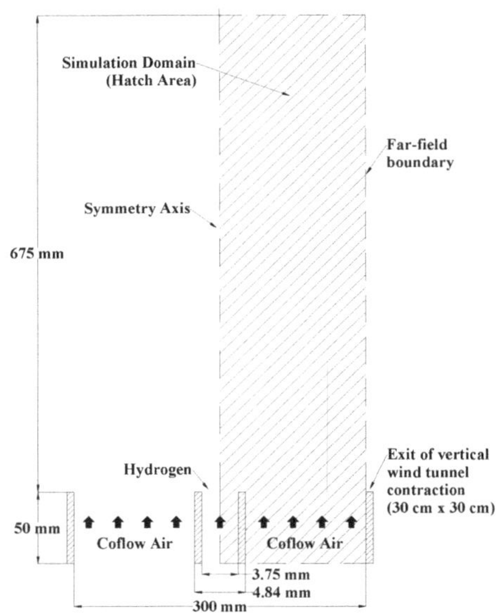

Figure 1. Schematic representation of the geometry of the experimental hydrogen jet flame [7-9], which was simulated in this work for model validation.

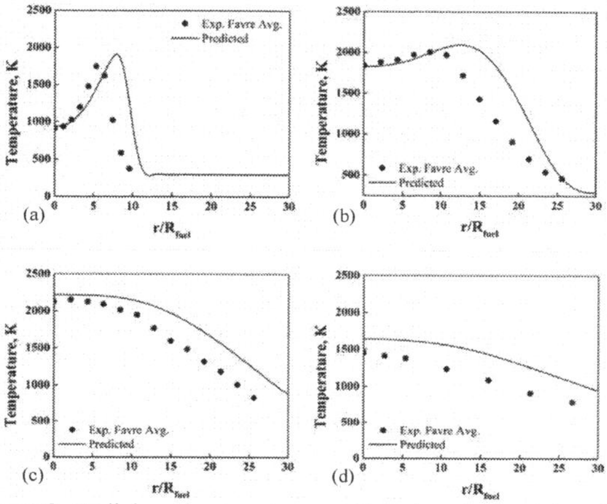

Figure 2. Radial gas temperature profiles at various axial positions.
z/L = (a) 1/8, (b) 3/8, (c) 5/8, and (d) 1.

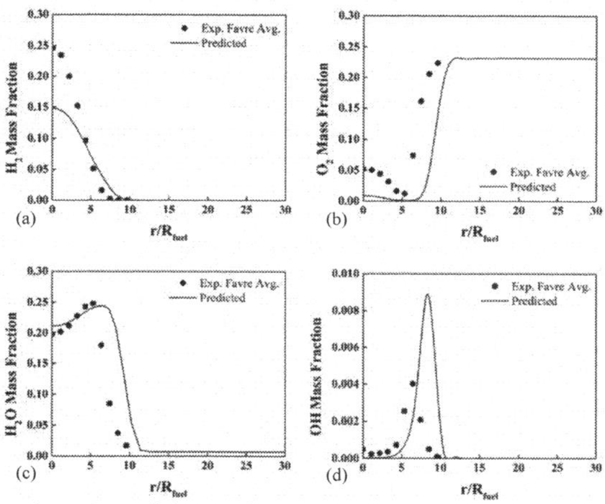

Figure 3. Radial mass fraction profiles at *z*/*L* equal to 1/8.
Species: (a) H_2, (b) O_2, (c) H_2O, and (d) OH radical.

Simulation of an Industrial Pilot Flash Reactor

The present computer model was used to simulate various design concepts of an industrial pilot flash reactor. The design objective is to determine the reactor volume (i.e., size) and configuration required to produce up to 7 tons of iron per hour (50,000 tpy) with a reduction degree of 92.5 %; that is, 90 % metallization. Hydrogen is the fuel/reductant.

In this work, the base reactor geometry (Figure 4) consists of a vertical cylindrical reaction shaft with four H_2-O_2 nonpremixed burners installed in the roof. Each burner is receded in the reactor roof and a refractory block that protects them from the high temperature inside the reactor is placed at each burner tip. The determination of the process flow rates and reactor dimensions (i.e., inner diameter and length) to be simulated involved material and energy balances, heat loss calculations through the composite reactor wall, and a nominal residence time of 3 seconds.

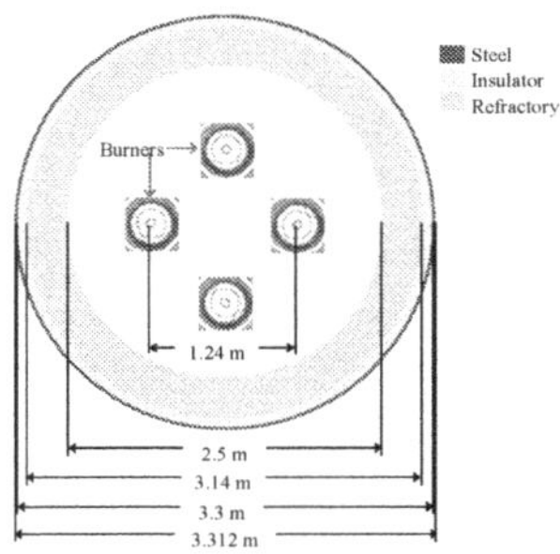

Figure 4. Cross-sectional view of the reactor roof perpendicular to the vertical direction.

Magnetite concentrate particles are fed at a rate of up to 10.4 ton h^{-1}, equivalent to 50,000 tons of iron per year. Thermal energy generated by the partial combustion of H_2 with O_2 maintains an off-gas target temperature of 1623 K and compensates for the reactor heat loss. Excess hydrogen is supplied to maintain an excess driving force in the off-gas of 0.5. Table III presents the operating conditions for two representative simulation runs of an industrial pilot flash reactor at a total pressure of 2 atm. The resulting reactor length was 20 m.

Table III. Operating Parameter for Representative Simulation Runs at a Total Pressure of 2 atm

Parameter	Run 143	Run 146
Production rate, tons of iron per year	50,000	25,000
Excess driving force	0.5	0.5
Flow rate of H_2, Nm^3 h^{-1}	23990	15800
Flow rate of O_2, Nm^3 h^{-1}	2250	1800
Amount of H_2 used/Fe, Nm^3 ton^{-1}	3460	4550
Amount of O_2 used/Fe, Nm^3 ton^{-1}	324	518

Figure 5 shows a profile of average conversion along the reactor axial position for a reactor producing 50,000 tons of iron per year operating at 2 atm. In this run, the iron concentrate was injected through the center of the reactor. The final conversion in this run was 56 %. Even though the nominal particle residence time under this condition was 3 seconds, the final reduction degree was much lower than the target value (i.e., 92.5 %).

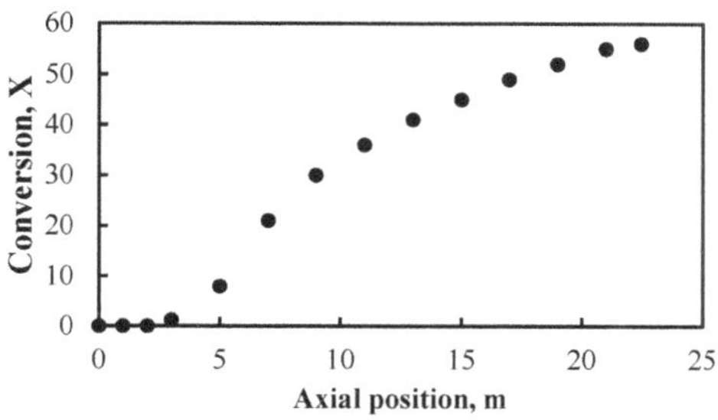

Figure 5. Computed axial profile of the average conversion for an industrial pilot flash reactor producing 50,000 tons of iron per year with an excess driving force of 0.5 at 1673 K.

To increase the conversion, we made the following changes: (1) increase the particle nominal residence time by decreasing the production rate and (2) inject the concentrate through the center of each burner. The latter change was intended to avoid the 3-m length of unreacted zone at axial positions close to the roof presented in the previous case [Figure (5)]. Figure 6 shows computed results for a production rate of 25,000 tons of iron per year in a reactor operating at 2 atm. As can be seen in Figure 6(a), there is a large temperature range with values between 1580 and 1722 K, which includes the target temperature of 1623 K. The concentrate particles start to react at shorter distances from the top of the reactor. A conversion higher than 90 % is reached at 7 m length and 99 % is reached up to the end of the reactor.

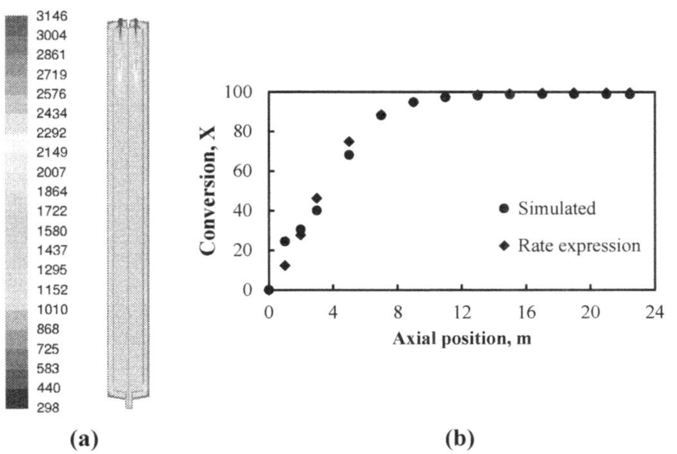

Figure 6. Computed results for an industrial pilot flash reactor producing 25,000 tons of iron per year with an excess driving force of 0.5 at 1623 K. Particles were injected through the centers of the burners. (a) Temperature contours (kelvins) and (b) axial profile of average conversion.

Concluding Remarks

A three-dimensional computational fluid dynamic model was developed for a novel flash ironmaking process. The model solves the turbulent governing equations of overall continuity, momentum, energy, and species transport with gas-phase and gas-solid reaction kinetics. The comparison of the model predictions with experimental data of a nonpremixed H_2 jet flame showed good agreement in terms of gas temperature and species concentrations. Distributions of velocity, temperature, and species concentrations, and reduction degree in a proposed industrial pilot flash reactor were computed and analyzed. The model was capable of simulating the interaction of the H_2-O_2 combustion with the presence of iron concentrate particles in a complex three-dimensional geometry. The importance of CFD modeling for predicting the reduction degree of concentrate particles realistically was demonstrated. The present model is a useful tool for further design and scale up of this novel flash ironmaking technology.

Acknowledgments

The authors acknowledge the financial support from the American Iron and Steel Institute (AISI) under the AISI's CO_2 Breakthrough Program. This material also contains the results of work supported in part by the US Department of Energy under Award Number DE-EE0005751.

References

1. ANSYS, Inc., *ANSYS FLUENT Theory Guide* (Canonsburg, PA: ANSYS, Inc., 2011).

2. B.F. Magnussen, "On the Structure of Turbulence and a Generalized Eddy Dissipation Concept for Chemical Reaction in Turbulent Flow" (Paper presented at the 19th AIAA Meeting, St. Louis, MI, January 12-15, 1981).

3. D.R. Eklund, J.P. Drummond and H.A. Hassan, "Calculation of Supersonic Turbulent Reacting Coaxial Jets," *AIAA J.*, 28 (9) (1990), 1633-1641.

4. J. Szekely, J.W. Evans, and H.Y. Sohn, *Gas-Solid Reactions* (New York, NY: Academic Press, 1976), 47.

5. H. Wang and H.Y. Sohn, "Hydrogen Reduction Kinetics of Magnetite Concentrate Particles Relevant to a Novel Flash Ironmaking Process," *Metall. Mater. Trans. B*, 44 (1) (2012), 133-145.

6. M.E. Choi and H.Y. Sohn, "Development of Green Suspension Ironmaking Technology Based on Hydrogen Reduction of Iron Oxide Concentrate: Rate Measurements," *Ironmaking Steelmaking*, 37 (2) (2010), 81-88.

7. R.S. Barlow, "Sandia H_2/He Flame Data – Release 2.0," http://www.ca.sandia.gov/TNF Sandia National Laboratories, 2003.

8. R.S. Barlow and C.D. Carter, "Raman/Rayleigh/LIF Measurements of Nitric Oxide Formation in Turbulent Hydrogen Jet Flames," *Combust. Flame*, 97 (3-4) (1994), 261-280.

9. R.S. Barlow and C.D. Carter, "Relationships Among Nitric Oxide, Temperature, and Mixture Fraction in Hydrogen Jet Flames," *Combust. Flame*, 104 (3) (1996), 288-299.

Celebrating the Megascale: Proceedings of the Extraction and Processing Division
Symposium on Pyrometallurgy in Honor of David G.C. Robertson
Edited by: Phillip J. Mackey, Eric J. Grimsey, Rodney T. Jones, and Geoffrey A. Brooks
TMS (The Minerals, Metals & Materials Society), 2014

APPLICATION OF THE COMBINED REACTORS METHOD FOR ANALYSIS OF STEELMAKING PROCESS

Simon N. Lekakh and D. G. C. Robertson

Missouri University of Science and Technology, 1400 N. Bishop, Rolla, MO 65409

Keywords: Steelmaking, Ladle metallurgy, Ar-mixing, Melt flow, Combined reactors

Abstract

A new integrated CFD-combined reactors approach is proposed for the description of processes in metallurgical vessels. CFD simulations were used to obtain the melt flow pattern in the vessels (ladle, tundish, and continuous caster mold). From these simulations, the characteristic curves were derived: (i) the residence time distribution curves (RTD) for flow-through systems (at tundish exit or at dendrite coherency surface in the mold) and (ii) the mixing curves for closed systems (ladle). In the next step, the melt flow was represented in a "combined reactors" system consisting of a combination of unit reactors (Plug Flow, Mixer, and Recirculated Volume). An inverse simulation was used to define the volumes of the reactor units and the melt flow rates between them by fitting to the characteristic curves from both methods (CFD and combined reactors). The suggested approach is demonstrated for multiple designs of Ar-stirred ladles, tundish, and SEN. This methodology can be used to enhance traditional post-processing CFD analysis and also as a tool for on-line process control.

Introduction

Physical modeling and CFD simulation of liquid metal processing are both carried out in order to achieve several major objectives, which can be classified into three levels:

Level 1: *to understand the process*, for example, the melt flow pattern, inclusion separation, and refining (chemical reaction) kinetics in metallurgical vessels;

Level 2: *to optimize existing and design new processes and equipment*, for example, to optimize the dam and weir geometry of an industrial tundish;

Level 3: *to control* the entire steelmaking process *on-line*, when several models need to be integrated in a fast-working algorithm, communicating with sensors and process control devices.

Knowledge of the melt flow phenomena is essential to meet these objectives. A broad review of the methods used to understand the nature of melt flow in the ladle, the tundish, and the mold (Level 1) can be found elsewhere [1-3]. For example, CFD simulations are used for visualization of melt flow; however this method alone is not effective in achieving Level 2 objectives and very complicated and time-consuming for on-line process control (Level 3 objectives). The combined reactors (CR) method was originally developed by Levenspiel [4] to analyze continuous chemical engineering processes as a set of interconnected unit reactors such as Continuous-Stirred-Tank Reactor (*Mixer*) and Plug Flow Reactor (*PF*). By solving the mass conservation equations, the flow can be characterized in the terms of the residence-time distribution (*RTD*) for continuous processes. The *RTD* curve in an actual process or simulations is obtained by short time dye injection into the in-flow stream and measuring the dye concentration in the out-flow stream. The combined reactors method was used by Sahai and Emi [5] for description of melt flow in a continuous casting tundish. The capabilities of different methods to solve different objective levels are ranked as high (H), low (L) or not applicable (N) in Table 1.

Table 1. Ranking the capabilities of the different methods

Objectives → / Methods ↓	Melt flow	Inclusion separation	Refining	Optimization	New design	Process control
Levels →	1	1	1	2	2	3
CFD + CR	H	H	H	H	H	H
Combined Reactors (CR)	N	N	H	L	L	N
CFD simulation	H	H	L	H	H	L
Physical modeling	H	H	L	L	L	N
Industrial measurements	H	H	H	L	L	N

In this paper, a new approach to analyze processes in metallurgical vessels is developed. The suggested method is based on CFD modeling of general melt flow in the real 3D-metallurgical vessel followed by inverse simulations using a virtual architecture of combined reactors to achieve a good match between the CFD and the CR characteristic curves (*RTD* or *Mixing*) by varying the volume of the unit reactors and the flow rates between them. This approach has a high capability to solve the different objectives shown in Table 1 and can combine high complexity with short simulation time (Figure 1).

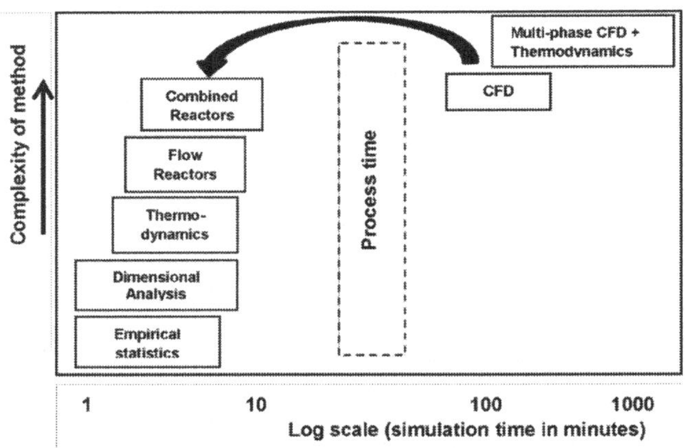

Figure 1. Complexity of methods versus simulation time.

Combined reactor design based on inverse optimization

CFD FLUENT 12.0 software [6] was used to solve 3-dimensional transient multiphase turbulent melt flow in a ladle with bottom *Ar*-plugs, tundish, and mold. Species transport was included to allow the derivation of the characteristic curves. A strategy of combined reactors (CR) design is suggested below for the achievement of a reasonable similarity between the CFD simulated characteristic curves with the characteristic curves from CR models. To represent the melt flow patterns, the CR architectures consist of combination of one plug flow reactor and two or three mixers, each loop-connected with recirculated volumes (RV). The specific architectures will be described below for ladle, tundish and mold. Mass conservation simulation tests showed [7] that these architectures were more suitable when compared to the set of unit reactors with dead volume [5].

In the CR architecture several parameters, such as the reactor volumes and the flow rates between mixers and recirculated volumes, needed to be calculated. The CR method used to represent "true" characteristic curves works as follows: (i) defining the architecture of the combined reactors by trial and error, based on general analysis of the CFD simulated melt flow pattern; (ii) building a CR mass conservation flow sheet with calculation of the characteristic curve for an arbitrary set of parameters of reactor volumes and flow rates; (iii) finding the best values of these parameters by fitting the calculated curve to the "true" curve by inverse optimization.

The flow process mass conservation equations for the different CR architectures were solved using Microsoft EXCEL. A function (φ) was minimized (Eq. 1) with the built-in EXCEL Solver:

$$\varphi = \Sigma(C_i^{true} - C_i^{calc})^2 \rightarrow min \qquad (1)$$

Combined Reactors Method for Different Metallurgical Vessels

Ar-Stirred Ladle. CFD FLUENT 12.0 software was used to solve isothermal transient multiphase turbulent melt flow in 100 ton ladle (2.0 m bottom and 2.4 m top diameters and 3.4 m height) with bottom *Ar*-plugs. Only the results for the central plug location are discussed here for reasons of brevity. To obtain the general geometry of the melt streams and their flow rate, iso-values of vertical (*Z*-direction) velocity were plotted at different levels from the bottom. Fig. 2a presents the negative (downward) melt *Z*-velocity and geometry of the rising plume (empty area). The *Ar*-flow rate affected the average downward vertical melt velocity (V_z) at different horizontal levels: increasing *Ar*-flow intensified melt downward flow at deep locations. The downward flow rate (tonne/s) was calculated based on the average V_z value over the area of downward flow (filled sections in Fig. 2a) and melt density. Downward flow intensity decreased from the top to the bottom of the ladle.

The generalized design of the 'architecture" of the combined reactors for the ladle was chosen as follows: the CFD simulation melt flow pattern was subdivided into four regions: the *Ar*-gas driven rising plume (*V0*), top horizontal layer (*V3*), central recirculated region (*V2*), and slow flow bottom layer (*V1*), as shown in Fig. 2b. The bottom volume (*V1*) was assumed to be plug flow and the others were assumed to be perfect mixers. The small top volume *V3* was assumed as 4% of the total ladle volume, the volumes *V1* and *V2* were included in the inverse simulation and volume *V0* was calculated as the remainder. Flow rates between reactors *V2* and *V1* (*F21*) and in the top of the rising plume *V0* (*F03*) were included in the inverse simulation based on the CFD data; all other flow rates were calculated based on mass conservation equations. The values of the four independent parameters included in the inverse simulation were varied to achieve similarity of mixing time between the CFD and the combined reactors.

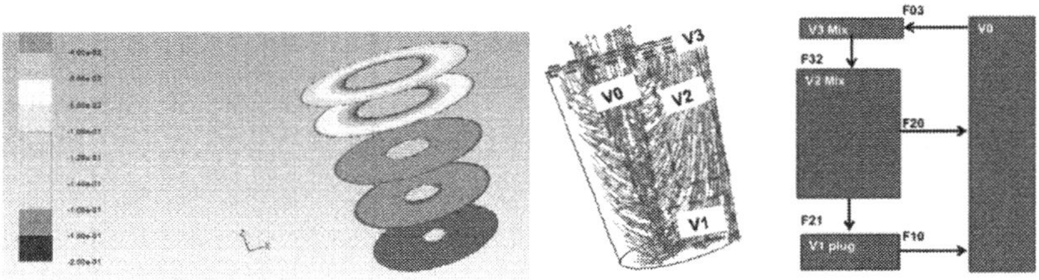

a) b) c)

Figure 2. Vertical velocity (V_z) values at different horizontal levels (a), vector velocity flow pattern from CFD simulation (b) and combined reactors architecture (c).

In the CFD simulation, a tracer was injected at four symmetrical points near the top of the melt surface of the *Ar*-stirred ladle. The relative tracer concentration was plotted as a function of time. Figure 3 illustrates CFD-simulated dye concentration curves for the bottom volume *V1* and the rising plume *V0*. The concentration curve in the bottom volume had a delay period - which is typical for plug flow. The mixing time was calculated based on 95% criterion. The *Ar*-flow rate affected the mixing time: 350 sec at 4 CFM and 220 sec at 20 CFM of *Ar* flow.

The similarity of melt flow in the CFD modeled *Ar*-stirred ladle and combined reactors was achieved by inverse simulation. The concentration curves in the bottom reactor *V1* were used for the inverse simulation. This region has a mixing delay when compared to the other ladle regions. Comparisons of the CFD and combined reactor mixing curves are given in Figure 3 for *Ar*-stirred ladle at 4 CFM and 20 CFM. After inverse simulation of the reactor volumes and flows between them, the combined reactors representation delivered identical to CFD mixing curve in the bottom volume *V1*. The agreement was not as good in other ladle regions, for example, in the rising plume (Figure 3); however, the mixing curves from both methods had similar averaged trends.

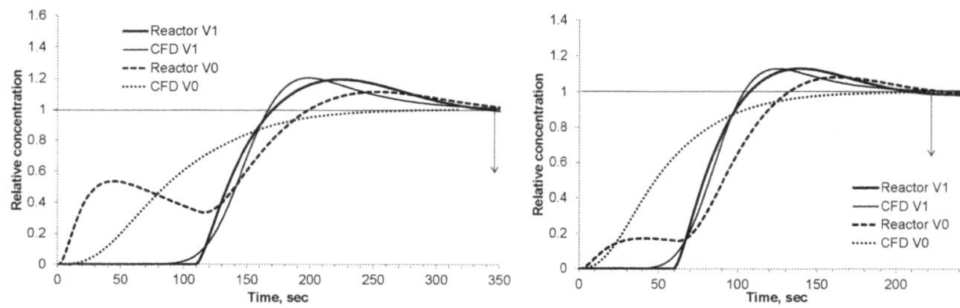

a) b)

Figure 3. CFD simulated and combined reactor modeled tracer concentration in the bottom region and plume stream of 100 t Ar-stirred ladle at 4 CFM (a) and 20 CFM (b). Arrows show mixing time.

Using the suggested combined reactors design (Figure 2c) the inverse simulation delivered a unique set of reactor parameters, including reactor volumes and melt flow rates. The results of inverse simulations are shown in Figure 4 for the centrally located *Ar*-plug at different *Ar* flow rates. Increasing *Ar* flow rate significantly decreased the bottom plug volume *V1*, had a minor effect on rising plume volume *V0*, and increased recirculated (*F20*) and plume (*F30*) melt flow rates. The combined reactor parameters describe the effect of increased *Ar*-flow rate by simultaneously increasing the overall melt recirculation flow rate and the volume of ladle regions involved in intensive mixing.

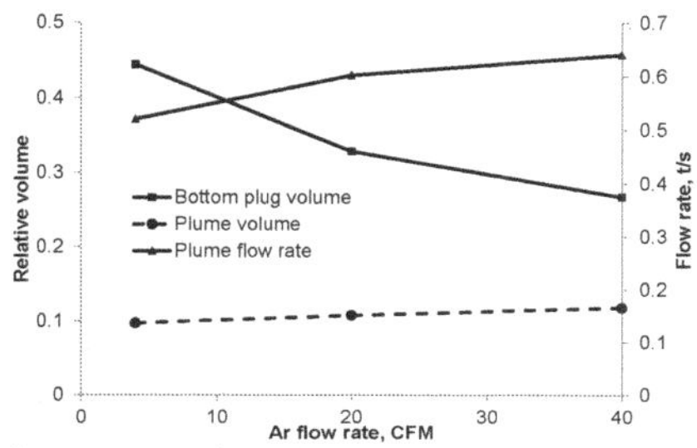

Figure 4. Effect of *Ar*-flow rate on the volumes of bottom "plug reactor" and plume "mixer reactor" and flow rate in plume.

Once the flow in the ladle is described then this can be used with a description of the kinetics of the metal-slag-gas reactions to calculate the refining and mixing in the ladle [8]. Figure 5 illustrates the predicted sulfur concentrations in the ladle in comparison to the experimentally measured values at the start and finish of refining [9]. There are large gradients in sulfur concentration in the metal between the different ladle regions initially, but these slowly disappear until the metal is more or less homogeneous at the end of the refining period.

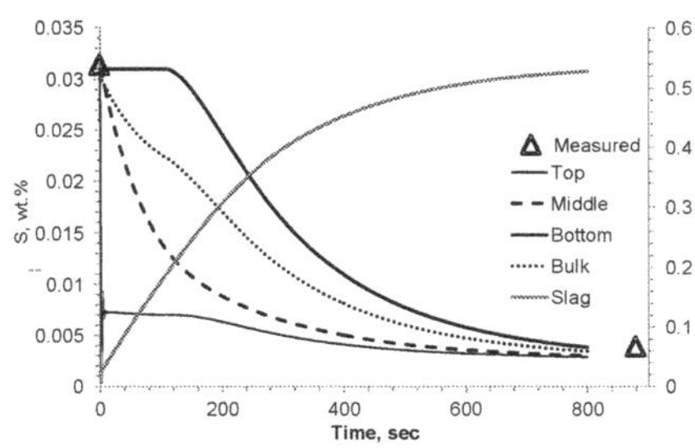

Figure 5. Comparison of predicted sulfur concentration in in the slag (right axis) and in the different ladle regions (left axis) with experimentally measured in the bulk steel at starting and finishing refining [9].

Tundish. The CR method is illustrated for one case: a single-strand tundish of 14 metric ton liquid steel capacity without flow control devices. The tundish had a trapezoidal shape with 10 degree vertical walls incline and 3 m long, 1 m wide and 0.8 m melt level (Figure 6a). The melt flow rate was varied from 2 t/min (slow) to 2.6 t/min (medium) and to 3.2 t/min (high). This gave 446, 340 and 270 seconds for the mean residence time values respectively. The tundish had a submerged entry nozzle (SEN) of 250 mm OD, and 150 mm ID submerged 300 mm into the melt, and a 200 mm OD stopper rod above a 70 mm diameter exit nozzle. CFD FLUENT 12.0 software was used to solve isothermal transient turbulent melt flow (Figure 6b); species transport was included to allow the derivation of the *RTD* curve (Figure 6c).

a) b) c)

Figure 6. Tundish design (a), vector velocity in central vertical section (b), RTD curves (c).

In the combined reactors architecture (insert in Figure 7 a), used by Sahai abd Emi [5] a plug flow reactor was followed by a mixer directly connected to dead volume. It can be seen that this simplified architecture of combined reactors was not in agreement with CFD simulated flow and the RTD_{CFD} curve at any combination of variables (reactor volumes and internal flow rate).

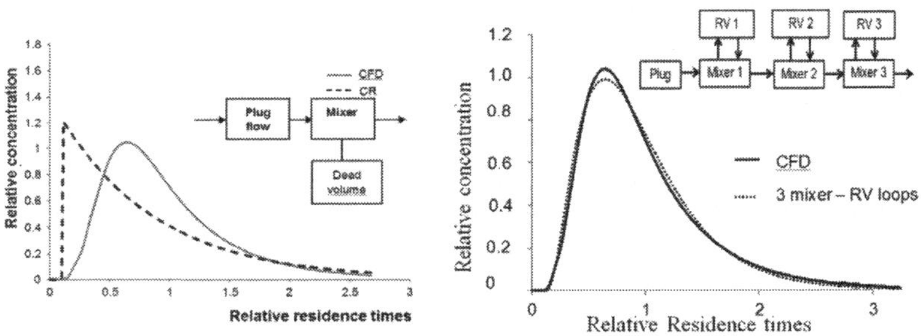

a) b)

Figure 7. Comparison of CFD simulated RTD_{CFD} curves with RTD_{CR} curves: a) obtained from the simplified combined reactor approach, and b) using the "plug plus 3-mixers-RV loops" architecture and inverse optimization.

Visual analysis of CFD simulated flow pattern (Figure 6b) indicated the existence of multiple recalculated zones in the tundish without flow control devices. To better represent this flow pattern, the combined reactor architecture contained a set of mixer reactors (Mixer) connected to recirculated volumes (RV) was suggested and optimized using inverse simulation (Figure 7b). Inverse simulation delivered a combination of the variables (Table 2) which indicated that RV volumes are negligible and can be omitted in this tundish design. The optimized structure consisted of a small plug volume connected to two or three in-line ideal mixers. This is a reasonable CR representation of the CFD-visualized flow pattern in the case of the tundish without flow control devices. The melt flow in tundish could be also characterized as an "eddy diffusion" or dispersed plug flow pattern [10]. In the case of a tundish with flow control devices and *Ar*-stirring, the CR approach well-represented the real flow pattern [7].

Table 2. Optimized parameters of combined reactors representing the tundish without flow control devices.

Reactor volume, part from tundish volume						
Plug	Mix 1	RV 1	Mix 2	RV 2	Mix 3	RV 3
0.14	0.20	-	0.41	0.01	0.2	0.01

Continuous casting mold. The SEN design, casting mold geometry, and casting speed all have a significant influence on steel quality through their effect on the flow pattern. The integrated CR approach was used to describe the melt flow in the mold. CFD simulations were performed to solve coupled transient multiphase turbulent melt flow and heat transfer in continuous casting. Modeling was done for a simplified geometry (½ of 1500 mm x 225 mm slab domain with square outport located on a central symmetry plane (Figure 8a). The parameter chosen to characterize the melt flow in liquid pool in the mold was the evolution of tracer concentration on the dendrite coherency surface (Figure 8b) assuming that mixing takes place only in the liquid pool. Inverse simulation was used to calculate the combined reactors volumes and flow rates by matching the residence time curves obtained from CFD simulation to those for the chosen combined reactors architecture (Figures 8c,d). Applying this approach, the effect of SEN design and casting speed on the melt flow and recirculation in the liquid pool was analyzed.

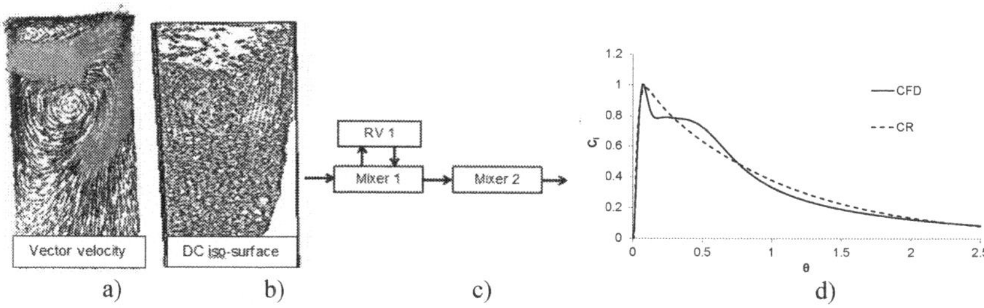

Figure 8. Continuous casting mold: a) CFD simulated vector velocity, b) dendrite coherency iso-surface, c) combined reactor architecture, and c) CFD simulated and inverse optimized RTD curves for combined reactor at dendrite coherency surface.

Conclusions

A combined reactors method is suggested for the "fast" and adequate representation of melt flow in metallurgical vessels. The examples described were the Ar-stirred ladle, the tundish and the continuous casting mold in steelmaking. CFD-modeling results were converted into melt flow in combined reactors, the architecture of which represented the different characteristic zones in the reactor. The flows between the zones were also specified. Similarity of melt flow between the CFD and combined reactors methods was achieved by matching the characteristic curves for: (i) melt mixing kinetics in the ladle, (ii) RTD curve at tundish exit and (iii) RTD curve at dendrite coherency iso-surface of the mold. The suggested combined reactors approach can be used for the description of the high level of complexity of pyrometallurical processes and implemented in an on-line fast-running algorithm for the control of such operations.

References

1. K. Chattopadhyay, M. Isac, and R. Guthrie: *Ironmaking and Steelmaking*, **8** (2010) 562.
2. L. Zhang, S. Taniguchi, and K. Cai: *Metallurgical and Material Transactions, B*, **31B** (2000) 253.
3. X. Huang and B. G. Thomas: *Metallurgical and Materials Transactions B*, **27B**, (1996) 617.
4. O. Levenspiel: Chemical Reaction Engineering, John Wiley & Sons, New York, third edition, 1999, 668.
5. Y. Sahai and T. Emi: *ISIJ International*, **6** (1996) 667.
6. Ansys Fluent 12.0. User's Guide, Ansys Inc., 2009.
7. S. N. Lekakh and D. G. C. Robertson: *ISIJ International*, **53** (2013) 622.
8. J. J. Peter, K. D. Peaslee, D. G. C. Robertson and B.G. Thomas: *AISTech Proceedings* (2005) 973.
9. L. Jonsson, D. Sishen, and P. Jonsson: ISIJ International. **38** (1998) 260.
10. S. P. Ferro, R. J. Principe, and M. B. Goldschmit: *Metallurgical and Materials Transactions B*, **12B** (2001) 118.

Celebrating the Megascale: Proceedings of the Extraction and Processing Division Symposium on Pyrometallurgy in Honor of David G.C. Robertson
Edited by: Phillip J. Mackey, Eric J. Grimsey, Rodney T. Jones, and Geoffrey A. Brooks
TMS (The Minerals, Metals & Materials Society), 2014

MODELLING OF SLAG FOAMING COUPLED WITH DECARBURISATION

M.A. Sattar, J. Naser, G. Brooks

Swinburne University of Technology

Hawthorn -3122 Australia

Keywords: CFD, Slag foam, Bubble break up, Coalescence, Decarburization

Abstract

CFD models have been developed and numerical simulations have been carried out to predict the formation of foam in oxygen steelmaking. Foam was considered as a separate phase comprising a mixture of gas and liquid. Bubble break up and coalescence models have also been incorporated in a CFD model to predict the number density of individual bubble classes. A population balance equation was used to track the number density of each bubble class. Decarburization with heat generation from chemical reactions was integrated in the process. User subroutines were written in FORTRAN to incorporate the foam formation, the bubble break up and coalescence rate and decarburisation in the main program. The model predicted the foam height, bubble number density, velocity of phases, decarburization, and turbulence. The result from the model has been compared with available data from literature and found to be in reasonable agreement with the experimental and plant data.

Introduction

The Basic Oxygen Steelmaking (BOS) process is widely used for the production of steel from pig iron. The production of steel from the BOS process was 67.1% of the total steel produced in the world using different processes in 2008 [1]. In the BOS process, a supersonic jet of oxygen is injected into the hot metal in a vessel that is protected by a basic inner liner. Oxygen is either supplied via a top lance on the liquid surface where the process is called top blown or through the bottom tuyeres and the process is called bottom blown [1]. The blowing of oxygen into the hot metal oxidizes carbon as well as impurities such as silicon (Si), manganese (Mn), Sulfur (S) and phosphorus (P) which are dissolved in the metal. These oxidation reactions are exothermic and at the same time the process generates large amounts of gas due to reaction of oxygen with carbon. Foam is formed in the process by the gas bubbles resulting from decarburisation. The formation of foam is important for the process because the large interfacial surface area between gas bubble and liquid phases in foam facilitate reactions in the process. Though foam is beneficial for the process, excessive foam has a detrimental effect on the process. Excessive foam can overflow the vessel; this is termed 'slopping' in the BOS. Slopping can reduce production and increase operation cost and sometimes can damage the vessel [2].

The formation of foam in oxygen steelmaking has been investigated by many researchers at laboratory scale [3-5]. In laboratory-scale experiments, the height of the foam is measured and the height is conveniently related to the superficial gas velocity, termed as the foaming index. Foam is formed by gas bubbles separated by a thin film of liquid, and the thin film is joined in a channel called the Plateau border channel. The drainage of liquid occurs throught the lamellae and the Plateau border channel. The drainage of liquid causes the coalescence and bursting of bubbles in foam. In an ordinary two-phase flow considering gas and liquid as working fluid, it is difficult to incorporate these characteristics of foam in a CFD model. In addition, foam is expected to accumulate at the top surface of liquid, but in the gas-liquid two-phase approach, it is not possible to keep the foam at the top surface of the liquid because gas diffuses into the gaseous atmosphere above the liquid. The above drawback is bypassed in this study by considering foam as a separate phase made up of gas and liquid.

In multiphase flow, gas bubbles split into smaller bubbles and aggregate to form bigger bubbles, due to interaction of bubbles with eddies or bubble-bubble interactions. The number density of each bubble class changes because of the aggregation of bubbles caused by coalescence and disintegration caused by break-up of bubble. The bubble break-up model of Luo and Svendsen [6] and the coalescence model of Prince and Blanch [7] are used in the present study. The bubble coalescence model based on film rupture by Tong et al. [8] has been used in the foam layer. Computational fluid dynamic (CFD) modeling, a type of numerical simulation, has been used for many years to investigate multiphase flow phenomena. Among the various techniques, the Euler-Euler and Euler-Lagrangian approaches are commonly used in multiphase flow. In the Euler-Lagrangian approach the dispersed phase is divided into particle or bubble and the particles or bubbles are tracked individually, whereas, in the Euler-Euler approach both the continuous and dispersed phases are considered to be interpenetrating continua, and bubbles or particles are introduced as a scalar in the gas phase. The time and cost of computation for tracking individual particles or bubbles in Euler-Lagrangian approaches is more than the Euler-Euler approach. Therefore, from computational considerations, the Euler-Euler approach is often more economical and commonly used [9, 10].

In this study, a thin slice of a 3D CFD model of a 6-tonne BOS converter has been developed, which is same as that of the pilot plant model of Millman et al. [11]. A 3D CFD model of a crucible, the same as that of the laboratory scale of Jiang and Fruehan [3], has also been developed. Numerical simulation was carried out based on the Euler-Euler approach. The model predicted foam height, bubble number density, decarburization and multiphase flow phenomena in the process with time. A comprehensive foaming model was incorporated into the model to account for foam formation and destruction. The model incorporated the bubble break-up and bubble coalescence model available in the literature [6-8]. In the present model, ten types of bubble class were considered and their number density was tracked using population balance modeling. The model incorporated the decarburization reaction with heat generation in the process.

Mathematical Model Used for the Modeling

The model was developed by writing user-defined subroutine (UDS) in FORTRAN language and incorporating these into the commercial CFD package AVL FIRE 2009.2. The Eulerian multiphase flow model solves the conservation of mass and momentum of each phase in each cell. The exchange of mass and momentum are dealt with by the interphase exchange term in the conservation equation. The mass conversion equation used for the present study is [11]:

$$\frac{\partial \alpha_k \rho_k}{\partial t} + \nabla . \alpha_k \rho_k V_k = \sum_{l=1, l \neq k}^{N} \Gamma_{kl} , \quad k=1,\ldots\ldots,N \tag{1}$$

Where α_k is the volume fraction of phase k, V_k is phase k velocity, and Γ_{kl} is the interfacial mass exchange between phases k and l. In the present study the gas liquid mixture is considered to be foam when the gas volume fraction is more than or equal to 0.75. The gas and liquid transformation mechanism is activated in the model when the gas fraction is 75% or above.

However, the amount of gas and liquid transformed into foam and vice versa are governed by equations (2)-(5). The mass transformation between gas and foam was calculated using the following equation [12]:

$$\Gamma_f = \frac{\frac{\pi}{6} d_i^3 \left(\rho_g \sum_i^n N_{i_g} - \rho_f \sum_i^n N_{i_f}\right)}{dt} = -\Gamma_g, \qquad N_{i_g} \leq N_{i_{av}} \tag{2}$$

Where N_{i_g} is the number of bubbles in gas phase transformed into foam phase, N_{i_f} is the number of bubble in foam transformed into gas phase, dt is time step and $N_{i_{av}}$ is the number of gas bubbles available in that cell. The number of gas bubbles transformed into foam bubbles was calculated from the equation [12]:

$$N_{i_g} = \frac{\alpha_l v_c}{n_p a_p l_i}, \alpha_g \geq 0.75 \tag{3}$$

Where α_l is volume fraction of liquid in the cell and v_c is the volume of that cell. The number of foam bubble transformed into gas bubble is calculated from the equation:

$$N_{i_f} = \frac{q_{PB}}{n_p a_p l_i}, \alpha_f > 0 \tag{4}$$

The mass transformation between liquid and foam was calculated using the following equation [13]:

$$\Gamma_f = \frac{n_p a_p l_i \left(\rho_l \sum_i^n N_{i_g} - \rho_f \sum_i^n N_{i_f}\right)}{dt} = -\Gamma_l, \qquad N_{i_g} \leq N_{i_{av}} \tag{5}$$

A population balance equation has been used as [12]:

$$\frac{\partial}{\partial t} \alpha_k \rho_k \emptyset_{ki} + \nabla . \alpha_k \rho_k V_k \emptyset_{ki} = \nabla . \alpha_k \rho_k D_{ki} \nabla_{\emptyset_{ki}} + S_i \tag{6}$$

Where S_i is the source term for bubble class i due to break up and coalescence in the foam and gas-liquid dispersion. The source term of break-up and coalescence of bubbles in the scalar transport equation has been updated using the daughter bubble distribution model with rectification [13]. The break up rate of bubble class i into k colliding with the eddy size λ_j can be written as [6]:

$$\Omega_B(\vartheta_k, \lambda_j, \vartheta_i) = \frac{\pi}{4} \left(d_i + \lambda_j\right)^2 \overline{u_\lambda} n_\lambda n_{d_i} \frac{12 C_f \sigma}{\beta \rho_c \varepsilon^{2/3} d_i^{5/3} \xi^{11/3}} \tag{7}$$

Where $\overline{u_\lambda}$ is the turbulent velocity of eddy λ_j, n_λ is the number of eddy and n_{d_i} is the number of bubble of class i. In the present study the turbulence and laminar shear collision rate is considered. The turbulent collision rate is given by Prince and Blanch [7]:

$$\theta_{ij}^T = 0.089 n_i n_j (d_i + d_j)^2 \varepsilon^{1/3} (d_i^{2/3} + d_j^{2/3})^{1/2} \tag{8}$$

The laminar shear collision rate is given by Prince and Blanch [7]:

$$\theta_{ij}^{LS} = \frac{4}{3} n_i n_j \left(r_i + r_j\right)^3 \left(\frac{d\overline{U_l}}{dR}\right) \tag{9}$$

The total coalescence rate Γ_{ij}:

$$\Gamma_{ij} = \{\theta_{ij}^{T} + \theta_{ij}^{LS}\} \times exp\left[-\frac{\left\{\frac{r_{ij}^{3}\rho_l}{16\sigma}\right\}^{1/2} \varepsilon^{1/3} \ln\frac{h_0}{h_f}}{r_{ij}^{2/3}}\right] \quad (10)$$

Where the equivalent radius $r_{ij} = \frac{1}{2}\left(\frac{1}{r_{bi}} + \frac{1}{r_{bj}}\right)^{-1}$, initial and final thickness for air-water is 10^{-4}m and 10^{-8}m. The coalescence rate $\Omega_{C_{i,j}}^{f}$ of a bubble of the ith group with bubbles of the jth group is [8]:

$$\Omega_{C_{i,j}}^{f} = n_{p_i} P_{i,j} \frac{n_{p_i} n_i}{\sum_{j=1}^{M} n_{p_j} n_j} \quad (11)$$

Where n_{p_i} is the number of films per bubble of the ith group, $P_{i,j}$ is the failure rate of films separating bubbles of the ith group. Amount of liquid per bubble, $\emptyset_i/d_i$ so the time required to drain out this liquid is $\tau = (\emptyset_i/d_i)/q_{PB}$. The liquid drainage due to gravity through Plateau border derived by Bhakta and Ruckenstein [14] is used in the present model which can be written as:

$$q_{PB} = \frac{3}{15} NR n_p a_p \frac{C_v a_p \rho g}{20\sqrt{3}\mu} \quad (12)$$

The cross section of the Plateau border $a_p = 0.161{r_p}^2$. The reaction of carbon with oxygen is:

$$2C + O_2 = 2CO \quad (13)$$

The decarburisation rate in each cell of the calculation domain was calculated from the following equation:

$$\frac{dC}{dt} = k\alpha_l v_c \rho_l (\emptyset_c - \emptyset_{eq}) \quad (14)$$

Here k is the rate constant, α_l is the volume fraction of liquid phase in the cell, v_c is the volume of the cell, ρ_l is the density of liquid, $\emptyset_c$ is the carbon fraction in the cell and $\emptyset_{eq}$ is the equilibrium carbon fraction. The equilibrium carbon fraction $\emptyset_{eq}$ is considered as equal to the carbon fraction at the end of the blowing time. The constant k was estimated from the slope of decarburization vs time graph of pilot-plant data [10] and was found to be 5.45e-03 %C/s. The authors are fully aware that decarburisation follows more complex kinetic equations but used this simple approach to demonstrate the general feasibility of the approach.

Model Geometry and Methodology

The simulations were carried out in a crucible of height 12cm and diameter of 4.5cm. Initially the crucible was filled up to 4cm of slag from the bottom and 8.7cm of gas above the slag as shown in the Figure 1(a). The properties of slag used for the CFD modeling were taken from Jiang and Fruehan's [3] laboratory model. In the present model the minimum diameter of the bubble was considered as 2.5mm and the maximum diameter of the bubble was considered as 20mm. For the pilot-plant model, the simulations were carried out in a thin slice of a 6-tonne BOS converter with height 2.905m and a diameter of 1.340m. Initially the BOS converter was filled up to 660mm from the bottom with the liquid iron and there was gas above the liquid iron, as shown in Figure 1(b). In this model, an unsteady state multiphase solution for momentum and continuity were used, and, for turbulence, a standard $k - \varepsilon$ turbulence model was used. A cell-centered finite volume approach was used to discretize the governing equations and the resulting discretized equations were solved iteratively using a segregated approach. In the pilot-plant model of Millman et al. [10], oxygen enters into the system through the lance tip with supersonic velocity. In this study, the exit diameter was considered as 268mm, at a distance 496mm from the original nozzle exit where the Mach number was 0.0322. Velocity boundary conditions were applied to match the flow rate of the pilot-plant model of Millman et al. [10]. The top surface of the BOS converter was specified as the outlet. Static pressure boundary conditions were applied at the outlet. The wall boundary type for this model is considered as a solid wall, and fixed temperature obtained from the pilot-plant model [10] was assigned.

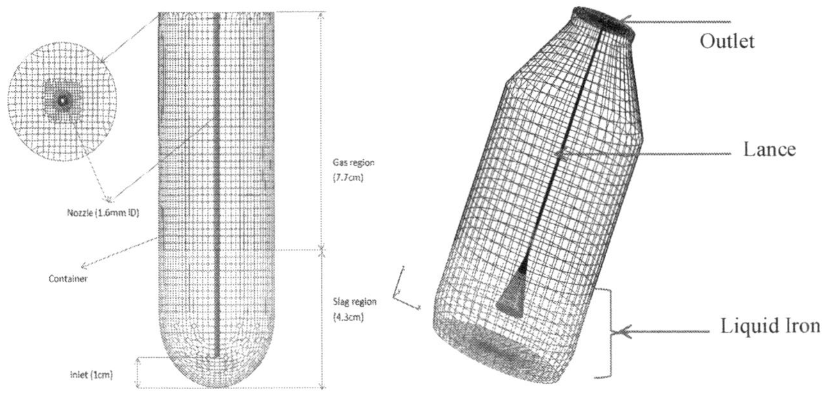

Figure 1. Mesh generated for the simulation (a) Crucible (a) Pilot-plant model

The density and viscosity of foam was calculated by the following equations:

$$\rho_f = \rho_l \alpha_l + \rho_g (1 - \alpha_l) \quad (14)$$

$$\mu_f = \mu_l \alpha_l + \mu_g (1 - \alpha_l) \quad (15)$$

Two different grids of the model were generated for both cases, and simulation was performed on both grids, and the convergence of the solution has been verified in both cases.

Table I. Properties of the fluids used for the pilot-plant model [15]

Properties	Fe	Oxygen
Density (kg/m^3)	7030	0.233
Viscosity (Ns/m^2)	1.875E-03	6.68E-05

Results and Discussion

The foaming index was evaluated and presented in Figure 2 from the CFD model of a crucible along with the data from Jiang and Fruehan [3] for a slag of CaO-SiO_2-FeO and CaO-SiO_2-Al_2O_3-FeO as well as the predicted data from Ito and Fruehan [4]. It can be seen from Figure 2 that in all cases the foaming index decreases with increase of FeO (for Jiang and Fruehan [3] when pct FeO > 2). The foaming index increases with viscosity and is inversely proportional to the square root of surface tension [3]. The net result for these slags is that the foaming index decreases with an increasing FeO content [3]. Increasing FeO in slag decreases the viscosity and increases surface tension. The above phenomena can be attributed to the fact that increasing FeO in slag decreases the viscosity of the slag. The drainage of liquid in foam increases with a decrease in viscosity. So a decrease in the height of the foam decreases the foaming index.

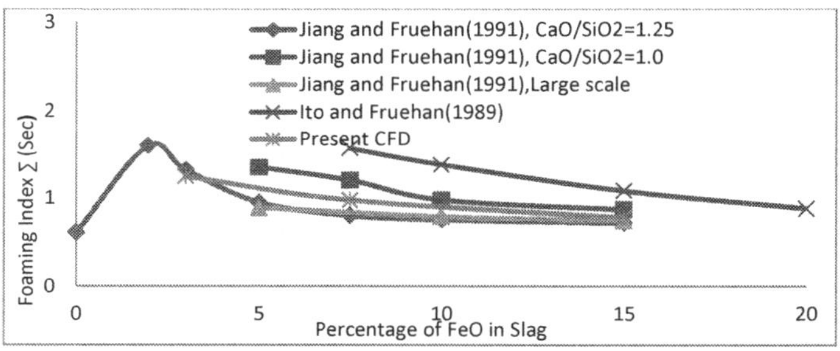

Figure 2. Foaming index of slag with different composition at 1773K.

The average height of foam for the pilot-plant model at different blowing times is presented in Figure 3. The figure shows that the foam height fluctuates but keeps on increasing in the first 7 minutes, then reaches a quasi-steady-state at around 9 minutes. The carbon content in liquid metal is higher at the beginning of blowing, therefore more oxygen reacts with the carbon dissolved in the liquid and produces a large volume of gas at the beginning of injection. The predicted foaming height is higher than that of the pilot-plant data for the first 4 minutes. The present CFD model over-predicted the foam height during the first 4 minutes of blowing. The reaction of Fe, Si, S, P, Mn with oxygen produces oxides, consuming the available oxygen. The decarburization reaction, which also requires oxygen, has to compete with the above-mentioned oxidation reactions for the availability of oxygen. In the present CFD model only the decarburization reaction is considered so the amount of CO gas produced is more than expected. As the production of gas (CO) is more than expected, the formation of foam and its height is also more than expected. At the initial stage, the flux added in the pilot-plant model does not dissolve completely in the liquid. Therefore, up to 4 minutes, the predicted foaming height is more than the pilot-plant data. On the other hand, the foaming height in the present CFD model is lower than the pilot-plant data during minutes 5-8. This can be attributed to the fact that with time the amount of carbon in the liquid decreases, as can be seen in Figure 4. In the present study, the reaction of carbon with oxygen was considered in gas and liquid phases only. However, in reality, the pure oxygen not only reacts with dissolved carbon in the bulk liquid, but CO_2 in the gas phase also reacts with dissolved carbon. There is also reaction between FeO in the slag and carbon droplets suspended in the emulsion. This reaction also produces CO gas, and this gas increases the volume of the foam due to expansion of the foam volume. These reactions were ignored in the present study.

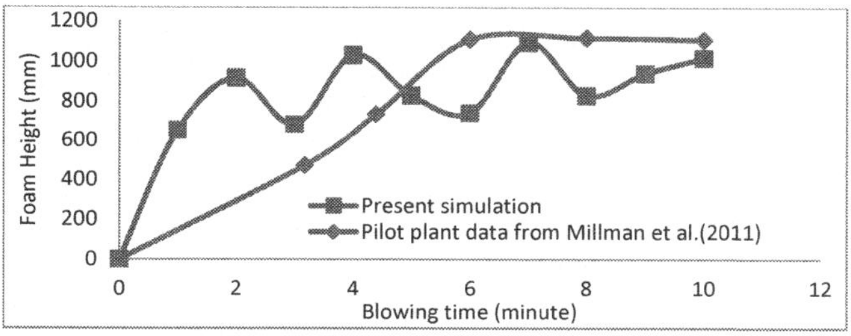

Figure 3. Foaming height as a function of process time

This simplistic approach still succeeded in providing a relatively accurate determination of foam height. A study is currently underway at Swinburne University of Technology to incorporate slag formation reactions and different mechanism of decarburization into the calculations.

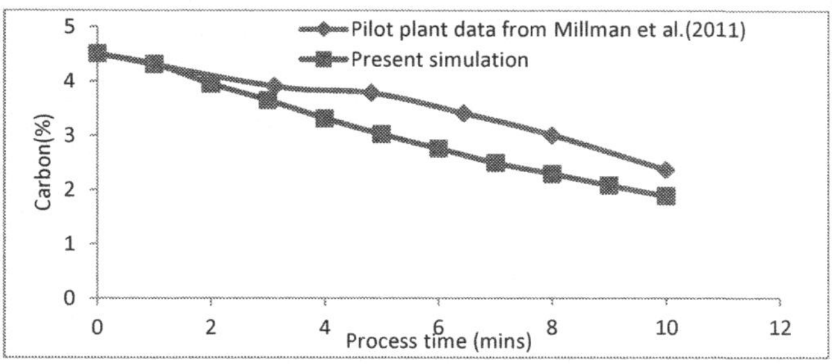

Figure 4. Carbon content as a function of process time

Conclusions

In this study, a CFD model of a pilot-plant converter (the same as that of the pilot plant model of Millman et al.[10]) and a CFD model of a crucible (the same as that of the laboratory model of Jiang and Fruehan [3]) have been developed. The results from the CFD model of the crucible show that the foaming index decreases with an increase of FeO content in slag. This phenomenon was attributed to the fact that increasing FeO in slag decreases the viscosity but increases the surface tension. The drainage of liquid in foam increases with the decreases of viscosity resulting in a decrease of the foaming index. The results from the CFD model of BOS show over prediction of the foaming height at the initial stage. This phenomenon was attributed to the fact that the present CFD model over-estimates the decarburization reaction that results in higher foam formation. The predicted results from the present study are in reasonably good agreement with the experimental data. The discrepancies between results of the present CFD model and plant data were due to the assumptions made to avoid complexity and numerical schemes and the turbulence model used to simplify the simulation.

Reference

[1] M. Brämming, "Avoiding Slopping in Top-blown BOS Vessels", Lulea, Sweden 2010.

[2] C. Nexhip, S. Sun, and S. Jahanshahi, "Physicochemical properties of foaming slags", *International Materials Reviews,* 49, (2004) 286-298.

[3] R. Jiang, and R. J. Fruehan, "Slag foaming in bath smelting", *Metallurgical Transactions B,* 22, (1991) 481-489.

[4] K. Ito, and R. J. Fruehan, "Study on the foaming of CaO-SiO2-FeO slags: Part II. Dimensional analysis and foaming in iron and steelmaking processes", *Metallurgical Transactions B,* 20,(1989) 515-521.

[5] H. Luo, and H. F. Svendsen, "Theoretical model for drop and bubble breakup in turbulent dispersions", *AIChE Journal,* 42, (1996) 1225-1233.

[6] M. J. Prince, and H. W. Blanch, "Bubble coalescence and break-up in air-sparged bubble-columns", *AIChE Journal,* 36, (1990)1485-1499.

[7] M., Tong, K. Cole, and S. J. Neethling, "Drainage and stability of 2D foams: Foam behaviour in vertical Hele-Shaw cells" *Colloids and Surfaces A: Physicochemical and Engineering Aspects,* 382, (2011) 42-49.

[8] R., Bannari et al., "Three-dimensional mathematical modeling of dispersed two-phase flow using class method of population balance in bubble columns", *Computers & Chemical Engineering,* 32, (2008) 3224-3237.

[9] M. R. Bhole, J. B. Joshi, and D. Ramkrishna, "CFD simulation of bubble columns incorporating population balance modeling", *Chemical Engineering Science,* 63, (2008) 2267-2282.

[10] M. S., Millman, et al., "Study of refining performance in BOS converter", *Ironmaking & Steelmaking,* 38, (2011) 499-509.

[11] AVL-FIRE 2008. *CFD Solver Manual,* Austria, AVL, Graz,

[12] M. A., Sattar, J. Naser, and G. Brooks, "Numerical simulation of creaming and foam formation in aerated liquid with population balance modeling", *Chemical Engineering Science,* 94, (2013a) 69-78.

[13] M. A., Sattar, J. Naser, and G. Brooks, "Numerical simulation of two-phase flow with bubble break-up and coalescence coupled with population balance modeling", *Chemical Engineering and Processing: Process Intensification,* 70, (2013b) 66-76.

[14] A. Bhakta, and E. Ruckenstein, "Decay of standing foams: drainage, coalescence and collapse, *Advances in Colloid and Interface Science,* 70, (1997) 1-124.

[15] Y.Cengel, and J. Cimbala, "Fluid Mechanics Fundamentals and Applications", McGraw-Hill Education. 2013

Celebrating the Megascale: Proceedings of the Extraction and Processing Division Symposium on Pyrometallurgy in Honor of David G.C. Robertson
Edited by: Phillip J. Mackey, Eric J. Grimsey, Rodney T. Jones, and Geoffrey A. Brooks
TMS (The Minerals, Metals & Materials Society), 2014

A Methodology for Modeling Electromagnetic Confinement Systems: Application to Levitation Melting

Nagy El-Kaddah[1], Thinium T. Natarajan[2]

[1]The University of Alabama
Dept. of Metallurgical and Materials Engineering; Tuscaloosa, AL 35487, USA
[2]United States Steel Corporation
Research & Technology Center; 800 E. Waterfront Drive; Munhall, PA 15120, USA

Keywords: Electromagnetic confinement, Modeling, Finite element method, Levitation melting.

Abstract

A modeling strategy is presented for computing the electromagnetic field and the shape of the molten metal in electromagnetic confinement systems. This strategy involves the use of a hybrid finite element/integral technique to calculate the electromagnetic field and force distribution in the melt. The free surface shape is determined from minimization of electromagnetic, gravitational and surface tension energies using the Lagrange method of multipliers. This approach was applied to model the electromagnetic levitation melting process. The model was found to accurately predict the measured shape of levitated droplets.

Introduction

The use of electromagnetic forces for containing and controlling the shape of molten metals is fairly widespread in the metallurgical industry. Applications include levitation melting, electromagnetic casting of aluminum, and edge confinement in twin roll casting of aluminum and steel. Melt confinement in these applications is accomplished by subjecting the metal to a high frequency oscillatory magnetic field generated by an induction coil of suitable geometry. The interaction between the induced currents and the associated magnetic produces a force field in a thin layer at the outer surface of the metal, commonly referred to as skin depth. The magnetic pressure differentials of the force field across the skin depth pinch and deform the free surface of the liquid metal until the magnetic pinch force balances the hydrostatic, dynamic, and surface tension pressures at every point on the free surface. In addition, the force field also produces a net upward force that may levitate and support the metal in space against gravity.

Modeling electromagnetic confinement systems involves the calculation of the electromagnetic field in the system and the resulting shape of the liquid metal. There are two basic approaches for determining the shape of the free surface. The first involves balancing the surface tension, gravitational, and electromagnetic pressure at every point on the free surface [1-8], while the second determines the shape of free surface of the liquid metal from minimization of an appropriate energy functional [9-11]. Although the first method is simpler to solve than the second one, it is does not ensure the conservation of the total volume of the metal. Calculation of the electromagnetic field is generally accomplished using well established finite difference/finite element techniques. However, for modeling electromagnetic confinement systems where the shape of the liquid metal is not known *a priori,* the field calculation must be carried out in every step during the search of the equilibrium shape. This makes conventional finite element/finite difference methods which require the solution of the field in free space computationally inefficient. This problem can be alleviated by using differential/integral formulation of the electromagnetic field [12-14] or using the boundary element method [15-17]. Although the boundary element method is far more efficient than other methods, it is prone to errors when the radius of curvature is comparable to the skin depth.

This paper describes an efficient algorithm for determining the shape of the liquid metal in electromagnetic confinement systems. The electromagnetic field problem in this algorithm is solved using a hybrid finite element/integral formulation of the field, while the shape of the free surface is determined using the Lagrange method of multipliers. This methodology was demonstrated in modeling levitation melting.

Formulation

Calculation of the Electromagnetic field

Consider a molten metal droplet placed in a time varying magnetic field generated by an induction coil characterized by axial symmetry shown in Fig. 1. For this system, the induced current has only azimuthal component and varies in r and z directions. For such a system, the field formulation can be considerably simplified by using the magnetic vector potential A, which has only θ component. For a time-harmonic field of frequency, ω, the induced current can be readily calculated from A_θ using:

$$J_\theta = -j\sigma\omega A_\theta \quad (1)$$

where σ is the electrical conductivity of the droplet, and j is $\sqrt{-1}$.

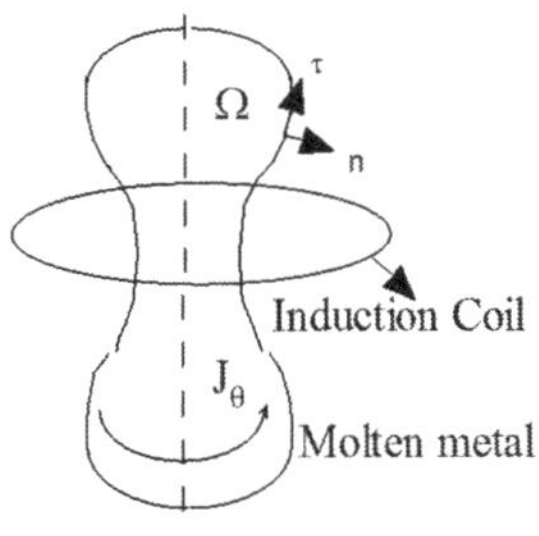

Figure 1. Sketch of liquid metal contained within an induction coil

The differential equation describing, A_θ, in cylindrical coordinates may be written as:

$$\frac{1}{r}\frac{\partial}{\partial r}\left(r\frac{\partial A_\theta}{\partial r}\right)+\frac{\partial^2 A_\theta}{\partial z^2}=\frac{A_\theta}{r^2}+j\sigma\mu_0\omega A_\theta \quad (2)$$

where μ_o is the magnetic permeability.

To avoid gridding and solution of A in free space, Biot-Savart law was used to provide A_θ on the boundary of the finite element domain, which may be expressed as:

$$A_\theta = \frac{\mu_0}{4\pi}\left(\sum_{k=1}^{inductor} I_k \oint \frac{dl}{|\mathbf{r}-\mathbf{r}'|^2} + \int_{\Omega'} \frac{J}{|\mathbf{r}-\mathbf{r}'|^2} dV'\right) \tag{3}$$

The first term in the right side of this equation represents the applied vector potential as result of the coil, while the second part represent the induced vector potential as result of induced currents in the melt.

Using the Bubnov-Galerkin method, the global finite element matrix equations of equation 2 with specified boundary condition can be written as:

$$\left[\sum_{j=1}^{Nel}([K_1]+[K_2]+[K_3])\right]\{A_\theta\} = 0 \tag{4}$$

where:

$$[K_1] = 2\pi \int_{\Omega_j} \left(\frac{\partial N^T}{\partial r}\frac{\partial N}{\partial r} + \frac{\partial N^T}{\partial z}\frac{\partial N}{\partial z}\right) r\, dr\, dz$$

$$[K_2] = 2\pi \int_{\Omega_j} \frac{N^T N}{r} dr\, dz$$

$$[K_3] = j2\pi\sigma\mu\omega \int_{\Omega_j} N^T N\, r\, dr\, dz$$

The solution of the finite element matrix equation yields the induced current distribution in the melt needed for computing the net lifting force, F_l, on the droplet, and the magnetic pressure, P_m, on the surface.

$$F_l = \int_{\Omega} \frac{1}{2}\mathrm{Re}\left(J \times \overline{B}\right) dV \text{, and} \tag{5}$$

$$P_m = \frac{B_s \cdot \overline{B}_s}{4\mu}. \tag{6}$$

In this work, the magnetic flux density, B, in these equations, is also calculated from the current distribution using Biot-Savart law. The bar in these equations denotes the complex conjugate.

Calculation of the Equilibrium Shape

The free boundary problem can be described by the variational statement of electromagnetic energy, E_m, surface tensional, E_γ, and gravitational, E_g, energies of the liquid is given by [9]:

$$\delta E = \delta\left(E_\gamma + E_g - E_m\right) = \oint_s \left(\gamma K + \rho g z + P_m\right)\delta l\, \delta S = 0 \tag{7}$$

together with the statement of conservation of the total volume of the metal:

$$\delta V = \oint_s \delta l\, \delta S = 0 \tag{8}$$

where γ is the surface tension, K is the curvature of the free surface, ρ is density, V is the volume of the melt, and l and S are displacement vector and surface area of the free surface, respectively.

Using the Lagrange method of multiplier, the augmented functional, W, of the energy functional, E, which satisfies the volume constraint to the first order may be expressed as:

$$W(\zeta,\eta) = E(\zeta) + \lambda\left(V(\zeta) - V_0\right) \tag{9}$$

where λ is the Lagrange multiplier, and V_0 is the volume of the liquid.

Minimization of this functional involves the determination of the Lagrange multiplier through the displacement of the free surface of the assumed shape of the droplet from the following equation

$$\delta l = \varepsilon \frac{\partial W}{\partial \zeta} \tag{10}$$

where ζ is the normal coordinate of the free surface, ε is a constant (step size), and its value depends on the difference in W between iterations. From Equations 7 to 10, the normal displacement vector is given by:

$$l(\zeta) = \varepsilon \frac{\left(P_m(\zeta) + \rho g z(\zeta) + \gamma K(\zeta)\right)}{\lambda} \tag{11}$$

and λ is given by:

$$\lambda = \frac{\int_s (P_m + \rho g z + \gamma K)\, dS}{\int_s dS} \tag{12}$$

Application to Levitation Melting

To test the accuracy of the present methodology, calculations were carried out using the system described in [10], where experimental data are available regarding the shape and location of the levitated aluminum droplet. The configuration and dimensions of the levitation coil are shown in Fig. 2. In essence, it consists of two opposing set of cylindrical coils; the lower four turns are for levitation, while the upper two turns are generally used for positioning the droplet within the coil. The field and shape calculations were carried out for 0.9740 gm droplet. The coil current and frequency were 168 A and 223 kHz, respectively. The physical and electrical properties used in the calculation are summarized in Table I.

In this work, the levitated droplet was meshed using 31 radial lines placed at equal angles from each other with at least five grids in the skin depth to ensure accurate field solution. For the 405 elements used in the calculation of the electromagnetic field, it took about 5 iterations to obtain a converged field solution. The value of the constant ε used in this study was 0.03.

TABLE I. PHYSICAL PROPERTIES OF ALUMINUM

Density	2360 Kg/m^3
Surface tension of liquid Aluminum	0.52 N/m
Electrical conductivity	4132000 mho/m

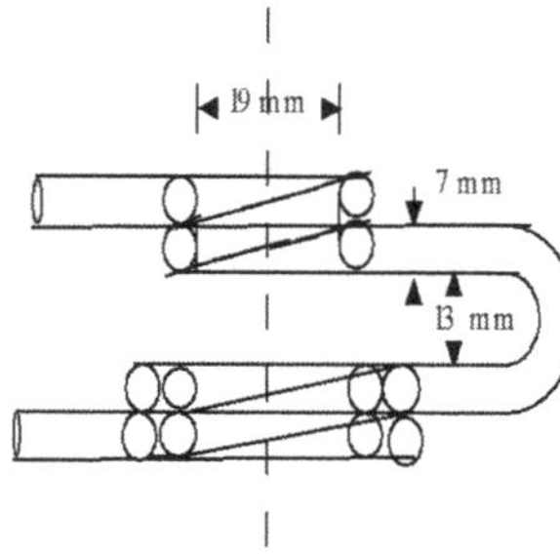

Figure 2. Sketch of the levitation coil used in this work

<u>Method of Resolution</u>

The following describes the iterative procedure developed for calculating the equilibrium shape of levitated liquid metal droplets. Starting with a guessed shape, typically a sphere, the solution begins by meshing and computing the electromagnetic field in the droplet. From the solution of the field, the location of the droplet within the coil is adjusted from the balance of the net lifting force and weight of the droplet. After finding the exact position of the droplet in the coil, the shape of the droplet at that location is calculated. These steps are repeated until the droplet assumes fixed location and shape.

To implement and efficiently execute this procedure in a computer code, meshing of the droplet was carried out along the radial lines connecting the geometrical center and free surface $\{\Gamma = f(R_o,\theta)\}$ of the droplet. The advantage of this scheme is that it utilizes computed coordinates of the displaced surface nodal points from shape calculation module to grid the droplet for field calculation in the next iteration. In addition, it preserves the initial fine grids in the skin depth during the search for the equilibrium shape as illustrated in Fig. 3.

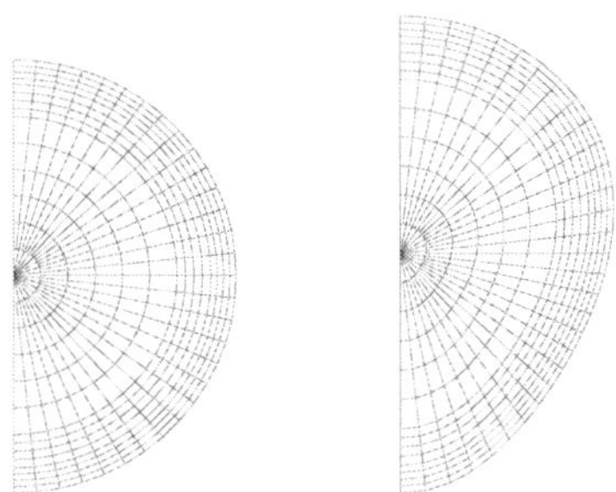

Figure 3. Finite element mesh of the droplet before and after deformation.

Computed Results

Figs. 4a and 4b show typical contour plot of the computed real and imaginary induced current distributions in the levitated droplet. It is seen that the induced current is predominantly in the lower half of the droplet, and penetrates to about one-third of its radius. It is interesting to see that decay of the induced current density deviates from this behavior over most of the induced current region

.

Fig. 5 shows a typical computed electromagnetic force distribution in the levitated droplet. The electromagnetic forces are essentially in the lower part of the droplet, as required for levitation, and confinement forces are normal and inward to the free surface. In noting that droplet deformation is caused by force gradient along the free surface, the lower part of the droplet will be pinched more than the upper part, resulting in the seen pearl shape.

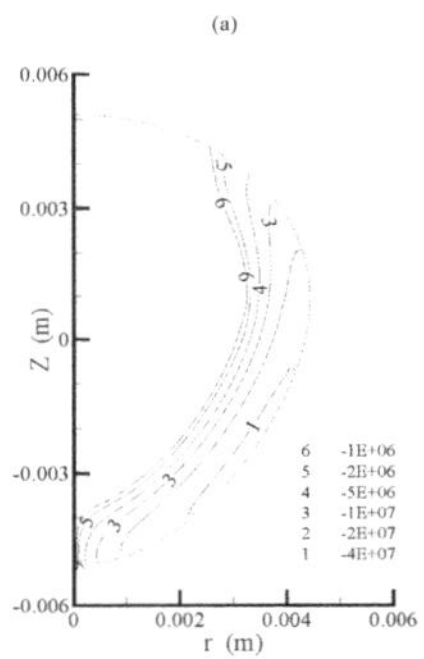

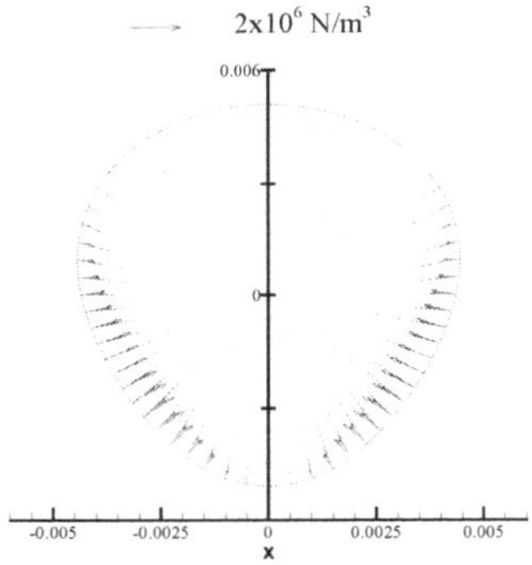

Figure 4. Computed induced currents: (a) Real, (b) Imaginary

Figure 5. Computed force field in the droplet

Fig. 6 shows the computed equilibrium shape and location of the aluminum droplet in the levitation coil. The computed position of aluminum sphere of the same volume used at the beginning of the shape calculation is shown in the same figure for comparison. This figure shows that the deformed droplet assumes a lower position in the coil than the spherical one. The computed distance between the bottom turn and the top of the droplet is 19.5 mm and this value compares well with the measured value of 19 mm.

A critical test of the solution methodology is to compare numerical predictions regarding the curvature of the free surface with measurements. Fig. 7 shows a comparison between the measured and the predicted shapes (solid line on the picture) of the quenched 0.9740 gm Aluminum droplet. It is seen that the computational method accurately predicts the curvature of the entire surface. The perfect agreement of both the location and shape demonstrates the accuracy of the computational methodology.

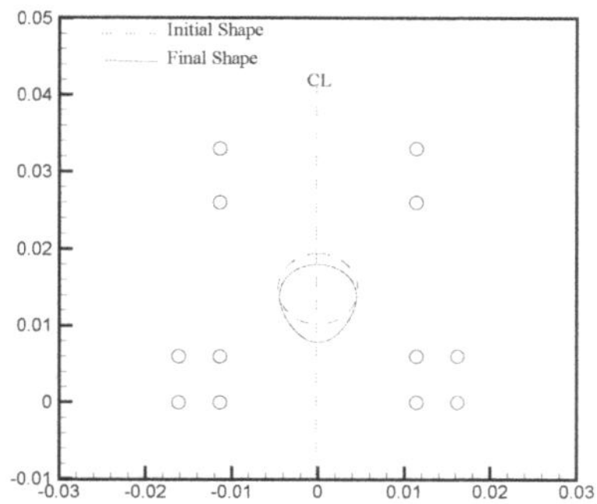

Figure 6. Computed location of the droplet within the coil.

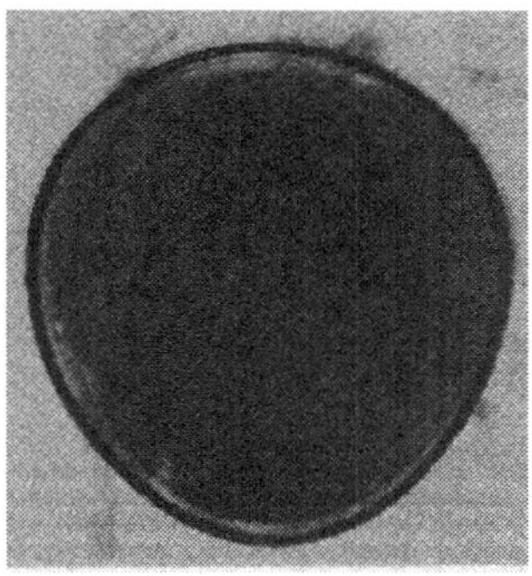

Figure 7. Comparison between measured and predicted droplet shape.

Conclusions

A finite element based methodology has been presented for computing the equilibrium shape and the location of the levitated metal droplets. The key feature of electromagnetic field formulation is the use of Biot-Savart's law as a boundary condition to limit the solution domain to the droplet. This feature not only reduces the computational time but also simplifies the regridding of the droplet during the search for the equilibrium shape. The accuracy of this method was demonstrated by comparing theoretical predictions against measurements.

References

1. A. Gagnoud, and M. Garnier, "Physical Analysis and Modelization of Phenomena in Electromagnetic Levitation in Conical Inductor," IEEE Transactions on Magnetics, 21(5) (1985), 1886-1888.

2. B. Q. Li and J. W. Evans, "Computation of Shapes of Electromagnetically Supported Meniscus in Electromagnetic Casters," IEEE Transactions on Magnetics, 25 (6) (1989), 4443-4448
3. J. R. Bhamidipati and N. El-Kaddah, "Calculation of Electromagnetic Field and Melt Shape in the Magnetic Suspension Melting Process," Magnetohydrodynamics in Process Metallurgy, ed. J. Szekely, J.W. Evans, K. Blazek and N. El-Kaddah (Warrendale, PA; The Metallurgical Society of AIME, 1992) 69-74.
4. D. W. Fugate, and J. F. Hoburg, "Shape and Stability Computations of Electromagnetically Confined Liquid Metal Boundaries," Metallurgical Transactions B, 24B (1) (1993), 171-178.
5. C. H. Winstead, P. C. Gazzerro, and J. F. Hoburg, "Surface-Coupled Modeling of Magnetically Confined Liquid Metal in Three-Dimensional Geometry," Metallurgical and Material Transcations B, 29(B) (1) (1998), 275-281.
6. S. Y. Hahn, Y. Sakai, T. Tsukuda, M. Hozawa, N. Imaishi, and S. Kitagawa, "Effect of Processing Conditions on Drop Behavior in an Electromagnetic Levitator," Metallurgical and Materials Transactions B, 29B (1) (1998), 223-228.
7. Y. Asakuma, S. H. Hahn, Y. Sakai, T. Tsukada, M. Hozawa, T. Matsumoto, H. Fujii, K. Nogi and N. Imaishi, "Equilibrium shape of a molten silicon drop in an electromagnetic levitator in microgravity environment," Metallurgical and Materials Transactions B, 31B (2) (2000), 327-329.
8. V. Bojarevics, G. Tinios, K. Pericleous and M. Cross, "CFD Modeling of Magnetic Levitation Casting," International Congress on Electromagnetic Processing of Materials" (Centre Francais de l'Electricite, Paris, 1997) 211-216.
9. A.D. Sneyd and H.K..Moffatt, "Fluid Dynamical Aspects of the Levitation Melting Process," J. of Fluid Mech., 117, (1982), 45-70.
10. N. El-Kaddah, and T. T. Natarajan, "The Influence of Coil Design on Melt Shape in Electromagnetic Levitation Melting," IISC. The Sixth International Iron and Steel Congress, Nagoya, Japan, 4 (1990), 380-387.
11. M. Pierre, and J. R. Roche, "Numerical Simulation of Tridimensional Electromagnetic Shaping of Liquid Metals," (Research Report RR-1687, Institut National De Recherche en Information et en Automatique, 1992).
12. S.J. Salon and J.P. Peng, "A Hybrid Finite Element-Boundary Element Formulation Of Poisson's Equation For Axisymmetric Vector Potential Problems," J. Appl. Phys., 53(11) (1982), 8420-8422.
13. J.L. Meyer, N. El-Kaddah and J. Szekely, "A New Method For Computing Electromagnetic Force Fields in Induction Furnaces," IEEE Trans. Magnetics, 23(1987), 1806-1810
14. T.T. Natarajan, and N. El-Kaddah, "A Methodology for Two-dimensional Finite Element Analysis of Electromagnetically driven flow in Induction Stirring Systems," IEEE Trans. on Magnetics, 35(3) (1999), 1773-1776
15. J.D. Lavers and M. Ramadan Ahmed, "Mathematical Modeling of Electromagnetic Confinement," Int. Symposium on "Casting of Near Net Shape Products," Ed. Y. Sahai, J.E. Battles, R.S. Carbonara and C.S. Mobley, (Warrendale, PA; The Metallurgical Society of AIME, 1988) 395-410.
16. M. R. Ahmed, and J. D. Lavers, "Boundary Element Analysis of the Electromagnet Casting Mould," IEEE Transaction on Magnetics, 25 (1989), 2843-2845.
17. K. Eppler, and H. Harbrecht, "Fast Wavelet BEM for 3D Electromagnetic Shaping," Applied Numerical Mathematics, 54 (2005), 537-554.

Celebrating the Megascale: Proceedings of the Extraction and Processing Division Symposium on Pyrometallurgy in Honor of David G.C. Robertson
Edited by: Phillip J. Mackey, Eric J. Grimsey, Rodney T. Jones, and Geoffrey A. Brooks
TMS (The Minerals, Metals & Materials Society), 2014

ELECTROCHEMICAL CHARACTERIZATION AND MODELING OF A SOLID OXIDE MEMBRANE-BASED ELECTROLYZER FOR PRODUCTION OF MAGNESIUM AND OXYGEN

Xiaofei Guan[1,2], Uday B. Pal[1,2,*], Srikanth Gopalan[1,2], Adam C. Powell[3]

[1]Division of Materials Science and Engineering, Boston University, MA 02446, USA
[2]Department of Mechanical Engineering, Boston University MA 02215, USA
[3]INFINIUM Inc.; 3 Huron Dr.; Natick, MA 01760, USA
[*]Email: upal@bu.edu

Keywords: Solid oxide membrane electrolysis, inert anode current collector, equivalent circuit, electronic transference number

Abstract

This paper reports a solid oxide membrane (SOM) electrolysis experiment using an LSM–Inconel inert anode current collector for production of Mg and O_2 at 1423K. The electrochemical performance of the SOM cell was evaluated by using various electrochemical techniques including electrochemical impedance spectroscopy, potentiodynamic scans, and potentiostatic electrolysis. The effects of Mg solubility in the flux on the current efficiency and the YSZ membrane stability during SOM electrolysis were discussed and examined through experiment and modeling. The electronic transference number of the flux were measured to assess the Mg dissolution in the flux during SOM electrolysis. A negative correlation between the electronic transference number of the flux and the current efficiency of the SOM electrolysis was observed.

Introduction

Solid oxide membrane (SOM) electrolysis process is used for the extraction of metals or alloys directly from their oxides or oxide compounds. It is both environmentally friendly and cost-effective. [1-6] The SOM process for magnesium (Mg) production employs a one-end-closed yttria-stabilized zirconia (YSZ) membrane to separate a liquid metal anode from a molten fluoride flux containing magnesium oxide (MgO). The YSZ membrane selectively conducts oxygen anions but blocks fluorine anions. During electrolysis, magnesium cations are reduced at the cathode, and oxygen anions are transported through the YSZ membrane to the anode where they are oxidized.

In order to produce pure O_2, a valuable byproduct, at the anode, it is critical to develop an inert anode current collector for use with the non-consumable liquid silver anode. Liquid silver is thermodynamically stable in O_2 and electrically conductive at the operating temperature, and therefore allows O_2 gas to evolve. The inert anode current collector is immersed in the liquid silver anode, and provides a good electrical connection between the liquid silver anode and the DC power supply. The inert anode current collector has the following properties: (a) good physical and chemical stabilities in O_2 and liquid silver at the operating temperature; (b) excellent thermal shock resistance and mechanical robustness; and (c) high electrical conductivity from ambient temperature (293 K) to the operating temperature (1423K). Guan *et al* [5] have recently reported a novel inert anode current collector that satisfies all the above requirements. It is not only cost effective and eco-friendly, but also shows high performance. This current collector incorporates a piece of sintered strontium-doped lanthanum manganite ($La_{0.8}Sr_{0.2}MnO_{3-\delta}$ or LSM) bar and an oxidation resistant Inconel alloy rod which are connected by a liquid silver contact.

In this work, SOM electrolysis experiment with the LSM–Inconel current collector is performed. The SOM experiment successfully produces Mg at the cathode and O_2 at the anode. During the experiment, the electrochemical performance of the SOM electrolyzer is evaluated. Prior work has reported that Mg has a small solubility in the flux used for SOM electrolysis of MgO. The dissolution of Mg imparts electronic conductivity to the flux. The electronic transference number of the flux between two steel electrodes are measured to assess the Mg dissolution in the flux. The correlation between the electronic transference number of the flux and the current efficiency of the SOM electrolysis is discussed. In addition, a new equivalent circuit is presented to model the SOM electrolysis process.

Experimental

Experimental Setup

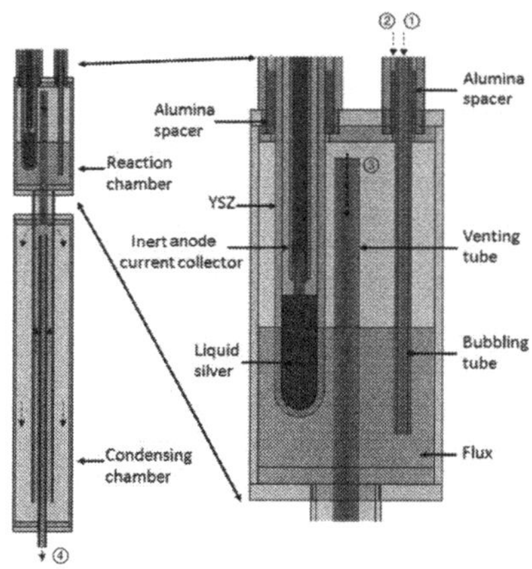

Figure 1. Schematic of the cross section of a SOM electrolysis setup. The components of the setup are indicated by the solid arrows. The dashed arrows show the forming gas flow: ① forming gas flow through the bubbling tube into the reaction chamber, ② forming gas flow through the annulus between the bubbling tube and its outside SS–304 tube, ③ forming gas and Mg vapor mixture flow through the venting tube leading to the condensing, and ④ forming gas flow out.

Figure 1 shows the schematic of a SOM electrolysis setup. It consisted of an upper reaction chamber heated to 1423K and a lower chamber (condenser) having a temperature gradient of 1373K to 473K. The setup was fabricated using grade 304 stainless steel (SS–304) and heated in industrial grade Ar (Airgas, USA). Inside the reaction chamber, 470g of powdered flux (45wt% MgF_2–55wt%CaF_2 containing 5wt%MgO and 2wt%YF_3) was used to form the molten electrolyte; 45wt%MgF_2–55wt%CaF_2 is a eutectic composition having a melting temperature of 1247K. [7] 5wt% MgO was added to the flux to reduce concentration polarization during SOM electrolysis of MgO, whereas the 2wt%YF_3 was added in the flux to prevent the diffusion of yttria into and out of the YSZ membrane during SOM electrolysis. [8] A 61-cm one-end-closed 6mol% YSZ tube (McDanel Advanced Ceramics, USA) separated the flux from 44g of liquid silver encased inside the YSZ tube. The LSM–Inconel inert anode current collector was disposed in the YSZ tube and the LSM bar of the current collector was submerged in the liquid silver anode. An SS–304 bubbling tube served as the cathode. Alumina spacers were used to electrically isolate the YSZ membrane and the bubbling tube from the reaction chamber. When an applied potential exceeded the dissociation potential of MgO, Mg vapor was produced at the cathode and O_2 evolved at the anode. The overall cell reaction is given as:

$$MgO = Mg(g) + 1/2O_2(g) \tag{1}$$

During the experiment, forming gas (95%Ar–5%H_2) was used to purge the reaction chamber. The gas flow rate through the bubbling tube was 180 cm^3/min. The bubbling gas stirred the flux to facilitate the transport of Mg vapor out of the flux and achieve chemical homogeneity in the flux. There was no gas flow through the annulus between the YSZ membrane and its outside SS–304 tube. The flow rate through the annulus between the bubbling tube and the SS–304 tube extending out of the top of the reaction chamber was 300 cm^3/min. This was done to lower the partial pressure of Mg vapor over the flux and to carry the Mg vapor into the condenser (see Figure 1). The Mg vapor in the condenser was cooled and condensed on a steel shim placed along the inner walls of the chamber.

Electrochemical Characterizations

Electrochemical impedance spectroscopy (EIS) scans were conducted to measure the ohmic resistance of the cell and the flux. The EIS scans were performed with a Princeton Applied Research 263A potentiostat and a Solartron 1250 frequency response analyzer. These scans were from 20,000 to 1 Hz with 20mV amplitude. The ohmic resistance of the measured cell was obtained from the value of the high frequency intercept on the real axis of the Nyquist plot.

Potentiodynamic scans at 5mV/s were performed between the LSM–Inconel inert anode current collector and the bubbling tube cathode to determine the dissociation potential of MgO. Electrolysis runs were performed by applying a constant electric potential greater than the dissociation potential of MgO. O_2 gas evolution was observed at the anode once the liquid silver anode became saturated with oxygen. The anode exit gas passed through a FMA–4305 digital flow meter (OMEGA Engineering, USA). The flow meter measured the flow rate of the anode exit gas to assist in calculating the current efficiency of the SOM cell. Current efficiency is defined as the ratio of the faradaic current for O_2 production to the total applied current.

Periodically, the electrolysis was halted and the electronic transference number of the flux between the bubbling tube and the reaction chamber wall was measured to assess the Mg dissolution in the flux. The electronic and ionic resistances of the flux can be modeled by two resistors in parallel. Therefore, the relationship between the total ohmic resistance ($R_{\text{flux,T}}$), the electronic resistance ($R_{\text{flux,e}}$), and the ionic resistance ($R_{\text{flux,i}}$) of the flux can be expressed by

$$1/R_{\text{flux,T}} = 1/R_{\text{flux,e}} + 1/R_{\text{flux,i}} \tag{2}$$

EIS scan was performed between the bubbling tube and the reaction chamber to determine $R_{\text{flux,T}}$. $R_{\text{flux,e}}$ was determined by performing a potentiostatic hold between the two electrodes. In each potentiostatic hold, a constant electronic potential (~0.1V) was applied and the current response was measured. $R_{flux,e}$ was obtained by dividing the applied potential by the current response. According to Equation (2), $R_{flux,i}$ was calculated from the measured values of $R_{flux,T}$ and $R_{flux,e}$. Using $R_{flux,e}$ and $R_{flux,i}$, the electronic transference number was calculated by [9]:

$$t_{flux,e}=R_{flux,i}/\left(R_{flux,e}+R_{flux,i}\right) \quad (3)$$

Equivalent Circuit Modeling of the SOM Electrolysis Process

Equivalent circuit is a useful tool to model the SOM electrolysis process. Impurity oxides (such as Fe_2O_3) in the flux contribute to the leakage current. In addition, the dissolved Mg in the flux imparts electronic conductivity to the flux. The electronic conductivity in the flux reduces the current efficiency of the SOM electrolysis and provides a pathway for the applied potential to dissociate zirconia in the YSZ membrane. Modeling of the SOM electrolysis process provides insights for increasing both the current efficiency and the YSZ membrane stability.

Dissociation of Oxides

In the SOM electrolysis, the oxides that could be dissociated include Fe_2O_3 (from the reaction chamber), ZrO_2, and MgO between the inert anode current collector and the bubbling tube cathode. The standard Nernst potential (E_N^0) for Fe_2O_3, ZrO_2, and MgO dissociations during SOM electrolysis with an inert anode current collector were calculated as shown in Table 1. [10] The absolute value of the standard Nernst potential for Fe_2O_3 dissociation is much smaller than those for MgO and ZrO_2; that is, the dissociation of Fe_2O_3 is the most favorable, assuming that all the three species behave ideally and given the same anode conditions in the SOM process. In actual SOM experiments, the dissociation of Fe_2O_3 contributes to the leakage current. A pre-electrolysis is usually performed to dissociate Fe_2O_3 by applying a DC electric potential, higher than the dissociation potential of Fe_2O_3 but lower than the dissociation potential of MgO.

From Table 1, the dissociation of ZrO_2 is more favorable than the dissociation of MgO assuming that both species behave ideally and given the same anode conditions in the SOM process. During the SOM process, however, the YSZ membrane is protected from electrochemical ZrO_2 dissociation due to the electric potential drop across the flux. Even when the flux has some electronic conductivity, the electric potential drop across the flux is such that the zirconia does not experience the full applied potential across the entire electrolysis cell. However, if the electronic conductivity in the flux and the applied potential across the electrolysis cell are both sufficiently high, then the potential drop across the YSZ membrane can be large enough to dissociate the zirconia.

Table 1. Standard Nernst potentials for relevant SOM reactions at 1423K.

Reaction	E_N^0 (V)
$Fe_2O_3 = 2Fe + 3/2O_2(g)$	−0.784
$MgO = Mg(g) + 1/2O_2(g)$	−2.275
$ZrO_2 = Zr + O_2(g)$	−2.153

Equivalent Circuit of the SOM Electrolysis Process

Based on current understanding of the SOM electrolysis process, a new equivalent circuit of the SOM electrolysis process is presented as shown in Figure 2. The symbols used in Figure 2 are defined in Table 2. Compared to the equivalent circuits in prior work by Krishnan *et al.* [2] and Gratz *et al.* [8], this equivalent circuit model is more comprehensive. It takes into account (1) the electronic conductivity of the flux caused by dissolved Mg, (2) the dissociations of Fe_2O_3 dissolved in the flux, and (3) the dissociation of zirconia in the YSZ membrane.

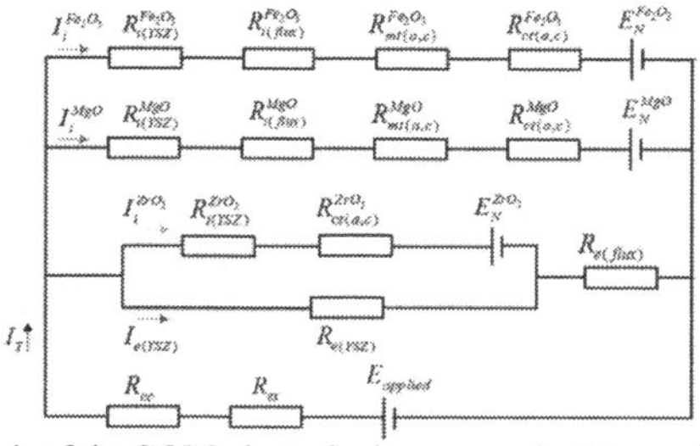

Figure 2. Equivalent circuit of the SOM electrolysis process for Mg production.

Table 2. Definitions of symbols in the SOM equivalent circuit shown in Figure 2.

Symbol	Definition	Symbol	Definition
$R_{i(YSZ)}^{Fe_2O_3}$	Ionic resistance of YSZ membrane involved for Fe_2O_3 dissociation	R_{cc}	Resistance of the inert anode current collector
$R_{i(YSZ)}^{MgO}$	Ionic resistance of YSZ membrane involved for MgO dissociation	R_{ex}	Resistance of external circuit elements
$R_{i(YSZ)}^{ZrO_2}$	Ionic resistance of YSZ membrane involved for ZrO_2 dissociation	$E_N^{Fe_2O_3}$	Nernst potential for Fe_2O_3 dissociation
$R_{i(flux)}^{Fe_2O_3}$	Ionic resistance of flux (between YSZ and bubbling tube) involved for Fe_2O_3 dissociation	E_N^{MgO}	Nernst potential for MgO dissociation
$R_{i(flux)}^{MgO}$	Ionic resistance of flux (between YSZ and bubbling tube) involved for MgO dissociation	$E_N^{ZrO_2}$	Nernst potential for ZrO_2 dissociation
$R_{mt(a,c)}^{Fe_2O_3}$	Mass transfer resistance at the anode and cathode for Fe_2O_3 dissociation	$E_{applied}$	Applied potential
$R_{mt(a,c)}^{MgO}$	Mass transfer resistance at the anode and cathode for MgO dissociation	$I_i^{Fe_2O_3}$	Ionic current for Fe_2O_3 dissociation
$R_{ct(a,c)}^{Fe_2O_3}$	Charge transfer resistance at the anode and cathode for Fe_2O_3 dissociation	I_i^{MgO}	Ionic current for MgO dissociation
$R_{ct(a,c)}^{MgO}$	Charge transfer resistance at the anode and cathode for MgO dissociation	$I_i^{ZrO_2}$	Ionic current for ZrO_2 dissociation
$R_{ct(a,c)}^{ZrO_2}$	Charge transfer resistance at the anode and cathode for ZrO_2 dissociation	$I_{e(YSZ)}$	Electronic current passing the YSZ membrane
$R_{e(YSZ)}$	Electronic resistance of the YSZ membrane	I_T	Total current
$R_{e(flux)}$	Electronic resistance of the flux between YSZ and bubbling tube		

The equivalent circuit first shows that impurity oxides (such as Fe_2O_3) with greater cation electronegativity than Mg, will dissociate before MgO. A pre-electrolysis is usually conducted to remove the Fe_2O_3. After the pre-electrolysis, an electric potential exceeding the dissociation potential of MgO is applied. Once the MgO dissociates, Mg is produced at the cathode. Some of the Mg produced dissolves in the flux, and the rest Mg evolves as vapor and is carried away by the forming gas. The dissolved Mg subsequently increases the electronic conductivity of the flux. The YSZ membrane has electronic conductivity at the operating temperature (1423K) and the other prevailing experimental conditions. A flux with electronic conductivity essentially serves as an extended cathode, and it allows for electronic current to pass through the flux and the membrane, decreasing the current efficiency. If the flux had no electronic conductivity, then the current passing through the flux would be purely ionic, and all of the energy input to the cell would be used in the process of dissociating MgO. The presence of an electronic current acts as an internal short circuit in the SOM process.

As more Mg dissolves into the flux, the electronic conductivity of the flux will further increase. The increased electronic conductivity also decreases the electronic potential drop across the flux, and then the potential drop across the YSZ membrane must correspondingly increase while a constant DC electric potential is applied across the entire cell. The increased potential drop across the YSZ can cause the dissociation of zirconia, which will degrade the YSZ membrane. To avoid ZrO_2 dissociation, the potential drop across the YSZ membrane should not exceed the dissociation potential of ZrO_2.

Results and Discussion

At the beginning, an EIS scan was performed between the inert anode current collector and the bubbling tube, and the results are shown in Figure 3. The high frequency intercept on the Nyquist plot corresponds to the cell ohmic resistance, which includes the resistances of the YSZ membrane, the flux, the electrodes, the external lead wires, and the contact resistances associated with all the interfaces. The cell ohmic resistance was measured to be 0.54Ω, close to earlier reported SOM electrolysis cell resistance with LSM–Inconel current collector. [5]

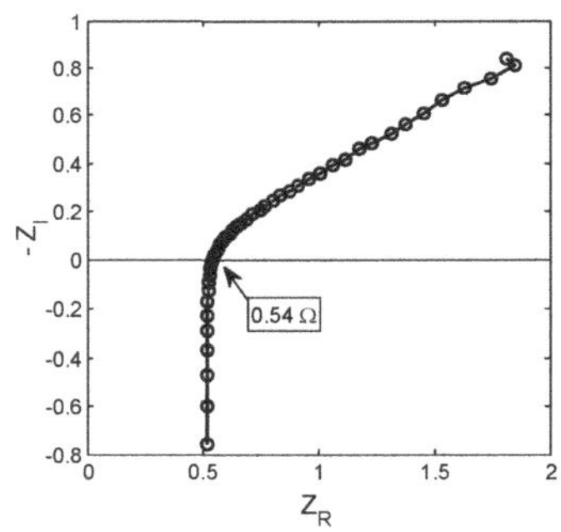

Figure 3. Electrochemical impedance spectroscopy results at the beginning of the electrolysis experiment at 1423K. The ohmic resistance of the cell was 0.54Ω.

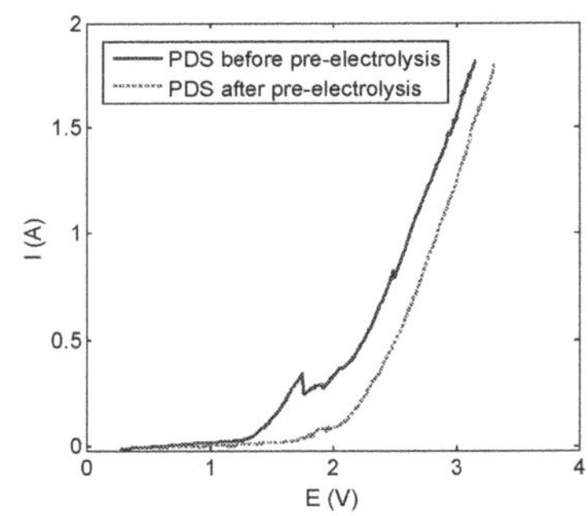

Figure 4. Potentiodynamic scans (5mV/s) showing the current–potential relationships before pre-electrolysis (solid line) and after pre-electrolysis (dashed line).

A PDS was conducted between the inert anode current collector and the bubbling tube after the EIS, and the current–potential relationship is shown in Figure 4. The dissociation potentials of the impurity oxides (such as Fe_2O_3) and MgO were identified to be approximately 1.3V and 2.17V, respectively. The leakage current caused by the impurity oxides dissociations is approximately 0.4A. To reduce the leakage current, a pre-electrolysis at 2.13V was performed for 70 mins to dissociate the impurity oxides. After the pre-electrolysis, another PDS was conducted between the inert anode current collector and the bubbling tube, and the current–

potential relationship is shown in Figure 4. The leakage current was significantly reduced after the pre-electrolysis. The electric potential for MgO dissociation was identified to be approximately 2.08V. At this point of time, the impurity oxides had been removed, and there was no Mg dissolved in the flux yet.

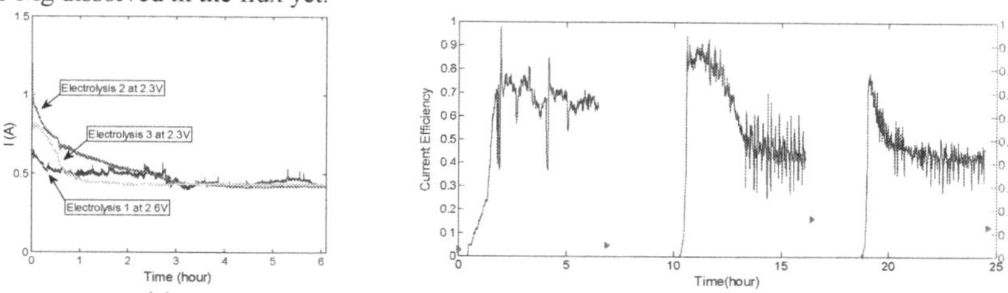

(a) (b)

Figure 5. (a) The current–time plots during the three electrolyses with potentials between 2.3–2.6V. (b) During electrolysis the current efficiency (solid line) decreases as the Mg dissolves in the flux and the electronic transference number (red triangle) increases. The asymptotic behavior indicates that the flux reaches the solubility limit for Mg dissolution.

Afterwards, electrolysis was performed thrice, each for 6 hours, between 2.3–2.6V, to dissociate MgO (see Figure 5(a)). The calculated current efficiency is shown in Figure 5(b). The steady state current efficiency decreased as more Mg was produced during SOM electrolysis. All the above electrochemical measurements showed that the LSM–Inconel current collector had low resistance and excellent stabilities during the experiment.

Before the first electrolysis, and immediately after each electrolysis, $t_{flux,e}$ between the bubbling tube and the reaction chamber was measured. The dependence of $t_{flux,e}$ as a function of time is shown in Figure 5(b). Initially $t_{flux,e}$ was low and the current efficiency was high, but with continued electrolysis the efficiency decreased and $t_{flux,e}$ increased until the flux reached the solubility limit for Mg dissolution. This indicates a negative correlation between the current efficiency of the SOM electrolysis and the electronic conductivity of the flux.

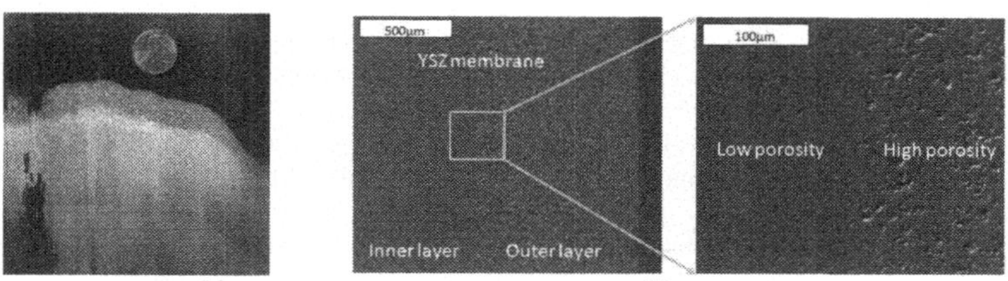

(a) (b)

Figure 6. (a) Mg collected on the stainless steel shim is shown in scale with a US one-cent coin. (b) SEM images of the membrane cross section after the SOM electrolysis.

After the SOM electrolysis, the furnace was cooled and the setup was disassembled. Figure 6(a) shows the Mg collected on the stainless steel shim lining the inner walls of the condenser. Figure 6(b) shows the SEM images of a piece of the YSZ membrane sample. The outer layer of the YSZ membrane in contact with the flux showed high porosity. This indicates that the YSZ membrane degradation took place during the electrolysis.

Conclusions

SOM electrolysis experiment was successful in producing Mg at the cathode and O_2 at the anode. The electrochemical measurements during SOM electrolysis indicated excellent stability of the LSM–Inconel current collector. The effects of Mg solubility in the flux on the current efficiency and YSZ membrane stability during SOM electrolysis were discussed and examined through experiment and modeling. Dissolved Mg metal was responsible for imparting electronic conductivity to the flux. $t_{\text{flux,e}}$ was measured to assess the Mg dissolution in the flux during SOM electrolysis. As more Mg was produced during SOM electrolysis, $t_{\text{flux,e}}$ increased and the current efficiency of the SOM electrolysis decreased. It suggests $t_{\text{flux,e}}$ should be reduced to achieve a high current efficiency of the SOM electrolysis.

Acknowledgements

This material is based on work supported by the Department of Energy under Award No. DE-EE0005547. The authors thank Dr. Eric Gratz for helpful discussion.

References

1. A. Krishnan, U.B. Pal, and X.G. Lu, "Solid Oxide Membrane Process for Magnesium Production Directly from Magnesium Oxide", *Metall. Mater. Trans. B*, 36 (4), (2005), 463-473.
2. U.B. Pal and A.C. Powell, "The Use of Solid-Oxide-Membrane Technology for Electrometallurgy", *JOM*, 59 (5), (2007), 44-49.
3. X. Guan et al., "Recycling of Magnesium Alloy Employing Refining and Solid Oxide Membrane (SOM) Electrolysis", *Metall. Mater. Trans. B*, 44 (2), (2013), 261-271.
4. X. Guan, U.B. Pal, and A.C. Powell, "An Environmentally Friendly Process Involving Refining and Membrane-Based Electrolysis for Magnesium Recovery from Partially Oxidized Scrap Alloy", *JOM*, 65 (10), (2013), 1285-1292.
5. X. Guan et al., "LSM ($La_{0.8}Sr_{0.2}MnO_{3-\delta}$)–Inconel Inert Anode Current Collector for Solid Oxide Membrane (SOM) Electrolysis", *J. Electrochem. Soc.*, 160 (11), (2013), F1179-F1186.
6. X. Lu et al., "Green Electrochemical Process Solid-Oxide Oxygen-Ion-Conducting Membrane (SOM): Direct Extraction of Ti-Fe Alloys from Natural Ilmenite", *Metall. Mater. Trans. B*, 43 (3), (2012), 503-512.
7. P. Chartrand and A.D. Pelton, "Thermodynamic Evaluation and Optimization of the LiF-NaF-KF-MgF_2-CaF_2 System Using the Modified Quasi-Chemical Model", *Metall. Mater. Trans. A*, 32 (6), (2001), 1385-1396.
8. E. Gratz, "Solid Oxide Membrane (SOM) Stability in Molten Ionic Flux for the Direct Electrolysis of Magnesium Oxide" (Ph.D. thesis, Boston University, January 2013).
9. A.V. Virkar, "Theoretical Analysis of the Role of Interfaces in Transport through Oxygen Ion and Electron Conducting Membranes", *J. Power Sources*, 147 (1-2), (2005), 8-31.
10. X. Guan, "Novel Process for Recycling Magnesium Alloy Employing Refining and Solid Oxide Membrane Electrolysis" (Ph.D. thesis, Boston University, September 2013).

Celebrating the Megascale: Proceedings of the Extraction and Processing Division Symposium on Pyrometallurgy in Honor of David G.C. Robertson
Edited by: Phillip J. Mackey, Eric J. Grimsey, Rodney T. Jones, and Geoffrey A. Brooks
TMS (The Minerals, Metals & Materials Society), 2014

PHENOMENOLOGICAL MODELS AND ANIMATIONS OF WELDING AND THEIR IMPACT

Tarasankar DebRoy[1]

[1]Professor of Materials Science and Engineering
115 Steidle Building, The Penn State University, University Park, Pennsylvania, USA, 16802

debroy@matse.psu.edu

Keywords: welding, phenomenological models, welding processes, metal structure

Abstract

Professor Robertson's recognized research on metallurgical thermodynamics and kinetics for over 40 years facilitated the emergence of rigorous quantitative understanding of many complex metallurgical processes. The author had the opportunity to work with Professor Robertson on liquid metals in the 1970s. This paper is intended to review the advances in the quantitative understanding of welding processes and weld metal attributes in recent decades. Over this period, phenomenological models have been developed to better understand and control various welding processes and the structure and properties of welded materials. Numerical models and animations of melting, solidification and the evolution of micro and macro-structural features will be presented to critically examine their impact on the practice of welding and the underlying science.

CELEBRATING THE MEGASCALE

Proceedings of the Extraction and Processing Division Symposium on Pyrometallurgy in Honor of David G.C. Robertson

Metallurgical Education

Celebrating the Megascale: Proceedings of the Extraction and Processing Division Symposium on Pyrometallurgy in Honor of David G.C. Robertson
Edited by: Phillip J. Mackey, Eric J. Grimsey, Rodney T. Jones, and Geoffrey A. Brooks
TMS (The Minerals, Metals & Materials Society), 2014

CURRENT AND SUGGESTED FOCUS ON SUSTAINABILITY IN PYROMETALLURGY

J.B. See[1], D.G.C. Robertson[2] and P.J. Mackey[3]

[1]Consultant, Millfield, NSW, 2325, Australia
[2] Missouri University of Science and Technology, 1400 N. Bishop, Rolla, MO 65409, USA
[3]P.J.Mackey Technology Inc, Kirkland, Quebec, H9J IP7, Canada

E Mail: mbbjbs@hunterlink.net.au

Keywords: Pyrometallurgy, Research, Sustainability, Education, Resource industry

Abstract

The production of iron and steel and non-ferrous metals by pyrometallurgical processes will remain a critical element in meeting the demand for materials in both developed and developing nations. Given the important need to reduce and minimise greenhouse gas emissions the technological focus of future pyrometallurgical R&D by universities and industry alike must concentrate on sustainability issues such as improved energy efficiency, recycling and waste minimization. Continued efforts are also needed on process optimization and new process development with a view to reducing capital and operating costs of the new large "mega" plants. Using the academic and industrial backgrounds of the authors, the present paper reviews the current status of R&D in pyrometallurgy in university departments with a particular emphasis on sustainability issues. The role of industry and government laboratories is also reviewed although primarily for developed countries. The paper also includes comments and suggestions on the future requirements for education and R&D in pyrometallurgy in developed countries to maximise sustainability. It is also suggested that future R&D in pyrometallurgy will be even more concentrated in developing countries – most notably China.

Introduction

The production of iron and steel and non-ferrous metals by pyrometallurgical processes will remain a critical element in meeting the demand for materials in both developed and developing nations. This paper first covers data on the production of metals to show what a large industry this is, and to show that it is still growing, with the growth being driven by development in the BRIC countries. We then discuss the important topic of increased efficiency in metals production and how this has been realized, among other factors, by increasing the size of plants up to capacities of the order of 1 to 20 Mtpa.

Efficiency/sustainability/recycling/resource conservation/and so on - these go hand in hand and require action across the industry. They also require action by society at large -

for example broad policy choices must be made by governments and utilities on energy generation. As engineers and academics we are particularly interested in how qualified metallurgical engineers and innovation can be fostered to serve the industry.

Current Production Capability

This section shows that the pyrometallurgical industry is large and projected to continue to grow. Pyrometallurgical processes are critically important for both industrialized and industrializing economies.

Data on the production of steel, copper, primary aluminium, and high-carbon ferro-manganese (HCFeMn), including examples of large plants producing these metals, are given in Tables I and II. Steel is the major (about 95% of all metals) product and world-wide these metals are still produced from ores/concentrates mainly by pyrometallurgical processes. Growth was calculated in Table I over the ten-year period from 2002 to 2011 (inclusive) and is unexpectedly robust, bearing in mind the global slow-down beginning in 2008. The vast majority of the growth in steel, aluminium and refined copper production has occurred in China.

Table I World metal production and examples of recent growth

Metal	World Production, Mtpa		Growth over the decade %	Compound annual %
	2002	2011		
Steel	888.0	1490.0	68	5.3
Aluminium	26.2	45.8	75	5.7
Copper	12.2	15.8	30	2.6
HCFeMn	2.83	4.4	55	4.5

Note: Copper output here is smelted copper from sulphide concentrate, roughly 80% of total copper output.

Table II Examples of large metallurgical plants

Plant, Location	Nominal capacity, Mtpa (metal output)	Ref.
POSCO, Gwangyang, South Korea (steel)	13.7	1
Dubai, Jebel Ali, Dubai (Al)	1.04	2
Guixi, China (Cu)	0.80	3
Metalloys, Meyerton, South Africa (FeMn)	0.40	4

Table III Examples of large capacity processes and corresponding annual output
(Annual capacity based on t/day rate, 365 days/yr and 95% on-stream time)

Metal	Unit process	Capacity t/day (metal)	Unit Process Annual Output, Mtpa
Steel	Blast furnace/BOF	12,000	4.20
Aluminum	Hall-Heroult cell	2.3 (per cell)	8E-04
Copper	ISASMELT™ *	1,000	0.35
HCFeMn	Electric Arc Furnace (81 MVA)	400	0.14

* Equivalent copper production from ISASMELT™ unit plus PS converters and anode facilities

Examples of the largest unit processes are listed in Table III. A single large blast furnace produces a large annual steel output whilst aluminum in contrast uses a very small capacity unit process and hence the largest aluminum plants have more than 1,000 units.

The intensity of use of a commodity (expressed as kg/capita) can be correlated with GDP per capita on a country-by-country basis [6]. This trend for steel is illustrated in Figure 1 [7] showing that at some point for countries such as the US, Germany and Japan there is a "saturation level", that is, a level where consumption per capita no longer increases with an increase in GPD per capita. The time frame for the approach to such world saturation levels comparable with the increase in world GDP per capita is illustrated in Figure 2 [8]. This indicates that demand for commodities produced by pyrometallurgical operations will continue to rise for quite some time to come.

Efficiency

The main reasons for the demise of smaller plants and the growth of mega plants are lower capital costs per unit of output, lower unit operating costs, together with increased overall efficiency. The "rule of thumb" is that capital costs increase with capacity to a power less than one (usually around 0.6), but this can vary from 0.1 on very small plants to 0.9 on very large plants. In the latter case, an increase in capacity above the largest unit process size can only be achieved by duplicating units, and so the scaling power must then be close to 1.

As noted, unit operating costs also decrease with scale, for a number of reasons including relatively lower fixed costs and lower manning costs per unit of output [9]. More recently, the use of robotic systems and advanced process control especially at mega plants can lower operating costs even further. Larger plants can generally be run more efficiently in terms of secondary objectives such as recycling, energy recovery, emission control and waste minimization, if only because the economic incentives are greater. This makes it easier to satisfy the triple bottom line of social, environmental and economic accounting.

Figure 1. Steel intensity of use (as kg/capita) as related to GDP per capita [7[.

Note: Stylistic representation
Source: Correlates of War, Maddison, Global Insight, Rio Tinto

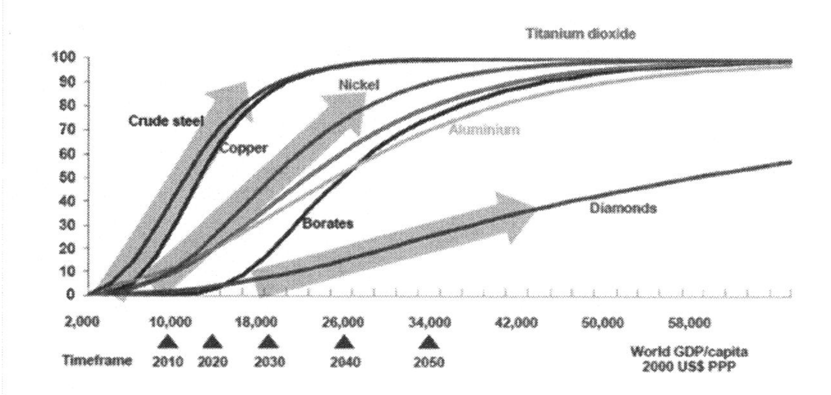

Figure 2. The approach to "saturation levels" for the intensity of use for a number of commodities as related to world GDP/capita [8]. Time frame estimates for attainment of increasing world GDP/capita are also indicated. It is evident that there will be continued growth in commodities.

Sustainability or Sustainable Development

Sustainable development is often defined by the following: "Sustainable development is development that meets the needs of the present without compromising the ability of future generations to meet their own needs" [10].

The extractive industries (water, coal, oil, gas, metals, fertilizers, etc.) cannot be fully sustainable in that valuable resources are inevitably being used up. So sustainability is not an absolute concept, and various indices are used to try and measure it [11]. Examples are indices calculated (a) at the country level by the World Economic Forum, and (b) on environmental and sustainability issues at the Yale Center for Environmental Law and Policy [12].

It is believed that, through recycling, any "Sustainability Index" for metals would be quite high. That is "metals are sustainable" since metals produced today can be used by "future generations". Key (and familiar) concepts that will contribute to reducing the environmental impact of meeting the need for metal are: (a) "End-of-pipe" approaches that convert emissions and wastes into benign materials for use, storage, or disposal, (b) Flow sheet changes to reduce emissions and wastes, (c) Both lower energy requirements and an improved efficiency of energy use, e.g. new ways for waste heat recovery, such as recovering heat from molten metallurgical slag, and (d) Improved recycling technologies [13]. All such changes must be subjected to cost-benefit analysis.

Waste Minimization

Waste minimization requires finding uses for waste streams, e.g. metallurgical slags, or the development of new technology to eliminate waste streams.

Recycling

Metals are highly recycled which means their production has a high "Sustainability Index". Life cycle assessment (LCA) is one method to quantify the benefits of recycling.

The resources to produce metals are not all used on a "one and done" basis – as is energy used for residential heating, lighting, transport, etc. Dr. Geoff Brooks of Swinburne University of Technology, Melbourne, Australia made this point regarding recycling, energy use, and the former Australian government's "carbon tax" as follows [14]:

> *"We need a more intelligent discussion about how to use our finite carbon, and it needs to be about metal. The recent furor about the carbon tax in this country (Australia) has not been a celebration of enlightened debate. I think much of the debate misses a vital aspect of carbon use, namely, that using carbon to make metal is an effective way of storing energy and that whatever approach we use to move industry to a lower carbon future should reflect this principle.*
>
> *There is also another major difference with this use of coal (for metal production) compared to power generation. Metals can be readily recycled. In fact, steel and aluminum (the world's two major metals) can be recycled many times over. When they are recycled, far less energy is used than that is initially used to make virgin metals. In the case of steel, this saving is about half. For aluminum, the savings are over 90%. Once you have made a metal, you have effectively embedded the energy associated with the coal into a new form. We should value metal more highly. Recent discussions and policies around carbon usage do not distinguish between these two activities. Our laws and policies don't differentiate between the impact of burning a tonne of coal to generate electricity for providing light for empty offices at night and using the coal to make some metal, which we can re-use over and over again."*

We expect that metal recycling will become even more important in future and for this reason would like to see more emphasis on recycling in R&D and economic analyses both by universities and natural resource companies.

Energy Efficiency, Renewable Energy and CO_2 emissions

Renewable energy (mainly solar and wind) is now in some cases roughly comparable in cost to coal and natural gas for the generation of electricity, even after allowing for the low capacity factor for renewables [15]. However, it is far cheaper (where possible, as in the iron blast furnace) to use coal or natural gas directly in pyrometallurgical processes since there is then the ability to utilize much more of the fuels' intrinsic energy, and so the cost of renewables is still not competitive in such cases. Also the application of electrical energy derived from renewables (which are not continuous) to any chemical or metallurgical process (that will be continuous) requires a full measure of back-up installed capacity.

CO_2 sequestration (essentially gas compression and capture in suitable geological formations) is a more distinct possibility with the so-called "oxygen blast furnace", since the top gas will be much richer in CO_2 than for a conventional blast furnace. Bio-fuels (e.g. switchgrass) could conceivably be used instead of coal or natural gas under special circumstances. Japan has announced a program to build a large blast furnace (10,000 t/day) fuelled by "hydrogen" – derived from coke oven gas [16]. The aim is reduce CO_2 emissions by 30% - a considerable and realistic improvement.

A more practical and realistic approach to minimizing the carbon footprint of the metallurgical industry will be to maximize recycling - as discussed previously. As is already the case in Europe and Japan [17] appropriate legislation will almost certainly be required to achieve this. Unrealistic mandates and failure to negotiate global policies will just cause the loss of economic activity and jobs in nations that insist on the use of renewable energy, to the benefit of those who do not.

As an example of improvements in the efficiency of energy utilization Figure 3 shows that there has been a significant drop in the energy consumption in the production of copper metal by smelting from feed material [18]. The drop around 1900 was achieved by both the introduction of mineral flotation and a significant increase in the size of the reverberatory furnaces then in use, while the continuing drop since about the 1950s is due to new tonnage-oxygen-driven processes introduced since that time; the 2011 data (11,403 MJ/tonne of anode copper) is the average calculated energy requirement for those processes now in operation.

It is noteworthy that the more intensive copper smelting processes used today were developed in the 1960s to 1970s by the major world non-ferrous metal companies existing in that period. Many of today's multinational metal companies have closed their research and development laboratories around the developed world and hence they are currently not investigating urgently needed options to "improve processes for tomorrow".

This begs the question of how the next improvements in energy consumption will be developed and applied to high temperature roasting, smelting and refining operations. In 2013 the industry seems to have abandoned the parallel "development" culture that previously existed. Inclusion of a suitable R & D program into the policies for sustainability that are already widely embraced by companies today is one recommendation arising out of the present study.

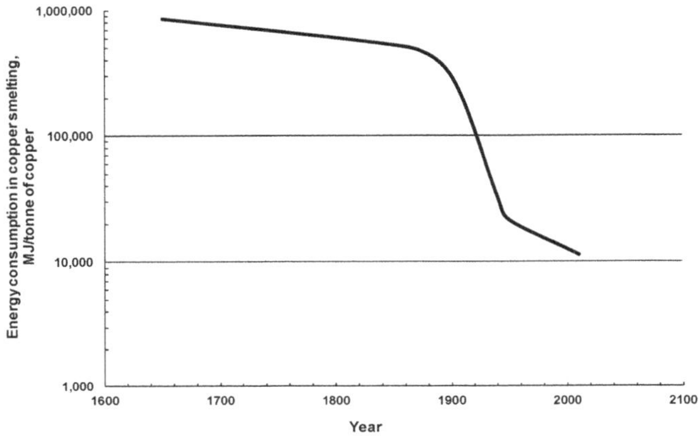

Figure 3. Trend in the energy consumption in copper smelting since about 1600 as MJ/tonne of copper, for feed to blister (prior to 1869) or later as anode copper [18]. Note: feed today is flotation concentrate but prior to about 1900, the feed was as-mined ore.

Cost-benefit Analysis

Cost-benefit analysis must be used by policy makers to avoid a waste of resources that could be better used elsewhere. In pyrometallurgy it makes no sense, for example, to insist on extraordinarily high air-quality around a smelter in a remote location (simply because the smelter is a fixed and easy target with funds available) while ignoring low air-quality in cities with millions of inhabitants. This has happened with very strict SO_2-in-air requirements for copper smelters in the western US while accepting much more air-pollution in some US cities. On the other side of the coin, industries have not been blameless, as evidenced by past opposition to what are now seen as sensible environmental policies - such as the removal of lead from gasoline and the elimination of the release of toxic wastes to the environment.

However new approaches are required to cost analyses to adequately capture costs created by effects of industrial processes on the sustainability of the environment and on society that are not normally currently measured by extractive industries, i.e. to capture in as quantitative a manner as possible the costs in achieving the triple bottom line. It is suggested that as part of the "sustainable policies" of companies today, this type of analysis should in fact be carried out and suitable R & D programs initiated to examine appropriate process changes.

How to Realize Sustainable Development in Pyrometallurgy

The realisation of sustainable development requires the direction and management by governments, industry, and individuals of a wide range of changes in the world's social, economic, political, environmental and technological environments. Further, it is essential that resource companies fully embrace sustainability as exemplified by the excellent Sustainability Report of Rio Tinto [19]. With respect to pyrometallurgy, it is essential that R&D be carried out by mining companies and other institutions on innovative approaches in areas like reduction in energy consumption and metal recycling. Improvements such as these would directly improve the sustainability of the operations of today's mining companies.

However, the terms sustainability and sustainable development mean many different things to different people, institutions and companies and are sometimes employed to consider "soft" topics. An example is the well prepared Rio Tinto Alcan Yarwun 2012 Sustainable Development Report [20] that reads in part:

> *"When we use the term 'sustainable development' within Rio Tinto, we are referring to our approach and performance in:*
> - *Demonstrating environmental stewardship;*
> - *Contributing to the region's social wellbeing;*
> - *Strengthening the economic prosperity of our business and community;*
> - *Maintaining strong governance and integrity systems."*

This statement does not include any direct or obvious link to technology, which is such a big part of Rio Tinto's present operations and their sustainable operations into the future. The Yarwun plant operates a mega 3.4 million tonne/year Bayer alumina refinery including a 160 MW co-generating power plant and it is therefore considered warranted to link the meaning and relevance of terms in a statement on sustainability to aspects of technical development. Due to the acknowledged importance of sustainability, companies share a responsibility to participate fully in technological development. Such participation would certainly have long term benefits to the industry and to society.

Education and R&D in Pyrometallurgy in University Departments

As scientists and engineers, pyrometallurgists have an important role to play in ensuring that this goal can be achieved and hence their current and future education must include a strong understanding of the concepts of sustainability relevant to the exercise of their professional skills.

Accordingly this section of our paper focuses on R&D and education in pyrometallurgy in university departments in different countries. Pyrometallurgy is the physical chemistry and process technology of the production of metals and alloys from ores and minerals by high temperature techniques and may therefore be considered as a specialised hybrid of science and engineering within the overall discipline of materials science and engineering. Education and training in pyrometallurgy is usually carried out within university departments of materials science and engineering and hence improvements in the sustainability of pyrometallurgical processing are directly affected by the professionals graduating from these institutions.

The successful and more sustainable operation of industrial plants that include pyrometallurgical processing also requires trained professionals from a wide range of other disciplines such as chemical engineering, mechanical engineering, electrical engineering and control engineering and it is worth noting in passing that in general these disciplines are placing increased emphasis on sustainability concepts [21].

As noted previously, pyrometallurgy is a specialised field that is critically important for the production of metals and alloys. Hence it is highly desirable firstly that there should be sufficient "Centres of Excellence" producing pyrometallurgists and secondly that such trained personnel should have some exposure to sustainability concepts like recycling, waste minimisation and improved energy efficiency.

To assess the availability of training and education in universities in pyrometallurgy and the associated extent of any exposure to sustainability concepts the authors carried out a survey of university departments active in extractive metallurgy in a number of countries. The responses from the surveys will now be summarised before commenting on the results. The total number of responses was 17 out of a total of 36 – an overall response rate of 47%.

University Departments in USA and Canada: Table IV summarises the responses received from universities in North America.

Table IV Sustainability in R&D and education in pyrometallurgy University departments in USA and Canada

General areas of effort in pyrometallurgy	- Calcination, roasting, smelting, refining, vapour phase extraction, ionic melts - Steelmaking, non-ferrous pyrometallurgy, refractories, recycling - EAF steelmaking, ironmaking, steelmaking, non-ferrous processing
Current topics in sustainability	- Recycling of magnet materials, Al and Mg alloys, refining of silicon - EAF steelmaking, biomass fuel and reductant - Reducing energy use, environmental footprint, materials for sustainable energy generation, values from wastes
Number of faculty (FTE):	Maximum number of academics involved in any one institution is three. Based upon funding levels they not totally engaged in sustainability research and/or education

In all the institutions surveyed, formal instruction in concepts of sustainability is less than 10% of course work for undergraduates and graduates, although there is some attention to this area as part of basic instruction in pyrometallurgy. One institution is exceptional in offering some specialised undergraduate and postgraduate courses that focus on sustainability from the viewpoint of energy balances, recycling and high temperature process systems.

University Departments in Australia and South Africa: Table V summarises the responses received from universities in Australia and South Africa.

Table V Sustainability in R&D and education in pyrometallurgy
University departments in Australia and South Africa

General areas of effort in pyrometallurgy	- Thermodynamics of melts, kinetics, reduction of oxides, process modelling, refractories - Iron ore sintering and reduction, PGM smelting, CVD applications - High temperature phase equilibria, slag/matte/metal systems, reaction kinetics and mechanisms in pyrometallurgical systems - Coke reactivity (mineral effect and strength), iron sinters, steel plant waste - Steelmaking fundamentals, aluminium production and refining, - e-waste treatment, magnesium production, silicon refining, ilmenite processing and solar thermal processing
Current topics in sustainability	- Sustainable processing, utilization of wastes (e.g. lignin) - Recycling waste plastics and rubber tyres - New technologies and reduction in energy and coke/carbon consumption for iron-making, ferroalloys, and aluminum production - Energy and exergy analysis - Reprocessing of slags to recover additional value - Use of leach residues as cement additives - Use of silicon-based reductants in ferro-alloy production to reduce CO_2 emissions. - Copper smelting and converting, iron ore sinter chemistry - Efficient use of carbon based materials in ironmaking, steel plant waste recycling - e-waste treatment and solar thermal processing
Number of faculty (FTE):	Maximum number of academics involved in any one institution: four - depending upon definition used for sustainability
Sustainability funding	Typically just staff salaries. Any other funding hard to determine but typically minimal. In one case could be significant but information not available

Funding in the Australian and South African universities surveyed is typically a blend of government and industry sources. Amongst these universities there are only a few that have a sufficient critical mass to continue to sustain an effort in pyrometallurgy.

In common with virtually all the universities surveyed in North America, formal instruction in concepts of sustainability is less than 10% of course work although there is some attention to this area as part of basic instruction in metallurgical engineering.

In general the approach in Australia is similar at all the universities surveyed. As one senior academic noted:

> *"Sustainability is dealt with generally in all courses, especially in energy balancing and waste minimisation, as well as waste reprocessing. It must be considered that our syllabus is already loaded with the essential subjects necessary for a balanced and adequate engineering education, and little time and effort is available for specialised courses in sustainability."*

Some senior academics also expressed concern about the continued viability of pyrometallurgical education and research as evidenced by the following quotes:

> *"I think it is highly likely that the peak of pyrometallurgical research in universities (and in CSIRO, as well) in this country at least may have passed. There is currently a dearth of academics working in pyrometallurgy in Australian Universities and the number is probably falling".*

> *"I just have a feeling that the sands of support within universities for research in the more experimentally difficult and less fashionable areas such as traditional laboratory-oriented pyrometallugy is shifting for the worst. Notwithstanding, I agree that a theme of sustainability as a present and future keystone of pyrometallurgical research in the sense that you are addressing it, is an important issue for the minerals industry, as is the inculcation of a culture of sustainability within undergraduate education."*

Although teaching and research in pyrometallurgy may be facing challenges within Australian universities, it is essential that both be continued in order to provide balanced technical education within programs which graduate extractive metallurgical engineers.

Some examples of world-class research centres and efforts in pyrometallurgy still prosper, such as at the Centre for Sustainable Materials Research and Technology at the University of New South Wales (SMaRT@UNSW), the Pyrometallurgical Group at the University of Queensland (UQ), Swinburne University of Technology and the University of Wollongong.

The overall aim of SMaRT is to develop innovative, sustainable materials and processes and a major accomplishment has been the development in conjunction with Onesteel (now Arrium) of a patented process using end-of-life plastics and rubber tyres in the electric arc furnace [22]. The UQ group are world leaders in the experimental study of non-ferrous phase equilibria and in their utilisation of FACTSage software.

The pyrometallurgical efforts at Swinburne and Wollongong are significant in ironmaking and steelmaking as one area and each institution has some projects with a sustainability focus.

University Departments in Other Countries: Survey forms were also received from Korea, Belgium and the UK. Projects included recycling of electronic scrap, slag utilization, waste minimization and the potential elimination of CO_2 emissions by using electrochemical processes.

R&D in Corporate and Government Laboratories

In some countries corporate and government laboratories make a significant contribution to sustainability in pyrometallurgy. Table VI lists some examples. However the picture is mixed in that many laboratories in the developed countries have closed in the past two decades.

Table VI Examples of selected major government and corporate laboratories active in pyrometallugy in various countries.

Country/region	*Government Laboratory*	*CorporateLaboratories*
Australia	CSIRO	Bluescope Steel
South Africa	MINTEK	Anglo American
Europe	IRSID	Eramet
Mexico	INSAMET	
Canada	CANMET	Barrick, Teck, Vale
Brazil		Vale
Sweden	MEFOS	
Norway	SINTEF	Hydro, Elkem
India	NIM	Tata

Closure of corporate R & D Laboratories

In introducing the CIM-Metsoc Centenary volume "The Canadian Metallurgical and Materials Landscape – 1960 to 2011:The Golden Anniversary of the Conference of Metallurgists", the editors commented on this era as a so-called "Golden Age" of mining and metallurgy in Canada and noted that the country's universities, research institutes and industrial laboratories played an invaluable role in those achievements.

Expressing a concern about the next fifty years and the widespread closure of industrial research and development laboratories in Canada over the period from the 1990s to the middle 2000s the editors lamented:

> "*Over the following 40 or 50 years* (since the 1960s), *these technological investments in the future were paid for many times over – the laboratories would bring enormous wealth to their owners. Each facility left a lasting legacy in terms of technology and technical processes. It is fair to conclude that without the technological developments carried out, Canada's mining and metallurgical industry would not have advanced in the way that it did, and it is suspected, would not be as vibrant as it is today. As further discussed below, the present editors fear the resulting competitive advantage enjoyed by Canadian companies for so long will slowly shrink due to closures of what were once world-class metallurgical development facilities. Granted, globalization as such brings specific changes and challenges to the established order, the twentieth century mining business models are different, but somehow the value of R&D today might be misjudged, seemingly overshadowed by the pursuit of short-term goals".*

In Canada the closure of key laboratories such as Noranda Technology Centre, Falconbridge Technology Centre, the Stelco laboratories and the Alcan laboratories is ultimately the industry and the country's loss. Interestingly, in their place are new laboratories like Kingston Process Metallurgy and expanded SGS facilities, however support from industry for research and development work is quite limited compared to specific engineering-type work related to a given mining/metallurgical project. These changes in the Canadian industrial scene have been accompanied by a significant curtailment of the mining and metallurgy activities of the government laboratory CANMET - evidently as a cost saving measure. The US closed the excellent laboratories of the US Bureau of Mines in 1996, and therefore the funding for the Centre of Pyrometallurgy at the then University of Missouri-Rolla (again evidently a cost saving).

There have been similar dramatic examples of the closure of industrial research and development laboratories across Australia. Since the 1990s, as an example, the country has seen the closure of the following laboratories: the BHPB laboratories in Melbourne, the BHPB laboratory in Newcastle, and the former CRA (now Rio Tinto) laboratories in Perth and Melbourne. However, unlike Canada, Australia continues to fund the CSIRO (Commonwealth Scientific and Industrial Research Organization) through initiatives such as the Minerals Down Under National Research Flagship that invests approximately A$80M yearly across the minerals value chain in exploration, mining, mineral processing and metal production [23,24].

South Africa, a country with a vast extractive metallurgy industry and a leading player in the international ferro-alloy industry, is another example of a country that has maintained a strong position in R &D in pyrometallurgy, primarily because of its strong continued funding of its government laboratory MINTEK (Council for Mineral Technology). MINTEK is a world leader in the development and implementation of DC arc furnace processes for smelting, fuming, remelting and slag cleaning. In late 2010 MINTEK's Pyrometallurgy Division employed 120 people including 21 engineering staff, and had a wide range of pilot plant equipment [25].

On a positive note non-ferrous metals companies in China have identified gaps in technology and are now carrying out leading edge research and development work in metallurgy, including pyrometallurgy, an example being the SKS process for copper and lead smelting. Further, a number of research laboratories and universities in China are also active in pyrometallurgical testwork on nickel - another major non-ferrous metal.

As noted above, by 2020 some 50% of global commodities will be consumed in China, with much of the metallurgical production carried out in the country as well. The corollary of this is that much of the research and development work in pyrometallugy will most likely be carried out there.

Possible future trends for R&D in pyrometallurgy

The conclusion that much future R&D in pyrometallurgy will probably be carried out in China suggests a strong linkage to the location of mega scale plants. However the scale of the pyrometallurgical industry in many developed countries (especially the steel industry, and in some cases, for example, Canada with a significant the non-ferrous industry) is still significant, and the decline of pyrometallurgical R&D in countries like the USA and Canada is to be regretted.

In countries where traditional pyrometallurgical processing is on the decline, incentives and funding for continuing research will likely be underpinned by sustainability challenges such as reduction of greenhouse gas emissions from blast furnaces. In a resource-rich developed country like Australia capital for investment in recycling must compete against capital for primary processing and shipping of minerals to the mega plants overseas in China. Recycling research is more attractive for the developed countries such as parts of Europe and Japan that are much less well endowed with minerals. Hence financial support for pyrometallurgical R&D may be difficult to come by in certain countries, but not necessarily in others, depending on their position in the minerals processing and metals production chain. In countries rich in certain resources, for example Chile which has large copper reserves, it is hoped that R&D in pyrometallurgy will continue to grow.

The present authors believe that due to the national importance of pyrometallurgy, it is in the interest of countries such as the USA, Europe, Australia and Canada to bolster pyrometallurgical research by industry and universities rather than just allow it to decline.

Conclusions

Pyrometallurgy is a key discipline in the production of many metals and plays an indispensible part in many of the mega-scale metallurgical plants today. This situation will continue into the foreseeable future. The paper showed that there will be continued growth in metal production and in particular, continuance of pyrometallugy. Further it was considered that the technological focus of future pyrometallurgical R&D by universities and industry alike must concentrate on sustainability issues such as improved energy efficiency, recycling and waste minimization.

In closing their research and development laboratories in a number of countries, multinational companies are implicitly admitting that the sustainability objectives they claim to support are not backed by R&D activities. While important work is presently carried out at numerous Government laboratories and universities in some developed countries, there has been a significant reduction in the pyrometallurgical R&D required to meet current and future needs. The recognition that sustainability is an important objective is not matched by adequate funding in developed countries.

Education in sustainability in pyrometallurgy is carried out in only a few Universities in the developed countries. Examples of the work going on in these Universities have been summarized.

It is highly likely that future effort in pyrometallurgical R&D will be concentrated in developing countries such as China and parts of South America because of the increasingly probable location of future mega plants in these countries. Such plants, especially for countries like China and India, will have strong domestic demand generated by increasing modernisation and urbanisation and this demand, plus the need for increased sustainability, will drive such R&D efforts.

As discussed in the paper, there is scope for the development of a sustainability index for metals production. In this way, improvements in the level of sustainability by a metal producer for example could be tracked. As described by Brennan [21], there is now a framework in place to attain this goal in the chemical process industry:

> *"Achieving higher levels of sustainability in industry demands focus across a range of activities. These include developing new technologies, improving existing technologies, designing new plants, re-vamping existing plants, and improved operation of existing plants. Cleaner production is an important concept for achieving more sustainable products and processes in these contexts".*

Notable amongst the themes here are those also discussed in the present paper regarding new technology development, technology improvement and new plant design as applied to the metals industry.

The new generation of pyrometallurgists will need training in all these areas. Achieving the specific goals identified here by the metallurgical industry will require investment in each of these activities.

It is hoped that delegates at this Symposium will further develop some of the ideas on sustainable development and the importance of these concepts in pyrometallurgy. The present authors are interested to discuss these topics further with delegates at the Symposium and with other interested parties after the Symposium.

Acknowledgements

The authors thank those faculty members who provided information and comments for this paper and the reviewer who provided some especially valuable insights. They also thank Dr John Rankin for his advice and support.

References

1. www.worldsteel.org/statistics/top-producers.html - accessed September 2013 globalblog.posco.com/posco-renovates-worlds-biggest-furnace-at-gwangyang-steelworks/
2. en.wikipedia.org/wiki/List_of_aluminium_smelters - accessed September 2013 www.dubal.ae
3. www.slideshare.net/PresentacionesVantaz/the-world-copper-factbook-2012-16401361 - accessed September 2013 www.nmm.jx-group.co.jp/english/industry/ind_02.html
4. minerals.usgs.gov/minerals/pubs/commodity/manganese/ - accessed September 2013 www.terrapacis.co.za/wp-content/uploads/2011/01/M14-09TPE02-Rev3.pdf
5. Report in: http://www.ey.com/Publication/vwLUAssets/, October 2013
6. W. Malenbaum, *World Demand for Raw Materials in 1985 and 2000*, E/MJ Mining Information Services (New York, NY, USA: McGraw Hill, 1978), 126 pages.
7. Rio Tinto Chartbook: www.riotinto.com/documents/investors_databook/chartbook.pdf
8. Rio Tinto-Presentation on Minerals and Diamonds, 12 September, 2012.
9. en.wikipedia.org/wiki/Economies_of_scale - accessed September 2013.
10. en.wikipedia.org/wiki/Sustainability - - accessed September 2013.
11. en.wikipedia.org/wiki/Sustainability_metrics_and_indices - Sept 2013.
12. YaleCentre for Environmental Law and Policy, http://epi.yale.edu/about/previouswork
13. W.J. Rankin, *Mineral, Metal and Sustainability-Meeting Future Material Needs*, (Melbourne, Victoria, Australia: CSIRO Process Science and Eng., 2011) 440 pages.

14. G. Brooks, "Treasure your metal: why we need to respect embedded energy", *The Conversation,* (14 May, 2012): https://theconversation.com/treasure-your-metal-why-we-need-to-respect-embedded-energy-6802
15. Lazard Levelized cost of energy: http://gallery.mailchimp.com/ce17780900c3d223633ecfa59/files/Lazard_Levelized_Cost_of_Energy_v7.0.1.pdf
16. K. Fujimoto and K. Suzuki, "Development of Technology for Advanced Utilization of Hydrogen from By-product Gas of Steelmaking Process", (Nippon Steel Technical Report, No. 101, November 2012), 203-207.
17. www.sae.org/events/green/reference/2010/End%20of%20Life%20 Vehicle%20Recycling.pdf
18. P. Coursol, P.J. Mackey and C.M. Diaz, "Noranda/Teniente Copper Bath Smelting Process Variations - Impact on Energy Requirements" (Paper presented at the Conference of Metallurgists, Montreal, Canada, October 2011).
19. Rio Tinto Sustainability Report: http://www.riotinto.com/sustainabledevelopment2012/
20. Rio Tinto Alcan Yarwun 2012 sustainable development report: http://www.riotintoalcan.com/documents/130726_FINAL_2012_Yarwun_SD_and_CF_report.pdf
21. D. Brennan, *Sustainable Process Engineering: Concepts, Strategies, Evaluation and Implementation* (Pan Stanford, 2012), 360 pages.
22. V. Sahajwalla, "Sustainable Processing", UNSW Science Research Highlights (2011), University of New South Wales, Sydney, pp. 62-63.
23. www.csiro.au/MDU
24. R. A. Shaw and W J. Rankin, "Research and Innovation in the Mining Industry", Australasian Mining and Metallurgical Operating Practices, Monograph No. 28, (Australasian Institute of Mining and Metallurgy, Melbourne, 2013), pp. 201-210.
25. http://www.pyrometallurgy.co.za/Mintek/capabilities.htm
http://en.wikipedia.org/wiki/Mintek

Celebrating the Megascale: Proceedings of the Extraction and Processing Division
Symposium on Pyrometallurgy in Honor of David G.C. Robertson
Edited by: Phillip J. Mackey, Eric J. Grimsey, Rodney T. Jones, and Geoffrey A. Brooks
TMS (The Minerals, Metals & Materials Society), 2014

TEACHING PROCESS SIMULATION IN ELEVEN EASY LESSONS USING EXCEL AND ITS TOOLS

Arthur E. Morris[1]

[1]Chief Scientist, Thermart Software, www.thermart.net

12102 Calle de Maria, San Diego, CA 92128 USA

Keywords: Processing, Energy Education, Combustion, Software, Simulation, Flowsheet

Abstract

The primary market driver for improving process technology is innovation, which requires a skilled and educated workforce. However, many Materials Science and Engineering departments have eliminated extractive metallurgy and chemical thermodynamics from their curricula, yet these topics contain the necessary fundamentals for process innovation. As a result, most MS&E students are ill-prepared for careers in processing. The dearth of process-oriented MS&E curricula has prompted some Universities to develop a "shared" effort to offer distance education between multiple institutions [1]. A target audience for a shared process simulation course would not only benefit students, but also be a basis for an on-line course for practicing engineers faced with new or changing career choices. To fill the gap, the basics of a process simulation course was developed in an abbreviated form as series of eleven articles and Excel workbooks published in *Industrial Heating* magazine between July 2012 and July 2013.

Introduction

For many years, course offerings in extraction and processing have declined owing to changes in research funding emphasis for Materials Science and Engineering (MS&E) faculty. As a result of changing emphasis and cost pressure, many MS&E departments have eliminated small-enrollment courses in extractive metallurgy and chemical thermodynamics from their curricula, yet these topics contain the necessary fundamentals for process innovation. As a result, most MS&E students are ill-prepared for careers in processing. In addition, graduated MS&E engineers often encounter situations where they must learn process technology on the job. The question discussed here is whether there exists an educational methodology for coping with the dearth of MS&E courses in extraction and processing, and the fallout in terms of continuing education for practicing engineers. In the case of MS&E students, the problem devolves into providing accredited educational course material offered to a small number of students with minimal expense and faculty participation. In the case of practicing engineers, the question devolves into the delivery of structured educational material using distance learning techniques.

For practicing engineers, distance learning has a long history, as do structured short courses and workshops in connection with TMS meetings. Courses were offered by correspondence, by videotape, and more recently by DVD. For the university student, courses delivered at the main university campus were televised by closed-circuit to satellite campuses. Where classroom space was limited, the lectures were rebroadcast later. The learning experience from these techniques was greatly enhanced by the development of satellite or web-connected classrooms, where participants could interact in real time. The latest development is Massive Open Online Courses (MOOCs), aimed at large-scale interactive participation and open access via the web [2]. The development of MOOCs has been chronicled by nearly every newspaper and newsmagazine in the US since its first mention in 2008. A list of MOOCs is obtainable on the web [3].

MOOC courses are generally selected to appeal to a large audience, and are therefore unlikely to be used in the MS&E arena. Nevertheless, as the acceptance of learning via on-line courses grows, a question arises: can TMS (and other professional societies), universities, students, and graduated engineers utilize some aspect of MOOCs and their imitators to the educational benefit of all concerned? Let's look in detail at a possible way to meet the needs of both practicing engineers and MS&E students for shared courses and distance learning. The suggested concept is to build competence in process engineering by treating one process topic in depth, and with computational intensity. The single topic should be one that is common to most everything a process engineer does. The hypothesis is that if a person can master the fundamentals of material/heat balances, and equilibrium thermodynamics for one topic, he can apply them to any topic. This concept is so far untested and unproven. But what are the alternatives?

At-Site Seminars, Short Courses, and Workshops

TMS (among other professional societies) has long sponsored these types of learning activities at the Annual Meeting, and at Specialty Conferences. The author of this paper, along with colleagues David Robertson and Eric Grimsey, have offered short courses and workshops on a variety of topics at Annual and Specialty TMS meetings. These offerings were intended to educate meeting attendees to the developing trends on software and calculational techniques for extractive and process metallurgy applications. However, there have been difficulties with this and similar types of offerings.

- Low enrollment. Many potential attendees cannot get permission to attend the meeting, or they have conflicts on coming early to attend a short course, or their company policy discourages course attendance unless it is specifically work-related.
- Unpredictable enrollment. TMS cannot afford to offer courses with enrollment below a certain level. A too-low enrollment is often not known until a couple of weeks before the meeting. Those who registered early and made travel plans are disappointed in a cancellation. The presenter, having spent considerable time on course development, is equally disappointed.

The situation is usually much different at specialty meetings, or when a major topical symposium is scheduled for the Annual meeting. Here, a course is usually well-attended when designed for the attendees coming for the meeting topic, and the presenters are well-known experts. The course is considered good value because the notes have lasting value to the attendee and his employer. But even here, there are drawbacks. The presenters expend a lot of effort to develop the course material, but it may be offered only once. Several years may elapse before the subject is again a meeting topic, and by then, the material may need updating, or the original presenters are no longer available. TMS may have no way to market the course notes through Wiley or Springer, so a very valuable resource expires.

Other professional societies and industry organizations have tackled the above-mentioned problems in a more organized way. First, they identify a course topic that is likely to have a long-term interest to professional engineers. Second, they develop a set of course notes that are professionally-organized, and a set of lecturer notes that go into depth on each course subtopic taught. Third, the presentation is rehearsed and the presenter evaluated and coached to do the best possible job. Fourth, the notes are structured to at least two levels of technical detail, so that the course can be tailored to either process engineers or process technicians. Fifth, the course includes a considerable industrial and practical content. And finally, very important, the course (structured in a slightly different way) could be offered in a distance-learning format, or as a shared university course. Might TMS have a role in promoting the development of such course content? Is this approach suitable for use by TMS and other professional societies?

The Connection between Materials Processing and Combustion

In mid-2012, the author embarked on a project to develop a set of magazine articles on the general subject of *Combustion Concepts*. Each article would occupy one page of *Industrial Heating* magazine [4], which is delivered free to those working in heat-related industrial processing. Each article had certain characteristics.

- Each article was accompanied by a downloadable Excel workbook that showed how to make the calculations described in the article.
- Each article built on the information presented in the previous course.
- Each article was envisioned to be one lesson for a process engineering course, using combustion as the main theme.
- The objective was teaching by simulation; each article used Excel-based software to simulate a combustion-related process [5].

Why combustion? After all, MS&E professionals do all sorts of things: leaching, precipitation, roasting, reducing, smelting, refining, alloying, etc. The first answer is that combustion is encountered so often in the processing of so many substances—every kind of material used by industry. Second, it is a familiar topic—who hasn't lighted a propane torch, watched a campfire, or turned on a gas-fired house furnace? People can build new knowledge faster from an existing base than starting from something new. Third, combustion involves fairly complex chemistry, and produces reaction products that interact in an interesting and complex way with the materials being heated. Finally, there is a wealth of readily-available information about the subject.

Why magazine articles? After all, why not write a treatise on the subject and be done with it. Well, that's been done [6], but by itself, it isn't suitable for the intended purpose. The main answer was to get feedback after each article, which would help develop the next one to better fit the interests of a wide audience of industrial people. Also, at the end of a year's articles, there would be nearly a dozen informational packages that might serve as a basis for a process course using combustion as an example. The series was given a theme: Combustion Concepts.

Eleven "Easy" Lessons on Combustion Concepts

The series started with simple fundamentals, with a first emphasis on units and stoichiometry. Additional articles showed how to use a thermodynamic database to calculate the heat of combustion of a fuel, and the heat content of combustion products. The next group of articles dealt with the calculational techniques of making material and heat balances, all dealing with the heating of solid aluminum. The concept of a system balance was introduced and expanded in two articles, which used Goal Seek and Solver to close the balances. The series concluded with articles on special aspects of combustion: available heat, adiabatic flame temperature, generating an exothermic atmosphere, and finally, a detailed burner analysis article. The theme in each article was that people could follow the procedures laid out in the worksheets, and construct heat and material balances for any process where sufficient data existed. Along the way, some useful Excel calculational tools were presented as a way to enhance the ordinary calculation techniques.

The general principles underlying the series content was based on a recently-published TMS book [6], and a set of lecture notes from a previous short course [7]. The technical level of the articles assumed familiarity with the content of a general college chemistry course. All of the calculations were Excel-based, so the workbook procedures require a modest competence in using Excel and some of its lesser-known tools.

In retrospect, the "lessons" were probably not as easy to understand as originally envisaged. The main difficulty was that the IH magazine article length was just one page, or about 700 words plus one small graphic. As the series went on, the article content became more technical and it was not really possible to do justice to the subject in the allotted space. A more useful collection of the material includes introductory text to better orient the reader, expansion of the articles to include more details on the examples, and longer explanatory text about the workbooks [8].

Article Descriptions

The Combustion Concepts articles started in July 2012, and continued through July 2013. You may download the articles and the Excel workbooks from the IH website [4]. An enhanced version of the series is available, along with a reference to useful Excel add-ins [8].

Standard Temperature and Pressure—July 2012

The article series began by discussing the need to take care when expressing flow in volumetric units [9]. There is no universally agreed-upon definition of STP, so the amount, or mass, of natural gas that occupies one volume unit cannot be determined without specifying the STP conditions. The worksheet contains a calculational table that calculates the actual and STP volumes for the SI and SES units system, as well as mass and molar amounts. Table I shows a portion of the worksheet table. Users enter values in the cross-hatched cells, and the table presents seven different unit values.

Table I. Portion of Excel Template Showing Conversion of Actual Gas Volume (Cross-Hatched Cells) to AES and SI Units, and STP.

Actual Values										STP Values			
AES			converted to		SI			converted to		AES STP			
V, ft^3	P, psia	T, °F	lb-mol		V, m^3	P, Pa	T, °C	g-mol		V, ft^3	P, psia	T, °F	lb-mol
1000	15.7	100	2.614		28.317	108,236	37.78	1185.7		992.0	14.696	60	2.614
SI			converted to		AES			converted to		IUPAC STP			
V, m^3	P, Pa	T, °C	g-mol		V, ft^3	P, psia	T, °F	lb-mol		V, m^3	P, Pa	T, °C	g-mol
360.6	101325	35	14,261		12734.4	14.70	95.00	31.443		323.88	100,000	0	14,261

Combustion Management—August 2012

The two principles of combustion management are [10]: (1) Provide more oxygen than you theoretically need to combust all of the fuel constituents. (2) Do not use too much oxygen. This led to a discussion of calculating the fuel-oxidant ratio for stoichiometric combustion, and the

$\%O_2$ in the combustion gas when using air in excess of stoichiometric. The calculation methodology was expanded to include gaseous fuels of any composition, and an oxidant based on humid air and oxygen enrichment. Since most combustion gas samples are dried before analysis, a correction was included for relating the % O_2 in dried offgas vs. the % excess oxidant. This article introduced the use of charts and Excel's Trendline tool to present calculated results.

Calculating the Heat of Combustion for Natural Gas—September 2012

There are a number of references where one can find the heats of formation of substances involved in combustion chemistry, and with a little effort, calculate the heat of oxidation of the substance [11]. For natural gas, it's common to see statements like "the heat of combustion is 1000 Btu/scf". But what is the STP here? What is the composition of the natural gas? This article describes the most useful internet sources for appropriate thermodynamic data, and organizes it into a template that calculates both the high heat value (HHV) and low heat value (LHV) for the combustion of all common fuel gas species at 25 °C.

The workbook calculations were built on the material presented in the previous two articles. Fuels may contain up to 10 different constituents, and the HHV and LHV is displayed in 10 different units. The workbook also shows that a fuel's LHV is relatively independent of the combustion temperature. One of the main information sources cited in the IH articles is Wikipedia. The entry for Combustion [12] needed work, along with related articles on enthalpy and air-fuel ratio. As a byproduct to writing for IH, these articles were edited to correct errors and omissions.

Making a Material Balance—November 2012

Material entering a process is consumed, used productively, and is either unaccountably lost, conveyed from the process for other uses, or vented. Making a material balance involves writing equations based on the law of conservation of mass, and writing chemical reaction equations to account for the fate of the instream constituents [13] The fate of "lost" mass or the values of any unknown stream properties is calculated by solving the material balance equation set. The calculations build on material developed in the first two IH articles.

This article required the invention of a process; here, the heating of aluminum ingots. The first step was the usual practice: stream flows were converted from the stated units to molar flows. Next, material balance equations were written around the process components, and closing the material balance by calculating the unknown leak air and stack gas flows. For novices, the article suggested making a ledger of the stream constituents to more clearly show the flow of each element of the process. Two results tables were prepared: Stream flows with constituents in absolute amounts and with constituents in composition units (volume %). The editable process cells and the resulting result tables are necessary for follow-on articles because subsequent workbooks can be linked to them. If properly linked, changes in the editable cell values in this worksheet will cascade through all subsequent process calculations.

Making a Heat Balance—December 2012

A heat balance is the primary tool for evaluating the thermal efficiency of a process [14]. 25°C was chosen as the reference temperature, with equations written in the standard heat balance format. Five calculational steps are involved:

1. Bring all instream substances to 25°C.
2. Carry out all reactions at 25°C.
3. Bring all outstream substances to their respective values.
4. Include a heat loss term.
5. Write equations for any subsidiary effects that occur among the stream constituents.
6. Close the balance by solving for any unknowns that bring the heat term sum to zero.

This article devoted considerable space to methods for using a thermodynamic database to find and manipulate heat content data. Some databases are expressed in table form only, while others present multi-term equations in a format that makes them difficult to use in Excel. The only Excel-based database is FREED [15], which is recommended because its format is very suitable for Excel-based calculations. The heat balance around the two process devices were used initially to calculate the heat loss from each device. In addition, the Goal Seek tool was introduced here as a way to conduct "What-If" calculations on other process variables.

Making a System Balance (Part 1)—February 2013.

Calculations for the previous two articles were relatively simple because the material and heat balances were *uncoupled*. That is, unknowns from one balance were not embedded as unknowns in the other. But when they are, unknowns from both balances must be solved at once, which is designated as a *system* balance [16]. Balance coupling is common when many different process variables are editable.

The example used in this article is the effect of changing the percent oxygen in the oxidant for the aluminum heating process. When $\%O_2$ is changed, Goal Seek can be used to make a system balance around the whole process, but the heat balance around the burner no longer closes. Goal Seek is used again on the burner heat balance to calculate the new hot combustion gas temperature, which then closes the entire system balance. This system balance example was rather simple, in that it could be closed with only two Goal Seek calculations.

Making a System Balance (Part 2)—March 2013.

One of the features that complicate making a system balance is when the flowsheet contains a recycle stream. For example, suppose that a heat exchanger is added to the aluminum ingot heating process to extract heat from the stack gas and deliver it to the oxidant stream [17]. The system balance for this flowsheet has the material and heat content equations well-coupled, so we end up with three separate equation sets to be solved at once. Here, Excel's Solver tool was employed for the task. This gives us a template for exploring (among other things) the effects of oxygen enrichment *and* preheating the combustion air. This example is probably the most computationally intense of all of the article examples.

Available Combustion Heat—April 2013

Much of a fuel's combustion heat is unavailable for productive use [18]. Hot offgas contains a considerable fraction of this heat, while some is lost to the surroundings by conduction and convection through walls and by radiation when a furnace door is opened. The difference between the heat entering a system and the heat discharged via the stack gas is called the "available heat" [19]. In some conventions, the heat loss may be added to the heat content of the stack gas. If so, this difference is the amount available for productive use, and is often expressed as a percentage of the entering heat. The lower the % available heat, the more fuel that must be burned to accomplish the objective.

An uncoupled material and heat balance calculation template was set up for a variable natural gas composition, with editable values for the % O_2 in the oxidant, the % excess oxidant, the offgas temperature, and the heat loss (per g-mol of fuel gas). This template includes features from the material and heat balance templates described in earlier articles. The template shows how to use Goal Seek to calculate the adiabatic flame temperature. The template can be modified to include the beneficial effects of combustion air preheat on % available heat.

Adiabatic Flame Temperature—May 2013

There are at least two ways to describe a furnace's thermal performance: the percent available heat (discussed above), and the adiabatic flame temperature (AFT), which indicates the intensity of heat transfer [20]. The AFT is mainly used to compare the changes caused by different combustion settings (i.e., type of fuel, % oxygen enrichment, % excess air, etc.). The flame temperature is important because most high-temperature heat transfer is by radiation, where thermal flux is a function of T^4. The computational techniques were based on material and heat balance calculations from earlier articles.

Exothermic Atmospheres—June 2013

When combustion occurs with less than stoichiometric air, the products contain CO_2, H_2O, N_2, CO and H_2, while O_2 is stoichiometrically absent. An exothermic atmosphere is one that contains a sufficiently large air-fuel ratio to provide sufficient combustion heat to raise the combustion gas to a temperature where it attains a near-equilibrium condition [21, 22]. The system balance calculations cannot be solved by using only material and heat balance equations. A Keq equation is also required to determine the tendency of the heat treat furnace to oxidize iron, and another one to satisfy the constraint that the gas phase is close to the equilibrium position. This is a common practice for this type of situation, and reasonable in view of the presence of a catalyst in the combustion furnace, and the presence of metallic parts in the heat-treat furnace. Here the water-gas shift (WGS) reaction and its associated equilibrium constant expression was employed as one of the system balance equations. T units are degrees K.

$$CO_2 + H_2 \rightarrow CO + H_2O;\ Keq = \frac{pCO \times pH_2O}{pCO_2 \times pH_2} = 10^{(-1654/T\ +\ 1.513)} \qquad [1]$$

Additional equilibrium constraint equations were employed to calculate the pO_2 in both furnaces, the activity of carbon in the heat-treat furnace gas, and the pH_2O in the cooler to determine the amount of water removed from the combustion gas. The results show the importance of water removal from the burner gas in preventing iron oxidation in the heat-treat furnace. The colder the gas cooler temperature, the lower the dew point and pO_2 in the heat-treat furnace.

Burner Analysis—July 2013

Many of the previous article's subjects are collected in this, the final article in the series. The workbook relates the factors involved in the performance of a burner operating with excess oxidant. In addition to the fuel composition, there are 13 more editable burner properties [23]. To avoid clutter, the calculational details are on a separate worksheet, as are the results.

However, it's easy to open the Data Entry and the Results worksheets on the same screen, so you can immediately view the results of a change in input. As with prior workbooks, experienced users can edit the location of selected cells, use a different calculational method for their own purposes, or add additional features.

This article introduced two new Excel add-ins for making repetitive calculations: Multicalc and Multigoal [24]. These two Excel utilities permit the user to make multiple worksheet calculations while changing one of the primary variables. The results are displayed in tabular and chart format. The worksheet contains examples of the use of these two tools.

Summary

Educational trends show increased use of on-line and other distance-learning techniques to accommodate the needs of graduated engineers, college students needing specialized training, or the learning of new skills by non-graduated technical professionals. A possible technique for developing such a course is to divide it into relatively short lessons, and apply the material to one common and easily understood processing operation (i.e., combustion). The teaching methodology is process simulation, using Excel and its tools. The technique is so far untested.

References

1. CourseShare, http://www.cic.net/projects/shared-courses/courseshare/introduction. September 2013.
2. Wikipedia contributors, "Massive open online course". Wikipedia, the Free Encyclopedia, (September 2013). http://en.wikipedia.org/wiki/Main_Page.
3. AboutEdu, "Directory of free Massively Open Online Courses providers", www.MOOCs.co. (September 2013).
4. http://www.industrialheating.com/articles/topic/2668?page=2. (September 2013).
5. Arthur Morris and Semih Perdahcioglu, "Software for Energy Education", *Energy Technologies and Carbon Dioxide Management*, TMS Annual Meeting & Exhibition, (2013).
6. Arthur Morris, Gordon Geiger, and H. Alan Fine, "*Handbook on Material and Energy Balance Calculations in Materials Processing*", TMS/Wiley, (2011).
7. Eric Grimsey, "*Basic Flowsheeting Principles with Examples*", www.thermart.net, (September, 2011).
8. www.thermart.net, "Combustion calculations" (September 2013).
9. Arthur Morris, "Standard Temperature and Pressure", VolCalc.xlsx, www.industrialheating.com, (July 2012), 22.
10. Arthur Morris, "Combustion Management", StoichCalc.xlsx, www.industrialheating.com, (August 2012), 24.
11. Arthur Morris, "Calculating the Heat of Combustion for Natural Gas", HeatCalc.xlsx, www.industrialheating.com, (September 2012), 28.
12. Wikipedia contributors, "Combustion". Wikipedia, the Free Encyclopedia, (September 2013). http://en.wikipedia.org/wiki/Main_Page.
13. Arthur Morris, "Making a Material Balance", MatBalCalc.xlsx, www.industrialheating.com, (Nov. 2012), 20.
14. Arthur Morris, "Making a Heat Balance", HeatBalCalc.xlsx, www.industrialheating.com, (December 2012), 22.
15. FREED database, www.thermart.net. (September 2013)
16. Arthur Morris, "Making a System Balance (Part 1)", SysBalBalCalc.xlsx, www.industrialheating.com, (February 2013), 22.
17. Arthur Morris, "Making a System Balance (Part 2)", SysBalBalCalc2.xlsx, www.industrialheating.com, (March 2013), 24.
18. Arthur Morris, "Available Combustion Heat", AvailHeatCalc.xlsx, www.industrialheating.com, (April 2013), 22.
19. http://www.gosolarcalifornia.ca.gov/tools/newCalcSys/Description/Calculate%20available%20heat.pdf. (May 2012).
20. Arthur Morris, "Adiabatic Flame Temperature", AFT-Calc.xlsx, www.industrialheating.com, (May 2013), 20.
21. Arthur Morris, "Exothermic Atmospheres", ExoCalc.xlsx, www.industrialheating.com, (June 2013), 22.
22. ASM Committee on Furnace Atmospheres, *Furnace Atmospheres and Carbon Control*, Metals Park, OH (1964).
23. Arthur Morris, "Burner Analysis", BurnerCalc.xlsx, www.industrialheating.com, (July 2013), 20.
24. www.thermart.net, "Software". (September 2013).

Celebrating the Megascale: Proceedings of the Extraction and Processing Division Symposium on Pyrometallurgy in Honor of David G.C. Robertson
Edited by: Phillip J. Mackey, Eric J. Grimsey, Rodney T. Jones, and Geoffrey A. Brooks
TMS (The Minerals, Metals & Materials Society), 2014

Enhancement of Pyrometallurgical Teaching Using Excel Simulation Models

Eric J Grimsey[1]

[1]Emeritus Professor, WA School of Mines, Curtin University

eric@wasm.curtin.edu.au

Keywords: pyrometallurgy teaching, Excel simulation, flowsheet, flash smelter, blast furnace

Abstract

Steady state Excel models for a copper flash smelter and an iron blast furnace are used to enhance the teaching of pyrometallurgical smelting principles within a fourth year level process engineering unit delivered at the Western Australian School of Mines. A lecture/workshop approach has been adopted in which student teams undertake process simulation assignments that illustrate the multifaceted responses of process outputs to variation of inputs, the objectives being to reinforce their understanding of smelting principles. The approach has proven to be popular with students, as evidenced by the consistently high ratings the unit has received through student feedback. This paper provides an overview of the teaching approach and process models used.

Introduction

The capstone pyrometallurgical processes unit at Curtin University's W A School of Mines in Kalgoorlie is delivered to a relatively diverse group of usually around 25 students, with approximately one third completing the final year of a 4 year metallurgical engineering degree, one third completing the 4th year of a 5 year double degree in chemical and metallurgical engineering, and one third of mainly international students (Chinese, African, Indonesian etc.) completing 1 or 2 year graduate metallurgy conversion programs, usually from an engineering background.

The technical objectives of the unit are to develop the student's analytical skills related to flowsheeting and steady state process simulation, and to use these skills to assist in understanding the principles of pyrometallurgical processing. The latter are exemplified for non-ferrous metals through the treatment of Ni and Cu sulfides and for ferrous metals through the smelting of iron oxides in a blast furnace, coupled with basic oxygen steelmaking. This relatively in-depth unit follows a much broader coverage of metal extraction processes undertaken in an earlier unit by the majority of students.

The personal development objective of the unit is to enhance the student's ability to work in teams through completion of three assignments, each in partnership with two other students, with assessment based on a team report. In addition to technical aspects, the report must include details of team meetings, a summary of team member contributions and a signed declaration that each member takes responsibility for the final content of the report, the assessment of which could include oral examination.

Feedback for the unit is provided anonymously through the Curtin University's on-line Evaluate system, through which students can "strongly agree, agree, disagree, or strongly disagree" that a unit has achieved positive educational outcomes divided into 11 categories. Over the past 5 years, 100% of students have either "strongly agreed or agreed" that this unit achieved the desired outcomes in an average of 10 categories. This is considered an excellent ranking. Students have also consistently reported that the use of Excel simulations to support learning and to provide a link to industrial practice is a major strength of the unit.

The reported problematic aspect is the requirement that students work in teams to complete their assignments, especially the expectation, clearly conveyed to all, that better prepared students are expected to assist other team members in understanding the material, such that all are able to take ownership of their team report. A controversial aspect is that team selection is not random. Each team is allocated one engineering student considered to be amongst the stronger academic performers, one considered to be amongst the weaker performers, and if numbers permit, one graduate student (usually non-Australian), with a view to maximizing the learning environment for all through peer interaction and support. The approach seems appreciated by the majority of mature age graduate students, many of whom have only recently arrived in the country and find peer interaction beneficial; not all undergraduate engineering students share this view, particularly the better academic performers.

Each year some students report feeling disadvantaged by the team approach, typified by comments such as: "not being able to choose my own group", "being forced to carry students who don't work", "not being able to make an honest report of team contributions", "not being able to learn since I have to spend time helping weaker team members" etc. Notwithstanding these attitudes, and to some extent because of them, there has been a determination to continue and hopefully to progressively improve student interaction within teams, especially given that the disadvantages perceived by some are not borne out in final grades. Team reports contribute 50% to the final mark, with 50% from a challenging examination; this combination usually results in a spread of grades consistent with those achieved by the students in other units.

Unit Coverage

The unit is divided into three sections, namely, (i) analysis and simulation of a basic steady state flowsheet, (ii) non-ferrous pyrometallurgy as illustrated by the treatment of Ni and Cu sulphides and (iii) ferrous pyrometallurgy as illustrated by the Fe blast furnace and aspects of basic oxygen steelmaking. Each section is normally taught over 3 weeks with class contact of 4 hours of lectures and workshops per week; this year for the first time the unit was taught as three short course modules. Students are provided with comprehensive notes on all topics and no textbooks are required. Theoretical coverage of the material is supported by three assignments which focus on process simulation through (i) a development and analysis of a simple steady state flowsheet using Excel, (ii) a process sensitivity analysis using a Cu flash furnace Excel model and (iii) a process sensitivity analysis using an Fe blast furnace Excel model.

Flowsheet Simulation

Flowsheet simulation is undertaken using a systematic approach based on well-established flowsheeting principles, as described in detail elsewhere [1, 2]. Topics covered include unit process and stream numbering, stream characteristics, material (mole and mass) balances, heat balances, basic unit process characteristics (mixers, splitters, separators, reactors, heat exchangers) and most importantly, unit process analysis. The latter involves recognition and development of a complete "zero degree of freedom" equation set which simulates the flowsheet as a steady state material and heat balance, solvable within Excel using matrix inversion for the linear equations, combined with Excel Solver for the non-linear equations. The focus is on the ability to develop the simulation, rather than on the characteristics of specific flowsheets, and the assignment is always designed to include at least a splitter, mixer, separator and reactor and to demonstrate the importance of bleeds and recycles.

Once the analytical and equation solving skills required for the first section are learned, they can be applied to the simulation of any basic flowsheet (notwithstanding the obvious convenience of a commercial simulation package when dealing with large flowsheets) or to the development of models to simulate individual processes such as the Cu flash furnace and Fe blast furnace as provided to students for the second and third sections of the unit.

As a result, students seem comfortable using these furnace models without any detailed explanation as to their structure and mechanics of operation, although detailed explanations are provided as reference material. Teaching in the rest of the unit can thus focus more on the physicochemical and engineering principles underlying non-ferrous and ferrous pyrometallurgy as illustrated through Ni, Cu and Fe processing, with some emphasis on the principles of operation of the flash and blast furnaces.

The Copper Flash Furnace Simulation

The use of Excel simulation models to assist pyrometallurgical teaching was inspired by the excellent book on copper flash smelting by Bill Davenport and co-workers [3], and the author's involvement in the development of process models for an industrial nickel smelter [4, 5].

The steady state Excel model of a copper flash furnace provided to students is based on a single stage, co-current reactor built around a linear equation set which contains 14 flow unknowns, matched by 9 component mass balances (Cu, Fe, S, O, C, H, SiO_2, CaO, Al_2O_3), 4 restriction equations (matte grade, Fe/SiO_2, oil/cons, Cu in slag) and 1 guessed flow (for N_2), to provide 14 independent equations. Matrix inversion in Excel is used to give a place-holder solution for all flows, which are then written into the heat balance. Data for sensible heats and enthalpies of formation are read from an adjacent thermodynamic table within which sensible heat values are linked to stream temperatures. The process heat loss is set to 500 MJ/t(cons) and all thermodynamic data within the table are imported from the FREED thermodynamic database, developed and made available by colleague Arthur Morris [1, 6].

Standard inputs for the model are concentrate flow rate and composition (%'s $CuFeS_2$, FeS_2, SiO_2, CaO, Al_2O_3), target matte grade, Fe/SiO_2 in slag, preheat temperature, oil addition and slag temperature, from which matte and gas temperatures are offset by -25°C and +75°C respectively. The simulation is run through Excel Solver by varying the guessed N_2 flow until the heat balance summation cell, set as the Solver objective, achieves a value of zero. In default mode, the required oxygen enrichment (OE, as vol% O_2 in air) for the blast is calculated by the model, along with all stream flows to meet process requirements. Alternatively, oil addition can be varied to meet a fixed value of OE, or off gas make, or matte grade etc.; indeed any value calculated within the model can be assigned a value within the Solver "constraints window" and then targeted by varying an appropriate input, to provide a very versatile simulation model.

The model has no dynamic aspects or physical restrictions or free energy calculations, with copper loss to slag determined from matte grade using a polynomial fit from industrial data and with matte taken as an ideal solution of Cu_2S and FeS. Notwithstanding, the simulation is able to reasonably predict the major heat and mass balance and chemically driven relationships between inputs and outputs for a Cu flash smelter as illustrated in Figures 1 and 2.

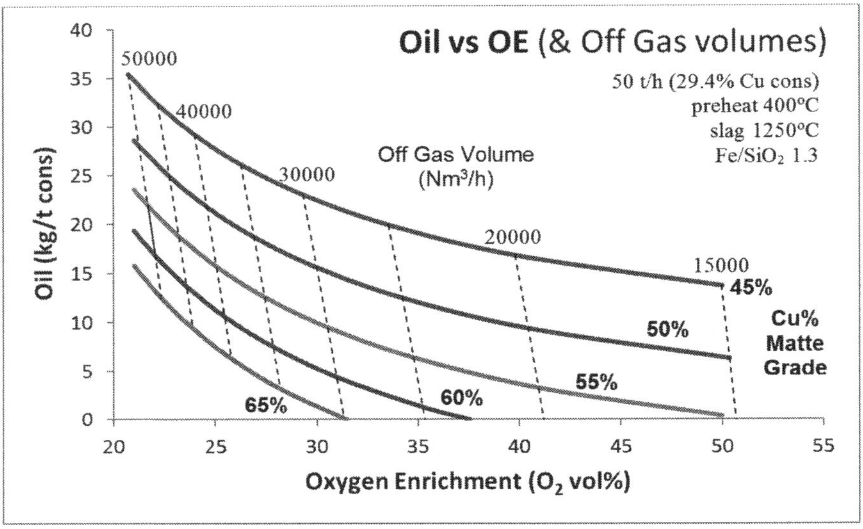

Figure 1. Simulation of oil addition and oxygen enrichment combinations, and resultant off gas make (volume), as necessary to produce specific Cu matte grades from a Cu concentrate containing 85% $CuFeS_2$, 5% FeS_2, 7%SiO_2, 1%CaO and 2%Al_2O_3, when using a 400°C air preheat while maintaining a 1250°C slag temperature.

Figure 1 shows the combinations of oil (0-35 kg/t cons) and oxygen enrichment (21-50 vol% O_2 in blast) available for the operation of a Cu flash smelter with an air preheat of 400°C, to produce matte grades of 45-65% Cu, from a 29.4% Cu concentrate, while maintaining the slag, matte and gas temperatures at 1250, 1225 and 1300°C respectively. Isochors for off gas make (15,000-50,000 Nm^3/h)[1] are also shown on the diagram.

Figure 1 confirms that the model reproduces the following important trends for student observation and explanation, (i) as oil is decreased at a given matte grade and preheat, the OE must be increased to maintain the heat balance, (ii) the smelter can operate without oil provided the matte grade and OE are high enough, (iii) OE rather than matte grade is the major influence on off gas make, (iv) the lowest off gas make (and potentially highest production capacity) is achieved at the highest OE and lowest oil input for a given matte grade, and finally, (v) matte grade must decreased as OE is increased (at fixed oil and preheat) in order to maintain the heat balance, with maximum grade being reduced to ~50% Cu if the OE is increased to 100% O_2 with no oil addition.

[1] Nm^3 is volume at STP, defined as 0°C and 1 atmosphere pressure.

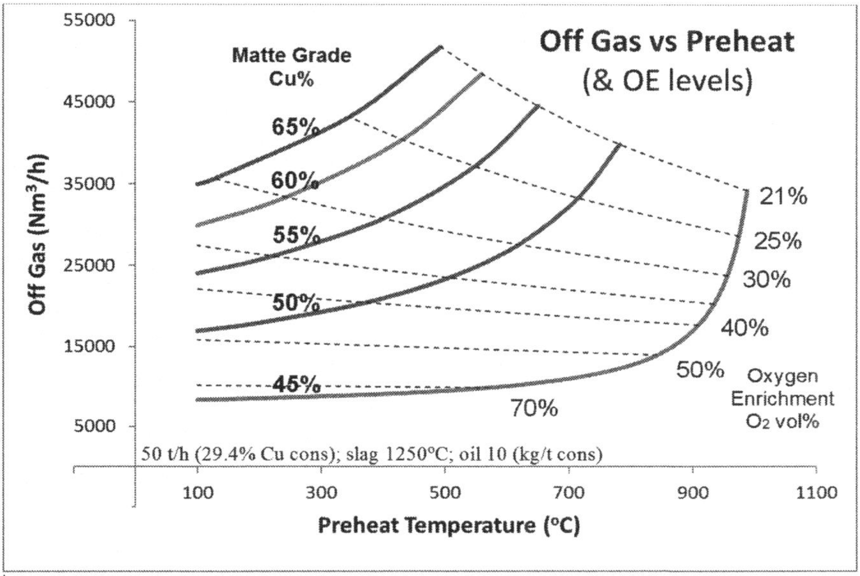

Figure 2. Simulation of off gas make and preheat temperature for a range of matte grades (45-65% Cu) and blast oxygen enrichment (21-70 vol%) when smelting a 29.4% Cu concentrate at a fixed oil addition of 10 kg/t cons, while maintaining a 1250°C slag temperature.

Figure 2 shows the effect on the off gas make (Nm^3/h) of increasing the air preheat temperature through the range 100°C to 950°C for matte grades of 45-65% Cu and OE levels of 21-70 vol% O_2, when the oil input is 10 kg/t cons and the slag temperature is 1250°C. The following additional trends are illustrated, (i) the required preheat temperature decreases with increase in matte grade at fixed OE, (ii) increasing the OE at a given matte grade decreases both the gas make and the required preheat temperature, (iii) the effect of OE on decreasing gas make becomes more significant at lower matte grades whereas the effect on preheat becomes less significant and, (iv) the most effective strategy to lower gas make (and potentially increase productivity) at a given matte grade is to lower the preheat and maximise the OE.

Explanation of the trends shown in Figures 1 and 2 provides reinforcement of the fuel value of Fe and S in the concentrate, of the need for additional oil to support lower grade matte smelting, of the difference between total oxygen, which dominantly controls the matte grade (via the oxygen/concentrate ratio), and oxygen enrichment, which dominantly controls gas volume and affects operating temperature, and the importance of nitrogen (through OE) as a heat sink and gas volume controller. Since the simulation also illustrates that a range of control options can produce a required smelter product, there is reinforcement of the need to consider other requirements such as gas cleaning and converter capacity and most importantly, input costs and market values, when choosing the most suitable option.

The Iron Blast Furnace Simulation

The steady state Excel model of an iron blast furnace provided to the students is developed using well-documented chemical and mass balance principles that underpin ideal furnace operation [7]. The furnace is modelled as a two stage (shaft and wustite reduction zone) counter-current, equilibrium limited, chemical and heat exchange reactor with no aerodynamic or physical restrictions or top gas pressure. The stages are separated at the assumed overlap of the chemical and thermal reserve zones where the default temperature is taken as 927°C (1200 K) for both gas and solids, and the CO/CO_2 and H_2/H_2O gas ratios are fixed by their assumed equilibrium with $Fe/Fe_{0.947}O$.

The ore is a mixture of hematite and magnetite with minor SiO_2, Al_2O_3 and moisture, the flux a mixture of CaO and $CaCO_3$ and the slag an ideal solution of Ca_2SiO_4, CaO and Al_2O_3, with no FeO content. The Si% in pig iron is calculated as a function of both the RAFT (raceway adiabatic tuyere flame temperature) and the hot metal temperature, while C% is calculated from the saturation solubility of C in liquid Fe, discounted by a factor which depends on the Si%; both equations are empirical. Currently, there are no inputs of Mn, P or S. Direct reduction (via the Boudouard reaction) is assumed to occur only in the lower wustite reduction zone with no CO disproportionation and C precipitation in the shaft zone; thus all solid C entering the shaft also enters the lower wustite reduction zone.

The model is built around a linear equation set which contains 21 flow unknowns, matched by 4 component mass balances for the shaft zone (C, O, Fe and H), 8 component mass balances for the wustite reduction zone (N_2, Al_2O_3, CaO, O, Fe, H, Si and C), 8 restrictions (3 gas ratios, slag basicity ratio, pig iron production rate, oil addition, pure O_2 addition and humidity in blast), which leaves 1 degree of freedom. This is accommodated by guessing the flow of normal air in the blast. The linear set is then solved by matrix inversion to give a place-holder solution for all flows which are used within the heat balances for the shaft and wustite reduction zones.

The required heats of formation and sensible heat data are again referenced from a table constructed using data from FREED [6] with the heat loss for each zone set at 8×10^3 MJ/h for each metre of hearth diameter, with 20% assigned to the shaft [7], resulting in a total heat loss of ~260 MJ/thm (thm: tonnes of hot metal) for the 8000 thm/day default model, in accord with industrial data [8]. In addition, a coupled mass and heat balance is written for the tuyere zone to calculate the mass of coke burnt in the flame and also the RAFT on the assumption that coke enters the raceway at 1800 K and none is vaporised [7].

Standard inputs for the model are pig iron production rate (thm/day), compositions (%'s) for the ore (Fe_2O_3, Fe_3O_4, SiO_2, Al_2O_3, H_2O), flux (CaO, $CaCO_3$, default 100% CaO), coke, slag basicity ($CaO/(SiO_2+Al_2O_3)$), relative humidity of air (default 25% @ 25°C), slag and metal temperatures (taken as equal) and blast preheat temperature.

The simulation is run through Excel Solver within which the guessed air flow is varied along with the top gas temperature (TGT) and the RAFT until all three heat balances are set to zero, as determined by the Solver "objective" and two "constraints". The simulation outputs all stream flows, including oil requirement (kg/thm), coke rate (kg/thm), oxygen addition to blast, normal air to blast, total blast rate and top gas make, all in Nm^3/thm, in addition to the TGT and RAFT. A summary of the percent reduction of FeO through indirect, direct and hydrogen reduction is also given. As with the Cu flash furnace model, it is possible to target any calculated values by varying appropriate inputs, such as setting the TGT, or the RAFT, or the top gas make (etc.) and then varying the oil input at fixed oxygen input, or vice versa.

Figures 3 and 4 illustrate that the model is able to simulate the major heat and mass balance and chemically driven relationships between inputs and outputs of an ideally operating Fe blast furnace. Figure 3 shows the normal air in blast (Nm^3/thm) versus oil addition (kg/thm) for an Fe blast furnace operating with various additions of pure oxygen (kg/thm), along with the associated coke rate (kg/thm). The figure essentially defines the operating window for an ideal blast furnace at constant preheat (1050°C), in which oil addition at fixed OE is limited by the need to maintain the RAFT temperature high enough for adequate heat transfer to melt the charge (~1900°C), and in which OE is limited at fixed oil by the need to maintain a TGT above ~130°C, to provide an adequate rate of moisture removal from the incoming charge.

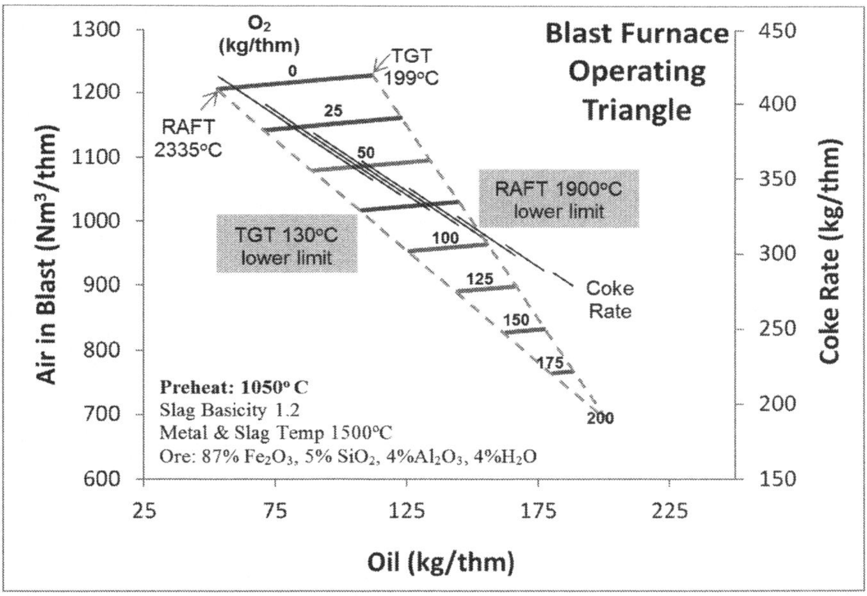

Figure 3. Simulation of the ideal operating window for a blast furnace with a constant preheat (1050°C), showing air in blast for various oil and pure oxygen additions, along with the associated coke rate, while maintaining a slag and metal temperature of 1500°C.

The figure illustrates the following important trends for student observation and explanation, (i) when oil addition is increased at fixed pure oxygen addition, the blast air must be increased, the RAFT temperature decreases and the TGT increases, (ii) when pure oxygen addition is increased at fixed oil, the blast air must be decreased, the RAFT temperature increases and the TGT decreases and (iii) when oil (or powdered coal etc.) is added to the furnace, the coke rate decreases almost linearly in response.

Figure 4 shows that the operating triangle shifts significantly downwards with increase in blast preheat temperature, in this case from 1050°C to 1250°C, as dictated by lower coke and air requirements. The decrease in coke rate with increase in oil is again clearly shown but also, at the higher preheat, the need to "step up" the coke rate with each increase in oxygen becomes clearer, as is necessary to compensate for a loss of preheat as hot air is replaced by cold oxygen.

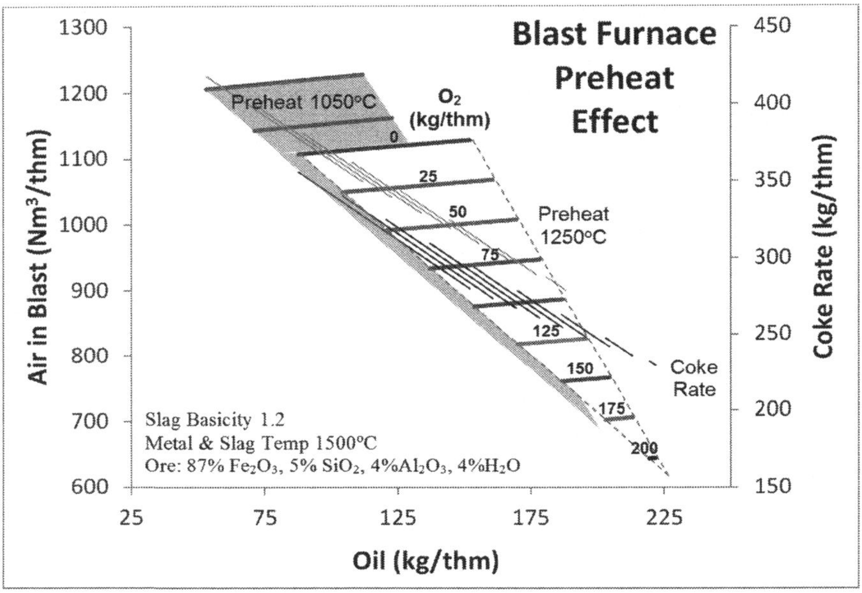

Figure 4. Simulation of the shift in the operating window for an ideal blast furnace with increase in the air preheat from 1050°C to 1250°C.

Observation and explanation of the above trends highlights to students, (i) the importance of "mass of gas" in addition to "available heat" in determining both the RAFT and TGT, (ii) the need for limits on oxygen and oil additions and preheat to maintain critical minimum values for these temperatures and (iii) the effectiveness of using hydrocarbon injection to reduce coke rate, with the resultant cooling of the RAFT able to be offset by increasing pure oxygen addition or preheat.

The Simulation Assignments

Although the base case and specifics of the assignment are varied each year, the main objectives remain the same, namely, to have the students engage in a team-based capstone hands-on activity which places the theory already covered into an operational context by highlighting through simulation, analysis and explanation (i) the multiple output effects created by the change in a single process input, (ii) the relative magnitude of those effects on the process control targets (iii) the combinations and ranges of multiple input changes which can and must be made to maintain a particular set of output targets and (iv) the potential impact of those combinations on the economical operation of the process, all of which are efficiently accomplished through use of relatively simple Excel simulation models as outlined here.

Teams are not necessarily expected to reproduce succinct comprehensive figures of the type shown above; rather they are expected to decide for themselves the best way to simulate, quantitatively illustrate and explain the important relationships between inputs and outputs and the implications for process control, as generally covered in the unit material. These relationships include those illustrated above, but extend, for example, to consideration of effects of slag mass and Cu grade on Cu recovery and the impact of oil addition on the extent of direct reduction of wustite within the Fe blast furnace and associated thermal implications.

Conclusions

The teaching of pyrometallurgical process principles to a relatively diverse group of later year undergraduate and graduate students has been enhanced through the use of relatively simple process models written in Excel, through which students are able to observe key relationships between process inputs and outputs for both a Cu flash furnace and Fe blast furnace, as confirmation of the basic principles which underlie these and other processes, as taught in the main body of the unit. The overall approach has proven very successful as evidenced by anonymous student feedback through Curtin University's online Evaluate system; however, the unwillingness of some students to positively engage in teamwork while completing assignments is yet to be resolved.

References

1. Eric J. Grimsey, "Basic Material and Heat Balances for Steady State Flowsheets", on CD insert in *Handbook of Material and Energy Balance Calculations in Materials Processing*, 3rd Ed., A. E. Morris, Geiger G. H. and H. A. Fine (New Jersey: Wiley, 2011).

2. Eric J. Grimsey, "Basic Flowsheeting Principles with Examples", www.thermart.net, (September, 2011).

3. W. G. Davenport et al., *Flash Smelting: Analysis, Control and Optimization, 2nd Ed.* (Warrendale PA: TMS-AIME, 2004).

4. Arthur G. Hunt, Eric J. Grimsey and Neil B. Gray, "A Simulation of the Kalgoorlie Nickel Smelter using METSIM", in *5th. Aus. I. M. M. Extractive Metallurgy Conference,* (Melbourne: AusIMM, 1991), 251 – 255.

5. A. G. Hunt et al., "Modelling for design and control at the Kalgoorlie Nickel Smelter", in *Proceeding of the Paul E. Queneau Symposium "Extractive Metallurgy of Copper, Nickel and Cobalt, volume 1: Fundamental Aspects,* eds. R. G. Reddy and R. N. Weizenbach, (Warrendale: TMS-AIME, 1993) 489-499.

6. Arthur E. Morris, FREED, A thermodynamic database in Excel using data from the U. S. Bureau of Mines and the U. S. Geological Survey, Ver. 7.8, www.thermart.net (June 13, 2010).

7. John G. Peacey and William G. Davenport, *The Iron Blast Furnace, Theory and Practice* (Oxford, England: Pergamon Press, 1979).

8. J. A. Burgo., "The Manufacture of Pig Iron in the Blast Furnace", in *The Making, Shaping and Treating of Steel, Ironmaking Volume*, ed. D. A. Wakelin (Pittsburgh: The AISE Foundation, 1999), p713.

Celebrating the Megascale: Proceedings of the Extraction and Processing Division Symposium on Pyrometallurgy in Honor of David G.C. Robertson
Edited by: Phillip J. Mackey, Eric J. Grimsey, Rodney T. Jones, and Geoffrey A. Brooks
TMS (The Minerals, Metals & Materials Society), 2014

THE ENGINEERING DESIGN SEQUENCE AND MATERIALS DEVELOPMENT: 990 GOLD–TITANIUM AS A CASE STUDY

Mark E. Schlesinger, P.E.
Department of Materials Science and Engineering
Missouri University of Science and Technology

Keywords: Engineering design, gold metallurgy, jewelry

Abstract

Engineering design follows a six–step sequence. Demonstration of this sequence in materials engineering is less common. A case study is presented, using the 1989 paper by Gafner on the development of 990 gold-titanium alloy. This paper identifies a need, develops the problem, identifies alternatives, demonstrates the use of metallurgical principles to identify alternative responses, and shows how more favorable options are selected. The paper can also be used to illustrate engineering design as an iterative process. The paper is also useful as a starting point for more in-depth examination of specific elements of the design process, and lends itself to the creation of student exercises.

Introduction

Figure 1, adapted from the text of Dieter,[1] describes the general engineering design process that is commonly taught to undergraduate students. It is a common element of mechanical engineering education programs,[2,3] but is less common to materials and metallurgical curricula. In developing a capstone design course for metallurgical engineering students at the Missouri University of Science and Technology, the author sought to introduce this process, and looked for examples relevant to students in the discipline. One example was the search for a means of responding to the problem of high levels of copper in steel scrap; the use of this problem as a case study has been described previously.[4]

This paper will describe the use of a second case study for introducing the general engineering design process to metallurgical students. This second case study relies on the 1989 paper by Gafner describing the development of 990 gold–titanium alloy.[5] This paper was written at a level that undergraduates can comprehend, and provides an excellent platform for a discussion of the separate steps of the design process.

Identification of Need

The first step in the design process is what Dieter calls a *statement of need.*[1] Statements of need have several potential sources: present or potential customers, organizational supervisors or management, industry–created roadmaps, or the intuition of the engineer.[3] Typically the need for design arises when something changes in the resources, processes, products, or markets of the organization.

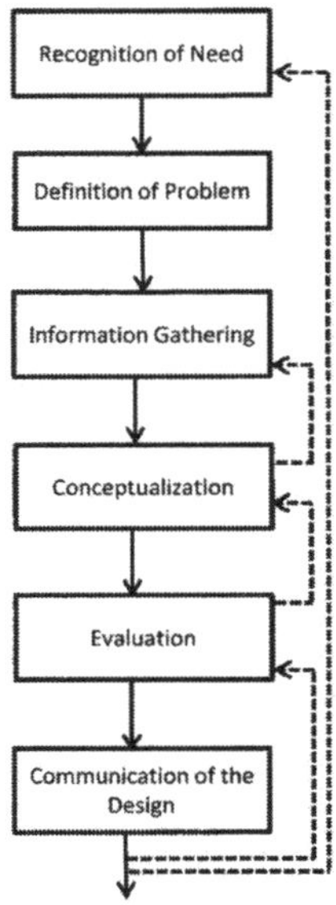

Figure 1. The general engineering design sequence, after Dieter.[1]

Figure 2 illustrates the need identified by Intergold of South Africa in 1983.[6] The item is a piece of gold jewelry in the Chuk Kam style. The phrase, which means 'pure gold' in Cantonese, denotes gold with a minimum millitesimal fineness of 990 (99.0 wt-%). Chuk Kam jewelry has been popular for several decades in Hong Kong and elsewhere in Asia, where its value as an investment is added to the normal attractiveness of gold jewelry. However, the Chuk Kam jewelry sold in Hong Kong before the 1990s had a negative attribute: the mechanical properties of gold, in particular the hardness and yield stress. Pure gold is soft, and easily damaged. Give the cost of items such as that shown in Fig. 2, improving the damage resistance of Chuk Kam jewelry was important to retaining an existing market as well as expanding it to new customers. As a result, Intergold was asked by its Hong Kong branch to 'develop a gold–based material with at least 990 fineness, a colour close to that of pure gold, and a durability as good as that of standard jewellery alloys'.[5] This is a statement of need, and the basis for the research program that followed.

Figure 2. Example of Chuk Kam jewelry. (DifferenceBetween.info, "Difference between 24k Gold and 22k Gold," http://www.differencebetween.info/difference-between-24k-gold-and-22k-gold).

Problem Definition

Defining a problem creates a basis for judging the options created later in the design sequence. There are two parts to a problem definition. The first is more information about the problem. In the case described by Gafner,[5] more information about the problem would include the size of the market (20 tons per year in the 1980s), the types of jewelry involved, the typical customer, the frequency of complaints, the available alternatives to Chuk Kam gold jewelry (gold coins or ingots, reduced–purity gold jewelry, platinum coins or jewelry, other investments), and reasons why the jewelry market might (or might not) be changing. Anticipating future changes is an important part of this step in defining a problem.

The second part of a problem definition is a list of *criteria* and *restrictions.* Criteria are the characteristics a response to the problem *should* achieve, while restrictions are the characteristics *required* of a solution. In this case, the need for 990 millitesimal fineness is a restriction, while the cost of processing is a criterion. (For most designs, cost is a restriction as well as a criterion, but this does not have the same level of importance in the gold jewelry business.) Government is often a source of criteria and restrictions as well as the marketplace, and should be included. (For some reason the use of radioactive materials in jewelry is frowned upon.) Gafner includes mechanical properties as a restriction, and color as a criterion.[5] Asking students to add to this list demonstrates the importance of visualizing the use of the product as well as its manufacture.

Information Gathering

The next step in the design sequence illustrated in Fig. 1 is the gathering of information needed to develop a response to the need identified in the first step. Some of this information is mentioned in passing in Gafner's paper,[5] but a more complete discussion of this step allows students to add to the discussion. The information required to deal with a need falls into two categories.

The first type of information is technical data. The type of technical data needed is often determined by the list of criteria and restrictions, and is an iterative process; as new criteria and restrictions are identified, the technical data needs grow as well. In the case of gold alloys, needed technical data would include standard testing techniques in the jewelry industry, phase diagrams, gold alloy compositions, physical properties data for gold alloys, corrosion characteristics of these alloys, measured color for different alloys, and costs for alloying materials and different types of processing. Gafner does not specifically list technical data needs;[5] however, students can be relied on to produce a list of their own.

The second type of information to be gathered is a set of existing solutions. It is not uncommon for a need to have been previously identified by others, and using a previous solution can save time and money. Gafner identifies one previous solution, the development of oxide composites,[5] but there may be others. Some can be obtained from formal sources such as patent databases and the technical literature. Others come from informal sources: customers, experts on the industry, and occasionally competitors (obtained ethically, of course). Previous solutions may have flaws, but knowing what they are can save time and effort later on. Most metallurgy seniors will be aware of patent and technical–paper databases, but this is a good opportunity to introduce these if it has not been previously done.

Conceptualization

Better known as *brainstorming* or the pretentious alternative *ideation,* this is the process of generating a list of possible responses to the need identified in the first step of the design process. There are four types of possible response:

The 'do-nothing' option. This includes present practice in the list of solutions. The present approach represents a benchmark against which other proposed concepts can be measured. While the do-nothing approach may occasionally turn out to be the best alternative, the organization that repeatedly reaches this conclusion will eventually find itself failing. The do-nothing option in this case was the present form of Chuk Kam jewelry.

A previous solution. This is the list of alternatives uncovered during the previous step. While most of these may have previously been found wanting, changing conditions and technology may make them worth looking at again. The use of direct reduced iron (DRI) as a source of iron units for electric–furnace steelmakers is an example. Lower natural gas prices have increased the economic viability of DRI manufacture in North America, improving its viability as a means of diluting impurities in steel scrap. A previously existing option for Chuk Kam gold jewelry, production of oxide composites, was abandoned in the 1980s because they "involved too high a level of technology for the usual goldsmith".[5] This may have changed over the 25 following years.

Tweaking. This is the minor adjustment of existing solutions, including the do-nothing option, to solve the problem with minimum cost and disruption. A 'tweak' can be put into operation quickly, and is likely to need less new equipment or capital expenditures than new processes or products. It also entails less risk. However, it also tends to produce the smallest benefits of proposed solutions, and can be an excuse for not thinking more boldly. Redesigning Chuk Kam jewelry to make it less damage–prone is an example of tweaking; producing a slightly different composite is another.

Inventing. This is often the highest–risk, highest–reward type of idea. The concept of 'thinking outside the box' can be introduced to students when describing this type of idea. It should be stressed that at this point no idea is too offbeat or unlikely to leave off the list. At the time of Gafner's article, the idea of producing harder Chuk Kam jewelry by putting a diamond coating on it would have been ludicrous; events since then have made it a potentially feasible option. Improving a gold alloy by having it bronzed also seems ridiculous, but the success of gold–colored copper–base coins suggests that this might not be dismissed out of hand. Students have vivid imaginations, and will come up with a long list of possibilities.

Table I, from Gafner's paper,[5] illustrates the importance of metallurgical knowledge in the invention process. The table shows the solid solubility of several elements in solid gold at 800 °C and at 400 °C. This facilitates the development of a precipitation hardened gold alloy as a potential response to the problem. Alloy systems where the solid solubility is at least 1 wt.-% include Au–Co, Au–Ru, Au–Ti, Au–Tl, Au–Zr, Au–Tb, Au–Dy, Au–Ho, and Au–Er. The difference between 1% and the solubility at 400 °C is the percentage of the alloy consisting of gold that can be precipitated into a second phase (H_A). If the second phase is an intermetallic with a low percentage of gold, a large fraction of the alloy (NH_A) can ultimately consist of that second phase, which could produce the desired improvement in mechanical properties. As Table I shows, titanium was clearly the best initial choice for this, with zirconium and several rare earth elements as less encouraging possibilities.

Table I
Hardening Indices of Solutes in Solid Gold (after Gafner[5])

Alloy	Solubility at 800 °C (wt.-%)	Solubility at 400 °C (wt.-%)	H_w (wt.-%)	H_A (wt.-%)	Hardening Phase	NH_A
Au-Co	2.2	0	1	3.3	Co	3.3
Au-Rh	0.6	0.2	0.4	0.8	Rh	0.8
Au-Ru	1	0	1	2	Ru	2
Au-Ti	1.2	0.4	0.6	2.5	Au_4Ti	12.5
Au-Tl	1	0.5	0.5	0.5	Tl	0.5
Au-U	0.7	0.1	0.6	0.5	Au_3U	2
Au-Zr	2	0.3	0.7	1.5	Au_4Zr	7.5
Au-Tb	1.2	0.3	0.7	0.8	Au_6Tb	5.6
Au-Dy	1.9	0.3	0.7	0.8	Au_6Dy	5.6
Au-Ho	3.2	0.4	0.6	0.7	Au_6Ho	4.9
Au-Er	4.8	0.4	0.6	0.7	Au_6Er	3.5

Evaluation

This fourth step in Dieter's design sequence is actually quite involved, with subprocesses that deserve discussion of their own.

Refinement is the process of narrowing the list of options generated during conceptualization. It begins with a *feasibility* analysis, which consists of comparing proposed solutions with the list of restrictions created earlier. Any proposed response which violates one of the restrictions can be eliminated from the list. A second refinement step involves comparing the remaining alternatives with the list of criteria

The use of semiquantitative tools as part of the refinement process is important. Two of the most accessible are the decision matrix and the decision tree.[1,3] Both of these involve quantifying the relative importance of the different criteria, determining how well the remaining alternatives satisfy these criteria, and calculating a 'score' for each alternative. The use of an objective approach for comparing alternatives eliminates bias and allows each alternative a fair chance. Discussion of these methods allows students to determine how degrees of satisfaction for different criteria can be quantified, and what assessment methods would work best.

A second part of the evaluation process is *testing and development*. The purpose of this is to (a) obtain a more accurate assessment of how well a proposed alternative meets the different criteria, and (b) optimize the specifications for the proposed solution. This step of the process is time–consuming and expensive, so only the 'survivors' of the refinement process are considered for this. The outcomes of development are measured against the criteria listed earlier; the importance of standardized testing methods is stressed here. The most significant source of testing procedures is ASTM, but individual companies often develop their own methods, and tests can adapted from other industries. In the case of 990 alloy, wear testing methods used by the coinage industry were used.[7] The use of experimental design in the design process is also important, and can be introduced here. The key is to obtain the optimal result for a given design from the fewest number of tests.

In the case of gold–titanium 990,[5] this meant an experimental program with different degrees of cold working before heat treatment. It was discovered that 500 °C was the best temperature for hardening, and that age hardening at 600 °C before cold working and precipitation hardening produced an even stronger alloy.[8,9]

The next subprocess of evaluation is optimization. This is the process of producing a 'best case' for a particular design alternative, based on the results of testing and development. Given the cost and time required for development, the optimization process is limited, and the engineer needs to understand when further development will yield diminishing returns. In the case of gold–titanium 990, replacing a small amount of the titanium with ruthenium and boron increased grain size, which improves castability for the alloy.

The final subprocess of evaluation is decision making, comparing the few remaining alternatives. Although quantitative tools such as those used for refinement are important for this process, this is to some degree an art as much as a science. The importance of risk in decision making is stressed to students at this point. In Intergold's case, the decision to sell gold–titanium 990 is now well known, as Fig 3. shows.[10]

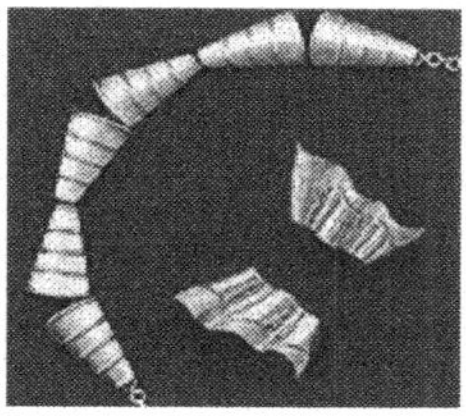

Figure 3. Chuk Kam jewelry produced from 990 Gold–Titanium (Gold Avenue, "Chuk Kam", http://info.goldavenue.com/info_site/in_glos/in_glos_chukkam .htm).

Communication of the Design

This is the last step in Dieter's version of the design sequence.[1] Communication can be internal (with an organization), or external (to the marketplace). At this point an introduction to intellectual property can be added to the discussion. In this case, Intergold's technical expertise was worth more than the specific invention, and publicizing their results helped to sell the material.[5,11]

Follow-Up

Design is an iterative process, and the production of a successful result often leads to new developments. In the case of 990 gold–titanium, this led in two directions.

The first was the development of new uses for the alloy. The availability of new materials often drives the design process, and the higher hardness and improved mechanical properties of 990 gold–titanium created new opportunities. In particular, this alloy has found use in wire bonding for electronic applications,[12] and is also sold for use in dental implants.[13] Because these applications have different criteria and restrictions than the jewelry business, the 990 millitesimal fineness restriction no longer applies, and a 1.7 wt.-% Ti alloy has become popular. Having students develop new sets of criteria and restrictions for the new potential applications, and reweight the criteria is a valuable exercise.

The second development is continuous improvement of the new alloy. 990 gold–titanium turned out to be less successful as a jewelry alloy than anticipated, due in part to the greater degree of technology needed to produce it.[14] In addition, the 990 fineness requirement in some countries for 'pure' gold jewelry is higher elsewhere. As a result, research began to find a hardenable gold alloy with even less alloy content. This resulted in the invention of an alloy with 0.3 wt.-% Sb and 0.2% Co or Ni.

References

1. G.E. Dieter, *Engineering Design: A Materials and Processing Approach (2nd ed.)*, McGraw–Hill, New York, 1983, pp. 6–11, 105–156.
2. E.V. Krick, *An Introduction to Engineering and Engineering Design*, John Wiley & Sons, New York, 1965, pp. 115–176.
3. A. Ertas and J.C. Jones, *The Engineering Design Process*, John Wiley & Sons, New York, 1993, pp. 2–11, 52–58.
4. M.E. Schlesinger, "Metallurgical Application of the Classical Engineering Design Sequence: Description and Examples," in *Challenges in Materials Education*, G.A. Irons and T.R. Meadowcroft, eds., The Metallurgical Society of CIM, Montreal, 1998, pp. 73–81.
5. G. Gafner, "The Development of 990 Gold–Titanium and Its Production, Use and Properties," *J. S. Afr. Inst. Min. Metall.*, 89 (1989), 173–181.
6. H. Schupak, "Hong Kong Jewelry: Blending Traditional and Modern," http://www.jckonline.com/article/283005-Hong_Kong_Jewelry_Blending_Traditional_And_Modern.php, accessed September 24, 2013.
7. H. Heidsiek and M. Clasing, "The Abrasive Wear of Gold Jewellery Alloys," *Gold Bull.*, 16 (1983), 76–81.
8. D. Heger, K. Jenkner, P. Klimanek, D. Bergner, and Ch.-J. Raub, "Zunderverhalten und Ausscheidungshärting von Hochkarätigen Gold–Titan–Legierungen," *Mat.-wiss., u. Werkstofftech.*, 26 (1995), 507–518.
9. D.M. Jacobson, M.R. Harrison, and S.P.S. Sangha, "Stable Strengthening of 990–Gold," *Gold Bull.*, 29 (1996), 95–100.
10. Gold Avenue, "Chuk Kam," http://info.goldavenue.com/info_site/in_glos/in_glos_chukkam.htm, accessed September 24, 2013.
11. F. Keller–Bauer, "990 Gold: An Unsung Alloy," *JQ Magazine*, 47 (1993), 94–96.
12. T. Nowicki and C. Carbonnaux, "Precious–Metal–Base Advanced Materials," *J. Phys. IV, Colloque C7*, 3 (1993), 509–518.
13. J. Fischer, "Mechanical, Thermal, and Chemical Analyses of the Binary System Au–Ti in the Development of a Dental Alloy," *J. Biomed. Mater. Res.*, 52 (2000), 678–686.
14. M. du Toit, E. van der Lingen, L. Glaner, and R. Süss, "The Development of a Novel Gold Alloy with 995 Fineness and Increased Hardness," *Gold Bull.*, 35 (2002), 46–52.
15. J. Fischer–Bühner, "Hardening of Low–Alloyed Gold," *Gold Bull.*, 38 (2005), 120–131.

Celebrating the Megascale: Proceedings of the Extraction and Processing Division Symposium on Pyrometallurgy in Honor of David G.C. Robertson
Edited by: Phillip J. Mackey, Eric J. Grimsey, Rodney T. Jones, and Geoffrey A. Brooks
TMS (The Minerals, Metals & Materials Society), 2014

The Challenges for Professional Metallurgical Education

Bob Hannah[1] and Peter C. Hayes[1]

[1]Metallurgical Engineering Program
School of Chemical Engineering,
The University of Queensland, Brisbane, Qld, Australia, Q4072

Keywords: metallurgical engineering, education, professional attributes, competency

Abstract

The changes that are happening in the world of metallurgy, and with it university education, are generational and require generational responses. Some of the key issues facing metallurgical education and ongoing professional training are explored.

What will be required of metallurgical engineering professionals in the coming decades? What knowledge and skills will be needed? How to provide the educational and learning pathways, and how to attract smart young people into this profession? These are issues more readily addressed by tertiary institutions in conjunction with industry; perhaps the greater challenges lie in building and maintaining competencies post-graduation. There is a need to establish and maintain a system that supports graduates and enables them to develop further into competent expert engineers able to successfully address complex engineering challenges.

Setting the Context

The organisers of this symposium have identified some of the key challenges facing the metallurgical engineering profession; a global industry working in a less certain social and economic environment, the need to genuinely embrace environmental impact and energy consumption, the increasing size and complexity of the projects and the need for skilled professionals. In terms of education and professional development post-graduation, these challenges arise at three levels of career development; undergraduate education, competent professional practitioner and the acquisition of specialist competencies needed to become expert in one or more areas of specialist practice. The challenges are sufficiently large and complex to warrant a holistic approach that more formally recognises the specialist level of professional practice. These concerns for the future are mirrored in the recent collection of papers commissioned by the International Mineral Processing Congress [1].

In this paper the responses by industry, the professions and Universities in Australia are explored. Underpinning the revival of undergraduate education for metallurgists, mining engineers and geoscientists in Australia is the strong mutual relationship between universities and industry [2]. This relationship has facilitated the development of a new metallurgical engineering program at The University of Queensland (a dual major in Chemical and Metallurgical Engineering) that meets the needs for a deeper and broader metallurgical education without compromising the existing strong industry focus. The issues of continuing professional development in its broadest sense are discussed with a focus on the development and recognition of competencies at the expert level.

Working in a World that has Greater Complexity and Uncertainty

One of the underpinning strengths of engineering as a profession is that we deal with complexity and change on a daily basis. Present engineering practice has various ways of dealing with this uncertainty; with risk management strategies, in their various manifestations, being central to understanding and minimising the risks. The existing engineering paradigm is one of determinism; i.e. every problem in engineering can be broken down into a series of sub-problems, each of which can be solved in a deterministic manner with an explicit solution. There is a growing volume of literature that suggests that this paradigm has limits when dealing with complex multidimensional problems where there is significant interaction between the sub-components; in particular when the interactions involve significant positive feedback [3]. The present paradigm remains valid for most engineering problems, those that are well defined. The literature does suggest, however, that we need to embrace the concept and implications of complexity, and develop engineering approaches and methodologies to better deal with large interactive problems and projects. Some progress has been made in this direction in the areas of plant design, process operations and risk management where the concepts and methodologies of resilience are being adopted.

Finding specific solutions that address the issues of complexity, size and uncertainty will need competencies at the expert level; based, of course, on a strong technical and professional foundation. An important driver for looking closely at the expert level is the increasing uncertainty and risk associated with professional practice. The changing environment with respect to professional liability means that professionals need to become more aware of the limits to their competency and have the capability to seek appropriate expert input from others. In particular, Duty of Care is becoming more explicitly defined in government statutes, with best practice being the minimum standard now expected. Professional engineering societies throughout the western world have made significant efforts to define the core competencies of engineers generally and have articulated this with generally good clarity at the graduate level [4]. There is not a similar clarity at the expert level.

The Warren Centre for Advanced Engineering at the University of Sydney has addressed some of these issues in its report titled *Professional Performance, Innovation and Risk* **[5].** The report suggests that there are generally accepted, well-defined ethical and competency standards; but we do not have a good definition of what constitutes professional performance. They summarise the key challenge as "*what should professional engineers expect of themselves and each other, and what should engineering professionals and clients expect of each other, in all key task-specific aspects of any engineering task.*" In effect, this report suggests that we need to move beyond a focus on the inputs, ethics and competence, and place more focus on the output, performance.

What Attributes and Competencies will Metallurgical Engineers Need?

Professional engineering societies have recognised that a narrow technical focus is no longer valid and, in their various ways have defined attributes and competencies that require both knowledge of the wider world and the skills to work interactively within this world [4]. All engineers, on graduation are now required to have minimum levels of team and communication skills to meet professional registration. The continued development of these skills, and in particular those related to team and project management, forms a significant component of tertiary education and also professional development programs within most of our major mineral companies.

All professional societies require that their members must meet minimum standards of professional behaviour in addition to technical competency. These behavioural standards are articulated in the various codes of ethics [6]. Within the minerals industry specifically, the International Congress on Mining and Metals (ICMM) has developed a set of ten principles related to sustainable development [7]. These principles are intended to provide guidance on responsible business practice and a broader vision of corporate responsibilities. Although not actually stated, there is an underpinning assumption that all planning and action will be based on well-founded knowledge and valid decision making, undertaken within an environment of complexity and change. The ten principles themselves relate to social, ethical and governance dimensions rather than just technical. In many ways, these ten principles provide a good overview of the range of capabilities required by practicing metallurgical engineers.

When expressed in terms of the three elements proposed in the Warren Report, ethics, competency and performance, it may be possible to build a better and more complete description of what is expected at different stages of an engineer's development; as a graduate, full professional and expert. This would be a significant contribution to helping engineers understand the full complexity of the tasks that they undertake; and, in particular, help them to minimise their professional risk.

Competencies Expected from Metallurgical Engineers on Graduation

Putting aside the need to embrace and incorporate complexity and resilience, most professional societies have undertaken a significant amount of excellent work on graduate competencies and now define the competencies very well, e.g. Engineers Australia. Perhaps the most significant step forward over the last 20 years has been to recognise that competency has several dimensions, and that all of the dimensions must be present for a person to be competent. Although there is some disagreement in the literature, most agree on a set of common dimensions that include Knowledge, Skills, Problem Solving Capability, Temperament and Experience.

These appear to represent a very sound articulation of the competencies (for the present) and therefore a very sound basis for the curriculum design of metallurgy programs. They will not be adequate in the future unless the concepts of complexity and resilience are embraced, and pathways found that integrate these concepts into the engineering paradigm that we use when solving problems and working with best practice.

In terms of the ability of students to embrace change, most of the students entering our universities already have all of the necessary personal and communication attributes needed to meet the challenges; strong intellects, flexibility of mind, natural ability to network, well developed skills to seek information where and when needed. The core requirements from Universities are the curriculum structures and learning experiences that encourage these students, and set the framework for their learning and development as engineers. Industry has an important role to play in helping to define the graduate outcomes and supporting the universities as they implement the learning programs.

Keeping Metallurgical Engineering Programs not just Alive but Growing and Thriving

When discussing the education and training for metallurgical engineers, we need to be particularly aware of the context in the English speaking world over the past 30 years, specifically

1. Declining enrolments
2. Schools closing
3. Changing competencies expected from graduate engineers

In short the profession is at risk of disappearing even from countries whose economies rely heavily on the minerals industry. The watershed in Australia came with the publication of a discussion paper prepared for industry on the future of Mining and Metallurgical education in entitled “Back from the Brink - reshaping minerals tertiary education”, published in 1998 [8]. This resulted in the establishment of the Minerals Tertiary Education Council (MTEC), funded by the Minerals Council of Australia (MCA). There were many important observations about the structural barriers to be overcome and the resources required to achieve the desired outcomes; essentially a world class tertiary education in the key professions mining, metallurgy and geosciences, and sufficient graduates to meet industry requirements for professions in these fields.

In spite of these initiatives and ongoing concerns it was not until 2004 that significant change took place in the state of Queensland, a state that relies heavily on the minerals industry. The state came within weeks of losing the only tertiary education capabilities in metallurgical engineering. The University of Queensland (UQ) and major companies in the minerals industry established a joint taskforce to address this crisis. The agreed outcome was the establishment of a new 4-year BE program a dual major in Chemical and Metallurgical Engineering (Chem Met program). The objectives of this initiative were to establish a sustainable program that provided graduates with the appropriate technical breadth and depth for employment in the minerals industry and significantly increased student numbers. Importantly, the even weighting in the program of conventional chemical engineering courses and specialist metallurgy courses means that that the program is truly a dual major between Chemical Engineering and Metallurgical Engineering [9].

An important feature of the ChemMet program is the flexible entry up until the start of year three of the program. Engineering students at UQ commence their selected specialist program in second year, following a common first year that includes courses on the underpinning mathematics, sciences and engineering sciences. For the ChemMet program, the second year follows the Chemical Engineering program with a focus on generic process engineering fundamentals; there is only one course taught fully by the metallurgy staff, a general introduction to physical and chemical processing of minerals. This introductory course is also taken as a compulsory course by students in a parallel BE Mining program. The delay in specifying the metallurgy major has been important factor in keeping opportunities open and allowing for articulation of students from related programs. Most students leaving high school are unaware of the metallurgical engineering discipline and the career opportunities it provides. Visits by 2nd year students to working process plants and integration of learning with these visits demonstrates to many students who had grown up in the city that there are exciting, challenging and promising career paths in the minerals industry. This structure not only allows students from other engineering programs within UQ, other tertiary institutions in Queensland or interstate to transfer without loss of credit for prior study into the program, but also provides opportunities to international students to join the program.

The dual major provides the core engineering science and process engineering knowledge and skills associated with conventional chemical engineering programs. The systems approach and with it the ability to analyse and model process flow sheets has added a new dimension to the program – one that is lacking from most conventional materials/metallurgy and mining/metallurgy programs. In addition the students are required to undertake a comprehensive suite of specialist courses in mineral processing, hydrometallurgy, pyrometallurgy and metallurgical plant design. Metallurgical case studies and examples are also included in chemical engineering courses and a number of key process engineering courses are delivered using a team teaching model in which academic staff, from a number of disciplines, contribute to supply contextual teaching to student cohorts in specialist fields. The dual major is the only metallurgy program accredited by I. Chem. E. London; I. Chem. E. assessors in 2012 commented that they were were particularly impressed with the design project for this program. The program is also accredited by Engineers Australia and AusIMM. The students are attracted to the wider career options on graduation and in their future careers but also view favourably the enhanced job opportunities on offer by specialisation in an identifiable industry sector.

All of the key objectives and targets of the UQ/industry taskforce have been achieved. During the first five years of the agreement, staff numbers in the program have increased from one to six continuing, full-time metallurgy academic staff - three of which are supported by industry. The students entering the program come from the top 10% of high school leavers. The program now regularly produces 20 to 30 four-year metallurgical engineering graduates annually, and has over 30 Research Higher Degree students enrolled.

Industry confidence in this program has been demonstrated by the initial financial commitment to UQ's strategic collaborations by Xstrata Copper and BMA and the extension of this commitment for a further five years; plus the development of a new partnership with RioTinto. In addition to the company specific support, a strategic collaboration has developed with the Minerals Council of Australia (through the Minerals Tertiary Education Council, MTEC) and has led to the establishment of the Metallurgy Education Partnership (MEP), a partnership between three Australian universities to teach a joint design capstone course in fourth year (see the paper by D. Ibana also presented at this symposium).

The reinvigoration of the UQ Metallurgical Engineering program has also provided the platform for development of research strengths in the core specialist areas of mineral processing (with the Julius Kruttschnitt Mineral Research Centre, JKMRC), hydrometallurgy (UQ RTA Bauxite and Alumina Technical Centre) and pyrometallurgy (PYROSEARCH). Researchers and postgraduate students from these Centres contribute to the high quality learning environment through teaching and tutoring in the undergraduate program.

It is not the intention to say this is the only way that all programs in metallurgical engineering programs should be structured. However, this approach has proven to be successful at The University of Queensland and incorporates many of the key features identified in the Back to the Brink report. Two critical factors in this success story were i) A strong industry–university partnership - industry got serious and became very involved with advice, support and money, and ii) the university got serious about retaining and rebuilding the program.

The Back to the Brink report concluded that "The opportunity exists for a true partnership between industry, government and academia." In Australia the missing element is government, which is struggling to address the broader issues of tertiary education and has been unable or unwilling to address the strategic needs for specialist programs. We suspect that this will continue to be the case and that the sustainability of metallurgical engineering programs will continue to rely on industry/University collaborations.

The experience at UQ suggests suggest that there are three vital elements to sustaining and growing metallurgical engineering education

- Industry leaders need to show the will to continue this specialist area of engineering – and provide the relevant universities with strong support for change as well as money.
- University metallurgy programs need to embrace the necessary change in structures. In an era of mass education small, single focused metallurgy programs are simply not sustainable on their own – they are not sufficiently attractive to students and are not financially self-sufficient. The ChemMet program at UQ is an exemplar that addresses both the issues of technical breadth and depth, and the development of generic professional skills without losing its industry focus. Also, against national trends [2] there are increased enrolments.
- There needs to be national leadership and co-ordination – the MTEC model is an exemplar for this.

Building and Recognising Expert Level Competencies

Shortages in personnel with metallurgical technical skills have been addressed in various ways **[1]**, some more successful than others. These encompass Coursework Masters general (certification); Coursework Masters focussed on particular industry sectors (certification); Conversion courses for engineers without metallurgy background (certification); Short course formats (attendance); Conference (certification). These approaches address to some extent the short term needs of industry and plant operations; but they do not necessarily go to the heart of developing expert skills and competency, especially in dimensions other than technical competency.

These approaches also rely on the long term sustainability of tertiary institutions to provide core knowledge, and high level skills and expertise. Research higher degrees provide high level technical competence and skills and provide career pathways, especially into process development and academe. A challenge for the profession is to provide, and encourage, career pathways that extend these skills and competencies to cover the non-technical competencies needed by metallurgical engineers at the expert level.

This need for well-structured development pathways extends to all areas of the profession. The very significant decline in experienced process engineers at operating sites is a serious constraint on the extent to which young professionals can develop under the guidance of experienced engineers, even though some of the larger corporations have well-structured generic graduate programs. The situation in consultancies may be more hopeful, where some very strong graduate programs exist.

However, the development of competencies from the professional level to expert still seems to be very unstructured and dependent on individuals moving between roles and employers to gain skills and experience based on what work is presently available. Whilst this has, to a significant extent, always been part of metallurgical engineering, the weakness is the absence any guidance as to what constitutes competence as performance, as identified in the Warren Centre Report.

Consultancies also provide the process, engineering and location diversity that has, to a significant extent, been lost from many operating companies that now focus, by necessity, on the core operations and financial resilience. These consultancies and targeted research centres may, in the future, provide the diversity of capability and approach to be the core providers of not only new technologies but of new ways to undertake metallurgical engineering generally.

Overall then, it would appear that the profession has a structure that can provide the new technical knowledge and ways to engineer this knowledge into process operations, even if many might argue that there is a very serious shortage of resources in terms of people, facilities and corporate will.

The significant gap appears to be in the structures and systems needed to identify and recognise the full dimensions of professional competence and performance.

Organisational Models for Building and Recognising Competencies

The Warren Centre Report identified the medical profession as a model worthy of study. This is also the experience of one of the authors (RH) who has children who have trained as medical specialists. In Australia, fully professional, but not expert, doctors train under the guidance and supervision of a certified expert (the specialist). It is expected that the specialist-in-training will refer his diagnoses and proposals for treatment to the specialist at a progressively reducing extent as his training progresses. It is the accountability of the specialist to set these boundaries and monitor the progress of the specialist in training.

This is backed by a system of certification (recognition) in which the trainee demonstrates competence in four dimensions i) Overall competence (knowledge, skills, experience) to make accurate diagnoses within the full range of medicine covered by that speciality, ii) A sound, but not fully expert, knowledge of all aspects of medicine covered by that speciality; and in particular, it is expected that the trainee be aware of the latest developments in all areas of the speciality and where to seek an expert level of knowledge when required, iii) An expert level of competence (knowledge, skills, experience) in one defined area of the speciality - e.g. the sub-speciality of infectious diseases within the speciality of medicine/physician, and iv) The ability to make advanced diagnoses and treatment plans in the expert area and take accountability for the decisions.

All metallurgical engineers work under various levels of supervision whilst their competencies are being developed. However, the system for medical specialists differs from that presently existing in the metallurgical profession in the following ways:

1. There is recognition that there is a third level of competence, the expert that differs from that of the graduate and professional engineer.
2. The practitioner moving from professional to expert level has a defined plan that addresses specific areas of competence.
3. There is no one single pathway, with many areas of sub-speciality being recognised.
4. The practitioner has one, or more, recognised specialists to act as mentors who monitor the development of competence and take final accountability, within limits, for advanced diagnoses during the training period.
5. There is a well developed system for identifying the competencies required and assessing whether the trainee has reached to necessary level of competence to act alone without the support of a certified specialist, e.g. entry into a college.

Although the metallurgical engineering profession differs significantly from that of medicine in terms of its organisation, in particular the hospital system for training doctors, moving to a system that acknowledges an expert level of competence and has a structure for developing and recognising these competencies would both enhance the reputation of the profession and significantly reduce the present exposure to professional risk.

This may seem a significant change from the present, but similar major changes have happened in the past; viz. the formation of Schools of Mines and professional societies to set base standards for technical competence, ethics etc. What may be required now is

- Recognition by industry, consultancies and the professional societies that something needs to be done and the will to do it.
- Leadership from the profession, with the consultancies maybe taking a leading role, to move from the present situation that identifies and certifies competencies at the graduate and professional level to include that of the expert level.

Summary

There is now plenty of evidence to demonstrate that metallurgical engineering professionals will be required into the future.

The key to the future of metallurgical education and professional development will bea sustained and active partnership between industry and tertiary institutions, to define what is required of future graduates and professionals, to build resilient structures that can deliver on these and to provide flexible pathways and attractive career paths for aspiring graduates. Now is perhaps the time for the profession to genuinely recognise a third level of competence – the expert.

Engineering can continue to provide value to companies and society as a whole as it has for centuries, with new skills and competencies they can reduce risk and maximise productivity as professional engineers.

References

1. *Special Symposium on Human Resource Development, held during XXVI International Mineral Processing Congress (IMPC2012), Sept., 2012, New Delhi, India*, The Indian Institute of Mineral Engineers.
2. G. Lind, Minerals industry engagement in metallurgical education in Australia, *Minerals Industry Education and Training: Special Symposium on Human Resource Development, XXVI International Mineral Processing Congress (IMPC 2012), September 24-28, 2012, New Delhi, India*, The Indian Institute of Mineral Engineers, pp. 87-104.
3. R.A. Meyers (Ed), Encyclopaedia of Complexity and Systems Science, 2009, Springer.
4. Engineers Australia 2011, Stage 1 competency standards for professional engineer, viewed on 30 September 2013 <http://www.engineers australia.org.au >.
5. Professional performance, Innovation and Risk in Australian Engineering practice, The Warren Centre for Advanced Engineering, Sydney, 2009,
6. Engineers Australia 2011, Code of Ethics, viewed on 30 September 2013 <http://www.engineers australia.org.au >
7. Sustainable Development Framework, 10 Principles, International Council for Mining and Metals, http://www.icmm.com/our-work/sustainable-development-framework/10-principles, 2012.
8. "Back from the Brink - Reshaping Minerals Tertiary Education" Discussion Paper, Minerals Council of Australia, Braddon, ACT, Australia, 1998.
9. Faculty of Engineering, Architecture and Information Technology (EAIT), The University of Queensland, Brisbane, Australia, http://www.eait.uq.edu.au/

Celebrating the Megascale: Proceedings of the Extraction and Processing Division Symposium on Pyrometallurgy in Honor of David G.C. Robertson
Edited by: Phillip J. Mackey, Eric J. Grimsey, Rodney T. Jones, and Geoffrey A. Brooks
TMS (The Minerals, Metals & Materials Society), 2014

SUSTAINABILITY EDUCATION FOR MINERALS AND MATERIALS INDUSTRY PROFESSIONALS

W. John Rankin[1]

[1]CSIRO Process Science and Engineering
Bayview Avenue, Clayton, VIC 3169, Australia
(Box 312, Clayton South VIC 3169)

Keywords: Sustainability, Sustainable Development, Minerals and materials industry, sustainability education

Abstract

The finite and non-renewable nature of mineral resources poses unique challenges for sustainability: How can mineral and material needs of society be satisfied indefinitely? How can the benefits of exploitation of minerals best be captured for present and future generations? How can the environmental problems caused by large scale mining and processing best be ameliorated? These issues are rarely addressed, or if so only incidentally, in traditional courses in geology, mining, metallurgy and materials yet addressing them will increasingly be a major role for professionals in the minerals and materials industry and related sectors and agencies. This paper describes a framework for examining these and other sustainability issues related to meeting future material needs from non-renewable resources. The framework may be useful for educational purposes, either as the basis of a dedicated subject or by including the elements of it throughout a traditional course in relevant subjects.

Introduction

The supply of the goods and services used by humans depends on access to materials obtained by exploiting non-renewable resources of the Earth's crust, particularly metallic and non-metallic minerals, rocks, coal, oil and gas. Mineral-derived-materials and fossil fuels form one major group of materials, the other being those derived from living matter. The latter, while not being derived from the crust, rely on the crust as the ultimate source of the nutrients they require for growth.

Perhaps the greatest challenge facing the world is to ensure that all people can have a good standard of living and quality of life without continuing to degrade the environment. The Earth's resources are finite and its land and water ecosystems have finite capacities to cope with the wastes produced by human production and consumption. In the United States, more than twenty-one billion tonnes of resources of all kinds are consumed each year. This amounts to about 80 tonnes per person per year and consists of 76 tonnes of non-renewable resources and 4 tonnes of biomass [1]. Of this only 19 tonnes are used as direct inputs to processing; the rest is waste. Further quantities of wastes are produced during processing of the direct inputs and during the use and ultimate disposal of the products made from them. Other developed countries have similar, though lower, patterns of consumption.

How to deal with the problems caused by this unprecedented level of production, and the associated consumption of resources and disposal of the wastes produced, is a global challenge. This challenge will only grow as living standards in China, India, Brazil and many other developing countries continue to rise.

The concept of Sustainable Development (SD), or more commonly sustainability, arose as an attempt to addresses these and other issues. The definition of SD is problematic as it is open to wide interpretation, and misinterpretation. Though vague, and providing little guidance on implementation, the Brundtland definition [2] has proved most enduring:

Sustainable development is development that meets the needs of the present without compromising the ability of future generations to meet their own needs.

A large body of theory and practice around sustainability has developed over the subsequent decades.

Sustainability principles have had growing influence on the development of environmental and social policy in most countries in recent decades, and SD has been adopted and promoted by major international organisations such as the United Nations, the International Monetary Fund (or IMF) and the World Bank.

Non-renewable resources pose a unique challenge for sustainable development, in particular: How global material needs can be satisfied sustainably when so many sources of materials are non-renewable and when their production and consumption results in huge quantities of wastes, many of which are environmentally harmful. How resource-rich countries and local communities can benefit from the exploitation of non-renewable resources and prosper beyond the time when they are exhausted.

The thesis of this paper is that minerals and materials professionals need to have a comprehensive understanding of the concept of sustainability and of the issues it raises for the production and use of materials and the implications it has for their jobs. This is supported by the findings of a recent survey undertaken by the TMS of members [3] which indicated that 88% of industry respondents thought that the importance of sustainability would increase in future. They also indicated that in all cases the actual level of proficiency in a range of sustainability areas in their companies was below the desired level. The purpose of this paper is to provide a "framework" for teaching sustainability and its implications for the minerals and materials industries.

The framework

It is assumed that technically-qualified professionals for the minerals and materials industries will continue to major in traditional disciplines such as geology/earth science, mining/civil/resource engineering, metallurgy, chemical engineering, mechanical engineering and materials science/engineering, or an environmental discipline such as environmental science/engineering. Most graduates from these programs will find employment somewhere within the materials cycle, either in industry or in government or non-government agencies. The materials cycle can be thought of as consisting of five interrelated components - primary production, secondary production, product design and manufacture, consumption and end-of-life management - as represented in Figure 1. Most minerals and materials technical vocations will fall fairly clearly within one of these areas (with the exception of consumption). Increasingly graduates working in these areas will need to understand, not only the scientific and technical aspects of their discipline, but how their disciplines, and the work they will do in industry, fits into the materials value chain and the contribution they, and their industry sector, can make to the transition to sustainability.

Further, the professionals of the future will need to understand the broader social, political and economic context within which the minerals and materials industries exist. Figure 2 illustrates, albeit simplistically, some of the complex interrelationship between socio-political and technological trends, basic human material needs and the potential problems caused. An exploration of these should be part of a professional's undergraduate education.

The proposed pedagogic framework is shown schematically in Figure 3. It attempts to address the issues raised in the Introduction based on the above assumptions. The individual topics, and some of the important issues in each, are listed in Table 1. Some familiarity with all these topics should be a prerequisite for the modern minerals and materials professional.

Just as important as the topics, however, is their interrelationships and the integration of the whole to provide a "sustainability narrative" within which each professional person can identify their role and potential contribution (or otherwise!).

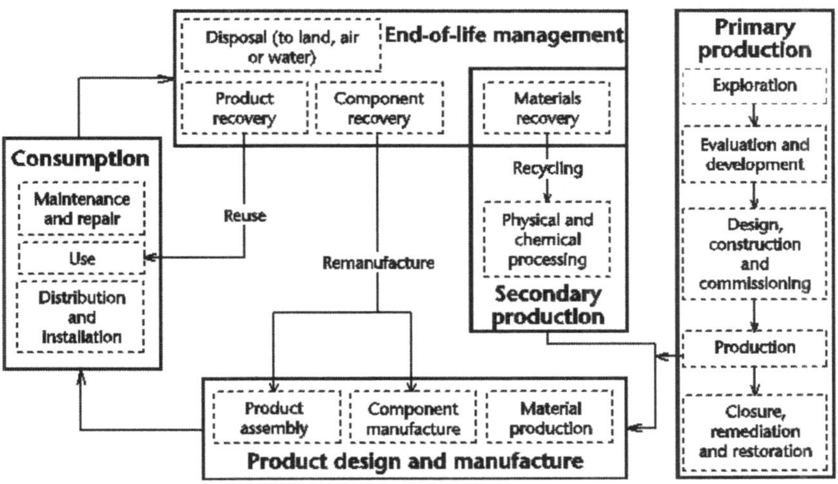

Figure 1. The flow of mineral and metallic commodities through the economy (after [4], with modifications).

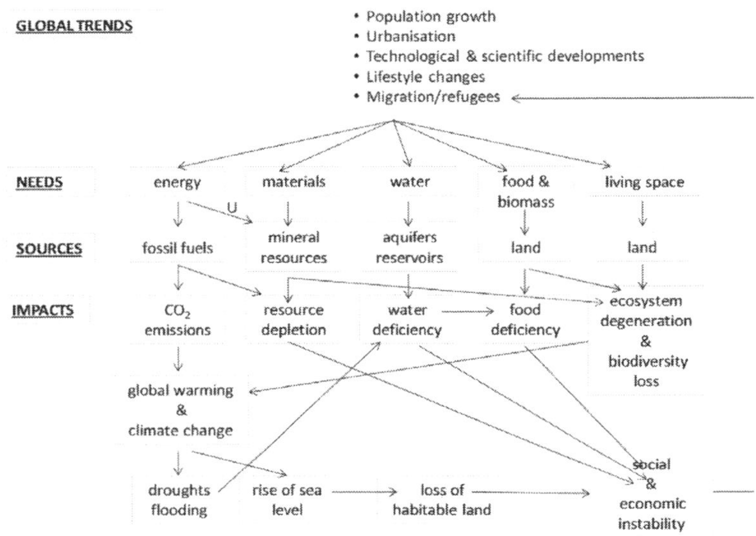

Figure 2. Schematic representation of the causes of global change and the possible resulting effects (after [5], with modifications and additions).

Implementation

The question of whether sustainability should be taught as a stand-alone subject within a degree program or be integrated throughout is moot. It is the author's opinion that ideally, the concept of sustainability should be introduced early in the program and the body of theory discussed (Topic 1, Table 1). The particular issues raised for the discipline area could also be identified at this stage (Topic 2, Table 1). Opportunities would then be found in individual subjects throughout the degree program to provide technical understanding of the issues and possible solutions. Thus, for example a mineral processing subject would include aspects of tailings handling, management and storage in addition to the traditional focus on recovery of the mineral(s) of value.

A follow-up subject, later in the course, would introduce the topics not considered to be part of the particular discipline and provide the integration and overview. Social and political aspects of sustainability could also be considered independently as a stand-alone subject or as part of a more general humanities subject.

Clearly, the emphasis on the various topics in the framework will vary from course-to-course depending on where the emphasis of the course is: ores in the ground (mining), treatment of ores (extractive metallurgy), materials and their properties/uses (materials science/engineering); etc.

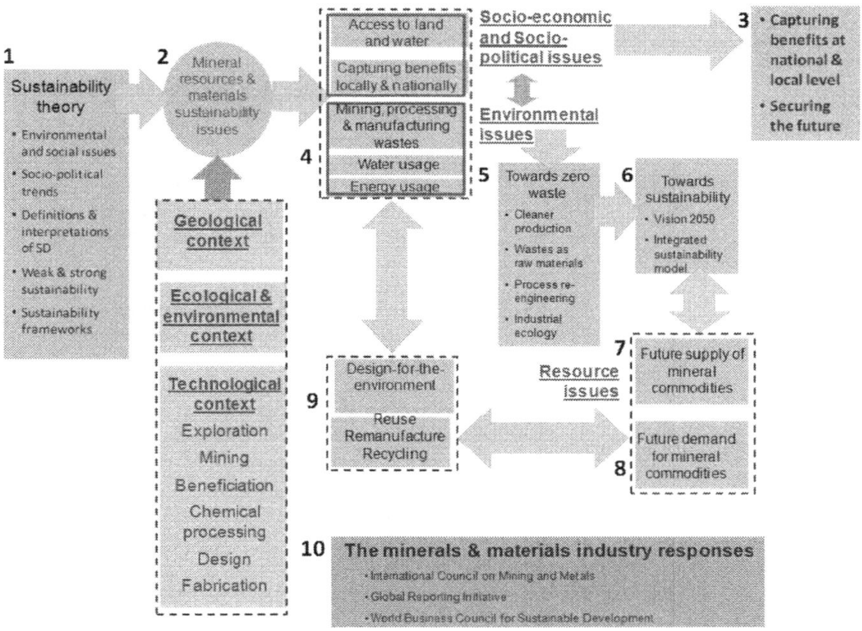

Figure 3. The proposed "framework' for introducing students to sustainability and its implications for minerals and materials professionals.

Table 1. The topics covered by the framework and the major issues in each relevant to the minerals and materials industries.

	Topic	Issues
1.	Sustainability theory and context	• The environmental context • The social and historical context • definitions of sustainability • Sustainability frameworks • A conceptual model of sustainability
2.	Mineral resources and materials sustainability issues	• Can mineral and metal needs of society be satisfied indefinitely into the future? • Can the benefits of exploitation be captured for present and future generations? • Can the environmental problems caused by large scale mining and processing be ameliorated?
3.	Socio-economic and socio-political issues	• Can resource-rich countries and local communities benefit from the exploitation of non-renewable resources and prosper beyond the time when they are exhausted?
4.	Environmental issues	• The materials cycle • Types and impacts of wastes • Management of solid, liquid wastes & gaseous wastes • Energy & water usage in the minerals & metals industries.
5.	Towards zero waste	• The waste hierarchy • Cleaner Production • Wastes as raw materials • Waste reduction through re-engineering • Industrial ecology (and examples) • Drivers and barriers to zero waste
6.	Towards sustainability	• Closing the materials cycle – the concept of stewardship • Stewardship models
7.	Resource issues - the future supply of minerals and materials	• Determinants of supply • Future potential sources: crustal resources, seawater, seabed; wastes
8.	Resource issues - the future demand for minerals and materials	• Determinants of demand • Substitution • Dematerialisation
9.	Secondary materials	• Recycling, reuse & remanufacturing • Potentials and limitations of recycling • Design for the environment • Recycling systems and technologies
10.	The minerals and materials industry responses to the sustainability challenge	• The International Council for Minerals & Metals • The Global reporting initiative • The World Business Council for Sustainable Development *Vision 2050*

This is quite appropriate - the common ground, however, is the overall narrative illustrated in Figure 3.

The aim of the framework is to put a theoretical basis to sustainability and provide an overview of the main issues so that the framework can provide a scaffold upon which further understanding can be built. Learning is a continuous process and graduates would continue to build on the framework while working and, importantly, learn the practical aspects of sustainability on the job and through on-going professional development activities (seminars, short courses, training programs, etc).

Concluding comments

The author has had an interest in the sustainability field, in particular how it relates to the primary production of mineral commodities and metals (finite resources), for nearly fifteen years. The views expressed here are a distillation of what has been learned over that period. Most recently, the author wrote a book *Minerals, Metals and Sustainability: Meeting Future Material Needs* [6] as an attempt to produce a concise, readable account which addresses sustainability issues at a level appropriate for professionals being prepared for the minerals and materials industries. Readers interested in the topics of the Framework will find these elaborated in some detail in the book in a relatively objective and non-polemical way. Extensive references to relevant literature provide an easy guide to further reading about a topic.

It is to be hoped that sustainability will become incorporated into all minerals and materials courses and ultimately become integrated into the thinking and actions of professionals in these industries. Perhaps an analogy is the way in which OHS has become integrated into the thinking, planning and activities of professionals in minerals and metals companies over the past couple of decades with resulting dramatic improvements in OHS performance. OHS performance improvement was very much a top-down driven approach. Sustainability performance improvement on the other hand will most likely result from a combination of bottom-up driven (from new graduates with different values) and top-down driven (from management reflecting changing government policies and public attitudes).

References

1. A. Adriaanse et al. "Resource Flows: The Material Basis of Industrial Economies," (Washington DC: World Resources Institute, 1997).

2. United Nations, Our Common Future Oxford (UK: Oxford University Press, 1987).

3. J. W. Fergus et al., "Sustainability in materials education," JOM, 65 (8) (2013), 935-938.

4. R. van Berkel, "Eco-efficiency in primary metal production: context, perspectives and methods," *Resources, Conservation and Recycling*, 51 (2007), 511–540.

5. P. A. Wilderer et al., "Sustainable Earth System Engineering: Incentives and perspectives," Handbook of Sustainable Engineering, ed. J. Kauffman and K-M. Lee (Springer Science+Business Media Dordrecht, 2013), 195-209.

6. W. J. Rankin, "Minerals Metals and Sustainability: Meeting Future Material Needs," (Leiden, The Netherlands: CRC Press/Balkema, 2011).

Celebrating the Megascale: Proceedings of the Extraction and Processing Division
Symposium on Pyrometallurgy in Honor of David G.C. Robertson
Edited by: Phillip J. Mackey, Eric J. Grimsey, Rodney T. Jones, and Geoffrey A. Brooks
TMS (The Minerals, Metals & Materials Society), 2014

DELIVERING A NATIONAL PROCESS DESIGN UNIT WITH INDUSTRY SUPPORT

Don Ibana[1]*

[1]Associate Professor, Department of Metallurgical and Minerals Engineering
WA School of Mines
Curtin University, Locked Bag 30, Kalgoorlie, WA, 6433, Australia
D.Ibana@curtin.edu.au

Keywords: Metallurgical Education, Metallurgical Process Design Unit, Teaching Metallurgical Process Design

Abstract

Supported by the Minerals Council of Australia (MCA) through the Minerals Tertiary Education Council (MTEC), three Australian universities–Curtin University, Murdoch University and the University of Queensland–have formed the Metallurgical Education Partnership (MEP) to jointly develop and deliver an engineering design capstone unit–Metallurgical Process and Plant Design–in their respective undergraduate programs in extractive metallurgy, in order to enhance the students' educational experience. A unique feature of the program is the close interaction of the students in all three universities and a significant involvement of industry professionals. Now in its sixth year, it is clear that this unit is achieving its objectives.

Introduction

Quality teaching of a capstone process and plant design unit in extractive metallurgical engineering can present significant challenges for academic staff given that few could expect to have the expertise and currency of experience to deliver such a unit at the standard of "best practice". This challenge however, can be efficiently and productively met through a partnership arrangement, in which the minerals industry financially supports the engagement of industry experts, to assist academics to deliver a commonly taught unit to final year students across multiple universities. Such is the partnership now in place between the Minerals Council of Australia (MCA) and the three universities offering extractive metallurgy programs within Australia, namely, Curtin University's Western Australian School of Mines (WASM), the University of Queensland (UQ) and Murdoch University (Murdoch). A common national process design unit has been offered by this partnership since 2008 through industry investment of approximately AU$500,000 per year (now totaling AU$3 million), and through a significant commitment and engagement by industry personnel and academic staff to ensure success, alongside the already established national MCA-university partnerships in mining engineering (including UQ, WASM and others) which deliver mining engineering as well as in minerals geoscience education [1]. The financial support includes the full cost of development and delivery of the design unit plus $100,000 funding per year to each of the partner universities that

*The author has been involved in this program since it its inception. He was one of Curtin University's representatives in the meeting called by MTEC that led to the creation of the MEP. He represented Curtin in the Implementation Committee in 2008 to 2009 and subsequently in the Steering Committee from 2010 to the present. Much of the information in this paper is based on his notes and minutes of meetings.

can be used on any endeavor that would enhance the quality of teaching of their respective metallurgy courses.

The current metallurgy partnership was initiated in 2007 when MCA, through their Minerals Tertiary Education Council subgroup (MTEC), convened a meeting with university representatives to explore a collaborative initiative with a view to increasing the number and quality of extractive metallurgy graduates and to ensure the sustainability of high quality teaching programs in this field, given the well-documented challenges such programs face within the university sector [2]. Further, in the face of rising graduate demand from the "mining boom" of the time, there was recognition from industry of the benefit of such investment, to offset future recruitment costs, and to enhance the retention of metallurgy staff within operating companies.

For some years prior to this crucial meeting, competitive issues between the two Western Australian universities (WASM and Murdoch), both competing within a relatively small local market to attract students, made it difficult to reach a satisfactory agreement to offer any common curriculum with shared resources, notwithstanding the financial incentives offered by industry to do so. Fortunately, the sharing of a capstone unit offered in the final year of their respective degree programs eliminated competitive concerns, and with the benefits of industry assistance obvious to all, the concept was enthusiastically endorsed by all three universities and the Metallurgical Education Partnership (MEP) was formed[2].

Within the partnership, the universities agreed to pool their resources to jointly develop, deliver and manage the national process design unit, with the continuing support of industry predicated on (a) a strong element of industry input and interaction, (b) a substantial component of student group interaction, (c) a substantial component of inter-university interaction, (d) endorsement by each university of the objectives, requirements, content and standards for the unit, and (e) issues of mineralogy as well as mineral and chemical processing to be addressed in the design.

In addition, it was necessary for the universities to align the delivery of their individual degrees such that all offered the same capstone design unit with the same credits within the same semester. Ultimately, it was agreed to place the unit in the final semester of all three 4-year degree programs, credited at 50 percent of a fulltime student's workload over the semester. The former allowed the completion of all necessary prerequisites while the latter accommodated sufficient time for intensive workshops for the engagement and input of industry professionals.

Once the initial agreement and course alignments were in place, the subsequent successful implementation and development of the program came about because academics from the three universities along with industry representatives were committed to work together to overcome the challenges of offering a national program across three widely spaced campuses with somewhat different academic environments. The purpose of this paper is not only to provide details of the program but also to recognise the efforts of those individuals who were largely responsible for its development. Their names are listed in Appendix 1.

[2] Notes on Meeting, MTEC Metallurgy Business Plan, Chamber of Minerals and Energy, 26-27 July 2007.

Management of the Program

Both Steering and Implementation Committees were formed to manage the program, each with members from MTEC, each partner university and the mining industry. University metallurgy department heads serve on the Steering Committee (SC), which is chaired by the Executive Director of MTEC and has a monitoring and decision making role, to ensure compliance of the program with the funding agreement. Academic unit coordinators from each university serve on the Implementation Committee (IC) with support from an MTEC project officer. The IC provides academic leadership and takes responsibility for the day to day teaching of the program; it is chaired by the representative of the "lead university" (currently WASM). The interrelationships between MTEC, partner universities, industry professionals and students in the program are shown schematically in Figure 1.

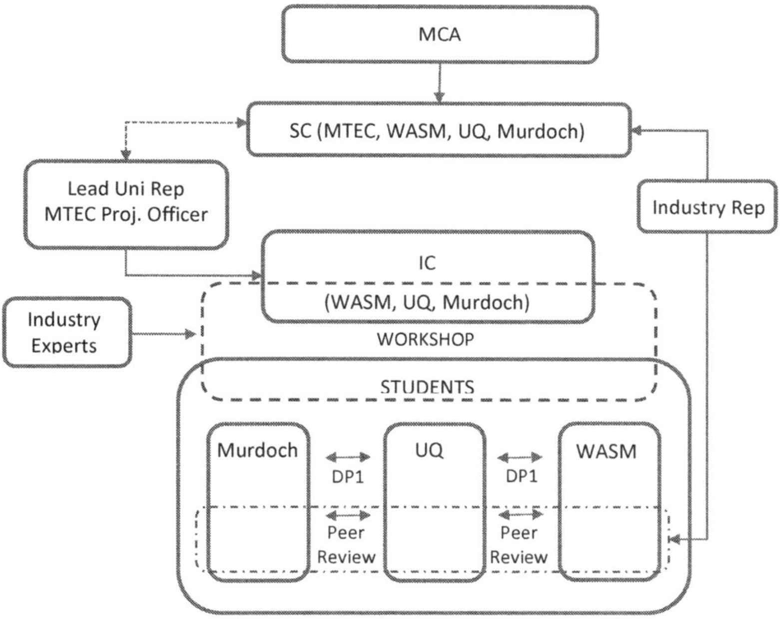

Figure 1. Interrelationship between MTEC, the partner universities and minerals industry professionals in the delivery of process design project (PDP).

The representative of the lead program and the MTEC Project Officer are non-voting members of the Steering Committee who provide an information link between the two committees and between the SC and the students in the program. The industry representative in the SC has direct interaction with students as facilitator of the students' peer review process, one of the key assessment activities in the program. Industry experts interact with students and teaching team as specialist lecturers, workshop facilitators and informal mentors. Student group interaction across universities is facilitated through the workshops, inter-university group work such as the completion of the first assessment item Design Package 1 (DP1), peer review and national project presentation via video conferencing.

Program Tasks

In summary, the program tasks are designed to facilitate the professional development of students as they prepare to enter the workforce, by engaging them with industry professionals and their peers, in a realistic development of a metallurgical process and plant design, which requires student teams to complete and deliver on the following requirements:

- Formulation and application of a project management plan
- Selection of the most appropriate processing option given a set of mineralogical and metallurgical data
- Development and optimisation of the selected process using sound metallurgical principles and design tools
- Development of an appropriate process flow diagram (PFD)
- Selection of appropriate process equipment
- Development of an appropriate plant layout
- Formulation of an appropriate process control scheme including control loops
- Assessment of the economic feasibility of the project
- Formulation of an environmental impact assessment and occupational health and safety (OH&S) plan for the project
- Demonstration of awareness of the importance of sustainability and social issues related to the project
- Composition of a written engineering report that demonstrates knowledge of correct presentation of data including appropriate use of tables and figures, and professional English including correct grammar, spelling and punctuation in conjunction with high-level structure and style
- Conducting of a professional-level technical presentation including the use electronic visual aids such as PowerPoint.

Program Features

<u>Intensive Workshops with Industry Experts</u>

A unique feature of the program, only possible through industry support, is that students and teaching staff from all three universities travel to a single venue to begin the program with an intensive session of specialist lectures and workshops with industry experts over a number of days. The aim is to provide a comprehensive review of core extractive metallurgy knowledge, as well as new knowledge related to process and plant design, including related professional skills such as teamwork, communications, time management, and project planning and management, to enable the team members to complete their design task at the required pre-feasibility degree of complexity.

In addition to providing students with the necessary skills to undertake the project, along with obvious networking opportunities with industry professionals and prospective employers, the intensive workshop facilitates an organized and detailed discussion and clarification of the project brief, which hopefully leaves the students in no doubt about the commitment required to successfully complete this significant undertaking. Specifically, the intensive workshop is intended to:

- Present students with technical information concerning their specific assigned commodity and the design process,
- Expose students to information sessions from a wide ranging group of technical, business and environmental experts provided by industry professionals,
- Provide cross-fertilisation of academic expertise and styles across the three universities, allowing students' access to a composite knowledge base not available to any individual university, and
- Develop a strong network of peers, on both a working and social level, which student participants can call on throughout the semester-long project and into their professional careers.

The workshop also provides a platform for initial group work by inter-university groups, focusing on consideration of process options and preliminary process selection, definition of the group's project scope and development of a project plan including a preliminary Gantt chart. The students therefore take away a good understanding of the tasks that they need to complete during the semester including project options, scope and milestones.

In the first two years of the implementation of the program, two workshops of 3 days with 12 hours per day were provided; one at the beginning of the semester at UQ and one in the middle of semester at Curtin University. This approach allowed the main topics to be considered in two parts, namely process selection and design in the first workshop and plant equipment selection and sizing, materials handling, process control and costing and financial analysis in the second workshop.

The conduct of two workshops was very well received by the students but proved to be too logistically demanding and expensive. So, in the following year, a single 5-day workshop was undertaken at the beginning of semester. A review showed this new approach to be successful also in achieving the required introductory objectives and it has been retained.

Student Teamwork

Teamwork is considered a vital part of the program and students are expected to work in teams of 4 or 5 to facilitate collaboration, development of teamwork skills, and inter-university interactions. Ideally, team members would be selected from within and across universities, with each group consisting of 2 to 3 students from two different institutions. This approach was adopted in the first two years of the program. However, some disagreements arose amongst the participants from different universities on the relative importance of various aspects of the design, the level of depth at which they should be considered, and the expectations of staff in relation to their assessment.

Although such disagreements are understandable, given the different degree programs of the students, with the focus at UQ on chemical and metallurgical engineering, at WASM on minerals and metallurgical engineering, and at Murdoch on mineral science and extractive metallurgy, they were an unnecessary distraction and difficult to manage given the physical separation of the campuses. Another concern arose because of the uneven distribution of students from amongst the universities, which necessitated the creation of intra-university as well as inter-university teams, with the former being perceived by students at least, to have a communications advantage.

As a result, from 2010 the student grouping was amended such that each group now consists of 4 to 5 students from within each institution only. However, to maintain a good degree of inter-university interaction, groups from different universities are paired into "super groups" for the purpose of mutual peer review of their output, facilitated by the industry representative to the IC. In addition, the first design project (DP1) is completed as super group work.

The project brief given to teams includes specific tasks that must be accomplished by the group as a whole such as process evaluation and selection of the metal extraction process for the defined commodity and ore body. There are also tasks that must be completed by individual members of the group such as taking charge of the design of one key unit operation within the process, as allocated within the group. The specific tasks involved in this aspect are, for example, development of a detailed process flow diagram, establishment of the mass and energy flows using flow-sheeting software, sizing of key equipment for this unit operation, establishing plant layout, etc. If the number of unit processes exceeds the number of members in a group, the excess can be black-boxed, unless considered of fundamental importance by the IC such as the design of the comminution circuit, which has always been a compulsory component of the design.

Industry Engagement

As anticipated, support from industry for the program has been very strong since the first year. Beginning with assistance on the development of the project brief to ensure that the ore definition, metallurgical data and all design parameters are realistic, and through to assistance as specialist lecturers, facilitators and mentors during and after the workshops, industry professionals approached by the teaching team have generously and enthusiastically provided their help. These include practitioners of metallurgical process and plant design from consulting companies and production professionals including principal process engineers and other senior people at their respective operations. In addition, mining companies have provided other ad hoc support including prize money for the best design report and funding for students' social activities to allow for positive and productive informal interactions amongst students and industry professionals.

Projects Undertaken

Each year, the Implementation Committee selects and recommends to the Steering Committee an economically important mineral to Australia's mining industry as the subject of the design project. In 2008, when the program was first implemented, the processing of nickel laterite was the topic. This was followed by zinc, alumina, gold and copper. Last year, it was decided that from 2013, these commodities will be revisited and in the same order and thus, this year, nickel processing is once again the focus of the project.

Monitoring and Assessment of Students Progress

Student groups complete project in three design phases, as described below:

- **Design Phase 1 (DP1)** builds on the preliminary group work during the intensive workshop sessions and involves three main components: (a) engineering methodology,

(b) professional skills, and (c) technical aspects. The engineering methodology involves clarification of project scope, identification of key project elements, principal deliverables, and critical project issues, and development of project plan. The professional skills involve project planning and analysis, ensuring that everyone in the group understands the project requirements and direction, and that the work is carried out in a professional manner. A code of conduct is prepared by the group setting out both inter and intra-university communication strategies, meeting schedules and protocols, document control, performance monitoring and conflict resolution strategies. Due consideration of OH & S, environmental, sustainability and financial issues, in addition to technical aspects, are involved in process selection.

- **Design Phase 2 (DP2)** involves flowsheet development, construction of a process flow diagram, determination of mass and energy balance for the plant, prime equipment selection, and specification of ancillary facilities and services such as storage and transport facilities for all saleable products, handling requirements for process waste and residues and source of power and water, identification of major on-site infrastructure and preliminary plant layout and analysis of operability, maintainability and constructability.
- **Design Phase 3 (DP3)** involves costing, plant layout, process control, and project viability.

Academic staff mentors assist and encourage their student teams throughout the project, not only technically, but also through overseeing their team interactions, communication with others and meeting of required deadlines. To assist students manage their time, provide staff an effective way of monitoring students' progress, and assess the quality of the students' performance, several assessment items are strategically scheduled throughout the semester and constitute important project milestones. The assessable items consist of group work and individual components to emphasize the importance of the development of teamwork skills while allowing differentiation of individual students' abilities. Full assessment details over the 14 weeks timetabled for the unit are shown in Table 1:

Assessment Item	Mark Breakdown	Due Date
Design Package 1 (DP1) Engineering Methodology such as outline of project scope, deliverables, preliminary project plan, etc.; Project Plan describing how project objectives will be addressed; Code of conduct; and Analysis of Process Options.	Group Mark: 15% Individual Mark: 0%	Week 4
Design Package 2 (DP2) Mass and energy balances; Overall plant flowsheet and layout; Major onsite infrastructure; Individual unit process flowsheet; Prime equipment selection and sizing; Ancillary process facilities and services; Individual process plant layout and analysis.	Group Mark: 7% Individual Mark: 8%	Week 7
Design Package 3 (DP3) Detailed plant design; Process area layout; Environmental and OH & S issues; CAPEX and OPEX.	Group Mark: 0% Individual Mark: 15%	Week 10
Group Presentation (National video link) Presentation of the groups' proposed design.	Group Mark: 5% Individual Mark: 0%	Week 11
Peer Review (PR) Mutual critical analysis of final designs by the paired groups.	Group Mark: 2.5% Individual Mark: 2.5%	Week 11
Final Report	Group Mark: 18%	Week 13

Consolidation of all the elements of the proposed design into a coherent design report document.	Individual Mark: 17%	
Oral Defense of the Final Design Group oral defense of the proposed final design with respective home university academic staff	Group Mark: 0% Individual Mark: 10%	Week 14
Total Group/Individual Mark	**Group Mark: 47.5%** **Individual Mark: 52.5%**	
Total Mark	**100%**	

Table 1. Assessment items and mark breakdown of the process design project.

DP1 is accomplished as combined work of the super groups, and is assessed and moderated by both unit coordinators of the groups of students' home universities, and thus, all students in the super group get the same mark. Peer Review (PR) is facilitated and assessed by the industry representative from the SC. All other assessment items are carried out by the unit coordinator of the students' home university but using common assessment rubrics to set a uniform standard.

Graduate Numbers

The enrolments and graduates from all the four-year metallurgy degree programs of the partner universities that include the process design unit are shown in Figure 2, along with predictions until 2016. Given the success of the program in enhancing the students' learning experience and industry readiness, it is hoped that it can be expanded to include other engineering students who seek employment in the minerals industry, such as the small but growing number of WASM students who undertake a 5-year double degree with chemical engineering, and students from a newly established metallurgical engineering degree at the University of South Australia, who study their core metallurgical units through WASM.

Student Feedback

Non-compulsory feedback is collected internally at each university and also collected by MTEC through their surveys. At WASM, feedback is obtained anonymously through the Curtin University's on-line eValuate system wherein students either "strongly agree, agree, disagree, strongly disagree or are "unable to judge" on whether 11 desired outcomes for a unit have been achieved. To date, feedback has been positive, with a 100% overall satisfaction rating in the first year and now with ratings of at least 80% each year, with many positive comments a recurring theme, such as: "the information at the (Perth) workshops provided excellent background knowledge on the tasks", "the workshop was definitely a good kick start to the unit", "the industry lectures were brilliant", "I found the collaboration with another institution to be very rewarding", "the group work with students from another university was a great idea", "the workshops were great for team bonding", "challenging but rewarding", "provided a good insight into the metallurgical design tasks in the real world", and "taught me how to consolidate and distill the relevant knowledge I gained over the years of study of metallurgy".

While the positive WASM feedback suggests that most of the objectives of the program are being consistently achieved, there is always room for adjustment and improvement, and comments from the students have also been particularly helpful in this regard. These include, for example, the need for the project brief to be clear and precise to avoid different interpretations; that the teaching team should have a unified view when parts of the brief could be interpreted in different ways; and the need for more structured guidance on effective communication and time management skills, conflict management and teamwork strategies including positive reinforcement, and better technology support, particularly through IT. Overall, the need for the teaching team to be always positive cannot be overemphasized as any negativity can be disproportionately amplified, particularly during periods when students are under immense pressure, which is almost always the case during their final semester of study.

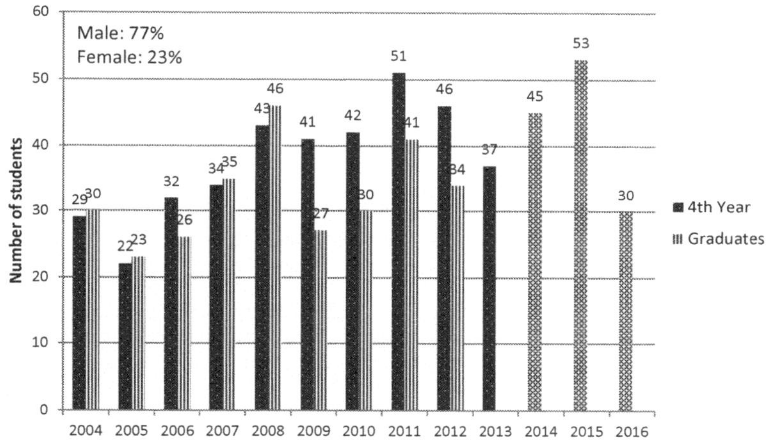

Figure 2. Actual and predicted graduations from MEP universities out to 2016 (reproduced with permission from Gavin Lind.[3]

The MTEC surveys have also been a very useful to the partnership. Before commencing the unit, the students are surveyed about their expectations of the tasks ahead and then after, about their perception of their learning experience. Consistently over the years, the results of the pre-course survey indicate that the students have a good understanding of what is expected of them, and they feel they are reasonably equipped for and optimistic about completing the upcoming tasks. The results of the survey after the course indicate that overall, the program provides a very positive and quality learning experience particularly through the substantial involvement of industry and the inter-institutional students' interactions.

Conclusions

Quality delivery of a capstone process and plant design unit in extractive metallurgical engineering requires the input of various experts including industry practitioners and thus, increasingly difficult to achieve within an individual university, given their progressively tightening teaching budget and the historically small cohorts of students in this discipline. To address this issue, the three main providers of metallurgical engineering education in Australia have collaborated, with on-going technical input from industry experts, and with funding support

[3] Gavin H. Lind, Executive Director of MTEC, Private communication with author, MTEC Australia, 11 November 2013.

through the Minerals Council of Australia, to jointly develop and successfully deliver a capstone unit in process and plant design as a national program delivered to final year metallurgical engineering students. Now in its sixth year, it is clear that the program is achieving its objectives.

References

1. G. Lind, "Minerals Industry Engagement in Metallurgical Education in Australia", presented at *XXVI International Mineral Processing Congress (IMPC 2012)*, 24-28 September 2012, New Delhi, India, The Indian Institute of Mineral Engineers), 87-104.

2. Minerals Council of Australia, "Back from the Brink, Reshaping Minerals Tertiary Education", A Discussion Paper, (Braddon ACT: Minerals Council of Australia 1998).

Appendix 1

Members of the MEP Steering Committee

The pioneering members of the Steering Committee were Eric Grimsey (Curtin WASM), Peter Hayes (UQ), Peter May (Murdoch) and Joe Pease (Industry Representative). Kevin Tuckwell (Executive Director of MTEC) was chair.

The current Steering Committee consists of Joe Pease (Industry Representative), Peter Hayes (UQ), Parisa Bahri (Murdoch), Steve Hall (Curtin WASM), and this author (Curtin WASM); Gavin Lind (Executive Director of MTEC) chairs the Committee.

The late Paul Dunn (Curtin WASM) also served in this Committee.

Members of the Implementation Committee

The pioneering members of the Implementation Committee were Bob Hannah (UQ), Nimal Subasinghe (Murdoch) and this author (Curtin WASM). Emmanuel Hondros (MTEC Project Officer) chaired the Committee and Dianna Drinkwater assisted the Committee as facilitator.

The current members of the Implementation Committee are Aleks Nikoloski (Murdoch), Yongjun Peng (UQ), Laurence Dyer (Curtin WASM), James Seaford (MTEC Project Officer), Chris Aldrich (Implementation Committee Chair and Representative of the Lead University Curtin WASM). The following have previously served on the Implementation Committee: Marc Steffens (Curtin WASM), James Didovich (Curtin WASM) and Boris Albijanic (WASM Curtin). Dan Churach and Nadine Smith (Parker Centre) also assisted the Committee previously as facilitators.

Celebrating the Megascale: Proceedings of the Extraction and Processing Division Symposium on Pyrometallurgy in Honor of David G.C. Robertson
Edited by: Phillip J. Mackey, Eric J. Grimsey, Rodney T. Jones, and Geoffrey A. Brooks
TMS (The Minerals, Metals & Materials Society), 2014

The MetSkill Program – Rapidly Developing Effective Young Engineers in the Workplace

Diana Drinkwater[1] and Nina Bianco[2]

1 Technical Training Manager, JKTech, Isles Road, Indooroopily, QLD Australia 4068
2 Senior CPD Designers, Centre for Innovation in Professional Learning (CIPL), The University of Queensland, QLD Australia 4067

Keywords: Education, professional development, skills shortage, graduate development

Abstract

MetSkill is a professional development program for metallurgical engineers that integrates with normal duties in their first one or two years of service. Graduates work together on a structured technical project, facilitated by specialists and supported by formal learning, and ultimately reported to their technical managers. The program enables graduates to "fill the gaps" in their undergraduate education, which is increasingly pertinent as engineering degrees become more general. Participants report that they enjoy the focus on more challenging (rewarding) aspects of their jobs and feel more confident about problem solving. Sponsor companies add that the relationships developed with external technical specialists enhances opportunities for innovation and development. MetSkill was delivered to two major resource companies in Australia in 2012. This paper provides an outline of the program and the reasons for its success, and demonstrates how the learning model could be applied to groups of graduates in other engineering disciplines.

Introduction

Metskill is a professional development program for graduate minerals processing engineers which integrates with their day to day activities during their first one or two years of service. The program aims to fast-track skills development and has been designed and developed by specialists at JKTech in Queensland, Australia, with support from researchers from the JKMRC and SMI Knowledge Transfer and teaching and learning specialists at The University of Queensland. Primarily focused on developing technical and analytical skills, the program also provides opportunities for participants to practice teamwork, communication, leadership and problem solving.

More than half of the graduates employed in Australia's mineral operations as metallurgists and mineral processors have degrees in general chemical engineering or materials science [1] and have encountered little or no mineral processing when they commence work. Once this skills gap would have been addressed by on-the-job training provided by senior professionals at the mine-sites, but this is less and less practical in today's lean workforce. This was the major drive behind development of MetSkill.

The program runs over one or two years and is delivered in-house to a mining company client. A site-based optimisation task is selected based on discussion between the JK specialists and the site sponsors. Activities throughout the training period include classroom style workshops on key topics and ongoing facilitation of project work and other activities via a purpose-built web platform. Much of the project work is scoped and scheduled by the program participants themselves, and project management and teamwork coaching are included in the curriculum. Additional on-line modules are being developed which will allow participants to select appropriate foundational material, such as geology, mineralogy and flotation chemistry, so that the face-to-face workshops can focus on group interactive experiences.

On completion of the program the graduates are expected to deliver a substantial project report, demonstrating their skills in problem solving and applying technical and analytical tools. A more important outcome, arguably, is the relationships developed within the graduate cohort and also between the graduates and the technical specialists both within their own organisation and from within the JK Teaching team that deliver the program.

Evaluating the learning experience, graduates report that they enjoy the focus on more challenging (rewarding) aspects of their jobs and feel more confident in their ability to make sound judgments and solve problems. There is also a feeling amongst the sponsor companies that the regular contact with external technical specialists increases the opportunities for innovation and development. In 2012 the MetSkill program was delivered to two 2 groups of Australian graduate metallurgists, and this paper reviews the outcomes and outlines the future directions of the program.

Background

The genesis of the MetSkill program can be traced back to a number of sources including the AngloPlatinum graduate development program described by Sweet et al. [2], and graduate development guidelines put in place in the 1970s and 1980s by Australian companies such as Mount Isa Mines, as well as the current guidelines by professional bodies such as the AusIMM and International Mineral Processing Congress (IMPC).

The aim of the MetSkill program is to give graduates an early and direct experience of a high quality metallurgical investigation, using best practice analytical methodologies and appropriate tools, facilitated by experienced specialists. The investigation is central to the learning experience, and for maximum benefit it should have the potential to make a real improvement to metallurgical performance.

The MetSkill Program

The learning objectives are summarised as follows: On completion of the program, the graduates should be able to successfully:

- Assess process plant data using fundamental metallurgical principles and appropriately selected analytical tools
- Recommend a day-to-day operating strategy that demonstrates knowledge of industry best practice for a selected area of plant
- Plan and execute a mineral processing plant survey, and conduct a technically competent optimisation study of a selected area of plant demonstrating sound judgement and using appropriate modelling software
- Apply mineralogical information from a variety of sources to a metallurgical problem
- Compose a written project report that demonstrates knowledge of correct presentation of data for a selected audience

- Work productively on a technical project both as an individual and a team member.

Implementation of the program requires commitment by the industry partner to a one or two year program, and agreement on an appropriate program of modules. Specific learning modules included in the package are negotiated with clients, based on a model of "core" and "elective" elements, as shown in Table 1.

Table I. Model of Core and Elective elements available in the MetSkill Program

Core Modules
1. Fundamentals of comminution, mineralogy, separation and ore characterisation testing (can be put online)
2. Working in high-performing teams, managing projects
3. Collecting good data, survey methodology, sample analysis, mass balancing and modelling - includes facilitated plant survey
4. Statistics and experimental design
5. Process optimisation

Elective Modules
6. Process control
7. Sustainable processing
8. Comminution circuit design and optimisation using JKSimMet
9. Flotation theory and JKSimFloat
10. Gold leaching and recovery
11. Geometallurgy
12. Metallurgical accounting

Next, a draft schedule needs to be established and an appropriate site must be selected for the technical investigation which forms the basis of the program. Note that the investigation should be only very broadly defined, as detailed scoping should be a part of the graduates' initial activities when they commence the program.

The benefits to the client companies go well beyond educational and training outcomes, as the investigation should provide high quality survey data for a selected operation, allowing operational problems to be assessed and solutions provided in-house. Because of the involvement of such a large group of metallurgists, the technical knowledge developed is likely to stay in-house for much longer than is often encountered with consultant-based technical work.

Generally no more than 3 or 4 graduates come from a single site, so in addition to technical skills development, this program encourages graduates to develop professional relationships with their off-site colleagues within the company and also with senior professionals from central office or supporting groups, such as mineralogical specialists or project managers.

Detailed Education Design

Underpinning Learning Model

The professional development model underpinning MetSkill is a small-group, collaborative, project-based model, which incorporates mentoring to enhance its effectiveness. MetSkill uses an authentic, work-situated project to develop the skills of early-career metallurgists, who are supported and guided by experienced metallurgists and consultants. The project is central to the learning and the underpinning knowledge and collaboration are built around it.

Project-based learning uses an authentic, defined, complex problem to trigger the learning process. It is similar to problem-based learning in that the "learning is organised around problems". In project-based learning, however, the problems are more structured and there is defined end-deliverable, e.g. a report [3]. The participants are motivated to engage with the learning by having an immediate work-situated problem to solve. A project-based model of learning uses an experiential learning or 'learning through action' approach.

A small-group, collaborative model was chosen for this program to situate the learning within a supportive community. Face-to-face workshops bring the participants (10 – 15 people) together at a mine site to work with each other, and with senior metallurgists, to execute their project. Outside the workshops, project collaboration and expert mentoring is supported by an online project space.

According to the research literature, guidance by workplace experts is a key feature of successful workplace learning - "individuals need guidance from experienced workers in the form of coaching and modelling that focuses on transferring knowledge to new situations" [4]. The use of ongoing coaching or mentoring can also increase the effectiveness of project-based models of learning. Mentoring is a particularly effective way to support novices or beginning professionals and has been used to provide support for professional development across a number of professions [5]. This support has ranged from career guidance to psychosocial support. In the MetSkill program, mentoring is used to provide skill development, coaching, feedback and networking opportunities.

The delivery model for MetSkill is blended. It combines face-to-face workshops and activities, with online collaboration and mentoring. The MetSkill learning model is made up of four main elements:

1. The project. This forms the focus of the learning. In MetSkill, the project is to investigate an aspect of plant processing and make recommendations for processing optimisation. By undertaking this project, the participants learn skills in applying key metallurgical tools and methodologies to improve the quality of plant operating practice.
2. Structured learning. This is in the form of face-to-face lectures in a workshop format which provide the underpinning evidence-based knowledge to support the project.

3. Collaboration. This occurs as two conversations around the project. One is between the early-career metallurgists who are collaborating with each other on the project deliverables. The other occurs between the teaching experts, the project mentors and the early-career metallurgists, where the experts and mentors provide project guidance and support to the early-career metallurgists.
4. Facilitation. The learning facilitator has an important role in MetSkill. They have a macro-view of the whole program, its different components and how they fit together. The facilitator is responsible for coordinating the MetSkill program. They ensure the participants are engaging with the learning, progressing their project deliverables and achieving the project milestones. They are also responsible for ensuring the experts and mentors understand their roles and are engaging with the participants.

Effective Professional Development Principles

There is considerable agreement within the research literature and across professions, about what constitutes effective professional development. MetSkill was designed using these best-practice, professional development principles. They are:

- Learning is ongoing and continuous (1-2 years), rather than one-off episodic short courses. Professional learning is a long-term process and effective professional development acknowledges this. The MetSkill program was designed to allow the participants to build metallurgical knowledge and skills over a 1-2 year period, by executing a long-term project.
- Learning is situated within the work context and is related to authentic work experiences. The most effective professional development occurs within the workplace, when professionals engage with authentic work experiences. The MetSkill program uses a work-based project that allows early-career metallurgists to work on process optimisation at a nominated mine site.
- Learning is social and collaborative. According to the research literature, "learning happens through social interaction and collaboration" [4]. In MetSkill, participants work collaboratively, with each other and with technical experts, to execute their project. This happens face-to-face in the workshops and site visits, and online using a dedicated web-space.
- Learning is learner-centred and self-directed. In MetSkill, early-career metallurgists determine the specific project they will work on, its planning and execution.
- Learning is active. MetSkill is an example of experiential learning, where participants learn by doing. Access to expert, evidence-based knowledge underpinned by research. MetSkill is underpinned by the latest research and by expert knowledge in mineral processing.

Effective professional learning also requires organisational leadership and support. Organisational support for MetSkill is negotiated in the initial planning phase and includes time-off for participants to attend the workshops, resources in the form of travel expenses, access to on-site data, equipment and personnel and support for, and involvement in, the program by management.

Online Support

One of the features of MetSkill design is the online support. The MetSkill website provides a flexible workspace where participants, experts and mentors can continue to collaborate on their project when they are no longer face-to-face. The website is designed to provide:

- An overview of the program and workshops, i.e. learning goals, timeline, calendar of events, deliverables and teaching faculty
- A space for filing project data and deliverables
- A place where experts, mentors and participants can collaborate with each other around the project
- A place where workshop resources could be stored and accessed

The project space is the focus of the website. It uses a wiki tool for filing project documents and facilitating participant communications about the project.

The website is designed using success factors for online groups. These are used to inform the design of the user interface, website structure, website content and access control. The design goal is to make explicit, the people, content, structure and processes that are critical to group success. Defining roles, responsibilities, communication pathways and permissions to access website content, were areas that required particular thought. The main challenge in designing the online support is predicting how participants and others are going to use the website and their preferred ways of communicating and accessing information.

Program Rollout 2012

In 2012 the MetSkill program was delivered to graduates at two Australian based mining companies, MMG, whose graduate group comprised a total of 11 from 5 different sites, and Newcrest mining, with 16 graduates from 6 different locations. Each group undertook a major survey at one of their sites with the support of consultants and research staff from the JKMRC and JKTech, and then worked collaboratively to manage the data processing, analysis and review, eventually producing process optimisation reports which were presented to senior metallurgical and technical staff in their organisations. In both cases the survey outcomes exceeded expectations in terms of data quality.

There were regular progress reports by the graduate groups to their own management, as well as discussion of opportunities available within the company to follow the current project with more activities. In both cases there has been spin-off work generated by the investigation. This is an important part of the learning process, as it provides the opportunity for graduates to embed newly learned skills into their everyday metallurgical practice.

A large team of technical experts has supported the data collection and analysis undertaken by graduates, drawn mostly from the JKMRC and JKTech but also including some external specialists. This team provided feedback and review on an ongoing basis, as well as specific instruction on tools and methodologies. The experience has been a rewarding one for many of these support staff, as many would not otherwise have many of these kinds of opportunities to pass on their skills to a large group of young professionals.

Reflections by the participants and their members were also very positive:

"For MMG the MetSkill program is all about front line metallurgy. The aim is to bring MMG's young metallurgists and chemical engineers "up to speed" in a relatively short time frame. We are grateful that JKTech have taken up the challenge." *Peter Tilyard, MMG Group Metallurgist 2012*

"The MetSkill course was a great opportunity not only to put what everyone has learnt at University into practice, but also to meet and form networks with the other metallurgy graduates within Newcrest." *Matthew Livingstone, Newcrest 2nd year graduate*

"The MetSkill course is a very enlightening package that brings together theory and practice providing an excellent foundation for working in the mineral processing industry. It provides the focus needed to heighten graduate skills that cannot be achieved at the University alone, by providing high level networking of JKTech staff that have invaluable knowledge in this industry." *Varun Patel, Gekko Graduate Process Engineer, participant in the Newcrest program.*

Amanda Rice (Cadia), Aaron Sitautau (Lihir), Sebastian Tello (JKTech) work on data analysis

Luke Vollert, Matt Livingstone and Aaron Sitautau take a SAG feed belt cut at Telfer

Future Directions

MetSkill is still a new and evolving program and there is certainly scope for streamlining resources and fine-tuning module content and schedules. However, the greatest opportunities are in extending the reach of the program by enhancing the capabilities of the supporting website, allowing some of the workshop material and activities to be delivered remotely, thereby providing access to larger or more remote groups of graduates.

There are also opportunities to apply this graduate development model to other professional groups within and without the mining industry, and JK Tech and The University of Queensland are actively exploring these options with other groups of professionals. We may well be looking at GeoSkill, EnviroSkill and MiningSkill programs in the future.

The other more direct expansion is regional, as the companies operating in the mining sector often have global development teams and want to run global professional development programs. The MetSkill development team is well aware of this and is working with its educational designers at The University of Queensland to develop a truly global program, drawing on experts around the world and underpinned by substantial on-line resources.

Conclusion

The MetSkill team at JKTech believes that they have the fundamental design right, and can deliver a program that directly addresses the minerals industry need for rapid, early career development of front-line professionals in the workplace. The focus on high quality educational design and best practice analytical techniques, delivered by world class specialists, is realistic and achievable, and delivers the expected results.

References

1. G. Lind, "Minerals industry engagement in metallurgical education in Australia", in *XXVI International Mineral Processing Congress (IMPC 2012)*, New Delhi, India, 1223-1229, September 24-28, 2012.

2. G. C. Sweet et al., "Industry taking the initiative in developing high caliber technical staff", in *Proceedings of XXIII International Mineral Processing Congress*, Istanbul, 3-8 September, 2011.

3. E. deGraaff and A. Kolmos, *Management of change: Implementation of problem-based and project-based learning in Engineering* (Rotterdam, The Nedlands: Sense Publishers, 2007).

4. C. Barker, "Embedding learning from formal training into sustained behavioral change in the workplace", in *National Vocational Education and Training Research and Evaluation Program* Occasional Paper 2011, NCVER: Adelaide.

CELEBRATING THE MEGASCALE

Proceedings of the Extraction and Processing Division Symposium on Pyrometallurgy in Honor of David G.C. Robertson

Pyrometallurgy Process Fundamentals I

Celebrating the Megascale: Proceedings of the Extraction and Processing Division Symposium on Pyrometallurgy in Honor of David G.C. Robertson
Edited by: Phillip J. Mackey, Eric J. Grimsey, Rodney T. Jones, and Geoffrey A. Brooks
TMS (The Minerals, Metals & Materials Society), 2014

OXIDATION OF FLASH REDUCED IRON PARTICLES IN VARIOUS GAS MIXTURES UNDER THE CONDITIONS OF A NOVEL FLASH IRONMAKING PROCESS

Zhixue Yuan[1], Hong Yong Sohn[1] and Miguel Olivas-Martinez[1]

[1]Department of Metallurgical Engineering, University of Utah;
135 S. 1460 E., Rm 412; Salt Lake City, UT 84112-0114, USA

Keywords: Oxidation kinetics of iron particles, Flash ironmaking, Reduction of iron oxide

Abstract

A novel flash ironmaking process that directly reduces iron oxide concentrate particles by gas is under development. The goal of this work was to study the possibility of reoxidation of iron particles in various gas mixtures. As the product iron cools down in the lower part of the flash reactor, conditions may become favorable for reoxidation because of equilibrium and high reactivity of iron particles. The effects of temperature (823 – 973 K) and H_2O partial pressure (40 – 100 pct., P_{total} = 86.1 kPa) on the reoxidation rate were examined. The pressure dependence was first order with respect to water vapor, and the activation energy was 146 kJ/mol. A complete rate equation that adequately represents the experimental data was developed. For oxidation in O_2-N_2 mixtures, the effects of temperature (673 – 873 K) and O_2 partial pressure (5 – 21 pct., P_{total} = 86.1 kPa) were determined. Reoxidation in pure CO_2 was also investigated at 873 – 1073 K for comparison.

Introduction

A novel flash ironmaking process is currently under development at the University of Utah as an alternative ironmaking process that would greatly reduce CO_2 emissions and energy requirement compared with ironmaking in a blast furnace [1-3]. In the temperature range of 1500–1900 K expected in the flash reduction process, H_2 is the main reductant even in the presence of CO when hydrocarbon gases are used.

Because of the equilibrium limitation of the H_2 reduction of wüstite, the possibility of reoxidation as the reaction mixture cools down needs to be considered. Iron product in the novel flash ironmaking process may also come into contact with air at moderate temperatures during transportation. Since direct reduced iron is usually pyrophoric and may ignite spontaneously due to the exothermic oxidation reaction [4], reoxidation of flash reduced iron particles in O_2-N_2 gas mixtures is also of interest. CO_2 will be present in the flash reactor when natural gas is used in the proposed process, and thus oxidation tests of the flash reduced iron by CO_2 were also carried out.

Experimental Work

Iron ore concentrate from ArcelorMittal (East Chicago, IN, U.S.A) was reduced to metallic iron particles at 1473 and 1623 K in a drop tube reactor, as described elsewhere [5, 6]. The reduced iron particles used in the reoxidation experiments had a reduction degree higher than 95 pct. The mean particle diameter was 17.5 μm. About 100 mg of iron particles were sprinkled thinly on top of the Kaowool bed in an alumina tray in order to eliminate the interparticle diffusion effect of the reactant gas in a tubular reactor. A thermogravimetric analysis unit was used to measure the oxidation rates in O_2-N_2 gas mixtures and pure CO_2. About 120 mg of iron particles were sprinkled thinly on the Kaowool bed in a cylindrical sample holder. Microstructures were characterized by an FEI NanoNova Scanning Electron Microscope (SEM) and a TOPCON SM-300 SEM. XRD analysis was carried out with Philips X'pert X-ray diffractometer with Cu-Kα source 40 kV.

Results and Discussion

In this study, the reoxidation degree was defined by the fractional oxygen content relative to Fe_xO_y, given by

$$F = (\Delta m + m_0) / \Delta m_{\text{Fe-Fe}_x\text{O}_y} \tag{1}$$

in which Δm is the weight gain caused by reoxidation, m_0 is the weight of oxygen in the starting sample, and $\Delta m_{\text{Fe-Fe}_x\text{O}_y}$ is the weight of oxygen in Fe_xO_y if all Fe in the starting sample is present as Fe_xO_y. For oxidation of iron in H_2-H_2O gas mixtures and CO_2, Fe_xO_y is Fe_3O_4, while for oxidation in O_2-N_2 mixtures, Fe_xO_y is Fe_2O_3. This is because in the appropriate temperature and gas partial pressure ranges, Fe could only be oxidized up to one stable oxidized form. This was also confirmed by the XRD patterns of the oxidized samples. In H_2-H_2O and CO_2, the oxidation of iron to wüstite [or magnetite below 833 K] was the dominant reaction investigated, because: the reactions are equilibrium limited; the reoxidation of iron to wüstite accounts for around 75% of the complete reoxidation process; and the early part of reoxidation is of significance as far as an ironmaking process is concerned. In O_2-N_2, however, the reaction studied was the oxidation of iron to hematite because of the large equilibrium constant for oxidation of iron in O_2.

Reoxidation in H_2O-H_2 Gas Mixtures

Different gas flow rates were tested at 923 K with a fixed sample size of 100 ± 5 mg and constant H_2O concentration of 80 pct (P_{total} = 86.1 kPa). Reaction rate was independent of gas flow rate above 6 L/min at experimental temperatures indicating elimination of the external mass transfer effect. The effect of interparticle diffusion of gaseous species was eliminated by using a thin sample layer with loosely scattered particles. The area and thickness of iron particles layer were, respectively, 9 cm^2 and 0.5 mm. Calculations indicated that the reaction of individual particles was not affected by interparticle diffusion. Iron particles with similar reduction degrees were produced at 1473 and 1623 K in the flash reduction process [7]. Reoxidation rates of these samples were similar, which was explained by their similar microstructures.

The proposed temperature range in the novel flash ironmaking process is 1500–1900 K, much higher than in the typical direct reduction processes [4]. Thus it is of interest to examine the differences of oxidation rates between the flash reduced iron particles and the conventional direct reduced iron particles. Fine magnetite concentrate was reduced by H_2 at 1073 K and by CO at 1223 K in a horizontal furnace to produce iron particles that are typical in conventional direct reduction processes. Reoxidation rates of these iron particles in H_2-H_2O were compared with that of the flash reduced iron particles produced at 1623 K. The results in Figure 1(a) show a slower oxidation rate for flash reduced iron particles. This is due to the lower specific surface area of iron particles produced at higher temperatures available in the novel flash ironmaking process. The data provided by the MICROMERITICS ASAP 2020 shows a specific surface area of 195 m^2/kg for flash reduced iron produced at 1623 K, whereas iron particles reduced by CO and H_2 at, respectively, 1223 and 1073 K are, respectively, 228 and 457 m^2/kg.

The influence of H_2O partial pressure on reoxidation was studied by varying the H_2O concentration from 40 to 100 pct in H_2O-H_2 mixtures (P_{total} = atmospheric pressure in Salt Lake City = 86.1 kPa), while temperature and the total gas flow rate were, respectively, maintained at a fixed rate of 873 K and 6 L/min and other conditions were all kept identical. The results shown in Figure 1(b) indicate that the rate and degree of reoxidation increased with H_2O partial pressure as expected. Temperatures were varied in the range of 823–973 K with the same $H_2O(g)$ concentration of 80 pct and other identical conditions (P_{total} = 86.1 kPa). As expected, the rise in temperature resulted in faster reoxidation and a larger reoxidation degree, as indicated in Figure 2. At even lower temperatures of 723 and 773 K, the reaction rates were too low to be measurable in the current system.

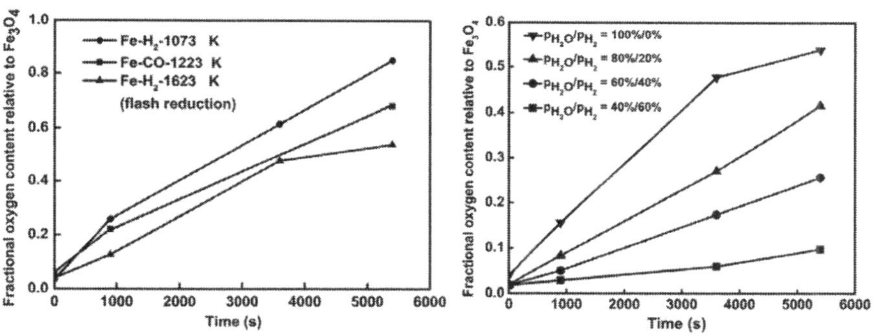

(a) (b)

Figure 1. Reoxidation rates of iron particles in H_2O-H_2 mixtures at oxidation temperature of 873 K. Effects of (a) reduction temperature for oxidation in 100% H_2O (g); and (b) H_2O/H_2 partial pressure ratio for flash reduced iron particles. P_{total} = 86.1 kPa (the atmospheric pressure in Salt Lake City).

The interpretation of the rate data was carried out using the nucleation and growth kinetics. In this model, the reoxidation degree of iron particles is related to the reaction time by [8]

$$\left[Ln\left(\frac{1-F_0}{1-F}\right)\right]^{1/m} = k_{app} \cdot t \qquad (2)$$

where m is the Avrami parameter, t the reaction time in seconds, F the fractional oxygen content relative to Fe_3O_4, F_0 the initial value of F at t = 0, and k_{app} the apparent rate constant. k_{app} is related to chemical reaction and varies with temperature and partial pressure, as follows [6]

$$k_{app} = b \cdot k \cdot f_p(p_{H_2O}, p_{H_2}) \qquad (3)$$

where b is a stoichiometry coefficient (number of moles of solid reacted per mole of gaseous reactant = 1 in the present study), and k is the intrinsic rate constant, p_{H_2O} and p_{H_2} are, respectively, the partial pressures of H_2O and H_2 in kPa, and f_p denotes the functional dependence on partial pressures. All data fit well with m = 1.26, as exemplified by Figure 3.

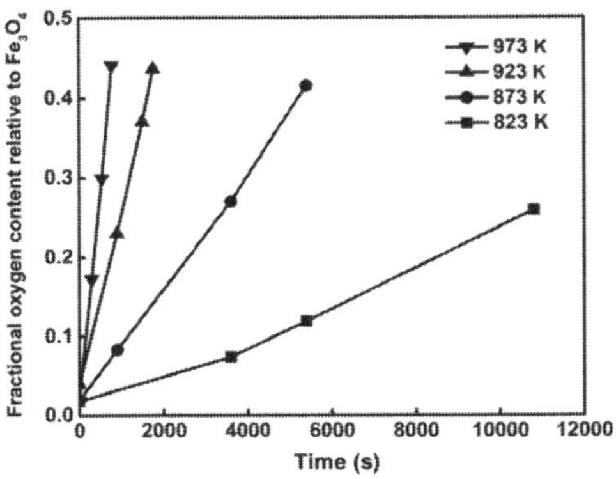

Figure 2. Effect of temperature on fractional oxygen content relative to Fe_3O_4; $p_{H_2O} / p_{H_2} = 80\% / 20\%$, P_{total} = 86.1 kPa.

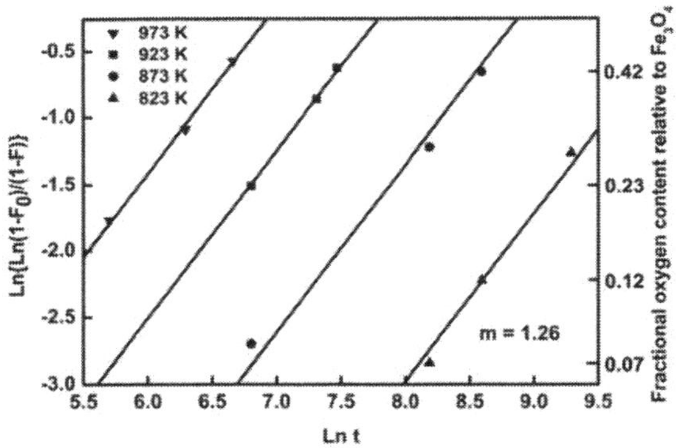

Figure 3. Relationship between $Ln\left[Ln\left(\frac{1-F_0}{1-F}\right)\right]$ and Ln t with $p_{H_2O} / p_{H_2} = 80\% / 20\%$, m = 1.26 and P_{total} = 86.1 kPa.

The reaction rate is dependent on the partial pressures of both gases, since the reduction of wustite by H_2 is significantly limited by equilibrium in the proposed temperature range of the flash ironmaking process. The rate dependence is expressed [6] by

$$f_p(p_{H_2O}, p_{H_2}) = p_{H_2O}^n - (p_{H_2} / K)^n \tag{4}$$

where K is the equilibrium constant of the oxidation of iron to wüstite [or magnetite below 833 K] by H_2O(g), and n is the reaction order with respect to H_2O and H_2 partial pressures. The value of n was determined to be 1. The activation energy was determined to be 146 kJ/mol. And the complete expression for the rate constant k (s^{-1} kPa^{-1}) was obtained as

$$k = 941.5 \cdot \exp(-146000 / RT) \tag{5}$$

At a constant temperature, the relationship between the fractional oxygen content relative to Fe_3O_4 (F) and time is given by:

$$\left[\ln\left(\frac{1-F_0}{1-F}\right)\right]^{1/1.26} = 941.5\cdot\exp\left(-\frac{146000}{RT}\right)\cdot(p_{H_2O}-p_{H_2}/K)\cdot t \quad (6)$$

where T is in K and p is in kPa. The instantaneous rate of reoxidation can thus be obtained as

$$\frac{dF}{dt} = 1.26\cdot k\cdot(p_{H_2O}-p_{H_2}/K)\cdot(1-F)\cdot\left[\ln\left(\frac{1-F_0}{1-F}\right)\right]^{1/4.85} \quad (7)$$

where $k = 941.5\cdot\exp(-146000/RT)$. The rate equation applies to iron particles produced from Fe_3O_4 in the temperature range of 1473–1673 K.

Reoxidation in O_2-N_2 Gas Mixtures

The effects of O_2 concentrations and temperatures were determined. Figure 4 show that the reaction rates are much faster than that in water vapor and quickly level off at a low extent. SEM pictures (not shown here) for starting iron particles and oxygen oxidized samples showed that the oxide grains develop on the iron particles after oxidation. It was also reported [9] that the oxidation of iron in oxygen starts as nucleation of iron oxides on the surface of iron. Therefore, the nucleation and growth model [Equation (2)] was again employed to fit the oxidation data till the level off point. Figure 5 shows that all data can fit very well with m = 1. The Avrami parameter m is again independent of temperature and pressure and m = 1 implies one dimensional oxide nuclei growth and explains the absence of a detectable induction period at the beginning of oxidation [10, 11].

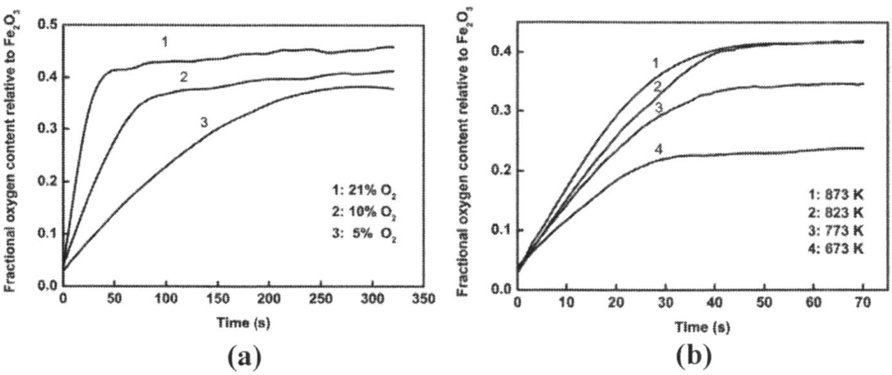

Figure 4. Reoxidation rates of iron particles in O_2-N_2 mixtures. Effects of (a) O_2 partial pressure at 873 K; and (b) temperature; $p_{O_2}/p_{N_2} = 21\%/79\%$. P_{total} = 86.1 kPa.

At the last stage, the oxidation rate became almost flat, as shown in Figure 4, probably due to the formation of a dense nonporous oxide layer causing a very slow diffusion of oxygen. Oxidation of iron in oxygen is strongly exothermic compared with in water vapor and carbon dioxide. Therefore, it is believed that the strong reaction heat results in local fusion, causing the rounding of surface morphology and impeding the diffusion of reactant oxygen.

The apparent rate constant in Equation (2) is related to the oxygen partial pressure as [12]

$$k_{app} = b \cdot k \cdot f_p(p_{O_2}) = b \cdot k \cdot p_{O_2}^n \quad (8)$$

where b is stoichiometry coefficient (number of moles of solid reacted per mole of gaseous reactant = 4/3 for iron oxidation in O_2), k the intrinsic rate constant, n the reaction order with respect to oxygen partial pressure, p_{O_2} oxygen partial pressure in kPa. k_{app} values from Figure 5(a) were plotted against $p_{O_2}^n$. Various values of n were tested and a first order relationship was found to be the most satisfactory as shown in Figure 6(a).

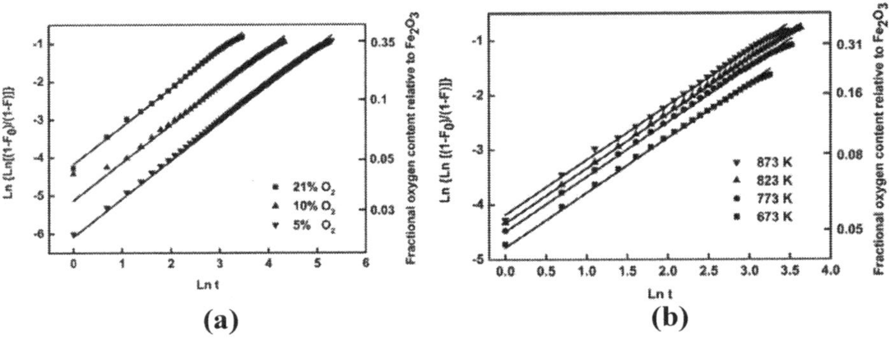

Figure 5. Relationship between $Ln\left[Ln\left(\frac{1-F_0}{1-F}\right)\right]$ and $Ln\, t$ at (a) 873 K and (b) $p_{O_2} / p_{N_2} = 21\% / 79\%$; $m = 1$ and $P_{total} = 86.1$ kPa.

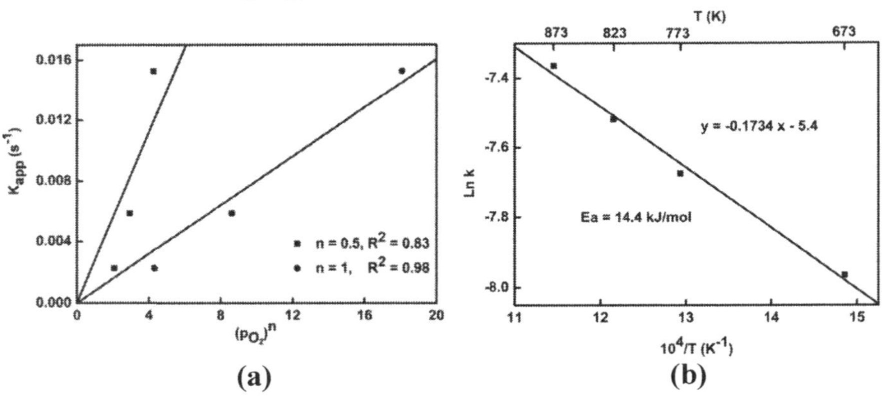

Figure 6. Kinetics parameters for flash reduced iron particles oxidized in O_2-N_2 mixtures. (a) Relationship between k_{app} and $p_{O_2}^n$, $n = 0.5$ and $n = 1$, p in kPa; (b) Arrhenius plot between $Ln\, k$ and $10^4/T$; k in kPa^{-1}.

From Equation (8), the intrinsic rate constant k is expressed as

$$k = 3/4 \cdot k_{app} / p_{O_2} \quad (9)$$

and the activation energy of the reoxidation reaction was obtained through the Arrhenius equation as 14.4 kJ/mol in Figure 6(b). The rate constant k (s^{-1} kPa^{-1}) is expressed as

$$k = 0.0045 \cdot \exp(-14400 / R / T) \tag{10}$$

where T is temperature in K, and R is the gas constant.

For the reoxidation until the level off point, the relationship between the fractional oxygen content and the time can be formulated as

$$\ln\left(\frac{1-F_0}{1-F}\right) = 0.006 \cdot \exp\left(-\frac{14400}{RT}\right) \cdot p_{O_2} \cdot t \tag{11}$$

where F is the fractional oxygen content relative to Fe_2O_3. After the leveling off point, the rate can be regarded as constant based on Figure 4, and thus the F values of leveling off is related to temperature in Figure 7 as

$$F = -3.05 \times 10^{-6} T^2 + 5.7 \times 10^{-3} T - 2.23 \tag{12}$$

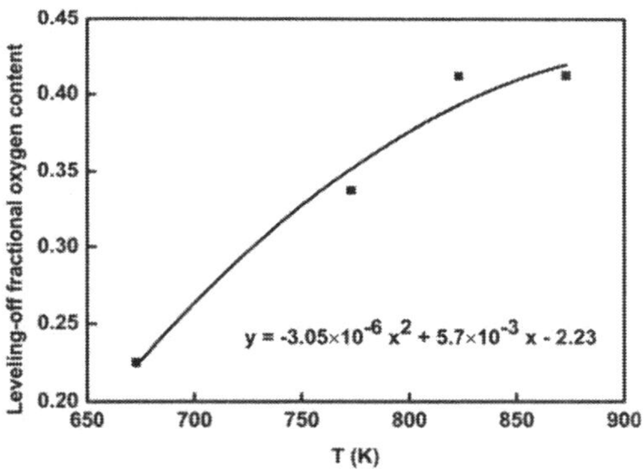

Figure 7. Relationship between leveling off F values and temperature.

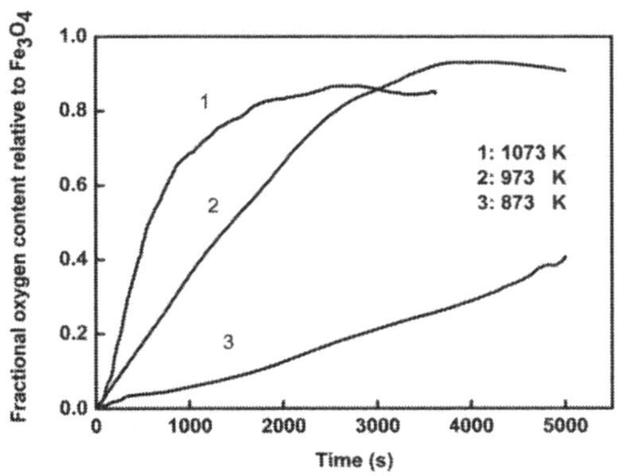

Figure 8. Effect of temperature on reoxidation rate; 100% CO_2, P_{total} = 86.1 kPa.

Reoxidation in CO_2

The reoxidation of flash reduced iron in pure CO_2 was carried out in the temperature range of 873–1073 K. The kinetics results in Figure 8 show that the oxidation in CO_2 is much slower than that in water vapor or oxygen. In addition, it can be concluded that both in the suspension reduction process, where the residence time of iron particles is typically less than 10 seconds, and in the collection bin, where the iron product may be kept for up to one hour at around 573 K, the oxidation by CO_2 is not of concern due to its slow reaction kinetics.

Concluding Remarks

A novel flash ironmaking process that directly reduces iron oxide concentrate particles is under development. As the product iron cools down in the lower part of the flash reactor, conditions may become favorable for reoxidation in H_2O and CO_2 because of equilibrium and high reactivity of iron particles. Flash reduced iron may also come into contact with air at moderate temperatures during transportation. The possibility of reoxidation of iron particles in various gas mixtures was investigated. The following conclusions are reached:

1. The nucleation and growth kinetics expression was found to describe the reoxidation of flash reduced iron particles in H_2O-H_2 and O_2-N_2 gas mixtures after detailed calculations and microstructure examination of the oxidized samples.
2. The oxidation reactions of flash reduced iron particles in H_2O-H_2 and O_2-N_2 gas mixtures both have first order dependence on the partial pressures of, respectively, water vapor and oxygen.
3. The activation energy of oxidation of flash reduced iron particles in H_2O-H_2 and O_2-N_2 gas mixtures are, respectively, 146 and 14.4 kJ/mol, which implies the former has greater temperature dependence than the latter.
4. Reoxidation of flash reduced iron particles in pure CO_2 was slow compared with in H_2O and O_2 and it will not be of concern, whether in the suspension reduction process or in the iron product collector, especially considering CO_2 accounts for less than 10% of the gas mixture from the partial combustion of natural gas.
5. Flash reduced iron particles are much less vulnerable to oxidation than conventional direct reduced iron particles, because the novel flash ironmaking process uses higher reduction temperatures, leading to lower specific surface area.
6. The findings in the present work indicate that within several seconds of residence time typically available in a flash reduction process the reoxidation degree of iron particles in water vapor is < 0.24 % at the temperature range of 823–973 K. In the collector where product iron particles may be kept for up to one hour at around 673 K, the expected reoxidation degree is around 0.02 %. To further reduce oxidation in the product collector, an inert gas device could be installed to dilute or prevent contact of oxidizing gas. And in long distance transportation, it is advisable to make briquettes of flash reduced iron particles and prevent contact with air or moisture as much as possible.

Acknowledgments

The authors acknowledge the financial support from American Iron and Steel Institute (AISI) under AISI's CO_2 Breakthrough Program. This material also contains results of work supported by the U.S. Department of Energy under Award Number DE-EE0005751.

References

1. H.Y. Sohn, M.E. Choi, Y. Zhang, and J.E. Ramos, "Suspension Reduction Technology for Ironmaking with Low CO_2 Emission and Energy Requirement," *AIST Iron & Steel Technology*, 6 (6) (2009), 158-165.

2. M.E. Choi, "Suspension Hydrogen Reduction of Iron Ore Concentrate" (Ph.D. dissertation, University of Utah, 2010).

3. H.Y. Sohn, "Suspension Ironmaking Technology with Greatly Reduced Energy Requirement and CO_2 Emissions", *Steel Times Int.*, 31 (4) (2007), 68-72.

4. R.L. Stephenson, *Direct Reduced Iron: Technology and Economics of Production and Use* (Warrendale, PA: The Iron and Steel Society of AIME, 1980) 64-99.

5. M.E. Choi and H.Y. Sohn, "Development of Green Suspension Ironmaking Technology Based on Hydrogen Reduction of Iron Oxide Concentrate: Rate Measurements," *Ironmaking Steelmaking*, 37 (2) (2010), 81-88.

6. H. Wang and H.Y. Sohn, "Hydrogen Reduction Kinetics of Magnetite Concentrate Particles Relevant to a Novel Flash Ironmaking Process," *Metall. Mater. Trans. B*, 44(1) (2013) 133-145.

7. H. Wang, "Reduction Kinetics of Iron Ore Concentrate Particles Relevant to a Novel Green Ironmaking Process" (Ph.D. dissertation, University of Utah, 2011).

8. M.E. Choi, H.Y. Sohn, Y.M.Z. Ahmed, F.M. Mohamed, G. Han, and M.E.H. Shalabi, "Effect of $CaSO_4$ Pelletization Conditions on a Novel Process for Converting SO_2 to Elemental Sulfur by Reaction Cycles involving $CaSO_4$/CaS: Part I. $CaSO_4$ Pellet Strength and Reducibility by Hydrogen," *Chem. Eng. Technol.*, 30 (5) (2007), 628-634.

9. W.E. Boggs, R.H. Kachik and G.E. Pellissier, "The Effect of Oxygen Pressure on the Oxidation of Zone-Refined Iron," *J. Electrochem. Soc.*, 112 (6) (1965), 539-546.

10. W.A. Johnson and R.F. Mehl, "Reaction Kinetics in Processes of Nucleation and Growth," *TMS-AIME*, 135 (1939), 416-458.

11. P.W.M. Jacobs and F.C. Tompkins, "Classification and Theory of Solid Reactions," *Chemistry of the Solid State*, W. E. Garner, Ed., (London: Butterworths, 1955), 184-212.

12. B.-S. Kim and H.Y. Sohn, "A Novel Cyclic Process Using $CaSO_4$/CaS Pellets for Converting Sulfur Dioxide to Elemental Sulfur without Generating Secondary Pollutants: Part II. Hydrogen Reduction of Calcium Sulfate Pellets to Calcium Sulfide," *Metall. Mater. Trans. B*, 33 (2002), 717-721.

Celebrating the Megascale: Proceedings of the Extraction and Processing Division Symposium on Pyrometallurgy in Honor of David G.C. Robertson
Edited by: Phillip J. Mackey, Eric J. Grimsey, Rodney T. Jones, and Geoffrey A. Brooks
TMS (The Minerals, Metals & Materials Society), 2014

A NEW APPROACH TO INVESTIGATING COKE REACTIVITY

Brian J Monaghan[1], Raymond J Longbottom[1], Mark Reid[1], Oluwatosin A Aladejebi[1], Apsara S Jayasekara[1] and Marc in het Panhuis[2]

[1]Engineering Materials Institute and School of Mechanical Materials and Mechatronics, University of Wollongong, Northfield Ave, Wollongong, NSW 2522, Australia
[2]Soft Materials Group, School of Chemistry and Intelligent Polymer Research Institute, ARC Centre of Excellence for Electromaterials Science, AIIM Facility, University of Wollongong, Northfield Ave, Wollongong, NSW 2522, Australia
E Mail: monagham@uow.edu.au

Keywords: Coke, Coke Reactivity, Coke minerals, Raman.

Abstract

The coke analogue is a new research tool that has been developed to quantify and characterize the effect minerals in coke have on coke reactivity. The analogue material is made from a number of different carbon or carbon containing materials and minerals to in part replicate real industrial coke's reactivity behavior. Elucidation of the effects of minerals on coke reactivity has proved difficult due to the complex and heterogeneous nature of industrial coke and the potential non-additive effects of minerals on coke reactivity. The coke analogue addresses this complexity and heterogeneity through control of its mineralogy, mineral size and mineral dispersion through the analogue, porosity and carbon structures. Initial studies using the analogue were focused on coke dissolution kinetics in liquid iron but recent work has focused in characterization of its porosity, carbon structure and rate of reaction.

Introduction

Coke is a key reagent in the iron blast furnace process. It is the furnace fuel for the high temperature process. In reaction with oxygen from the hot blast, it is the source of the CO reductant for the iron oxide bearing materials. Further, given that it is the only solid charge material in the lower zone of a blast furnace, it is a critical material in maintaining the furnace permeability required for the high production rates needed for profitable operation in a modern blast furnace [1]. The principal measure of coke quality in an iron blast furnace is its resistance to degradation [2]. This is typically measured using the coke reactivity CRI-CSR test [3,4], where CRI and CSR denote Coke Reactivity Index and Coke Strength after Reaction respectively. Coal characteristics such as coal rank, ash %, various components in the ash and maceral reflectograms can account for ~70% of the variation seen in the CRI-CSR tests of Australian coals [2]. It is difficult to meaningfully characterize what accounts for the other 30%, but it is known that mineral type and ash chemistry in general are implicated [2].

The difficulty in addressing this last 30% is related to the nature of the coke material itself. Coke is a complex heterogeneous material comprised of different forms of carbonaceous material, mineral components and a pore structure dependent on the volatile matter in the source coal (and coking conditions). When exposed to high temperatures and reactive atmospheres, the aforementioned compositional and structural features, inherent in a given coke, render isolating the specific component effects on the behavior of a coke difficult.

There have been a number of approaches to addressing the mineral effects on coke reactivity; these have involved both experimental and statistical studies. The general approach used in the statistical studies was to correlate coke mineralogy with coke reactivity using large industrial data sets [5,6,7,8]. While these studies did allow some degree of ranking of the relative effect of specific minerals, they produced data of limited accuracy and usefulness when predicting a coke's reactivity. Experimental measurement of the reactivity of cokes made from coals doped with specific minerals prior to coking [9,10] found that the effect of mineral cations present in coke could be ranked as follows:

$K_2CO_3 > Na_2CO_3 > CaCO_3 > MgCO_3 = MgO > FeCO_3 > FeS_2 > Al_2O_3 = SiO_2$ (little or no change).

While this work represented a significant improvement in the understanding of key mineral effects on reactivity there were some limitations with the approach, as the ranking was only semi quantitative, not readily translatable to mineral combinations (more than one mineral effect) and did not account for the porosity effects on reactivity or the different carbonaceous types in the coke.

A new approach, originally developed for evaluating the effects of minerals on the dissolution behavior of coke in iron [11,12], is the coke analogue. The analogue is a laboratory-grade material made from graphite, bakelite, novalac and minerals. In the analogue the ash mineralogy, mineral dispersion (distribution within the analogue), particle size and porosity can be controlled to isolate specific effects of the mineralogy on reactivity.

In the dissolution studies [11,12] it was shown that even though the analogue was made from a mixture that could contain up to around 50% graphite, its dissolution properties more closely represented that of coke rather than graphite.

In reactivity studies the analogue was reacted with CO_2 in a small scale experiment designed to replicate the CRI test [13,14]. In these studies the effects of single minerals and of binary combinations of selected minerals were assessed in terms of their effect on the coke analogue's reactivity. The results of this approach for single mineral additions are given in Figure 1 [14]. This figure shows a plot of the effect of added single minerals to the analogue on the reactivity, measured as a fractional weight change (FWC) of the analogue with time. The mineral concentration in the analogues was 0.1 mol. of cations per 100 g of carbonaceous material after firing. This approach allowed clear delineation of the relative effects of the single minerals on the reactivity of the analogue. For binary systems, mixtures of quartz, soda feldspar or troilite with magnetite showed complex reactivity behavior, which was consistent with what might be expected from industrial coke [14].

While it is clear from the reported work on the coke analogue that it can test effects of specific minerals (and perhaps combination of minerals), there are still many issues associated with the analogue that need addressing.

- Does the analogue have similar rate-controlling kinetics to that of industrial coke?
- Given significant quantities of graphite are used in the production of the analogue, why does its reactivity more resemble that of coke than graphite?
- Is the porosity of the analogue controllable, is it of a similar order to coke and is it less variable than coke?

Studies are underway to address these issues. What is presented in this paper represents some preliminary results/findings from these studies.

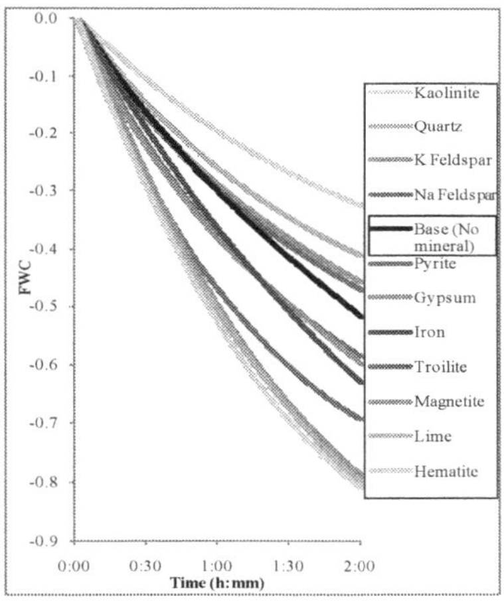

Figure 1. Changes in FWC with time for single mineral-containing coke analogues in CO_2 gas at 1100°C [14]. Note the legend lists the minerals in their order of reactivity from lowest (top) to highest (bottom). The mineral concentration in the analogues was 0.1 mol. of cations per 100 g of carbonaceous material after firing.

Experimental

The general approach was to produce coke analogues doped with selected minerals and assess the effect of the addition of these minerals had on the analogue porosity and carbon structure. Coke analogues were prepared with no minerals, quartz and magnetite minerals and characterised by using optical and scanning electron microscopy (SEM), energy dispersive spectroscopy (EDS) and Raman spectroscopy. The optical microscopy results were used for general observation of the analogue micro structure and in combination with image analysis to measure porosity. SEM/EDS analysis was used for general observation of the analogue microstructure, and for establishing mineral phase fractions and any changes in mineral compositions as a result of firing. Raman spectroscopy was used to analyze the carbon structure of the analogue. Porosity and Raman measurements were also made on a metallurgical coke to facilitate comparison with the analogue. A Jobin Yvon Horiba HR800 Raman spectrometer and an SEM Jeol JSM-6490LA were used. The Raman spectra were measured between 500 and 3500 cm^{-1} using a 638.5 nm laser source with a 950-line diffraction grating.

Making the Coke Analogue

The coke analogue is an extruded cylinder manufactured from mixtures of graphite and phenolic resin that is then fired in an argon atmosphere at 1200°C. The stepwise firing procedure involved curing at 170°C for 24 hours, heating to 950°C and holding for one hour then cooling to 50°C at 5°/min. It is then heated again to 1200°C and held for one hour and then cooled to 50°C at 5°/min. The proportions of the different components used in the preparation of the coke analogue samples are given in Table I. The preparation procedure results in an analogue of cylindrical shape ~8 g in mass 18 mm in diameter and 30 mm in height. After firing, the analogue was sectioned, mounted in resin and prepared for optical and electro optical analysis and Raman spectroscopy. All samples were polished to a 0.25 μm finish and SEM samples were carbon coated.

Table I. Components used to produce coke analogue

Material	Purity	Size	Addition	Primary Function(s)
Graphite Powder 1	99.99+%	< 45μm	28% of dry base	Carbon source
Graphite Powder 2	99.99+%	< 150μm	28% of dry base	Carbon source
Phenolic resin (Bakelite)	N/A	+100μm -250μm	44% of dry base	Porosity control
Liquid 45% Novolac Resin in propylene glycol	N/A		Ratio of 0.52 liquid / dry mix	Binder Porosity control
hexamethylenetetramine (HTMA)	N/A		3mass% of liquid resin	Curing agent

Minerals

Minerals were either purchased from laboratory suppliers or produced in-house using methodologies developed in previous research [15]. In general, these minerals were crushed and particles of size fractions in the range 125–150 μm were used to dope the analogue. The phases of the minerals were confirmed using X-ray diffraction. The diffractometer was a Phillips PW1730 with molybdenum radiation and a zirconium filter.

Porosity Measurement

The analogue porosity was measured using an image analysis type method developed in-house. Using a Leica DM/RM optical microscope images were obtained at 50x magnification. Using ImageJ analysis software version 1.46r, 12 fields were taken giving a scale of one pixel to 1.0157 microns. These were stitched together and cropped to give an area of ~4 mm × 4 mm of the analogue. Images for porosity measurement were prepared by imaging the surface of the sample with high intensity green filtered light and subsequently using the LASv4 software to select only the green channel of the red, green and blue input from the digital camera. In essence this method only allows light from the highly reflective polished coke analogue material, as opposed to the low reflectivity mounting material and pores, to form an image. Figure 2 gives an example of an image prepared for porosity measurement. The ImageJ software was used to exclude all porosity with an equivalent diameter less than 10 μm (considered unreliable) and greater than an equivalent diameter of 500 μm (considered as fissures rather than porosity).

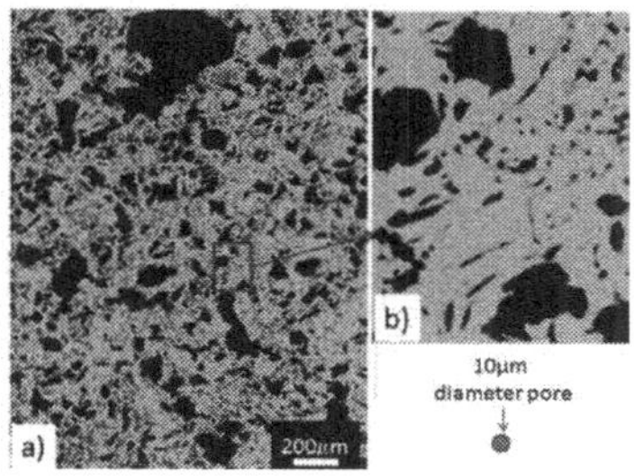

Figure 2 a) An example of an optical image configured for porosity measurement and b) an increased magnification of the square indicated in a). Included is a scaling element of what would be an equivalent 10 μm diameter pore in b).

Extensive porosity measurements were made on coke analogue discs obtained from positions along the length of the coke analogue. The surfaces of these discs were then analyzed for porosity. A similar approach was attempted for porosity measurement on industrial coke, though in this case the position of top and bottom was arbitrary.

Results/Discussion

Porosity The analogue had an average porosity of 29% with a standard deviation of +/- 2.3% (absolute) throughout the analogue. These values account for effects of position along the length of the sample, across the diameter of the sections and batch variation. On comparable measurements on an industrial coke a value of 24% with a standard deviation +/-4.3% (absolute) was obtained. This standard deviation from this value is likely an underestimate, as it does not adequately represent the inherent variation in a single batch of coke where it is known that there are significant porosity changes with respect to coke position in the coke oven. A comparison of pore distribution of the coke analogue with that of industrial cokes is given in Figure 3.

Cokes 1 to 6 are from the literature [16] and cokes 7 to 9 were measured using a technique similar to that for the coke analogue. To facilitate comparison with published literature values size fractions > 500 μm were also considered. From this figure it can be seen that the majority of the porosity of the coke analogue is biased towards values of <200 μm and occurs over a narrow range. The <200 μm fractions of the analogue are similar to that of industrial coke. In the industrial coke there is a greater variation in the porosity of the ranges quoted [16] and measured in this study. Larger pores are also more numerous.

The ability to control the porosity and porosity distribution will be critical in any heterogeneous gas-solid kinetic study, as it is a factor in defining contact area. In the work so far on using the coke analogue it is possible to produce analogues with reproducible, representative porosities and porosities over a small (relative to industrial coke) size range. What has not yet been demonstrated is how readily the porosity can be varied.

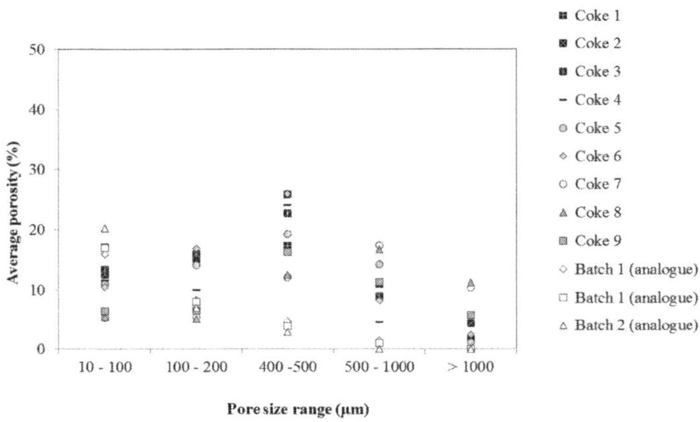

Figure 3. A comparison of pore distribution of the coke analogue with that of industrial cokes. Cokes 1 to 6 are from the literature [16] and cokes 7 to 9 were measured using a technique similar to that for the coke analogue.

<u>Carbon Structure.</u> A Raman and SEM image of the same area of a coke analogue containing no mineral matter is given in Figures 4 a) and b) respectively. In Figure 4 a) areas of the coke analogue identified using Raman spectroscopy to have sp^2 (typically a light grey/white phase) and sp^3 carbon bonding are indicated. sp^2 bonding is typical of graphite and sp^3 indicates other carbon structures. The sp^2-sp^3 bonding in the analogue is important as it shows that the analogue material has a mixed carbon structure, similar to industrial coke [17]and not simply graphitic. In Figure 4 b) the SEM image, there are a number of white areas shown. Using EDS analysis these have been identified as oxides. The oxide matter shown in Figure 4 b) represents a problem for the use of the coke analogue material. The source of the oxide is the bakelite used in the analogue preparation. It represents an oxide addition of ~1 mass % to the analogue. At present no suitable substitute for bakelite has been identified and this oxide fraction represents a maximum level of impurity oxide that will have to be tolerated in any assessment of specific mineral doping of the analogue.

Preliminary measurements of the effect of mineral additions on carbon structure of the analogue as assessed by the ratios of intensities of their Raman D and G bands (I(D)/I(G)) as a function of the ratios of the minimum between the bands (I(V)/I(G)) are given in Figure 5. This figure is for an analogue with no minerals and for analogues with magnetite and quartz added. The G band is representative of graphite (sp^2 bonding) and the D band is representative of a defected graphitic structure (sp^3 bonding). Therefore low I(D)/I(G) values indicate graphite, and high values indicate a more disordered structure.

While these are only preliminary results it appears that the addition of the minerals changes the carbon structure of the analogue and that the size of the effect differs for quartz and magnetite.

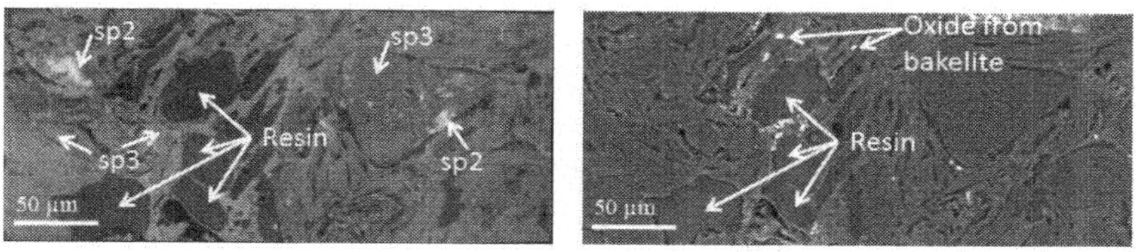

a) Raman image b) Backscattered SEM

Figure 4. a) A Raman image of a typical coke analogue containing no mineral addition and showing the type of carbon bonding, and b) an SEM image of the same material and same area showing an oxide phase is present.

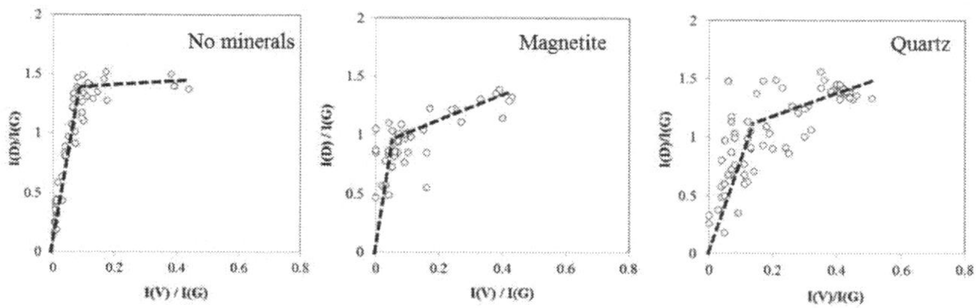

Figure 5. Raman I(D)/I(G) ratios vs I(V)/I(G) for coke analogues containing no mineral, magnetite mineral and quartz mineral additions. Dotted lines are added to guide the reader's eye.

More work is required to confirm and explain these results. The minerals were added to the pre-fired analogue mixture. Their structure has not been confirmed post-firing. It is known from previous work that the magnetite reduces to metallic iron [14] during firing; therefore care will be needed when expressing the mineral component and component effects of the oxide. Depending on what is being studied it may be instructive to quote the mineral structure both pre- and post-fining.

Conclusions

A material called coke analogue has been developed and characterised that should allow for a quantitative evaluation of effects of minerals on coke reactivity. Results have been presented that show the porosity levels in the analogue are similar to that of industrial coke and that the variability of the porosity in the analogue is much less than that of coke. Further Raman and microscopy characterization of the analogue carbon structure indicates that there is both graphitic and non-graphitic carbon in the analogue and that there appears to be an effect of mineral type in the analogue on the resulting carbon structure. The porosity and carbon structure results are encouraging, as they indicate that the analogue is a suitable material for evaluating the effects of mineralogy on both carbon structure of the analogue and its reactivity.

References

---

1. A..K. Biswas: *Principles of Blast Furnace Ironmaking*, (Brisbane, Australia: Cootha Publishing House, 1981).
2. P. Bennet, A Reifenstein, G. O'Brien and B. Jenkins, "Coke Reactivity and Characterisation" (Report C12057, ACARP, 2008).
3. ASTM D5341 - 99(2010) Standard Test Method for Measuring Coke Reactivity Index (CRI) and Coke Strength After Reaction (CSR).
4. ISO 18894:2006 Coke - Determination of coke reactivity index (CRI) and coke strength after reaction (CSR).
5. D. French, R. Sakurovs and M. Grigore, "Mineral Matter Reactions in Cokes", (Report C14074, ACARP, 2009).
6. M. Grigore, R. Sakurovs, D. French and V, Sahajwalla, "Influence of Mineral Matter on Coke Reactivity with Carbon Dioxide", *ISIJ International*, 46 (2006), 503-512.
7. B-C. Kim, S. Gupta, D. French, R. Sakurovs, and V. Sahajwalla, "Effect of Thermal Treatment on Coke Reactivity and Catalytic Iron Mineralogy", *Energy & Fuels*, 23 (2009), 3694-3702.
8. R. Sakurovs, D. French and M. Grigore, "Quantification of Mineral Matter in Commercial Cokes and their Parent Coals", *International Journal of Coal Geology*, 7 (2007), 81-88.
9. W.W. Gill, N.A. Brown, C.D.A. Coin, and M.R. Mahoney, "Influence of Ash on the Weakening of Coke", (Paper presented at the 44th ISS-AIME Ironmaking Conference, 1985), pp. 233-238.
10. W.H. Van Niekerk and R.J. Dippenaar, "The Influence of Potassium on the Reactivity and Strength of Coke, with Special Reference to the Role of Coke Ash", *J. S. Afr. Inst. Min. Metall.*, 86 (1986), 25-29.
11. B.J. Monaghan, S.A. Nightingale and M. Chapman, "Carbon Transfer in the Lower Zone of a Blast Furnace", *Steel Research International*, 81 (2010), 829-833.
12. R.J. Longbottom, B.J Monaghan, M.W. Chapman, S.A. Nightingale, J.G. Mathieson, and R.J. Nightingale, "Techniques in the Study of Carbon Transfer in Ironmaking", *Steel Research International*, 82 (2011), 505-511.
13. R. Longbottom, B.J. Monaghan, O. Scholes, and M.R. Mahoney, "Development of a Metallurgical Coke Analogue to Investigate the Effects of Coke Mineralogy on Coke Reactivity",(Paper presented at Scanmet IV, 4th International Conference on Process Development in Iron and Steelmaking, Luleå, Sweden, 2012), 147-156.
14. M.H. Reid, M.R. Mahoney and B J. Monaghan, A Coke Analogue for the Study of the Effects of Minerals on Coke Reactivity, *ISIJ International*, in print.
15. M.W. Chapman, "Insoluble Oxide Product Formation and its Effect on Coke Dissolution in Liquid Iron", (Ph.D. Thesis, University of Wollongong, 2009), 63-99.

16. P. Loison, P. Foch, and A. Boyer, *Coke Quality and Production*. (London, UK: Butterworth & Co Press. 1989), 129.
17. M. Kawakami, H. Kanba, K. Sato, T. Takenda, S. Gupta, R. Chandratilleke, and V. Sahajwalla, "Characterisation of Thermal Annealing Effects on the Evolution of Coke Structure Using Raman Spectroscopy and X-Ray Diffraction," *ISIJ International*, 46, (2006), 1167.

Celebrating the Megascale: Proceedings of the Extraction and Processing Division Symposium on Pyrometallurgy in Honor of David G.C. Robertson
Edited by: Phillip J. Mackey, Eric J. Grimsey, Rodney T. Jones, and Geoffrey A. Brooks
TMS (The Minerals, Metals & Materials Society), 2014

THE USE OF NATURAL GAS FOR REDUCTION OF METAL OXIDES: CONSTRAINTS AND PROSPECTS

Oleg Ostrovski[1]

[1]The University of New South Wales, Sydney 2052, Australia

Keywords: Metal oxides, Reduction, Natural gas

Abstract

Industrial pyrometallurgical processes in ferrous metallurgy are based on carbothermal reduction of metal oxides. Carbothermal reduction of stable oxides requires high temperatures. Low-temperature reduction can be implemented by using methane-containing gas with high carbon activity, or by carbothermal reduction under reduced CO partial pressure. Under standard conditions, methane is thermodynamically unstable above 550 °C and decomposes to solid carbon and hydrogen. At appropriate CH_4/H_2 ratio and temperature, carbon activity in the methane-containing gas phase can be well above unity relative to graphite, which provides favorable thermodynamic conditions for reduction. To maintain these conditions, the rate of reduction/carburisation should be higher than the rate of solid carbon deposition. The paper discusses reduction of pure manganese and chromium oxides at relatively low temperatures, and constraints in reduction of manganese and chromium ores. Reduction of metal oxides by carbon in hydrogen as an alternative use of natural gas is also discussed.

Introduction

Carbothermal reduction of stable metal oxides such as MnO, Cr_2O_3, SiO_2, TiO_2 and Al_2O_3 requires high temperatures. Low temperature reduction can be implemented by using methane-containing gas with high carbon activity, or by carbothermal reduction under reduced CO partial pressure. Advantages of gaseous reduction over carbothermal reduction are: a) lower operating temperatures, b) ability to process fines, c) faster kinetics, and d) elimination of coke.

Under standard conditions, methane is thermodynamically unstable at temperatures above 550°C and cracks (decomposes to solid carbon and hydrogen) in accordance with reaction (1).

$$CH_4 \rightarrow C_{(s)} + 2H_2 \quad (1)$$

At appropriate CH_4/H_2 ratio and temperature, carbon activity in the methane-containing gas phase can be well above unity relative to graphite, which provides favorable thermodynamic conditions for reduction to occur at relatively low temperatures.

Potential and constraints in reduction of metal oxides by the methane-containing gas can be illustrated considering reduction of iron ore. Reduction of hematite by methane-containing gas can be presented by the following overall reaction:

$$3Fe_2O_3 + 11CH_4 = 2Fe_3C + 9CO + 22H_2 \quad (2)$$

Cementite is unstable and decomposes to iron and carbon:

$$Fe_3C \rightarrow 3Fe + C_{gr} \tag{3}$$

Equilibrium carbon activity for reaction (3) as a function of temperature is shown in Figure 1. Thermodynamic instability of cementite increases with decreasing temperature. Figure 1 also shows activity of carbon in the CH_4-H_2-Ar gas mixture (35 vol% CH_4 and 55 vol% H_2). It can be concluded from Figure 1 that formation of cementite by reduction of iron oxides using CH_4-H_2-Ar gas mixture (35 vol% CH_4 and 55 vol% H_2) occurs at T>571 °C.

The reduction process starts with adsorption of methane on the active sites of the oxide surface and its decomposition. The overall reaction of methane adsorption and cracking can be presented as

$$CH_4 = \ldots = C_{ad} + 2H_2 \tag{4}$$

where C_{ad} represents active carbon species adsorbed on solid surface, which is substantially different from deposited solid carbon. This active adsorbed carbon provides higher reducibility and reduction rate compared with those of carbothermal reduction. Adsorbed carbon is consumed by reduction/carburization reactions.

The key factor is a high rate of the reduction reaction compared with the rate of carbon deposition. If solid carbon is formed on the sample surface, it blocks the access of reducing gas to metal oxides, and the carbon activity at the oxide surface is decreased to unity, regardless of the high $P_{CH4}/(P_{H2})^2$ ratio in the inlet gas.

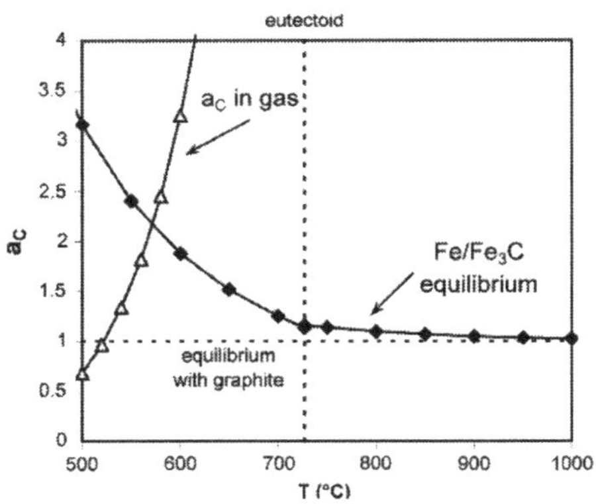

Figure 1. Equilibrium carbon activity in cementite as a function of temperature (from [1]).

Therefore, the following fundamental condition for reduction of metal oxides by methane containing gas can be formulated: the rate of reduction/carburisation should be higher than the rate of solid carbon deposition. Reduction and carbon deposition rates are affected by temperature and gas composition (methane to hydrogen ratio), oxide surface area, which depends on particle size, porosity and pore size. Reaction of solid carbon deposition is strongly affected by catalysts.

Reduction of manganese oxides and ores by methane-containing gas

Manganese oxide MnO is reduced to manganese carbide by methane by the following reaction:

$$MnO + 10/7CH_4 = 1/7Mn_7C_3 + CO + 20/7H_2;\ \Delta G° = 377\ 682 - 314.44T\ \text{(J/mol)} \quad (5)$$

At standard conditions, reaction (5) proceeds spontaneously at temperatures above 928 °C, while carbothermal reduction of MnO starts at 1340 °C (standard conditions).

Results of a temperature programmed reduction of MnO_2 by the CH_4 - H_2- Ar gas mixture (10 vol % CH_4, 20 vol % H_2, 70 vol % Ar) are shown in Figure 2 [2]. Reduction started with conversion of MnO_2 to lower manganese oxides Mn_2O_3, Mn_3O_4 and MnO by hydrogen (only H_2O was evolved in the reduction process). This reaction was completed at 610-620 °C. Formation of manganese carbide by reduction of MnO started at 760 °C and was completed at about 1200 °C.

Anacleto and Ostrovski studied reduction of Wessels (South Africa) and Groote Eylandt (Australia) manganese ores using methane-containing gas [3]. Manganese ores had significant differences in chemical composition: Wessels ore had high iron oxide and calcia content, lower concentration of silica and only traces of potassium; Groote Eylandt (GE) ore contained high potassium concentration.

The reaction rate of Wessels ore increased with temperature. At 1200 °C, the extent of reduction was close to 100% after one hour reduction. The rate of $(Mn,Fe)_7C_3$ formation increased rapidly with increasing temperature to 1150 °C. Further increase in temperature to 1200 °C affected reduction rate quite marginally. Excessive methane cracking at higher temperatures hindered reduction due to carbon deposition.

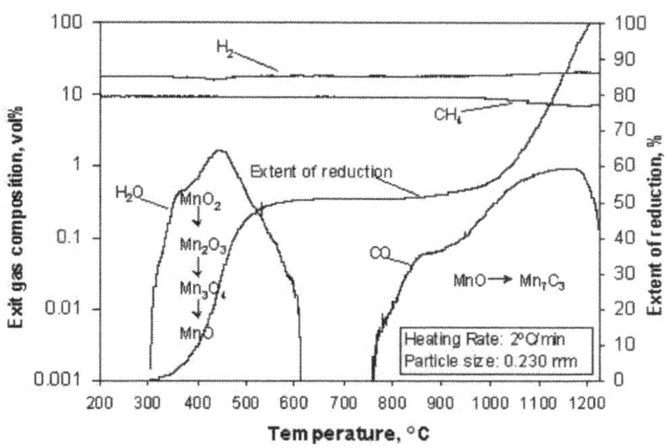

Figure 2. Temperature-programmed reduction of MnO_2 by methane–hydrogen mixture with 10 vol % CH_4, 20 vol % H_2, 70 vol % Ar (from [2]).

Solid carbon deposition had a particularly strong effect on the reduction of the GE ore. Reduction curves obtained in reduction of the GE ore at different temperatures are shown in Figure 3. The reduction rate of this ore increased with increasing temperature from 1000 to 1050 °C. It was practically independent of temperature in the temperature range 1050-1100 °C, and decreased with further increase in temperature. Above 1050 °C there was a rapid increase in pores size and a significant decrease in total pores surface area with increasing temperature. Examination of a variation of average pore diameter and total pores area with reduction temperature showed that fine pores were eliminated with increasing reduction temperature.

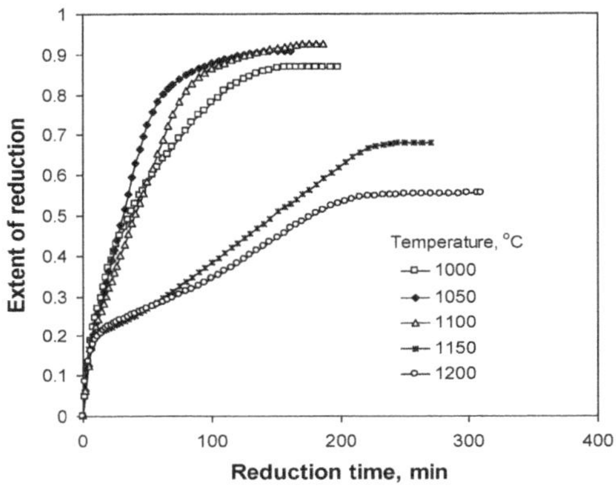

Figure 3. Reduction of Groote Eylandt ore by the methane-hydrogen-argon gas mixture at different temperatures (from [3]).

Relatively low softening temperature of GE ore and predisposition to sintering are related to higher silica content in comparison with the Wessels ore and formation of low melting silicates. An addition of 10-15 wt% CaO to the Groote Eylandt ore significantly increased the rate and extent of reduction at elevated temperatures.

Reduction of chromium oxides and ores by methane-containing gas

Reduction of chromium oxide by methane follows the reaction:

$$Cr_2O_3(s)+13/3CH_4(g)=2/3Cr_3C_2(s)+3CO(g)+26/3H_2(g);\ \Delta G^{o}=1{,}097{,}519-989.8T\ (J) \qquad (6)$$

Reaction (6) proceeds spontaneously at temperatures above 841 °C, which is much lower compared to the carbothermal reduction of chromium oxides (1113 °C).

CO evolution and the extent of Cr_2O_3 reduction in the temperature-programmed experiments are shown in Figure 4 (from [4]). The reduction of chromium oxide to chromium carbide, Cr_3C_2, started at approximately 770 °C. Cr_2O_3 was completely reduced to Cr_3C_2 at 910 °C. This was confirmed by oxygen analysis (LECO) and XRD analysis.

Reduction of chromium oxide in isothermal experiments was completed in about 100 minutes at 900 °C, and 22 minutes at 1200 °C. The increase of rate of reduction with temperature was found to be most significant between 900 °C and 1000 °C. When the temperature was higher than 1000 °C, this increase slowed down with the increase of temperature. At high temperatures, the increase in temperature has a double effect: (1) increase in the reduction rate constant and (2) decrease in the reaction gas-solid interface due to sintering. As a result, the increase in the reduction rate with increasing temperature in the range 1000 °C to 1200 °C was marginal.

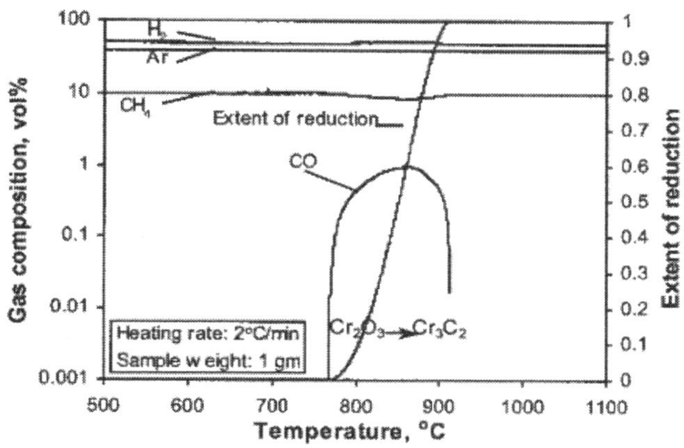

Figure 4. Temperature-programmed reduction of Cr_2O_3 by 10 vol % CH_4–50 vol % H_2–40 vol % Ar mixture (from [4])

However, a study of reduction of chromium ores by methane containing gas in [5] gave negative results. N. Anacleto [5] examined reduction of Coobina (Australia) and Chromtau (Kazakhstan) ores. The original ores consisted of chromite spinel and cementing magnesium silicate gangue. The chromite spinel is a solid solution formed by $FeO.Cr_2O_3$ (chromite), $MgO.Cr_2O_3$ (magnesiochromite), $MgO.Al_2O_3$ (spinel) and $FeO.Fe_2O_3$ (magnetite).

In the temperature-programmed experiments, iron oxides were reduced to metallic iron; carbide phase started to form at 1000 °C in the iron-rich area. The final extent of Coobina ore reduction in the temperature-programmed experiment (up to 1200 °C) reached 47%, while the extent of Chromtau reduction was below 25%. Low extent of reduction was also observed in the isothermal experiments at temperatures 1000-1200 °C.

Chromtau ore contained much less chromite and significantly more magnesiochromite than Coobina ore. Both ores were composed of large dense grains and had low surface area. It can be concluded that thermodynamic and kinetic conditions for reduction of chromium oxides by methane-containing gas are not appropriate to have acceptable degree and rate of reduction.

Carbothermal reduction of metal oxides and ores in hydrogen

Cracking of methane with carbon deposition poses a serious obstacle in the use of natural gas in reduction reactions. Alternative use of the natural gas may include its reforming with formation of carbon black and hydrogen, which can be employed for reduction reactions. It was demonstrated that carbothermal reduction of metal oxides in hydrogen atmosphere has significant advantages over conventional carbothermal reduction [6-9].

It is well recognised that carbothermal reduction of metal oxides in the solid state goes through the gas phase. Reduction of an oxide MO to a carbide MC can be presented by reactions (7) and (8):

$$MO + 3CO = MC + 2CO_2 \quad (7)$$

$$CO_2 + C = 2CO \quad (8)$$

In these reactions, carbon and oxygen are transferred between solid phases by CO and CO_2, respectively. Gas species are directly involved in the carbothermal reactions; the composition of the gas phase affects the reaction rate.

Reduction of manganese oxides and ores

Carbothermal reduction of MnO in H_2, He and Ar at 1300 °C is shown in Figure 5 [6]. Reduction in argon and helium at 1275 °C and below was slow and not completed in 120 min. Reduction in hydrogen was close to completion even at 1200 °C. Reduction in helium was faster than in argon at all temperatures. The effect of gas atmosphere on the MnO reduction depended on the temperature. The difference in the reduction behavior in different gases decreased with increasing temperature. This effect was particularly strong on reduction in helium and hydrogen; reduction curves obtained in hydrogen and helium at 1400°C were almost identical; reduction in argon was slower.

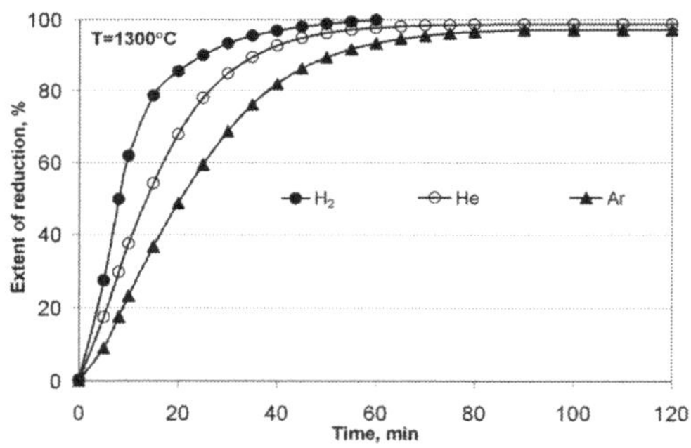

Figure 5. Carbothermal reduction of MnO at 1300°C in argon, helium and hydrogen (from [6])

Carbothermal reduction of Wessels and Groote Eylandt ores was also faster in hydrogen than in inert gas [7]. Extent of reduction of Wessels ore at 1100°C reached 55% in argon and 60% in helium in 160-min, while reduction in hydrogen achieved about 90% for 1 hour. Increasing the temperature from 1100 °C to 1200 °C significantly accelerated the reduction process. Reduction in hydrogen at 1200 °C was above 90% for 20 min and close to completion when reduction time approached 1 hour. At 1200 °C, the degree of reduction of Wessels ore was also significant in helium and argon; it was above 80% in helium and 70% in argon within 1 hour.

Reduction of rutile and ilmenites

The rate of carbothermal reduction of rutile in hydrogen was much higher than in inert gases; the rates of reduction in helium and argon were about the same.

Evolution of CO in the temperature programmed reduction of rutile in a temperature range 350 to 1600 °C, with a heating rate of 3°C/min in different gas atmospheres is shown in Figure 6 [8]. Generally, reduction of titania in hydrogen started at a lower temperature and the reduction rate reached a higher peak at a lower temperature than in argon and helium.

Carbothermal reduction of ilmenites was also faster in hydrogen than in inert atmosphere [9].

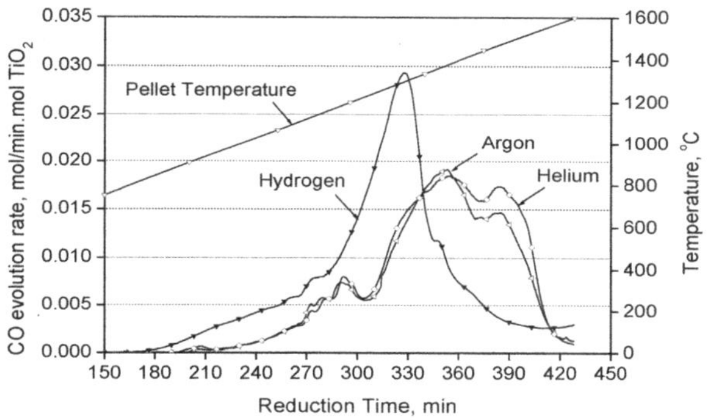

Figure 6. Temperature programmed reduction of rutile in different gas atmospheres. The furnace temperature was ramped from 350 to 1620°C at 3°C /min (from [8]).

Reduction mechanisms

Carbothermal reduction of metal oxides in solid state proceeds via the gas phase, as shown by reactions (7) and (8). The rate of the overall reduction of stable metal oxides MnO, Cr_2O_3 or TiO_2 is suggested to be limited by the interfacial Boudouard reaction, or by mass transfer of CO_2 within the porous bed of the solid reactant mixture from oxide to carbon particles [10].

The strong effect of gas atmosphere on the rate of manganese oxides reduction indicates that external and internal mass transfer of gaseous species involved in the reaction play an important role in the reduction kinetics. The difference between Ar and He is the degree of resistance that they impose onto the diffusivity of other gaseous species. Diffusivity of CO in helium and argon, calculated using the Chapman-Enskog formula at 1400 °C and 1 atmosphere is 16.9 cm^2sec^{-1} and 3.6 cm^2sec^{-1} respectively [6]. The difference is a factor of 4.7. Reduction of titanium oxides in helium and argon had about the same rate; this indicates that external and internal mass-transfer was not significant in the rate control in reduction of rutile or ilmenite.

Diffusivity of CO in hydrogen and helium is about the same, therefore internal and external mass transfer is not a factor distinguishing carbothermal reduction of manganese and titanium oxides in hydrogen and helium. However, hydrogen is involved in the reduction process. Carbothermal reduction of metal oxide in hydrogen atmosphere includes formation of methane:

$$C + 2H_2 = CH_4 \quad (9)$$

Although partial pressure of methane in the gas phase is low, activity of carbon in the gas phase in equilibrium with graphite is 1.0. Methane may act as a means for carbon transportation, which accelerates the overall reduction process. In the reduction of manganese and titanium oxides, partial pressure of methane is significantly higher than partial pressure of CO_2. Therefore, mass transfer of methane from carbon to oxide is expected to be much faster than mass transfer of CO_2 from oxide to graphite. Further reaction of intermediate methane is also expected to be faster than that of CO_2 as the reaction rate depends on the partial pressure of a reactant.

Conclusions

In the reduction of metal oxides by methane-containing gas, the high carbon activity of the reducing gas is the key to achieve high rates and extents of reduction of metal oxides by methane-containing gas. However, it also causes deposition of solid carbon, which has a strong retarding effect on the reduction processes. The rate of reduction/carburisation should be higher than the rate of solid carbon deposition. This condition is held in reduction of pure Mn, Cr and Ti oxides; however, in reduction of ores, the rate of carbon black formation is higher than the reduction rate, which is an inherent limitation in industrial implementation of reduction processes using natural gas.

Alternative approach can be based on the use of natural gas as a source of carbon black and hydrogen for carbothermal reduction of metal oxides in hydrogen. Carbothermal reduction in hydrogen can be implemented at much lower temperatures in comparison with reduction in CO or inert atmosphere. Low-temperature reduction of metal oxides occurs in the solid state with potential advantages over conventional reduction of oxides from molten slags.

References

1. R. Longbottom, O. Ostrovski, J. Zhang, and D. Young, "The Stability of Cementite formed from Hematite and Titanomagnetite Ore," *Metall. Mater. Trans. B*, 38 (2007), 175-184.
2. N. Anacleto, O. Ostrovski, and S. Ganguly, "Solid State Reduction of Manganese Oxides by Methane-Containing Gas," *ISIJ International*, 44 (2004), 1480-1487.
3. N. Anacleto, O. Ostrovski, and S. Ganguly, "Reduction of Manganese Ores by Methane-Containing Gas," *ISIJ International*, 44 (2004), 1615-1622.
4. N. Anacleto, and O. Ostrovski, "Solid State Reduction of Chromium Oxide by Methane-Containing Slag," *Metall. Mater. Trans. B*, 35B (2004), 609-619.
5. N. Anacleto, "Solid State Reduction of Manganese and Chromium Ores by Methane-containing Gas" (PhD Thesis, UNSW, 2002), 144-174.
6. R. Kononov, O. Ostrovski, and S. Ganguly, "Carbothermal Reduction of Manganese Oxide in Different Gas Atmospheres," *Metall. Mater. Trans. B*, 39B (2008), 662-668.
7. R. Kononov, O. Ostrovski, and S. Ganguly, "Carbothermal Solid State Reduction of Manganese Ores: 2. Non-isothermal and Isothermal Reduction in Different Gas Atmospheres," *ISIJ International*, 49 (2009), 1107-1114.
8. M. A. R. Dewan, G. Zhang, and O. Ostrovski, "Carbothermal Reduction of Titania in Different Gas Atmospheres," *Metall. Mater. Trans. B,* 40B (2009), 62-69.
9. M. A. R. Dewan, G. Zhang, and O. Ostrovski, "Carbothermal Reduction of a Primary Ilmenite Concentrate in Different Gas Atmospheres," *Metall. Mater. Trans. B*, 41B (2010), 182-192.
10. W. J. Rankin and J. R. Wynnycky, "Kinetics of Reduction of MnO in Powder Mixtures with Carbon," *Metall. Mater. Trans. B*, 28B (1997), 307–319.

Celebrating the Megascale: Proceedings of the Extraction and Processing Division Symposium on Pyrometallurgy in Honor of David G.C. Robertson
Edited by: Phillip J. Mackey, Eric J. Grimsey, Rodney T. Jones, and Geoffrey A. Brooks
TMS (The Minerals, Metals & Materials Society), 2014

DECOMPOSITION OF METHANE DURING OXIDE REDUCTION WITH NATURAL GAS

Halvor Dalaker[1] and Pål Tetlie[1]

[1]SINTEF Materials and Chemistry, N-7465, Trondheim, Norway.

Keywords: Metal production, oxide reduction, methane, natural gas, cracking, thermal decomposition.

Abstract

The decomposition of methane has been studied experimentally in contact with various oxides at elevated temperatures. For each oxide, the temperature at which the cracking starts has been determined. No correlation was found to exist between the temperature for the onset of cracking for a given oxide and the affinity for carbon of the oxide's metallic component. In terms of potential for using methane as a reducing agent, calculations are presented that indicate that the suitability of an oxide for such reduction depends on the desired methane+oxide reaction being more energetically favourable than the methane+methane=ethylene/acetylene reactions.

Introduction

Replacing coal and coke with methane in metal production would lead to energetic and environmental benefits. An ambitious method would be to use the methane directly as a reducing agent at temperatures where it is metastable. Ostrovski and Zhang has demonstrated that at such temperatures the carbon activity is much higher than for thermodynamically stable carbon, and that this metastability allows reduction to take place at lower temperatures than predicted by equilibrium calculations [1].

The problem is that the increased carbon activity depends on there being no decomposition of methane resulting in carbon depositions. The interactions between the oxide to be reduced and the methane are therefore critical to how any reactions proceed, and whether carbon deposition will dominate reduction or if the metastability of methane can be beneficially exploited.

As shown in an equilibrium consideration by Ostrovski and Zhang [1], in a gas containing hydrogen and metastable methane, the activity of carbon in methane relative to that of carbon as graphite is given by equation (1):

$$\frac{a_{C^*}}{a_{C_{Gr}}} = \frac{\left(p_{CH_4} / p_{H_2^2} \right)}{\left(p_{CH_4} / p_{H_2^2} \right)_{Gr}} \tag{1}$$

Following the notation of Ostrovski and Zhang, a_{C^*} is the activity of the "active carbon", or the carbon from the metastable methane; $a_{C_{Gr}}$ is the activity of carbon in its thermodynamically stable state, graphite; p_{CH_4} and $p_{H_2^2}$ are the partial pressures of methane and hydrogen respectively, and the subscript "*Gr*" used in the denominator means that the ratio in brackets represent the value of this ratio when methane and hydrogen are in equilibrium with graphite. Using thermodynamic software one can calculate the equilibrium ratio as a function of temperature, in this work, FactSage is used.

For a given gas it is then straightforward to calculate the relative carbon activity as a function of temperature. In the current work, the gas mixture used is 10%CH_4-10%H_2-80%Ar, and Figure 1 shows a plot of the carbon activity in this gas mixture. As is seen, the carbon activity increases dramatically with temperature if cracking can be suppressed.

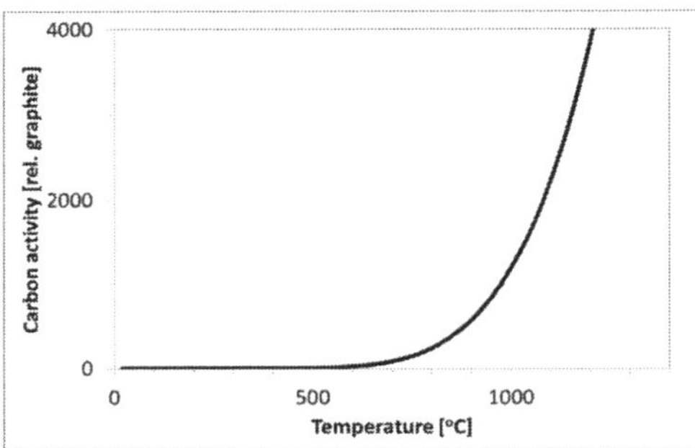

Figure 1. (left) Carbon activity in metastable methane in a 10%CH_2-10%H_2-80%Ar gas mixture, relative to graphite, as a function of temperature.

In the current work, the thermal decomposition of methane is studied experimentally in contact with different metal oxides (Al_2O_3, MgO, SiO_2 and CaO), and the results are discussed in light of thermodynamic calculations with respect to oxide reduction.

Experimental Details

The experiments were performed in a tube furnace with a sealed alumina inner tube. The crucibles were also made from alumina, and the process gas was supplied via a hole in the crucible bottom. The sample charge rested on an alumina disk atop the gas inlet, with slits cut to allow good gas exposure. The experimental set-up was designed by Ksiazek et al. [2]

In each experiment the furnace was flushed with argon at room temperature to drive out oxygen, then heated under a constant stream of 10%CH_4-10%H_2-80%Ar, total gas flow 2.5 l/min. The gas was supplied by an Environics Series 4000 gas mixing unit, and the out gas was monitored by a mass-spectrometer scanning for argon, hydrogen, methane, CO, H_2O and CO_2 over the charge to molecular weight ratio lines of 2, 12, 13, 16, 18, 28, 29, 40. The spectrographic signals were calibrated using known mixtures of H_2, CH_4, Ar, and CO over a range of concentrations, and each experiment was preceded by a single-point calibration with a single known gas mixture (7.5%CH_4-7.5%H_2-3.5%CO-81.5%Ar).

The temperature in the charge was measured by a type C thermocouple in the middle of the charge, protected by an alumina sheath. A charge of only 4 grams was chosen in order to reduce the volume of sample, and thus any temperature gradients within the charge material. The materials used were in pure powder form, pelletized with no other binder than water, which was driven out during drying of the pellets. The pellets were not made to be perfectly spherical or of a given size distribution, the aim was simply to obtain the raw materials in a form that was not ejected from the crucible by the gas flow, as would likely be the case with a powder. The size of the "pellets" was 2-3mm.

The heating rate was not fixed to a specific ramp, but corresponded to approximately 10°C/min. The temperature was steadily increased until the output gas analysis showed significant signs of methane decomposition, at which point the process gas was replaced with 100% Ar and the heating shut down. This temperature was generally around 1100°C.

In order to avoid the presence of any deposited carbon, a new crucible set-up was used in each experiment, and the inner furnace tube was removed from the furnace and treated in air at 1100°C to burn off all carbon as CO/CO_2 before the next experiment.

Results

The out-gas signals for the different experiments are presented in this section on an oxide-by-oxide basis, with a brief discussion of the interpretation of the different responses. In the next section the results are discussed with respect to a comparison between oxides, and with respect to theoretical considerations regarding reduction potential. Although other signals besides H_2/CH_4 were monitored, these bear no relevance to the following discussions and are not included.

<u>Experiment one: empty crucible</u>

The first experiment was done on an empty furnace in order to evaluate the behaviour of methane with no oxide-pellets present. Figure 2 shows the output graph for the hydrogen and methane content in the off-gas. At low temperatures, there is reasonable correlation between the two gases, as is expected since they are present in identical concentrations (10%) in the input gas. At higher temperatures however, the signals deviate as CH_4 decompose into H_2 and C.

The temperature at which cracking of methane sets in should thus be associated with a simultaneous increase in hydrogen, decrease in methane, and loss of correlation between the H and CH_4 signals. Due to the noise in the signals, determining exactly where this point occurs is challenging, and an attempt must be made to interpret where there is a "clear deviation from the trend" on a case by case basis. For the empty crucible, the hydrogen-signal starts to lift predominantly above the methane signal at a temperature in the region of 700°C. However, this is also true in the region around 400°C, so cannot alone be taken as an indication of cracking, and the methane signal is more or less completely flat between 700°C and 900°C. If the slight upwards drift of the hydrogen signal is the trend, then at 950°C this trend breaks, as the increase in the hydrogen signal starts to accelerate. This also coordinates with a start of a drop in the methane signal, and is taken as the temperature for the onset of cracking in this case.

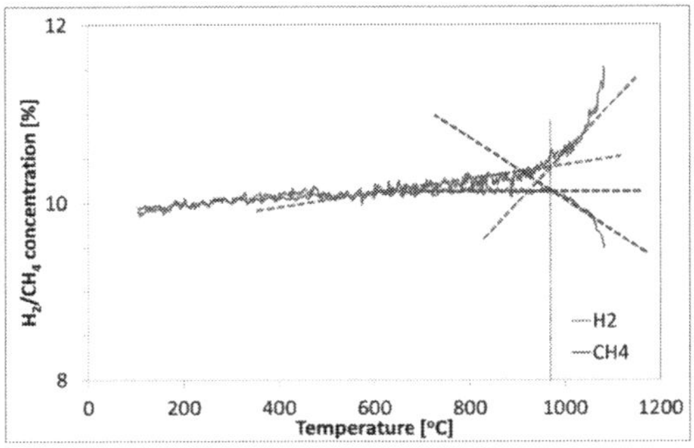

Figure 2. Out-gas signals for experiment one: empty crucible. The dashed lines are a guide to the eye to identify the break in trend at the onset of cracking.

Experiment two: quartz, SiO_2

The gas-curves from the experiments with SiO_2 are shown in figure 3. In this system, there is a very clear departure of the two signals at around 770°C. At the same temperature, a slight downwards trend in the hydrogen signal is inflected to a slight upwards trend. The expected downwards deviation in the already dropping methane signal is likely masked by the noise and 770°C is taken as the onset of cracking.

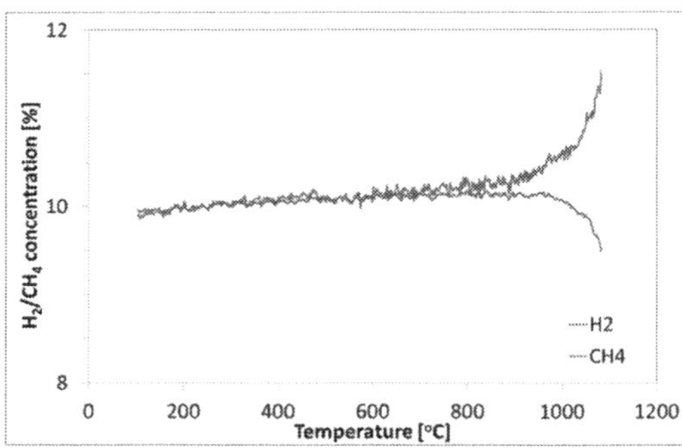

Figure 3. Out-gas signals for experiment two: SiO_2

Experiment three: MgO

Figure 4 shows the gas-curves for the MgO-test. Much as for the SiO_2 case, there is a reasonably easily recognizable point of departure where the two curves deviate; in this case at 820 °C. From this point the hydrogen signal is increasing, although that trend started somewhere around 780.

However, as the methane signal is also increasing between 780 °C and 820 °C—more or less proportionally to the hydrogen signal—this cannot be attributed to cracking. The methane signal also starts to drop off at 820 °C, and the decrease accelerates from this point onwards. The conclusion is that cracking is initiated at 820 °C.

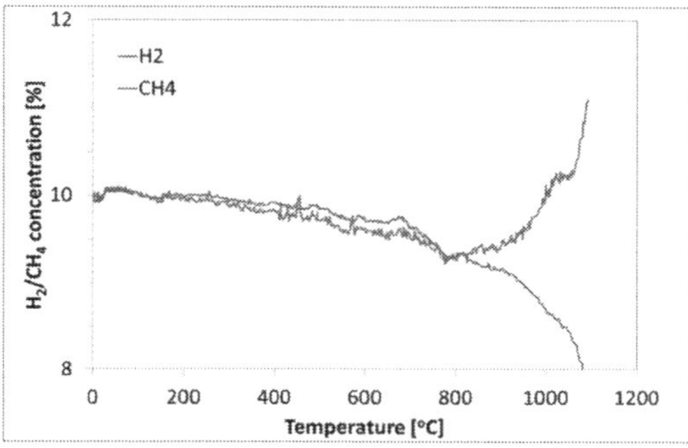

Figure 4. Out-gas signals for experiment three: MgO

Experiment four: Al_2O_3 in pellet form

An experiment was performed with Al_2O_3 in pellet form, mainly in order to check the influence of available surface area. Since the interior of the furnace is made from aluminium oxide, the chemical system would basically be the same as for the empty furnace. The difference being that in this test the alumina was present in pelletized form and thus represented a vastly greater surface area. Unfortunately, the temperature logger had an error during this experiment, so the temperature readings above 970°C are not trustworthy. Nevertheless, the available data are included in the paper for information. Figure 5 shows the off-gas analysis up to this point. The lack in higher temperature data makes it impossible to see the clear breakdown of methane content which is evident in all other curves at the top-end of the temperature plot, and this also makes it difficult to identify a spot where the two curves start to diverge. Based on the curves however, there does not seem to be a clear deflection point indicating the onset of cracking. The fact that the curves deviate somewhat from each other from around 600°C cannot be due to cracking, as the deviation shows the opposite behaviour of cracking: Increase in methane and decrease in hydrogen. These deviations are therefore likely due to an artefact of the measurements. From around 920°C it appears as if the curves converge, and this could be the first step towards crossing and deviating sharply, i.e. the onset of cracking. Without the curves after 970°C however, it is hard to make any conclusion.

Based on the limited information available, the tentative conclusion must be that there is nothing to indicate that the extra surface area available due to the pelletized form of the oxide has lowered the temperature for the onset of cracking, but neither can it be concluded that there definitely is no such connection.

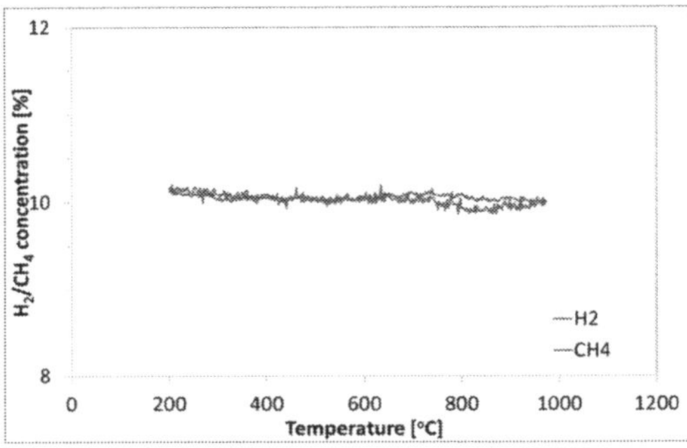

Figure 5. Out-gas analysis for experiment four: Al_2O_3,

Experiment five: CaO

The off-gas chart for the CaO-experiment shows a situation very similar to that for the MgO-case. Again there is a clear point of departure between the two signals, this time at 810°C, and it appears in a slight downwards trend. The acceleration in the downward-trending methane signal is much less obvious however, and any deflection in the hydrogen signal is also much more slight. Nevertheless, taken together it all point to 810°C as the best candidate for the onset of cracking.

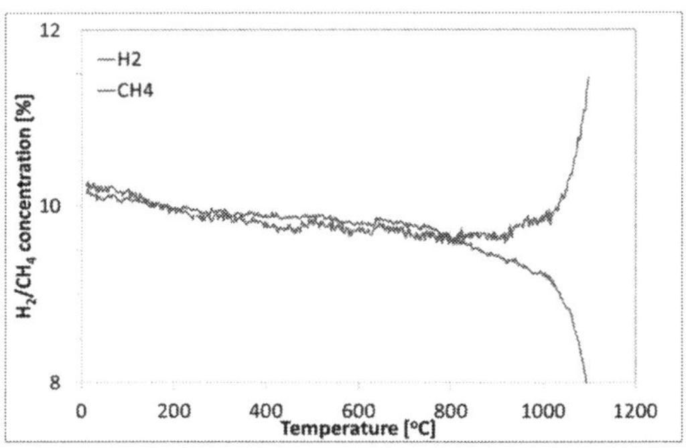

Figure 6. Out-gas analysis from experiment five: MgO.

Discussion

The temperature for the onset of cracking varies across the oxides from 770 °C for SiO_2, to 810-20 for CaO and MgO, to 950°C for the empty Al_2O_3 crucible. It is interesting that the oxides of Mg and Ca, both being alkaline earth metals, show comparably similar results. Table I summarises these results together with other parameters that will be referred to in the discussion.

In an investigation of different metals' ability to catalyse carbon nano-tube growth, Esconjauregui et al. [3] found a correlation between a metal's catalytic abilities and its affinity for carbon. It does not seem far-fetched that an oxide's interactions with carbon in methane are correlated to the enthalpy of formation of the respective metallic carbide. In the present work, the carbides in question are SiC, CaC_2, and MgC_2. When organising these carbides by enthalpy of formation [4], from the strongest interaction with carbon to the weakest, the order is the same as when organising the oxides in order of increasing temperature for onset of cracking: SiC (-65kJ/mol); CaC_2 (-60kJ/mol); MgC_2 (88kJ/mol). However, aluminium has a very strong affinity for carbon, as represented by the enthalpy of formation of Al_4C_3 (-209kJ/mol). Thus if the metal – carbon-affinity was the only determining factor, the onset of cracking in the experiment with Al_2O_3-pellets should be expected to take place at a temperature below the 770°C observed for SiO_2. And while the exact temperature for the onset of cracking could not be determined for the Al_2O_3-pellets, it was certainly higher than 800°C. Thus an oxide's ability to crack methane cannot be predicted from its metallic component's affinity for carbon.

Table I. Temperature for the onset of cracking (T_{crack}) for all examined oxides based on the experiments in the present work. In addition, FactSage calculations of the carbon activity and distribution of carbon at T_{crack}, and the temperatures for 0.1% and 10% methane-to-carbide conversion.

Compound	T_{crack} [°C]	a_C (at T_{crack}) [Relative to graphite]	Theoretical Distribution of Carbon (T_{crack})	T for 0.1% Methane-to-carbide conversion. [°C]	T for 10% methane-to-carbide conversion [°C]
SiO_2	770	176.3	97.6% CH_4, 0.63% SiC	740	840
CaO	810	259.4	98.1% CH_4, 0.43% CaC_2	800	860
MgO	820	284.5	100% CH_4	990	1080
(Al_2O_3)	950	827.5	70% CH_4, 18% C_nH_m, 0.12%Al_4C_3	950	1110

The theoretical carbon activities at the temperatures of the onset of cracking, using the same calculation as shown in figure 1, are listed in table I in the case of each of the oxides. Using this activity at this temperature, FactSage equilibrium calculations were performed on the Ar-CH_4-H_2-MeO systems. As the degree of conversion of the metal oxides in equilibrium will depend on the amount of methane available, the calculations are performed with an excess of oxide, and are discussed in terms of the conversion/distribution of carbon. Table I shows the calculated distribution of carbon between CH_4 and carbide at T_{crack} assuming no cracking, i.e. the carbon activity is as given in figure 1. Where the distributions do not add up to 100%, this is mainly because higher order hydrocarbons, C_nH_m are predicted to form in the calculations.

For SiO_2, CaO and Al_2O_3 the calculations predict the transformation of some oxide to carbide before the onset of cracking, while this is not the case for MgO. The levels of carbide are quite low however, and considering a system with a constant gas-flow, the amount of gas necessary to perform reduction near T_{crack} would likely be very large. The table shows the temperatures at which carbon has shifted from methane to carbide at levels of 0.1% and 10%. These levels are somewhat arbitrarily chosen as examples of: a) a lower bound for an operating temperature and b) a temperature at which reduction could be expected to proceed more favourably if it could be maintained without cracking

For the three oxides where carbide formation is predicted thermodynamically at or before T_{crack}, T_{crack} is much closer to the 0.1% limit than the 10% limit. For SiO_2 and CaO, the 10% limit would be achieved if the system could be pushed a further 70 or 50 degrees respectively while repressing cracking. Without discussing the feasibility of such a task, it can be noted that for Al_2O_3 the required temperature increase would be much greater: 160 degrees. In terms of carbon activity, for SiO_2 and CaO the increase is a factor 1.5-2, while for Al_2O_3 it is closer to 3. Part of the reason for this is that for Al_2O_3, the breaking of the oxide and creation of the carbide becomes less energetically favourable than the chaining together of carbon atoms from methane to molecules such as C_2H_2 and C_2H_4.There is thus less metastable methane available to react to form carbide.

The situation is similar for magnesium, where the calculations also predict large amounts of C_nH_m-formation. Probably due to a lower stability of the oxide however, it turns out that the temperature for 10% carbide conversion is lower for magnesium than for aluminium, even though the temperature for first carbide formation is higher.

Figure 7 shows the ΔG (given in J per mole of CH_4, calculated using FactSage) as a function of temperature for the reactions for carbide formation from oxide and methane:

$$SiO_2 + 3\,CH_4 \rightarrow SiC + 2\,CO + 6\,H_2 \quad (2)$$
$$CaO + 3\,CH_4 \rightarrow CaC_2 + CO + 6\,H_2 \quad (3)$$
$$MgO + 3\,CH_4 \rightarrow MgC_2 + CO + 6\,H_2 \quad (4)$$
$$2\,Al_2O_3 + 9\,CH_4 \rightarrow Al_4C_3 + 6\,CO + 18\,H_2 \quad (5)$$

and the reactions for formation of acetylene and ethylene from methane:

$$2\,CH_4 \rightarrow C_2H_2 + 3\,H_2 \quad (6)$$
$$2\,CH_4 \rightarrow C_2H_4 + 2\,H_2 \quad (7)$$

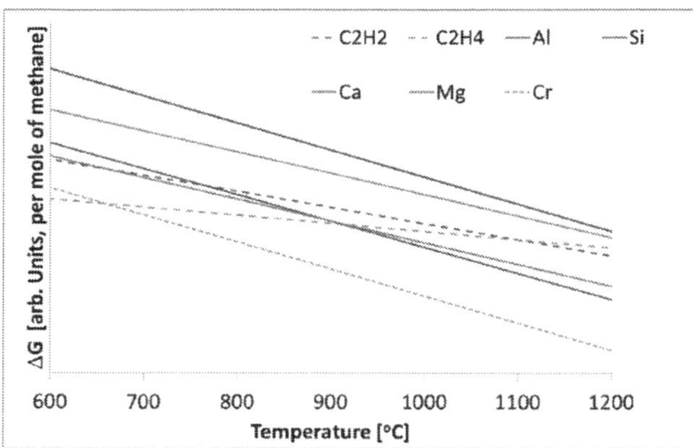

Figure 7. ΔG for the oxide+methane=carbide reactions for the metallic oxides and carbides corresponding to the elements named on the curves, together with ΔG for the methane→ethylene/acetylene reactions. The energy is given in units per mole of methane.

As can be seen, it is more energetically favourable for methane to combine into acetylene and ethylene than to form carbides from aluminium or magnesium oxides. Thus, even if the process gas can be pushed far from equilibrium, and the metastability of methane resulting in very large carbon activities, a reaction with the oxides is still not favourable.

The problem is also there for silicon and calcium, as the formation of ethylene is favourable over the formation of carbide for both oxides up until around 910°C, which means that it is favourable in the temperature regime up to T_{crack} found in this work.

When selecting an oxide for methane reduction then, it could be advised to choose a metal for which ΔG for carbide formation is below ΔG for ethylene/acetylene formation in the temperature range one aims to do the reduction. For example, looking at chromium, for which the ΔG of carbide formation from oxide is also plotted in figure 7, carbide formation is favoured over ethylene formation for temperatures above 660°C. Indeed, chromium oxide has been shown to be successfully carburized in methane without soot formation [1].

In closing it should be emphasized that the discussions based on the equilibrium calculations do not take the kinetics of the problem into account. Kinetic factors are likely what will determine, for example, if it is possible to push the SiO_2-CH_2-H_2-system the additional 70 degrees beyond the T_{crack} found in this work which would correspond to 10% carbide conversion. The kinetics would also have a determining impact on the real conversion that would occur at such conditions.

Nevertheless, equilibrium calculations can be instructive in their own way; the amount of superheating necessary beyond T_{crack} will for example give an indication on the likelihood for success. Furthermore, observations such as those made in the present work regarding competing reaction mechanisms to those desired will be of use in planning and evaluating new experiments.

Conclusions

The onset of cracking of methane in contact with oxides does not happen immediately upon reaching metastable conditions for the methane molecule. How far from equilibrium the system can be pushed—and to what degree the metastability of methane can be taken advantage of in terms of increased carbon activity—varies between different oxides.

The degree of metastability that can be achieved is *not* correlated to the affinity for carbon of the metallic component of the oxide.

The metastability of a methane-containing gas may lead to the formation of ethylene and acetylene. When designing experiments for reduction experiments with methane, choice of oxides and temperature range should be made so that the desired reaction is energetically favoured over the gas-phase reactions forming acetylene and ethylene.

Acknowledgements

This work is financed by the Norwègian Research Council through the KiSelRox project (grant # 224942). The authors also wish to express gratitude to our colleagues at SINTEF and NTNU for help with experiments: Delphine Leroy, Michal Kziachek, Edit Thomassen, Ole Kjos, Thor Anders Aarhaug.

References

1. Oleg Ostrovski and Guangqing Zhang, "Reduction and Carburization of Metal Oxides by Methane-Containing Gas", *AIChE Journal*, 52 (2006), 300-310.
2. Michal Ksiazek, Halvor Dalaker, Eli Ringdalen, and Merete Tangstad, "Synthesis of SiC by direct reduction of SiO_2 by using a methane-gas experimental approach" , to be published in Celebrating the Megascale: An EPD Symposium in Honor of David G. C. Robertson, TMS (2014).
3. S Esconjauregui, CM Whelan and K Maex,"The reasons why metals catalyze the nucleation and growth of carbon nanotubes and other carbon nanomorphologies", *Carbon* 47 (V) 659 – 669.
4. G Aylward & T Findlay, *SI Chemical Data, 4th edition*, (Brisbane: JP Wiley and Sons, 1998)

Celebrating the Megascale: Proceedings of the Extraction and Processing Division Symposium on Pyrometallurgy in Honor of David G.C. Robertson
Edited by: Phillip J. Mackey, Eric J. Grimsey, Rodney T. Jones, and Geoffrey A. Brooks
TMS (The Minerals, Metals & Materials Society), 2014

Reduction of Mixtures of Ni-Oxide and Ti-Oxide Powder in a Natural Gas Containing Atmosphere

Casper van der Eijk[1] and Kai Tang[1]

[1]SINTEF Materials and Chemistry
7465, Trondheim, Norway

Keywords: Natural gas, Reduction, Ni-oxide, Ti-oxide, Ni3Ti

Abstract

Natural gas has been used as reductant in the production of iron for several decades. Recently, the use of natural gas as reductant for other metals has been drawing attention because of the low price of natural gas, as well as the reduction of CO_2 emissions compared to the use of coke and coal. In this work, the reduction of a mixture of Ti- and Ni-oxides by natural gas has been studied. Within the Ni-Ti system, the intermetallic compound Ni_3Ti is thermodynamically the most stable one. Therefore, powders of Ni-oxide and Ti-oxides were mixed in an atomic ratio of three to one and subsequently pelletized. Reduction was performed in a TGA furnace at constant temperature with varying reducing gas atmospheres.

In a natural gas containing atmosphere, the Ni-oxide was easily reduced to metallic nickel while the Ti was reduced to Ti-oxycarbides or Ti-oxynitride when nitrogen was present. Depending on the process parameters, several percent of Ti was dissolved in the nickel-based solid phase.

Introduction

The Ni-Ti based materials are interesting because of their shape-memory and super-elastic properties. The production of Ni-Ti based alloys is however expensive since it requires pure titanium metal. Titanium has a high affinity with oxygen and the costs of titanium production by the intricate Kroll process are high.

Direct reduction of Ti-oxides to metallic titanium by hydrogen is thermodynamically not possible. In principle, however, reduction of the activity of product may help to enhance the reduction process. Since the activity of titanium is expected to be much lower in an intermetallic compound, it is possible to reduce the Ti oxide to Ni-Ti alloy. There are several M_3Ti compounds with a large negative enthalpy of formation, as shown in Figure 1. In this work, it is attempted to produce Ni_3Ti by direct reduction of Ni and Ti oxide mixture, since this is the M_3Ti compound with most engineering relevance [1-3].

Experimental

Chemically pure NiO (Nickel(II)-oxide from Alfa Aesar) and TiO_2 (Titan(IV)-oxide from Merck) powders were mixed in an atomic ratio of three to one and subsequently pelletized. The pellets had a diameter between 3 and 5 mm and were sintered at 1000°C in air. Reduction was performed in a TGA (ThermoGravimetric Analysis) furnace at a temperature of 1200°C with varying reducing gas atmospheres. In the TGA measurements, the green pellets are placed in an alumina crucible with slots in the bottom in order to allow the gas to reach the sample material. The gas is injected in the bottom of the vertical tube furnace, close to the crucible.

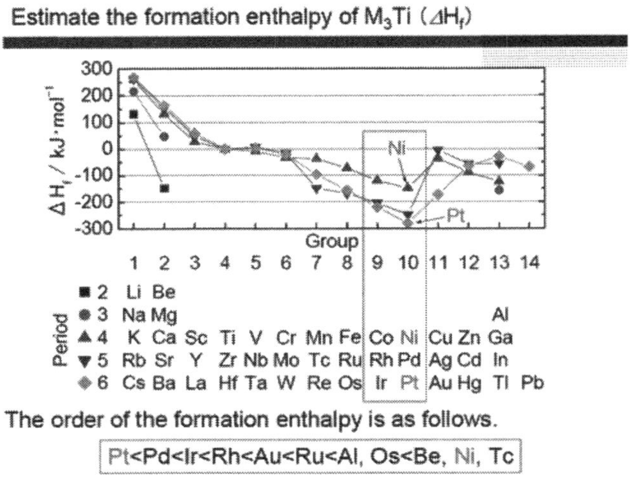

Figure 1. An estimation of the formation enthalpies of M_3Ti compounds (ΔH_f)[1]

All samples were heated to 1200°C in 100% hydrogen atmosphere with a heating rate of 10°C/minute. The heating program up to 1200°C for all specimens is identical. When the temperature reached 1200°C, the atmosphere was changed to the gas mixtures listed Table I and ran for 4 hours at 1200°C. In every experiment, 50 grams of pellets were used and placed in an alumina crucible. This means that there were 13 grams of oxygen: 8 grams bound to NiO and 5 grams bound to TiO_2. After reduction the pellets were analyzed by microprobe (EPMA).

Table I. Gas mixtures used during the experiments

o.	Gas composition
1.	100% H_2
2.	50% H_2, 50% N_2
3.	90% H_2, 10% CH_4
4.	35% H_2, 50% N_2, 15% CH_4
5.	35% H_2, 50% Ar, 15% CH_4
6.	85% H_2, 15% CH_4

Results and Discussion

The reduction of the nickel in pure hydrogen is completed during heating at a temperature between 750 and 760°C. The weight loss is then 8 grams and the rate of weight change decreases sharply. Above 800°C, the reduction of the Ti-oxide by hydrogen starts. A summary of the experimental results is listed in Table II. Reduction of the pellets in 100% H_2 atmosphere (Exp. No. 1) leads to a weight loss of 9.1 grams. 8 grams of oxygen are from the reduction of NiO. Hence 1.1 grams is from the reduction of Ti-oxide. Some of the Ti dissolves into the Ni-based solid solution, while the TiO_2 is reduced to the more Ti-rich oxides, such as Ti_3O_5 or Ti_4O_7. Reduction in a 50% H_2 and 50% N_2 mixture (Exp. No. 2) gives similar results as Exp. No. 1. The only difference is a slightly lower degree of reduction due to the lower H_2 activity. Nitrogen behaves like an inert gas in this case.

Table II. Summary of the experimental results

No.	Weight loss (grams)		O/Ti atomic ratio for the Ti containing phase	At.% Ti dissolved in Ni phase
1.	9.1		1.7	2.0
2.	8.9		1.7	1.7
3.	10.3	Top	1.6	7.6
		Bottom	0.3	5.7
4.	9.7*	Top	1.7	2.3
		Bottom	0.1	3.7
5.	9.6*	Top	1.7	5.5
		Bottom	0.2	4.4
6.	8.5*	Top	1.6	7.4
		Bottom	0.2	4.1

* Cracking of CH_4 leaves carbon black deposition on the crucible (Figure 5).

Introducing 10 vol.% CH_4 (Exp. No. 3) to the gas stream results in a much higher degree of reduction. There is also a distinct difference between the pellets that were on the bottom of the crucible, where the gas has first contact with the oxide material, and those on the top of the stack of pellets in the crucible. The reduced pellets from the top of the crucible contain a nickel and a Ti-oxide phase, as shown in Figure 2a. This is similar to what is found in Exp. No. 1 and 2. There is however much more Ti in Ni-based solid solution, which indicates a higher degree of reduction. It appears that the Ti-oxides on the bottom of the crucible that have been in contact with CH_4 were reduced to Ti-oxy-carbides, as shown in Figure 2b. The CH_4 probably cracks into hydrogen and carbon when it comes into contact with nickel. The free C and H atoms have a high activity.

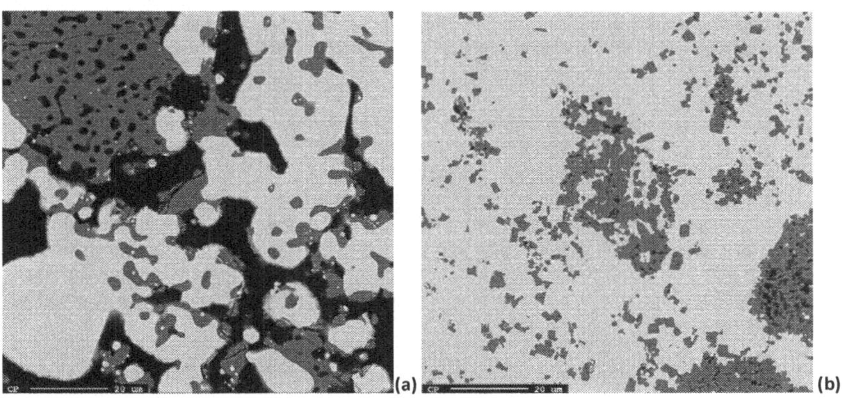

Figure 2. Experiment No. 3 backscattered image, a) top, Ni phase bright with Ti-oxide phase dark, b) bottom Ni phase bright with Ti-oxy-carbide phase dark.

In Exp. No. 4, nitrogen and CH_4 were both present in the gas stream. The pellets that had been on the bottom had a gold-like color, as shown in Figure 3. This is the typical color for Ti-nitrides. EPMA analyses show the formation of Ti-oxy-nitrides in the pellets in the bottom of crucible, which had first contact with the gas stream. In this case, nitrogen has played a role in the reaction. The pellets that were on the top of the stack of pellets in the crucible possess a low amount of Ti in Ni-based solid solution (Figure 4). Exps No. 5 and 6 confirmed the results from Exp. No. 3 with the formation of Ti oxycarbides in the pellets which were in contact with CH_4.

Figure 3. Experiment No. 4 Pellets after reduction in a Gas atmosphere containing H_2, CH_4 and N_2. The color indicates the presence of Ti-nitrides.

In the experiments No. 4 to No. 6, carbon was deposited on the outside of the crucible. This is shown in Figure 5. Also some graphite phase was observed in the Ni-based phase. The weight of the carbon flakes on the outside of the crucible is about 0.2 grams. This means that the weight of the oxygen removed is slightly higher then what is listed in Table II. However, the Ni_3Ti phase was not observed in the all measurements.

Figure 6 shows the equilibrium relation in the same condition as for the laboratory investigations in experiment No. 3. It shows that the Ni-oxide in pellets is first reduced to Ni. After all the Ni-oxide has been reduced, TiO_2 begins to be reduced to $Ti_{20}O_{39}$, Ti_9O_{17}, ..., Ti_3O_5. The final stable Ti-oxide is Ti_3O_5. Further reduction of Ti_3O_5 results in producing Ti oxycarbide, Ti_2OC. It should be noted that the stoichiometric of TiO_xC_y varies from 0-0.5, so that Ti_2OC in the FACTSAGE calculation is only used to represent the Ti oxycarbides here. Meanwhile small amounts of TiC also begin to form. The amount of TiC increases tardily. A further increasing in gas reduction will produce carbon black substantially and thus hinder the reduction of Ti oxycarbides.

The results from the thermodynamic calculations agree well with the experimental observation. The calculation predicts the formation of Ti_3O_5 as the final oxide. This agrees with well the observation that the lowest O/Ti is 1.6. Also the observation of the Ti-oxy-carbides and the deposited carbon are as foreseen by the calculations.

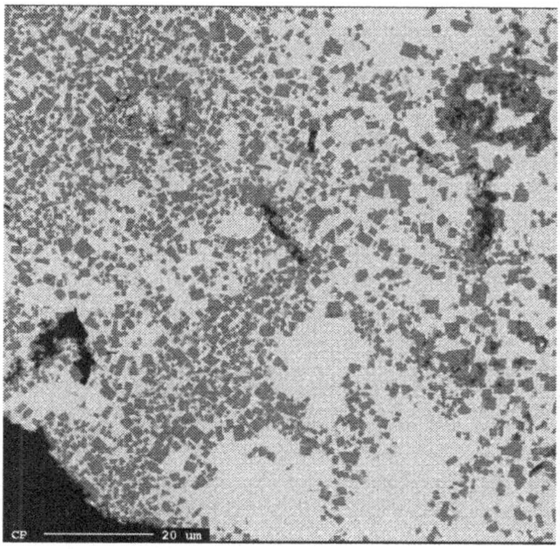

Figure 4. Exp. No. 4 Bottom, Ti-oxy-nitrides in a nickel matrix from the pellets on the bottom after experiment No. 4.

Figure 5. Carbon deposition on the outside of the crucible after experiment No. 5.

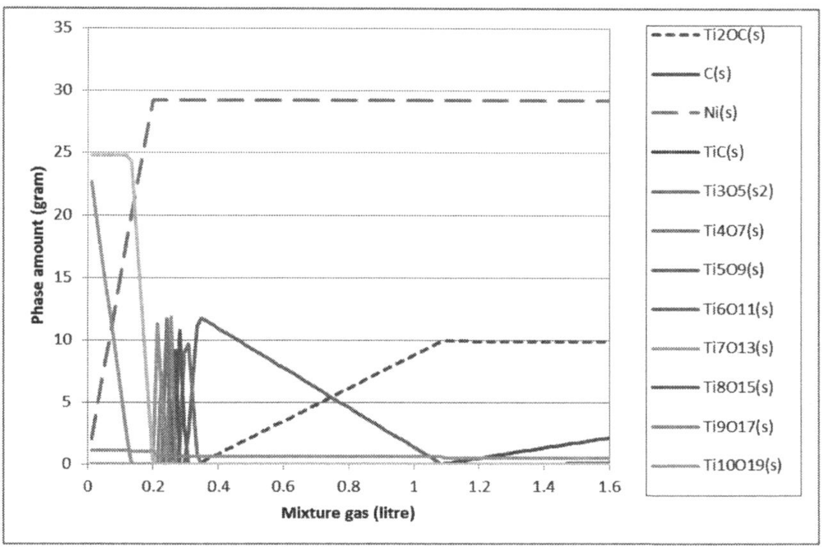

Figure 6. Amount of phase in equilibrium with gas containing 50% H_2-15% CH_4-35%Ar.

Conclusions

- The reduction of a NiO/TiO_2-mixture in a hydrogen atmosphere is enhanced when CH_4 is introduced into the gas mixture.
- The addition of CH_4 in the gas mixture leads to the formation of Ti oxycarbides in the pellets that are in direct contact with the CH_4.
- The combination of nitrogen and CH_4 in the gas mixture leads to the formation of Ti-oxy- nitrides in the pellets that are in direct contact with the CH_4.
- When the CH_4 concentration is 15% in the gas mixture, deposition of carbon on the crucible and sample material is observed. This is not observed when the CH_4 concentration in the gas is 10%.

References

1. H. Sekimoto et al., "Reduction of Titanium Oxide to Titanium Alloy by Hydrogen" (Paper presented at the 3rd Workshop on Reactive Metal processing, March 2-3, 2007, Massachusetts Institute of Technology, Cambridge, MA, USA).

2. H. Sekimoto et al., "Reduction of Titanium Oxide in the Presence of Nickel by Non-equilibrium Hydrogen Gas", *Journal of Material Research*, 24 (7) (2009), 2391-2399.

3. A. Qiu et al., "Calculation of phase diagram of Ti-Ni-O system and application to deoxidation of TiNi alloy", *Trans. Nonferrous Met. Soc. China*, 21 (2011), 1808-1816.

Celebrating the Megascale: Proceedings of the Extraction and Processing Division Symposium on Pyrometallurgy in Honor of David G.C. Robertson
Edited by: Phillip J. Mackey, Eric J. Grimsey, Rodney T. Jones, and Geoffrey A. Brooks
TMS (The Minerals, Metals & Materials Society), 2014

KINETIC AND THERMODYNAMIC ANALYSIS OF THE REDUCTION OF OXIDES OF Cu AND Co IN A SiO_2-CaO-$(Al,Fe)_2O_3$ SLAG

Yotamu Stephen Rainford Hara and Animesh Jha

The Institute for Materials Research, Houldsworth Building
Clarendon Road, Leeds University, Leeds LS2 9JT (UK), England

Key words: Cu-Co slag, Carbothermic reduction, Kinetics, Thermodynamics, Sulphidation

Abstract

The investigation focuses on a low temperature recovery of Cu and Co from a 40 wt% SiO_2-(30 wt%Fe,6 wt%Al$)_2O_3$-10wt% CaO-7wt%CuO-7wt%CoO slag over a temperature range of 1173K to 1323K. The alloy phases containing Cu-Co and Fe-Co alloy formed via the carbothermic reduction of oxides: MO + C = M + CO(g), where M represents metallic copper, cobalt or iron. In the direct reduction of oxides, the recovery of metallic phase was well below 90% at 1323K, due to the kinetic barrier which was analysed and attributed to oxygen and metal-ion transport in the slag. This barrier was overcome by adding $CaSO_4$ and carbon, which yields a matte (MS) phase via MO + $CaSO_4$ + 4C = MS + CaO + 4CO reaction. Lime thus produced in situ participates in metal oxide/metal sulphide reduction reactions, which are analysed with the help of X-ray powder diffraction, scanning electron microscopy, and thermogravimetric analysis.

Introduction

The pyrometallurgical extraction of copper from the mineral sulphide concentrates is achieved via selective oxidation of iron and sulphur. The iron and sulphur are typically oxidised in two stages: (i) smelting and (ii) converting. A metal-rich sulphide or matte (Cu_2S,FeS) and slag molten phases are produced during the smelting stage. The remaining iron and sulphur from the matte phase are removed during converting stage and, blister copper and molten slag phases are produced. The details of copper smelting and converting are found in reference [1]. Because copper smelting and converting are oxidising processes, part of copper and nearly all cobalt are oxidised [1] and dissolve in the molten slag phase. The Cu and Co content in the Nkana slag dump in Zambia can be as high as 3.9 wt-% [2] and 2.6 wt-% [3] respectively. Microscopic examinations of copper smelting and converting slag have revealed that the Cu and Co oxides are either dissolved in the oxide (fayalite, glassy) phases or exist as small spinel phases enclosed within the oxide phases [2].

A majority of Cu and Co in the slag phase are in the oxide state such that carbothermic reduction (equation 1) may be employed to produce metallic/alloy phases (M) of Cu, Co and Fe. Thermodynamic prediction indicates that Cu,Co and Fe oxides can be reduced to the metallic state at temperatures as low as 1073K [4, 5].

$$MO + C = M + CO(g) \qquad 1$$

However, the solid state carbothermic reduction of metal (Cu,Co and Fe) oxides from the slag phase may not be straight forward. This is due to lack of contact between the metal oxides and the solid carbon. It may be suggested that the reduction of the metal oxides may require different mechanism(s) other than direct contact between metal oxides and solid carbon and the reactions may proceed as follows and, these are; (i) the diffusion of the O^{2-} and metal (M^{2+}) ions from the metal oxide (MO) to the slag/carbon interface (Figure 1a). The rate of reaction may be determined by the diffusion of O^{2-} and metal (M^{2+}) ions through the slag matrix and/or gasification of carbon. (ii) Diffusion of carbon in the slag matrix so that the metallic phase is formed within the slag phase (Figure 1b). This mechanism may be undesirable for solid state reduction as a result of the lack of separation of the metallic phases from the slag phase.

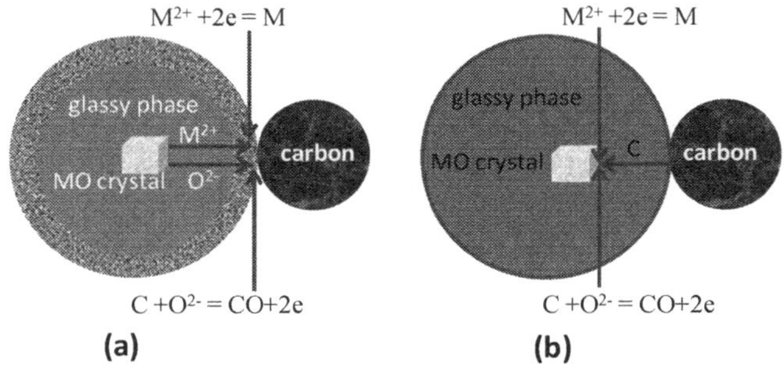

Figure 1. Illustration of reduction mechanism of the metal oxide crystal in slag.

Thermodynamic Considerations

The Cu-Co-Fe-Ca-O-S system at 1273K is shown in Figure 2 and a number of observations can be made:

i. The metal oxides can be reduced to metallic state (Cu, Co and Fe) at low partial pressure of O_2 gas ($\log_{10}P(O_2) < -15$) and SO_2 gas gas ($\log_{10}P(SO_2) < -9$).
ii. The oxides of copper and cobalt may be preferentially reduced in preference to the oxides of iron at low partial pressures of SO_2 gas ($\log_{10}P(O_2) < -7$) and $\log_{10}P(O_2)$ -6.2 to -14.9.
iii. Sulphidation of the metal (Cu, Co and Fe) oxides may be achieved at low and high partial pressures of O_2 and SO_2 gases, respectively.
iv. Preferential sulphidation of Cu and Co may be possible at slightly low partial pressures of SO_2 and O_2 gases (areas A and B).

Based on the Cu-Co-Fe-Ca-O-S system in Figure 2, carbon and SO_2/S_2 are required for the sulphidation of metal oxides and, this is possible by using $CaSO_4$ as the reduction calcination of $CaSO_4$ yields SO_2 gas (equation 2a) [6]. The SO_2 gas from the reduction calcinations can react with the metal oxide (MO) to give metal sulphide (MS) (equation 2b). The summation of equations 2a and 2b gives the overall reaction in equation 2c. The use of $CaSO_4$ is attractive because: (i) $CaSO_4$ is readily available; (ii) sulphidation may be achieved with minimal or no generation of SO_2 gas via reaction 2c. (iii) A calcium silicate (Ca^{2+}=0.99 Å) slag may be produced which may be a sink for transition metal ions [7] e.g. Fe^{2+}(0.77 Å), Fe^{3+} (0.65 Å), Cu^{1+} (0.96Å), Cu^{2+} (0.73 Å) and cobalt (Co^{2+}=0.74 Å) ions, and depending upon its basicity of the silicate slag, the relative concentrations of the transition metal ions may be present.

$$CaSO_4 + 0.5C = CaO + SO_2 + 0.5CO_2(g) \qquad 2a$$
$$MO + SO_2(g) + 1.5C = MS + 1.5CO_2(g) \qquad 2b$$
$$MO + CaSO_4 + 2C = MS + CaO + 2CO_2(g) \qquad 2c$$

The Cu-Co-Fe-Ca-O-S system in Figure 2 further suggests that the metal sulphides can be converted/reduced to the metallic state by decreasing the partial pressures of SO_2 and O_2 gases. Therefore, metallisation of the sulphidised slag or matte phase can be achieved via carbothermic reduction in the presence of CaO (equation 3) because, CaO is thermodynamically stable sink for sulphur under reducing condition.

$$MS + CaO + C = M + CaS + CO(g) \qquad 3$$

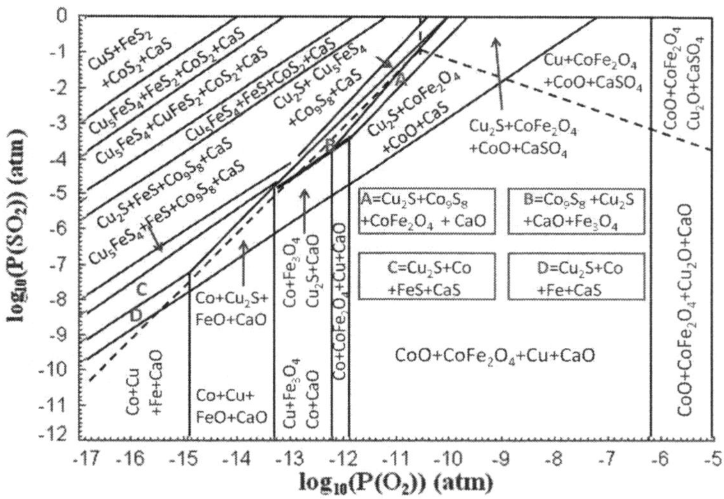

Figure 2. Cu-Co-Fe-O-S system (continuous line) and Ca-O-S system (dotted line) at 1273K, computed using FactSage software [4].

Experimental

The Cu–Co slag was prepared by mixing metal oxides, Al_2O_3, CaO and SiO_2 according to Table I. The composition in Table I is typical of Nkana copper smelting slag [7] except that CuO and CoO were added in higher quantities for easy analysis and for understanding of metal oxide during recovery. The mixed sample powders were melted in air for 2 hours at 1523K. The molten sample was cooled down to room temperature in air, crushed and ground down to < 200μm.

Table I. Composition of synthetic slag with the oxides of cobalt, copper and iron

Constituent	SiO_2	CaO	Fe_2O_3	Al_2O_3	CoO	CuO
Weight (%)	40	10	30	6	7	7

For carbothermic and sulphidation experiments, the samples were mixed as follows:

i. *Carbothermic reduction* - 1.4g of the slag sample was mixed with 0.44g carbon (carbon black - X72 or activated charcoal), which is equivalent to MO:C = 1:3.
ii. *Sulphidation of slag* – 0.7g of slag was mixed with $CaSO_4$ and graphite in a molar ratio of $MO:CaSO_4:C$ = 1:1.1:0.9.

All carbothermic reduction and sulphidation experiments were carried out isothermally, in the thermogravimetric equipment, in order to study the weight loss as a function of time. Details of the thermogravimetric equipment can be found in our previous studies [8, 9]. The reaction vessel was purged with argon gas at a flow rate of 660 litres $(minute)^{-1}$, for maintaining inert atmosphere and for sweeping off the product gas [8]. The weight loss of the sample was recorded every 10 seconds, throughout the experiments.

The phases in the reacted samples were analysed by X-ray diffraction (XRD) techniques, using the Philips X'pert machine with a copper Kα radiation (λ = 1.5417 Å.), operated at 40kV and 30mA. The phases in the powder diffraction patterns were identified by using X'pert high score plus software which is supported by the JCPDS PDF-4+ data base (RDB 2012) [9]. For microscopy examination, a portion of the sample was mounted in a mixture of 3 parts epoxy hardener and 1 part epoxy hardener and cured for 8 hours. The samples were gently ground on different grades of silicon carbide paper and polished down to 3μm. The polished samples were placed inside the Agar high resolution sputter coater for coating a 6nm thick layer of platinum to minimise charging during SEM analysis. The samples were examined for structural and chemical changes over a large area in order to understand the distribution of the phases.

Results and Discussion

The X-ray powder diffraction pattern for the synthetic slag is shown in Figure 3. The only crystalline phases were $Co_{0.62}Fe_{2.38}O_4$ (Co-Fe spinel) and silica (SiO_2). There are no peaks for Ca or Al containing compound(s) in the XRD pattern and this can be due to formation of amorphous glassy phase [10]. The presence of amorphous phase is evident as there is a broad hump between $2\theta = 20$ and 34° [11]. The SEM-EDX analysis of the slag showed three phases, SiO_2, a multicomponent silicate glassy (Ca-Al-Si-Cu-Fe-O) phase and Co-Fe spinel crystals which formed within the glassy matrix.

<u>Carbothermic Reduction of Slag</u>

The plots of % reduction versus t curves for the carbothermic reduction of slag with carbon black and activated charcoal are shown in Figures 4a and 4b, respectively. It is evident from Figure 4a that the extent of reduction/metallisation is less sensitive to the reduction temperature in the temperature range between 1173K and 1273K.

For two samples reduced at 1173K and 1223K isotherms, the increase in % reduction (%R) was only 5%. However, the % reduction increased by about 35%R in 1.5 hours as the reduction temperature was increased from 1273K to 1323K. From the results in Figure 4a, it can be concluded that the completion of carbothermic reduction of slag with carbon black requires temperature higher than 1323K.

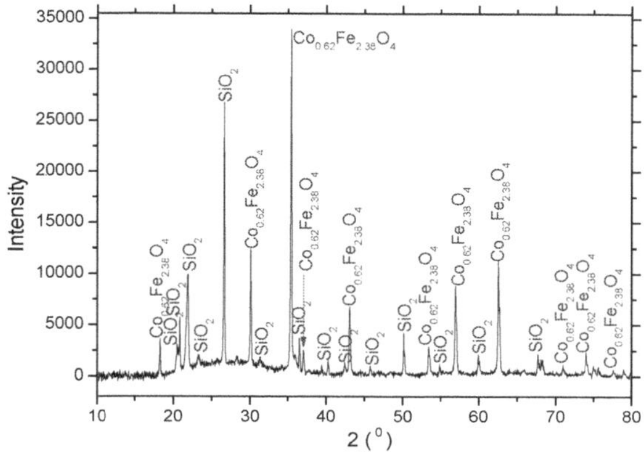

Figure 3. XRD pattern (2θ = 10 and 70°) for slag, synthesised at 1523K.

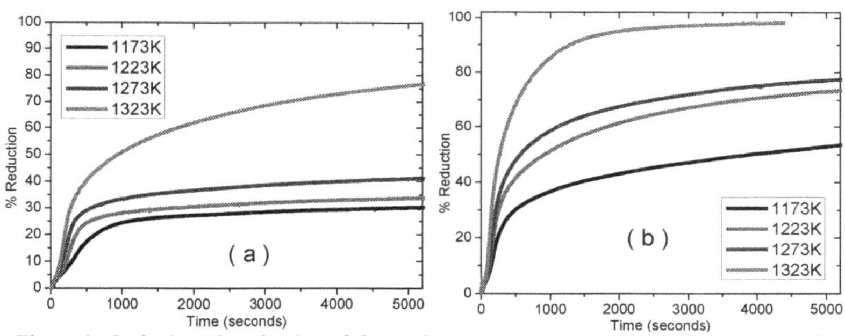

Figure 4. Carbothermic reduction of slag under argon atmosphere at MO:C = 1:3 (a) with carbon black and (b) activated charcoal.

The extent of reduction/metallisation was higher for the carbothermic reduction of slag with activated charcoal as shown in Figure 4b. Furthermore, from this figure it is apparent that the carbothermic reduction of slag with activated charcoal is more sensitive to the reduction temperature than that observed for carbon black. The apparent sensitivity to the reduction temperature may be deduced from the apparent 93%R (Fig.4b) achieved in 1.5 hours at 1323K. This comparative observation in Figures 4a and 4b between carbon black and activated charcoal, respectively, suggest that the activated charcoal is a better reducing agent than carbon black.

The XRD patterns for the carbothermically reduced slag at 1173K and 1323 K are shown in Figures 5a and 5b. Similar phases were obtained at 1173K and 1223K, for the carbothermic reduction of slag with carbon black, which is in good agreement with the TGA results in Figure 4a. The XRD results in Figure 5a show that the reduction of copper was higher than the reduction of cobalt and iron at 1173K. In other ways, there was a preferential reduction of Cu-oxides at 1173K with carbon black.

The higher or preferential metallisation of copper in the presence of carbon black at 1173K suggests that the partial pressure of $Log_{10}(PO_2)$(atm) was less than -11.9 (see Figure 2).

The TGA results showed that the extent of reduction was higher for the carbothermic reduction of slag with activated charcoal and this can be confirmed by the XRD results in Figure 5b, as the XRD peak intensities for Cu and Fe are high. There are no peaks for metallic Co in the XRD patterns in Figure 5b due to dissolution of Co in Fe [8, 9]. The iron oxide phases (FeO, Fe_3O_4) can be observed in the XRD patterns at 1323K and it is for this reason that reduction did not reach 100 %R. The XRD peak intensities for SiO_2 were high at all temperatures for the carbothermic reduction of slag with carbon black. However, the low relative intensities of XRD peaks for SiO_2 is apparent at 1323K, with activated charcoal, because of high fractions of metallisation. It appears CaO produced in situ during the reduction reaction 2a-reacted with SiO_2 and yielded a $CaAl_2Si_2O_8$ phase.

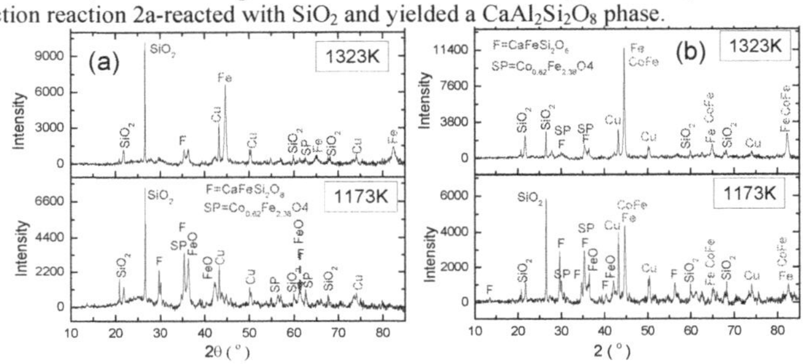

Figure 5. XRD patterns after carbothermic reduction of Cu-Co slag under argon atmosphere at MO:C = 1:3 (a) with carbon black and (b) activated charcoal.

Reaction Kinetics for Carbothermic Reduction of Slag

The rate of reduction, derived from %R versus time plots in Figures 4a and 4b,were fitted to the chemical (equation 4a), diffusion (equation 4b) and mixed kinetic (equation 4c, the sum of equation 4a and 4b) models in order to establish the reaction mechanism which might be governing the overall rate. Although the experimental data fitted well individually each model, for comparative purpose here we have selected the mixed model based on the assumption that the slag is in liquid state above 1273K and that both the diffusion barrier and chemical reaction might be relevant in determining the overall rate of reduction. The cross-sectional SEM analysis of the reduced samples, showed the presence of metallic beads at the periphery of the slag particles thereby confirming that there is migration/diffusion of the metallic (M^{2+}) and O^{2-} ions, as illustrated in Figure 1a. The diffusion of O^{2-} ions (10^{-10} to 10^{-15} cm^2 s^{-1}) in crystalline solids (e.g. alumina, silica) and glassy states tend to be slow with activation barrier varying between 350 and 450 kJ mol^{-1}) [12] above 1573K. In the temperature discussed herein, the diffusion barrier might be larger due to lower temperatures than 1573K, which is why analysis of data using mixed reaction control model may be relevant [13].

$$1-(1-X)^{\frac{1}{3}} = kt \qquad \text{4a}$$

$$1-\frac{2}{3}X-(1-X)^{\frac{2}{3}} = kt \qquad \text{4b}$$

$$2-\frac{2}{3}X-(1-X)^{\frac{1}{3}}+(1-X)^{\frac{2}{3}} = kt \qquad \text{4c}$$

Where, X is the fraction reacted at time t and k is the rate constant.

The plots of ln k vs. 1/T lines are shown in Figure 6 and the calculated value of overall activation energy is shown against each fitted line. For carbon black in Figure 6, the derived value of activation energy is 814 ± 83 kJ mol^{-1}), which compares with the structural relaxation energy of silicates due to shear flow in the glass transition regime, Kingery et al. [12]. It should be noted that the derived value of 814 ± 83kJ mol-1 compares much better with molar enthalpy of formation of silica which is of the order of 900 kJ mol^{-1}. By comparison the activation energy for the gasification of carbon black was found to be 333 ± 21kJ by Rao and Jalan [14], which then suggests that the rate controlling step in the reduction of slag might have been controlled by the diffusion of cations and anions in the glassy phase.

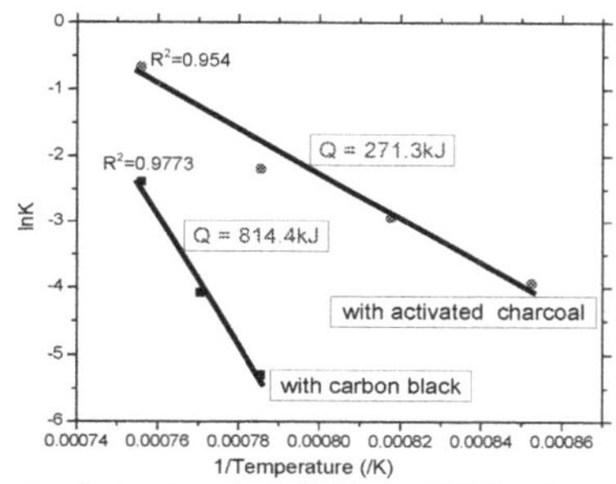

Figure 6. Arrhenius plots (lnK vs. 1/T) for the carbothermic reduction of slag, the rate constant k was calculated using equation 4c. The activation energy is shown in each line.

The value of apparent activation energy decreased to about one third (271 ± 34kJ) when carbon black was replaced with activated charcoal. This value is in a reasonable agreement with the values (252-268 kJ mol^{-1}) reported in the literature for the gasification of activated charcoal [15]

Sulphidation of Metal Oxides in the Presence of $CaSO_4$ and Graphite

As stated above, the reduction calcination of $CaSO_4$ can yield CaO and SO_2 gas. However, carbon with a low reactivity is required for the calcination of $CaSO_4$ to CaO and SO_2 gas [5]. To maximise the yield of SO_2 gas from the calcination of $CaSO_4$, the sulphidation of the slag was carried out in the presence of graphite. The plots of %conversion vs. time curves showing the effect of reaction temperature at a molar ratio of MO:CaO:C = 1:1:0.9 are shown in Figure 7a. It is evident from Figure 7a that sulphidation reaction is very sensitive to the reaction temperature. For example, the reaction was incomplete after 2 hours at 1173K, however it reached completion within 15 minutes at 1323K.

The XRD patterns in Figure 8 show the effect of reaction temperature. The powder diffraction patterns at 1173K and 1273K in this figure show that: (i) the reduction calcination of $CaSO_4$ was incomplete due to presence of residual $CaSO_4$ and a part of $CaSO_4$ was calcined to CaS. It is worth noting that the phases obtained at 1173K agree broadly with the Ca-Fe-O-S system [4] as there is an equilibrium between CaS + Fe_3O_4 and $CaSO_4$ + $Ca_2Fe_2O_5$ phase fields, at low partial pressure of O_2 gas. However, the $CaSO_4$ + $Ca_2Fe_2O_5$ phases become unstable as the partial pressure of O_2 gas reaches $\log_{10}(PO_2)$(atm) < -11.2 at 1223K. The presence of $CaSO_4$ in the XRD pattern in Figure 8 suggests that the sulphidation reaction was limited by gasification of graphite.

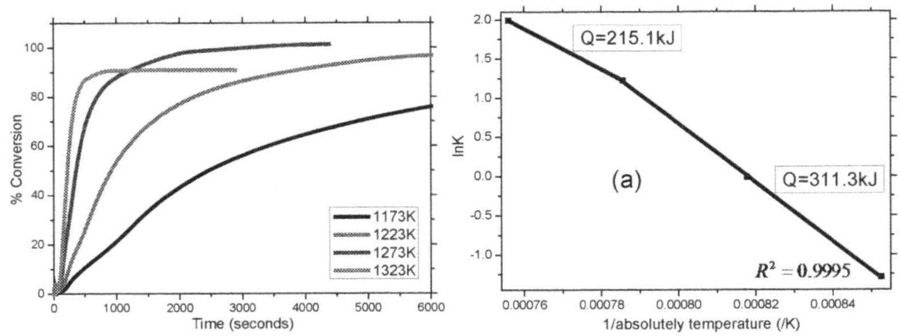

Figure 7. (a) Plot of conversion vs. time curves for the sulphidation of slag at a molar ratio of MO:$CaSO_4$:C = 1:1:1, with graphite and (b) Arrhenius plot of ln K vs.1/T.

The relatively strong intensities of XRD peaks in Figures 8a and 8b for the metal sulphides ($CuFeS_2$ and Cu_2S) suggest that the metal sulphide formation tendency increased from 1173K to 1323K, and that these sulphides co-exist with Ca_2SiO_4 and $Ca_2Al_2SiO_7$ phases as result of the decomposition of $CaSO_4$ to CaO which then reacts with SiO_2 present in situ. However, the residual presence of Fe_3O_4 suggests an incomplete reaction between iron oxides and SO_2 gas. In other ways, part of the SO_2 gas escaped without reacting with the metal oxides.

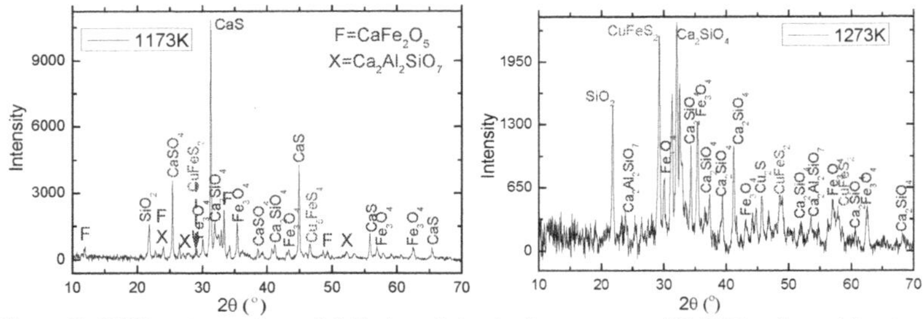

Figure 8. XRD patterns after sulphidation of slag in the presence of CaSO4 and graphite at a molar ratio of MO:$CaSO_4$:C = 1:1:0.9.

Estimation of Activation Energy for Sulphidation of Slag

The experimental data for the sulphidation of slag in the presence of graphite fit well with the mixed reaction model, discussed above. The plot of the lnk vs. 1/T line is shown in Figure 7b and it is evident that the slope of the line changes at 1273K. The change in the slope of the line agreed very with the XRD results (Figure 8) as similar phases were obtained at 1173K/1223K and 1273K/1323kJ. The activation energy of 311 ± 5 kJ in the lower region (between 1173K and 1273K) is very comparable to the activation of 305kJ [6] (1123 and 1373K) and 342kJ for the gasification of CO_2 gas with graphite, determined by Turkdogan et. al. [6] and Karatas [16, 17] respectively. The activation energy in the upper region (215kJ) is also comparable to activation energy of 239 ± 18 kJ, for the reaction between CaS and $CaSO_4$ [18].

Carbothermic Reduction of Matte in the Presence of CaO

The sulphidised slag was mixed with CaO and activated chalcoal and, was reduced between 1173K and 1323K. The plots of the % reduction against time curves are compared in Figure 9a, from which it is evident that the carbothermic reduction of sulphidised slag (matte) was much faster than the sulphide formation reaction. The implication of this observation is that as long as the metal sulphide phase is present in the slag via the decomposition of $CaSO_4$, discussed above, it will reduce to metals and alloys, as can be seen in, Figure 9b at 1173K.

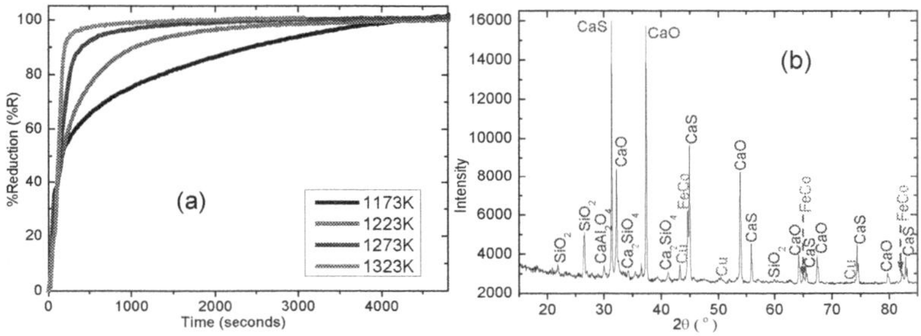

Figure 9. Carbothermic reduction of sulphidised slag the presence of CaO with activated charcoal, for MS:CaO: = 1:2:2 (a) TGA results and (b) XRD pattern at 1173K.

Conclusions

1. Complete solid state reduction of metal oxides from slag via carbothermic reduction requires higher reduction temperature, T ≥ 1323K and a very reactive carbon such as activated charcoal as shown in Figures 5a and b.
2. The activation energy of 271 ± 34kJ and 814 ± 83kJ were determined for the carbothermic reduction of slag with activated charcoal and carbon black respectively. The activation energy for the carbothermic reduction with activated charcoal is comparable to the gassification of CO_2 gas with coal char.
3. Metal sulphides ($CuFeS_2$, Cu_2S, Cu_5FeS_4) were obtained during sulphidation of slag with $CaSO_4$ and graphite at a molar ratio of $MO:CaSO_4:C$ = 1:1.1:0.9
4. The sulphidation of slag in the presence of $CaSO_4$ and graphite is limited by the gasification of graphite at T < 1273, with an activation energy of 311kJ
5. Complete metallisation was achieved via sulphidation of the slag, followed by carbothermic reduction of sulphidised slag (matte) at 1173K (Figure 9a and b).
6. There is a broad agreement between the experimental results and the thermodynamic prediction if Figure 2 as preferential reduction and sulphidisation Cu and Co were achieved at low temperatures.

Acknowledgements

The authors would like to thank the Institute of Materials, Minerals and Mining (IOM[3]) and the Copperbelt University in Zambia for the financial support.

References

1. Davenport, W.G.L., et al., *Extractive Metallurgy of Copper (4th Edition)*. Chemical, Petrochemical & Process, (2002), Elsevier, 1-452.
2. Imris, I., Cobalt distribution in Rokana smelter. *Transactions of the Institution of Mining and Metallurgy, Section C: Mineral Processing and Extractive Metallurgy*, (1982), **91,** p. C153-C161.
3. Jones, R.T., et al., Recovery of cobalt from slag in a DC arc furnace at Chambishi, Zambia. *Journal of The South African Institute of Mining and Metallurgy*, (2002). **102**(Compendex), 5-9.
4. Bale, C.W., et al., FactSage thermochemical software and databases. *Calphad: Computer Coupling of Phase Diagrams and Thermochemistry*, (2002). **26**(2), 189-228.
5. Turkdogan, E.T., *Physical chemistry of high temperature technology, (*Academic Press, 1980).
6. Turkdogan, E.T., et al., Rate of oxidation of graphite in carbon dioxide. *Carbon*, (1968), **6**(4), 467-484.
7. Cutler, C.J., et al. Phasing out reverbatory furnace operations at KCM Nkana. in *Southern African Pyrometallurgy 2006*. 2006. Johannesburg, South Africa.
8. Stephen, Y., R. Hara, and A. Jha, *A Novel Low-Energy Route for the Extraction of Copper and Cobalt Metals/Alloys from the Zambian Sulphide Concentrates.* Characterization of Minerals, Metals, and Materials, (2012), 77-87.
9. Hara, Y. and A. Jha, Carbothermic reduction of Zambian sulphide concentrates in presence of lime, *Mineral Processing and Extractive Metallurgy*, (2013), **122**(3), 146-156.
10. Manasse, A. and M. Mellini, Chemical and textural characterisation of medieval slags from the Massa Marittima smelting sites (Tuscany, Italy), *Journal of Cultural Heritage*, (2002). **3**(3), 187-198.
11. Ward, C.R. and D. French, Determination of glass content and estimation of glass composition in fly ash using quantitative X-ray diffractometry, *Fuel*, (2006), **85**(16), 2268-2277.
12. W.D, K., B. H.K, and U. D.R., *Introduction to ceramics (*2nd Ed., 1976, New York, Chichester, John Wiley).
13. Habashi, F., *Kinetics of metallurgical processes,* (Metallurgie Extractive, Quebec, Canada, 1999).
14. Rao, Y.K. and B.P. Jalan, A study of the rates of carbon-carbon dioxide reaction in the temperature range 839 to 1050 C, *Metallurgical Transactions*, (1972), **3**(9), 2465-2477.
15. Matsui, I., D. Kunii, and T. Furusawa, Study of char gasification by carbon dioxide. 1. Kinetic study by thermogravimetric analysis. *Industrial & Engineering Chemistry Research*, (1987), **26**(1), p. 91-95.
16. Karata and C., Catalysis of the graphite–CO2 reaction by iron from in-situ reduction wustite over the range 870–1100°C, *Mineral Processing and Extractive Metallurgy*, (2001), **110**(1), 7-13.
17. Karatas, C., Catalytic enhancement of graphite - CO2 reaction by in situ reduced chromium and nickel from their oxides Cr2O3 and NiO over the temperature range 850-1100°C, *Mineral Processing and Extractive Metallurgy*, (2004), **113**(1), 19-24.
18. Kutsovskaya, M.L., M.T. Hepworth, and J.R. McGaa, Recovery of Lime, Sulfur, and Iron from Gypsum and Pyrite Wastes. *Ind. Eng. Chem. Res.*, (1996), **35**(5), 1736-1746.

Celebrating the Megascale: Proceedings of the Extraction and Processing Division
Symposium on Pyrometallurgy in Honor of David G.C. Robertson
Edited by: Phillip J. Mackey, Eric J. Grimsey, Rodney T. Jones, and Geoffrey A. Brooks
TMS (The Minerals, Metals & Materials Society), 2014

CARBOTHERMIC REDUCTION OF ILMENITE CONCENTRATE WITH COKE ASSISTED BY HIGH ENERGY BALL MILLING

Bing Song[1], Kai Zhang[1], Fei Xi[1] and Xuewei Lv[1]*

[1]College of Materials Science and Engineering, Chongqing University, China, 400044
*Corresponding author - Email: lvxuewei@cqu.edu.cn

Key words: Ilmenite concentrate, High energy ball milling, Carbothermic reduction

Abstract

The electric arc furnace process is a common method to produce high TiO_2 slag from ilmenite. The pre-reduction process was proposed in recent years in order to reduce electricity consumption. In order to enhance the pre-reduction degree, high energy ball milling was proposed to improve the specific surface area of the ilmenite prior to reduction. In this study, the carbothermic reduction behavior of milled ilmenite concentrate was investigated. It was found that the ball milling method can result in a higher reduction rate and improved metallization of iron. The metallization of iron increased with increasing the time of ball milling. The metallization of iron with milling reached 89.49% when the time of ball milling was 70 min., while the FeO content decreased with 70 min. of milling from 11.84% to 4.76%.

Introduction

The development of the titanium industry requires large amounts of high quality titanium ore materials. As the resources of natural rutile and high-grade titanium mineral decrease worldwide, the low grade titanium ores should be paid more attention [1]. Although ilmenite is a type of low grade titanium ore, it has a significant utility value in the production of titanium dioxide (TiO_2) because of its abundant reserves [2-3]. Titanium dioxide is considered as a raw material for industrial processes related to pigment or other titanium products. However, extraction of the iron from the ilmenite has been, and still is, an expensive undertaking [4].

During the past several decades, a large number of investigations [5-8] have been published regarding the reduction of ilmenite ores. However, research on the improved reduction of ilmenite is relatively limited. The technical features for improved reduction of ilmenite mainly have pre-oxidation, addition of additives and mechanical activation treatment, etc. For example, Zhang and Ostrovski [9-10] investigated the pre-oxidation and sintering of ilmenite concentrate. Ilmenite concentrate with a high iron content had a higher reduction rate at the initial stage. Gupta et al. [11] examined the effect of ferric chloride addition on the reduction of ilmenite. The addition of ferric chloride to ilmenite-graphite mixtures significantly increased the rate. No reduction occurred in the absence of ferric chloride, while with 10 pct $FeCl_3$ the reaction occurred rapidly at 1273 K. Huang et al. [12] and Ranganathan et al. [13] reported on the effect of ferrosilicon addition to the reduction of ilmenite. The metallization ratio of iron and the particle size of iron in the reduced samples increased with increasing Fe-Si amount and reduction time.

If molten Fe-Si alloy was present during reduction, it would be possible enhance the rate of reduction and the segregation of iron. Chen et al. [4, 15] and Welham et al. [14] investigated the carbothermic reduction of ilmenite by a mechanical activation method. It was found that a longer milling time (exceeding 100 hr) resulted in a lower reduction temperature and a higher reduction rate. Tian et al. [16] reported that ball milling resulted in a crystal lattice distortion of ilmenite and an increase in the specific surface area. The oxidation rate of ilmenite can be significantly accelerated and the time of oxidation reaction and phase transition was reduced. Welham et al. [17] reported that ball milling is a non-equilibrium process in which physical energy is transferred into a powder by crystalline damage, the formation of defects and localized heating during impact, giving an increase in the enthalpy and entropy of the system.

A long reduction time, slower reduction rate and higher energy consumption are the major problems in the existing electric furnace processes. The purpose of the present investigation was to study the effect of ball milling time on the solid state reduction of ilmenite concentrate, while offering a theoretical foundation. These results are expected to be valuable to the industry.

Experimental Work

For the present experiments, an ilmenite concentrate was supplied by the Panzhihua Iron and Steel (Group) Co. The chemical composition and particle size distribution of the ilmenite concentrate is listed in Tables I and II respectively. The X-ray diffraction of ilmenite concentrate results showed that $FeTiO_3$, Fe_3O_4 and $MgTiO_3$ were main phases. Coke was used as reducing agent; its composition is shown in Table II. The average particle size of the coke was approximately 74 μ m. The stoichiometric amount of carbon to reduce iron oxides to iron was 9.01%. The amount of coke addition in the tests was 14%, and this wt-% was used with 200 gm of ilmenite concentrate in each experiment. It was found previously[18] that it was important to provide sufficient carbon for full reduction of ilmenite concentrate, hence a small excess was used. High energy ball milling was performed using a ball mill, the diameter of the balls was 10 mm and the volume of the cell was 250 ml. The ball movement was controlled by adjusting the rotational speed. The rotational speed was 300 r/min in the present experiments. A schematic representation of the ball milling action is shown in Figure 1. The ball milling times for mixtures of ilmenite concentrate and coke under Ar atmospheres were 10, 30, 50 and 70 min. at room temperature.

After ball milling, the particle size was determined and the mixture of powdered ore and coke was then pelletized using a mould press; each pellet was approximately 30 mm in diameter. The pellets were dried at 393 K for 2 hr to remove residual water before used in the experimental run. The samples were placed in a corundum crucible positioned in a vertical tube furnace, and then introduced into the constant temperature zone of the main furnace once the desired temperature had been reached. The samples were reduced for 30 min at 1653 K under an Ar atmosphere. A schematic diagram of the reduction experimental apparatus is shown in Figure 2. It is noted that Chen et al. [4] reported that oxygen gas should be excluded from the milling step. In the present work, an Ar purge gas was used; however it is still possible that some oxygen could have been absorbed within the particles which may have affected the reduction process to a small degree. After solid state reduction, the sample was sent for wet chemical analysis of TFe, Fe^{2+} and metallic iron (MFe).

The metallization of the solid state reduction was defined as the following equation:

$$\text{Metallization} = \frac{MFe}{TFe} \times 100\% \quad (1)$$

Where, TFe and MFe are the weights of total iron and metallic elemental iron in the solid state reduction sample respectively.

Table I. Main chemical components of ilmenite concentrate (wt. percent)

TiO_2	Fe_2O_3	FeO	CaO	MnO	MgO	SiO_2	Al_2O_3	V_2O_5
45.64	6.53	36.45	1.12	0.855	3.22	3.65	1.02	<0.10

Table II. Particle size distribution of ilmenite concentrate (raw material)

Particle size (μm)	+150	-150~+110	-110~+75	-75~+44	-44~+37	-37
Content (wt %)	0.9	2.4	23.8	44.5	12.8	15.6

Table III. Main chemical components of coke(wt. percent)

Fixed carbon	Ash	Volatile	S	P
83.66	14.12	2.22	0.65	0.125

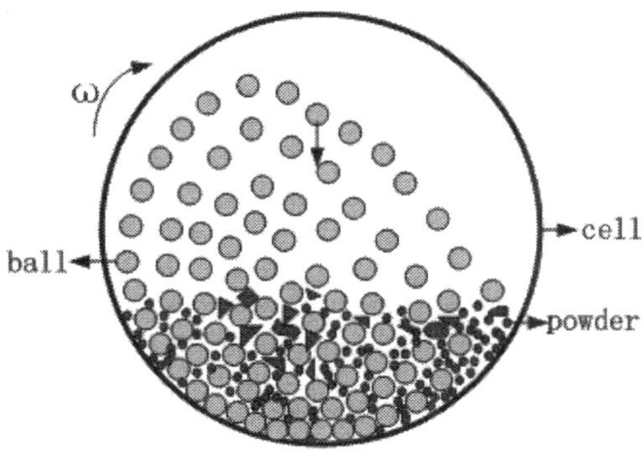

Figure 1. A schematic illustration of the ball milling action (ω = rotational speed)

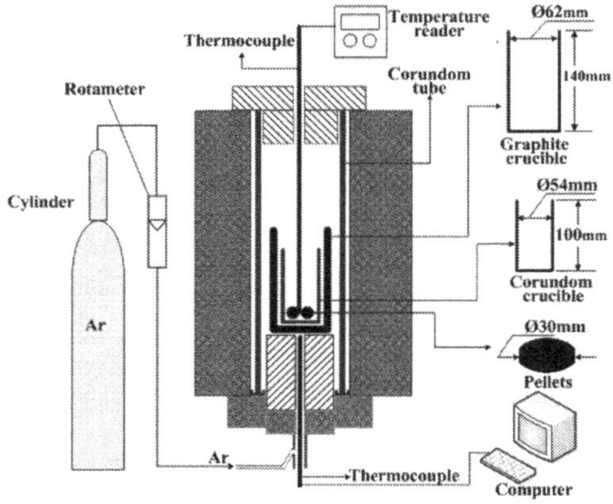

Figure 2. Experimental apparatus for the pellet reduction tests

Results and Discussion

Behavior during Ball Milling - Average Grain Diameter and Surface Area

The average grain diameter and the change in specific particle surface area of the samples ball milled at different times are shown in Figure 3. This shows that the average grain diameter decreased with increasing the time of ball milling. The average grain diameter was reduced from 50.13 to 5.81 μm after 70 min. of ball milling. Because the milling arrangement used in the present experiments provided high energy impact milling rather than a type of abrasive milling, the small size obtained is considered reasonable. This result suggests that nanometer-sized mixtures of ilmenite concentrate and coke could be produced by ball milling for extended times. During the present milling tests, the specific surface area of the material increased from 0.23 to 2.57 m^2/g with increasing the time of ball milling; this increase in specific surface area corresponded to the smaller particle size. The increase in the surface area of ilmenite and coke particles produced by ball milling was considered to increase the contact area between ilmenite concentrate and coke particles. The degree and homogeneity of coke mixing with ilmenite concentrate thus also increased with ball milling time.

Powder Morphology

The SEM pictures taken of the original mixture of ilmenite and coke powder and typical milled mixtures are shown in Figure 4. The particle size of the raw material is mostly less than 100 μm as shown in Figure 4 (at 0 min.). Figure 4 also shows that the particles have clear edges and corners when the ball milling time was 10 or 30 min, but the particles of the former are larger than that of the latter. The largest particles were reduced in size with increasing time of ball milling. While the size of the small individual particles stabilized during further milling, the smaller particles have a tendency towards agglomeration. These results suggest that ilmenite concentrate and coke powders were likely first reduced to a sub-micron size by ball milling.

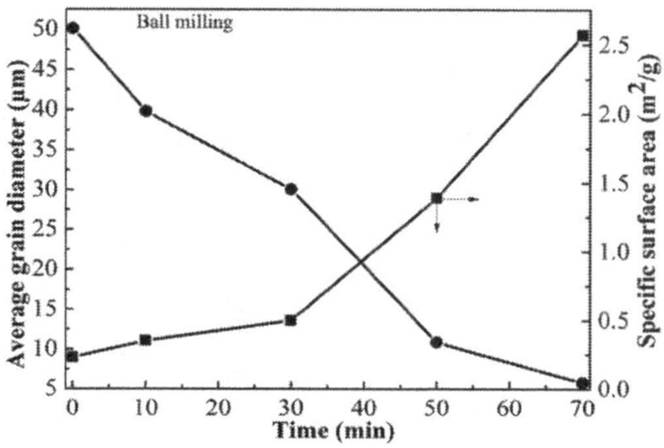

Figure 3. The average grain diameter and specific surface area of test mixtures after ball milling.

Reduction Experiments - Metallization of Iron and the FeO Content in Solid State Reduction Samples

The MFe and FeO content of solid state reduction samples were analyzed by wet chemical methods. The results of the chemical analyses are shown in figure 5. It is indicated that the metallization of iron increased with increasing the time of ball milling, the metallization of iron can reach up to 89.49% with 70 min. milling time.

The metallization of iron significantly improved when the time of ball milling was from 0 to 10 min. This suggests that high energy ball milling can likely improve the subsequent reduction of ilmenite concentrate during the carbothermic reduction process.

This is believed to occur because the particle surface has some active sites, and the active energy decreased after ball milling treatment. On the other hand, Figure 5 also shows that the content of FeO decreased from 11.84% to 4.76% with 70 min. of ball milling. This is because the surface area of particle was increased, accelerating the reduction rate of iron oxides in the ilmenite concentrate. The reduction reaction was such that the FeO content was reduced. Compared with the metallization of iron, as expected, the FeO content showed a reverse trend.

Figure 4. SEM of ilmenite-carbon mixtures after ball milling at different times.
Times for top set, L to R are: 0, 10 and 30 min.
Time for lower set, L to R are: 50, 70, and 70 min. but at a higher magnification.

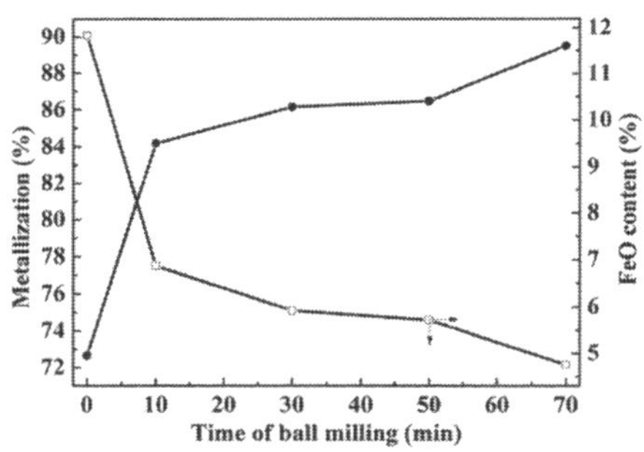

Figure 5. The metallization of iron and the FeO content in the sample after different times of ball milling.

Thermogravimetric Analysis

The TG curve of reduced ilmenite following different ball milling times is shown in Figure 6. This shows that the mass loss was nearly unchanged from the start, up to about 1100^{0}C. This is considered because the reduction reaction is a slower solid-solid reaction in this temperature range, illustrated by equation (2). The Boudouard reaction proceeded at higher temperatures, with carbon monoxide being the main constituent in the gas phase. Therefore, the reaction was considered mainly a gas-solid reaction at this stage and the mass loss rate significantly improves; the process is illustrated by the equations represented in (3). The effect of ball milling time was more noticeable when the temperature approached 1100^{0}C; the mass loss increased with the increasing time of ball milling. Due to the smaller particles size increasing the contact area between ilmenite concentrate and coke, the reduction reaction can proceed more easily.

$$2FeTiO_3 + C = 2Fe + 2TiO_2 + CO_2 \quad (2)$$

$$C + CO_2 = 2CO \text{ and, } FeTiO_3 + CO = Fe + TiO_2 + CO_2 \quad (3)$$

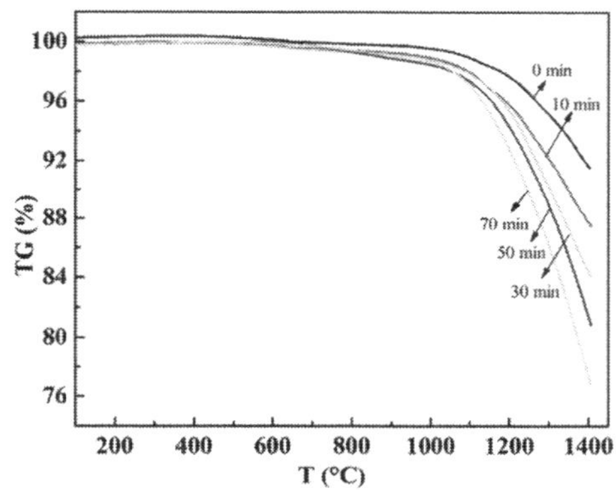

Figure 6. TG curve of reduced ilmenite concentrate at different milling times.

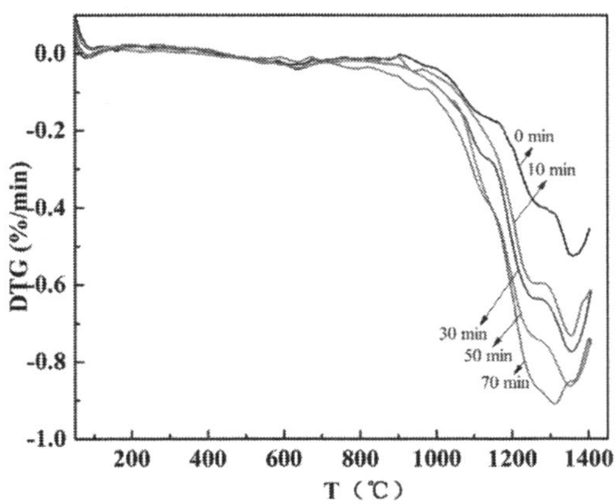

Figure 7. DTG curves of reduced ilmenite concentrate at different milling times.

The relationship between the instantaneous mass loss as a percent and temperature during the reaction process is shown in Figure 7. This shows that the temperature needed to achieve a given maximum mass loss can be reduced by milling; the extent of the temperature reduction was 3^0C, 9^0C, 18^0C and 53^0C respectively for the four milling times given in Figure 7. It is seen that at a given temperature, the mass loss increased with an increasing time of ball milling; this result suggests that high energy ball milling can improve the rate of the reduction reaction.

Conclusions

After high energy ball milling of coke-ilmenite mixtures at room temperature, ilmenite concentrate was reduced during subsequent carbothermic reduction process. The enhanced reduction process in pre-milled samples was attributed mainly to intimate mixing of coke with disordered ilmenite nanocrystallites during milling. The conclusions are summarized as follows:

1. The high energy ball milling can help to improve the reduction rate of ilmenite concentrate during subsequent carbothermic reduction process. The metallization of iron in the solid state reduction sample generally increased with increasing the time of ball milling, while the content of FeO decreased with increasing the time of ball milling.

2. The maximum mass loss rate increased with the increasing time of ball milling. It was found that the temperature needed to achieve a given maximum mass loss can be reduced by milling; the extent of the temperature reduction was 3^0C, 9^0C, 18^0C and 53^0C for the milling times of 10, 30, 50 and 70 mins. respectively. The period of reduction can be shortened after high energy ball milling.

References

[1] P.C. Pistorius, and C. Coetzee, "Physicochemical aspects of titanium slag production and solidification", *Metallurgical and Materials Transactions B,* (2003), 34 (5), 581-588.

[2] H. Elstad, J.M. Eriksen, A. Hildal, T. Rosenqvist and S. Seim, "Equilibrium between titania slags and metallic iron", *Heavy Minerals Conference*, (2007), 35-42.

[3] M. Pourabdoli, Sh. Raygan, H. Abdizadeh and K. Hanaei, "Production of high titania slag by electro-slag crucible melting (ESCM) process", *International Journal of Mineral Processing,* (2006), 78 (3), 175-181.

[4] Y.Chen, T. Hwang, M.Marsh and J.S. Williams, "Mechanically activated carbothermic reduction of ilmenite", *Metallurgical and Materials Transactions A*,(1997), 28 (5), 1115-1121.

[5] M.Tripathy, S. Ranganathan and S.P. Mehrotra, "Investigations on reduction of ilmenite ore with different sources of carbon", *Mineral Processing and Extractive Metallurgy,* (2012), 121 (3), 147-155.

[6] Y.M.Wang, Z.F. Yuan, Z.C Guo, Q.Q. Tan, Z.Y. Li and W.Z. Jiang, "Reduction mechanism of natural ilmenite with graphite", *Transactions of Nonferrous Metals Society of China*, 2008, 18

(4), 962-968.
[7] C.S. Kucukkaragoz and R.H. Eric, Solid state reduction of a natural ilmenite, *Minerals Engineering,* (2006), 19 (3), 334-337.
[8] S.K. Gupta, V. Rajakumar and P. Grieveson, "The influence of weathering on the reduction of Ilmenite with carbon", *Metallurgical Transactions B*, (1989), 20 (5), 735-745.
[9] G.Q. Zhang and O. Ostrovski, "Reduction of ilmenite concentrates by methane containing gas, Part II: Effects of preoxidation and sintering", *Canadian Metallurgical Quarterly*, (2001), 40 (4), 489-497.
[10] G.Q.Zhang, "Reduction of Ilmenite Concentrates By Methanecontaining Gas: Part I. Effects of Ilmenite Composition, Temperature and Gas Composition", *Canadian Metallurgical Quarterly*, (2001), 40 (3), 317-326.
[11] S.K. Gupta, V. Rajakumar and P. Grieveson, "Kinetics of Reduction of Ilmenite with Graphite at 1000 to 1100° C", *Metallurgical Transactions B,* (1987), 18 (4), 713-718.
[12] H. Run, X.W. Lv, C.G Bai, K. Zhang and G.B. Qiu, "Enhancement Reduction of Panzhihua Ilmenite Concentrate with Coke and Conglomeration of Metal with Ferrosilicon", *Steel Research International*, (2013).
[13] S.Ranganathan, K.K. Bhattacharyya, A.K. Ray and K.M. Godiwalla. "Investigations on the reduction and growth of particles of iron from ilmenite ore", *Mineral Processing and Extractive Metallurgy*, (2012), 121 (1), 55-63.
[14] N.J. Welham, "A parametric study of the mechanically activated carbothermic reduction of ilmenite", *Minerals Engineering,* (1996), 9 (12), 1189-1200.
[15] Y.T.Chen, T. Hwang and J.S. Williams, "Ball milling induced low-temperature carbothermic reduction of ilmenite", *Materials Letters,* (1996), 28 (1), 55-58.
[16] W.Tian, L. Lv, B. Liang, P. Wu and C. Li, "Effect of Mechanical Activation on High Temperature Oxidation of Ilmenite", *Chemical Reaction Engineering and Technology*,(2011), 27 (6), 537-542.
[17] N.J.Welham and J.S. Williams,"Carbothermic reduction of llmenite (FeTiO3) and rutile (TiO2)", *Metallurgical and Materials Transactions B*, (1999), 30 (6), 1075-1081.
[18] H. Run, X.W. Lv, C.G. Bai, Q.Y. Deng and S.W. Ma, "Solid state and smelting reduction of Panzhihua ilmenite concentrate with coke", *Canadian Metallurgical Quarterly*, (2012), 51 (4), 434-439.

CELEBRATING THE MEGASCALE

Proceedings of the Extraction and Processing Division Symposium on Pyrometallurgy in Honor of David G.C. Robertson

Pyrometallurgy Process Fundamentals II

Celebrating the Megascale: Proceedings of the Extraction and Processing Division Symposium on Pyrometallurgy in Honor of David G.C. Robertson
Edited by: Phillip J. Mackey, Eric J. Grimsey, Rodney T. Jones, and Geoffrey A. Brooks
TMS (The Minerals, Metals & Materials Society), 2014

ELECTRICITY-INDEPENDENT GENERATION OF SILICON BASED ON THE USE OF RICE HUSK: A CONCEPT PROCESS

Mansoor Barati

University of Toronto
Department of Materials Science and Engineering
184 College Street, Toronto, Ontario, Canada, M5S 3E4,
(mansoor.barati@utoronto.ca)

Keywords: Silicon production, rice husk ash, process evaluation

Abstract

The present paper is based on a term project for final year undergraduate students, requiring them to evaluate a process for producing silicon from rice husk. The traditional method for production of silicon involves carbothermal reduction of quartz in electric arc furnaces at temperatures exceeding 2000 °C. The technology consumes a considerable amount of electrical energy as well as fossil fuels, resulting in significant CO_2 emissions. A flowsheet for production of silicon metal from rice husk is laid out in this paper. The technology will consist of combustors, which will generate steam using the rice husk biomass, power generator, and electric furnace for reduction of the rice husk. Materials and energy balance calculations presented here indicate that such process can be operated independent of external electricity.

Introduction

This paper is presented in honor of Professor David Robertson who is a prominent educator and researcher in the field of pyrometallurgy. His contributions in developing process concepts, evaluation of metallurgical processes using materials and energy balance tools, and training students and engineers to use such tools and techniques are numerous. This article was prepared based on a final year undergraduate term project, in which the students are required to evaluate an alternative process for production of silicon, using materials and energy balance techniques. The aim of this exercise was to (a) encourage creative thinking, with sustainability aspects in mind, for improving existing technologies, (b) strengthen the students' abilities in energy/materials balance calculations as well as flowsheet design and evaluation, and (c) promote the use of engineering tools (HSC Chemistry in this case). The theme of the paper intersects with two primary academic interests of Professor Robertson, giving students hands–on training on the use of computational tools, and process evaluation.

Silicon Production Process and the Proposed Method

Conventional Silicon Metal Production Method

Silicon metal, also known as metallurgical grade silicon, is conventionally produced in submerged electric arc furnaces, by reduction of quartz with carbon from various sources including coal, coke, charcoal, and woodchips. The overall reaction may be presented as following.

$$SiO_2 + C = Si(l) + CO(g) \tag{1}$$

In the operating temperatures of the furnace (1900–2200°C), side reactions including formation of silicon monoxide (SiO) and silicon carbide (SiC) complicate the process. For example, it has been shown that dissociation of SiO gas at lower temperatures (<1900 °C) is responsible in part for the generation of Si

$$2SiO(g) = Si + SiO_2 \tag{2}$$

where SiO is a product of SiO_2 reduction with SiC in the inner hot zone of the furnace (1900–2200 °C) according to Reaction (3). SiC is also generated outsize the hot zone of the furnace via Reaction (4).

$$2SiO_2 + SiC = 3SiO(g) + CO(g) \tag{3}$$
$$SiO(g) + 2C = SiC(g) + CO(g) \tag{4}$$

Proposed Process

The primary source of silica, SiO_2, in the existing silicon making process is lumpy quartz or quartzite rocks. In the proposed process, the material is replaced with rice husk ash (RAH), that is silica-rich by-product of rice husk (RH) combustion. Also, the electrical energy to the EAF is supplied through combustion of RH, rendering the process independent of grid electricity.

RH is a by–product of the rice milling that has traditionally been used as a low quality biomass fuel. It constitutes on average 20 wt% of the rice paddy [1] although the range can be as wide as 14 [2] to 33 wt% [3]. The mineral content of RH is approximately 20 wt%, the remaining being combustible matter and water. Although the combustible matter of RH is composed of hemicellulose, cellulose, lignin, and extractive matter, in the present evaluation it is assumed to be 100% cellulose, noting that the atomic ratio of C/H in an analysis [4] corresponded to that of cellulose. Table I presents a typical analysis of rice husk, Table II provides the composition of RHA produced from different sources and under various combustion temperatures.

The industrial applications of RHA to date are as a pozzolan in the high quality cements and as an insulator in the steel industry. Several attempts[5-9] have been made to generate solar grade silicon from RHA, noting the high silica content and low concentration of photovoltaic-sensitive impurities such as B and P. However, a commercial scale operation has not yet been developed because of two reasons: (a) the high cost of reductants (Ca and Mg) used for low temperature reduction of RHA and (b) the quality of produced silicon that is below photovoltaic Si specifications. The new process concept involves the use of RHA as feedstock for production of silicon metal in the conventional EAFs, therefore both above problems would be eliminated as carbon is used as the reductant, also the impurity level specifications are more relaxed.

Table I. Ultimate analysis of rice husk[4]

Composition (%)	
Carbon	28.08
Total hydrogen	4.89
Nitrogen	0.68
Sulphur	0.21
Oxygen	38.71
Moisture	8.21
Ash	19.22
Gross calorific value	13 $MJ.kg^{-1}$

Table II. Chemical analysis of several RHA (different sources or treatment conditions)[10]

SiO_2	94.6	93.7	91.6	89.1	94.0
Al_2O_3	0.3	0.3	0.5	0.1	0.1
Fe_2O_3	0.3	0.2	0.4	0.04	0.1
CaO	0.4	0.6	0.6	0.6	0.4
MgO	0.3	0.4	0.5	0.5	0.4
Na_2O	0.2	0.2	0.3	0.2	0.1
K_2O	1.3	1.4	1.6	1.0	0.9
P_2O_5	0.3	0.4	0.4	0.9	<0.01
TiO_2	0.03	0.03	0.04	0.01	0.30
SO_3	–	–	–	–	1.5
Carbon	1.0	1.4	2.6	5.1	
LOI	1.8	2.3	3.7	7.0	2.7

The heating value of RH is in the range of 12.1-15.2 MJ/kg[2]. Established technologies for efficient recovery of the RH energy include direct combustion for cogeneration of power and heat, as well as gasification for generation of synthetic gas. Several plants in China and Thailand operate using RH as fuel, with the power generation capacities up to 10 MW [11] which is restricted by the availability of RH within the perimeter of the facility and transportation costs. In this work, the feasibility of using RH as both the source of electrical energy and silica (as RHA) is investigated. As Figure 1 depicts, in this process, rice husk is combusted in a reactor, generating hot gases and RHA. The off–gas is directed to boiler for steam generation and RHA is fed to a briquetting machine to prepare the EAF feed. Electricity from the generator is supplied to the EAF. A detailed materials and energy balance is expected to answer the question of whether the electrical energy produced by combustion of RH is sufficient for carbothermal reduction of the produced RHA. Also, it is of interest to know the overall energy efficiency of RH combustion (fuel to electricity), and the CO_2 footprint of the process.

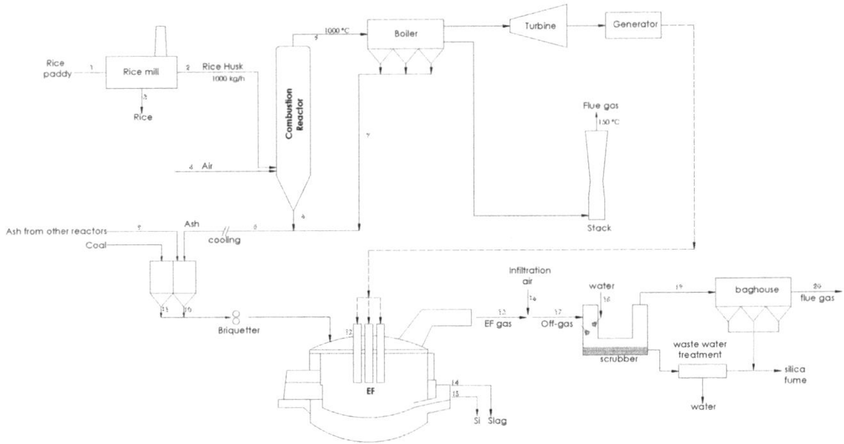

Figure 1. Proposed process flowsheet

Materials and Energy Balance

The materials and energy balance model was set up based on the process flowsheet in Figure 1. Certain assumptions have been made corresponding to typical process conditions pertaining to silicon EAF and RH combustion. These assumptions and some data are provided in the appendix. It is to be noted that the capacity of the combustor is 1 t RH/h whereas the EAF productivity is 1 t Si/h. Therefore, several combustors need to operate in parallel to supply sufficient RH to the EAF.

As the first step, an overall materials balance may be completed around the combustor using Eqs. (5) through (11). In these equations, m_i^j represents the mass of constituent j in stream i.

Carbon balance: $$0.382 \times 1000 = m_5^{CO2} \times \left(\frac{12}{44}\right) + \frac{1.8}{100}\left(m_6 + m_5^{ash}\right) \quad (5)$$

Hydrogen balance: $$\frac{\left(5.5 + 8.3 \times \frac{2}{18}\right)}{100} \times 1000 = m_5^{H2O} \times \left(\frac{2}{18}\right) \quad (6)$$

Mineral Balance: $$0.165\, m_2 = 0.982\left(m_6 + m_5^{ash}\right) \quad (7)$$

Ash distribution: $$m_6 = m_5^{ash} \quad (8)$$

N_2 balance: $$0.767\, m_4 = m_5^{N2} \quad (9)$$

Off-gas O_2: $$\frac{\frac{m_5^{O2}}{32}}{\frac{m_5^{O2}}{32} + \frac{m_5^{N2}}{28} + \frac{m_5^{H2O}}{18} + \frac{m_5^{CO2}}{44}} = 0.04 \quad (10)$$

Overall mass balance: $$m_2 + m_4 = m_5^{N2} + m_5^{O2} + m_5^{H2O} + m_5^{CO2} + m_5^{ash} + m_6 \quad (11)$$

The above calculation gives mass of streams 1-6. The mass of Streams 7 is equal to the mass of the ash recovered from the off-gas, i.e. $m_7 = m_5^{ash}$.

Second, materials and energy balance is established around the EAF. If the system boundary for this calculation is drawn around the EAF and prior to infiltration air, 15 unknowns are identified as following

- Mass of rice husk (1 unknown)
- Mass of coke (1 unknown)
- Mass of silicon metal constituents(Si, Al, Fe, C) (4 unknowns)
- Mass of slag components (CaO- SiO_2, Al_2O_3, Na_2O, K_2O) (5 unknowns)
- Mass of EAF gas components (CO, H_2, SiO_2) (3 unknowns)
- Electrical energy (1 unknown)

The basis of calculations is 1000 kg of silicon metal (this represents one hour of operation and is consistent with the basis of calculations for the rice husk combustor).

Overall mass balance:	$m_{10} + m_{11} + 140 = m_{13} + m_{14} + 1000$ (12)
Si metal mass	$m_{15}^{Si} + m_{15}^{Fe} + m_{15}^{Al} + m_{15}^{C} = 1000$ (13)
% C in Si metal	$\frac{m_{15}^{C}}{m_{15}^{Si} + m_{15}^{Fe} + m_{15}^{Al} + m_{15}^{C}} = \frac{0.1}{100}$ (14)
Fe balance	$\frac{0.5}{100} \times \left(\frac{112}{160}\right) m_{10} = m_{15}^{Fe}$ (15)
Al Balance	$\frac{0.3}{100} \times m_{10} \times \left(\frac{54}{102}\right) + \frac{2}{100} \times m_{11} \times \left(\frac{54}{102}\right) = m_{15}^{Al} + m_{15}^{Al2O3} \times \left(\frac{54}{102}\right)$ (16)
Al distribution	$m_{15}^{Al} = m_{15}^{Al2O3} \times \left(\frac{54}{102}\right)$ (17)
Si Balance	$\frac{95.5}{100} \times m_{10} \times \left(\frac{28}{60}\right) + \frac{3.5}{100} \times m_{11} \times \left(\frac{28}{60}\right) = m_{15}^{Si} + m_{14}^{SiO2} \times \left(\frac{28}{60}\right) + m_{13}^{SiO2} \times \left(\frac{28}{60}\right)$ (18)
CaO Balance	$\frac{0.4}{100} \times m_{10} + \frac{1}{100} \times m_{11} = m_{14}^{CaO}$ (19)
Na_2O Balance	$\frac{0.3}{100} \times m_{10} = m_{14}^{Na2O}$ (20)
K_2O Balance	$\frac{1.2}{100} \times m_{10} = m_{14}^{K2O}$ (21)
C balance	$\frac{89.6}{100} \times m_{11} + \frac{1.8}{100} \times m_{10} + 140 = m_{15}^{C} + m_{13}^{CO} \times \left(\frac{12}{28}\right)$ (22)
H balance:	$\left(\frac{2.5}{100} \times \left(\frac{2}{18}\right) + \frac{0.3}{100}\right) \times m_{11} = m_{13}^{H2}$ (23)
Si distribution between metal and silica fume:	$\frac{m_{15}^{Si}}{m_{13}^{SiO2} \times \left(\frac{28}{60}\right)} = 8.5$ (24)
Si distribution between metal and slag:	$\frac{m_{15}^{Si}}{m_{15}^{SiO2} \times \left(\frac{28}{60}\right)} = \frac{85}{5} = 17$ (25)
Energy balance	$(m_{13}H_{13} + m_{14}H_{14} + m_{15}H_{15}) + H_{loss} - E - (m_{10}H_{10} + m_{11}H_{11} + 140H_{12}) = 0$ (26)

In the energy balance equation, H_{loss} is the energy loss, E is the electrical energy input and based on the given data, $H_{loss} = 0.3E$.
Materials flow for streams 10 – 15, as well as total electricity input are calculated by solving this set of equations using Matrix inversion method in MS–Excel. The enthalpies of the streams (shown as H in Eq. 26) are calculated per kg, using HSC Chemistry functions imported to Excel. Streams 8 and 9 are simply calculated as $m_8 = m_6 + m_7$ and $m_9 = m_{10} - m_9$.

Results

Materials and Energy Balance Results

Solving Eqs. (5)–(11) yield the following:

Table III. Mass balance results for the RH combustor

Stream #	Stream Name	Mass Flow (kg/h)	Volumeric flow/ % (Nm^3/h)	Temp. (°C)
1	Rice paddy	5000	–	25
2	Rice	4000	–	25
3	Rice husk	1000	–	25
4	Combustion air	9755	7581	25
		O_2	21 vol%	
		N_2	79 vol%	
5	Combustor off-gas	9771	7621	1000
			4 vol%	
			79 vol%	
		CO_2	8 vol%	
		H_2O	9 vol%	
6	RHA - 1	84		1000
7	RHA - 2	84		

Solving the second set of equations, streams 8-15 and the EAF power requirement are calculated. The mass balance results are provided in Table IV . The electrical energy consumption in the EAF is 47,000 MJ/t Si or 13.1 MWh/t. This is consistent with typical values of 12–14 MWh/t, although the best performing operations can achieve energy rates as low as 10–11 MHh/t [12]. Since the production rate is 1 t/h, the power requirement would be 13.1 MW.

As seen in in Table IV, a combustor will produce 168 kg/h of RHA (M_8) by processing 1t/h of RH. The amount of RHA required in the EAF, based on 1tSi/h production, is 2569 kg. Therefore:

$$n_{combustors} = \frac{2569}{168} = 15.3 \cong 15 \tag{27}$$

This is the number of combustors that need to work at the same time to provide sufficient RHA feed for the EAF to operate at full capacity. Also, each combustor will generate 0.9 MW power. For 15 combustors, the total power generation is 13.8 MW. This indicates that there is a surplus electricity of ~ 0.7 MW.

Table IV. Mass balance results for streams 9–20

Stream #	Stream Name	Mass Flow (kg/h)	Volumeric flow/ % (Nm^3/h)	Temp. (°C)
8	RHA from one combustor	168	–	25
9	RHA from other combustors	2401	–	25
10	RHA to EF	2569	–	25
11	Coke	763	–	25
12	Electrode	140	–	25
13	EF gas	2281	1673	1600
		CO	97 vol%	
		H_2	3 vol%	
14	Slag	192		1950
		SiO_2	64.6 wt%	
		Al_2O_3	6.0 wt%	
		CaO	9.3 wt%	
		K_2O	16.1 wt%	
		Na_2O	4.0 wt%	
15	Metal	1000		1950
		Si	98.4 wt%	
		Fe	0.9 wt%	
		Al	0.1 wt%	
		C	0.6 wt%	
16	Infiltration Air	16952	13170	25
		O_2	21 vol%	
		N_2	79 vol%	
17	Off-gas	19233	14006	1200
		CO_2	11.6 vol%	
		H_2O	0.4 vol%	
		O_2	13.8 vol%	
		N_2	74.3 vol%	
18	Water	9631.4	9.6	25
19	Cooled off-gas	28873	25999	80
		CO_2	6.2 vol%	
		H_2O	46.3 vol%	
		O_2	7.4 vol%	
		N_2	40.0 vol%	

Energy Distribution around EAF

An energy analysis on the EAF was conducted to obtain the energy distribution on the unit. In these calculations, the chemical energy from combustion of C and H in RHA, electrode, and coke as well as electricity were considered as the input. The energy is distributed between metal, slag, and off–gas, satisfies the endothermic reactions, and also compensates for the heat losses. As seen, the off–gas contains chemical energy because of incomplete combustion (CO).

Table V. Energy balance results for the EAF

INPUT ENERGY

	Thermal	Chemical	Total	%
Rice Husk	0	362	362	0.7
Coke	0	5357	5357	9.9
Electrode	0	1096	1096	2.0
Electricity	–	–	47041	87.3
		Total	53856	100

OUTPUT ENERGY

	Thermal	Chemical	Total	%
Metal	3570	0	3570	6.6
Slag	508	0	508	0.9
Off-gas	3883	2040	5923	11.0
Loss	–	–	14112	26.2
Reactions	–	29743	29743	55.2
		Total	53856	100

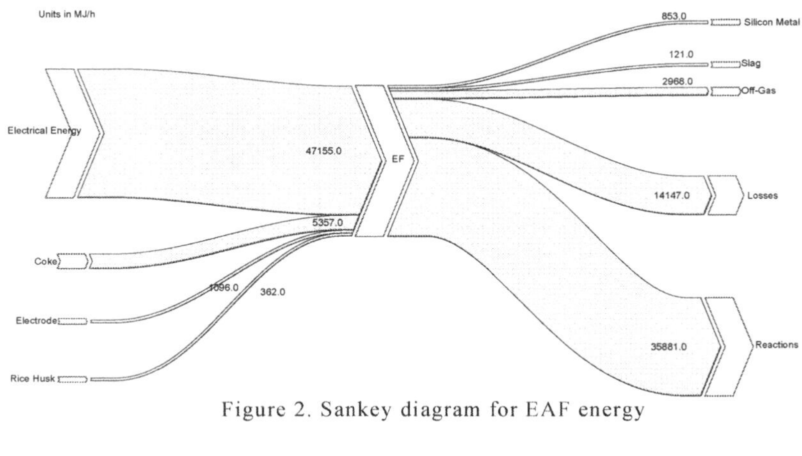

Figure 2. Sankey diagram for EAF energy

The Sankey diagram provided in Figure 2 shows that the silicon EAF is a rather efficient reactor, consuming a large part of the input energy for the reduction reactions.

Carbon Footprint of the RH–based Method

The carbon dioxide footprint of the process is estimated as following, using the mass balance results.

- CO_2 from each combustor = 1192 kg/hr
- CO_2 from all 15 combustors = 18 t/hr
- CO_2 from EF = 3 t/hr
- Total CO_2 = 21 t/hr

In order to compare this value with the grid-dependent EAFs, the indirect CO_2 emission associated with the use of electricity must be taken into account. For example, considering a CO_{2_e} rate of 0.5 kg/kWh (world average), the total CO_2 for a typical silicon metal EAF will be approximately $3 + 13100\times0.5/1000 = 9.6$ t/h. The value will be smaller, if cleaner sources of electricity are used. For example, powered by hydroelectricity (0.015 kg CO_{2-e}/kWh), the emission rate will be $3 + 13100\times0.015/1000=3.2$ t/h. Both these values are smaller than the CO_2 footprint of the RH based process. However, it must be noticed that RH is a neutral CO_2 fuel, *i.e.* the emitted CO_2 as a result of its combustion has already been captured from the environment during the rice plant growth. Therefore, the true CO_2 footprint of the proposed method is ~ 3 t/h.

Overall Efficiency of Power Generation based on RH

The lower heating value of the given RH is 11,360 MJ/t. Once combusted, it yields 0.9 MWh or 3250 MJ of electricity. Therefore, the overall fuel to electricity efficiency is 28.6%.

Challenges

An important challenge for the use of RHA in the conventional silicon EAFs is ensuring that the material bed remains permeable. Generally, lumpy quartz in the range of 15–100 mm[13] is preferred over small particles or fines, as these could reduce the bed porosity and result in incomplete and/or inefficient reactions. Also, smaller particles melt faster that leads to sintering the solids and formation of large cavities that trap gases. Sudden release of the gases creates operational problems, as well as loss in effective use of the reducing power of the gases. Briquetting the RH particles that are generally fine (<150 μm[9]) with coal/coke may solve this problem.

Another challenge could be adequate supply of RH from the plant's vicinity as transportation of large volumes of RH could be cost prohibitive.

Lastly, the economics of the method need a careful analysis, as this could be extremely sensitive to the geographical location of the operation and local market conditions for RH. A highly rudimentary analysis is given below for the operating costs.

For production of one tonne silicon through the proposed method, approximately 15 tonnes of RH is required based on the materials balance presented. Assuming a cost of ~ 30$/ t for RH (price in India and Thailand, where there is an active market for RH), the cost of silica source together with electricity, all generated from RH, is 15×30=$450 /t Si. On the other hand, one can estimate the cost of grid electricity for a conventional process as ~ (13000 kWh/t Si) × (0.07 $/kWh) = $910. Therefore, the cost of electricity alone in a traditional operation could be twice as large as the cost of electricity and silica source in the proposed method, making the latter a more attractive approach. However, a fair comparison should include many more elements including the capital costs, the cost of transporting RH to the plan and the cost of quartz for the conventional method, the value of the by–products (silica fume), local prices, etc.

Summary and Conclusions

A concept process was put forward for generation of silicon metal from RHA, independent of grid electricity. In this method, RHA is reduced in EAF using the electricity generated in–situ by combustion of RH. A materials and energy balance analysis shows that the capacities of the combustion unit and EAF match perfectly so that both the RHA and electric power generated in the combustion plant can satisfy the EAF feedstock and electricity requirements.

References

[1] E. Natarajan, et al., Overview of Combustion and Gasification of Rice Husk in Fluidized Bed Reactors, *Biomass and Bioenergy*, 14 (1998), 533-546.

[2] E.C. Beagle, *Rice Husk Conversion to Energy*, in, FAO, Rome, 1981.

[3] T. Kapur, et al., Electricity generation from rice husk in Indian rice mills: Potential and financial viability, *Biomass and Bioenergy*, 14 (1998), 573-583.

[4] K.K. Larbi, et al., Use of Rice Husk for Neutral Emission Energy Generation and Synthesis of Solar Grade Silicon Feedstock, *Biomass Conversion and Bioenergy*, 2 (2012), 149-157.

[5] R. Singh, B.K. Dhindaw, *Production of High Purity Silicon for Use in Solar Cells*, in: F. De Winter, M. Cox (Eds.) Sun, Mankind's Future Source of Energy: Proceedings of the International Solar Energy Congress, Pergamon Press, Oxford ; New York, 1978, pp. 776-781.

[6] H.D. Banerjee, et al., Investigations on the Production of Silicon from Rice Husks by the Magnesium Method, *Materials Science and Engineering*, 52 (1982), 173-179.

[7] D.N. Bose, P.A. Govindacharyulu, *Progress in Solar-Grade Silicon from Rice Husk*, in, Pergamon Press, Perth, Aust, 1984, pp. 2735-2739.

[8] N. Ikram, M. Akhter, X-ray Diffraction Analysis of Silicon Prepared from Rice Husk Ash, *Journal of Materials Science*, 23 (1988), 2379-2381.

[9] K.K. Larbi, et al., Reduction Behaviour of Rice Husk Ash for Preparation of High Purity Silicon, *Canadian Metallurgical Quarterly*, 50 (2011), 341-349.

[10] M. Nehdi, et al., Performance of rice husk ash produced using a new technology as a mineral admixture in concrete, *Cement and Concrete Research*, 33 (2003), 1203-1210.

[11] L. Jeng Shiun, et al., A review on utilisation of biomass from rice industry as a source of renewable energy, *Renewable and Sustainable Energy Reviews*, 16 (2012), 3084-3094.

[12] A. Luque, S. Hegedus, *Handbook of Photovoltaic Science and Engineering*, in, John Wiley & Sons, West Sussex, UK, 2003.

[13] A. Schei, et al., *Production of High Silicon Alloys*, (TAPIR FORLAG), Trondheim, 1998.

Appendix: Assumptions and Data

- Silicon metal (stream # 15) is an alloy of Si, Al, Fe and C, with C = 0.1 wt%.
- 50% of the total ash produced in the combustor reports to the gas phase. The ash is later recovered (stream # 7).
- In EAF:
    - 50% of the total alumina input is reduced to aluminum (and reports to Si), the remaining reports to slag
    - Iron oxide is completely reduced.
- The EAF gas (stream # 13) contains only CO, H_2 and silica fume (solid SiO_2). The gas is later mixed with infiltration air (# 16), to reduce the temperature to 1200 °C (stream # 17). The gas is fully oxidized in the process as well (will contain CO_2, H_2O, O_2, N_2, silica fume)
- Silicon in the EAF is distributed between slag (as liquid SiO_2), metal (as liquid Si), and gas (as solid silica fume). The distribution of Si between the three phases is metal: slag: gas = 85:5:10
- Energy loss in the EAF is equal to 30% of the total electrical energy input
- The EAF slag contains CaO, SiO_2, Al_2O_3, Na_2O, K_2O (all of iron oxide in the ash is reduced to metal)
- Graphite electrode is 100% Carbon
- The EF off-gas is scrubbed with water to lower its temperature. In the scrubber, the gas is saturated with water (temperature of the gas reaches Adiabatic Saturation Temperature)

Ambient conditions:

Temperature	25 °C
Air composition	21% O_2 and 79% N_2 (volume %)

Analysis of rice husk ash, in wt%

SiO_2: 95.5 Al_2O_3: 0.3 Fe_2O_3: 0.5 CaO: 0.4 Na_2O: 0.3
K_2O: 1.2 C: 1.8

Coke analysis (for EAF) in wt%

CaO: 1.0 SiO_2: 3.5 Al_2O_3: 2.0 H_2O: 2.5 H_2: 0.3
O_2: 1.1 C: 89.6

Crude Si analysis (from EAF) in wt%

Fe: ? Al: ? C: 0.1 Si: ?

Combustion Reactor:

Capacity:	1000 kg/hr of Rice Husk
Off-gas Temp.:	1000 °C
Off-gas O_2:	4 vol%

Electric Furnace

Production:	1000 kg/h (silicon metal)
Off-gas Temp.:	1600 °C
Diameter	0.5 m
Height	4.0 m
Off-gas O_2:	4 vol%
Si Metal Temp.:	1950 °C
Graphite electrode consumption	140 kg/ton silicon

Efficiencies of the power generation system

Boiler efficiency:	85% (thermal energy to steam energy)
Turbo-generator:	33% (steam energy to electricity)

Celebrating the Megascale: Proceedings of the Extraction and Processing Division
Symposium on Pyrometallurgy in Honor of David G.C. Robertson
Edited by: Phillip J. Mackey, Eric J. Grimsey, Rodney T. Jones, and Geoffrey A. Brooks
TMS (The Minerals, Metals & Materials Society), 2014

ELECTRICALLY ENHANCED METAL PURIFICATION USING SLAG

Md Saiful Islam, M Akbar Rhamdhani, Geoffrey A. Brooks
Swinburne University of Technology, Melbourne, Australia

Keywords: silicon; metal purification, solar grade silicon; electrochemistry; kinetics

Abstract

A novel electrically-enhanced metal purification using slag has been developed. This was achieved by applying DC potential between liquid metal and slag phases. The concept was demonstrated through small-scale experiments at temperatures of 1500-1600°C for the removal of boron from Si-B melt using $CaO\text{-}SiO_2\text{-}16wt\%Al_2O_3$. Silicon (containing B) and slag were reacted in an alumina crucible in a resistance tube furnace. A circuit was produced by immersing graphite rods into the silicon bath and the slag layer. Electrical potential differences of up to 3.5 V were applied during the reactions. The reaction rates, the open-circuit voltages and short-circuit currents (in the case when the potential was applied) were then measured and determined for various slag to silicon ratio. It has been shown from this study that both the apparent rate and the boron partition ratio were increased by a factor of approximately 1.6.

Introduction

Slag-metal reactions are of significant importance in the refining and extraction of many metals in metallurgical processes. In a steelmaking process, slags have been commonly used to refine steel, for example in the dephosphorization, desiliconization, and desulfurization processes. A similar process using slag has also been used for refining of liquid silicon. Slags can be reacted with liquid silicon to absorb impurities such as B and P. In such processes, the extent of impurities removal will be limited by the equilibrium, i.e. by the impurities partition ratio. Some of the slag-metal reactions in the refining of the common metals (such as iron and steel) are electrochemical in nature because the components in a slag are predominantly in the ionic state, e.g. when the slag has high basicity. The electrochemical nature of slag-metal reaction has been demonstrated by a number of investigators, for example in the case of the desulphurisation of cast iron and Fe-Cr-S alloy by $CaO\text{-}SiO_2$ and $CaO\text{-}SiO_2\text{-}CaF_2$ slags [1], the transfer of oxygen through $PbO\text{-}SiO_2$ slag to liquid lead [2], the reduction of MnO from molten $MnO\text{-}SiO_2\text{-}CaO\text{-}Al_2O_3$ slag [3], and reaction between Fe-C alloys and $CaO\text{-}SiO_2\text{-}Al_2O_3$ slags [4-5].

Considering the electrochemical nature of the above processes, one would expect that external potential difference and/or electrical field would have an effect on the reactions. In this study a novel method has been developed to improve the refining of metals using slag. This was achieved by applying DC potential across the two liquid phases to alter the kinetic and to some extent the thermodynamic limit of the process. This has been demonstrated in the refining of liquid silicon using slag. In the first set of experiments, the kinetics and boron partition ratios in the case of reaction between Si-B and $CaO\text{-}SiO_2\text{-}16wt\%Al_2O_3$ slag were investigated. In the second set of experiments, DC voltage difference was applied between the two liquid phases. The enhancement of the kinetics and the increase in the boron partition ratio were then compared with the first set of experiments.

Existing Processes for Refining of Silicon

The production of polycrystalline solar grade silicon (SOG-Si) mainly depends on the off-spec scrap of electronic grade silicon from the semiconductor industry. With the increase demands of photovoltaic power generation, it is important to have alternative processes with more cost effective and lower environmental impact. Refinement of metallurgical grade silicon (MG-Si) and the electrochemical routes could be alternative to produce low-cost production route. A list of electrochemical and other alternative refining of MG-Si is shown in Table 1.

Table 1. Production routes of electrochemical and Metallurgical refining

Process Route	Raw Material	Chemical Reactions/Temperature/Pressure	Developer
Refinement Methods of MG-Si			
Refinement of MG-Si: Etching/Acid [6]	MG-Si	Crushing → Sieving → Acid Leaching (HCl, HF, HNO_3) with Stirring (mechanical or Ultrasonic) at 50^0C for 8 hrs → Washing & Drying	Hemlock, Elkem ASA, Heliotronic, etc
Refinement of MG-Si :Slag process [7]	MG-Si	$[B] + (3/2)O^{2-} + (3/4)O_2 \rightarrow (BO_3^{3-})$;$1500^0C$ $[P] + (3/2)O^{2-} + (3/4)O_2 \rightarrow (PO_4^{3-})$	Nippon Steel, Crystal Systems, etc
Refinement of MG-Si : electron beam and Plasma [8]	MG-Si	Electron beam melting (Phosphorus removal) → Plasma($Ar\text{-}O_2$, $Ar\text{-}H_2O$ etc) treatment (Boron removal); (2027^0C) → Directional solidification	JFE, Univ. Tokyo, Kawasaki steel Corporation, UNICAMP
Refinement of MG-Si : Solidification from Si-Al alloy [9]	MG-Si	Directional solidification of Si-Al alloy to remove Boron (1273-1473K)	Univ. Tokyo
Electrochemical Approach			
Three layer Electrorefining [10]	MG-Si	Me (Na, K, Ca, Ba, etc) → $Me^{n+} + ne^-$;(1700K) (Anode) $Si^{4+} + 4e^- \rightarrow Si(l)$; (Cathode)	NTNU, SINTEF
Direct electrolytic reduction of SiO_2 [11]	Pure silica	$SiO_2\,(s) + 4e^- \rightarrow Si(s) + 2O^{2-}$(1123K)	

Boron is one of the most important impurities to remove from molten silicon because of the relatively high segregation coefficient in silicon and low partial pressure. The slag refining for removal of boron from MG-Si has been studied by several investigators. One of the limitation of the process is the limit of the boron partition ratio. There have been a number of studies on the thermodynamics of silicon melt in equilibrium with different slag systems. In this method, liquid silicon is treated with $CaO\text{-}SiO_2$ based slag with addition of CaF_2, Al_2O_3, MgO [12-13]. The slag's boron capacity is often represented by the boron partition ratio, defined as $L_B = (wt\%B)/[wt\%B]$ where (wt% B) and [wt% B] represent the concentration of boron in slag and silicon phase, respectively. Teixeira and Morita [14-15] investigated the use of $CaO\text{–}SiO_2$ and $CaO\text{–}SiO_2$–(25% and 40%) CaF_2 with changing basicity ratio 0.55-1.21 at 1550°C and the highest value of L_B of 5.5 was attained for basicity of 1.21. Their results also suggested a strong dependence on both the oxygen partial pressure and the basicity. Johnston and Barati [7] also confirmed that L_B increases through a local maximum with increasing basicity and oxygen partial pressure for $Al_2O_3\text{–}CaO\text{–}MgO\text{–}SiO_2$ and $Al_2O_3\text{–}BaO\text{–}SiO_2$ slags at 1773K. Furthermore, the general observation was that L_B reached a maximum when the CaO/SiO_2 ratio close to 2 for the slag systems $CaO\text{-}SiO_2$ with additions of CaF_2, BaO, and MgO and $CaO\text{-}SiO_2\text{-}Al_2O_3$ respectively [12-13]. Luo et al. [16] claimed that the value of L_B increased with increasing temperature due to reduced viscosity, enhanced mobility and increased silica activity of the $CaO\text{-}SiO_2\text{-}Al_2O_3$ slag. It is useful to assess the removal rate of boron along with the L_B value. A list of selected previous studies on kinetics of boron removal using slag is shown in Table 2.

Table 2. Selected previous studies on kinetics of boron removal

Investigators	Initial Slag & Alloy Composition	Experimental Conditions	Mass Transfer Coefficient (m/s)	Boron Partition Ratio, L_B
Krystad et al. [17]	SiO_2-CaO CaO/SiO_2=1 Si with 250 ppm B	Electric resistance furnace T=1873K; Gas=Ar; Slag/Si = 1, 2 Equilibrium time=3hr	2.1×10^{-6} (S/M=1)	3.3
			1.2×10^{-6} (S/M=2)	2.6
	SiO_2-CaO-20%MgO CaO/SiO_2=1 Si with 250 ppm B	Electric resistance furnace T=1873K; Gas=Ar, Slag/Si = 2 Equilibrium time=3hr	3.2×10^{-6}	2.3
Nishimoto and Morita [18]	SiO_2-CaO CaO/SiO_2=1.22 Si with 300 ppm B	Electric resistance furnace T=1873K;Gas=Ar; Slag/Si = 2.5 Equilibrium time=3hr	1.4×10^{-6}	3.0

Electrically Enhanced Silicon-Slag Reactions

Since the reactions in the refining of silicon using slag are electrochemical in nature, one could expect that disturbances to the electrical field and charge distributions in the phases (e.g. by applying external potential difference) would affect the overall reactions. From the kinetics perspective, this potential difference could either enhances or slows down the diffusion or mass transfer of charged species involve in the reactions. Once the kinetics barrier is overcame (i.e. the controlling step is shifted from diffusion or mass transfer of reacting species to chemical reaction) or when the B concentration reaches its original equilibrium concentration (with no applied potential), the additional external applied potential could also affect the overall chemical reaction equilibrium. Consider, a process that is a chemical reaction controlled or close to equilibrium. The overall reaction of boron removal from silicon into the slag with probable anode and cathode reactions are

$$[B] + 3/2\ O^{2-} + 3/4\ O_2 = (BO3^{3-}) \tag{1}$$

$$\text{Anode:}\quad \underline{B} + 3\ O^{2-} = BO_3^{3-} + 3e^- \tag{2}$$

$$\text{Cathode:}\quad \tfrac{3}{4}\ Si^{4+} + 3e^- = \tfrac{3}{4}\ Si \tag{3}$$

The central idea is to enhance this process electrochemically, by using the Si-B liquid as an anode. Let us consider a liquid silicon / molten slag two-phase system, with at least two components in both phases (they can contain more components), namely Si-B and SiO_2-B_2O_3. Let us assume first chemical equilibrium occur in the system, with an equilibrium partition ratio of boron, L_B^o. By applying an anodic potential, ΔE to the Si-B alloy, the new equilibrium state can be related to the original using the following equation based on Nernst equation:

$$\Delta E \cong \frac{R \cdot T}{3 \cdot F} \cdot \ln\left(\frac{L_B}{L_B^o}\right) \tag{4}$$

where L_B is the new equilibrium partition ratio, R = 8.3145 J/mol.K, F = 96485 C/mol, T is temperature in K. The sign of approximation is used to accommodate the assumptions of dilute solution of B in liquid silicon and dilute solution of B_2O_3 in slag i.e. that is the activity coefficients are assumed to be constant. The Equation (4) can be re-written as:

$$L_B \cong L_B^o \cdot \exp\left(\frac{3 \cdot F \cdot \Delta E}{R \cdot T}\right) \tag{5}$$

Teixeira and Morita [14] suggested that at equilibrium the partition ratios of boron between $CaO-SiO_2$ slags and liquid Si are between 2 and 5. By assuming a value of initial boron partition ratio of 3 and also assuming that there are no additional resistances across the circuit and phases (i.e. the anodic potential is equal to the applied potential through electrodes), one can evaluate the change in the boron partition with applied electrical potential difference. The calculated values are shown in Figure 1. It can be seen that the partition ratio increases exponentially with applied potential. By applying potential difference of 100mV, theoretically, the partition ratio can be increased by seven times (from 3 to 20). For small applied potential, one could only expect small increase in the boron partition ratio.

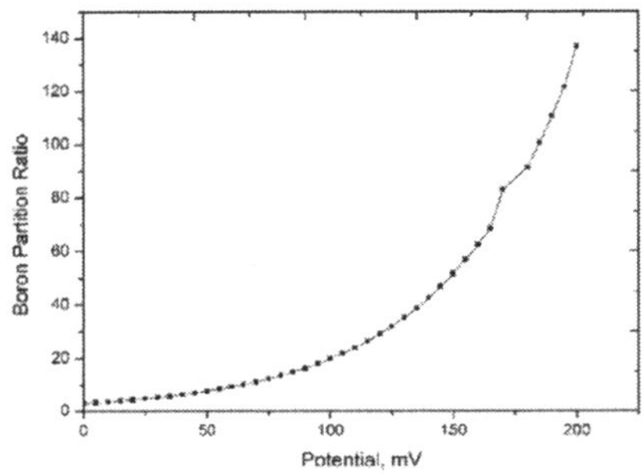

Figure 1. The dependence of the partition ratio of boron on the applied potential in $CaO-SiO_2$-16wt% Al_2O_3 slag using $L_B{}^{o}$ = 2.76, T = 1550°C

Experimental Methodology

To demonstrate the concept of electrically enhanced refining process using slag, experiments have been undertaken using $CaO-SiO_2$-16wt% Al_2O_3 slag. The rate of boron removal from silicon was evaluated by tracking the change in boron concentration in the silicon with reaction time. The bulk chemical analyses were carried out using the inductively coupled plasma (ICP-AES) technique. FactSage 6.4 thermodynamic package was used to assist with the selection of experimental parameters. Relatively small silicon-slag masses and crucible geometries were used for the kinetics experiments to allow better quenching after completion of the experiments. High purity chemicals, obtained from Sigma Aldrich (SA) and Alfa Aesar (AA), were used in the study, as shown in Table 3. Master slags were prepared from powdered oxides of $CaCO_3$, SiO_2, and Al_2O_3. The $CaCO_3$ was dried at 500°C for 5 hours, followed by heating for successive 2-hour periods at 1000°C and 1250°C to decompose the carbonate to produce CaO. The appropriate powders were weighed and mixed thoroughly in the desired ratio using ball mill for 24 hours. Slag was prepared by melting a batch composition in a graphite crucible using induction furnace. The slags were melted two times to allow homogenization. The master slag was analysed using ICP-AES to confirm the composition, as shown in Table 4. The intended composition was $CaO-SiO_2$-16wt%Al_2O_3 slag with basicity of 1.2. High-purity silicon and was doped with boron by mixing crushed silicon with powdered boron and melted in an alumina crucible using a vertical tube furnace. Table 5 shows the initial master silicon alloy composition analysed by ICP-AES. Alumina crucibles with inner diameter 16 mm were used to contain slag and silicon for the kinetic experiments. Crushed silicon alloy and slag were placed in the crucible inside a resistance heated vertical tube furnace. After evacuating the chamber, the furnace was flushed with high purity argon with flow rate of 250 ml/min. Figure 2(a) shows the setup of the furnace during the experiment. The crucible was placed in a cold zone inside the furnace at the beginning of the experiment. The crucible was then moved to the hot zone using a movable pedestal when the temperature reached 1550°C. The start of the reaction (t = 0) was defined as the time when the crucible was placed in the hot zone. For every reaction time, at least one experiment was carried out. When the pre-determined time had elapsed, the crucible was withdrawn from the hot zone, quenched by flushing argon gas into the water cooled copper coil. The experimental parameters used in the study are shown in Table 6.

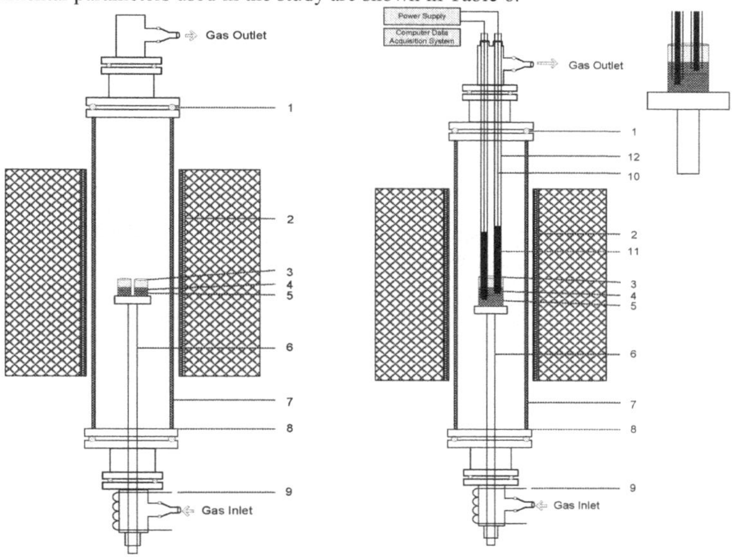

Figure 2: Furnace set up for kinetics experiments (a) with and (b) without applied potential. Legend: 1-Silicone O-ring, 2- $MoSi_2$ heating Element, 3-Crucible, 4-Silicon, 5-Slag, 6-Alumina Pedestal, 7-Alumina Tube, 8-water cooled flanges, 9-copper coil, 10-Kanthal wire, 11-Graphite electrode.

Table 3. Summary of chemicals used

Chemicals	**SiO_2**	**$CaCO_3$**	**Al_2O_3**	**Si**	**B**
Purity (wt. %)	99.5	99.5	99.5	99.9999	>95-97
Shape	Powder	Powder	Powder	Granular	Powder

Table 4. Master slag composition analysed by ICP-AES

Type	**SiO_2**	**Al_2O_3**	**Fe_2O_3**	**CaO**	**MgO**	**Na_2O**
Slag#1 (wt. %)	37.57	15.9	0.24	46.50	0.07	0.09

Table 5. Initial master silicon alloy composition analysed by ICP-AES

Elements	**B**	Ca	Al	Fe	Mg	Na	K	P	Ti	Zn
Concentration (ppm)	**371**	56	30	53	<10	<10	<10	<10	15	15

Table 6. Experimental parameters for kinetics experiments

Slag type & alloy details	**Parameters**	**Time (mins)**			**Slag/Silicon ratio**
Slag#1 *Alloy:* Si with 350 ppm B	Temp: 1550°C Gas: Argon Crucible: Alumina	9	45	120	1.5 (slag: 4g.; Si: 2.66g)
		15	60	180	2 (slag: 4g; Si: 2g)
		30	90	210	2.5 (slag: 4g; Si: 1.6g)

The schematic of the set up and the details of the furnace for the electrically enhanced kinetics experiments are shown in Figure 2(b). The top and bottom part of the furnace were designed so that they can suspend a larger crucible and also allow electrodes to be introduced from the top part of the furnace. The crucible was positioned in the furnace using an alumina pedestal from the bottom. A silicone seal was used on the furnace end cap to hold the electrodes. An external circuit was formed by connecting one high purity graphite electrode to the silicon bath and a second graphite electrode to the slag layer. These two electrodes were sheathed by thin alumina tubes and were connected by kanthal wire to a power supply and a computer data acquisition system. A DC power supply was connected in order to apply the potential difference to the slag-metal cell. In this study, external potential of 0V, 2.5V and 3.5V were applied to the system. A computer based data acquisition card (National Instrument DAQ Card 9250) was used to record the voltage and current data (measured through a current shunt) during the experiments. The voltage and data were stored in a computer through the LabView 11.0 software. The start of the reaction (t = 0) was defined as the time when the electrodes immersed into the two phases. When the pre-determined time had elapsed, the crucible containing the silicon-slag system was quenched by lowering it to the cool zone at the bottom of furnace where cool Ar gas was flowing. This cooling zone was cooled by circulating water through copper coil. At the end of the experiments for each reaction time, the slag and silicon phases were physically separated from the crucible, then ground to a powder for analysis. The boron content in all the samples was analysed by ICP-AES technique. Selected repeated experiments at particular reaction times were carried out to estimate the error and variation of results.

Results and Discussion

Figure 3(a) shows the change in the boron concentration during the slag-silicon reaction. Selected repeated experiments were carried out to evaluate the variation of the concentration, and these were presented in the Figure 3(a) as vertical bars. The experimental results show that the rate of boron removal was high at the beginning of the reaction and it dropped to zero after 7000 seconds of the reaction, depending on the ratio of the amount of silicon to slag. In this study, the rates of boron removal were found to increase with increasing slag to silicon ratio. To quantify the increase in the apparent rate, further kinetic analysis was carried out. In high temperature metallurgical processes it can be expected that chemical reactions are sufficiently fast and rarely to be the rate limiting step. It was pre-assumed that the kinetic was controlled by mass transfer, thus the kinetic could be described by the Equation (6) [19].

$$Y_i \ln Z = -k_i t \tag{6}$$

where Y_i is a term that depends on the particular rate-determining step, i.e. $Y_i = Y_{Si}$ and $Y_i = Y_S$ for silicon and slag mass transfer controlled kinetics, respectively. Where,

$$Y_{Si} = \frac{V_{Si}}{A}\frac{(wt\%B)_o - (wt\%B)_e}{(wt\%B)_o},\ Y_S = \frac{V_S}{A}\frac{(wt\%B)_e}{(wt\%B)_o},\ \text{and } Z = \left[\frac{(wt\%B) - (wt\%B)_e}{(wt\%B)_o - (wt\%B)_e}\right] \tag{7}$$

In these equations, V_{Si} and V_S are volume of the Si and slag phases, A is the reacting interface area. The $(wt\%B)$, $(wt\%B)_o$, $(wt\%B)_e$ are the boron concentration at time t, at initial and at equilibrium, respectively.

First order integrated rate plots assuming both silicon and slag mass transfer controlled were developed. The plot of the data assuming a slag phase mass transfer controlled resulted in a straight line for all the data with different slag to silicon ratio, as shown in Figure 3(b). This may suggests that the kinetics were controlled by mass transfer in the slag phase. The slope of the line shown in Figure 2(b) corresponds to the mass-transfer coefficient in the slag (k_S), which is 1.33×10^{-6} m/s for CaO-SiO_2-16wt%Al_2O_3 slag. Detailed kinetics analyses are currently being carried out and these are beyond the scope of the present paper.

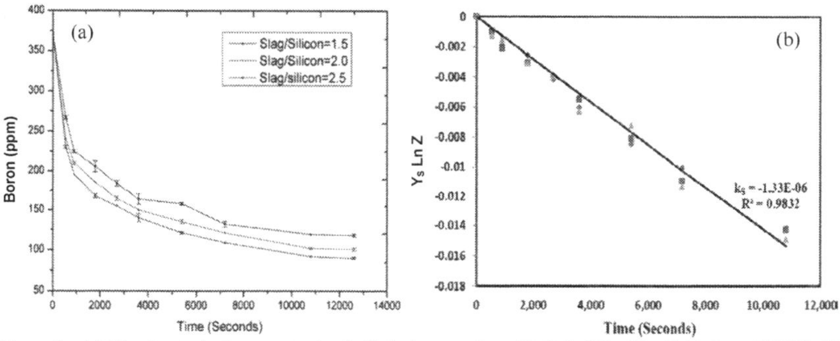

Figure 3 – (a) The change in B concentration in Si during reaction with CaO-SiO_2-16wt%Al_2O_3 at 1550°C, (b) first order integrated kinetics of different slag to silicon ratio assuming slag mass transfer controlled model

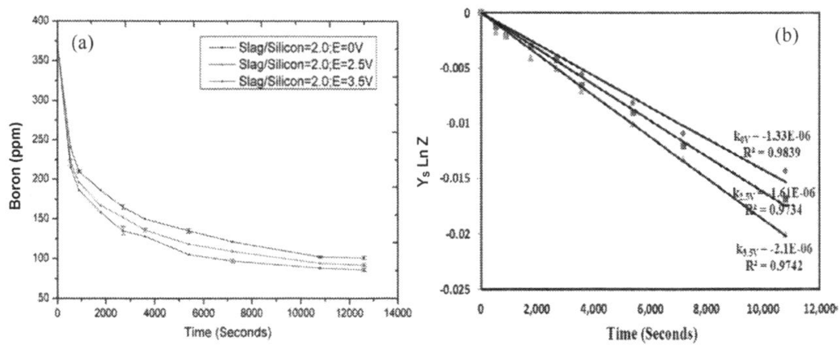

Figure 4 – (a) The change in B concentration in Si, and (b) the integrated kinetic plots, during reaction with CaO-SiO_2-16wt%Al_2O_3 slag at 1550°C with applied potential of 0V, 2.5V and 3.5V

A preliminary measurement of an open circuit voltage showed a value of approximately 0.25V. This was carried out by dipping a graphite electrode into the slag melt contained in a graphite crucible which connected to the circuit. In the main experiments, Figure 4(a) shows the change in the boron concentration during the experiments for CaO-SiO_2-16wt%Al_2O_3 slag with and without applied DC voltage for slag/silicon ratio 2.0. By applying a 3.5V potential difference, the apparent equilibrium boron concentration was decreased by approximately 15%. That is the apparent boron partition ratio was increased by a factor of 1.6. From the trend in the Figure 4(a), the rate of boron removal was found to increase with increasing the applied anodic potential into silicon bath. To quantify the increase in the apparent rate, kinetic analysis was carried out using Equation (6). In this analysis, the apparent equilibrium B concentrations were used for the value of $(wt\%B)_e$, i.e. concentrations at time 3.5h (12,600s) as shown in Figure 4(a). Therefore, when evaluating the kinetics of the case of 2.5V and 3.5V experiments, the new equilibrium values shown in the Figure 4(a) were used. It should be noted that the predicted equilibrium B concentration calculated using FactSage 6.4 was 65ppm for the case of equilibrium reaction with no applied potential, which was lower by approximately 30ppm than the apparent equilibrium values measured for 0V experiments in our study.

The slope of the Figure 4(b) represents the mass transfer coefficients in the slag (k_s). The mass transfer coefficient in the case of 0V, 2.5V and 3.5V applied potential were 1.33 x 10^{-6}, 1.61 x 10^{-6}, and 2.1 x 10^{-6} m/s, respectively. From this, it can be clearly seen that the mass transfer coefficient was increased by 1.6 when a potential difference of 3.5V was applied across the two liquid phases. Since, preliminary experiments showed that the slag used could be considered as ionic melt. The reaction mechanism will involve the diffusion or mass transfer of the ionic species in the slag phase. The application of external potential may provide electrical field thus enhancing the diffusion and mass transfer of the ionic species in the slag leading to the increase in the overall kinetics of the reaction.

Conclusion

This study shows that refining of metal using slag can be enhanced by applying external potential across the two liquid phases. This has been demonstrated in the refining of silicon using CaO-SiO_2-16wt%Al_2O_3 slag. It appeared that both the kinetics and the equilibrium were affected by the application of the potential. The results showed that by applying a 3.5 V potential, the apparent rate and the boron partition ratio were increased by a factor of 1.6. The observed increase in the apparent boron partition ratio was found to be much lower compared to the theoretical limit, e.g. as calculated using Equation (5). It should be re-emphasized that Equation (6) does not take into account additional resistances across the liquid slag, silicon and the circuits; as well as additional voltage to overcome the kinetics barrier. Currently, complete thermodynamic and kinetic studies are being carried out to understand the detailed mechanism of the process. It is hoped that this study stimulates and provides the basis for further works in the area. All the limitations aside, the technique has the potential to be used for improving the existing metals refining processes using slag.

References

1. N. Sen et al., "Desulphurisation of high sulphur cast iron using an electrochemical technique," Scandinavian journal of metallurgy 28(6) (1999), 249-253.
2. U. Pal, T. Debroy, and G. Simkovich,"Electronic and ionic transport in liquid PbO-SiO_2 systems," Metallurgical Transactions B 16(1) (1985), 77-82.
3. X. Pan,"Electrochemical effects of direct current on the reduction of manganese oxide from molten MnO-SiO_2-CaO-Al_2O_3 slag," J of the South African Inst of Min Metall 95(5) (1995), 221-224.
4. G. G. Krishna Murthy, A. Hashem, and U.B Pal, "Reduction rates of FeO in CaO-SiO_2-Al_2O_3-X slags by Fe-C droplets," Ironmaking and Steelmaking 20(3) (1993), 191-200.
5. D. Woolley and U. Pal, "Experimental evidence for electrochemical nature of the reaction between iron oxide in calcia-silica-alumina slag and carbon in liquid iron," Metall Mater Trans B 30(5) (1999), 877-889.
6. X. Ma et al., "Hydrometallurgical purification of metallurgical grade silicon," Rare Metals 28(3) (2009), 221-225.

7. M.D. Johnston and M. Barati, "Distribution of impurity elements in slag-silicon equilibria for oxidative refining of metallurgical silicon for solar cell applications," Solar Energy Materials and Solar Cells 94(12) (2010), 2085-2090.
8. N. Yuge et al., "Purification of metallurgical grade silicon up to solar grade," Progress in Photovoltaics: Research and Applications 9(3) (2001), 203-209.
9. T. Yoshikawa and K. Morita, "Boron removal in the solidification refining of Si with Si-Al melt," Sohn International Symposium on Advanced Processing Metals and Materials,1 (2006), 673-684.
10. E. Olsen and S. Rolseth, "Three-Layer Electrorefining of Silicon," Metallurgical and Materials Transactions B 41(2) (2010), 295-302.
11. K. Yasuda et al., "Direct electrolytic reduction of solid SiO_2 in molten $CaCl_2$ for the production of solar grade silicon," Electrochimica Acta 53(1) (2007), 106-110.
12. K. Suzuki et al., "Thermodynamics for removal of boron from metallurgical silicon by flux treatment," Journal of the Japan Institute of Metals 54(2) (1990), 168-172.
13. H. Fujiwara et al., "Reducing removal of phosphorous from calcium containing silicon alloys," Materials Transactions, JIM 37(4) (1996), 923-926.
14. L. A. V. Teixeira et al., "Removal of Boron from Molten Silicon Using $CaO–SiO_2$ Based Slag," ISIJ international 49(6) (2009), 783-787.
15. L. A. V. Teixeira et al., "Behavior and state of boron in $CaO-SiO_2$ slags during refining of solar grade silicon," ISIJ international 49(6) (2009), 783-787.
16. D. W. Luo et al., "Removal of boron from metallurgical grade silicon by electromagnetic induction slag melting," Transactions of Nonferrous Metals Society of China 21(5) (2011), 1178-1184.
17. E. Krystad, K. Tang and G. Tranell, "The Kinetics of Boron Transfer in Slag Refining of Silicon," JOM Journal of the Minerals, Metals and Materials Society, (2012), 1-5.
18. H. Nishimoto, H. and K. Morita, "The Rate of Boron Elimination from Molten Silicon by Slag and Cl_2 gas Treatment," Supplemental Proceedings: Materials Processing and Energy Materials, Vol 1(2011), 701-708.
19. M.A. Rhamdhani, G.A. Brooks, and K.S. Coley, "Kinetics of metal/slag reactions during spontaneous emulsification," Metallurgical and Materials Transactions B, 36 (B) (2005), 219-27.

Celebrating the Megascale: Proceedings of the Extraction and Processing Division Symposium on Pyrometallurgy in Honor of David G.C. Robertson
Edited by: Phillip J. Mackey, Eric J. Grimsey, Rodney T. Jones, and Geoffrey A. Brooks
TMS (The Minerals, Metals & Materials Society), 2014

CRYSTALLIZATION BEHAVIOR OF MOLTEN BLAST FURNACE SLAG USING CONFOCAL SCANNING LASER MICROSCOPE

Liu Lu, Hu Meilong, Bai Chenguang
College of Materials Science and Engineering, Chongqing University, China

Key words: Crystallization, Perovskite, CSLM, High Ti-bearing slag.

Abstract

Extracting perovskite from high titanium-bearing blast furnace (BF) slag is a green and potential method to recover the second titanium resource. The non-isothermal crystallization process of perovskite in synthesized high titanium-bearing slag was studied in situ by confocal scanning laser microscope (CSLM) with cooling rate of 20 K/min. The results showed that perovskite was the primary phase formed during cooling process in the titanium-bearing synthesized slag (TiO_2=23%). Perovskite started to precipitate at 1719 K and finally presented an orthogonal straight lines morphology.

Introduction

China has abundant titanium resources, 92.4% of which is vanadium-titanium bearing magnetite, deposited mostly around the Panzhihua-Xichang area in the southwestern part of China. At present, smelting the iron concentrate ores in blast furnace is the main process to use these resources. However, titanium resource is not used effectively through BF process because the 54% of titanium in the raw ores finally ends up in blast furnace slag (TiO_2=22-23 mass %) and it is hard to extract titanium from the high titanium-bearing BF slag [1]. Moreover, the high titanium-bearing BF slag is unsuited to be cement [2]. As a result, more than 60 million tons of slag has been stacked in the open from the beginning of BF smelting, which not only brings environmental pollution but also wastes resources. How to extract the valuable secondary titanium resources and use waste slag is a serious problem in China.

The comprehensive utilization of titanium-bearing BF slag has aroused wide concern since 1980 but none of the proposed techniques is easy to be put into industrial and commercial use [3-12]. However, green extractive process of titanium component shows a bright prospect [1,12-14]. In this technique, making titanium gather in perovskite crystals and making these grow up to big sizes is essential for following mineral separation [15]. Many researches on kinetics of precipitation and crystal growth process of perovskite phase have been done [13,16-17] but the crystallization process of perovskite was rarely reported.

In this work, the non-isothermal crystallization process of perovskite in synthesized high titanium-bearing slag based on the industrial slag (the composition is shown in **Table I**) was studied by confocal scanning laser microscope (CSLM) in order to provide information for extracting titanium from slag. Compared with the conventional methods of studying crystallization, such as quenching followed by microscopy, or hot-stage microscopy, CSLM is a new advanced technology which has higher resolution and enables real time in-situ observation of high temperature (up to 1973 K) transient phenomena [19-20]. The CSLM has not been used extensively for studying the complex melting and solidification behavior of slag.

Table I. The main chemical composition of the industrial slag and the experimental synthesized slag (mass %).

	TiO_2	CaO	SiO_2	Al_2O_3	MgO	MoO_3	Total
Industrial slag	22.34	26.96	24.21	14.35	8.32	0	96.18
Aim composition	23	28.8	26.2	14	8	0	100
XRF analyzed composition	22.65	28.68	25.47	14.21	7.54	1.45	100

*In XRF result, the minor impurities (Fe, K, P, Cr, Sr, Zr, Nb) were ignored and the major species were normalized to give a total of 100%. It was assumed that MoO_3 had little effect on the experiments.

Experimental

The synthesized slag was prepared by pre-melting 100 g chemical reagent powder mixture, with the aim composition as shown in **Table I.** The powders used were reagent grade CaO (98 pct), MgO (98 pct), SiO_2 (99 pct), TiO_2 (99 pct) and Al_2O_3 (98 pct). The chemical reagent powders were dried at 373 K for 6h.The mixed powder was put into a molybdenum crucible (50 mm in inner diameter and 80 mm in height) and melted under an argon atmosphere at 1773 K for 1h in $MoSi_2$ furnace so that the slag was homogenized. After melting, the slag was quenched in water and was cut into a small slag sample in a wafer shape of 7 mm in diameter and 3 mm in height for running the experiment. The slag was also analyzed by X-Ray fluoroscopy, and the result in **Table I** showed that there was small deviation from the aim composition.

In the present experiment, the crystallization behavior was observed in-situ through a confocal scanning laser microscope (CSLM) with an infrared furnace (Lasertec, VL2000DX). The melting and solidification of the synthesized slag sample took place in a platinum crucible (8 mm in inner diameter and a height of 5 mm) under an argon atmosphere in the CSLM (Ar>99.999%). The moisture and oxygen were removed by purifier provided with the CSLM. The infrared furnace was evacuated and back-filled with argon three times before heating. The temperature schedule in the experiment is shown in **Figure 1.** The experiment focussed on the crystallization process of slag from 1723 K to 1373 K at cooling rate of 20 K/min. The temperature accuracy was confirmed by melting experiments of pure copper (melting point: 1356 K) and pure nickel (melting point: 1726 K).

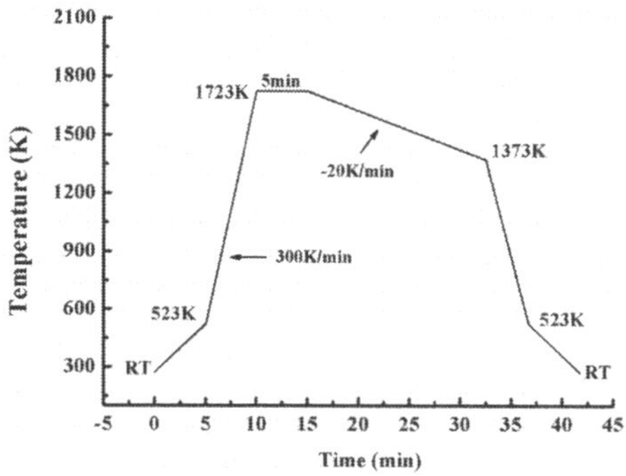

Figure 1. Temperature profile of the CSLM experiment.

After the CSLM experiment, the images captured and digitized from the video recordings of experiment were analyzed to obtain information on the crystallization process of slag. In addition, the quenched sample was examined by X-ray Diffraction (D/max 2500PC).

Results and Discussion

The surface of liquid slag fluctuated because of surface tension driven flow. So the crystals moved in and out of the viewing field with the moving liquid slag. **Figure 2** shows the crystallization process of slag at a cooling rate of 20 K/min. The slag remained totally liquid above 1723 K and the liquid phase presented as bright white field. The pictures a, b and c in **Figure 2** were of a different location because the movement of the slag. Pictures c to h are for the same location.

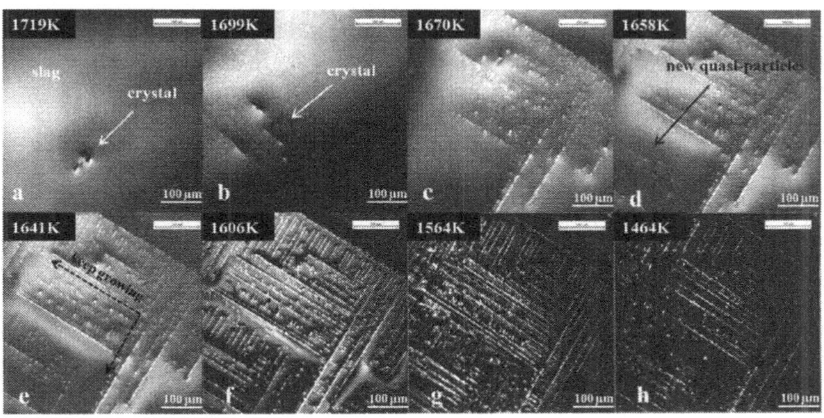

Figure 2. The CSLM images of crystallization process of slag cooled at 20 K/min.

The primary phase precipitates as three quasi-particles at 1719 K (**Fig.2(a))** Then larger pieces of crystals appear below 1703 K. The crystals present mutually perpendicular closely-spaced straight lines which consist of the quasi-particles. As the temperature decreases, crystals keep growing and extending in mutually perpendicular directions in a way that new quasi-particles precipitate one by one in alignment, forming closely-spaced structure – presumably made up of dendrites (**Fig.2(b-h)**). Meanwhile, crystals keep coarsening and the amount of liquid phase reduces gradually with decreasing temperature. It is shown that the dendrites present orthogonal straight lines morphology at the cooling rate of 20 K/min. The dendrites grow in a way that new dendrites precipitate successively in orthogonal directions.

Figure 3 illustrates the XRD patterns of the sample. As can be seen, perovskite is the main crystalline phase during experiments which coexists with another phase clinopyroxene - $CaMgSi_2O_6$. Besides, an amorphous peak is also present. In order to better understand the crystallization process, the equilibrium phases at different temperatures of the synthesized Ti-bearing slag were calculated by FactSage 6.2 shown in **Figure 4.** The liquidus temperature of the slag is 1705 K and perovskite is the primary phase during cooling process. There is small deviation between the theoretical calculation and experimental result. Combining the analytical and calculated results, it can be concluded that the crystals observed in CSLM experiments are perovskites, which crystallized along a specific path mentioned earlier. The growth process of perovskite is shown in **Figure 5.**

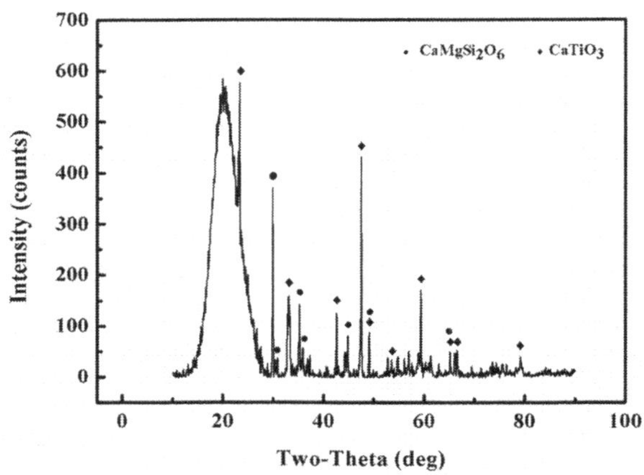

Figure 3. XRD patterns of the sample cooled at 20 K/min.

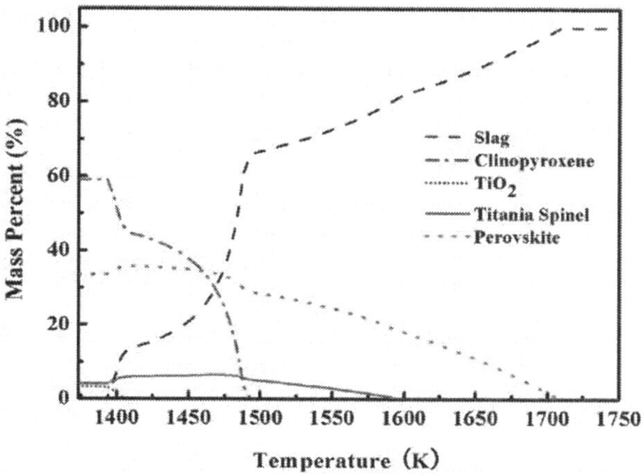

Figure 4. Theoretical isothermal phase composition of 28.8%CaO, 26.2%SiO_2, 14%Al_2O_3, 8%MgO, 23%TiO_2 slag during cooling.

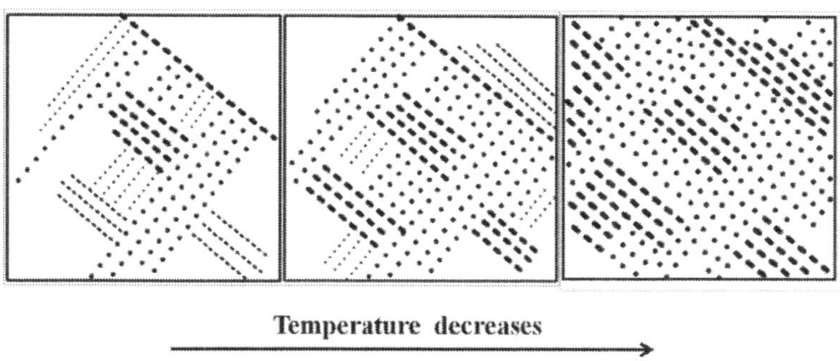

Figure 5. Schematic representation illustrating the growth process of perovskite during CSLM experiment at a cooling rate of 20 K/min.

Conclusions

Perovskite was the primary phase formed during cooling process of the Ti-bearing synthesized slag (TiO_2=23%) at cooling rate of 20 K/min. Perovskite started to precipitate at 1719 K and finally presented as orthogonal dendrites. Perovskite grew along a specific growth path with new dendrites precipitating successively.

References

1. Fu Nianxin, Lu Ling and Sui Zhitong, "Precipitating Behavior of Perovskite Phase in the Blast Furnace Slag Bearing Higher Titanium," *Journal of Iron and Steel Research*, 10(3) (1998), 70-74.
2. Weng Qingqiang, "High Ti-bearing Slag Comprehensive Utilization Outline and Forecast," *Sichuan Metallurgy*, 31(6) (2009), 40-43.
3. Wu Yucheng, Song Linyun, Li Yun, Li Yong, Li Guanghai and Zheng Zhixiang, "Preparation and Photocatalystic Activity of Ce-doped TiO_2 Nanoparticles," *Journal of Synthetic Crystals*, 37(2) (2008), 427-430.
4. Yang He, Xue Xiangxin, Zuo Liang and Yang Zhongdong, "Photocatalytic Degradation of Brilliant Red X-3B With Blast Furnace Slag Containing Titanium and Rare Earth Oxides," *Journal of the Chinese Ceramic Society*, 31(9)(2003), 896-899.
5. Xiao Xingcheng, Jiang Weihui and Wang Yonglan, "Study on Crystallization Process of Titanate - Slag Glass Ceramics ," *Glass&Enamel*, 27(2)(1999), 7-11.
6. Liu Songli, Yang Shaoli and Gao Shizhong, "The Development of Titanium Extraction Technology in Panzhihua High Titanium-bearing Blast Furnace Slag," *Panzhihua Sci-Tech&Information*, (4) (2006), 10-12.
7. Fang Rongli, Jin Chengchang and Xu Anyi, "Highly Valuable Utilization of Ilmenite Slag of Panzhihua Iron and Steel Company," *Cement*, (2) (1998).
8. Sun Xiwen, Zhang Jiantao and Yang Zhiyua, "The Research of Slag Brick Using High Titaniferous Blast-Furnace Slag," *New Building Materials*, (3)(2003), 5-7.
9. Wang Peng, Han Binqiang, Han Yanlei, Ke Changming and Li Nan, "Research on Hydration Capability of Panzhihua Iron and Steel Co. BF Slag after Extracting Silicon titanium Alloy," *Bulletin of the Chinese Ceramic Society*, 27(6) (2008), 1208-1210.
10. Xiong Yao, Liang Bin and Li Chun, "Extraction and Separation of Titanium in Natural-cooling BF Slag," *The Chinese Journal of Process Engineering*, 8(6) (2008), 1092-1097.
11. Liu Xiaohua, Sui Zhitong, "Pressure Acid Splitting of Ti-bearing BF Slag," *The Chinese Journal of Nonferrous Metals*, 12(6) (2002), 1281-1284.
12. Zhou Zhiming, Zhang Binghuai and Zhu Zizong, "A Test of Titania Separation From High Titania Bearing Blast Furnace Slag," *Iron Steel Vanadium Titanium*, 20(4) (1999), 35-38.
13. Lou Taiping, Li Yuhai, Li Liaosha and Sui Zhitong, "Study on Kinetics of Perovskite Phase Precipitate in Slag Bearing Titanium," *Journal of the Chinese Ceramic Society*, 28(3) (2000), 255-258.
14. Z.Z. Guo, T.P. Lou, L.Zhang, L.N. Zhang and Z.T.Sui, "Precipitation and Growth of Perovskite Phase in Titanium Bearing Blast Furnace Slag," *Acta Metall.Sin.(Engl.Lett.)*, 20(1) (2007), 9-14.
15. L.Zhang, L.N.Zhang, M.Y.Wang, G.Q. Li and Z.T. Sui, "Recovery of Titanium Compounds from Molten Ti-Bearing Blast Furnace Slag under the Dynamic Oxidation Condition," *Minerals Engineering*, 20(2007), 684–693.
16. Wang Mingyu, Zhang Li, Zhang Linan and Sui Zhitong, "Characteristics Change of Ti-bearing Molten Blast Furnace Slag under Dynamic Oxidation Condition,"

Nonferrous Metals, 60(2) (2008), 59-61.
17. Wang Mingyu, Zhang Li, Zhang Linnan, Sui Zhitong and Tu Ganfeng, "Study on Selection of the Best Titania Enrichment Phase in Titanium-Bearing Slag," *Journal of Materials and Metallurgy* ,4(3) (2005), 175-177.
18. Joo Hyun Park, Jin Gyun Park, Dong Joon Min, Young E. Lee and Youn-Bae Kang, "In Situ Observation of the Dissolution Phenomena of SiC Particle in CaO–SiO_2–MnO Slag," *Journal of the European Ceramic Society*, 30(2010), 3181–3186.
19. J. Liu , M. Guo, P.T. Jones, F. Verhaeghe, B. Blanpain and P. Wollants, "In Situ Observation of the Direct and Indirect Dissolution of MgO Particles in Cao–Al_2O_3–SiO_2-based Slags," *Journal of the European Ceramic Society*, 27(2007), 1961–1972.

Celebrating the Megascale: Proceedings of the Extraction and Processing Division Symposium on Pyrometallurgy in Honor of David G.C. Robertson
Edited by: Phillip J. Mackey, Eric J. Grimsey, Rodney T. Jones, and Geoffrey A. Brooks
TMS (The Minerals, Metals & Materials Society), 2014

Viscosity-Structure Relationship in the CaO-SiO_2-MnO-CaF_2 Slag for the Production of Manganese Ferroalloys

Joo Hyun Park[1,*], Kyu Yeol Ko[2]

[1]Department of Materials Engineering, Hanyang University, Ansan 426-791, Korea
[2]Metals and Materials Research Department, LS-Nikko Copper, Ulsan 70, Korea
*email: basicity@hanyang.ac.kr

Keywords: Viscosity, CaO-SiO_2-MnO-CaF_2 slag, Raman spectra, Structure, Depolymerization

Abstract

The viscosity of the CaO-SiO_2-$xMnO$-$yCaF_2$ slags (C/S=1.0; x=10, 40%; y=0 to 15%) was measured to clarify the effect of CaF_2 on the viscous flow of molten slags at high temperatures. Furthermore, the Raman spectra of the quenched glass samples were quantitatively analyzed to investigate the structural role of CaF_2 in a depolymerization of silicate networks. The critical temperature of the slags abruptly increased at 15%CaF_2, which was confirmed to originate from a crystallization of cuspidine using XRD analysis. The viscosity of the slags continuously decreased by CaF_2 addition in the 10%MnO system, whereas the viscosity of the 40%MnO system was not significantly affected by CaF_2 addition. The effect of CaF_2 on the viscosity of the slags was quantitatively analyzed using micro-Raman spectra of quenched glass samples accompanying with a concept of silicate polymerization index, Q^3/Q^2 ratio. A polymerization index continuously decreased with increasing content of CaF_2 in the 10%MnO system, whereas it was not affected by CaF_2 in the 40%MnO system. Consequently, the bulk thermophysical property of the CaO-SiO_2-MnO-CaF_2 slags was quantitatively correlated to the structural information.

Introduction

Viscosity of silicate melts at high temperatures has been widely investigated, because of the close relationship between their macroscopic thermophysical properties and the microscopic molecular (or ionic) structure, which strongly affect the reaction efficiency in process metallurgy.[1-5] Recently, we thoroughly evaluated the novel structure-viscosity relationship of CaO-SiO_2-MnO slag system using micro-Raman spectroscopy due to its importance in smelting and refining processes for the production of manganese ferroalloys as well as high manganese steels.[5]

The slag-metal reaction kinetics should be enhanced in view of high production rate and low cost in smelting and refining processes of Mn ferroalloys as well as high Mn steels. In order to meet these requirements, it is necessary to consider and quantify the effect of CaF_2 addition on the viscosity of MnO-containing slags at high temperatures, because the CaF_2 is well known to decrease the viscosity of silicate melts.[2-4,6,7] The experimental study of the effect of CaF_2 on the viscosity of various silicate melts has been investigated by several researchers for CaO-SiO_2(-MgO)-CaF_2,[2,3,7] FeO(-CaO)-SiO_2-CaF_2,[2] CaO-SiO_2-Na_2O-Li_2O-CaF_2,[8] and CaO-Al_2O_3-SiO_2-MgO-CaF_2 systems.[6]

From the above studies, the viscosity itself and the activation energy of Newtonian flow of silicate melts generally decrease with increasing content of CaF_2 due to depolymerization reactions of silicate networks, indicating that the addition of CaF_2 is more effective in higher silica content region. However, the effect of CaF_2 on the viscosity of the FeO-SiO_2 slags was not found to be as significant as in the case of the CaO-SiO_2 slags.[2] In MO (M=Mg, Na_2)-containing slags, the addition of CaF_2 has limited effect on viscosity within a specific content due to additional contribution of basic oxides such as MgO and Na_2O besides CaO in the silicate depolymerization.[3,8]

Also, for the highly basic slags, of which silica content is lower than about 10%, CaF_2 suppresses the precipitation of solid phases at lower temperatures, whereas the effect of CaF_2 on viscosity itself is negligible when the slags are completely liquid at high temperatures.[6] From solid-state ^{19}F NMR spectroscopic analyses of CaO-SiO_2-CaF_2 (-Na_2O) systems, it was reported that F^- was dominantly coordinated with Ca^{2+} in alkali-free ternary system, whereas it predominantly bonded with Na^+ rather than Ca^{2+} in quaternary system. In both cases, however, F^- ions were commonly believed not to break the Si-O-Si bonds.[9]

Alternatively, there is no experimental work investigating the effect of CaF_2 on the viscosity of MnO-containing silicate melts, even though there are some experimental data for the viscosity of the CaO-MnO-SiO_2 ternary slag.[1] Therefore, in the present study, the viscosity of the CaO-SiO_2-MnO-CaF_2 slags has been measured to clarify the effect of CaF_2 on the viscous flow of molten slags at high temperatures. In addition, the Raman spectra of the quenched glass samples were quantitatively analyzed to investigate the structural role of CaF_2 in a depolymerization of silicate networks with a reference to our previous works.[5]

Experimental

All of slag samples were prepared using reagent-grade chemicals such as MnO, SiO_2, CaF_2 and CaO calcined from $CaCO_3$ at 1273 K for 10 hours and the pre-melted slags were used. The pre-melting was carried out using a Pt-10%Rh alloy crucible under a purified Ar atmosphere in a super kanthal electric furnace. The rotating cylinder method was used in the present study. A rotating viscometer was set on the super kanthal electric furnace. The temperature was measured using an R-type thermocouple. The experimental temperature range was about 1523 to 1873 K.

The measurements of viscosity in the present study were made using a Pt-10%Rh alloy spindle, crucibles, and suspending wire. The details of experimental procedure is available in the literature.[10] In addition, the CaF_2 content of the slag samples was analyzed using XRF after experiments, because evaporation of the fluoride species from the melts was expected.[11] It was confirmed that the weight loss of CaF_2 by fluoride emission during viscosity measurements was within about 5% of the initial content, which is similar to previous works.[3,8] Hence, the composition in the present article refers to initial content.

The procedures for the preparation of glass samples and the quantitative analysis of Raman spectra are already described in detail in a previous article.[5]

Results and Discussion

Effect of CaF_2 Addition on the Viscosity and Critical Temperature of CaO-SiO_2-MnO Slags

The viscosities of the CaO-SiO_2-MnO(-CaF_2) slags (C/S=1.0) are shown as a function of temperature in Fig. 1 for 10% and 40%MnO systems. The viscosity values for CaF_2-free system at 1673, 1773 and 1823 K measured by other researchers are also shown in Fig. 1 for comparison.[1,12,13] One can see that the literature values and the present results are in good agreements within experimental scatters.

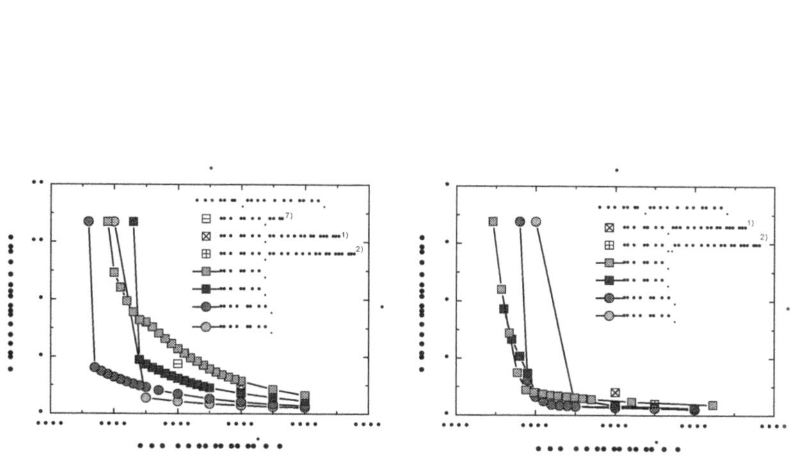

(a) (b)

Figure 1. Viscosities of the $CaO-SiO_2-CaF_2-xMnO$ (C/S=1.0) slags for (a) x=10% and (b) x=40% as a function of temperature at different CaF_2 contents.

The viscosity of 10%MnO system is generally higher than that of 40%MnO system at a given CaO/SiO_2 (=1.0) ratio and CaF_2 content. Also, it gradually increases with decreasing temperature and this tendency is significant as the CaF_2 content decreases when MnO content is 10% (Fig. 1(a)), whereas the viscosity does not change so much as temperature decreases to about 1723 K when MnO content is 40%. For the latter in Fig. 1(b), the viscosity drastically increases at a specific temperature, which has been called critical (or crystallization) temperature (T_{CR}).[3,8]

A temperature dependency of the viscosity for the 10 and 40%MnO systems are compared in Fig. 2 at different CaF_2 contents. The critical temperature of 10%MnO system decreases from about 1623 K to 1543 K as the content of CaF_2 increases from 0 to 10% (Figs. 2(a)-(c)), followed by a rebound to about 1623 K when the CaF_2 content reaches 15% (Fig. 2(d)). For the 40%MnO system, there is no significant change in critical temperature (1673 (±5) K) when the CaF_2 content varies from 0 to 10% (Figs. 2(a)-(c)). However, it abruptly increases to about 1723 K when the CaF_2 content changes from 10 to 15% (Fig. 2(d)).

An increase in the critical temperature at 15%CaF_2 composition in both 10 and 40%MnO systems was found to originate from a crystallization of cuspidine ($Ca_4Si_2O_7F_2$), which was confirmed by XRD analysis as shown in Fig. 3. The slag samples attached to the surface of spindle was quenched when the viscosity of slags sharply increased. It was found that 10%CaF_2 system exhibited a crystallization of Mn oxides, viz. MnO and Mn_3O_4 as shown in Fig. 3(a), whereas Mn oxides and cuspidine were detected in 15%CaF_2 system in Fig. 3(b). Here, the Mn_3O_4 is believed to form by oxidation of MnO in slag samples attached in a spindle during cooling after viscosity measurement. Therefore, even though a reliable phase diagram of the $CaO-SiO_2-MnO-CaF_2$ quaternary system is not available at this time, the cuspidine is expected to primarily crystallize in the $CaO-SiO_2-MnO-15\%CaF_2$ system (C/S=1.0, MnO=10-40%).

Furthermore, in Fig. 2, the effect of MnO content on the viscosity of slags at temperatures greater than T_{CR} increases by decreasing the content of CaF_2. This means that MnO is also relatively strong network modifier as likely as CaO in the CaF_2-free silicate melts. The structural role of Ca^{2+} and Mn^{2+} ions in a depolymerization process of the $CaO-SiO_2-MnO$ ternary system was systematically discussed in our previous articles.[5]

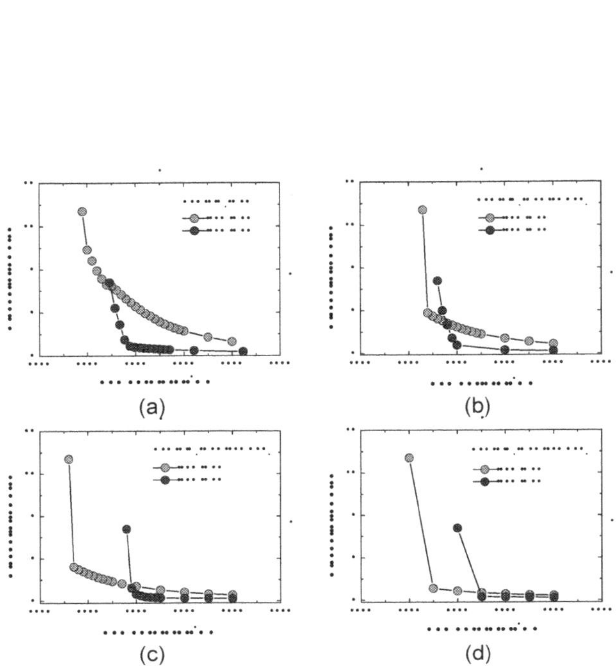

Figure 2. Effect of CaF_2 and MnO content on the viscosities of the CaO-SiO_2-MnO-CaF_2 (C/S=1.0) slags. (a) CaF_2-free, (b) 5%, (c) 10%, (d) 15% CaF_2.

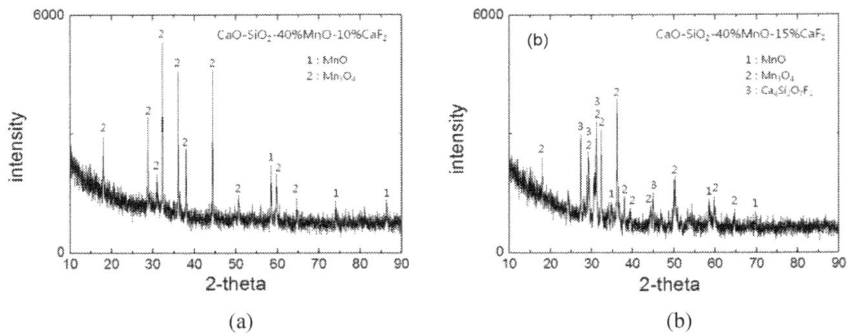

Figure 3. XRD pattern for the solidified (a) CaO-SiO_2-40%MnO-10%CaF_2 and (b) CaO-SiO_2-40%MnO-15%CaF_2 system.

The viscosities of the CaO-SiO_2-CaF_2-10%MnO system at various temperatures are plotted against the content of CaF_2 in Fig. 4. The viscosity of slags continuously decreases as the content of CaF_2 increases at a given temperature and the effect of CaF_2 addition is more dominant at lower temperatures, whereas CaF_2 did not significantly affect the viscosity of slags at higher temperatures. However, the viscosity of slags containing 40%MnO is very low, viz. lower than 0.1 Pa.s even in the CaF_2-free system at temperatures greater than 1673 K.

Therefore, CaF_2 is not so useful in the 40%MnO system in view of slag fluidity and furthermore the combination of high MnO (e.g. 40%) and high CaF_2 (e.g. 15%) promotes the precipitation of solid phase (e.g. cuspidine), which is undesirable in the viewpoint of not only smelting and refining processes but also post-treatment of skimmed slags. Hence, the addition of large amounts of CaF_2 is not recommended at high MnO-containing slags because the larger the amounts of CaF_2 added, the severe corrosion of refractory materials expected.[14]

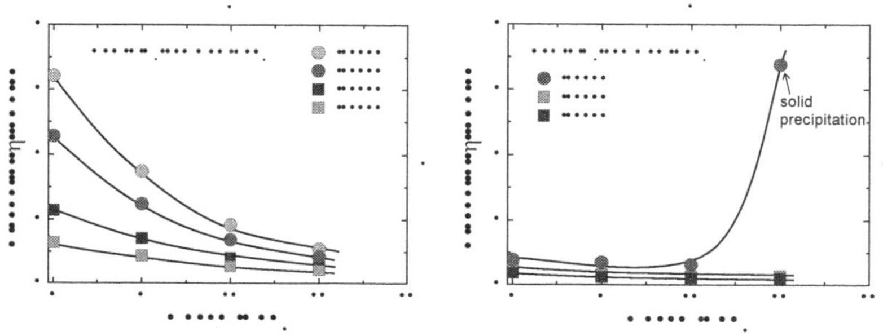

(a) (b)

Figure 4. Effect of CaF_2 addition on the viscosities of the CaO-SiO_2-xMnO (C/S=1.0) slags for (a) x=10% and (b) x=40% at temperatures from 1623 to 1873 K.

Structural Changes of Silicate Melts by CaF_2 Addition ; micro-Raman Spectra Analysis

Raman spectroscopy was originally used to probe the local anionic structure of silicates.[5] The bands assigned to antisymmetric stretching of Si–O^- (non–bridging oxygen; NBO) and Si–O^0 (bridging oxygen; BO) bonds occur in the 850–1200 cm^{-1} region, whereas Si–O–Si bending modes are found between about 500 and 700 cm^{-1}.[5] The frequencies of the stretching modes decrease with decreasing degree of polymerization, viz. increasing NBO/Si. There are several types of units such as SiO_2 (fully polymerized), Si_2O_5 (sheet), SiO_3 (chain), Si_2O_7 (dimer), and SiO_4 (monomer) groups.[15-19] In the study of NMR spectra of silicates,[20] the stoichiometric notations for each unit were replaced by so–called Q^n concept, wherein they are equivalent to Q^4 (NBO/Si=0), Q^3 (NBO/Si=1), Q^2 (NBO/Si=2), Q^1 (NBO/Si=3), and Q^0 (NBO/Si=4), respectively.

The Raman spectra of the CaO-SiO_2-CaF_2-10%MnO system quenched from 1873 K are shown in Fig. 5 as a function of wavenumbers (Raman shifts) range from 700 to 1300 cm^{-1}. The intensity of high frequency Q^3 (1050-1100 cm^{-1}) band is very weak and it is not easy to catch the effect of CaF_2 on an intensity change of this band. However, it seems that the relative intensity of the Q^2 (970(±5) cm^{-1}) band decreases with increasing CaF_2 content and becomes week shoulder at 15%CaF_2 composition. Alternatively, the Q^0 (850(±5) cm^{-1}) and Q^1 (910(±5) cm^{-1}) bands increase in intensity by increasing the content of CaF_2. The more quantitative analysis of structural changes by CaF_2 addition is as follows.

The relative fractions of the silicate anionic units obtained from a Gaussian deconvolution of Raman bands shown in Fig. 5(a) are plotted against CaF_2 content in Fig. 5(b). The details for the Gaussian deconvolution process are available in our previous works.[5] The fraction of Q^3 (Si_2O_5-sheet) unit is less than 10% and continuously decreases with increasing content of CaF_2 and the fractions of Q^0 (SiO_4-monomer) and Q^1 (Si_2O_7-dimer) units increase as the content of CaF_2 increases.

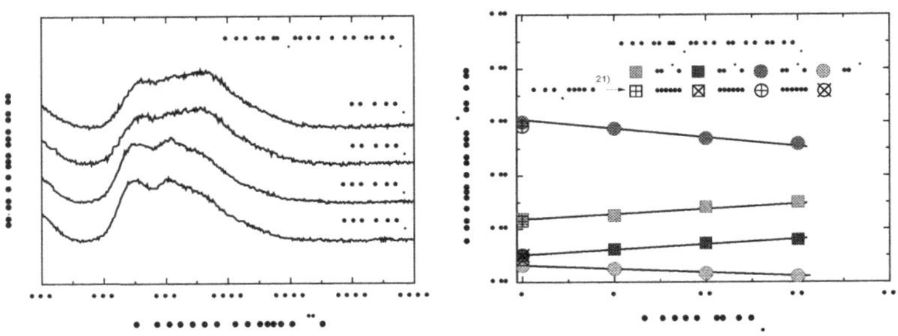

(a) (b)

Figure 5. (a) Raman spectra of the quenched CaO-SiO_2-CaF_2-10%MnO (C/S=1.0) slags and (b) relative fractions of Q^n (n=0, 1, 2, 3) structural units with CaF_2 content ranges from 0 to 15%.

The CaO-SiO_2-10%MnO system is mainly composed of Q^2 (SiO_3-chain) structure (~60%). The relative fraction of Q^2 unit decreases as the content of CaF_2 increases possibly based on the depolymerization reactions shown in Eqs. [1] and [2]. In Fig. 5(b), the fractions of each Q^n unit for the CaF_2-free ternary system exhibit a good agreement with the previous measurements,[5] indicating that the present Raman analysis is highly consistent.

$$2[SiO_3]^{2-} \text{ (chain)} + F^- = [Si_2O_6F]^{5-} \text{ (dimer)} \quad [1]$$

$$[Si_2O_6F]^{5-} \text{ (dimer)} + F^- = 2[SiO_3F]^{3-} \text{ (monomer)}$$

[2]

The Raman spectra of the CaO-SiO_2-CaF_2-40%MnO system as a function of wavenumbers (Raman shifts) range from 700 to 1300 cm^{-1} are shown in Fig. 6. The intensity of high frequency Q^3 (1050-1100 cm^{-1}) band is negligible through the composition range investigated. There is no remarkable change in the relative intensity of Q^0, Q^1, and Q^2 bands through the CaF_2 content from 0 to 10%, although the apparent height of Q^2 band seems to slightly increase at 15%CaF_2. However, the more quantitative analysis of structural changes by CaF_2 addition is as follows.

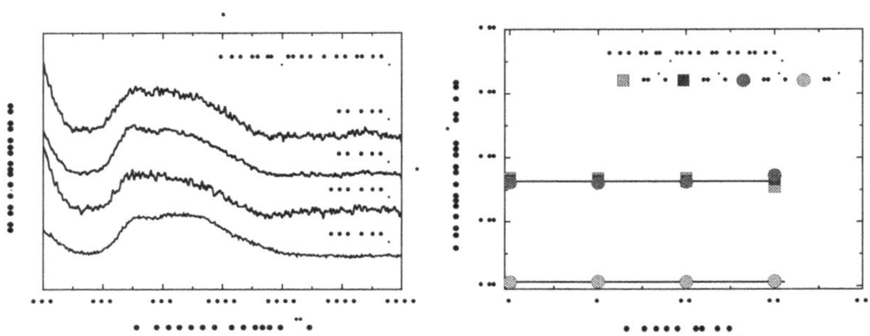

(a) (b)

Figure 6. (a) Raman spectra of the quenched CaO-SiO_2-CaF_2-40%MnO (C/S=1.0) slags and (b) relative fractions of Q^n (n=0, 1, 2, 3) structural units with CaF_2 content ranges from 0 to 15%.

The relative fractions of the silicate anionic units obtained from a Gaussian deconvolution of Raman bands shown in Fig. 6(a) are plotted against CaF_2 content in Fig. 6(b). The fraction of Q^3 (Si_2O_5-sheet) unit is negligible through the composition range investigated, i.e. C/S=1.0 and MnO=40%, irrespective of CaF_2 content. Also, it is very interesting that the relative fractions of Q^0, Q^1 and Q^2 units are nearly the same and do not change as the content of CaF_2 increases to 15%.

The Q^3/Q^2 ratio, which was originally proposed as a quantitative polymerization index of silicates in our previous works,[5] is plotted against CaF_2 content in Fig. 7. The Q^3/Q^2 ratio continuously decreases with increasing CaF_2 content in 10%MnO system, whereas it does not change as CaF_2 content increases in 40%MnO system. This means that CaF_2 effectively contributes to a depolymerization of silicate networks in the former, while it does not in the latter.

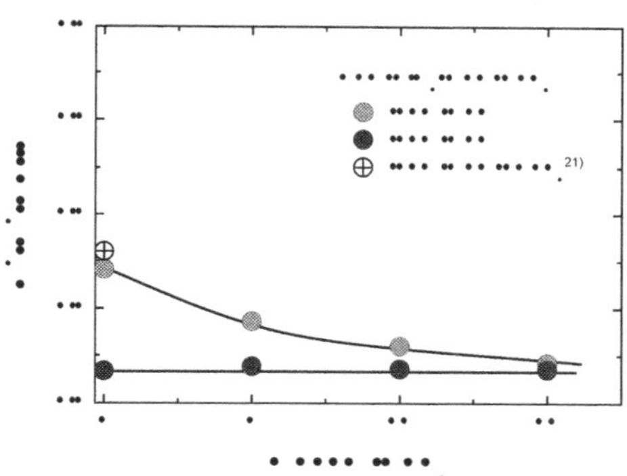

Figure 7. Effect of CaF_2 on the polymerization index, viz. Q^3/Q^2 ratio in the CaO-SiO_2-MnO-CaF_2 slags quenched from 1873 K.

In summary, the present experimental results for the viscosity measurement and the structural analysis for the CaO-SiO_2-MnO-CaF_2 slags quenched from 1873 K using Raman spectroscopy are consistent with those of CaF_2-free ternary system, which was previously reported.[5]

Conclusions

The addition of CaF_2 in the CaO-SiO_2-MnO (CaO/SiO_2=1.0) slag affected not only the critical (crystallization) temperature (T_{CR}) but also the viscosity at high temperatures and its influence to slag properties were strongly dependent on the content of MnO in the slags. The T_{CR} of the 10%MnO system slightly decreased with increasing amount of CaF_2 to 10%, followed by a rebound at 15%CaF_2. The T_{CR} of the 40%MnO system did not change until CaF_2 was added to 10%, but sharply increased at 15% CaF_2. An increase in the T_{CR} at 15%CaF_2 systems in both cases was confirmed to originate from a crystallization of cuspidine as primary phase using XRD analysis. The viscosity of the slags continuously decreased by CaF_2 addition in the 10%MnO system, whereas the viscosity of the 40%MnO system was very low, viz. lower than 0.1 Pa.s, even without CaF_2 and it was not significantly affected by CaF_2 addition at temperatures greater than 1773 K. The effect of CaF_2 on the viscosity of the slags was quantitatively analyzed using micro-Raman spectra of quenched glass samples accompanying with a concept of silicate polymerization index, Q^3/Q^2 ratio. A polymerization index continuously decreased with increasing content of CaF_2 in the 10%MnO system, whereas it was not affected by CaF_2 in the 40%MnO system. Consequently, the bulk thermophysical property of the $CaO-SiO_2-MnO-CaF_2$ slags was quantitatively correlated to the structural information.

References

1. L. Segers, A. Fontana, and R. Winand: *Electrochim. Acta* 24 (1979) 213.
2. F. Shahbazian, Du Sichen, and S. Seetharaman: *ISIJ Int.* 39 (1999) 687.
3. J.H. Park, D.J. Min, and H.S. Song: *Metall. Mater. Trans. B* 35B (2004) 269.
4. J.H. Park, H. Kim, and D.J. Min: *Metall. Mater. Trans. B* 39B (2008) 150.
5. J.H. Park: *ISIJ Int.* 52 (2012) 1627.
6. L. Wu, J. Gran and Du Sichen: *Metall. Mater. Trans. B* 42B (2011) 928.
7. O. Takeda, T. Okawara, and Y. Sato: *ISIJ Int.* 52 (2012) 1544.
8. H.S. Park, H. Kim, and I. Sohn: *Metall. Mater. Trans. B* 42B (2011) 324.
9. M. Hayashi, T. Watanabe, H. Nakada and K. Nagata: *ISIJ Int.* 46 (2006) 1805.
10. K.Y. Ko and J.H. Park: *ISIJ Int.* 53 (2013) 958.
11. J.H. Park and D.J. Min: *Steel Res. Int.* 75 (2004) 807.
12. M. Kawahara, K. Mizoguchi, and Y. Suginohara: *Bull. Kyushu Inst. Technol.* 43 (1981) 53.
13. F.Z. Ji: Metall. *Mater. Trans. B* 32B (2001) 181.
14. J.H. Park, M.O. Suk, I.H. Jung, M. Guo, and B. Blanpain: *Steel Res. Int.* 81 (2010) 860.
15. B.O. Mysen and P. Richet: *Silicate Glasses and Melts: Properties and Structure*, Elsevier, Amsterdam, Netherlands, (2005).
16. B.O. Mysen: *Eur. J. Mineral.* 15 (2003) 781.
17. T. Furukawa, K.E. Fox, and W.B. White: *J. Chem. Phys.* 75 (1981) 3226.
18. P. McMillan: *Am. Mineral.* 69 (1984) 622.
19. Y. Iguchi, S. Kashio, T. Goto, Y. Nishina and T. Fuwa: *Can. Metall. Quart.* 20 (1981) 51.
20. J.F. Stebbins, J.B. Murdoch, E. Schneider, I.S.E. Carmichael, and A. Pines: *Nature* 314 (1985) 250.

Celebrating the Megascale: Proceedings of the Extraction and Processing Division
Symposium on Pyrometallurgy in Honor of David G.C. Robertson
Edited by: Phillip J. Mackey, Eric J. Grimsey, Rodney T. Jones, and Geoffrey A. Brooks
TMS (The Minerals, Metals & Materials Society), 2014

RECOVERY OF VANADIUM FROM A HIGH Ca/V RATIO VANADIUM SLAG USING SODIUM ROASTING AND AMMONIA LEACHING

Song Xu, Mujun Long*, Dengfu Chen, Helin Fan, Yuting Chen, Xue Sun
Laboratory of Metallurgy and Materials, College of Materials Science and Engineering,
Chongqing University, Chongqing 400044, P.R. China.
*Email: longmujun@cqu.edu.cn

Keywords: Vanadium slag; Extraction of vanadium; Sodium salt roasting; Ammonia leaching

Abstract

In order to seek an effective extraction process for vanadium, the recovery of vanadium from a high Ca/V ratio vanadium slag was studied by sodium roasting and ammonia leaching. In the present paper, the oxidation and leaching process of vanadium slag was investigated by X-ray diffraction (XRD), scanning electron microscopy and energy dispersive X-ray spectrometry (SEM/EDS) techniques. The effects of ammonium carbonate concentration, leaching temperature and leaching time on the leaching ratio of vanadium were discussed. As indicated in the experimental result, the optimal $(NH_4)_2CO_3$ concentration was 120g/L, leaching temperature was 60°C and leaching time was 20 min. Approximately 92% of the vanadium was recovered under the optimal conditions. Furthermore, by means of X-ray diffraction analysis, the phase transformations of the vanadium slag during roasting and leaching processes were analyzed and discussed.

Introduction

Vanadium is considered as an important rare metal and has been widely used in many fields due to its physical properties such as high tensile strength, hardness, and fatigue resistance [1]. About 85% of vanadium is used in the iron and steel industry [2]. The other 15% is mainly employed in non-ferrous metal industry, chemical industry , electronic industry and so on [3, 4].

Vanadium bearing slag plays a dominant role in the sources of vanadium which accounts for 58% for the production of vanadium [5]. Currently, the most common process to extract vanadium [6,7,8] is to smelt and reduce the vanadic titano-magnetite and produce vanadium-bearing hot metal. Then, the vanadium and iron are separated into vanadium slag and semi-steel by blowing oxygen in LD converter, as shown in Figure 1. Vanadium slag can be treated by several processes such as:

1. Sodium salt roasting-water leaching[8-10],
2. Calcium salt roasting followed by acid or alkaline leaching [11],
3. Acid leaching-solvent extraction [12, 13].

Among these, sodium salt roasting-water leaching technology is the most commonly used for the extraction of vanadium from vanadium slag.

Leaching process is known as a key stage in the process of extracting vanadium from the vanadium slag. The role of leaching process is to dissolve the roasted slag and separated the leachate into solid residues and leaching solution. What's more, the aim of leaching process is to recover as much as possible the vanadates that had been generated during the roasting process.

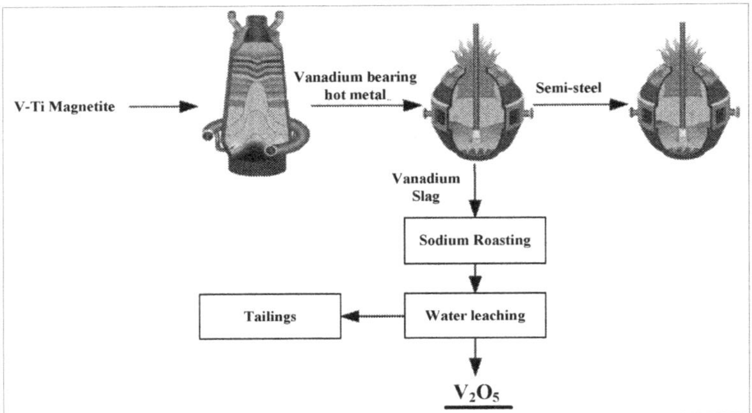

Fig 1. Schematic diagram of vanadium extraction process [8, 14]

Compared with the normal converter vanadium slag, the vanadium slag in the present work has a high Ca/V ratio. In general, high CaO content in the vanadium slag may induce water-insoluble calcium vanadates or calcium-containing vanadium bronze during roasting process [15], leading to difficulties in extraction of vanadium from the high Ca/V ratio slag. The present paper focuses on the feasibility of vanadium recovery from high Ca/V ratio vanadium slag using roasting followed by ammonia leaching (instead of conventional water leaching). Thus, it is necessary to study the ammonia leaching process and to investigate the factors influencing the vanadium extraction. The phase transformations of vanadium slag during the leaching process were also studied with XRD techniques.

Experimental

Vanadium Slag Specification

The prepared vanadium slag samples were ground and screened through 74μm sieve prior to leaching. The chemical composition was analyzed using X-ray fluorescence spectroscopy (XRF, model XRF1800, SHIMADZU, Japan), and the results arre presented in Table I, where the vanadium oxide was analyzed as V_2O_5 for convenience. Compared with the common converter vanadium slag [8], the vanadium slag in this work has a lower content of vanadium and a higher Ca/V ratio.

Table I. Chemical composition of vanadium slag (main fraction, %)

V_2O_5	FeO	SiO_2	MnO	TiO_2	CaO	MgO	Cr_2O_3	Al_2O_3	P_2O_5
13.60	36.80	18.30	8.60	8.30	4.40	3.85	4.02	1.34	0.45

The mineralogical analysis of the samples was investigated using X-ray diffraction (XRD), which was performed with a Rigaku D/MAX 2500PC using Cu *Kα* radiation. An X-ray diffraction pattern of the vanadium slag is presented in **Figure 2**. It can be observed that the major phases in this slag were vanadium-containing spinel $(Mn,Fe)(Cr,V)_2O_4$, olivine phase $(Fe,Mg)_2SiO_4$, pyroxene phase $Ca(Fe,Mg)Si_2O_6$, and some iron oxide Fe_3O_4.

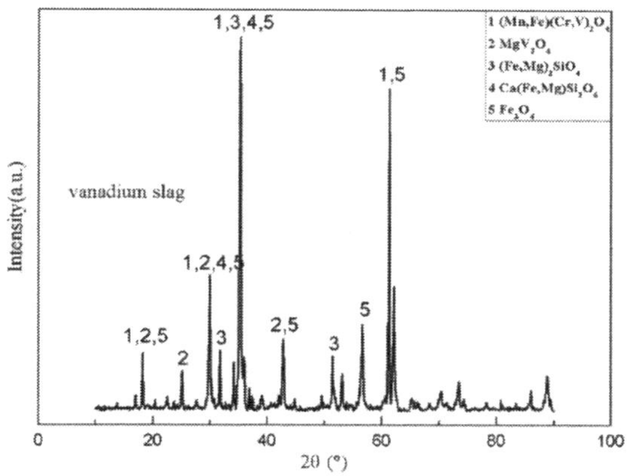

Fig 2. XRD pattern of raw vanadium slag

Sample Preparation

The sample that would be used for leaching experiments was prepared by the following sodium roasting process. The ground vanadium slag was thoroughly mixed with a prior determined mass of sodium carbonate. Then the mixture was put into an electric muffle furnace where the roasting experiments were carried out. A proportional-integral-differential (PID) controller (±5K) was employed to control the temperature, and the furnace was kept with free access to air in order to maintain an oxidizing atmosphere during roasting process so as to obtain a maximum efficiency in the conversion of the vanadium. After a determined roasting time, the samples were taken out and cooled down to the room temperature rapidly. Then, the roasted samples were ground to the size of -74μm and saved for further processing.

Experimental Leaching Procedure

The prepared roasted samples were leached with an ammonia carbonate leaching solution which was prepared with analytical reagent $(NH_4)_2CO_3$ and distilled water. The leaching procedure was performed in a stirred three-neck flask which was placed in a water bath. At the end of the leaching time the leachate was separated from the solid residues by means of vacuum filtration. Vanadium concentration in the leachate was analyzed by the ammonium ferrous sulfate titration, which is a standard analytical method.
Leaching ratio (η) of sodium salt roasting process was defined as Eq.(1).

$$\eta = \frac{[V_s]}{[V_t]} \tag{1}$$

where $[V_t]$ is the mass of total vanadium in the raw slag before roasting, and $[V_s]$ is the mass of vanadium in the leaching solution.
(1) Characterization of the sample
The crystalline phases of raw vanadium slag and roasted slag were characterized by powder X-ray diffraction. Scanning electron microscopy (SEM) and energy disperse X-ray spectrometry (EDS) were also used, to investigate the surface morphologies and surface compositions.
(2) Optimization of ammonia leaching parameters
The roasted slag used for the leaching experiments was prepared under optimal roasting conditions determined in a previous study [16]: addition amount of sodium carbonate of 19%, roasting temperature of 800°C, roasting time of 2 hour. Effects of $(NH_4)_2CO_3$ amount, leaching temperature and leaching time on the leaching ratio of vanadium were also investigated by single-factor experimental analysis. During the leaching process, the liquid to solid ratio was kept at a constant 8.0 ml/g. Scheme of the single-factor experiments is illustrated in Table II.

Table II. Scheme of the single-factor leaching experiments

Factors studied	$(NH_4)_2CO_3$ concentration, g/L	Leaching temperature, °C	Leaching time, min
$(NH_4)_2CO_3$ concentration	80,100,120,140,160	50	20
Leaching temperature	120	30,40,50,60,70	20
Leaching time	120	50	10,15,20,25, 30,40

(3) Phase transformation during roasting and leaching process
It is of importance to understand the leaching mechanism by investigating the phase transformations occurring during the leaching process. The phases present in the roasted slag and leaching residue under optimized conditions were studied using XRD.

Results and Discussion

Characterization of the Sample

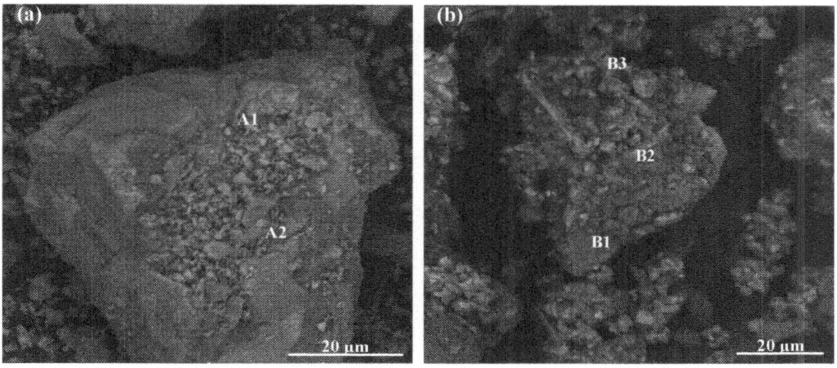

Fig 3. SEM images of slag samples: (a) raw vanadium slag, (b) roasted slag

The SEM images of slag samples are presented in Fig.3. **The EDS analyses of the areas marked in Fig. 3 are given in Table III.** As shown in Fig. 3(a), there were some smaller and brighter grains (shown in area A1) which were dispersed irregularly in the matrix phase. Based on the results of XRD and EDS analysis, these grains were the vanadium spinel (see the analysis data for area A1 in Table III). The matrix phase was olivine. It can be seen from Table III that V, Ti and Cr were mainly concentrated in the spinel grains (A1), while Al, Si and Ca existed mainly in the olivine (silicate) phase (A2). Fe and Mn existed in both phases.

As shown in Fig. 3(b), it can be seen from the SEM image of the roasted sample that some particles cohered together and presented a flower pattern. Results from EDS and XRD analysis demonstrate that the oxidized vanadium spinel had reacted with the Na_2CO_3 to produce sodium vanadates (shown in areas B1, B2 and B3, with the corresponding analyses in Table III). It was worth noting that there was a certain content of Mg and Ca in areas B1 and B3.

Figure 4 illustrates the XRD patterns of the sample before and after roasting process. Compared with the XRD pattern of raw vanadium slag (which has also been presented earlier in Fig. 2), the peaks of Fe_2O_3, acmite ($NaFeSi_2O_6$) and sodium pyroxene ($NaTiSi_2O_6$) appeared in the XRD pattern of roasted slag, while the characteristic peaks of $(Mn,Fe)(Cr,V)_2O_4$ disappeared. Meanwhile, the water-soluble sodium vanadates such as $NaVO_3$ and NaV_3O_8 were generated owing to the oxidizing reaction between the sodium carbonate and vanadium oxide. Some minor material phases such as MgV_2O_4 and $Mg_3(VO_4)_2$ were also generated which are water-insoluble substances. As a consequence, it is worthwhile to carry out experiments to recover the vanadium from these water-insoluble compounds by the ammonia leaching process.

Table III. EDS analysis of raw vanadium slag and roasted samples (mass fraction, %)

Area	O	Na	Mg	Al	Si	Ca	Ti	V	Cr	Mn	Fe
A1	26.15	-	-	-	-	-	7.31	21.79	9.06	5.74	29.95
A2	48.93	-	-	3.38	23.88	8.81	-	-	-	5.00	10.01
B1	44.47	18.84	-	-	2.75	1.59	1.90	13.59	2.85	3.21	10.79
B2	41.37	20.91	-	-	2.47	-	-	11.29	-	3.25	20.71
B3	38.19	13.37	2.69	-	8.55	7.62	2.99	9.21	-	2.25	15.13

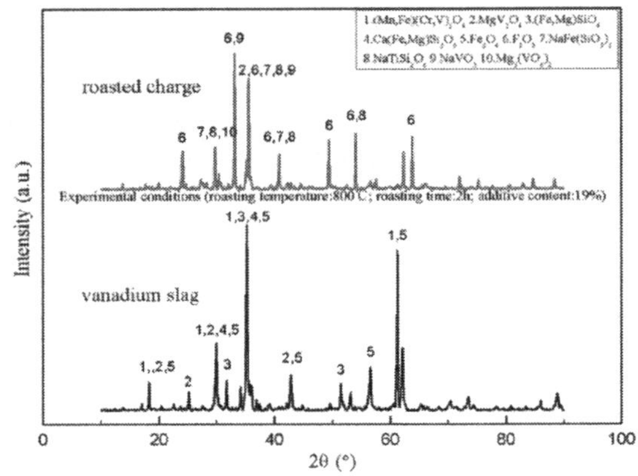

Fig 4. XRD patterns of raw vanadium slag and roasted slag

Parameters of Ammonia Leaching Process

The main aim of leaching process is to extract vanadates from the roasted charge to the greatest extent possible. Hence, in present paper, the effects of $(NH_4)_2CO_3$ concentration, leaching temperature and leaching time on the leaching ratio of vanadium were investigated by single-factor experimental analysis.

Effect of Ammonia Carbonate Concentration

Figure 5 illustrates the effect of $(NH_4)_2CO_3$ concentration on the leaching ratio of vanadium. This increased with the increment of $(NH_4)_2CO_3$ concentration at low concentrations, however it decreased slightly when the concentration of $(NH_4)_2CO_3$ exceeded 120 g/L. This phenomenon may be attributable to carbonate precipitation which occurred at high $(NH_4)_2CO_3$ concentrations, and tended to block the reaction.

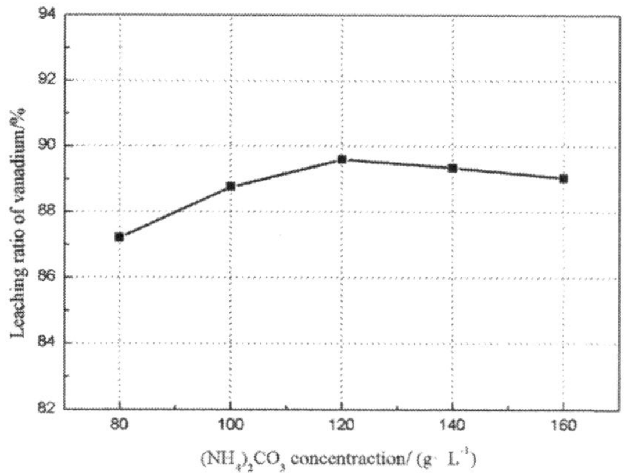

Fig 5. Effect of $(NH_4)_2CO_3$ concentration on the leaching ratio of vanadium
Temp 50°C, time 20 min

Effect of Leaching Temperature

The leaching ratio of vanadium versus leaching temperature is shown in Figure 6. Increasing leaching temperature played a significant role in the recovery of vanadium. However, the extraction of vanadium increased only slowly as the leaching temperature varied from 60 to 70°C. A maximum extraction of 91.2% was achieved at 70°C, yet there was only a 0.3% differential with the temperature at 60°C. Therefore, considering that higher temperature would accelerate the hydrolization of the $(NH_4)_2CO_3$ solution, 60°C was chosen as the optimum leaching temperature in present work, also saving energy.

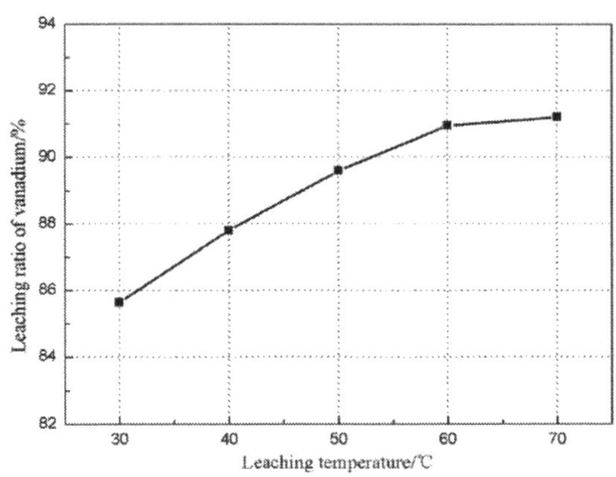

Fig 6. Effect of leaching temperature on the leaching ratio of vanadium.
Conc 120 g/L, time 20 min

Effect of Leaching Time

Figure 7 shows the profile of the leaching ratio of vanadium against the leaching time. It can be observed that the ratio gradually increased with the increasing leaching time up to 25 minutes. From 25 to 40 min there was a small decrease. The vanadates cannot be sufficiently dissolved when the leaching time was too short, nevertheless, it will be a waste of time if the leaching time was too long. Therefore the optimal leaching time was 20 min.

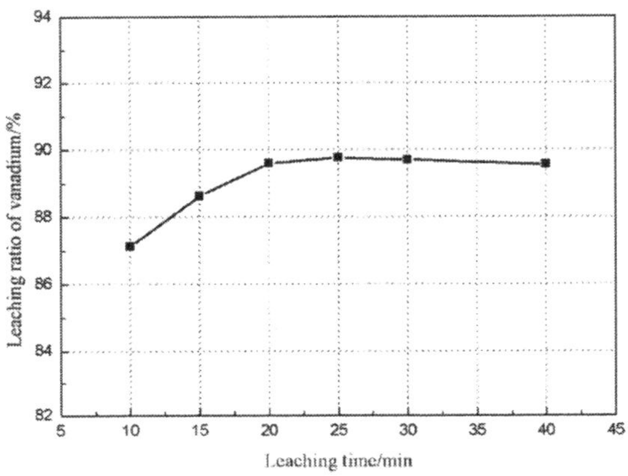

Fig 7. Effect of leaching time on the leaching ratio of vanadium
Conc 120 g/L, temp 50°C

Phases Present Before and After the Ammonia Leaching Process

Under optimized leaching conditions, the leaching residue was analyzed by means of XRD analysis. Figure 8 shows the XRD patterns of roasted slag before and after ammonia leaching process. The results indicated that the major phases of the leaching residue are basically the same as that of the roasted sample except that the sodium vanadates had dissolved away. It is worth mentioning that the minor phase manganese vanadate was not detected in the leaching residue, indicating that the water insoluble vanadates were also leached along with the water soluble sodium vanadates in the ammonia carbonate leaching solution. Therefore the ammonia leaching process is an efficient way to deal with the roasted high Ca/V ratio vanadium slag.

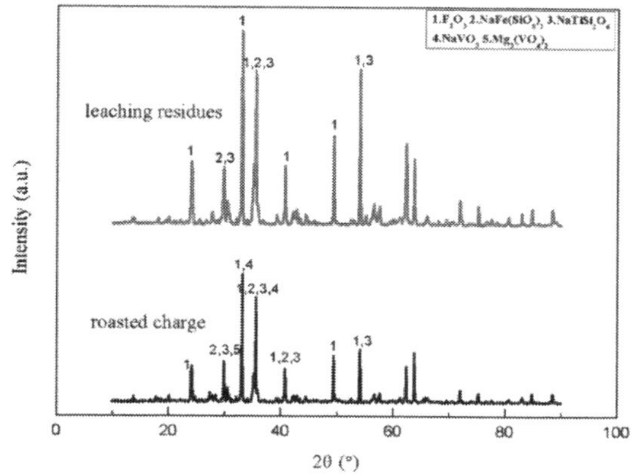

Fig 8. XRD pattern of the roasted slag and leaching residue

Summary

Experiments on extraction of vanadium from a high Ca/V ratio vanadium slag by sodium roasting and ammonia leaching process were carried out in the current study, indicating that the sodium roasting and ammonia leaching process was a practicable method to extract vanadium from a high Ca/V ratio vanadium slag. As shown in the experimental study, the optimal ammonia carbonate concertraction was 120g/L, the leaching temperature was 60°C, and the leaching time was 20 min. By means of X-Ray diffraction (XRD) techniques, the phases present in vanadium slag during leaching process under the optimal leaching conditions were analyzed and discussed. Under the ammonia leaching process, most of the vanadates (>90%) were leached to the leaching solution.

Acknowledgements

The work was financially supported by the geological survey project from Ministry of Land and Resources of P.R. China (project no. 12120113087400).

References

1. R.R. Moskalyk and A.M. Alfantazi. Processing of vanadium: a review[J]. *Minerals Engineering*, 2003, 16(9): 793-805.
2. P.R. Taylor, et al. Extractive metallurgy of vanadium-containing titaniferous magnetite ores: A review[J]. *Minerals and Metallurgical Processing*, 2006, 23(2): 80-86.
3. A. Parasuraman, et al. Review of material research and development for vanadium redox flow battery applications[J]. *Electrochimica Acta*, (0).
4. W. You. The Resources of Vanadium and Its Applications, Development and Outlook[J]. *Rare metals and cemented carbides*, 1996, 01: 51-55. (in Chinese)

5. B. Raja. Vanadium market in the world[J]. Steelworld, 2007, 13(2): 19-22.
6. B.C. Jena, W. Dresler, and I.G. Reilly. Extraction of Titanium, Vanadium and Iron from Titanomagnetite Deposits at Pipestone Lake, Manitoba, Canada[J]. *Minerals Engineering*, 1995, 8(1-2): 159-168.
7. D. Chen, et al. A novel process for recovery of iron, titanium, and vanadium from titanomagnetite concentrates: NaOH molten salt roasting and water leaching processes[J]. *Journal of Hazardous Materials*, 2013, 244–245(0): 588-595.
8. H. Daoxin. Vanadium Extraction and Steelmaking[M]. Beijing,China: *Metallurgical Industry Press*, 2000. (in Chinese)
9. A. Mahdavian, et al. Recovery of Vanadium from Esfahan Steel Company Steel Slag; Optimizing of Roasting and Leaching Parameters[J]. *International Journal of Iron and Steel Society of Iran*, 2006.
10. L. Shiming. The metallurgy of vanadium abroad[M]. Beijing,China: *Metallurgical Industry Press*, 1985. (in Chinese)
11. H.S. Chen. Extraction of vanadium pentoxide from vanadium slag by lime calcination[J]. *Iron Steel Vanadium Titanium*, 1992, 13(6): 1-9. (in Chinese)
12. Y. JingLing. A new way of recovering vanadium from iron/vanadium slag[J]. *Journal of Beijing University of Chemical Technology*, 2007, 34(3): 254-257. (in Chinese)
13. Z.L. LI, et al. Acid Leaching of Calcined Vanadium Titanomagnetite with Calcium Compounds for Extraction of Vanadium[J]. *The Chinese Journal of Process Engineering*, 2011, 11(4): 573-578.
14. X. Zhang, et al. Nucleation and growth kinetics of spinel crystals in vanadium slag[J]. *Ironmaking & Steelmaking*, 2012, 39(2): 147-154.
15. H.A. Barolin. Oxidation of Vanadium Slag[M]. Beijing,China: *Metallurgical Industry Press*, 1982.
16. Song Xu, et al. Study on Roasting Parameters for Extraction of Vanadium from a High Ca/V Ratio Vanadium Slag[J]. *Metalurgia International*, 2013, 18(8): 38-40.

Celebrating the Megascale: Proceedings of the Extraction and Processing Division Symposium on Pyrometallurgy in Honor of David G.C. Robertson
Edited by: Phillip J. Mackey, Eric J. Grimsey, Rodney T. Jones, and Geoffrey A. Brooks
TMS (The Minerals, Metals & Materials Society), 2014

Sintering Process for Limonitic Nickel Laterite

Enguang Guo[1], Mei Liu[1], Cheng Pan[1], Qiugang Yuan[1], Xuewei Lv[1]

[1]College of Materials Science and Engineering, Chongqing University, Chongqing, 400044

Keywords: Sintering; Limonitic nickel laterite; Binder phase

Abstract

Land-based nickel resources include nickel sulfide and nickel laterite. With the consumption of high grade nickel sulfide, use of nickel laterite has received more and more attention. The mineralogy and sintering behavior of limonitic nickel laterite with high iron and low nickel and silica was studied to offer technical support for producing ferronickel through sintering–blast furnace route. The mineralogy results showed that the main phases in this kind of nickel laterite are goethite (FeO(OH)), gibbsite ($Al(OH)_3$) and $NiFe_2O_4$. The TGA (Thermogravimetric Analysis) and DSC (Differential Scanning Calorimetry) revealed hydroxide minerals in addition to absorbed water. After the sintering experiments the chemical composition, phases present (XRD analysis) and physical properties of the sinter were studied. The content of FeO in the sinter increased with increasing basicity up to 1.3, and then decreased with further increase in basicity. The yield of sinter increased (from 73% to 80%) with increase in the basicity from 1.1 to 1.7, and then the yield decreased with the further increase in basicity. $Mg(Ni)(Fe,Al)_2O_4$ is the main phase in the sinter while $MgCaSiO_4$ and Fe_2SiO_4 is the main binder phase when the basicity is 1.1~1.5, and $MgCaSiO_4$, Fe_2SiO_4 and SFCA is the main binder phase when the basicity is 1.7~2.1.

Introduction

As a commercially valuable metal, nickel is largely used in production of stainless steel or high temperature alloys [1-3]. In the last decade, the rapid increase of demand for stainless steel has led to a significant rise in ferronickel production. In 2011, about 60% of the nickel was extracted from sulfide ores, even though laterite accounts for 70% of world nickel resources [3, 4]. However, with the continuous depletion of economical sulfide ores, much more attention has been drawn to laterite processing.

A variety of flowsheets are used to process laterite ores. They generally fall into two categories – pyrometallurgy and hydrometallurgy. In recent years the pyrometallurgical processes have been commonly applied in China, including nickel matte and ferro-nickel processes. For the latter, one process is the rotary kiln-electric furnace (RKEF) process, which is largely used to produce high grade nickel pig iron and has good adaptability for various nickel content laterites [5-7]. The other process is the blast furnace process.

The sintering process is necessary for nickel laterites before smelting in the blast furnace, because the laterites have poor strength and high water content. The yield and strength of laterite sinter currently produced is so low that the sinter performs poorly in the blast furnace and therefore the cost of the blast furnace process is high. There are few previous studies on sintering of nickel laterites. In this work the properties and sintering of nickel laterite were studied. The effect of carbon dosage (as fuel) and flux additions (to help with the formation of a binder phase) on the sinter process were part of the sintering study.

Materials

The chemical composition of the nickel laterite and the lime flux were analyzed by the standard chemical method (GB/T13551-1995, China national standard) (**Table I** and **Table II**). Coal was analysed by industry analysis (**Table III**). The result shows that this kind of nickel laterite contains low nickel and high ferrous. It illustrates that this kind of nickel laterite comes from the zone of limonite.

Table I. Chemical composition of dried limonite - %(mass)

	Ni	TFe	CaO	SiO_2	MgO	Al_2O_3	P
Ni laterite	1.2	46.26	0.96	3.3	1.51	8.77	0.0068

Table II. Chemical composition of limestone - %(mass)

	CaO	other
Limestone	75.98	24.02

Table III. Proximate analysis (%(mass) and calorific value of the coal

				Bomb calorific value
Fixed carbon	Ash	Volatile	moisture	J/g
72.00	20.86	7.14	5.1	24717

The phase composition of the limonite and the sinter were examined with XRD (X-ray diffraction) analysis which was carried out using D/MAX 3C (Co Kα). XRD analysis of nickel laterite (**Fig. 1**) shows that the main phases in the nickel laterite are goethite ($FeO(OH)$), gibbsite ($Al(OH)_3$) and $NiFe_2O_4$.

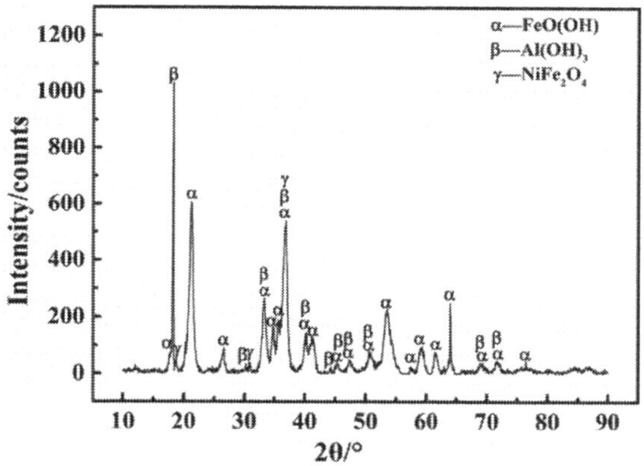

Figure 1. XRD pattern of limonitic nickel laterite

The thermal behavior such as DSC and TGA measurements were carried out using a NETZSCH STA 449C. The maximum temperature was fixed at 1400°C, and the heating rate was 20K/min. All the powder of limonite used was smaller than 0.074μm. During the TGA experiment, the mixture was loaded into the Pt crucible in air. The TG and DSC curves for the nickel laterite up to 1400°C are given in **Fig. 2**. and verified the XRD results, as the endothermic peaks at 117 and 305℃ are respectively caused by free water desorption and hydroxide decomposition. The XRD analysis of nickel laterite was carried out at 1000°C and 1260°C respectively to study the reason for the endothermic peak appearing at 1244 °C on the TGA-DSC curve (**Fig. 2**) of nickel laterite. The result is shown in **Fig. 3**, which indicates that there is some hydroxide present at 1000 °C, where the greater height of the second large peak for Fe_2O_3 is considered to be due to the presence of hydroxide. This indicates that the reason of the endothermic peak appearing at 1244°C on the TGA-DSC curve is the decomposition of residual hydroxide.

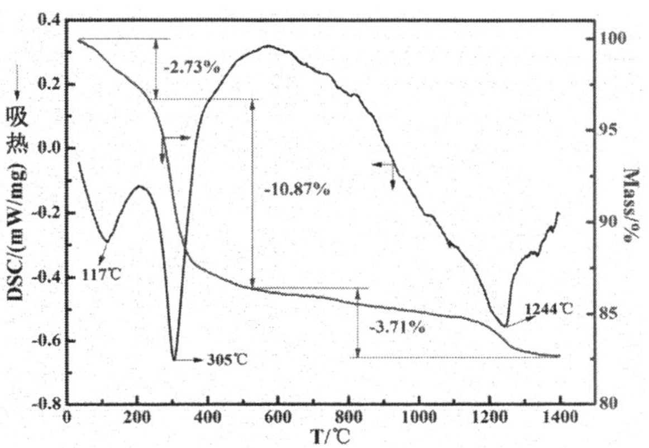

Figure 2. TG-DSC curve of nickel laterite

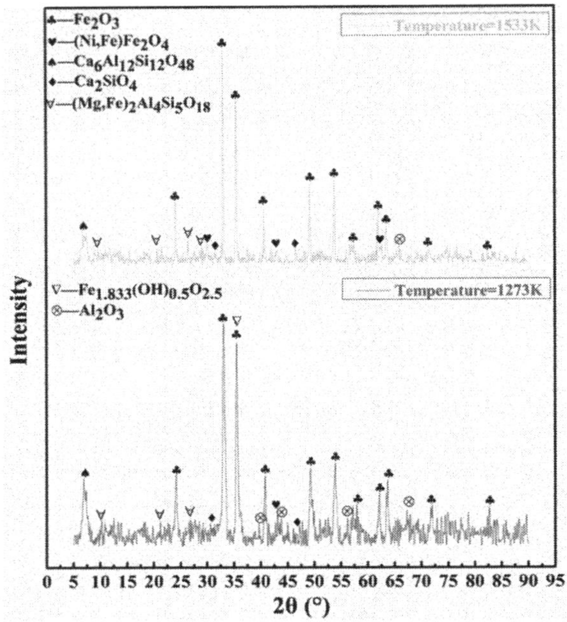

Figure 3. XRD pattern of nickel laterite at 1273K and 1533K

Experimental procedure for the sintering experiments

The composition of the sinter mixture is one of the factors that determine the amounts of liquid oxide phase and other mineral phases formed at temperature during the sintering process. These in turn affect the strength and yield of sinter. One of the most important compositional variables is the basicity (CaO/SiO_2). The effect of basicity on sinter properties was studied in this work. The basicity was varied from 1.1 to 2.1 in steps of 0.2. No other variables were changed during the series of sintering experiments. The sintering temperature is determined by the fuel addition (maintained constant in the experiments) and the nature of the starting raw materials.

For 100 units of limonite (after the addition of limestone to adjust the basicity) 12.5 units of coal (9 units of fixed carbon) and 18.0 units of moisture were added. The granulation process was then carried out in a rotating drum, and the sinter mixture was charged into the laboratory sintering pot (**Fig. 4**). The mass of the mixture used in each experiment was 28~30 kg.

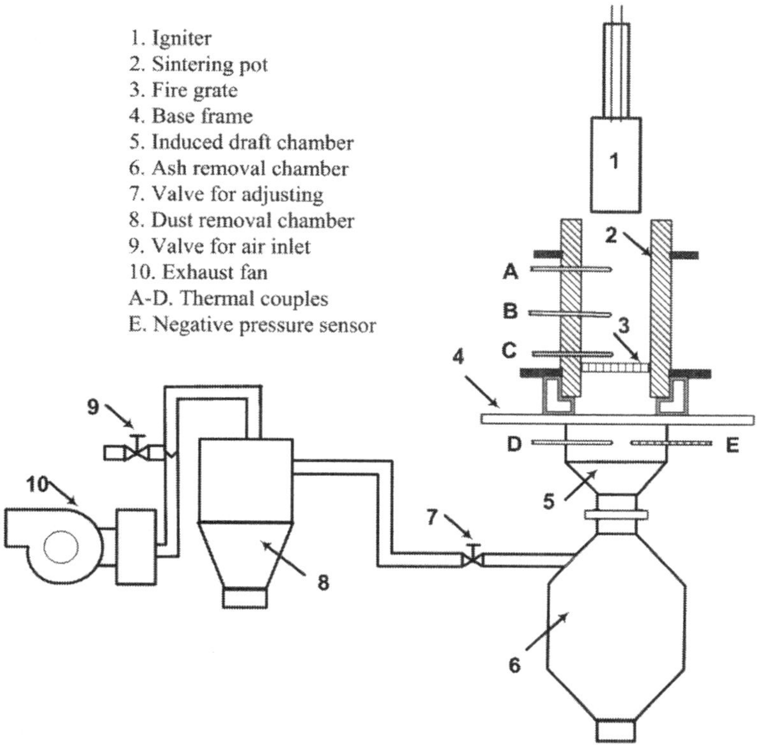

Figure 4. Schematic of sintering experimental apparatus

Results and discussion

The change of the content of TFe (total iron) and FeO and yield of sinter with basicity is shown in **Fig. 5**. As expected the content of TFe dosage did not change when the basicity increased. The basicity had an obvious effect on the content of FeO, as the content of FeO increased from 38.9% to 40.2% with increasing basicity from 1.1 to 1.3, and then decreased when the basicity continued to increase. The yield increased from 72.8% to 80.2% with the basicity increasing from 1.1 to 1.7, and then the yield decreasing with the basicity increasing.

As shown in **Fig. 6**, the shatter strength was highest at the intermediate basicities. A similar behavior was found for the tumbler strength, the highest value being is 66.3% when the basicity is 1.5. The tendency of abrasion was opposite with the lowest value when the basicity is 1.9.

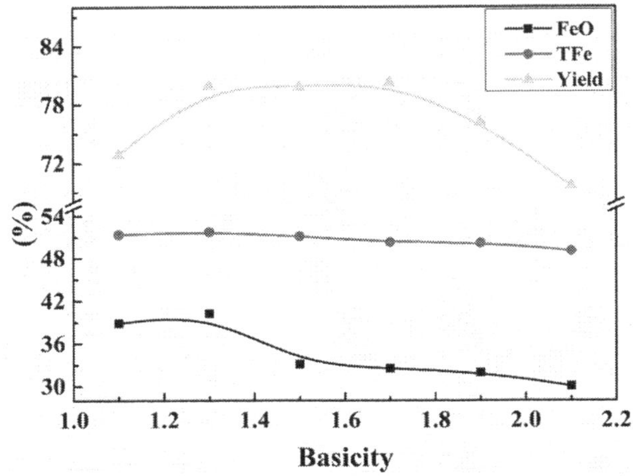

Figure 5. Dependency of content of TFe and FeO, yield on the basicity

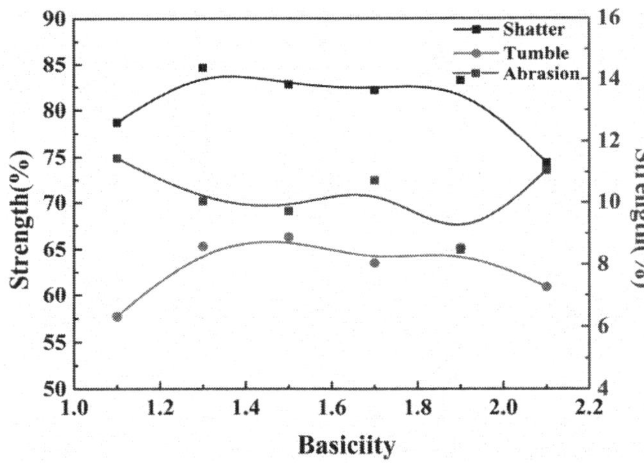

Figure 6. Dependency of tumbler strength, shatter strength, abrasion index on the basicity

The strength of sinter is influenced by the binder phases in sinter. In order to investigate the binder phases, the XRD analysis of sinter was also studied. The results are shown in **Figs. 7** and **8**. The results shows that the binder phases in the sinter were $MgCaSiO_4$ and Fe_2SiO_4 when the basicity is 1.1~1.5, $MgCaSiO_4$, Fe_2SiO_4 and SFCA (silico-ferrite of calcium and aluminum) when the basicity is 1.7~2.1. The Fe_2SiO_4 peak was strongest with a basicity of 1.3. SFCA appeared in the sinter when the basicity increased up to 1.7. The strength of SFCA is worse than the strength of Fe_2SiO_4 so the strength of sinter decreased with the basicity increasing up to 1.7.

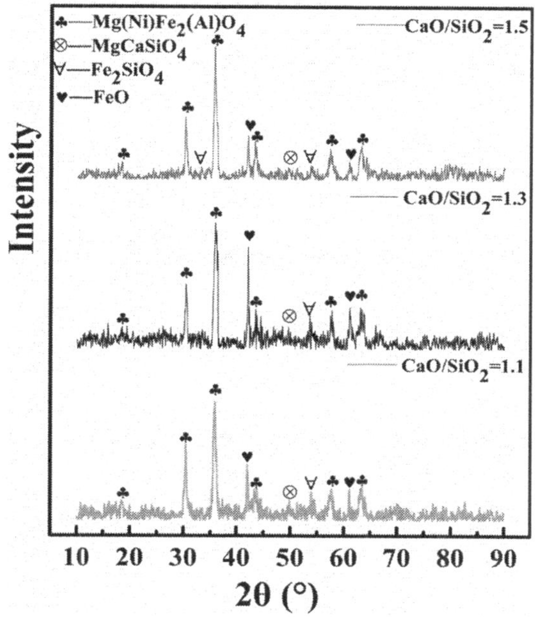

Figure 7. The XRD phase of nickel sinter

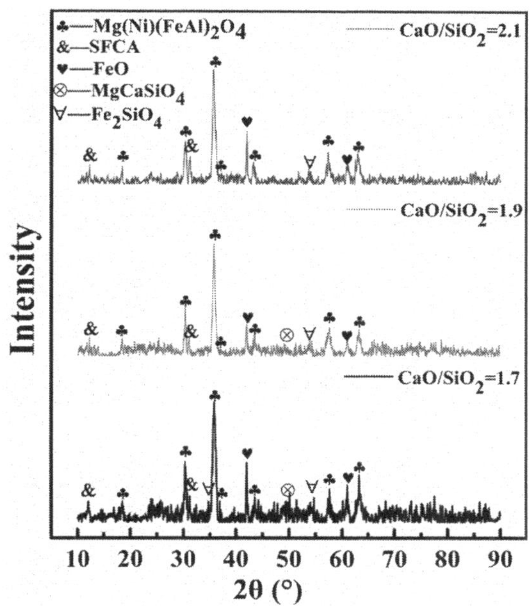

Figure 8. The XRD phase of nickel sinter

Conclusions

1. The main phases in the limonitic nickel laterite are goethite (FeO(OH)), gibbsite ($Al(OH)_3$) and $NiFe_2O_4$.
2. The content of TFe dosage not change when the basicity increases. The content of FeO increased with increasing basicity up to 1.3, and then decreased when the basicity continued to increase. The yield increased from 72.8% to 80.2% with the basicity increasing from 1.1 to 1.7, and then the yield decreased with further increases in basicity.
3. $Mg(Ni)(Fe,Al)_2O_4$ was the main phase in the sinter while $MgCaSiO_4$ and Fe_2SiO_4 were the binder phases when the basicity was 1.1~1.5, and $MgCaSiO_4$, Fe_2SiO_4 and SFCA were the binder phases when the basicity was 1.7~2.1 .

References

1. N. Voermann et al., "Furnace Technology for Ferro-Nickel Production – an Update" in International Laterite Nickel Symposium, March 18, (Minerals, Metals and Materials Society, Charlotte, NC, 2004) 563-577.

2. R.A. Bergman, "Nickel production from low-iron laterite ores: Process descriptions", CIM Bulletin, 96 (2003), 127-138.

3. C. Panet al., "Gaseous Reduction of Pellets of Laterite Ore Containing Carbon", *Metalurgia International*, 16(1) (2011), 5-9.

4. J. Kim et al.,"Calcination of low-grade laterite for concentration of Ni by magnetic separation", *Minerals Engineering*, 23(4) (2010), 282-288

5. T. Hu et al., "Effect of Coal Blending Ratio on Sintering of Philippine Nickeliferous Laterite', *Metalurgia International*, 16(8) (2011), 32-35.

6. X.W.Lv et al.,"Mineral Change of Philippine and Indonesia Nickel Lateritic Ore during Sintering and Mineralogy of Their Sinter", ISIJ International, 50(3), (2010), 380-385.

7. X.W. Lv et al., "Dehydrating and sintering of Philippine nickel laterite", *Canadian Metallurgical Quarterly*, 50(1), (2011), 20-27.

CELEBRATING THE MEGASCALE

Proceedings of the Extraction and Processing Division Symposium on Pyrometallurgy in Honor of David G.C. Robertson

Pyrometallurgy Process Fundamentals III

Celebrating the Megascale: Proceedings of the Extraction and Processing Division Symposium on Pyrometallurgy in Honor of David G.C. Robertson
Edited by: Phillip J. Mackey, Eric J. Grimsey, Rodney T. Jones, and Geoffrey A. Brooks
TMS (The Minerals, Metals & Materials Society), 2014

Development of a Thermodynamic Database for Mold Flux Applications to the Continuous Casting Process for Steelmaking

[1]Marie-Aline Van Ende and [1]In-Ho Jung

[1]Department of Mining and Materials Engineering, McGill University
3610 University Street, Montréal, Québec, CANADA H3A 0C5

Keywords: Thermodynamic database, mold flux, continuous casting, steelmaking, kinetic model

Abstract

A thermodynamic database for the oxyfluoride system CaO-MgO-Al_2O_3-SiO_2-Na_2O-K_2O-Li_2O-MnO-FeO-F has been developed based on the critical evaluation and optimization of all available experimental thermodynamic and phase diagram data. The developed database can be used for phase diagram and equilibrium solidification calculations for multicomponent systems and for understanding the solidification behavior of mold fluxes. Based on this thermodynamic database, a simple kinetic model was developed to simulate the interactions between the mold flux and molten steel using effective equilibrium reaction volumes combined with the thermodynamic database.

Introduction

Mold fluxes play crucial roles during the continuous casting of steel: prevention of steel reoxidation, absorption of non-metallic inclusions, control of horizontal heat transfer and lubrication between mold and solidified steel. Typical mold fluxes being used in continuous casting process of steel belong to the CaO-SiO_2-Al_2O_3-Na_2O-F system with small addition of MgO, K_2O, Li_2O, MnO, etc. Controlling the primary solidification temperature of the flux, which is mostly synchronized to the bcc → fcc phase transformation temperature of steel during solidification, is one of the key parameters in the design of the mold flux composition. Typical primary solidified phase is the well-known cuspidine ($Ca_4Si_2F_2O_7$) crystal. A wrong choice of mold powder can lead to defects such as longitudinal cracking and sticker breakouts, along with process problems. A small amount of FeO, Fe_2O_3 and MnO can be picked up in the mold slag through reaction with liquid steel.

Although the general effects of each mold flux component on the relevant mold flux properties are known, determining the optimum mold flux composition still requires a trial-and-error approach. Casting a new steel grade will usually require the development of a new mold flux, which can become expensive and time consuming with such approach. In spite of their industrial importance, the phase equilibria of fluorine containing oxide systems have not been well investigated due to the complexity of their chemistry and the difficulties in performing experiments. No systematic thermodynamic modeling of the mold flux systems has been conducted to date, either.

In order to assist in the design of mold flux for the continuous casting process of steel and fluxes for special refining processes, a thermodynamic database for the oxyfluoride system containing CaO-MgO-Al_2O_3-SiO_2-Na_2O-K_2O-Li_2O-MnO-FeO-F has been developed by our research group for many years. Such database gives precious information on the solidification behavior of a given mold flux, which is of great help to understand its performance and provide candidates for new mold flux design, thereby surpassing the conventional trial-and-error approach. In the present study, the recent thermodynamic database for the oxyfluoride system will be overviewed. Then, a kinetic model, developed to calculate the change in composition in the mold slag during continuous casting by considering the reactions between molten mold flux and steel, and the absorption of non-metallic inclusions, is presented.

Thermodynamic Database Development for Mold Flux

A thermodynamic database for the mold flux system CaO-MgO-Al_2O_3-SiO_2-Na_2O-K_2O-Li_2O-MnO-FeO-F [1-3] has been built based on the critical evaluation and thermodynamic optimisation of all reliable experimental thermodynamic and phase diagram data found in literature. When the oxide slag system is expanded to fluoride, the slag structure becomes more complex. In the case of oxide slag system, the short-range-ordering of cations connected by oxygen anion as common anion should be considered. This has been well described by the Modified Quasichemical Model. The addition of fluoride to the oxide slag can create short-range-ordering between both cations and anions. For example, the CaO-SiO_2-CaF_2 system is, in reality, a reciprocal system of $(Ca^{2+},Si^{4+})[O^{2-},F^{-}]$, which involves four binary systems.

In order to describe the Gibbs energy of the reciprocal system, the Gibbs energy parameters for each binary system should be determined first and, if necessary, Gibbs energy parameters for the reciprocal system can be added to obtain a better fit of all the experimental data in the multicomponent system. The database for the oxide system has been already well developed in the FToxid database [4], which is further expanded toward Na_2O, K_2O and Li_2O systems. Then, the oxide system is expanded toward the fluoride system. In order to develop the large thermodynamic database for oxyfluoride slag, all reciprocal systems in the given oxyfluoride system should be modeled. The oxyfluoride liquid phase was described using the two-sublattice Modified Quasichemical Model [5, 6], which allows taking into account the short-range ordering and reciprocal exchange reaction of cations and anions in the oxyfluoride slag.

Equilibrium Solidification Calculation of Mold Flux

The database can be applied to the accurate calculation of the equilibrium solidification of mold fluxes. An example of equilibrium solidification calculation is provided in Fig. 1, where a typical mold flux with composition (in wt%) 32.0 SiO_2, 36.7 CaO, 0.7 MgO, 6.8 Al_2O_3, 8.4 Na_2O and 15.4 CaF_2 was employed. The calculation results predict that cuspidine ($Ca_4Si_2F_2O_7$) is the first compound to crystallize in the mold flux at approximately 1230 °C, which is in good agreement with experimental observations [7, 8] and predicted solidification and breaking temperatures. The predicted solidification behavior of many other commercial mold fluxes and new mold flux candidates were consistent with experimental data. The database can be effectively used to design new mold fluxes.

Development of a Kinetic Model for Steel/Mold Flux Interactions

A kinetic model was developed to calculate the change in composition in molten slag during continuous casting by considering the reactions between molten slag and steel, and the absorption of non-metallic inclusions. The kinetic model is based on the Effective Equilibrium Reaction Zone Model that was successfully applied to the RH degassing process [9]. In this approach, a complex process is divided in a finite number of reaction zones in which equilibria are calculated. For instance, in the simplified case of a slag/metal reaction (Fig. 2), the metal phase would be divided in a bulk volume (V1 in Fig. 2) and a smaller volume near the slag/metal interface (V2 in Fig. 2). The slag phase would be divided in a similar way (V3 and V4 in Fig. 2). In the Effective Equilibrium Reaction Zone Model, the equilibrium would be first calculated between V2 and V3, followed by equilibrium homogenization reactions in the metal phase (between V1 and V2) and in the slag phase (between V3 and V4).

Kinetics are taken into account by varying the reaction zone volumes depending on the process conditions, based on physical descriptions of the different reaction mechanisms. Simplified mathematical functions and empirical relations derived from simulations, experimental studies and plant data can be used to describe the effective reaction zone volumes. This approach allows easy linkage of the thermodynamic database to the kinetic simulation.

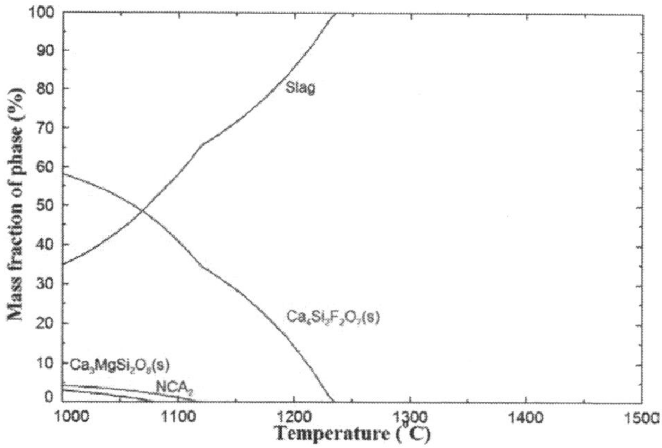

Figure 1. Equilibrium solidification calculations for the mold flux with 32.0 SiO_2, 36.7 CaO, 0.7 MgO, 6.8 Al_2O_3, 8.4 Na_2O and 15.4 CaF_2 (in wt%).

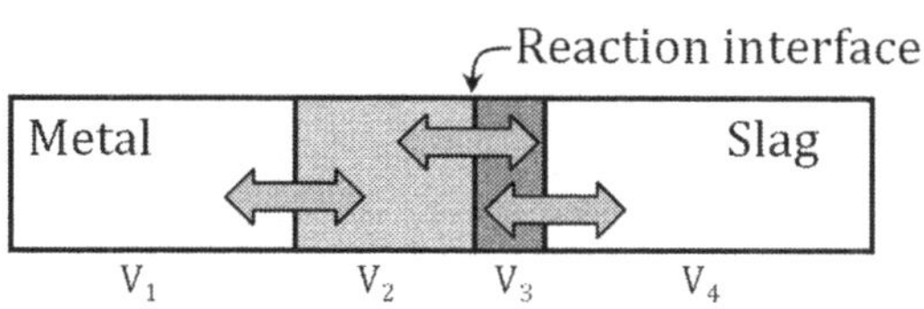

Figure 2. Concept for Effective Equilibrium Reaction Zone Model.

The schema of the effective reaction zones in the mold flux model is represented in Fig. 3. The mold slag pool was divided into three layers of different thickness and temperature to consider a temperature profile through the mold slag pool. The continuous fresh mold flux input into the upper mold slag layer (slag layer 3) and materials output from each mold slag layer were modeled by streams. A steel layer, which is continuously refreshed by steel from the steel pool, was reacted with the bottom slag layer (slag layer 1). The dissolution of non-metallic inclusions was conducted in the bottom slag layer (slag layer 1). In total, six reactions (R1 to R6 in Fig. 3) were created to simulate transport of matter and equilibrium reactions between the layers at each time step and are briefly described below:

- Reaction 1 (R1 in Fig. 3) represents the reaction between the entire steel layer, the entire slag layer 1 and part of the inclusions. The amount of inclusion dissolved in slag layer 1 is defined by the inclusion removal fraction. The steel layer is also partially renewed by Fresh steel.
- Reactions 2 to 4 (R2 to R4 in Fig. 3) correspond to the downward material flow between and out of each slag layer. As Fresh mold flux is continuously added on top and the slag layer thicknesses are constant, material must enter and exit each slag layer at the same rate. In the present study, the amount of material exiting a given slag layer was assumed to be proportional to its thickness.
- Reactions 5 and 6 (R5 and R6 in Fig. 3) denote the partial homogenization between the slag layers. In the present study, it was assumed that 50% of a given slag layer is equilibrated with 50% of the neighboring layer.

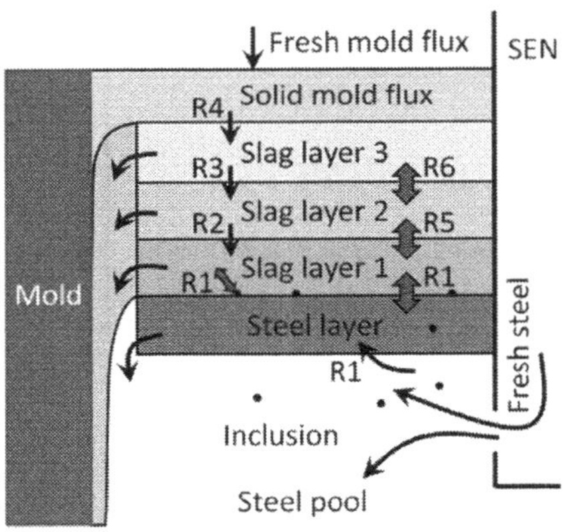

Figure 3. Schematic of a kinetic model for mold flux in continous casting process.

The details of the model are presented elsewhere [10]. Thermodynamic equilibrium at each reaction zones and each cycle was calculated using FactSage software [4] (version 6.3) with the accurate oxyfluoride database presented above. The Equilib files describing the reactions occurring in each reaction zone were linked and controlled by a large macro file running in FactSage software.

Simulation of Mold Flux Composition During the Casting of High Al Steel

Although an optimized mold flux is used for each steel grade, the molten mold flux will take more time to reach its steady state composition than steel during the transition between two different steel grades. Thus, the surface quality of the steel slab during the transition period between two steel grades is poor compared to that in steady state casting conditions. In order to simulate such transient period, the kinetic model for the mold flux has been tested. In particular, the transition between a low carbon steel to a high Al TRIP steel was calculated in Fig. 4. The molten steel/molten mold flux interactions were first simulated for the low-carbon steel with composition and casting conditions listed in Table I. After 15 min of casting, the low-carbon steel was replaced by a high Al TRIP steel (Table I). The mold flux presented in Fig. 1 was employed for both steel grades. The thickness of the mold slag pool was assumed to be 7 mm: the thickness of the low (layer #1), intermediate (#2) and top layer (#3) is assumed to be 1, 2 and 4 mm, respectively. The temperature of each layer is fixed at 1550, 1500 and 1450 °C, respectively. The thickness of slag layer 1 was made smaller than the two others to obtain a steep temperature gradient in the mold slag pool near the steel surface. The steel layer is assumed to be 10 mm to have a reaction volume ratio between slag and metal of 1:10. Both steel grades were assumed to contain about 20 ppm Al_2O_3 inclusions with a fixed removal rate of 15%.

Table I. Steel composition and casting parameters for the simulation.

Mold dimensions (mm)		Simulation parameters						
Length	900		Thickness (mm)	Temperature (°C)		Thickness (mm)	Temperature (°C)	
Width	1500	Slag layer 1	1	1550	Slag layer 3	4	1450	
Thickness	250	Slag layer 2	2	1500	Steel layer	10	1550	
Steel grade	Casting time (min)	Casting speed (m min^{-1})	Casting temperature (°C)	Powder consumption (kg m^{-2})	Steel composition (wt%)			
					Mn	Si	C	Al
Low C steel	0-15	1.1	1550	0.4	0.9	0.04	0.01	0.05
High Al steel	16-30	0.8	1550	0.5	2.0	0.9	0.01	1.5

The composition of slag layer #1 is changing drastically by the reactions with the TRIP steel. In particular, Al in the molten steel can readily reduce SiO_2 in mold flux to increase Al_2O_3 in mold flux, suggesting the global reaction between steel and slag to be given by Eq. (1).

$$3/2\ (SiO_2) + 2\ \underline{Al} = (Al_2O_3) + 3/2\ \underline{Si} \quad (1)$$

After the change in mold flux composition in slag layer #1, the mold flux composition in slag layers #2 and #3 are continuously changed due to diffusion and mixing between mold flux layers. Fig. 4 shows the change in the mold flux composition (averaged composition of layers # 1 to 3) with casting time. The Al_2O_3 content is increasing up to about 15 wt% within 10 min following the transition from low C steel to Al TRIP steel. The solidification behavior of the altered mold flux is largely modified, resulting from the interactions with the high Al TRIP steel.

The equilibrium solidification of the modified mold flux is presented in Fig. 5. As can be seen, the primary phase is changed from cuspidine to $Na_2CaAl_4O_8$ and Ca_2SiO_4 phases. The primary solidification temperature of the mold flux is increased drastically. The evolution of the solid and liquid fractions of the mold flux with temperature is also largely changed from that of the original flux. This is a well-known phenomena in high Al steel production. Because of this reason, mold flux for high Al steel has not been well optimized yet in real production line, which causes low surface quality of steel products.

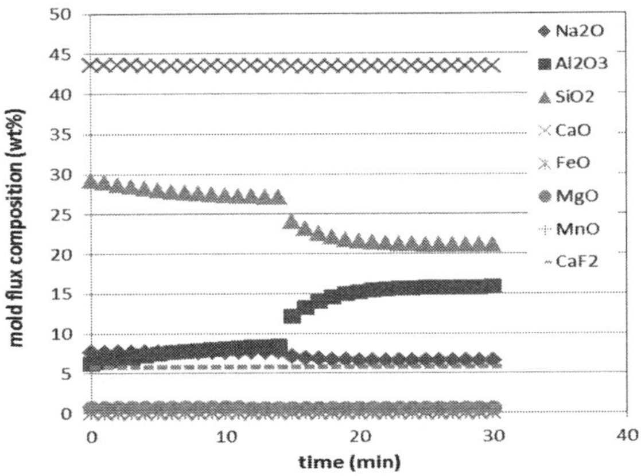

Figure 4. Evolution of the calculated composition of mold flux with casting time.

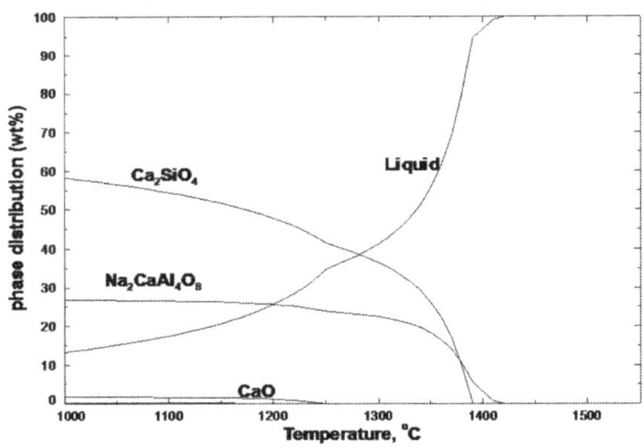

Figure 5. Equilibrium solidification calculation of the average mold slag obtained after casting of high Al TRIP steel for 15 min (total casting time =30 min).

Summary

An accurate and consistent thermodynamic database for the CaO-MgO-Al_2O_3-SiO_2-Na_2O-K_2O-Li_2O-MnO- FeO-F mold flux system was built. The liquid oxyfluoride phase is described using the Two-sublattice Modified Quasichemical Model, which can accurately represent its thermodynamic properties. All reliable experimental data could be reproduced within experimental error, as was illustrated with the $CaO-SiO_2-CaF_2$ system in the present study. Accurate equilibrium solidification calculations for a wide range of mold flux can be performed. In the future, the thermodynamic database will be expanded to ZrO_2, TiO_x, B_2O_3, S, H, etc. to cover a wider range of casting applications and also development of new mold flux system. A simple kinetic model based on the reaction zone approach was developed to simulate the changes in the molten mold flux composition with casting conditions. The model can be tuned to represent the variation of mold flux composition during the continuous casting process.

Acknowledgements

Financial support from Hyundai Steel, JFE Steel Corporation, Nippon Steel & Sumitomo Metal, Nucor Steel, Posco, QIT, RHI, RIST, Tata Steel Europe, Voest Alpine and the Natural Science and Engineering Research Council of Canada (NSERC) is gratefully acknowledged. The authors would like to express their gratitude to Prof. Jung-Wook Cho, GIFT, Postech, for the valuable discussions during the progress of this study.

References

1. I.-H. Jung, "Thermodynamic modeling of CaF_2 containing system and its applications to steelmaking process" (Invited paper presented at Asia Steel Conference, The Korean Institute of Metals and Materials, Seoul, South Korea, 2009), S3/30.

2. I.-H. Jung, "Thermodynamic modeling of CaF_2 containing slag and its applications to steelmaking process", *Proc. AISTech 2010* Association for Iron & Steel Technology, Warrendale, PA, (2010), 1211-1220.

3. I.-H. Jung, M.-A. Van Ende and D.-G. Kim, "Thermodynamic modeling of the F containing oxide system and applications to refractory corrosion" *Proc. UNITECR 2011*, The Technical Association of Refractories, Japan, Tokyo, Japan, (2011), 582.

4. C.W. Bale et al., "FactSage thermochemical software and databases - recent developments", *Calphad,* 33, (2009), 295-311.

5. A.D. Pelton et al., "The Modified Quasichemical Model – I Binary Solutions", *Metall. Mater. Trans. B*, 31 (2000), 651-660.

6. A.D. Pelton and P. Chartrand, "The Modified Quasichemical Model II - Multicomponent Solutions", *Metall. Mater. Trans. A,* 32 (2001), 1355-1360.

7. M. Hanao, M. Kawamoto and T. Watanabe, "Influence of Na_2O on Phase Relation Between Mold Flux Composition and Cuspidine" *ISIJ Int.*, 44 (2004), 827-835.

8. R.G. Hill, N. Da Costa and R. V. Law, "Characterization of a mould flux glass", *J. Non-Cryst. Solids,* 351 (2005), 69-74.

9. M.-A. Van Ende et al., "A kinetic model for the Ruhrstahl Heraeus (RH) Degassing process", *Metall. Mater. Trans. B,* 42 (3) (2011), 477-489.

10. M. A. Van Ende and I.-H. Jung, *ISIJ Int.* (2013), accepted.

Celebrating the Megascale: Proceedings of the Extraction and Processing Division
Symposium on Pyrometallurgy in Honor of David G.C. Robertson
Edited by: Phillip J. Mackey, Eric J. Grimsey, Rodney T. Jones, and Geoffrey A. Brooks
TMS (The Minerals, Metals & Materials Society), 2014

THERMODYNAMIC OPTIMIZATION OF Mn-Si-C SYSTEM

Min-Kyu Paek[1], Youn-Bae Kang[2] and Jong-Jin Pak[1]

[1]Department of Metallurgical and Materials Engineering, Hanyang University, Ansan, 426-791 Korea, jjpak@hanyang.ac.kr

[2]Graduate Institute of Ferrous Technology, Pohang University of Science and Technology, Pohang, 790-784, Korea

Keywords: Mn-Si-C, Thermodynamic database, Modified Quasichemical model, Phase diagram

Abstract

The critical evaluation and thermodynamic optimization of Mn-Si-C and its sub-binary systems have been carried out over the whole composition range from room temperature to above the liquidus temperature. The solution properties of the liquid systems have not been clearly described because the liquid solution exhibits a high degree of short-range-ordering. In order to predict the thermodynamics of the liquid solution, the liquid phases were optimized by the modified quasichemical model. The model parameters of the solid phases were also optimized by the compound energy formalism to best reproduce the phase diagram and important thermodynamic properties in Mn-Si-C system. Using the model parameters, various thermodynamic calculations were carried out. The present database will be a part of larger thermodynamic database for the ferromanganese alloy database.

Introduction

Recently, the demand for the production of high purity ferromanganese alloys has been increasing with the development of high Mn steels such as TWIP (Twining Induced Plasticity) steels. Ferroalloy companies in South Korea have been attempting to produce high purity ferromanganese alloys by refining processes of ferroalloys in preparation for the depletion of high purity raw materials in the future. In order to determine the refining procedure of ferromanganese alloys, the C content should be controlled first due to its strong effects on the behavior of other impurities [1].

However, although the Mn-C system is the most fundamental system for the refining of carbon in ferromanganese alloys, the thermodynamic properties of the liquid solution have not been clearly described. Moreover, no reliable thermodynamic optimization in Mn-Si-C ternary system has been reported. This is mainly due to the experimental difficulties caused by the high vapor pressure of Mn and because the liquid solution exhibits a high degree of short-range-ordering (SRO) by the strong interactions among the alloying elements. Several studies on the thermodynamic assessment of these systems have been reported using the Bragg-Williams (BW) random mixing model [2-5]. However, it is very difficult to describe the thermodynamics of liquid solution over the whole composition range by the random mixing model.

In the present study, the Modified Quasichemical Model (MQM) in the pair approximation [6] was used to consider the SRO for the optimization in Mn-Si-C and its sub-binary systems. The liquid Mn-C system was modeled using the available thermodynamic data such as the C solubility, activity of C and Mn, and the excess partial enthalpy data. For the liquid Mn-Si system, Shukla et al.'s optimization by the MQM [7] was adopted, but slightly revised to improve the accuracy at low silicon content region. The Si-C binary system was optimized by MQM for consistency. Combining the optimization of the binary systems, thermodynamic properties of the liquid solution in the Mn-Si-C ternary system were reproduced well. Using the optimized model parameters of the liquid solution, the phase diagrams of Mn-Si-C and its sub-binary systems were also calculated with a reassessment of all solid phases such as solid solutions and intermediate compounds by the Compound Energy Formalism (CEF).

Thermodynamic models

The MQM in the pair approximation was used to model the liquid alloys. In the pair approximation of this model, atoms A and B distribute over the sites of a quasilattice. The pair exchange reaction can be expressed as:

$$(A-A)+(B-B)=2(A-B) \qquad \Delta g_{AB} \tag{1}$$

where *(i-j)* represents a first-nearest neighbor (FNN) pairs. The nonconfigurational Gibbs energy change for the formation of two moles of (A-B) pairs is Δg_{AB}. Δg_{AB} is expanded in terms of the pair fractions:

$$\Delta g_{AB} = \Delta g^{\circ}_{AB} + \sum_{i\geq 1} g^{i0}_{AB} X^{i}_{AA} + \sum_{j\geq 1} g^{0j}_{AB} X^{j}_{BB} \tag{2}$$

where Δg°_{AB}, g^{i0}_{AB} and g^{0j}_{AB} are the model parameters of MQM which can be functions of temperature. A description of the MQM and its associated notation are discussed elsewhere [6, 8]. As Δg_{AB} becomes progressively more negative, reaction (1) is shifted progressively to the right, and the calculated enthalpy and configurational entropy of mixing assume, respectively, the negative "V" and "m" shapes characteristic of SRO [6].

The Gibbs energy of the ternary Mn-Si-C liquid solution is then estimated using Toop-like interpolation method [8,9] with manganese as asymmetric component. Thermodynamic properties of the ternary solution estimated in this way could reproduce available experimental data as close as possible without addition of the ternary model parameter.

All solid solutions were modeled using the Compound Energy Formalism (CEF) [10]. For example, solid solution phases in Mn-C system treated with a two-sublattice model, $Mn_a(C,Va)_b$, with manganese atoms occupying substitutional sites and others occupied by the interstitial sites. Its Gibbs energy in the CEF is given by:

$$\begin{aligned} G^{\varphi} &= y_{Va}\,{}^{\circ}G^{\varphi}_{Mn:Va} + y_{C}\,{}^{\circ}G^{\varphi}_{Mn:C} + bRT(y_{Va}\ln y_{Va} + y_{C}\ln y_{C}) \\ &+ y_{Va}y_{C}\left[{}^{0}L^{\varphi}_{Mn:C,Va} + {}^{1}L^{\varphi}_{Mn:C,Va}(y_{C}-y_{Va})\right] + {}^{mag}G^{\varphi} \end{aligned} \tag{3}$$

where y_i represents the site fractions of component *i* on each sublattice, ${}^{\circ}G^{\varphi}_{Mn:Va}$ are the Gibbs energy of the pure manganese with structure of α–Mn, β-Mn, γ–Mn, δ–Mn and ε–Mn, ${}^{\circ}G^{\varphi}_{Mn:C}$ is the Gibbs energy of a hypothetical compound. $L^{\varphi}_{Mn:C,Va}$ are interaction parameters between carbon and vacancy on one sublattice when the other sublattice is occupied only by manganese.

Contribution of magnetic properties to Gibbs energy of mixing was described by Huang [2] employing a phenomenological equation suggested by Hillert and Jarl [11].

Results and Discussion

Mn-C binary system

This system has been evaluated previously by Huang [2] and Djurovic et al. [3]. They have extensively reviewed the available thermodynamic data and developed gradually the Mn-C binary system. According to their compilation, there are six solution phases (liquid, α-Mn, β-Mn, γ-Mn, δ–Mn, and ε–Mn) and four carbide phases as stoichiometric compounds ($Mn_{23}C_6$, Mn_3C, Mn_5C_2, and Mn_7C_3). However, the available thermodynamic data such as activity coefficients of C and Mn are not well reproduced by their calculations as shown in Fig. 1.

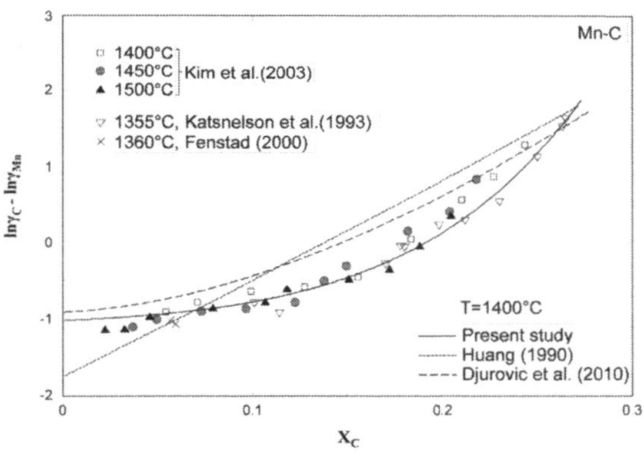

Figure 1. Ratio of activity coefficient of C and Mn in Mn-C melts.

Katsnelson et al., [12] Fenstad, [13] and Kim et al. [14] measured the equilibrium carbon content in Mn-C melts under the different CO partial pressures in the presence of pure solid manganese oxide MnO, according to the following reaction:

$$C(s) + MnO(s) = Mn(l) + CO(g) \tag{4}$$

$$K_4 = \frac{a_{Mn} P_{CO}}{a_C} = \frac{X_{Mn} P_{CO}}{X_C} \cdot \frac{\gamma_{Mn}}{\gamma_C} \tag{5}$$

where K_4 is the equilibrium constant for reaction (4), and a_{Mn} and a_C are the activities of Mn and C, γ_{Mn} and γ_C are the activity coefficients of Mn and C, respectively, and P_{CO} is the CO partial pressure. The standard states for Mn and C are liquid Mn and graphite, respectively. From the equilibrium experiments, the ratio of the γ_C and γ_{Mn} can be directly obtained by Eq. (5). As shown in Fig. 2, the $\ln\gamma_C - \ln\gamma_{Mn}$ values remain almost constant at $X_C < 0.15$ and then increase sharply, but the temperature effect is not so significant.

This tendency becomes more obvious on the partial excess enthalpy in the liquid Mn-C system as shown in Fig. 3. Witusevich et al. [15] measured the excess partial enthalpy of mixing in Mn-C melts at 1400°C by a high-temperature isothermal calorimetry. The excess property of C was re-evaluated for a hypothetical supercooled liquid carbon (HSLC) using the database of Scientific Group Thermodata Europe (SGTE) [16]. As shown in the figure, their result also shows a concentration dependence with a horizontal portion at $X_C < 0.15$, while the partial excess enthalpy of HSLC, $\Delta h_{C(l)}^{mix}$ significantly increases at $X_C > 0.15$ and the value near the C saturation is much higher than 0.

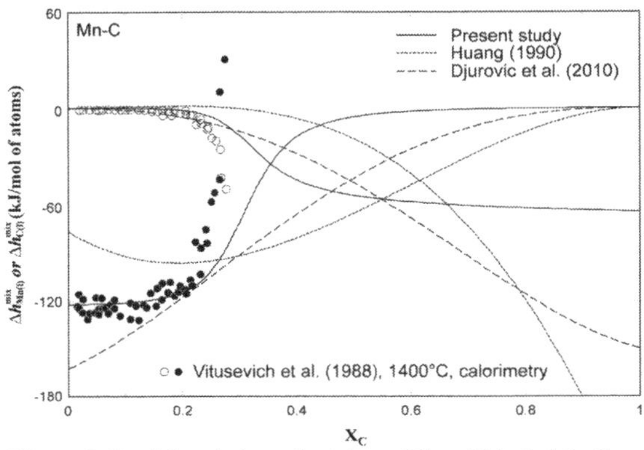

Figure 2. Partial enthalpy of mixing of C and Mn in Mn-C melts.

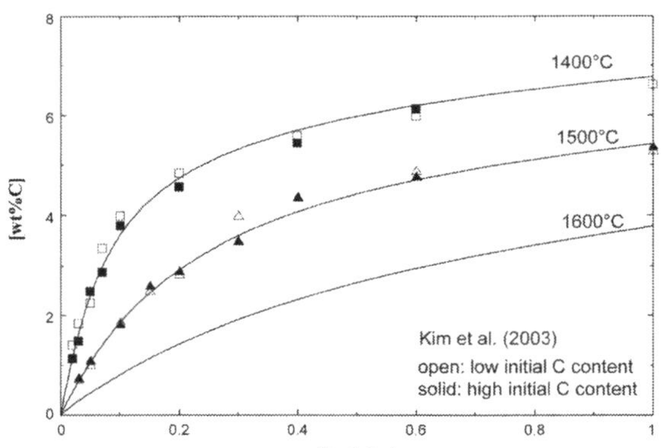

Figure 3. Refining limit of C in Mn-C melts.

As pointed out by Witusevich et al., [15] it is very difficult to obtain the accurate enthalpy data at high concentration of C due to the delay of the dissolution by adding a large amount of graphite. Therefore, the significant increase of the excess partial enthalpy of HSLC at $X_C > 0.15$ could be attributed to the experimental errors in the calorimetric measurement. In order to obtain one set of model parameters for the C solubility, partial enthalpy of mixing and the activity coefficient ratio of C and Mn in liquid Mn-C system, the Gibbs energy of pair exchange reaction was determined as:

$$\Delta g_{Mn-C} = -42677+4.50T+(2092+1.75T)X_{MnMn} \quad (6)$$

Using the optimized model parameters of liquid solution, the refining limit of C can be calculated as functions of temperature and CO partial pressure in Mn-C melts. The predicted lines agreed very well with the authors' previous results [14] at 1400 and 1500°C as shown in Fig. 3. The equilibrium C content significantly decreases with a reduction of CO partial pressure especially at the low pressure range. Combining this optimization with the database of Fe-C and Mn-Fe binary systems, the decarburization process of ferromanganese alloys can be predicted.

Figure 4 shows the optimized enthalpy of formation of the intermetallic compounds along with the measured enthalpy data in Mn-C system. The present optimization puts more weight on the directly measured values by Meshcel and Kelppa [17] rather than on the derived values [18] from the Gibbs energy measurements. As shown in the figure, the present result is also in good agreement with the enthalpy of formation of $Mn_{23}C_6$ measured by Dawson and Sale [19] using the adiabatic oxygen combustion calorimetry.

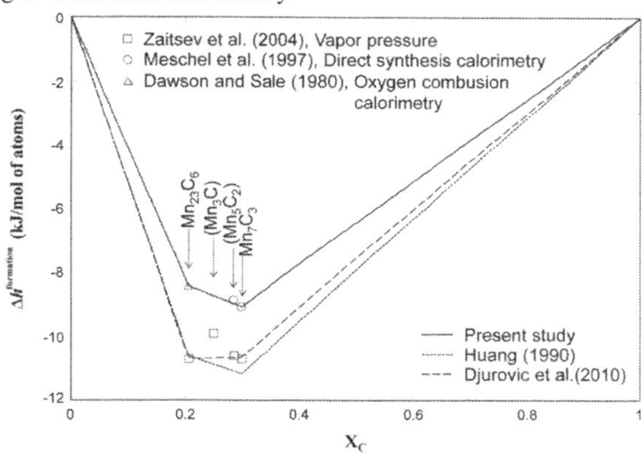

Figure 4. Enthalpy of formation of carbides in Mn-C system.

The optimized phase diagram of the Mn-C system is shown in Fig. 5. The main difference between the present and previous optimization [2, 3] is that of the two phase region of γ and ε phases. The two phase region became much larger than the previous optimization and this result agreed very well with Benz et al.'s equilibrium phases [20] measured in Mn-C alloys at 1100°C.

Mn-Si-C ternary system

There were no reports on the Mn-Si-C ternary system. This is attributed to the poor prediction of the ternary system at high Si concentration by the strong interactions among the alloying elements. In addition, the liquid Mn-C and Mn-Si systems showed considerable negative deviation, while the liquid Si-C system exhibits positive deviation from the ideal solution behavior. Therefore, in the present study, the asymmetric approximation [8,9] with Mn as the asymmetric component was considered for the optimization of the liquid solution in the Mn-Si-C system.

Figure 6 shows the C solubility data measured by various investigators [21-25] at 1500℃. As shown in the figure, the C solubility is significantly decreases with increasing the silicon content. Then, at a critical C and Si contents, the melt is simultaneously saturated with graphite and β-SiC. When the Si content exceeds the critical value, the melt is saturated with pure solid β-SiC. Although the C solubility was significantly changed by the strong interaction with Si in Mn-Si-C melts, the various experimental results can be reproduced well without the additional ternary parameter of the MQM.

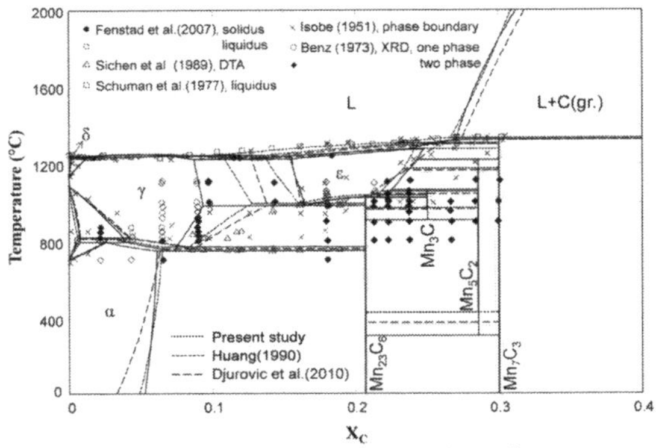

Figure 5. Optimized Mn-C phase diagram.

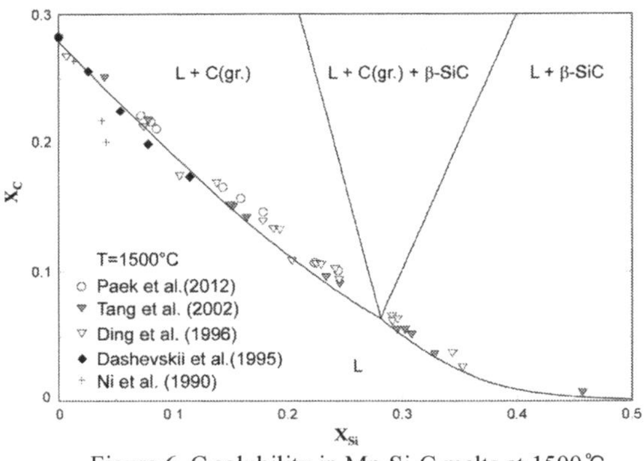

Figure 6. C solubility in Mn-Si-C melts at 1500℃.

The present optimization was also compared with the a_{Mn} data measured by Tanaka [26] in Mn-Si-C melts below the C saturation. Although he drew the squiggles as the isoactivity curves of Mn in his study, his experimental results corresponded well along the Mn isoactivity curve predicted in the present study as shown in Fig. 7.

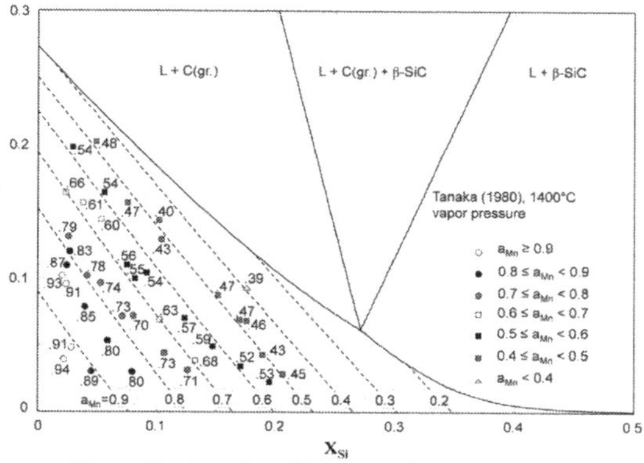

Figure 7. Activity of Mn (a_{Mn}) in Mn-Si-C melts.

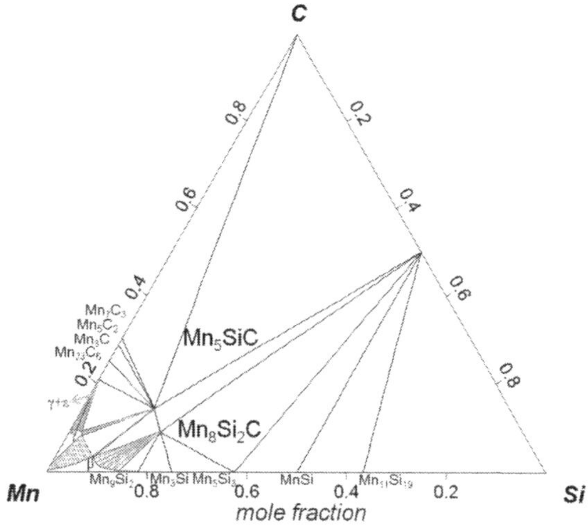

Figure 8. Isothermal section in Mn-Si-C system at 1000°C

According to Spinet et al., [27,28] there are two ternary compounds of Mn_8Si_2C and Mn_5SiC in the Mn-Si-C system. Later, Shuster [29] measured the equilibria among Mn_8Si_2C, β-SiC, Mn_5Si_3Cx, and the intermediate compounds of Mn-Si system. However, the phase equilibria involving Mn_5SiC and Mn carbides are not known. Therefore, the Gibbs energies of formation of the ternary compounds were evaluated based on the Shuster's [29] isothermal section at 1000°C in the Mn-Si-C system. Figure 8 shows the optimized isothermal section at 1000 °C in the Mn-Si-C system. The ternary compounds coexist with the solid solutions and intermediate compounds in Mn-rich side can be calculated.

Conclusions

The critical evaluations and thermodynamic optimizations of Mn-Si-C systems have been presented. The liquid solution has been optimized using the Modified Quasichemical Model (MQM) to account for the short-range-ordering (SRO). The available thermodynamic data of the liquid solution were reproduced within the experimental error limits without excessive entropy terms of the MQM. Combining the model parameters of the binary liquid solution, the thermodynamic properties of the liquid Mn-Si-C system were predicted well without the additional ternary parameter. The solid phases of the solid solutions and intermediate compounds in Mn-Si-C and its sub-binary systems were also reassessed using all available thermodynamic and phase diagram data. The optimized model parameters will be a part of larger thermodynamic database for the ferromanganese alloy database by extending the system to the quaternary or higher order system containing iron.

References

1. B.D. You, B.W. Lee and J.J. Pak, "Manganese loss during the oxygen refining of high-carbon ferromanganese melts", *Metals and Materials*, 5 (1999), 497-502.

2. W. Huang, "Thermodynamic assessment of the Mn-C system", *Scand. J. Metall.*, 19 (1990) 26-32.

3. D. Djurovic et al., "Thermodynamic assessment of the Mn-C system", *Calphad*, 34 (2010) 279-285.

4. P.Y. Chevalier, E. Fisher and A. Rivet, "A Thermodynamic evaluation of the Mn-Si system", *Calphad*, 19 (1995) 57-68.

5. J. Gröbner, H.L. Lukas and F. Aldinger, "Thermodynamic calculation of the ternary Al-Si-C", *Calphad*, 20 (1996) 247-254.

6. A.D. Pelton et al., "The modified quasichemical model I – Binary solutions", *Metall. Mater. Trans. B*, 31B (2000) 651-659.

7. A. Shukla, Y.B. Kang and A.D. Pelton, "Thermodynamic assessment of the Si-Zn, Mn-Si, Mg-Si-Zn and Mg-Mn-Si systems", *Calphad*, 32 (2008) 470-477.

8. A.D. Pelton and P. Chartrand, "The modified quasichemical model: Part II – Multicomponent solutions", *Metall. Mater. Trans. A*, 32A (2001) 1355-1360.

9. A.D. Pelton, "A general geometric thermodynamic model for multicomponent solutions", *Calphad*, 25 (2001) 319-328.

10. M. Hillert, "The Compound Energy Formalism", *J. Alloys Comp.*, 320 (2001) 161-176.

11. M. Hillert and M.Jarl, "A model for alloying effects in ferromagnetic metals", *Calphad*, 2 (1978) 227-238.

12. A. Katsnelson, F. Tsukihashi and N. Sano, "Determination of manganese and carbon activities of Mn-C melts at 1628K", *ISIJ Int.*, 33 (1993) 1045-1048.

13. J. Fenstad, Ph.D. Thesis, Norwegian University of Science and Technology, Trondheim, Norway (2000).

14. E.J. Kim, B.D. You and J.J. Pak, "Thermodynamics of carbon in liquid manganese and ferromanganese alloys", *Metall. Mater. Trans. B*, 34B (2003) 51-59.

15. V.T. Witusiewicz, A.K. Biletski and V.S. Shumikin, "Thermodynamic properties of liquid Mn-C alloys", *Metally*, 6 (1988) 26.

16. A.T. Dinsdale, *Calphad*, "SGTE data for pure elements", 15 (1991) 317-425.

17. S.V. Meschel and O.J. Kleppa, "Standard enthalpies of formation of some 3d transition metal carbides by high temperature reaction calorimetry", *J. Alloys Comp.*, 257 (1997) 227-233.

18. A.I. Zaitsev et al., "Thermodynamic studies of manganese carbides", *Dokl. Phys. Chem.*, 395 (2004) 632-636.

19. W.M. Dawson and F.R. Sale, "Enthalpy of formation of manganese carbides, $Mn_{23}C_6$", *Metall. Mater. Trans. A*, 11A (1980) 1849-1852.

20. R. Benz, J.F. Elliott and J. Chipman, "Solid phases of the Mn-C system", *Metall. Trans.*, 4 (1973) 1449-1452.

21. M.K. Paek et al., "Thermodynamic interactions among carbon, silicon and iron in carbon saturated manganese melts", *Korean J. Met. Mater.*, 50 (2012) 45-51.

22. K. Tang, V. Olso and S.E. Olsen, "Manganese and silicon activities in liquid carbon-saturated Mn-Si-C alloys", *Steel Research*, 73 (2002) 77-81.

23. W. Ding and S.E. Olsen, "Reaction equilibria in the production of manganese ferroalloys", *Metall. Mater. Trans. B*, 27B (1996) 5-17.

24. V.Y. Dashevskii et al., "Solubility of carbon in liquid Mn-Si system", *Dokl. Akad. Nauk*, 345 (1995) 75-78.

25. R. Ni, Z. Ma and S. Wei, "Thermodynamics of Mn-Fe-C and Mn-Si-C system", *Steel Resarch*, 61 (1990) 113-116.

26. A. Tanaka, "Activities of manganese in Mn-Fe-C, Mn-Si-C and Mn-Fe-Si-C melts at 1673K", *Trans. JIM*, 21 (1980) 27-33.

27. P. Spinat et al., "Characterization of two isotypic phases Mn_8Si_2C and Fe_8Si_2C", *Compt. Rend. Acad. Sci. Paris Sér. C*, 274 (1972) 1159-1162.

28. P. Spinat and P. Herpin, "Neutron diffraction studies of the Mn_5SiC phase and the solid solutions $(Mn_{1-x}Mo_x)_5SiC$ and $(Mn_{1-x}Fe_x)_5SiC$", *Bull. Soc. fr. Minéral. Crystallogr.*, 99 (1976) 13-20.

29. J.C. Schuster, "Silicon carbide and transition metals", *Int. J. Refract. Met. Hard Mater.*, 12 (1994) 173-177.

Celebrating the Megascale: Proceedings of the Extraction and Processing Division
Symposium on Pyrometallurgy in Honor of David G.C. Robertson
Edited by: Phillip J. Mackey, Eric J. Grimsey, Rodney T. Jones, and Geoffrey A. Brooks
TMS (The Minerals, Metals & Materials Society), 2014

REMOVEL OF NON-METALLIC INCLUSIONS FROM MOLTEN STEEL USING A HIGH FREQUENCY MAGNETIC FIELD

Shengqian Wang, Lifeng Zhang, Yue Tian

Beijing Key Laboratory of Green Recycling and Extraction of Metals (GREM)
School of Metallurgical and Ecological Engineering
University of Science and Technology Beijing
Beijing 100083, China
Email: zhanglifeng@ustb.edu.cn

Keywords: Electromagnetic separation, Non-metallic inclusions, Molten steel

Abstract

This paper is to discuss the removal of non-metallic inclusions from molten steel using a high frequency magnetic field. In experiments, a certain amount of aluminum was added to the molten steel in a crucible to generate aluminum inclusions. A layer of aluminum clusters were observed close to the wall of the crucible after the electromagnetic (EM) field was imposed. Three dimensional EM field and fluid flow simulation were performed and showed that over 91% inclusions transported to the skin depth (2.31 mm) of the molten steel. The calculation of inclusion removal agreed well with the experimental observation. The separation mechanisms of inclusions from the molten steel by the high frequency EM field were discussed.

Introduction

The application of electromagnetic (EM) technologies provides a considerable potential to control the fluid flow in the molten steel. The technologies have been used successfully to achieve a better product quality and a higher productivity in the industry, such as EMS and EMBR [1]. Whereas, as it has been reported [2] that this potential technology has been imposed a limitation to the removal of inclusions due to the skin depth effect of high frequency electromagnetic system. The fluid velocity severely induced by the EM system in the molten metal resulted in EM stirring. In the course of the application of AC magnetic fields for inclusion removal, the effect of EM stirring has been reported to be negative because the strong fluid velocity induced close to the wall on which particles are entrapped, resulting in the re-entrainment of these trapped particles back to the melt [3]. For another, because the induced fluid flow by the EM field can carry inclusions from the melt to the vicinity of the wall of the crucible in which the EM Archimedes force is effective at trapping the inclusions, there are also some positive effects. The study by Shu et al [4] showed that the induced secondary flow may enhance separation efficiency with high frequency resulting in stronger secondary flow. It was also reported that EM separation is efficient to remove particles as small as 1μm-2μm in a short time [5]. Studies to remove inclusions from molten silicon and aluminum using EM separation have been reported and attained satisfying results [5-12]. However, there were very few reports for the purification of molten steel through EM separation. The current work is to investigate 1) the possibility to apply EM separation to remove inclusions from molten steel under high frequency magnetic Field, 2) the motion and removal of inclusions using laboratory experiments and numerical simulation.

Theoretical Background of EM Purification

When the EM field is applied to the molten metal, the magnetic field and current density combined to create the EM force known as the Lorentz force at orthogonal angles to each other. Strong flow is induced as a result of Lorentz force acting on the molten steel. The mechanism of inclusion removal from molten metal by EM field employs the difference in conductivity of the metal and inclusion particle to drive its motion in the melt under an applied external force. The moving direction of particles also depends on the electrical conductivity difference between the melt and the inclusions.

Leenov and Kolin [13] proposed a theoretical expression of the EM force acting on a single sphere particle in a conductive liquid. The EM force can be expressed by

$$F = J \times B \tag{1}$$

where, F is the EM force (N), J is the induction current intensity (A) and B is the magnetic flux density vector (T).

When the EM force is applied on the liquid metal which contains non-conductive nonmetallic inclusions, it only exists in the conductive metal and can be ignored for the non-conductive inclusion. So, the EM Archimedes force acting on the inclusion is in the opposite direction of the EM force that acts on the liquid. According to Leenov and Kolin [13], the separation force acting on a sphere particle can be expressed by

$$F_P = -\frac{3}{2}\frac{\sigma_L - \sigma_P}{2\sigma_L + \sigma_P} V_P F \tag{2}$$

where, σ_L is the electrical conductivity of the liquid metal (S/m); σ_p is the electrical conductivity of the sphere particle (S/m); V_p is the volume of the sphere particle (m^3). As to most of nonmetallic inclusions, the value of σ_p is approximately zero in Eq.(2). So the following equation is given

$$F_P = -\frac{3}{4} V_P F \tag{3}$$

The negative sign in Eq.(3) shows that the EM separation force is inverse to the imposed EM force and the force is independent of the kinetic viscosity. If the EM force acts on the horizontal direction, the gravity and the magnetic force get away from each other. Inclusion should move in the opposite direction of the EM force. If inclusion particles are small enough to obey the Strokes law of viscous resistance, the EM repulsion is equal to the Stokes viscous force, and Eq.(3) can be modified into

$$u_P = -\frac{d_P^2}{24\mu} F \tag{4}$$

where, μ is kinetic viscosity of liquid metal (kg/m·s).

Experimental Methodology

The experimental set-up included induction coil powered by a 100 kW capacity power supply with a frequency range of 50~200 kHz. A crucible with induction pure iron was put inside the coil (Figure 1). High temperature insulation cotton was put in the space between the crucible and the coil to achieve enough heat to remelt the steel sample.

The experimental materials and parameters were listed in Table 1. After the induction pure iron was melted, aluminum pieces were added to the melt to generate alumina inclusions and a ceramic rod was used to stir the molten steel. A recirculation flow was observed on the surface of the molten steel. The separation time was set to 30~50 seconds. The temperature at the wall of the crucible was measured using a thermocouple immediately after the power supply was shut off. The frozen steel together with the crucible was cut and polished, inclusions were observed and analyzed using the optical microscope and SEM-EDS.

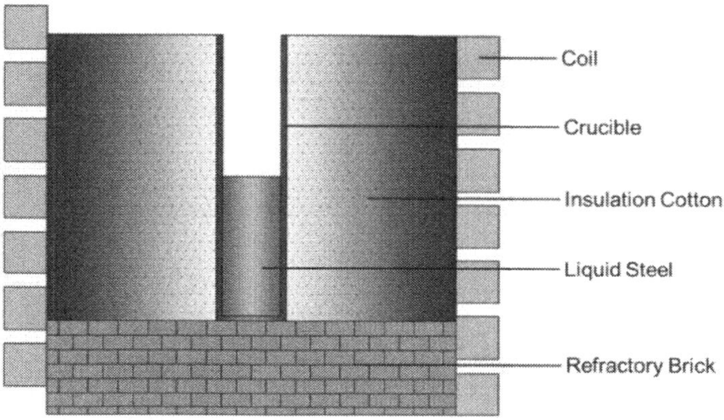

Fig. 1 Experimental set-up for EM separation of inclusions from steel

Table 1 Experimental materials and parameters

No.	Crucible material	Crucible diameter (mm)	Steel mass (g)	Aluminum mass (g)	Frequency (kHz)	Coil current (A)	Separate time (s)
S-EM-1	Al_2O_3	20	90	0	63	333	30
S-EM-2	Al_2O_3	20	90	4	63	333	45

Experimental Results

It has been reported that when EM force was imposed on the melt of silicon and aluminum, nonmetallic inclusions were entrapped near the wall of the crucible container due to the induced pressure gradients in the melt under alternating magnetic field [14]. It has been verified that inclusions were mainly pushed to the wall by the EM Archimedes force which generate due to the existence of EM Field [15]. Figure 2 gives complete view for the cross-section of 15 mm from the bottom of complete assembly of optical images of the edge of sample S-EM–1 (without aluminum addition) and sample S-EM–2 (with aluminum addition). The experiment S-EM-1 was designed to compare with the experiment S-EM-2. Few Al_2O_3 particles were observed in the matrix of the steel and close to the wall. No alumina were observed, which also implies that there was no dissolution of alumina lining refractory materials to the molten steel. This phenomena have been verified by Wang and Zhang [16, 17]. Again, without aluminum addition there no inclusions near the wall of the crucible since no alumina inclusions generated in the steel. With aluminum addition, many alumina clusters and the thickness of the inclusion layer was approximately 0.5~1.5 mm.

The horizontal cutting section of the steel sample S-EM-2 after EM separation in Figure 3, indicating that Al_2O_3 particles were observed close to the wall of the crucible and implying the effect of the EM force on the motion of particles. The gap in black was due to the shrinkage during cooling and solidification. Figure 4 gives a close view of the inclusion layer close to the wall on a horizontal cross section of sample S-EM–2, clearly indicating pure alumina inclusions.

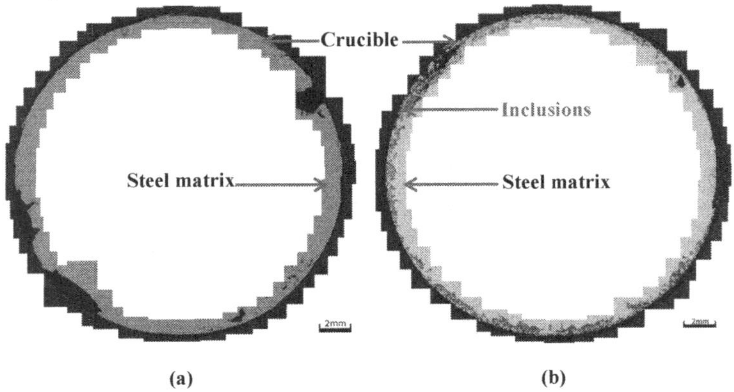

Fig. 2 Optical macrographs of horizontal cross-section 15mm from the bottom of (a) sample S-EM–1 (no aluminum addition), (b) sample S-EM–2

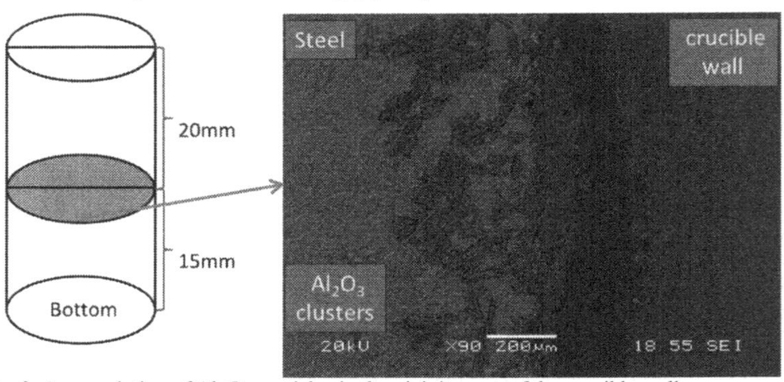

Fig. 3 Accumulation of Al_2O_3 particles in the vicinity area of the crucible wall

Fluid Flow and Inclusions Entrapment in Steel under EM Field

The dimension and parameters for the current EM separation process were shown in Fig. 1 and Table 1. The current simulation results were for the experiment S-EM-2. Figure 4 indicates that there were two recirculation eddies with big speed at bottom and top of the crucible. There was other four recirculations at the middle height but much weaker than the top and the bottom recurculations. This kind of recirculation has been reported by Oeters. The sedimentation velocity was calculated to be 6.6×10^{-6} m/s by Stokes law. It is several times smaller than the velocity of fluid.

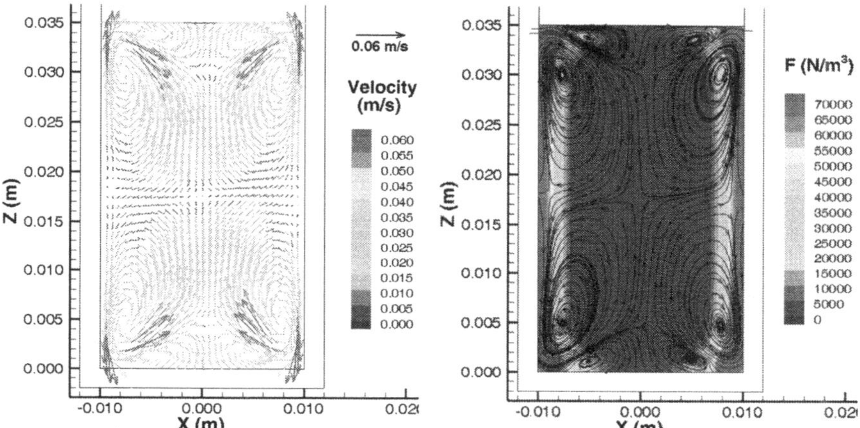

Fig. 4 Vector and the streamline with the background is contour of the EM force (N/m^3) in the molten steel

In the current simulation, 20000 spherical inclusions (d=20 μm, ρ=5000 kg/m^3) were assumed uniformly distributed in the interior domain at the beginning. Figure 5 gives the distribution of particles under EM and without EM. It is can be seen that the particles distributed in solid steel irregularly without EM. But the particles have moved upward because the density of particles is smaller than the surrounding steel. So few particles are found at the bottom of the steel (about 2mm from the bottom) after being completely solidified. Then the same parameters were used with EM to calculate. It is found that the distribution of particles has changed observably. About 28.2% and 38.6% of the particles were captured by the wall of crucible at the height between 0 mm to 6 mm and 29 mm to 35 mm respectively where the fluid moves toward the central of the steel. There are over 91% inclusions can be removed by the EM separation within 2.31mm (skin depth) from the wall and there are a few particles in the center area of the steel.

Figure 6 show the EM force, the fluid flow and the distribution of particles on the vertical-section and cross-section. It can be seen that the distribution of particles has a close relationship with EM force and fluid flow. The molten steel is moving toward to the central in the Area 1 and Area 3. The particles have the inverse moving direction toward to the wall under pressure gradient from the molten steel and then were conditioned to these areas. Finally, over 65% of the particles were entrapped by the wall of crucible in these areas. There are more particles than the other part in the center due to the motion of molten steel toward to the center and the strongest EM force. At the same time, only 13.2 % of the particles were entrapped in the area2. A conclusion can be drawn that the distribution of particles is determined by EM force and fluid flow. The molten steel moving toward to the wall contribute to migrating of particles toward the wall and stronger EM force is helpful to entrap the inclusions. It can be seen that the distribution of the particles on the horizontal cross-section of the steel in Area 1, Area 2 and Area 3. The particles entrapped by the wall in Area 2 is less than Area1 and Area3 while the mount of particles entrapped is more than area1 and area3 in the central area. The experimental result also shows that most of the particles were entrapped near the wall within 2.31 mm (the skin depth). It is very clearly that the calculated result coincides with the experimental result as shown in Fig.7.

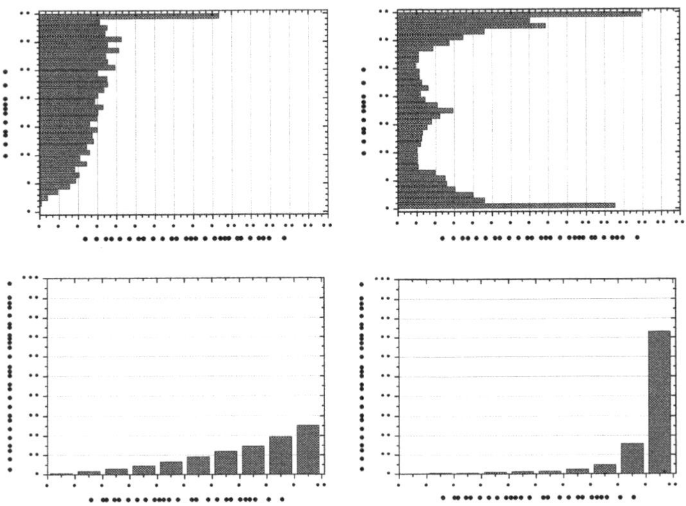

No EM With EM

Fig.5 Percentage of particles entrapped in the vertical and horizontal steel

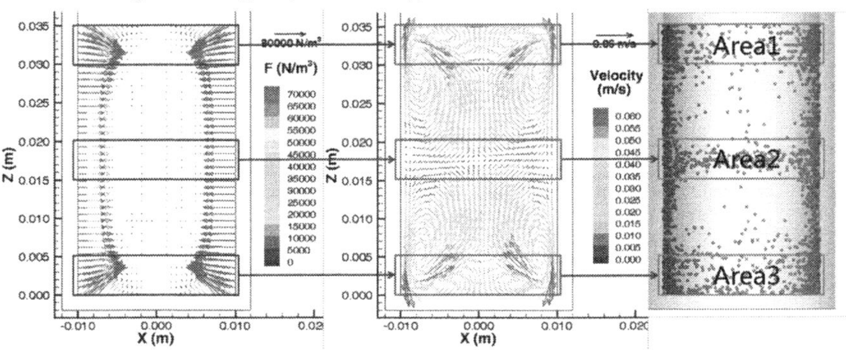

Fig.6 Comparison for the molten steel flow and the distribution of particles on the vertical-section

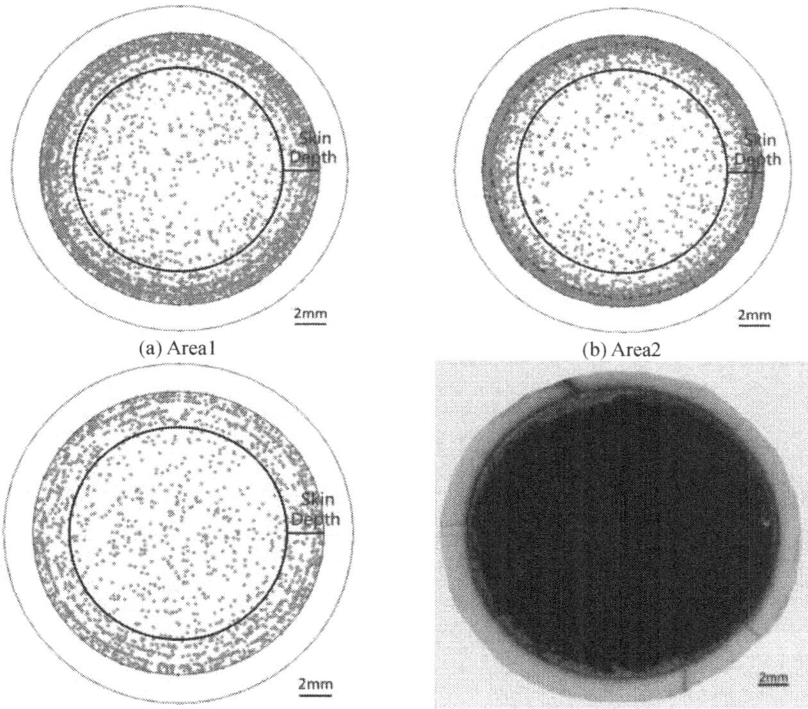

(a) Area1 (b) Area2

(c) Area3 (d) Experimental result

Fig.7 Distribution of particles on the cross section for calculated result and experimental result

Conclusions

The removal mechanisms of inclusions from the molten steel under high frequency EM separation were investigated by experimental investigation and numerical simulations. Aluminum was added into the molten metal to generate alumina inclusions. In the experiments, clusters of the Al_2O_3 inclusions were observed close to the wall of the crucible. In the simulation, eight recirculation eddies were formed due to the flow pattern induced by the EM field and a non-flat free surface at the top were formed because the velocity there is not uniform. The current experimental work and theoretical simulation shows that the high frequency EM separation is feasible and high-efficiency method to separate inclusions from the molten steel. Over 91% particles can be captured within 2.31mm (the skin depth) and only a few distributed in the central region. Most of particles were transported to the vicinity of crucible wall by the fluid flow, and then entrapped to the inside surface of the crucible wall by the strongest EM Archimedes force and the viscous force.

Acknowledgements

The authors are grateful for support from the National Science Foundation China (Grant No. 51274034 and Grant No. 51334002), Beijing Key Laboratory of Green Recycling and Extraction of Metals (GREM), the Laboratory of Green Process Metallurgy and Modeling (GPM^2) and the High Quality steel Consortium (HQSC) at the School of Metallurgical and Ecological Engineering at University of Science and Technology Beijing (USTB), China. Correspondence author: Lifeng Zhang, E-mail: zhanglifeng@ustb.edu.cn.

References

1. G. Gerbeth and S. Eckert, "The LIMMCAST Progrom at HZDR: Modelling of Fluid Flow and Transport Phenomena in the Continuous Casting Process," *Iron and Steel Research*, 19 (s) 2012, 1-8.
2. M. Sergey, L. Reinhold, and A. Diran, "Electromagnetic Separation Techniques in Metal Casting. I. Conventional Methods," *Ieee Transactions on Magnetics*, 36 (4) 2000, 2015-2021.
3. K. Takahashi and S.J. Taniguchit, "Electromagnetic Separation of Nonmetallic Inclusion from Liquid Metal by Imposition of High Frequency Magnetic Field," *ISIJ International*, 43 (6) 2003, 820-827.
4. D. Shu, B. Sun, and Y. Zhou, "Effects of Secondary Flow on the Electromagnetic Separation of Inclusions from Aluminum Melt in a Square Channel by a Solenoid," *ISIJ Inter.*, 42 (11) 2002, 1241-1250.
5. D. Wang, Z.Y. Li, and L.J. He, "Research on Purifying Emulsion Using Electromagnetic Technique and Its Control System," *Industrial Lubrication and Tribology*, 53 (1) 2001, 155 - 161.
6. L.F. Zhang and A. Ciftja, "Recycling of Solar Cell Silicon Scraps through Filtration, Part I: Experimental Investigation," *Solar Energy Materials and Solar Cells*, 92 (11) 2008, 1450-1461.
7. B.W. Zhang, Z.M. Ren, and J.X. Wu, "Continuous Electromagnetic Separation of Inclusion from Aluminum Melt Using Alternating Current," *Transactions of Nonferrous Metals Society of China*, 16 (1) 2006, 33-38.
8. D. Shu, T. Li, and B. Sun, "Study of Electromagnetic Separation of Nometallic Inclusions from Aluminum Melt," *Metallurgical and Materials Transactions A*, 30 (11) 1999, 2979-2988.
9. K. Takahashi and S. Taniguchi, "Electromagnetic Separation of Nonmetallic Inclusion from Liquid Metal by Imposition of High Frequency Magnetic Field," *ISIJ International*, 43 (6) 2003, 820-827.
10. J.H. Kim, J.P. Park, and E.P. Yoon, "A Study on Elimination of Alumina Particles in Molten Aluminum Using Direct Electromagnetic Force," *Metals and Materials International*, 5 (4) 1999, 339-343.
11. N. El-Kaddah, A. Patel, and T. Natarajan, "The Electromagnetic Filtration of Molten Aluminum Using An Induced-current Separator," *JOM*, 47 (5) 1995, 46-49.
12. A. Ciftja, L.F. Zhang, and A. Kvithyld, "Purification of Solar Cell Silicon Materials through Filtration," *Rare Metals*, 25 (6s) 2006, 180-185.
13. D. Leenov and A. Kolin, "Theory of Electromagnetophoresis. I. Magnetohydrodynamic Forces Experienced by Spherical and Symmetrically Oriented Cylindrical Particles," *The Journal of Chemical Physics*, 22 (4) 1954, 683-688.
14. L.N.W. Damoah and L.F. Zhang, "High Frequency Electromagnetic Separation of Inclusions from Aluminum," *Light Metals 2012 Edited by: Carlos E. Suarez*, 2012.
15. S.Q. Wang, L.F. Zhang, and Y.L. Li, "Electromagnetic Separation of Nonmetallic Inclusions from Liquid Steel by High Frequency Magnetic Field," *Journal of Iron and Steel Research*, 19 (s) 2012, 866-878.
16. Y.F. Chen, J.L. Wang, and L.F. Zhang, "Transient Interaction between Molten Iron, Alumina Lining Refractory and Slag," *The 9th International Conference on Molten Slags, Fluxes and Salts (MOLTEN12), (Beijing), CSM*, 2012.
17. J.L. Wang, L.F. Zhang, and S.F. Yang, "Interaction Between Molten Steel, Lining Refractory and Slag Phase," *The 8th International Conference on May 14-16*, 2012

Celebrating the Megascale: Proceedings of the Extraction and Processing Division Symposium on Pyrometallurgy in Honor of David G.C. Robertson
Edited by: Phillip J. Mackey, Eric J. Grimsey, Rodney T. Jones, and Geoffrey A. Brooks
TMS (The Minerals, Metals & Materials Society), 2014

FLUID FLOW, ALLOY DISPERSION AND INCLUSION MOTION IN ARGON-STIRRED STEEL LADLES

Yanlong Li, Lifeng Zhang, Ying Ren

Beijing Key Laboratory of Green Recycling and Extraction of Metals (GREM)
School of Metallurgical and Ecological Engineering,
University of Science and Technology Beijing, Beijing 100083, China
Email: zhanglifeng@ustb.edu.cn
Office Phone: +86-010-62332267

Keywords: Fluid Flow, Ladle Refining, Gas Stirring, Alloy Dispersion

Abstract

In this work, a multiphase mathematical model for the fluid flow in argon-stirred steel ladles was established and validated by experimental measurement of wood's metal. Alloy dispersion in the molten steel was calculated. The melting and mixing of alloy was faster with smaller size of the alloy particles, larger superheat and bigger slip velocity. The current mixing time is approximately 300 s with 30 mm alloy diameter and 91.5 K superheat.

Introduction

Demands for higher productivity, higher quality, and lower production costs in steelmaking have led to the development of ladle metallurgy, Gas stirring has been used in steelmaking ladles to enhance the removal of inclusions, homogenize the melts and increase the rate of mass transfer between the metal and the top slag such as the desulfurization reaction[1]. Therefore, flow features of the molten steel during ladle refining, particularly the interfacial phenomenon in the contact region between molten steel and slag, are of importance to understand the comprehension of various phenomena such as mixing, inclusion removal, slag emulsification and desulphurization reactions between phases in a gas-stirred ladle system[2].

Due to recent rapid advances in computer technology and developments in numerical calculation techniques, mathematical modeling is now widely applied in engineering and science practices. Mathematical models describing fluid flow in gas-stirred ladle systems can be broadly classified into three groups, namely: (1) the quasi-single phase of single-phase variable density formulation procedures[3]; (2)the Eulerian-Eulerian two-phase model[4-7] and (3) the Eulerian-Lagrangian two-phase approach[2, 8].

Over the past three decades, most of the models usually assumed a flat and frictionless free surface at the top of molten melts, neglecting the level fluctuation of steel and the interfacial behavior among slag/steel/gas phases. However, melts contained in refining ladles are typically covered with a protective slag layer to minimize radiation heat losses, prevent reoxidation and capture inclusions. Consequently, most model studies fall short of realistically representing industrial ladle refining operations.

In the current study, a full scale, three-phase, three dimensional and transient mathematical model was established to predict the fluid flow phenomenon in a real industrial ladle. In this model, three continuous phases are used to track interfacial behavior of steel, slag and argon and a discrete phase is used to simulate the gas injection by a stream of argon bubbles into the continuous phase. The breakup and coalescence of bubbles are ignored while the bubble size distribution is taken into account with logarithmic normal function[9, 10]. The lift force and turbulence source are considered in this model and the simulation was validated by the measurement of Wood's metal done by Xie and Oeters[11]. Next, the addition, transport, melting and mixing of ferroalloy are studied based on the melting model proposed by Zhang and Oeters[12, 13].

Flow Model

A set of Navier-Stokes equations is solved for the steel, slag and argon phases and the Lagrangian form of Newton's law equation for the motion of each bubble. The continuous phases' equations are as follows:

Continuity equation:

$$\frac{\partial \rho}{\partial t} + \nabla \cdot u = 0 \tag{1}$$

Momentum equation:

$$\frac{\partial}{\partial t}(\rho \cdot u) + u \cdot \nabla(\rho \cdot u) - \nabla\left[(\mu_l + \mu_t)\nabla u\right] = -\nabla p + (\rho - \rho^*)g + f_\sigma - S_b \tag{2}$$

where u is the velocity shared by three continuous phases, ρ is the averaged density, p is the pressure, μ_l and μ_t are laminar viscosity and turbulent viscosity respectively, g is the gravitational acceleration rate, ρ^* is modified density, f_σ is volume force source defined by Continuum Surface Force (CSF) model[14] of VOF and S_b is momentum source exchanged with injected bubbles.

$$\rho = \alpha_l \rho_l + \alpha_s \rho_s + \alpha_g \rho_g \tag{3}$$

$$f_\sigma = \sigma_{ij} \frac{\rho\left(-\nabla \cdot \frac{\nabla \alpha_i}{|\nabla \alpha_i|}\right)\nabla \alpha_i}{\frac{1}{2}(\rho_i + \rho_j)} \tag{4}$$

where α is volume fraction calculated by VOF model and subscript l, s and g stand for steel, slag and gas and σ_{ij} is surface tension between phase i and j.

The injected argon bubbles are calculated in each time step given by:

$$\frac{du_{b,i}}{dt} = \frac{3\mu C_{D,i}\,\mathrm{Re}_i}{4d_{b,i}^2 \rho_{b,i}}(u - u_{b,i}) + \frac{\rho_{b,i} - \rho}{\rho_{b,i}} g + C_{VM}\frac{\rho}{\rho_{b,i}}\frac{d}{dt}(u - u_{b,i}) + C_L \frac{\rho}{\rho_{b,i}}(u - u_{b,i}) \times \nabla \times u + \frac{\rho}{\rho_{b,i}} u \cdot \nabla u \tag{5}$$

The terms of right hand side represent the drag force, buoyancy force, the virtual mass force, the lift force and the pressure gradient force. The coefficients C_{VM} and C_L are the virtual mass and lift coefficients, which are set to 0.5 and 0.1 respectively. The drag coefficient is from Kuo and Wallis[15], which takes bubbles change into account according to its diameter.

$$\begin{aligned}
C_D &= \frac{16}{\mathrm{Re}} && \mathrm{Re} < 0.49 \\
C_D &= \frac{20.68}{\mathrm{Re}^{0.643}} && 0.49 < \mathrm{Re} < 100 \\
C_D &= \frac{6.3}{\mathrm{Re}^{0.385}} && 100 < \mathrm{Re} \\
C_D &= \frac{We}{3} && \mathrm{Re} > \frac{2065.1}{We^{2.6}}
\end{aligned} \tag{6}$$

$$C_D = \frac{8}{3} \qquad We > 8$$

The momentum source term of equation (2) S_b is described as follow:

$$S_b = \sum_i^{N_{b,cell}} \left[\begin{array}{l} \frac{3\mu C_{D,i} \mathrm{Re}_i}{4 d_{b,i}^2 \rho_{b,i}} \left(u - u_{b,i} \right) + C_{VM} \frac{\rho}{\rho_{b,i}} \left(\frac{D}{Dt} u - \frac{D_b}{Dt} u_{b,i} \right) \\ + C_L \frac{\rho}{\rho_{b,i}} \left(u - u_{b,i} \right) \times \nabla \times u + \frac{\rho}{\rho_{b,i}} u \cdot \nabla u \end{array} \right] Q_{b,i}^S \Delta t \tag{7}$$

where $N_{b,cell}$ is the number of bubbles in the computational cell, $u_{b,i}$ is velocity, d is diameter, μ is viscosity, and ρ is density. The subscript b,i means the "i^{th} bubble". $Q_{b,i}^S$ is the flow rate (kg/s) of the injected bubble stream, and Δt is the time step of bubble trajectory calculation. D/Dt and D_b/Dt are total derivatives with respect to the liquid and bubble phases. *Re* and *We* are the particle Reynolds number and Weber number, which are defined in terms of the relative velocity between the fluid and the particle as follows:

$$\mathrm{Re} = \frac{\rho d_{b,i} \left| u - u_{b,i} \right|}{\mu} \tag{8}$$

$$We = \frac{\rho d_{b,i} \left| u - u_{b,i} \right|^2}{\sigma_{lb}} \tag{9}$$

It should be pointed out that in ANSYS-FLUENT the particle bubble is treated as "a point source" and the properties are assigned only to the cell which includes the bubbles center. This procedure is valid only if the bubble volume is smaller than the computational cell volume. Otherwise, excess value may by assigned to the central cell which can lead to non-physical results. To solve this problem, user-defined subroutines are used to distribute the bubbles' concentration, the exchange force and the source terms of turbulence into the group of computational cells whose centers are within the effective radius from the bubble center. The gas fraction calculated by Discrete Particle Model in a given cell is calculated as follows:

$$\gamma_g = \frac{1}{\Delta V_{cell}} \sum_i^{N_{b,cell}} \frac{Q_{b,i}^s \Delta t}{\rho_{b,i}} \tag{10}$$

As illustrated in Figure 1, the bottom of ladle is full of steel and the middle is slag by default while the γ_g acting as gas phase should be taken into account by momentum equation (2). In order to fix this problem, a negative gravity component source $-\rho^* g$ is introduced into momentum equation when $\gamma_g > \alpha_g$, which expresses as follow:

$$\rho^* g = \left(\gamma_g - \alpha_g \right) \rho_l g \qquad \alpha_l > 0, \alpha_s = 0 \ \text{(steel zone)}$$

$$\rho^* g = \left(\gamma_g - \alpha_g \right) \left(0.5 \times \rho_l + 0.5 \times \rho_s \right) g \qquad \alpha_l > 0, \alpha_s > 0 \ \text{(mixture zone)}$$

$$\rho^* g = \left(\gamma_g - \alpha_g \right) \rho_s g \qquad \alpha_l = 0, \alpha_s > 0 \ \text{(slag zone)}$$

For turbulence equations, the modified *k-ε* model with turbulence source caused by bubbles motion and "random walk" model for discrete phase were illustrated more detailed elsewhere[7].

Melting Model of Ferroalloy Model

According to Zhang and Oeters' model[12], a solidified steel shell is formed around the surface of the ferroalloy particle when it first enters into the molten steel. This initially prevents any alloy mixing. Then, as the ferroalloy particle is heated by the surrounding liquid, the steel shell melts and finally disappears. Because the melting point of Mn-Fe (1215°C) is much lower than that of steel, the ferroalloy particle should be fully liquid when the steel shell disappears. Thus, the molten ferroalloy is suddenly introduced into the liquid where it starts to diffuse. The time when this occurs matches the duration of the solidified steel shell.

Geometrical model and Operational conditions

The ladle studied in this work is a 135 ton in capacity. Input conditions such as geometrical, thermo-physical properties and operating parameters are shown in Figure 1 and Table 1.

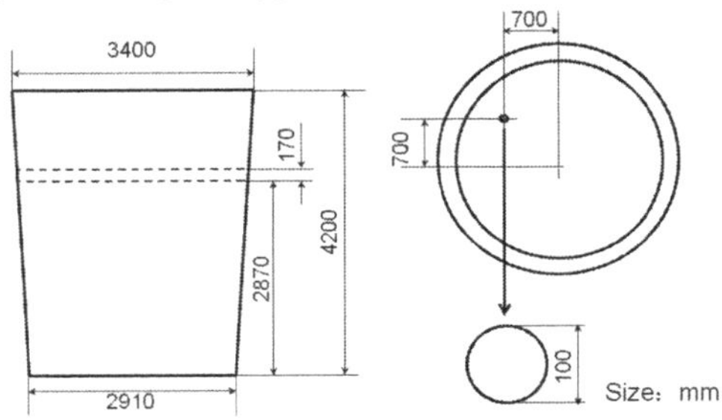

Fig. 1 Geometry of ladle

Table 1 Physical Properties and Operating Parameters

Parameters	Value	Unit	Ref.
Density of molten steel	7020	kg/m^3	[7]
Viscosity of molten steel	0.0067	Pa·s	[7]
Density of slag	3500	kg/m^3	[5]
Viscosity of slag	0.03	Pa·s	[4]
Density of argon(STP)	1.784	kg/m^3	[5]
Viscosity of argon	8.285×10^{-5}	Pa·s	-
Surface tension of metal/gas	1.82	N/m	[2]
Surface tension of slag/gas	0.58	N/m	[2]
Surface tension of metal/slag	1.15	N/m	[2]
Density of alloy	6120	kg/m^3	-
Thermal conductivity of alloy	4.184	w/(m·K)	-
Thermal conductivity of steel	43.095	w/(m·K)	-
Diffusion coefficient of alloy	5.5×10^{-9}	m^2/s	[16]
Latent heat of steel	2.47×10^{5}	J/kg	[16]
Mass content of alloy	97.46%	-	-
Solidifying temperature of steel	1717	K	-
Superheat	91.5	K	-
Gas blow rate	220	NL/min	-

Note: the alloy used in the current model is Mn97

Initial and Boundary Conditions

For the fluid flow, the no-slip boundary condition is used at the bottom wall and side wall with standard "wall functions" in order to capture the steep gradients with reasonable accuracy on a coarse grid. There are 20 groups of argon particles with different diameters and gas flow rates injected into the domain at random circle faces of the porous plug. The top surface is set as pressure outlet boundary condition. For the species transport of alloy dispersion, the zero flux boundary condition is used at walls.

Numerical Methodology

The solutions of governing equations with boundary conditions and source terms are obtained using the commercial fluid dynamics package ANSYS-FLUENT with added user-defined subroutines. The calculation domain divided by the 645 000 hexahedral cells. The calculations are conducted in the unsteady solution mode using PISO algorithm to solve the three phase flow problem. A criterion for convergence in all cases simulated in the present study is established when the sum of all residuals (normalize and un-scale) for the dependent variables is less than 10^{-5}.

Results and Discussion

Validation

The experimental data of the Wood's metal (low melting point) flow by nitrogen injection did by Xie and Oeters[11]are used to validate the current model. Nitrogen was injected into wood's metal (consists of 50% bismuth–25% lead–12.5% tin–12.5% cadmium) from a nozzle located at the bottom of the ladle by gas flow rate of 200 Ncm^3/s.

The calculated flow field in a vertical section of the ladle is shown in Figure 2. A comparison between experimental and calculated time-averaged liquid velocity along the vertical direction is plotted in Figure 3, which is in reasonably good agreement.

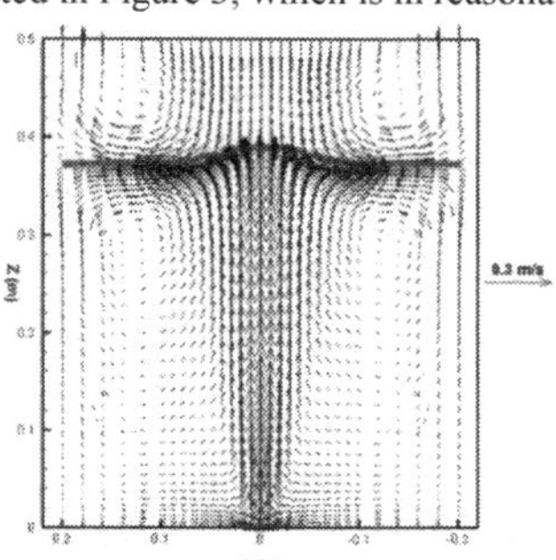

Fig. 2 Time-averaged flow field of Wood's metal experiment

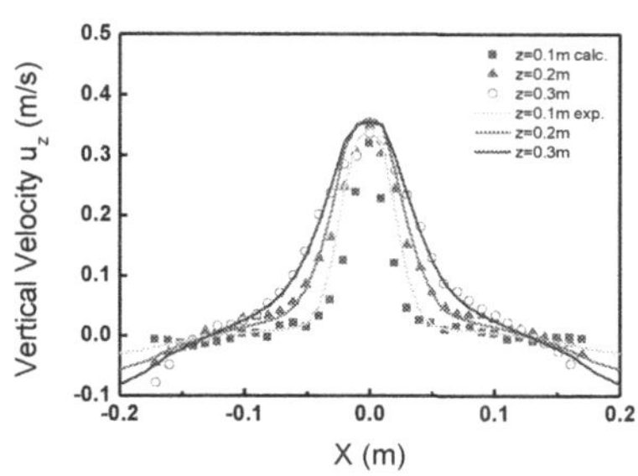

Fig. 3 Comparison of vertical velocity between calculation and measurement[11]

Fluid Flow

The calculated velocity field in the 125 ton steel-refining ladle with off-centered bottom gas injection is shown in Figure 4. A large recirculation pattern in the ladle is formed and it is good to homogenize the melt. Figure 5 is the velocity field of slag layer where the top of blow point is marked as black circle. There are two large eddies formed at the other side, which are helpful for mixing and may help trap inclusions. In Figure 6, the path line of magnitude velocity sheds light on the motion of steel, slag and inclusions.

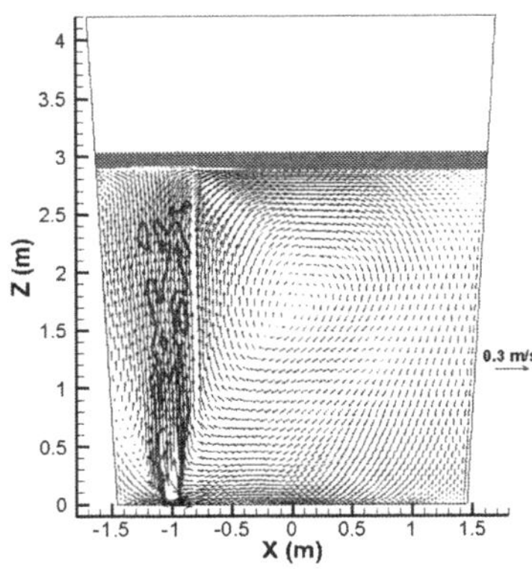

Fig. 4 Velocity field of Ladle (220 L/min)

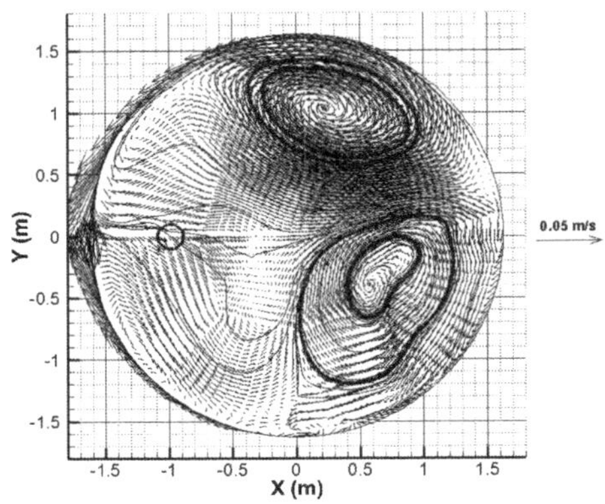

Fig. 5 Velocity field of slag layer (Z = 3 m)

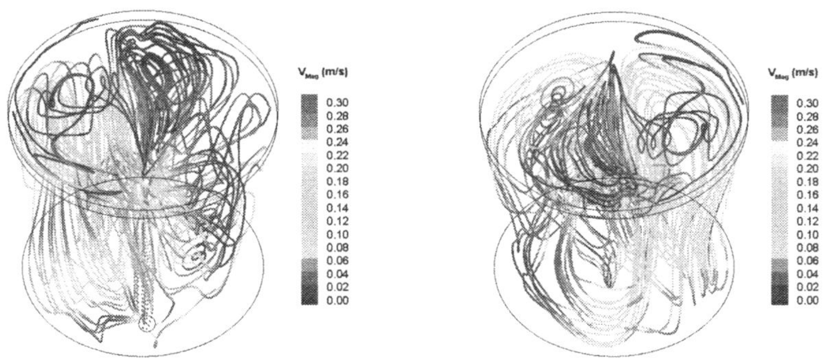

Fig. 6 Path line of Magnitude velocity (view from left and right side of Figure 5 respectively)

Alloy Dispersion

The alloy particles are assumed as spheres with 30 mm diameter and dropped into the molten steel at the top of gas blow point. Figure 7 shows that the alloy particles float up to the top after drop and sink into the melt. As mentioned above, when alloy first enter into the molten steel, a solidified shell formed around the surface then begin to melt because of the heating by surrounding liquid. Within this time, the alloy is heated up to its melting point and turns to liquid. The liquid alloy starts to diffuse when the shell disappears then homogenizes after a few minutes. It should be pointed out that the alloy dispersion model ignores its burning loss. The calculated mixing time is approximately 300 s.

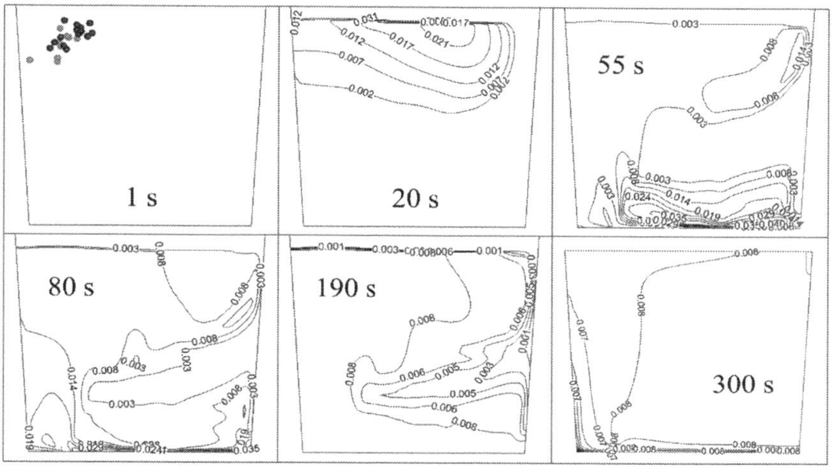

Fig. 7 Melting and dispersion of alloy particles with time (showing concentration of alloys in kg/m^3)

The concentration of alloy is shown in Figure 8 as a function of time and the monitor point is half height (molten bath) of centerline. Figure 9 shows that the calculated shell existence time decreases when with the increase of steel's superheat and with the decrease of alloy diameter.

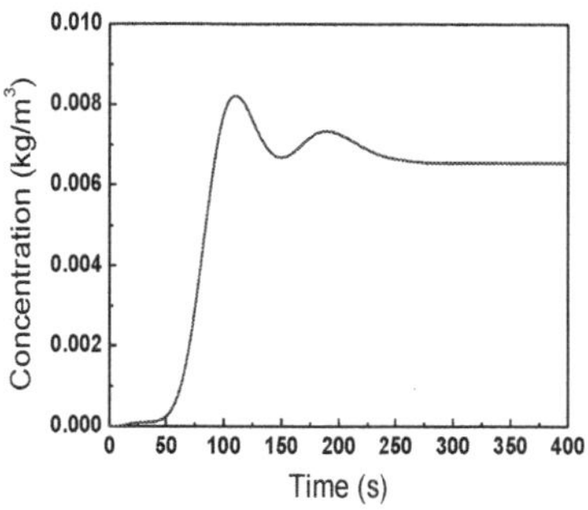

Figure 8 Computed concentration at monitor point

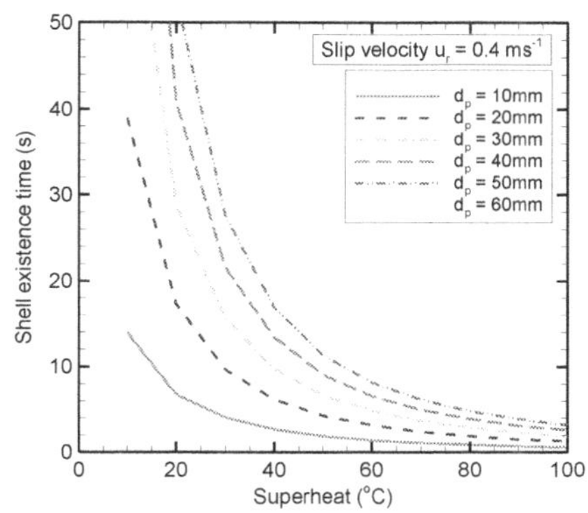

Figure 9 Shell existence time with steel's superheat and the particle diameter of alloy particles

Summary

A multiphase mathematical model of argon-stirred steel ladles is established and validated by experimental measurement of wood's metal. VOF model is used to track the free surface of phase (*i.e.* steel, slag, gas) and discrete particles act as gas bubbles and are modeled by calculating their trajectories. The melting and mixing time of alloy particles are sensitively depends on their diameter and liquid superheat. The melting and mixing time of alloy are reduced with decreasing the size of the alloy particles and increasing steel superheat.

Acknowledgements

The authors are grateful for support from the National Science Foundation China (Grant No. 51274034 and Grant No. 51334002), Beijing Key Laboratory of Green Recycling and Extraction of Metals (GREM), the Laboratory of Green Process Metallurgy and Modeling (GPM^2) and the High Quality steel Consortium (HQSC) at the School of Metallurgical and Ecological Engineering at University of Science and Technology Beijing (USTB), China. Correspondence author: Lifeng Zhang, E-mail: zhanglifeng@ustb.edu.cn.

References

1. S. Kim and R. Fruehan, "Physical modeling of gas/liquid mass transfer in a gas stirred ladle," *Metallurgical and Materials Transaction B*, 18 (4) 1987, 673-680.
2. H.P. Liu, Z.Y. Qi, and M.G. Xu, "Numberical Simulation of Fluid Flow and Interfacial Behavior in Three-phase Argon-Stirred Ladles with One Plug and Dual Plugs," *Steel Research International*, 82 (4) 2011, 440.
3. J.S. Woo, J. Szekely, and A.H. Castillejos, "A Study on the Mathematical Modeling of Turbulent Recirculating Flows in Gas-Stirred Ladles," *Metallurgical and Materials Transaction B*, 21B (4) 1990, 269.
4. J.H. Wei and L. Zeng, "Numberical Simulation of Fluid Flow in Bath during Combined Top and Bottom Blowing VOD Refining Process of Stainless Steel," *Steel Res. Int.*, 83 (11) 2012, 1053.
5. B.K. Li, H.B. Yin, and C.Q. Zhou, "Modeling of Three-phase Flows and Behavior of Slag/Steel Interface in an Argon Gas Stirred Ladle," *ISIJ Int.*, 48 (12) 2008, 1704.
6. H. Turkoglu and B. Farouk, "Mixing Time and Liquid Circulation Rate in Steelmaking Ladles with Vertical Gas Injection," *ISIJ Int.*, 31 (12) 1991, 1371.
7. L.F. Zhang, S.J. Taniguchi, and K.K. Cai, "Fluid flow behaviour of liquid in cylindrical vessels stirred by one or two air jets," *Steel Res. Int.*, 7 (3) 2000, 170.
8. S. Johansen and F. Boysan, "Fluid dynamics in bubble stirred ladles: Part II. Mathematical modeling," *Metallurgical and Materials Transaction B*, 19 (5) 1988, 755-764.
9. Y. Xie, S. Orsten, and F. Oeters, "Behaviour of bubbles at gas blowing into liquid Wood's metal," *ISIJ Int.*, 32 (1) 1992, 66-75.
10. J. Davidson and B. Schüler, "Bubble formation at an orifice in a viscous liquid," *Chemical Enginering Research and Design*, 75 1997, S105-S115.
11. Y. Xie and F. Oeters, "Experimental Studies on the Flow Velocity of Molten Metals in a Ladle Model at Centric Gas Blowing," *Steel Res. Int.*, 63 (3) 1992, 93.
12. L.Y. Zhang and F. Oeters, "Mathematical Modeling of Alloy Melting in Steel Melts," *Steel Res. Int.*, 70 (4+5) 1999, 128.
13. L.Y. Zhang, "Modelling on melting of sponge iron particles in iron-bath," *Steel Res. Int.*, 67 (11) 1996, 466.
14. J. Brackbill, D.B. Kothe, and C. Zemach, "A continuum method for modeling surface tension," *Journal of computational physics*, 100 (2) 1992, 335-354.
15. J. Kuo and G. Wallis, "Flow of bubbles through nozzles," *Int. J.Multiphase Flow*, 14 (5) 1988, 547-564.
16. J. Aoki, L. Zhang, and B.G. Thomas, "Modeling of inclusion removal in ladle refining," *Urbana*, 51 2005, 319-331.

CELEBRATING THE MEGASCALE

Proceedings of the Extraction and Processing Division Symposium on Pyrometallurgy in Honor of David G.C. Robertson

Poster Session

Celebrating the Megascale: Proceedings of the Extraction and Processing Division
Symposium on Pyrometallurgy in Honor of David G.C. Robertson
Edited by: Phillip J. Mackey, Eric J. Grimsey, Rodney T. Jones, and Geoffrey A. Brooks
TMS (The Minerals, Metals & Materials Society), 2014

NITROGEN SOLUBILITY IN LIQUID Mn-Fe-Si-C ALLOYS

June-Yong Eom, Jung-Mock Jang, Min-Kyu Paek and Jong-Jin Pak

Department of Metallurgical and Materials Engineering, Hanyang University, Ansan, 426-791 Korea, jjpak@hanyang.ac.kr

Keywords: Ferromanganese, Nitrogen solubility, Silicon, Interaction parameters

Abstract

The nitrogen solubility in liquid Mn-Si, Mn-Si-Fe, Mn-Si-C and Mn-Si-Fe-C alloys has been measured by the gas-liquid metal equilibration technique in the temperature range of 1673-1773 K. The additions of silicon, iron and carbon significantly decreased the nitrogen solubility in liquid manganese alloys. The experimental results were thermodynamically analyzed by the Wagner's formalism to determine the first- and second-order interaction parameters of silicon, iron and carbon on nitrogen in liquid manganese. The thermodynamic parameters can be used to predict the nitrogen solubility in ferromanganese and silicomanganese alloy melts as functions of the melt composition and temperature at given nitrogen partial pressures.

Introduction

Ferromanganese and silicomanganese alloys are the main sources of manganese for various grades of steels. Recent development of high manganese steels such as Twinning Induced Plasticity (TWIP) steels prompted the research on the production of high purity manganese alloys.[1-2] The nitrogen control in manganese alloys became also important for the production of low nitrogen containing high manganese steels. The thermodynamics of nitrogen in manganese alloys containing iron, silicon and carbon over wide ranges of melt composition and temperature is very important to control the nitrogen content in commercial manganese alloys.

The nitrogen solubility in pure Mn[3-7] and Mn-Fe alloy melts[7-10] has been studied by various investigators. In the authors' previous study, the thermodynamics of nitrogen in Mn-Fe-C melts has been also investigated.[7] Typical manganese alloys such as ferromanganese and silicomanganese alloys contain silicon up to 2 and 20 mass%, respectively, however, the studies on the effect of silicon on nitrogen in liquid manganese alloys are very limited. Baratashvili *et al.*[10] measured the nitrogen solubility in liquid Mn-Si alloy containing silicon up to 8 mass% at 1683 and 1933 K. There is no nitrogen solubility data in ternary or quaternary manganese alloy melts containing silicon.

In the present study, the nitrogen solubility in Mn-Si, Mn-Si-Fe, Mn-Si-C and Mn-Si-Fe-C melts was measured by the gas-liquid metal equilibration technique in the temperature range from 1673 to 1773K. From the analysis of all the data obtained by the present study and the authors' previous study,[7] the first- and second-order interaction parameters of silicon, iron and carbon on nitrogen in liquid manganese were determined to establish the database for the prediction of nitrogen solubility in ferromanganese and silicomanganese alloy melts as functions of the melt composition and temperature at a given nitrogen partial pressure.

Experimental Procedures

The gas-liquid metal equilibration experiments were carried out using an electric resistance furnace heated by the silicon carbide with a mullite reaction tube at 1673 and 1773 K as shown in Fig. 1. Desired portions of master alloys were charged to make an aimed melt composition for each experiment. Six grams of alloys were placed in a high purity alumina or graphite crucibles (OD: 13 mm, ID: 10 mm, H: 23 mm). Six crucibles containing different compositions were tied up by Mo wire and placed in the outer alumina crucible. The equilibration time was determined as 5 hours in the preliminary experiments.

After each experiment, the crucible was pulled out of the furnace and quenched rapidly in a helium gas stream on the water-cooled copper plate. The nitrogen and carbon contents in the metal samples were measured by the nitrogen/oxygen analyzer and the carbon/sulfur analyzer, respectively. The iron and silicon contents in the metal samples were measured by ICP-AES.

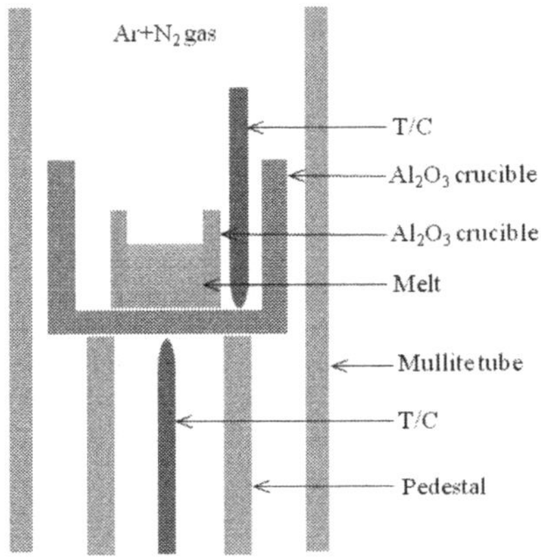

Figure 1. A schematic diagram of the experimental apparatus.

Results and Discussion

The dissolution of nitrogen in liquid manganese can be written as

$$\frac{1}{2}N_{2(g)} = [N]_{1\ \text{mass\% in Mn}} \tag{1}$$

$$\Delta G^{\circ} = -67{,}222 + 30.32T\ J/g \cdot atom^{[10]}$$

$$K_N = \frac{f_N[\%N]}{P_{N_2}^{1/2}} = f_N K_N^{'} = exp(-\frac{\Delta G^{\circ}}{RT}) \tag{2}$$

where K_N is the equilibrium constant for Reaction (1), [%N] is the equilibrium nitrogen content in mass% and, f_N is the Henrian activity coefficient of nitrogen for which the reference state is the infinitely dilute solution, i.e., $f_N \rightarrow 1$ when $[\%N] \rightarrow 0$. P_{N_2} is the nitrogen partial pressure in atm over the melt surface, K_N' is the apparent equilibrium constant, ΔG° is the Gibbs free energy change for the dissolution of one g-atom of nitrogen in liquid manganese, and R and T are the universal gas constant, 8.314 J/mol·K and the absolute temperature in K, respectively.

Sieverts' law was followed at low nitrogen contents below 1 mass% in liquid manganese where the interaction between nitrogen atoms in the melt could be neglected.[7] Therefore, in the present study, the nitrogen solubility in manganese alloy melts was measured under reduced nitrogen partial pressures of 0.01 and 0.03 atm to keep the nitrogen content below 1 mass%.

Manganese-Silicon Alloys

The nitrogen solubility in Mn-Si melts was measured to determine the effect of silicon on nitrogen in liquid manganese. Figure 2 shows the apparent equilibrium constants, $\ln K'_N(=\ln([\%N]/P_{N_2}^{1/2}))$ plotted against the silicon content in manganese-silicon melt at 1673 and 1773K. The $\ln K'_N$ value decreases with silicon content, and the effect is more significant at higher silicon content in the melt. It is interesting to see that the temperature dependence of $\ln K'_N$ value changes with silicon content, and it becomes opposite at silicon contents above 15 mass% as shown in the figure.

Figure 2 also compares the K'_N value determined at 1683 K by Baratashvili *et al.*.[10] They measured the nitrogen solubility at a nitrogen partial pressure of 0.97 atm in liquid Mn-Si alloys containing silicon up to 8 mass% using a high-frequency vacuum tube oscillator. Their lower K'_N values may be attributed to the interaction between nitrogen atoms in the melt at their high nitrogen contents. Therefore, the specific effect of silicon on nitrogen in manganese-silicon melt could not be measured in their study.

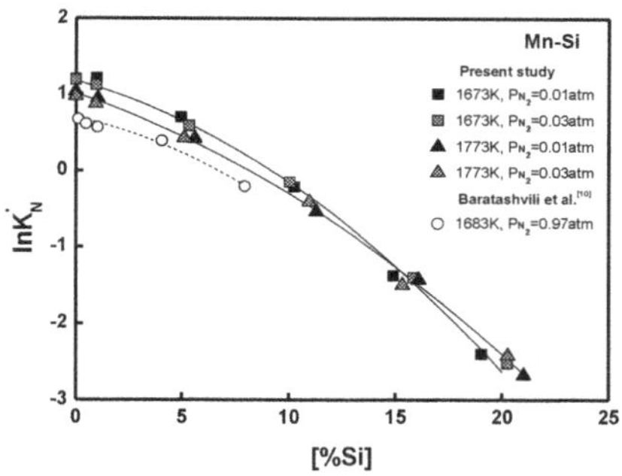

Figure 2. K'_N values with Si content in Mn-Si melts.

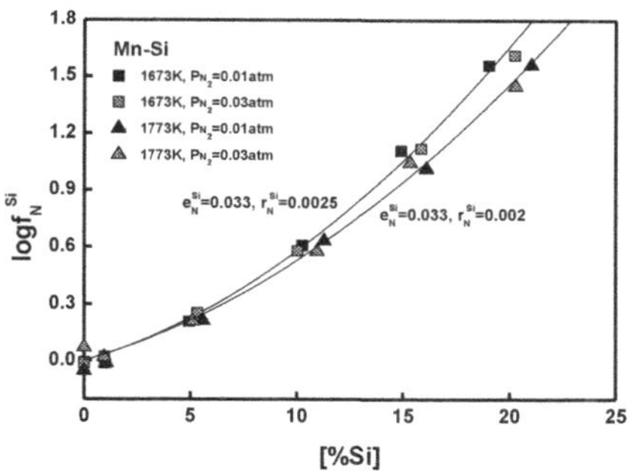

Figure 3. activity coefficient of nitrogen with Si content in Mn-Si melts.

The activity coefficient of nitrogen, f_N in manganese-silicon melts can be calculated using Eq. (2) as a function of silicon content. The resulting values of $\log f_N$ are plotted versus silicon content in Fig. 3. The activity coefficient of nitrogen increases sharply with increasing silicon content. The data determined at different nitrogen partial pressures show an excellent correlation with silicon content at each experimental temperature. In other words, the Sieverts' law of the nitrogen dissolution was obeyed at reduced nitrogen partial pressures in the present study.

The relationship can be analytically expressed using the first- and second-order interaction parameters:[14]

$$\log f_N = e_N^{Si}[\%Si] + r_N^{Si}[\%Si]^2 \quad (3)$$

where the values of e_N^{Si} and r_N^{Si} at 1673 and 1773K are determined as 0.033 and 0.0025, and 0.033 and 0.002, respectively, by the multiple regression analysis of the data in Fig. 3. The temperature dependence of r_N^{Si} can be expressed as 14.83/T - 0.00636.

Manganese-Silicon-Iron Alloys

The nitrogen solubility in Mn-Si-Fe melts was measured to determine the simultaneous

effect of silicon up to 10 mass% and iron up to 20 mass% on nitrogen in liquid manganese at 1673 and 1773 K. The activity coefficient of nitrogen in Mn-Si-Fe melts can be calculated from Eq. (2) using the nitrogen solubility data at given nitrogen partial pressures.
Then the following relationship can be expressed using the interaction parameters.

$$logf_N = e_N^{Si}[\%Si] + r_N^{Si}[\%Si]^2 + e_N^{Fe}[\%Fe] + r_N^{Fe}[\%Fe]^2 + r_N^{Si,Fe}[\%Si][\%Fe] \quad (4)$$

Eq. (4) can be rearranged as

$$r_N^{Si,Fe}[\%Si][\%Fe] = logf_N - e_N^{Si}[\%Si] - r_N^{Si}[\%Si]^2 - e_N^{Fe}[\%Fe] - r_N^{Fe}[\%Fe]^2 \quad (5)$$

where $r_N^{Si,Fe}$ is the second-order interaction parameter of silicon and iron on nitrogen, the values of e_N^{Si} and r_N^{Si} were determined in the preceding section, the values of e_N^{Fe} and r_N^{Fe} were previously determined as 0.015 and 0, respectively, in the temperature range from 1623 to 1823 K.[7]
Figure 4 shows the plot for the relation of Eq. (5) to determine the $r_N^{Si,Fe}$ value. The data determined over a wide range of melt composition at different nitrogen partial pressures and temperatures show an excellent linear relationship to obtain the $r_N^{Si,Fe}$ value as zero. No temperature dependence of the $r_N^{Si,Fe}$ value was observed.

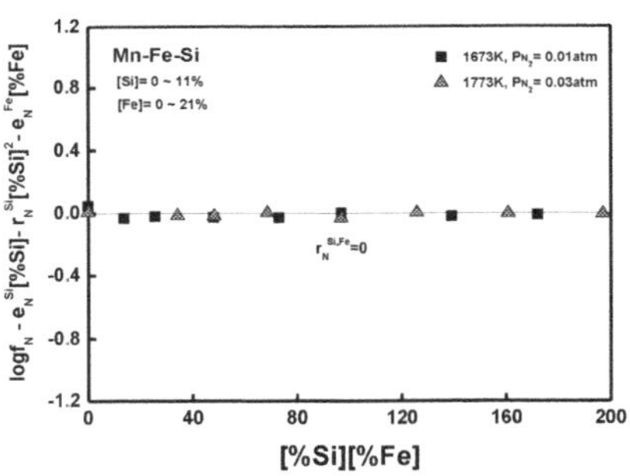

Figure 4. Relation to determine the $r_N^{Si,Fe}$ value in Eq. (5).

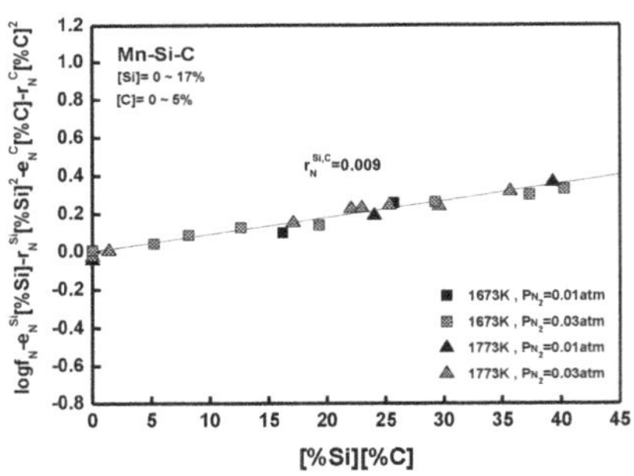

Figure 5. relationship to determine the $r_N^{Si,C}$ value in Eq. (6).

Manganese-Silicon-Carbon Alloys

The nitrogen solubility in Mn-Si-C melts was also measured to determine the simultaneous effect of silicon up to 15 mass% and carbon up to its saturation on nitrogen in liquid manganese at 1673 and 1773 K. The activity coefficient of nitrogen in Mn-Si-C melts can be obtained from Eq. (2), and it can be expressed as the following relationship using relevant interaction parameters.

$$logf_N = e_N^{Si}[\%Si] + r_N^{Si}[\%Si]^2 + e_N^{C}[\%C] + r_N^{C}[\%C]^2 + r_N^{Si,C}[\%Si][\%C] \quad (6)$$

where the values of e_N^C and r_N^C were previously determined as 0.09 and 0.013, respectively, in the temperature range from 1623 to 1823 K.[7]
Using the similar method described in the preceding section, the $r_N^{Si,C}$ value can be determined from the plot shown in Fig. 5. The data determined over a wide range of melt

composition up to carbon saturation show an excellent linear relationship. The $r_N^{Si,C}$ value can be determined as 0.009 by the linear regression analysis of the data shown in the figure. No temperature dependence of the $r_N^{Si,C}$ value was observed in the temperature range from 1673 to 1773 K.

Prediction of Nitrogen Solubility in Manganese-Silicon-Iron-Carbon Alloys

Table I summarizes the first- and second-order interaction parameters of elements on nitrogen in liquid manganese alloys determined in the present study and the authors' previous study.[7]

Table I. Interaction parameters of elements on nitrogen in liquid manganese.

System	i	e_N^i	r_N^i	Temp. (K)	Ref.
Mn-N	N	0.005	0.029	1623 ~ 1823	7
Mn-Fe-N	Fe	0.015	0	1623 ~ 1823	7
Mn-C-N	C	0.09	0.013	1623 ~ 1823	7
Mn-Si-N	Si	0.033	$\frac{14.83}{T} - 0.00636$	1673 ~ 1773	Present study
Mn-Si-Fe-N	Si, Fe		0	1673 ~ 1773	Present study
Mn-Si-C-N	Si, C		0.009	1673 ~ 1773	Present study

Using these interaction parameters, the activity coefficient of nitrogen in Mn-Si-Fe-C alloy melts can be evaluated. Then the nitrogen solubility in commercial manganese alloy melts can be calculated at given nitrogen partial pressures and melt temperatures from the following relation.

$$\begin{aligned} logK_N &= logf_N + \log[\%N] - \tfrac{1}{2}logP_{N_2} \\ &= e_N^N[\%N] + r_N^N[\%N]^2 + e_N^{Si}[\%Si] + r_N^{Si}[\%Si]^2 + e_N^{Fe}[\%Fe] + r_N^{Fe}[\%Fe]^2 \\ &\quad + e_N^C[\%C] + r_N^C[\%C]^2 + r_N^{Si,Fe}[\%Si][\%Fe] + r_N^{Si,C}[\%Si][\%C] \\ &\quad + \log[\%N] - \tfrac{1}{2}logP_{N_2} \end{aligned}$$

In order to check the validity of the interaction parameters determined in the present study, the nitrogen solubility measurement was carried out for Mn-Si-Fe-C melts over a wide range of composition under the nitrogen partial pressures of 0.01 and 0.03 atm at 1673 and 1773 K. Figure 6 compares the experimental results of nitrogen solubility with the calculated values using Eq. (7). They are in excellent agreement, and it suggests that the interaction parameters determined in the present study can be used to predict the nitrogen content for ferromanganese as well as silicomanganese alloys over wide range of melt composition at given nitrogen partial pressures in the temperature range from 1673 to 1773 K.

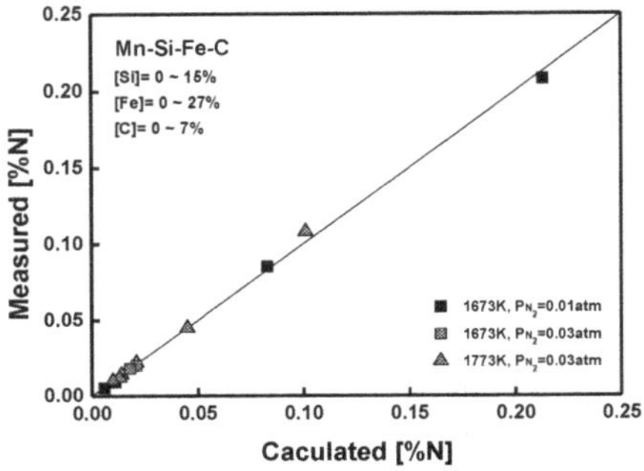

Figure 6. Correlation between calculated and measured nitrogen solubility in Mn-Si-Fe-C melts.

Conclusions

The interaction parameters of silicon, iron and carbon on nitrogen in liquid manganese have been determined from the nitrogen solubility in liquid Mn-Si, Mn-Si-Fe, Mn-Si-C and Mn-Si-Fe-C alloys over a wide range of composition at 1673 and 1773K. These parameters could be used to accurately predict the nitrogen solubility in ferromanganese and silicomanganese alloy melts at given nitrogen partial pressures.

References

1. M. K. Paek, W. K. Lee, J. Jin, J. M. Jang and J. J. Pak, "Thermodynamic Interactions Among Carbon, Silicon and Iron in Carbon Saturated Manganese Melts," *Korean. J. Met. Mater.*, 50 (2012), 45-51.

2. Y. E. Lee and D. S. Min, "Development of Mn alloys for steelmaking process" (Paper presented at the 1st Int. Conf. on High Manganese Steels, Seoul, Korea, 17 May 2011), 125.

3. N. A. Gokcen, "Solubility of Nitrogen in Liquid Manganese," *Trans.Metall. Soc. AIME*, 221 (1961), 200-201

4. I. B. Baratashvili, V. P. Fedotov, A. M. Samarin and V. M. Berezhiani, The Solubility of Nitrogen in Liquid Manganese," *Dokl. Akad. Nauk SSSR*, 139 (1961), 1354-1355.

5. Y. Tsu and T. Saito, "Absorption of N by Molten Fe and Mn at High Temperatures" (Paper presented at the Proc. Int. Conf. on the Science and Technology of Iron and Steel, Tokyo, Japan, 1971), 461-466.

6. S. Z. Beer, "Solubility of Nitrogen in Molten Iron Manganese Alloys," *Trans. TMS-AIME*, 221 (1961), 2-8.

7. E. J. Kim, B. D. You and J. J. Pak, "Nitrogen Solubility in Liquid Manganese and Ferromanganese Alloys," *Metall. Mater. Trans. B*, 32B (2001), 659-668.

8. R. A. Dodd and N. A. Gokcen, "Solubility of Nitrogen in Liquid-Iron Manganese Alloys," *Trans. TMS-AIME*, 221 (1961), 233-239.

9. C. Qiu, "A Thermodynamic Evaluation of the Fe-Mn-N System," *Metall. Trans. A*, 24A (1993), 629-649.

10. I. B. Baratashvili, V. P. Fedotov, A. M. Samarin and V. M. Berezhiani, "The Solubility of Nitrogen in Manganese-Iron and Manganese-Silicon Melts," *Dokl. Akad. Nauk SSSR*, 140 (1961), 131-132.

11. C. Wagner, *Thermodynamics of Alloys*, (Cambridge, MA: Addison-Wesley Press, 1952), 47.

Celebrating the Megascale: Proceedings of the Extraction and Processing Division
Symposium on Pyrometallurgy in Honor of David G.C. Robertson
Edited by: Phillip J. Mackey, Eric J. Grimsey, Rodney T. Jones, and Geoffrey A. Brooks
TMS (The Minerals, Metals & Materials Society), 2014

STUDY ON DEZINCIFICATION AND DE-LEAD OF BLAST FURNACE DUST BY FLUIDIZED REDUCTION EXPERIMENT

Shufeng Yang[1,2], Chengsong Liu[1,2], Xiaojie Gao[1,2], Jingshe Li[1,2]

[1]State Key Laboratory of Advanced Metallurgy
University of Science and Technology Beijing; Beijing 100083, China

[2]School of Ecological and Metallurgical Engineering
University of Science and Technology Beijing; Beijing 100083, China

Keywords: Blast furnace dust, reduction experiment, de-zinc, de-lead.

Abstract

In the blast furnace process, the dust entrained in the blast furnace gas enters into the down-comer, flows through the gravity dust separator (to eliminate coarse particles) and then is collected in a bag-house. The powder collected by the baghouse is called bag dust, while both fractions are called blast furnace dust whose main components are C and Fe. The dust also contains small amounts of nonferrous metals such as Zn and Pb, which have some value. Also, due to the small particle size and low density the dust is easily suspended in air and so can endanger human health. Therefore it is necessary to develop a process to both treat the dust to recover the metal values and to dispose of the residue – preferably by recycling to the blast furnace itself via the sinter strand. These objectives will result in good economic, environmental and social benefits [1].

Introduction

Due to regional mineral resources, there is a factory in Xinjiang whose main resources of iron ore used in sintering and iron making contain Zn and Pb. Some of the ores have quite high zinc content. The Zn and Pb in the iron ore are recirculated and enriched continually in the blast furnace. The zinc and lead vapor are oxidized to fine particles in the upper shaft. A fraction of these adhere to the ore and recirculate and a fraction report with the gas flow to the dust removal system.

Using the blast furnace dust as the raw material, reduction experiments were conducted to study the effect of several factors such as the reaction time, temperature, basicity, and the content of $CaCl_2$ addition on the de-Zn and de-Pb of the dust. The proposed process converts the zinc and lead oxide to volatile species so that they can be separated from the dust. This paper describes the theoretical foundation and an effective technical approach for utilizing blast furnace dust, thereby saving resources, and reducing the pollution of the environment.

Methodology

Materials

The experimental materials were obtained from a steel plant in Xinjiang and included the blast furnace dust, reducing coal, burnt lime, and powder of $CaCl_2$. The main components of these materials are shown in Table 1. After repeated testing for the contents of Zn and Pb in the dust, the average content of Zn was 7.3% and Pb was 0.8%. In the reduction experiments, the average values were adopted as the initial values.

Table 1. Main components in experimental materials

Experimental material	Main components
Blast furnace dust	Fe: 48~52%, Zn: 6~9%, Pb: 0.6%~1.1%
Reducing coal	C>60%; ash<10%; S<1%
Burnt lime	CaO:>80%
Powder of $CaCl_2$	$CaCl_2$>90%

Experimental procedures and apparatus

The high-temperature apparatus used is shown in Fig. 1. Some reducing coal was spread on the bottom of the volatilization furnace in advance (hearth layer). The proportion of blast furnace dust, reducing coal and burnt lime in the charge was 1:0.5:0.05. After balling and drying, the pellets were placed in the high-temperature volatilization furnace. ZnO and PbO werr reduced at 1000~1300°C. The Zn, Pb, $ZnCl_2$ and $PbCl_2$ vapor produced was draw into the oxidation chamber by means of an induced draft fan (not shown in Figure 1) and reacted with air producing ZnO and PbO which was cooled down in W-type cooling tubes. When the temperature was below 100°C, the ZnO was captured in the baghouse. The residue was stored and analyzed. Fig. 2 illustrates the procedures of the experiment.

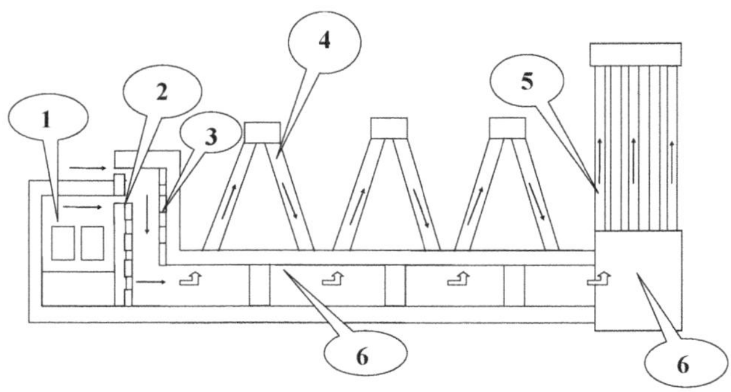

1- High-temperature volatilization chamber; 2- oxidation chamber; 3- Water cooling chamber; 4- W-type cooling tubes; 5- bag-house dust collector; 6- Zn dust collection chamber

Figure 1. Schematic of high-temperature boiling furnace

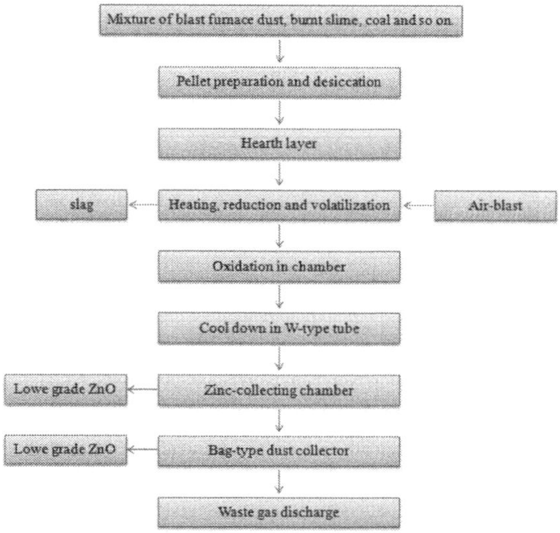

Figure 2. Schematic of experimental procedures

Based on the apparatus and method described above, this work investigated and analyzed the effect of reduction time, reduction temperature, basicity and the addition of $CaCl_2$ on de-Zn and de-Pb of blast furnace dust. After the experiment was finished, the mass of dust accumulated in Zn-collecting chamber and bag-type collector was measured. The contents of Zn and Pb in in the dust and the residue were measured by chemical analysis.

Results and Discussions

Effect of reduction time on the dezincification and de-lead

Fig. 3 shows the change of the content of Zn and Pb in the residue with the reduction time when the temperature was 1100℃, the basicity 0.8, and there was no $CaCl_2$ addition. According to Fig. 3, we can see that in the first 30 minutes, the contents of Zn and Pb in the residue were falling, especially Zn. The speed of volatilization for lead oxides was slower than that for zinc. After 30 minutes, the curve fluctuated slightly. The main cause of this was small differences between the different experiments.

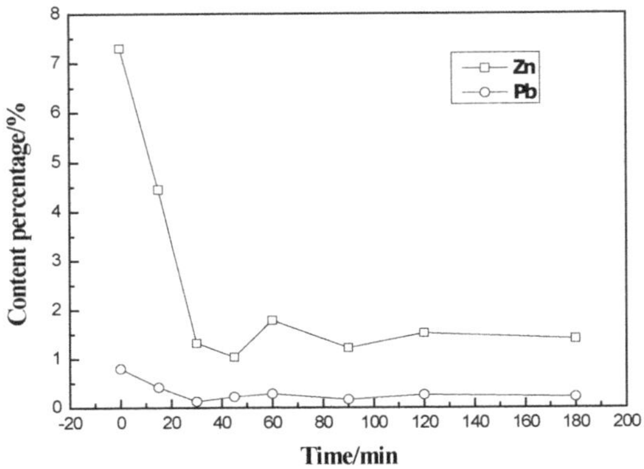

Figure 3. Change of Zn and Pb during reduction

In conclusion, when the temperature was 1100°C and the basicity 0.8, and no $CaCl_2$ addition was used, the reaction time required for best effect of de-Zn and de-Pb was 30 minutes, or perhaps 45 minutes to ensure complete reaction. Under this condition, the amounts of Zn and Pb in the residue were about 1.5% and 0.2% respectively. These values were not low enough for the residue to be recycled to the blast furnace where Zn and Pb should both be below 0.1%.

Effect of temperature on the dezincification and de-lead

Figs. 4 and 5 show that, with the same reduction time and basicity, and without $CaCl_2$ addition, the removal rate of both Zn and Pb from the dust increased gradually with increase of the reaction temperature. As the temperature was increased (from 1100 ℃ to 1300 ℃) the percent of zinc and lead removed also increased. The rate controlling step of de-Zn reaction is the chemical reaction that CO reduces zinc oxide to form CO_2 and zinc vapor. The apparent activation energy of this reaction is large at around 79 kJ/mol.

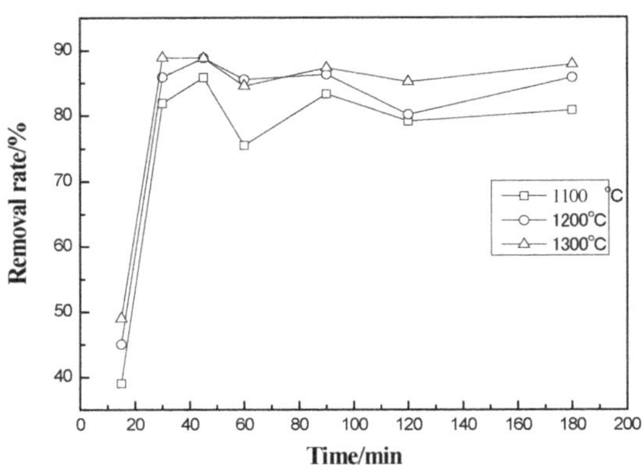

Figure 4 Relationship between temperature and the removal ratio of de-Zn

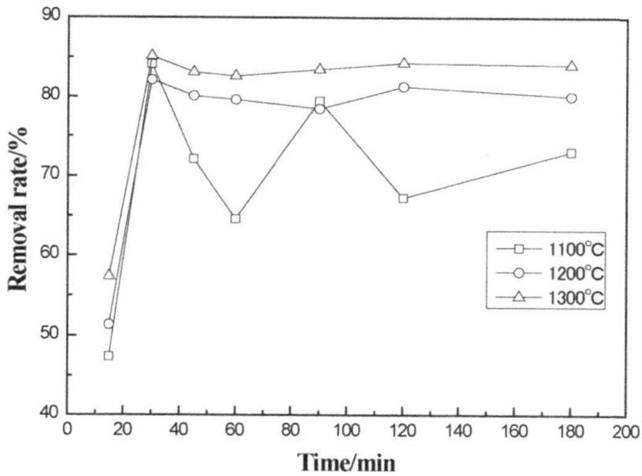

Figure 5. Relationship between temperature and removal ratio of de-Pb

The rate controlling steps of de-Pb are the evaporation reaction of the lead liquid and the diffusion of the lead vapor through the gaseous boundary layer [3], with an apparent activation energy of 88 kJ/mol.

Effect of basicity on the dezincification and de-lead

The blast furnace dust contains a large amount of total iron (48% ~ 52% content), and so the processes of de-Zn and de-Pb were accompanied by iron oxide reduction. Increased lime addition increased the porosity of the reduced iron, which was conducive to the further penetration of reducing gas and to the easier escape of the volatile reduction products [7]. Fig. 6 shows the trend of % Zn remaining in the residue with basicity, when the temperature is 1100°C and without $CaCl_2$ addition.

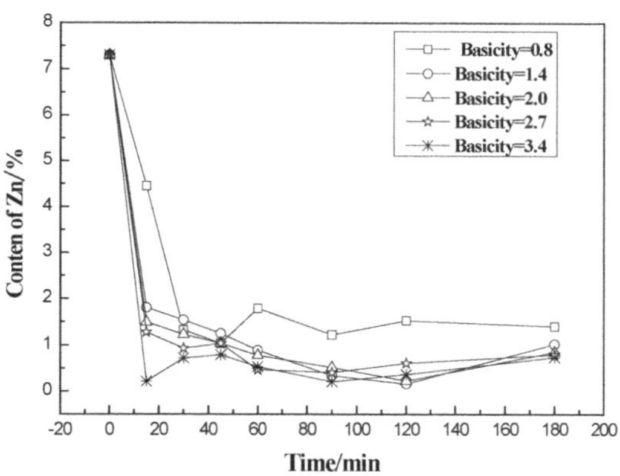

Figure 6. % Zn in the residue with different basicities

According to the Fig. 6, increasing the basicity is beneficial to the transmission of CO and the reduction and volatilization of Zn. Final zinc contents in the residue were around 0.4%. The reaction time required was in the range of 80- 120 minutes.

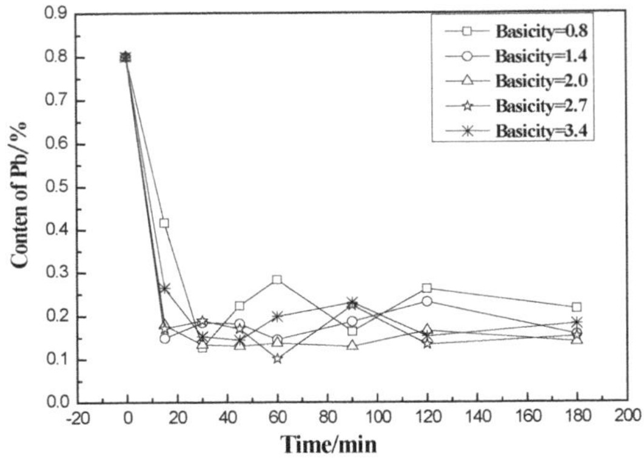

Figure 7. Changing of content of Pb with different basicities

Fig. 7 shows the trend of the lead content change in the residue with basicity, when the temperature was 1100°C, and no $CaCl_2$ additions were made. Since the reduction ratio of lead oxide is not controlled by the CO chemical step there was little if any effect of basicity on the lead volatilization.

Fig. 8 shows the trend with change in basicity, when the temperature was 1100°C and there were no $CaCl_2$ additons. As shown in Fig. 9, with the increase of basicity from 0.8 to 1.4, the removal ratio of Zn is improved significantly. With further increase in basicity to 2.7 there was a small improvement. To sum up, considering the reduction time and removal ratios of Zn and Pb, the best basicity range lies in the range of 2 ~ 2.7, and best reduction time is around 90 minutes.

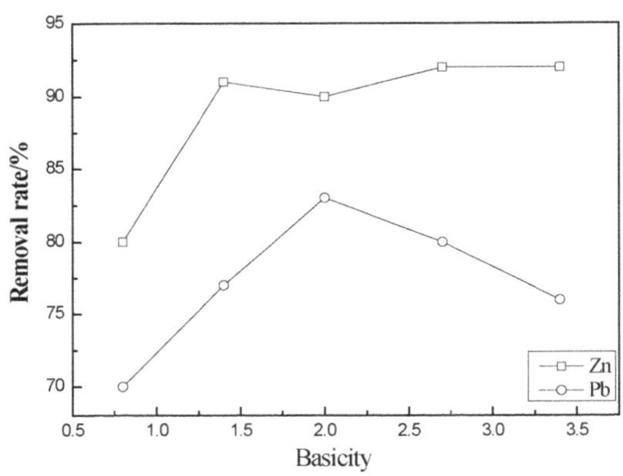

Figure 8. Removal ratio of Zn and Pb with basicity

Effect of additions on de-Zn and de-Pb

The amount of $CaCl_2$ addition was determined by the following reaction equations.

$$ZnO+CaCl_2=CaO+ZnCl_2$$

$$PbO+CaCl_2=CaO+PbCl_2$$

Assuming that x is the amount of blast furnace dust, a_1 and a_2 are theoretical addition amount of $CaCl_2$ used to remove the ZnO and PbO in dust, then the total calculated amount $a=a_1+a_2=0.073x+0.008x=0.081x$, namely the percentage of $CaCl_2$ addition should be about 8.1%. The amount of addition selected in this work was 1 times, 1.2 times and 1.5 times as large as the theoretical addition amount – giving 8%, 10% and 12% respectively. Figs. 9 and 10 indicate the effect of additions of $CaCl_2$ on de-Zn and de-Pb when the temperature is 1100°C and the basicity is equal to 0.8.

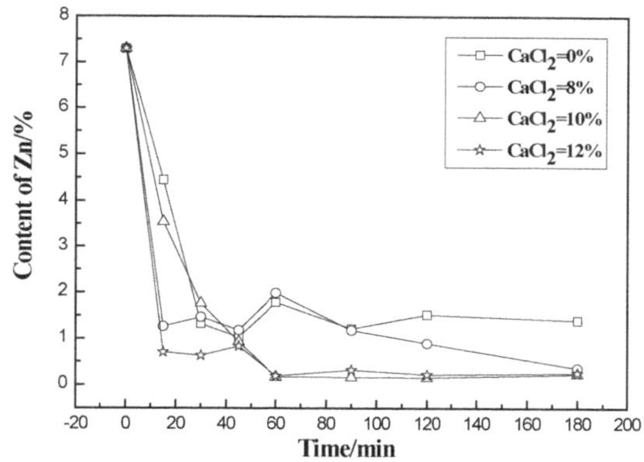

Figure 9. Changing of content of Zn with different amounts of $CaCl_2$

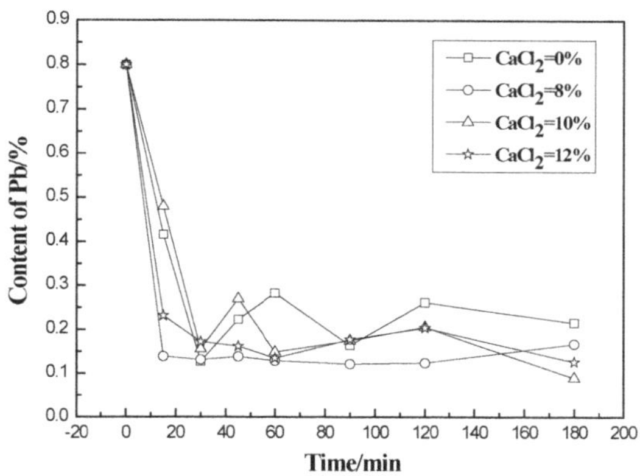

Figure 10. Changing of content of Pb with different amounts of $CaCl_2$

The final content of Pb in the residue is much lower than that of case without $CaCl_2$ additions, and was in the range of 0.1 to 0.2%.

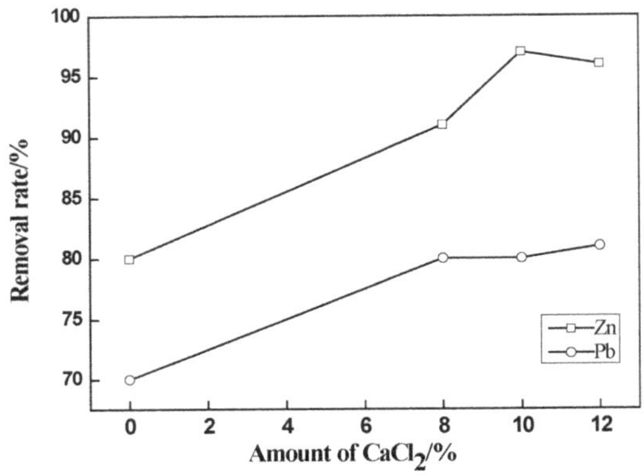

Figure 11. Changing of removal ratio of Zn and Pb with different amounts of $CaCl_2$

The removal ratio of Zn and Pb improved greatly after adding $CaCl_2$ as indicated in Fig. 11, especially with the amount of $CaCl_2$ is in the range of 10%~12%, when the removal ratio of zinc reached about 96%. The removal ratio of Pb was always just above 80% after adding $CaCl_2$. The best amount of $CaCl_2$ addition was 10%, and the best reduction time was 60 to 90 minutes. 96% removal of zinc represents a value of around 0.3% Zn in the residue and 80% removal of lead represents around 0.2% lead in the residue. So by using the CaCl2 addition the residue from reduction was probably low enough in Zn and Pb to be returned to the blast furnace, probably via the sinter strand. Ideally the values for Zn and lead in the residue would be less than 0.1%.

Overall, it is totally possible treat the blast furnace dust in the high-temperature volatilization furnace. By adjusting the experimental conditions, the removal ratio of both Zn and Pb in the blast furnace dust could almost reach the ideal requirement. What we need to do is to find the balance point between the removal ratio of Zn and Pb and the cost of the process, and finally optimize the removal effect which will create benefits for the environment, economy and society.

Conclusions

(1) Under the actual condition of reduction in the high-temperature volatilization furnace, ZnO and PbO could be reduced and chloridized to produce volatile species - elemental Zn and Pb, and $ZnCl_2$, $PbCl_2$

(2) The effects of increasing reaction time, reaction temperature, basicity, and the addition of $CaCl_2$ on de-Zn and de-Pb of blast furnace dust were all positive.

(3) Compared with lead oxide, the zinc oxide in blast furnace dust wase removed in a shorter time.

References

1 Liaosha Li, Kaiyuan Li. *Recovery of iron and zinc from blast furnace sludge.* The

Chinese Journal of Process Engineering, 2009, Vol. 19(3), pp. 468-473.

2 Huichao Yin, Jianliang Zhang. *Study on the law of zinc absorption and change of metallyrgical property of sinter and pellet.* Iron and Steel, 2010, Vol. 45(2), pp. 15-18.

3 Yingsheng Wang, Shengli Wang, Zhixing Zhao, Wen Pan. *Study on application of MGF during sintering process.* Iron and Steel, 2010, Vol. 45(9), pp. 15-17+61.

4 Lv X W, Bai C G, Deng Q Y, et al. *Behavior of Liquid Phase Formation during Iron Ores Sintering.* ISIJ Int, 2011, 51(5): 114.

5 Zhimin Liu, Shengli Wu, Jun Jin, Canyang Song, Juan Zhu. *Study on optimizing the sintering proportioning in Mansteel.* Sintering and Pelletizing, 2012,Vol. 37(2), pp. 13-18.

6 Shengli Wu, Yuming Dai, Dauter Oliveira, Yuandong Pei, Jian Xu, Hongliang Han. *Optinization of ore blending during sintering based on complementation of high temperature properties.* Journal of University of Science and Technology Beijing, 2010,Vol.32(6), pp. 719-724.

7 Jian Chang, Buxin Shu, Jianliang Zhang, et al. *Sintering proportioning optimization model based on high temperature properties of iron ore fines.* Sintering and Pelletizing, 2011,Vol. 36(5), pp. 12-14.

8 Shengli Wu, Yu Liu, Jianxin Du, Kun Mi, Hong Lin. *Experiment study of assimilation ability between iron ores and CaO.* Journal of University of Science and Technology Beijing, 2002,Vol.24(3), pp. 258-261.

9 Shengli Wu, Jianxin Du, Hongbin Ma, Junqing Tian, Haifa Xu. *Fluidity of liquid phase in iron ores during sintering,* Journal of University of Science and Technology Beijing, 2005,Vol.27(3), pp. 291-293.

10 Zhixing ZHao, Yuandong Pei, Wen Pan, Haibing Jiang, Hongge Li. *Influencing factors on high temperature properties of iron ore in Shougang*, Iron and Steel, Vol.45(12), pp. 12-16.

11 Yang L X, Davis L. *Assimilation and Mineral Formation During Sintering for Blends Containing Magnetite Concentrate and Hematite/Pisolite Sintering Fines.* ISIJ Int, 1999, Vol.39(3), pp. 239-245.

AUTHOR INDEX
Celebrating the Megascale: Proceedings of the Extraction and Processing Division Symposium on Pyrometallurgy in Honor of David G.C. Robertson

K

L

M

N

O

P

R

S

T

V

W

X

Y

Z

SUBJECT INDEX
Celebrating the Megascale: Proceedings of the Extraction and Processing Division Symposium on Pyrometallurgy in Honor of David G.C. Robertson

The manufacturer's authorised representative in the EU is Springer Nature Customer Service Centre GmbH, Europaplatz 3, 69115 Heidelberg, Germany. If you have any concerns regarding our products, please contact ProductSafety@springernature.com

Printed and bound by CPI Group (UK) Ltd, Croydon, CR0 4YY
25/09/2026
02222881-0012